ASTROPHYSICS IN THE EXTREME ULTRAVIOLET

Astrophysics
in the Extreme Ultraviolet

Proceedings of Colloquium No. 152 of the
International Astronomical Union,
held in Berkeley, California,
March 27–30, 1995

Edited by

STUART BOWYER

Professor in the Graduate School,
University of California, Berkeley, U.S.A.

and

ROGER F. MALINA

Director, Center for EUV Astrophysics,
University of California, Berkeley, U.S.A.

KLUWER ACADEMIC PUBLISHERS
DORDRECHT / BOSTON / LONDON

A C.I.P. Catalogue record for this book is available from the Library of Congress

ISBN 0-7923-3908-8

Published by Kluwer Academic Publishers,
P.O. Box 17, 3300 AA Dordrecht, The Netherlands.

Kluwer Academic Publishers incorporates
the publishing programmes of
D. Reidel, Martinus Nijhoff, Dr W. Junk and MTP Press.

Sold and distributed in the U.S.A. and Canada
by Kluwer Academic Publishers,
101 Philip Drive, Norwell, MA 02061, U.S.A.

In all other countries, sold and distributed
by Kluwer Academic Publishers Group,
P.O. Box 322, 3300 AH Dordrecht, The Netherlands.

Printed on acid-free paper

Printed in the Netherlands

ORGANIZED BY

The Center for EUV Astrophysics, University of California at Berkeley

SCIENTIFIC ORGANIZING COMMITTEE

S. Bowyer, co-chair
B. Haisch, co-chair
P. C. Agrawal
A. Dupree
G. Fontaine
E. Jenkins
Y. Kondo
R. F. Malina
R. Mewe
J. Sahade
J. Schmitt
A. Vidal-Madjar
K. Yamashita

LOCAL ORGANIZING COMMITTEE

S. Bowyer
J. Hinchman
S. Lilly

SUPPORTING INSTITUTIONS

The Center for EUV Astrophysics
Vice Chancellor for Research at the University of California at Berkeley
California Space Institute

Contents

I. Summaries of Recent Missions

II. Extragalactic Sources in the EUV

III. Coronae of Cool Stars

IV. White Dwarf Structure/Evolution

V. The Interstellar Medium and Diffuse Background

VI. Cataclysmic Variable Stars

VII. Photospheres and Winds of Early-Type Stars

VIII. Novae, X-ray Binaries

IX. Neutron Stars

X. Solar System Observations

XIII. Plasma Diagnostics

XIV. Future Opportunities in EUV Astronomy

Index

Editors' Preface

The new field of EUV astrophysics has come of age. Since the first IAU sponsored conference on EUV astrophysics in 1989, the *EUVE* and *ROSAT* missions have completed the first all-sky EUV surveys. *EUVE* spectroscopy with $1 - 4$ Å resolution has been obtained on about 100 objects in the EUV sky. Some observers have commented that EUV spectroscopy of other stars is now more advanced than EUV spectroscopy of the Sun.

When the current EUV missions were originally proposed, many reputable scientists argued that spending government funds on EUV astronomy was pointless; there would be nothing to see. There are now over 800 sources in the first all-sky EUV catalogues—more sources than existed in either the first-generation X-ray or gamma-ray catalogues. *EUVE* is currently discovering a new EUV source every few days.

EUV spectroscopy has now been carried out at 100 Å for a number of extragalactic objects, and EUV spectra at 700 Å have been obtained unexpectedly for B stars at 200 pc. Recent papers report EUV observations of the neutron star Geminga at 150 pc. And observations of EUV emission from stellar atmospheres and coronae reveal that no theoretical models fit the new data—our fundamental understanding must be revised.

As the papers in this book make clear, observational EUV astronomy is now providing intriguing insights and discoveries on a wide range of astrophysical phenomena—probing accretion disk physics, the energetics of stellar coronae, and the structures of hot stellar atmospheres in normal and compact objects.

Simultaneous observations are now being carried out with *EUVE* and the *Hubble Space Telescope*, X-ray observatories, and ground-based observatories, providing multi-wavelength data which elucidate the underlying emission mechanisms.

EUV observations of the local interstellar medium provide a detailed picture of the thermal and ionization conditions in the local ISM. At this time no consistent model of the local ISM can explain available EUV data coupled with data at other wavelengths.

A particular success of EUV astronomy has been the development of new observational probes to study helium in stellar atmospheres and in the ISM. Helium's primary absorption features, for both neutral and ionized helium, are in the EUV. *EUVE* has rightfully been called a "helium measuring" machine for the string of first reports of helium densities and ionization in Mars, the Comet Shoemaker-Levy impact on Jupiter, stellar coronae, cataclysmic variables, white dwarfs, B stars, and the ISM.

In the proceedings of the first EUV conference, we stated in our preface, "Will the current optimism reflected in these pages be borne out by the data soon to be obtained? Only the data will reveal the answer." And the answer, as shown in these pages, is a resounding *yes*. Yet ironically, at the time of writing, no space agency has new missions under development that would build upon these first-generation EUV missions. Whereas the fields of X-ray and infrared astronomy are about to embark on specialized fourth-generation follow-up missions, EUV astronomy may be entering an observational hiatus following the end of the *EUVE* and *ROSAT* missions. Perhaps this book will help emphasize the need for new EUV missions. We hope so.

We would like to thank all those who contributed to the success of the IAU Colloquium on Astrophysics in the Extreme Ultraviolet. We particularly want to thank Sharon Lilly and Jennifer Hinchman of the Local Organizing Committee. We also thank the Scientific Organizing Committee and Dr. Bernhard Haisch, who co-chaired the conference. We are grateful to Andrea Frank and her staff for their professional and timely preparation of this book.

Stuart Bowyer and Roger F. Malina, Editors
Center for EUV Astrophysics
University of California at Berkeley
March 1995

Welcoming Remarks by the Chancellor of the University of California

Welcome to IAU Colloquium 152 on EUV Astrophysics.

Many of you have come a long way to participate in this conference; from Russia, Japan, England, Germany, Italy and many other countries. You are the experts but I would like to remind you of some of the milestones in this field, and in particular, Berkeley's role in the development of EUV astrophysics. Twenty years ago we had no EUV sources. From 1959 when the first papers were published on this subject until 1974, everyone assumed that interstellar space was filled with gas that would be entirely opaque to EUV radiation. In 1974, a Berkeley group, led by Stu Bowyer, published an article in the Astrophysical Journal. The article predicted that conventional wisdom was wrong and that space was not nearly as opaque in the EUV as predicted.

The only way to know was to look. Within a year, Bowyer and his group submitted a winning proposal to NASA for an EUV mission. Unfortunately, the mission did not get very far before it was terminated; fortunately, Stu was not easily discouraged. In less than a year he found a way to fly a small EUV telescope on the joint US–Soviet Apollo-Soyuz mission. On that mission, the Berkeley group discovered four EUV emitting stars. Based on that success, the group sent another proposal to NASA. That proposal was selected for development in 1976. But it did not fly until June of 1992 when the *Extreme Ultraviolet Explorer* mission finally got off the ground.

Was it worth the wait? I think so. Here you are today, about to start the first major international conference dedicated to exploring this new spectral window. The window was opened by that long-awaited launch.

Let me congratulate Stu Bowyer for never giving up and thereby bringing about today's proceedings. His work is indicative of the quality of research that goes on here at Cal.

It is with great pride that I welcome you to the University of California at Berkeley. Thank you and I wish you well in your discussions!

Chang-Lin Tien, Chancellor
University of California, Berkeley

Participants

Ken Anderson	Center for EUV Astrophysics, UCB (USA)
Aldo Altamore	University of Roma III (Italy)
Victoria Alten	University of Colorado (USA)
Thomas Ayres	University of Colorado (USA)
Colin Barber	Leicester University (UK)
Martin Barstow	Leicester University (UK)
Manuel Bautista	Ohio State University (USA)
Thomas Berghoefer	Max Planck Institute for Extraterrestrial Physics (Germany)
Jeffrey Bloch	Los Alamos National Lab (USA)
Stuart Bowyer	Center for EUV Astrophysics, UCB (USA)
Nancy Brickhouse	Harvard-Smithsonian Center for Astrophysics (USA)
Alex Brown	University of Colorado (USA)
Fred Bruhweiler	Catholic University (USA)
Eric Brunet	Magistere Interuniversitaire de Physique (France)
Matthew Burleigh	Leicester University (UK)
Paul Callanan	Harvard-Smithsonian Center for Astrophysics (USA)
Joseph Cassinelli	University of Wisconsin, Madison (USA)
Supriya Chakrabarti	Boston University (USA)
Pierre Chayer	Center for EUV Astrophysics, UCB (USA)
K.P. Cheng	California State University Fullerton (USA)
Damian Christian	Center for EUV Astrophysics, UCB (USA)
Carol Christian	Center for EUV Astrophysics, UCB (USA)
David Cohen	University of Wisconsin, Madison (USA)
Donald Cox	University of Wisconsin, Madison (USA)
Nahide Craig	Center for EUV Astrophysics, UCB (USA)
Scott Cully	Space Sciences Laboratory, UCB (USA)
Michael Dahlem	Space Telescope Science Institute (USA)
Marthijn deKool	Max Planck Institute for Astrophysics (Germany)
Rosa Izela Diaz	Instituto de Astronomia UNAM (Mexico)
George Doschek	Naval Research Laboratory (USA)
Jeremy Drake	Center for EUV Astrophysics, UCB (USA)
Stephen Drake	Goddard Space Flight Center (USA)
Andrea Dupree	Harvard-Smithsonian Center for Astrophysics (USA)
Jean Dupuis	Center for EUV Astrophysics, UCB (USA)
Jerry Edelstein	Center for EUV Astrophysics, UCB (USA)
Richard Edgar	University of Wisconsin, Madison (USA)
Bradley Edwards	Los Alamos National Lab (USA)
David Finley	Center for EUV Astrophysics, UCB (USA)
George Fisher	Space Sciences Laboratory, UCB (USA)
Roger Foster	Naval Research Laboratory (USA)
Matthias Frank	Lawrence Livermore National Laboratory (USA)
Antonella Fruscione	Center for EUV Astrophysics, UCB (USA)
Marc Gagn	University of Colorado (USA)
Peter Gaposchkin	City of San Francisco (USA)
Ricardo Génova	Center for EUV Astrophysics, UCB (USA)
Mark Giampapa	AURA, National Optical Astronomy Observatories (USA)
Alvaro Gimenez	Institute Nationale de Technica Aerospacial (Spain)
Don Goldsmith	Interstellar Media (USA)

Caroline Greer	The Queen's University of Belfast (UK)
Cecile Gry	Laboratoire d'Astronomie Spatiale (France)
Edward Guinan	Villanova University (USA)
Bernhard Haisch	Lockheed Palo Alto Research Laboratory (USA)
Doyle Hall	Johns Hopkins University (USA)
Gregory Hanson	Harvard Smithsonian Center for Astrophysics (USA)
Peter Hauschildt	Arizona State University (USA)
Isabel Hawkins	Center for EUV Astrophysics, UCB (USA)
Adrienne Herzog	Planetary Science Institute (USA)
Larry Hiller	Lawrence Livermore National Laboratory (USA)
Jay Holberg	University of Arizona (USA)
Steve Howell	Planetary Science Institute (USA)
Ivan Hubeny	NASA Goddard Space Flight Center (USA)
Mark Hurwitz	Center for EUV Astrophysics, UCB (USA)
Chorng Hwang	Center for EUV Astrophysics, UCB (USA)
Carole Jordan	University of Oxford (UK)
Stefan Jordan	Institut für Astronomie und Astrophysik (Germany)
Jelle Kaastra	SRON Laboratory for Space Research (Netherlands)
Menas Kafatos	George Mason University (USA)
Mariko Kato	Keio University (Japan)
Maria Katsova	Sternberg State Astronomical Institute (Russia)
Francis Keenan	The Queen's University of Belfast (UK)
Detlev Koester	University of Kiel (Germany)
Arieh Königl	University of Chicago (USA)
Joachim Krautter	Landessternwarte Königstuhl (Germany)
Jacek Krelowski	Torun University (Poland)
Hideyo Kunieda	Nagoya University (Japan)
Simon Labov	Lawrence Livermore National Laboratory (USA)
Martin Laming	Naval Research Laboratory (USA)
Michael Lampton	Center for EUV Astrophysics, UCB (USA)
Denis Leahy	University of Calgary (Canada)
Jim Lewis	Center for EUV Astrophysics, UCB (USA)
Peng Li	Space Sciences Laboratory, UCB (USA)
James Liebert	University of Arizona (USA)
Duane Liedahl	Lawrence Livermore National Laboratory (USA)
Richard Lieu	Center for EUV Astrophysics, UCB (USA)
Jeremy Lim	Institute of Astronomy & Astrophysics (China)
Mark Lindeman	Lawrence Livermore National Laboratory (USA)
Jeffrey Linsky	JILA, University of Colorado (USA)
Moissei Livshits	Russian Academy of Sciences (Russia)
Knox Long	Space Telescope Science Institute (USA)
James MacDonald	University of Delaware (USA)
Joseph MacFarlane	University of Wisconsin, Madison (USA)
Roger Malina	Center for EUV Astrophysics, UCB (USA)
Herman Marshall	Eureka Scientific Inc. (USA)
Helen Elizabeth Mason	Cambridge University (UK)
Mihalis Mathioudakis	Center for EUV Astrophysics, UCB (USA)
Christopher Mauche	Lawrence Livermore National Laboratory (USA)
Bruce McCollum	NASA Goddard Space Flight Center (USA)
Kelley McDonald	Center for EUV Astrophysics, UCB (USA)

Rolf Mewe	SRON Laboratory for Space Research (Netherlands)
Warren Miller III	Instituto de Astronomia UNAM (Mexico)
Jonathan Mittaz	Mullard Space Science Laboratory (USA)
Richard Monier	Obsevatoire de Strasbourg (France)
Brunella Monsignori-Fossi	Arcetri Astrophysical Observatory (Italy)
Luisa Morales	Center for EUV Astrophysics, UCB (USA)
Dermott Mullan	Bartol Research Institute (USA)
James Neff	Penn State University (USA)
Harrie Netel	Lawrence Livermore National Laboratory (USA)
Ron Oliversen	NASA Goddard Space Flight Center (USA)
Frits Paerels	Physics Department, UCB (USA)
Jianfang Peng	Ohio State University (USA)
Geraldine Peters	University of Southern California (USA)
Timothy Pfafman	Los Alamos National Laboratory (USA)
Anil Pradhan	Ohio State University (USA)
Thomas Preibisch	Universität Würzburg (Germany)
John Pye	Leicester University (UK)
Mohan Rajagopal	Stanford University (USA)
Elio Ramos-Colon	George Mason University (USA)
Saul Rappaport	Massachusetts Institute of Technology (USA)
Ted Rodriguez-Bell	Space Sciences Laboratory, UCB (USA)
Simon Rosen	University of Leicester (UK)
Diane Roussel-Dupre	Los Alamos National Laboratory (USA)
Slavek Rucinski	David Dunlap Observatory (Canada)
Pedro Safier	Astronomy Department, UCB (USA)
Edwin Salpeter	Cornell University (USA)
Wilton Sanders	University of Wisconsin, Madison (USA)
Daniel Wolf Savin	Physics Department, UCB (USA)
Jürgen Schmitt	Center for EUV Astrophysics, UCB (USA)
Karel Schrijver	Lockheed Palo Alto Research Labs (USA)
Edward Sion	Villanova University (USA)
Martin Sirk	Center for EUV Astrophysics, UCB (USA)
Barham Smith	Los Alamos National Laboratory (USA)
Sumner Starrfield	Arizona State University (USA)
Robert Stern	Lockheed Palo Alto Research Laboratory (USA)
Guy Stringfellow	Pennsylvania State University (USA)
Gaghik Tovmassian	Instituto de Astronomia UNAM (Mexico)
John Vallerga	Eureka Scientific Inc. (USA)
Olaf Vancura	Harvard-Smithsonian Center for Astrophysics (USA)
Stéphane Vennes	Center for EUV Astrophysics, UCB (USA)
Alfred Vidal-Madjar	Institut d'Astrophysique de Paris CNRS (France)
Richard Wade	Penn State University (USA)
Fred Walter	Earth & Space Sciences (USA)
John Warren	Space Sciences Laboratory, UCB (USA)
Volker Weidemann	University of Kiel (Germany)
Barry Welsh	Eureka Scientific Inc. (USA)
Klaus Werner	University of Kiel (Germany)
Stephen White	University of Maryland (USA)
Kenneth Widing	Naval Research Laboratory (USA)
Tod Woods	Lawrence Livermore National Laboratory (USA)

Xiaoyi Wu   Center for EUV Astrophysics, UCB (USA)
Koujun Yamashita   Nagoya University (Japan)
Wei-Hong Yang   Northwest Polytechnic University (USA)
Peter Young   Cambridge University (UK)

Results from the *ROSAT* EUV Wide Field Camera

JOHN P. PYE

Department of Physics & Astronomy, Leicester University, Leicester, LE1 7RH, UK

I review some of the major achievements of the *ROSAT* extreme-ultraviolet all-sky survey, including the results of the recently completed 2RE source catalogue and associated study of temporal variability in EUV sources.

1. Introduction

In January 1989, at the first Berkeley colloquium on extreme-ultraviolet (EUV) astronomy, I reviewed, prelaunch, the expected performance and calibration parameters of the *ROSAT* EUV Wide Field Camera (Pye et al. 1991), and predicted numbers of sources of different classes that should be detectable in the all-sky survey. So it is timely now, six years later at this second Berkeley colloquium on EUV astronomy, to summarise the achievements of the WFC, especially those from the all-sky survey.

During 1990–1991 the WFC performed the first all-sky survey at EUV wavelengths. The survey was conducted in two 'colours' using broad-band filters to define wavebands covering the ranges 60–140 Å and 112–200 Å. It was fully imaging, with effective spatial resolution of about 3 arcmin FWHM, and point source location accuracy of typically better than 1 arcmin. From an initial analysis, Pounds et al. (1993) published the WFC Bright Source Catalogue (BSC) of 383 sources. They also assessed why, for the two main classes of EUV source, there were substantially fewer detections than predicted prelaunch. Briefly, for white-dwarf stars (WDs), the hotter objects ($T_{\rm eff} \gtrsim 4 \times 10^4$ K) have their EUV (and soft X-ray) luminosities greatly reduced by the opacity of trace metals radiatively levitated in the WD atmospheres (Barstow et al. 1993). For the coronal emission from late-type (F–M) stars the explanation appears to lie, at least in part, with over-estimation of the high-luminosity tails in early X-ray luminosity functions, especially for dM stars. The sensitivity of flux/count-rate conversion factor to the assumed (and rather ill-defined) source spectra, may also be a significant cause of error (Hodgkin & Pye 1994).

2. 2RE Catalogue

Reprocessing of the complete survey database has resulted in a new catalogue of EUV sources, designated '2RE' (Pye et al. 1995). The 2RE Catalogue contains 479 sources (see figure 1), 120 of which were not reported in the BSC; of these, 97 also do not appear in the first *EUVE* Catalogue (Bowyer et al. 1994) and hence have not previously been reported at EUV wavelengths. There are 387 2RE sources detected in both survey wavebands, a significant advance on the BSC (80 percent versus 60 percent). Improvements over the original BSC include: (i) better rejection of poor aspect periods, and smaller random errors in the aspect reconstruction; (ii) improved background screening; (iii) improved methods for source detection; (iv) inclusion of a time-variability test for each source; (v) more extensive investigation of the survey sensitivity. Most sources (444 out of 479) have proposed identifications; these include 75 with late-type stars and 6 with hot white-dwarf stars, amongst the 120 new, 'post-BSC' sources. Overall, as with the BSC,

1

S. Bowyer and R. F. Malina (eds.), Astrophysics in the Extreme Ultraviolet, 1–4.
© 1996 *Kluwer Academic Publishers. Printed in the Netherlands.*

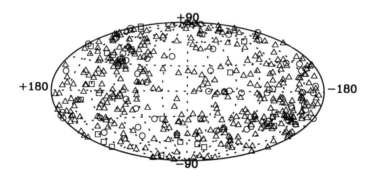

FIGURE 1. Map of the sky in galactic coordinates, showing the locations of the EUV sources in the 2RE Catalogue. The projection is Hammer equal area. Sources detected in both S1 and S2 wavebands are indicated by triangles, while those detected only in S1 or S2 are marked with circles and squares respectively.

the main constituents of the 2RE Catalogue are, in order, late-type stars (F–M), hot white-dwarf stars, and cataclysmic variables (see figure 2).

3. Temporal Variability

Light-curves for each 2RE source were generated and tested for constancy on timescales ~ 1.5 h (the scanning-survey sampling interval). As a result, 31 sources were classified as variable (McGale et al. 1995a,b). Cataclysmic variables (CVs) were found to be the most active group with about 50 percent of them displaying variability, compared with about 10 percent of late-type stars. All the variable G-type stars are known or possible RS CVn-type binaries; ~ 40 percent of the 2RE G-type RS CVn systems are found to vary. Other variable sources are the high-mass X-ray binary Her X-1, the eclipsing binaries Algol (B8 V + K0 IV) and V471 Tau (DA + K2 V) and the active galaxy Mkn 478.

One of the most extreme examples of EUV variability seen to date, and certainly a highlight of the WFC pointed observation programme, is the EUV transient source RE J1255+266 (see Dahlem et al. 1995a; also Dahlem & Kreysing 1994; Dahlem et al. 1995b). It has been identified optically as a binary system consisting of a DA white dwarf with a low-mass companion (Watson et al. 1995).

4. Optical Identification Programme

Mason et al. (1995) have reported the results of the WFC survey optical identification programme. This programme, which concentrated on BSC sources without secure catalogue counterparts, produced identifications in 195 of the 216 EUV source fields observed. The newly identified EUV emitters include 69 new white dwarfs, 114 active stars, 7 new magnetic cataclysmic variables and 5 active galaxies, 4 of which are newly

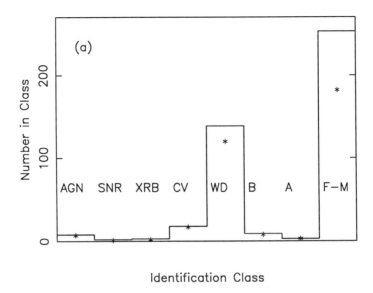

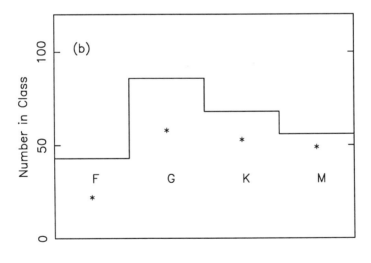

FIGURE 2. (a) Distribution of optical counterparts in the 2RE Catalogue. The star symbols show the same information, but for the BSC (figure 4a of Pounds et al. 1993). (b) Distribution of identifications in the main subgroup of late-type stars.

identified. Several of the white dwarfs identified are in non-interacting binaries, one CV (2RE J0751+144) uniquely appears to share characteristics of the both the Polars and the DQ Her-type CVs, and one of the AGN (2RE J1034+393) has the highest EUV/UV flux ratio known.

5. Other Survey Studies

Based on the BSC or 2RE, and on the associated data products (images, light-curves, photon event lists, etc.), there have been various detailed investigations of specific topics and classes of source. Many references to papers on such studies and on individual sources, can be found in Pounds et al. (1993) and Pye et al. (1995). Descriptions of the instrument, its performance and calibration can be found in Barstow & Willingale (1988), Sims et al. (1990), Pye et al. (1991), Wells et al. (1990), Pounds et al. (1993) and Pye et al. (1995), and references therein.

6. On-Line Access to the Survey Archive

The 2RE Catalogue, and other survey data products such as images, are available on-line via Internet, at the Leicester University Database and Archive Service (LEDAS). LEDAS can be accessed by telnet (ledas.star.le.ac.uk), anonymous ftp (ledas-ftp.star.le. ac.uk) and WWW (URL: http://ledas-www.star.le.ac.uk). The WFC survey also has its own WWW page (accessible from the LEDAS home page): http://ledas-www.star.le.ac.uk/rosat_euv.

ROSAT is a collaborative project between Germany, the UK and USA. I am happy to acknowledge the great efforts of many colleagues in the *ROSAT* Project in ensuring the successful operation of the WFC. The UK *ROSAT* Project is funded by the Particle Physics and Astronomy Research Council.

REFERENCES

BARSTOW, M. A. & WILLINGALE, R. 1988, JBIS., 41, 345

BARSTOW, M. A., ET AL. 1993, MNRAS, 264, 16

BOWYER, S., LIEU, R., LAMPTON, M., LEWIS, J., WU, X., DRAKE, J. J., & MALINA, R. F. 1994, ApJS., 93, 569

DAHLEM, M. & KREYSING, H. -C. 1994, IAU Circular No. 6085

DAHLEM, M., ET AL. 1995a, these proceedings

DAHLEM, M., KREYSING, H. -C., WHITE, S., ENGELS, D., CONDON, J. J., & VOGES, W. 1995b, A&A, in press

MCGALE, P. A., PYE, J. P., BARBER, C. R., & PAGE, C. G. 1995a, these proceedings

MCGALE, P. A., PYE, J. P., BARBER, C. R., & PAGE, C. G. 1995b, MNRAS, in press

MASON, K. O., ET AL. 1995, MNRAS, 274, 1194

POUNDS, K. A., ET AL. 1993, MNRAS., 260, 77

PYE, J. P., WATSON, M. G., POUNDS, K. A., & WELLS, A. 1991, in Extreme Ultraviolet Astronomy, ed. R. F. Malina & S. Bowyer, 409, New York: Pergamon

PYE, J. P., ET AL. 1995, MNRAS, 274, 1165

SIMS, M. R. ET AL. 1990, Opt. Eng., 29, 649

WATSON, M. G., MCMAHON, R. G., & PAGE, M. J. 1995, IAU Circular No. 6126

WELLS, A., ET AL. 1990, Proc. SPIE, 1344, 230

Results from the Second *EUVE* Source Catalog

M. LAMPTON, S. BOWYER, J. LEWIS, X. WU, P. JELINSKY, R. LIEU, AND R. F. MALINA

Center for Extreme Ultraviolet Astrophysics, 2150 Kittredge Street,
University of California, Berkeley, CA 94720–5030, USA

We present the results of the Second *EUVE* Source Catalog, including all detections from the *EUVE* all-sky survey, the *EUVE* deep survey, and sources detected during dedicated instrument pointings. Where available, we furnish identifications of these objects and statistics with regard to type of stellar or extragalactic object.

We summarize the results of the Second *Extreme Ultraviolet Explorer* (*EUVE*) Source Catalog (Bowyer et al. 1995). The data include (a) all-sky survey detections from the initial six-month scanner survey phase, (b) additional scanner detections made subsequently during specially programmed observations designed to fill in low-exposure sky areas of the initial survey, (c) sources detected with Deep Survey telescope observations along the ecliptic, (d) objects detected by the scanner telescopes during targeted spectroscopy observations, and (e) other observations. The Second Catalog employs an improved all-sky survey source detection method that offers better detection sensitivity and reliability than achieved in the First Catalog. The Second Catalog lists three classes of source detections separately: the all-sky survey detections, the deep survey detections, and sources detected during other phases of the mission. Each list gives positions and intensities in each waveband. The total number of objects listed is 734, of which 67% have plausible identifications.

TABLE 1.

Objects	Number
White Dwarfs	104
Early-type stars (A, B)	24
Late-type stars (FGKM)	273
Cataclysmic variables	14
Low-mass X-ray binaries	2
Extragalactic objects	37
Other	35
No identification	245
TOTAL	734

The full Second Catalog will appear in *The Astrophysical Journal Supplement Series*, and is available via the World Wide Web at http://www.cea.berkeley.edu.

This work has been supported by NASA Contracts NAS5-30180 and NAS5-29298.

5

S. Bowyer and R. F. Malina (eds.), Astrophysics in the Extreme Ultraviolet, 5–6.
© *1996 Kluwer Academic Publishers. Printed in the Netherlands.*

REFERENCES

BOWYER, S., LAMPTON, M., LEWIS, J., WU, X., JELINSKY, P., LIEU, R., & MALINA, R. F.
1996, The Second EUVE Source Catalog, ApJS, in press

EUV Astrophysics with *ALEXIS*: The Wide View

JEFFREY BLOCH

Los Alamos National Laboratory, Astrophysics and Radiation Measurements Group, Group NIS-2, Mail Stop D436 Los Alamos, NM 87545 USA

The *Array of Low Energy X-ray Imaging Sensors* (*ALEXIS*) satellite is Los Alamos' pathfinding small space mission achieving low cost and rapid development time for its technology demonstration and science goals. The *ALEXIS* satellite contains the *ALEXIS* telescope array, which consists of six EUV/ultrasoft X-ray telescopes utilizing normal incidence multilayer mirrors, microchannel plate detectors, and thin UV rejecting filters. Each telescope is tuned to a relatively narrow bandpass centered at either 130, 171, or 186 angstroms. Each telescope has a 33° field-of-view, and a resolution of $\sim 0.25°$. With each 50 s rotation of the satellite, the telescopes scan most of the anti-solar hemisphere of the sky. The spacecraft is controlled exclusively from a ground station located at Los Alamos.

This paper discusses the characteristics and performance of the *ALEXIS* telescopes and the results from the mission in spite of the damage incurred to the spacecraft at launch.

1. Introduction

The *ALEXIS* small satellite contains an ultrasoft X-ray or extreme ultraviolet (EUV) monitor experiment that consists of six compact normal-incidence telescopes operating in narrow bands centered on 66, 71, and 93 eV (186, 176, and 130 Å). The satellite also contains a VHF broadband ionospheric survey experiment called BLACKBEARD. Los Alamos National Laboratory (LANL) is the project lead and where the experiments were built and integrated with the spacecraft bus. Sandia National Laboratory (SNL) provided the payload data processors and detector high voltage supplies, and the University of California Space Sciences Laboratory (UCB SSL) built and calibrated the detectors. The spacecraft bus, built by AeroAstro, Inc., is a custom, low-cost, miniature satellite, made to be compatible with several expendable launch vehicles. The Air Force Space Test Program provided the launch for *ALEXIS* via a Pegasus air-launched booster into a 400 × 450 nautical mile orbit on April 25, 1993. The satellite and experiments are controlled entirely from a small groundstation at LANL. The project, excluding launch, was funded entirely by the Department of Energy (DOE) Office of Non-proliferation and National Security as part of an advanced technology development program for potential future uses in Comprehensive Test Ban Treaty (CTBT) monitoring.

The *ALEXIS* experiment takes up 100 pounds of the 242 pound total satellite mass, draws 45 watts, and produces an orbit average of 10 kbits s^{-1} of data. Position and time of arrival are recorded for each detected photon. The satellite is spin stabilized, with magnetic torque coils providing attitude control authority. *ALEXIS* is always in a survey-monitor mode, with no individual source pointings. It is well-suited for simultaneous observations with ground-based observers who prefer to observe sources at opposition. Coordinated observations need not be arranged before the fact, because most sources in the anti-Sun hemisphere will be observed and archived. A single ground station in Los Alamos tracks and controls *ALEXIS*. Between ground station passes, experiment and spacecraft data are stored in 78 Megabytes of solid state memory in the spacecraft bus.

S. Bowyer and R. F. Malina (eds.), Astrophysics in the Extreme Ultraviolet, 7–14.
© *1996 Kluwer Academic Publishers. Printed in the Netherlands.*

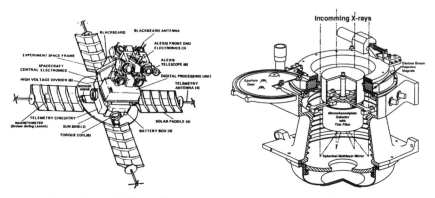

FIGURE 1. Left: The *ALEXIS* satellite in its on-orbit configuration. The damaged solar panel hinge and magnetometer are indicated. Right: Cross sectional view of one of the six *ALEXIS* telescopes.

2. Telescope Design

Normal incidence multilayer X-ray/EUV telescopes have been used for solar observations, but *ALEXIS* is the first time that they have been used successfully on-orbit for non-solar cosmic ultrasoft X-ray/EUV measurements. Figure 1 shows a cross sectional view of an *ALEXIS* telescope. It is an extremely simple $f/1$ optical design, consisting of an annular entrance aperture, a spherical mirror, an optical and UV rejecting filter, and a curved, microchannel plate detector with wedge and strip anode readout. The curved front of the microchannel plate follows the curvature of the focal surface so that the spatial resolution of the system is approximately constant over the entire 33° field-of-view. Spherical aberration limits the system's spatial resolution to about 0.25°.

As shown in Figure 1, the six EUV telescopes are arranged in three co-aligned pairs and cover three overlapping 33° fields-of-view. During each 50 s rotation of the satellite, *ALEXIS* scans most of the anti-solar hemisphere (see Figure 2). The geometric collecting area of each telescope is about 25 cm^2. Analysis of the preflight X-ray throughput calibration data indicates that the peak on-axis effective collecting area for each telescope's response function ranges from 0.25 to 0.05 cm^2. The peak area-solid angle product response function of each telescope ranges from 0.04 to 0.015 cm^2-sr. In one twelve hour data collection period, the brightest EUV sources in the sky can be detected (see Figure 3).

The spacing of the molybdenum and silicon layers on each telescope's mirror is the primary determinant of the telescope's photon energy response function. The *ALEXIS* multilayer mirrors also employ a "wavetrap" feature to significantly reduce the mirror's reflectance for He II 304 angstrom geocoronal radiation which can be a significant background source for space borne EUV telescopes. These mirrors, produced by Ovonyx, Inc., are highly curved yet have been shown to have very uniform multilayer coatings and hence have very uniform EUV reflecting properties over their entire surfaces. Our efforts in designing, producing and calibrating the *ALEXIS* telescope mirrors have been previously described in Smith et al. (1990), and Smith et al. (1989).

The left portion of Figure 3 represents an estimate of each telescope's on-axis effective area vs. input photon energy. These curves are currently being refined based on actual count rates observed from different EUV bright white dwarfs whose spectra have been measured with the *EUVE* spectrometers.

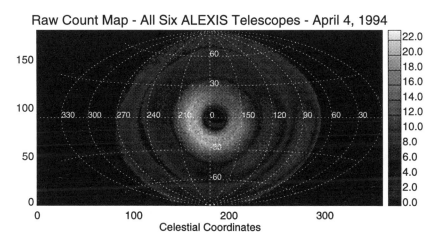

FIGURE 2. Raw count sky map (0.25 degree pixels) summing the data from all six *ALEXIS* telescopes demonstrating the wide area coverage that *ALEXIS* affords. The apparent variations in signal are entirely due to exposure and vignetting effects. The scanning motion of each telescope pair produced each of the annular ring regions on the map.

TABLE 1. Filter and photocathode makeup for each of the six *ALEXIS* telescopes.

Telescope	Look Direction Offset to Spin Axis (°)	Multilayer Mirror Bandpass	Filter	Photocathode
1A	87.5	93 eV / 130 Å	Lexan/boron	MgF_2
1B	87.5	71 eV / 172 Å	Al/Si/C	NaBr
2A	56	93 eV / 130 Å	Lexan/boron	MgF_2
2B	56	66 eV / 186 Å	Al/Si/C	NaBr
3A	31.5	71 eV / 172 Å	Al/Si/C	NaBr
3B	31.5	66 eV / 186 Å	Al/Si/C	NaBr

Table 2 describes the makeup of each of the telescopes' filters and detector photocathodes. Mechanically, each of the telescopes are identical.

3. Launch, Loss, and Scientific Recovery

ALEXIS was launched by a Pegasus booster on 1993 April 25 into a 844×749 km orbit with an inclination of 70°. The Pegasus dropped from the wing of a B-52 bomber at 13:56 UT. Initial reports from the launch site indicated a perfect, nominal launch. Initial attempts to contact the *ALEXIS* satellite after the launch were unsuccessful. Video taken

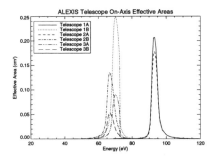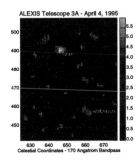

FIGURE 3. Left: Preflight on-axis effective area estimates for each *ALEXIS* telescope. Right: Detection significance map constructed from a 12 hour dataset from Telescope 3A (176 angstroms) on April 4, 1995. The scale is minus the base 10 log of the probability that the number of counts in a given source region is a fluctuation above the surrounding background. The bright source HZ 43 is unambiguously detected and a second white dwarf (WD1254+223) to the south is marginally detected. The horizontal grid lines are 10 degrees apart in Declination.

from the second stage of the Pegasus booster showed the *ALEXIS* +Y solar paddle to have prematurely unstowed.

During the weeks after launch, data gathered from external assets indicated that all four solar paddles had deployed. This implied that *ALEXIS* was not dead on arrival in orbit, but turned on and stayed on long enough to deploy the three undamaged solar paddles. After a 15 s transmission on June 2, on June 30, 1993 *ALEXIS* transmitted a strong signal and stayed in contact for 4 minutes. The telemetry data showed that *ALEXIS* was spinning about an axis nearly 90° from the Sun. All systems appeared functional except the magnetometer, which had failed. By July 5, 1993, commands had been successfully sent to *ALEXIS* to conserve power usage and the batteries became fully charged. Once regular contact with *ALEXIS* had been established, we devised a manual method of controlling the magnetic torque coils that did not depend on the broken magnetometer to control the orientation of the satellite spin axis (see Bloch et al. (1994a) for more details). By the end of July 1993, the solar panels were finally facing the Sun. Telescope doors were opened, and scientific operations began with the telescopes.

For several months that followed, telescope data were collected blindly in the hope that we could eventually devise an attitude solution algorithm that took into account the modified mass properties of the spinning satellite, the missing magnetometer, and the possible motion of the broken solar paddle. The first significant success with a new algorithm (a Kalman filter/backsmoother dynamics estimator) was achieved in April of 1994, when it was used to produce an image of the Moon and the EUV bright white dwarf HZ43 (Bloch et al. (1994b)). This first version appeared to have pointing accuracy of about one degree, still short of the desired 0.25° original goal. By November of 1994, a new revision of the attitude solution software was producing solutions with accuracies less than 0.5°, at which time three months of archival data were co-added to show the detection of several bright EUV sources, as well as the Cataclysmic Variable VW Hyi in super outburst from a dataset collected in late May and early June of 1994. These and subsequent source detections are detailed in Roussel-Dupré et al. (1995a). A further refinement of the attitude solution software became available in February of 1995 which allowed for solutions to be generated during spacecraft maneuvers which had not been possible previously. Since we have to perform maneuvers to correct for the drift away from the Sun of the spacecraft spin axis every 2−5 days (which would last up to 8 hours), we can now recover a significant additional portion of the data for analysis.

4. Science Goals

As an astronomical instrument, *ALEXIS* is like a set of wide-angle fish-eye lens cameras with narrow bandpass filters as opposed to observatory-class telescopes. With its wide fields-of-view and well-defined wavelength bands, it complements the scanners on *EUVE* and the *Rosat* WFC, which are sensitive, narrow field-of-view, broad-band survey experiments. The 66 and 71 eV *ALEXIS* bandpasses are tuned to the Fe IX–XII emission line complex, characteristic of million degree optically thin plasmas which exist in the coronae of stars and which are thought to fill a large fraction of interstellar space around the sun, creating the soft X-ray background. While the maximum effective areas are small compared to *EUVE* or WFC (which have peak effective areas of several cm^2), each *ALEXIS* telescope has a significant area-solid angle product, which is the true figure of merit for sky survey/monitor experiments. As a result of the telescopes' fast optics, *ALEXIS* can also excel in diffuse background studies, where large area-solid angle products are needed. Now that we know about the instrument's on-orbit performance, we can comment in detail on the status of the preflight science goals.

4.1. *The Diffuse EUV Background*

Measurements of the diffuse EUV cosmic background on degree angular scales would produce valuable information about the structure of the million degree gas that produces the soft X-ray background. Comparisons of *ALEXIS* data with the *Rosat* 0.25 keV sky maps would help set limits on low neutral column density structures that may exist in the local cavity, due to the large difference in photon absorption cross sections between the *Rosat* soft X-ray and *ALEXIS* bandpasses. The absolute fluxes observed in the *ALEXIS* 176 and 186 angstrom bandpasses could help set limits on the Fe abundance in the hot phase of the local interstellar medium (Bloch et al. (1991)).

We have only recently undertaken the first efforts at understanding the diffuse EUV cosmic flux in the *ALEXIS* data (Smith et al. (1995)). The analysis is very dependent on understanding all of the non-cosmic background sources. Despite great efforts in the design of the telescopes to reduce or eliminate a variety of on-orbit backgrounds, several unwanted diffuse signals are apparent in the *ALEXIS* data. These must be carefully cataloged and modeled before any diffuse maps with believable structure can be produced. In particular, an intense anomalous background is seen in the data that appears to be correlated (when it is present), with the spacecraft velocity direction vector ((Roussel-Dupré et al. 1995d),Bloch et al. (1994b)). This background may be similar to a background seen with the *Rosat* WFC (West et al. (1994)). There are other background trends that are related to high energy particle populations at different orbital positions, as well as geocoronal contributions. In spite of these non-cosmic background issues, *ALEXIS* can already place interesting upper limits on the flux from Fe lines in the million degree gas that is supposed to produce the soft X-ray background (Smith et al. (1995)) by looking at the difference between Earth and sky pointing counting rates.

4.2. *Narrow Band Source Survey*

Over 18 steady EUV sources have been detected in the first attempt at analyzing a year's worth of *ALEXIS* archival data (Roussel-Dupré et al. (1995a)) These sources, mostly white dwarfs, have been cataloged already by WFC and *EUVE* and for most have had detailed EUV spectra obtained with the *EUVE* spectrometers. In this regard, *ALEXIS* can add little to what is known about these sources, save that they provide convenient on-orbit calibration lamps to check for on-orbit instrument sensitivity changes. In addition, the *ALEXIS* telescopes and individual components (detectors, filters, and mirrors) underwent extensive calibrations prior to flight, so that when the final versions of *ALEXIS* preflight responses are available, an independent absolute calibration check can be performed with other EUV instruments.

4.3. *Transient Phenomena*

Bright transient EUV source detections have turned out to be the most significant science topic that *ALEXIS* can immediately address in the first months after the scientific mission recovery. The interest in this area of study intensified after the serendipitous discovery by the WFC of an extremely bright EUV transient that brightened by more than a factor of 4000 from its quiescent state (Dahlem et al. (1995)).

Several EUV outbursts from cataclysmic variables (CVs) have been observed with *ALEXIS*, namely VW Hyi, U Gem, and AR Ursae Majoris (=1ES1113+432) (Remillard et al. 1994). In addition, *ALEXIS* has detected at least two bright EUV transients lasting 24 hrs or less with (at the time of this writing), no obvious counterparts (Roussel-Dupré et al. 1995b), (Roussel-Dupré et al. 1995c). An *EUVE* target of opportunity observation was performed for each of these transients, but preliminary indications are that the sources had faded completely away by the time *EUVE* was pointed at the source location.

Before launch, it was hoped that *ALEXIS* would be able to gather statistics on the frequency of flare star outbursts using the 130 angstrom bandpass telescopes that include the intense Fe XX–XXIII lines, as seen in *EUVE* observations of AU Mic (Cully et al. (1994)). It is not clear as of this writing how much this goal can be achieved due to the fact that *ALEXIS'* actual observing efficiency is somewhat less than predicted preflight.

Since January 1995, we have been operating an automated software procedure to notify the *ALEXIS* team by e-mail if there are any sources or transients in the last 12, 24, or 48 hrs of telescope data. This procedure is usually completed within 2 hrs of the time that the data is downloaded to Los Alamos groundstation from the satellite.

4.4. *Gamma Ray Bursts*

Current theories to explain the isotropic distribution of Gamma Ray Bursts (GRBs) over the sky as seen with the Compton Gamma Ray Observatory (CGRO) now tend to put their source locations at cosmological distances. Because *ALEXIS* scans over such a wide area of the sky with every 50 s rotation of the satellite, it is natural to look for possible EUV counterparts to GRBs. If such a counterpart emission is observed, the source would have to be within $\approx$ 100 pc of the earth due to the opacity of the ISM at EUV wavelengths. Such an observation would be extremely significant to the study of GRBs.

We have begun to search the archival *ALEXIS* data for good datasets to use for this GRB study. For the 39 GRBs examined thus far, 10 were not in any *ALEXIS* telescope scan area, 12 occurred behind the earth from *ALEXIS'* vantage point, and 3 occurred during satellite shutdown periods. The remaining 14 bursts are good candidates to search for pre- or post-event EUV emission, and 4 of these events are good candidates for near-simultaneous observations, i.e., an *ALEXIS* telescope may have been scanning over the GRB error box at the exact moment that the GRB occurred.

4.5. *Lunar EUV Observations*

ALEXIS has collected a significant number of Lunar EUV flux measurements in the last two years. In fact, back in November of 1993, before we had developed any real spacecraft attitude solution capability, the Moon was the first celestial object identified in the *ALEXIS* data using the first rough analysis tools. These observations are useful for Solar EUV variability measurements, as well as Lunar surface composition studies (Edwards et al. (1995)).

5. Current Status

As of this writing (May, 1995), the *ALEXIS* satellite and payloads continue to function nominally. The difficulties caused by the launch damage have been mostly overcome through new procedures and software. The telescopes appear not to have lost any significant sensitivity over the two years that they have spent on orbit (Roussel-Dupré et al. 1995a).

ALEXIS would not have been possible without the dedication of a great many people over the last six years. We would like to particularly thank those who lent their hearts and minds to the launch and rescue, and those that continue to work tirelessly on flight operations and data analysis tasks. Past and present, these include Ron Aguilar, Frank Ameduri, Tom Armstrong, Mark Bibeault, Doug Ciskowski, Don Cobb, Jim Devenport, Bryan Dunne, Brad Edwards, Don Enemark, John Gustafson, Amy Hodapp, Mark Hodgson, Dan Holden, Irma Gonzales, Dave Guenther, Meg Kennison, Phil Klingner, Cindy Little, Lisa May, Carter Munson, Greg Nunz, Greg Obbink, Tim Pfafman, Mick Piotrowski, Bill Priedhorsky, Keri Ramsey, Diane Roussel-Dupré, Sean Ryan, Ernie Serna, April Smith, Barry Smith, Steve Smoogen, Steve Stem, Ralph Stiglich, and Steve Wallin of Los Alamos; Bob Dill, Frank McLoughlin, Robert Miller, Mark Psiaki, Richard Warner, and Chris Wright of AeroAstro; David Bullington, Jim Griffee, Jim Klarkowski and Harvey Temple of Sandia National Laboratories; David Hastman and Jean Floyd of Orbital Sciences, Inc.; Lt. Frank Dement and Capt. Kurt Hall of the United States Air Force, and the staff of the Vandenberg Tracking Site. The *ALEXIS* detectors were built and calibrated at the University of California-Berkeley Space Sciences Laboratory by Oswald H. Siegmund, Scott Cully, and John Warren. This work was supported by the US Department of Energy.

REFERENCES

BLOCH, J. J., PRIEDHORSKY, W. C., ROUSSEL-DUPRÉ, D., EDWARDS, B. C., & SMITH, B. W. 1990, Design, Performance, and Calibration of the ALEXIS Ultrasoft X-Ray Telescopes SPIE Proc., EUV, X-Ray, and Gamma-ray Instrumentation for Astronomy, 154, 1344

BLOCH, J. J., & SMITH, B. W. 1991, The ALEXIS Project and the Local Interstellar Medium, in Extreme Ultraviolet Astronomy, ed. R. Malina & S. Bowyer, New York: Pergamon, 1991

BLOCH, J. J., ARMSTRONG, W. T., DINGLER, R., ENEMARK, D., HOLDEN, D., LITTLE, C., MUNSON, C., PRIEDHORSKY, W. C., ROUSSEL-DUPRÉ, D., SMITH, B. W., WARNER, R. W., DILL, B., HUFFMAN, G., McLOUGHLIN, F., MILLS, R., & MILLER, R. 1994, The ALEXIS Mission Recovery, American Astronautical Proceedings, Advances in the Astronautical Sciences, 86, 505

BLOCH, J., EDWARDS, B., PRIEDHORSKY, W., ROUSSEL-DUPRÉ, D., SMITH, B. W., SIEGMUND, O. W. H., CARONE, T., CULLY, S., RODRIGUEZ-BELL, T., WARREN, J., & VALLERGA, J. 1994, On Orbit Performance of the ALEXIS EUV Telescopes, SPIE Proc., EUV, X-ray, and Gamma-Ray Instrumentation for Astronomy V, 2280, 297

CULLY, S. L., FISHER, G. H., ABBOTT, M. J., & SIEGMUND, O. H. W. 1994, A Coronal Mass Ejection Model for The 1992 July 15 Flare on AU Microscopii Observed by the Extreme Ultraviolet Explorer, ApJ, 435, 449

DAHLEM, M., KRYSING, H. -C., WHITE, S., ENGELS, D., CONDON, J. J., & VOGES, W. 1995, RE J1255+266: Detection of an extremely bright EUV transient, A&A, 295L, 13

EDWARDS, B. C., BLOCH, J. J., DUPRÉ, D., PFAFMAN, T. E., & RYAN, S. 1995, ALEXIS Lunar Observations, these proceedings

LAMPTON, M. 1994, Two Sample Discrimination of Poisson Means, ApJ, 436, 784

PRIEDHORSKY, W. C., BLOCH, J. J., SMITH, B. W., STROBEL, K., ULIBARRI, M., CHAVEZ, J., EVANS, E., SIEGMUND, O. W. H., MARSHALL, H., VALLERGA, J., & VEDDER, P. 1988, ALEXIS: An Ultrasoft X-Ray Monitor Experiment Using Miniature Satellite Technology, in X-Ray Instrumentation in Astronomy II, 982

PRIEDHORSKY, W. C., BLOCH, J. J., CORDOVA, F. A., SMITH, B. W., ULIBARRI, M., SIEGMUND, O. W. H., MARSHALL, H., VALLERGA, J., & VEDDER, P. 1991, ALEXIS: A Narrow Band Survey/Monitor of the Ultrasoft X-ray Sky, in Extreme Ultraviolet Astronomy, ed. R.F. Malina & S. Bowyer, New York: Pergammon Press, 464

PRIEDHORSKY, W. C., BLOCH, J. J., HOLDEN, D. H., ROUSSEL-DUPRÉ, D. C., SMITH, B. W., DINGLER, R., WARNER, R., HUFFMAN, G., MILLER, R., DILL, B., & FLEETER, R. 1993, The ALEXIS Small Satellite Project: Initial Flight Results, SPIE Proc., EUV, X-ray, and Gamma-Ray Instrumentation for Astronomy, 2006, 114

REMILLARD, R. A., SCHACHTER, J. F., SILBER, A. D. & SLANE, P. 1994, 1ES 1113+432: Luminous, Soft X-ray Outburst from a Nearby Cataclysmic Variable AR Ursae Majoris, ApJ, 426, 288

ROUSSEL-DUPRÉ, D., BLOCH, J. J., CISKOWSKI, D., DINGLER, R., LITTLE, C., KENNISON, M., PRIEDHORSKY, W. C., & RYAN, S. 1994, On-Orbit Science in a Small Package: Managing the ALEXIS Satellite and Experiments, SPIE Proc., Advanced Microdevices and Space Science Sensors, 2267

ROUSSEL-DUPRÉ, D., BLOCH, J. J., RYAN, S., EDWARDS, B. C., PFAFMAN, T., RAMSEY, K., & STEM, S. 1995, The ALEXIS Point Source Detection Effort, these proceedings

ROUSSEL-DUPRÉ, D., BLOCH, J. J., EDWARDS, B. C., PFAFMAN, T. E., PRIEDHORSKY, W. C., RYAN, S., SMITH, B. W., SIEGMUND, O. H. W., CULLY, S., RODRIGUEZ-BELL, T., VALLERGA, J., & WARREN, J. 1995, ALEXIS J1139, IAU Circ., #6152

ROUSSEL-DUPRÉ, D., & THE ALEXIS TEAM 1995, ALEXIS J1644, IAU Circ., #6170

ROUSSEL-DUPRÉ, D., & BLOCH, J. J. 1995, in The Proc. of TAOS X, Workshop on the Earth's Trapped Particle Environment, in press

SMITH, B. W., BLOCH, J. J., & ROUSSEL-DUPRÉ, D. 1989, Metal multilayer mirrors for EUV/ultrasoft X-ray wide-field telescopes, SPIE Proc., X-ray/EUV Optics for Astronomy and Microscopy, ed. R.B. Hoover, 1160

SMITH, B. W., BLOCH, J. J., & ROUSSEL-DUPRÉ, D. 1990, Metal multilayer mirrors for EUV/ultrasoft X-ray wide-field telescopes, Opt. Eng., 29, 6, 592

SMITH, B. W., PFAFMAN, T. E., BLOCH, J. J., & EDWARDS, B. C. 1995, ALEXIS Observations of the Diffuse Cosmic Background in the Extreme Ultraviolet, these proceedings

WEST, R. G., SIMS, M. R., & WILLINGALE, R. 1994, Evidence for a far-ultraviolet spacecraft glow in the ROSAT Wide Field Camera, Planet. Space. Sci., 42(1), 71

Temporal Behaviour of Sources in the *ROSAT* Extreme-Ultraviolet All-Sky Survey

PAUL McGALE, J. P. PYE, C. R. BARBER, AND C. G. PAGE

X-Ray Astronomy Group, Department of Physics and Astronomy,
University of Leicester, Leicester LE1 7RH

From a total catalogue of 479 2RE sources, 31 have been found to be variable. All the variable sources are optically identified, with a breakdown by source type as follows: cataclysmic variables 9, late-type (F–M) stars 18, the high-mass X-ray binary Her X-1, the eclipsing binaries Algol and V471 Tau, and the active galaxy Mkn 478. The most highly variable objects in the EUV band are cataclysmic variables. The survey was sensitive to timescales from ~ 1.5 h to ~ 5 d, and observed variability ranges from flare-like events lasting < 1.5 h to irregular and periodic flux changes over ~ 0.5–2 d. With the exception of the cataclysmic variables, and possibly a few late-type dK-dMe stars with large flares, the observed variability levels should not substantially affect EUV luminosity functions

1. Introduction

The Wide Field Camera (WFC) on the *ROSAT* satellite performed the first all-sky survey at extreme-ultraviolet (EUV) wavelengths over a continuous six-month period starting in 1990 July. The survey was conducted in two wavebands, S1 and S2, covering the ranges 60–140 Å and 110–200 Å respectively. Detailed descriptions of the instrument, survey, and analysis procedures can be found in Sims et al. (1990), Pounds et al. (1993) and Pye et al. (1995).

From an initial analysis of the data a bright source catalogue (BSC), containing 383 sources, was published by Pounds et al. (1993). Recently, an updated and expanded version, the '2RE' Catalogue, has been produced (Pye et al. 1995), taking advantage of better knowledge of the survey data base and improved analysis procedures. The 2RE Catalogue contains 479 EUV sources. Active, late-type (F–M) stars (LTSs) and hot, white-dwarf stars (WDs) account for most of the optical identifications, 251 and 140 respectively. Other identifications are with cataclysmic variables (CVs), active galactic nuclei (AGNs), B-type stars, X-ray binaries (XRBs) and supernova remnants (SNRs) contributing 18, 8, 10, 3, and 2 counterparts each. A total of 34 sources has, so far, remained unclassified.

During the data processing associated with the 2RE Catalogue, light-curves were produced for all sources. The information from these light-curves has been summarised in the Catalogue in the form of a "time-variability flag." Of the 31 2RE sources reported as variable, all have probable optical identifications, with a breakdown as follows: LTSs 20, CVs 9, high-mass X-ray binaries (HMXB) 1 (Her X-1), AGNs 1 (Mkn 478). Hence, 50 percent of the CVs and ~ 10 percent of the LTSs in 2RE are classed as variable.

S. Bowyer and R. F. Malina (eds.), Astrophysics in the Extreme Ultraviolet, 15–19.

2. Variability Analysis

Each 2RE light-curve was 'χ^2-tested' against the hypothesis of a constant source at the measured mean count rate

$$\chi^2 = \sum_{i=1}^{n}(C_i - \bar{C})^2 / \sigma_i^2.$$

Here, C_i is the value of the ith time bin ($C_i \geq 0$), in counts per second, and σ_i^2 its variance. The mean of the C_i's is denoted by $\bar{C}$. This yielded a probability P($> \chi^2$).

A source was only flagged as variable if P < 0.001 in one or both wavebands. Also, to reduce further potential spurious variability due to small errors in correcting the raw counts to count rates, the χ^2 per degree of freedom (χ_ν^2) was required to be > 2.0. Thus 31 2RE sources have been identified as variable by Pye et al. (1995). Simulations show that we should expect no more than ~ 1 source in the 2RE Catalogue to be falsely flagged as 'variable'.

For those 2RE sources which passed the variability criteria, two other statistics were computed: (i) the ratio of maximum to mean count rate, as a measure of flare strength; (ii) a 'normalised', rms variability (ϵ_{rms} e.g., Pallavicini et al. 1990), to measure the overall variability level.

χ_ν^2 is plotted in Fig. 1. Fig. 2 summarises the EUV variability as a function of source class. For a full discussion on the methods of light–curve production and the test statistics applied see McGale et al. (1995).

3. Discussion

As can be seen from Fig. 1, the ability to detect variability in the 2RE sources is clearly flux limited, though this manifests itself in several different ways. For example, a source that is very weak, even undetectable, most of the time, may suddenly flare briefly, for say 1 or 2 *ROSAT* orbital scans, e.g., 2RE J2047−363 (= HD 197890 = SAO 212437, Matthews et al. 1994). On the other hand, less dramatic variability can be seen in sources with higher 'quiescent' flux levels, e.g. 2RE J0308+405 (= Algol). In terms of timescales, the survey is sensitive to variability in the range ~ 1.5 h to ~ 5 d (with some drop in detectability for timescales near 1 d — the nominal interchange frequency of the EUV filters.) In principle, the WFC is also sensitive to timescales from ~ 40 ms (set by the on-board storage of the individual detected events) to ~ 80 s (corresponding to one scan of a source through the centre of the detector). However in practice, the source needs to be either very strong or periodic to utilise this information, see e.g. the analysis of Her X-1 (= 2RE J1657+352) by Rochester et al. (1994). Owens et al. (1993) have searched the WFC all-sky survey data for rapid transients (< 50 s duration) excluding sky locations near known BSC sources; there were no detections.

3.1. *Cataclysmic Variables*

A striking feature of the results (Figs. 1 and 2) is how clearly the CVs stand out from the other source classes, both in the fraction of sources that are seen to be variable and in the level of their variability. Much follow-up work has already been undertaken on the CVs newly-discovered in the WFC all-sky survey. Watson (1993), has summarised the properties of the new EUV-selected sample. 13 out of the total of 18 2RE CVs are AM Her-type systems (ie. 'polars'); 8 out of 9 of the variable 2RE CVs are AM Her systems. The other variable CV is the dwarf nova SS Cyg, one of two such systems in the 2RE Catalogue.

S1 Waveband   S2 Waveband

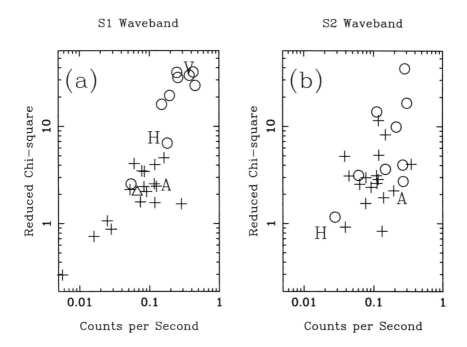

FIGURE 1. Light-curve variability test: χ^2 per degree of freedom (χ^2_ν) versus source mean count rate $\bar{C}$ for reported variable sources only. The symbols $+$, $\bigcirc$, $\triangle$ represent, in turn, F–M stars, CVs and AGNs. Her X–1 is denoted by H, Algol by A and V471 Tau by V. For the S2 waveband V471 Tau is out of scale bounds, at (1.16, 288.0). (a) S1 band, (b) S2 band.

3.2. Late-Type Stars

More than half of the 2RE Catalogue is made up from coronally active stars of spectral types F–M. Of the 18 (or 19 if we include Algol) classified here as EUV-variable, most are also known to be highly variable at optical and/or X-ray wavelengths. There are examples of dMe flare stars, single rapidly rotating dK stars, and RS CVn binary systems. Although it is difficult to make definitive statements due to the limited numbers of objects, the fraction of F-type EUV-variable stars (0/44) appears to be rather lower than for G–M stars (see Fig. 2). Fleming et al. (1995) have found a similar result in the long-term ($\sim$ 10 years) behaviour of an X-ray-selected (*Einstein* Extended Medium Sensitivity Survey) stellar sample. All the G-type EUV-variable stars (9) are either known (7), or possible (2, 2RE J0458+002 = BD+00 908, Strassmeier et al. (1988); 2RE J0106−225 = SAO 166806†) RS CVn systems. Thus $\sim$ 40 percent of the 2RE G-type RS CVn systems are found to vary. RS CVn systems were also found to be the major identified class of object in the *Ariel V* sky survey of fast-transient X-ray sources (Pye & McHardy 1983).

† We suggest 2RE J0106−225 = SAO 166806 as a possible RS CVn system, and certainly as a very active star, on the following basis. From the G5 V spectral type we estimate a distance of $\approx$ 33 pc, and hence, from the mean S1-band count rate of 0.078 count s^{-1}, an EUV luminosity $L_{\rm EUV}$ (erg s^{-1}) of log $L_{\rm EUV} \sim$ 29.8 (see Hodgkin & Pye 1994, Fig. 2a). Such a high value strongly suggests that SAO 166806 is an active spectroscopic binary (see Hodgkin & Pye 1994, Figs. 5, 6 and 8).

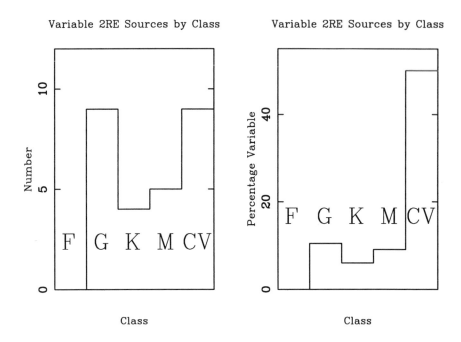

FIGURE 2. Variable 2RE sources by identification class.

Similar variability levels in LTSs have been reported from soft X-ray pointed observations from *Einstein* and *EXOSAT* (see e.g. review by Pallavicini 1993, and references therein).

3.3. *Other Source Types*

The eclipsing binary systems V471 Tau (2RE J0350+171, DA + K2 V, detached) and Algol (2RE J0308+405, B8 V + K2 IV, semi–detached) clearly show periodic behaviour in their EUV light-curves, with eclipses seen in both systems (V471 Tau: Barstow et al. 1992, Algol: M. Barstow, private communication). The WFC survey observations of the HMXB Her X-1 have been reported by Rochester et al. (1994). Gondhalekar et al. (1992) have presented WFC survey observations of AGNs, including Mkn 478.

3.4. *Long-Term Variability*

Two years after the WFC survey, the EUV sky was re-examined by NASA's *Extreme Ultraviolet Explorer* (EUVE, Bowyer et al. 1994). A comparison of the two surveys has been performed by Barber et al. (1995), who find high levels of variability (factors > 10) for CVs, and more modest variability (factors ~ 2) in the LTSs, during the two years between the surveys. Thus, albeit from this rather limited evidence, the 'short-term' and 'long-term' EUV variability appear similar, though the fraction of LTSs exhibiting variability from the EUVE/WFC comparison may be rather higher, at ~ 35 percent, than the ~ 10 percent reported here for the shorter timescales. However, this conclusion is tentative given the added uncertainty of cross-calibration between the two surveys.

The UK *ROSAT* project is funded by the Particle Physics and Astronomy Research Council.

REFERENCES

BARBER, C. R., WARWICK, R. S., McGALE, P. A., PYE, J. P., & BERTRAM, D. 1995, MNRAS, 273, 93

BARSTOW, M. A., SCHMITT, J. H. M. M., CLEMENS, J. C., PYE, J. P., DENBY, M., HARRIS, A. W., & PANKIEWICZ, G. S. 1992, MNRAS, 255, 369

BOWYER, S., LIEU, R., LAMPTON, M., LEWIS, J., WU, X., DRAKE, J. J., & MALINA, R. F. 1994, ApJS, 93

FLEMING, T. A., MOLENDI, S., MACCACARO, T., & WOLTER, A. 1995, ApJS, in press

GONDHALEKAR, P. M., POUNDS, K. A., SEMBAY, S., SOKOLOSKI, J., URRY, C. M., MATTHEWS, L., & QUENBY, J. J. 1992, in Physics of the Active Nuclei, W.J. Duschl & S.J. Wagner, Springer–Verlag, 52

HODGKIN, S. T., & PYE, J. P. 1994, MNRAS, 267, 840

MATTHEWS, L., BROMAGE, G. E., KELLET, B. J., SIDHER, S. D., ROCHESTER, G. K., QUENBY, J. J., SUMNER, T. J., O'DONOGHUE, D., & WILLOUGHBY, G. 1994, MNRAS, 266, 757

McGALE, P. A, PYE, J. P., BARBER, C. R., & PAGE, C. G. 1995, MNRAS, in press

OWENS, A., PAGE, C. G., SEMBAY, S., & SCHAEFER, B. E. 1993, MNRAS, 260, L25

PALLAVICINI, R. 1993, in Physics of Solar and Stellar Coronae: G.S. Vaiana Memorial Symposium, ed. J.F. Linsky & S. Serio, Dordrecht: Kluwer, 237

PALLAVICINI, R., TAGLIAFERRI, G., & STELLA, L. 1990, A&A, 228, 403

POUNDS, K. A. ET AL. 1993, MNRAS, 260, 77

PYE, J. P. & McHARDY, I. M. 1983, MNRAS, 205, 875

PYE, J. P. ET AL. 1995, MNRAS, in press

ROCHESTER, G. K., BARNES, J., SIDHER, S., SUMNER, T. J., BEWICK, A., CORRIGAN, R., & QUENBY, J. J. 1994, A&A, 283, 884

SIMS, M. R. ET AL. 1990, Opt. Eng., 29(6), 649

STRASSMEIER, K. G., HALL, D. S., ZEILIK, M., NELSON, E., EKER, Z., & FEKEL, F. C. 1988, A&AS, Ser., 72, 291

WATSON, M. G. 1993, Adv. Space Res., 13, No 12, 12,125

EUV Observation with Normal Incidence Multilayer Telescopes

HIDEYO KUNIEDA,[1] KOUJUN YAMASHITA,[1]
TAKASHI YAMAZAKI,[1] KAZUAKI IKEDA,[1]
KAZUTAMI MISAKI,[1] YOSHIYUKI TAKIZAWA,[2]
MASATO NAKAMURA,[2] ICHIRO YOSHIKAWA,[2]
AND ASAMI YAMAGUCHI[3]

[1] Department of Physics, Nagoya University, Nagoya 464-01

[2] Department of Earth and Planetary Physics, University of Tokyo, Tokyo 113

[3] National Astronomical Observatory, Mitaka 181

The EUV emission from hot interstellar plasmas is observed by normal incidence telescopes on board a sounding rocket. It was performed on January 29, 1995, to observe the sky area around the HZ 43 close to the north galactic pole. The wave bands ($\delta\lambda \sim 10$ Å) are provided at 130 and 170 Å by the multilayer coating on the reflectors of 20 cm in diameter and of 30 cm in focal length. The focal plane images are detected by CsI coated MCP's. The observed flux of HZ 43 is 1.5 counts s^{-1} at 130 Å and 3 counts s^{-1} at 170 Å. The diffuse emission is 27 c/s/deg^2 at 130 Å and 20 c/s/deg^2 at 170 Å. Those preliminary numbers are subjects to change along with the data analysis.

---

1. Introduction

Interstellar field is assumed to be filled with cold gas with dust and thin hot plasmas of about 10^5 to 10^6 K. The latter component has been recognized in soft X-ray observations below 1 keV. The plasmas of 10^6 K is characterized by emission lines from O VII and OVIII at 0.56 and 0.65 keV, respectively, while EUV emission lines are expected from plasmas of lower temperature in 10^5 K range.

Thinner windows were developed to enhance the detection efficiency in longer wavelength range for the proportional counters. However, strong contamination was found due to the He II (304 Å) emission lines from geocorona. The diffuse EUV emission is difficult to separate from such strong geocorona emission, because spectrometer is not available but filtering, which allows only factor of ten reduction of He II lines even around the absorption edges.

A breakthrough comes from the progress of X-ray optics. Multilayer coatings, thin layer pairs of light and heavy elements, constructed on mirror surfaces efficiently reflect X-rays of selected wavelength. The first mission to measure celestial objects is performed with *ALEXIS* (see the paper in the same issue). In order to improve sensitivity, large normal incidence telescopes are prepared for a Japanese rocket mission in two wave bands at 130 and 170 Å to observe the sky region around the brightest EUV source HZ 43. The flight was successful and some complex structure of EUV emission is observed together with a bright point source, HZ 43.

2. Emission Lines from Hot Interstellar Medium

Kato (1976) calculated radiation of a hot thin plasmas from 1 to 250 Å. The lower temperature plasmas than 10^6 K are expected to emit EUV lines. Figure 1 shows the line emission power in various wavelength bands as a function of temperature (quoted

S. Bowyer and R. F. Malina (eds.), Astrophysics in the Extreme Ultraviolet, 21–26.

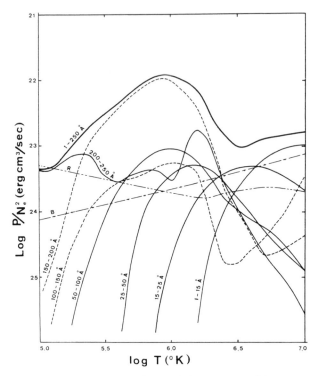

FIGURE 1. Line emission power vs. T.

from Fig. 4 of Kato 1976). The plasmas below 10^6 K are characterized by strong lines in 150–200 Å. Covering 10–100 Å, very soft X-ray detectors only detect the small portion of energy flux from cooler gas. The calculated spectrum of 8×10^5 K is shown in Fig. 2 in wave length range from 100 through 250 Å (Fig. 3 of Kato 1976). The strongest lines are found at around 170 Å, all emerged from ionized iron ions (Fe IX 171.06 Å, Fe X 170.9 Å). Below 140 Å, no lines are seen more than 1/100 of the strongest one at 170 Å.

Our strategy is to provide one pass band at 170 Å to get those strong lines and another band at 130 Å to put a limit of continuum flux without line emission. The lines at 170 Å have strong temperature dependence, which is clear in the Fig. 1.

3. Instrumentation

3.1. Multilayer Normal Incidence Telescopes

Since the two wave bands at 130 and 170 Å are favorable to decide the physical situation of hot plasmas at the temperature of $10^5 - 10^6$ K, 2d of multilayer coatings are tuned to be 140 Å and 184 Å with 20 and 25 layer pairs of Mo/Si, respectively. The reflectivity measured at a synchrotron facility (UVSOR at the Institute for Molecular Science: Okazaki, Japan) are shown in Fig. 3a and 3b. Peak reflectivity is about 50–60% and the width of pass band is about 10 Å in both cases.

Multilayer structure is coated on a spherical mirror of 20 cm in diameter and 30 cm in focal length. Polishing and coatings were done by Nikon. It is placed at the bottom

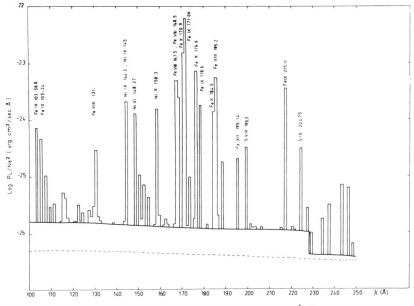

FIGURE 2. Spectrum of plasma of 8×10^5 K.

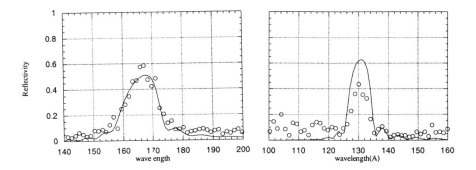

FIGURE 3. (*a*) Mirror reflectivity (NIT-1). (*b*) Mirror reflectivity (NIT-2).

of a housing, whose cross section is shown in Fig. 4. A micro-channel plate of 28 mm in diameter (corresponding field of view of 4 degree) is put on the focal plane. Images are obtained by a registive plate after the two stage MCP. The CsI coating on the MCP yields detection efficiency of about 20% at 130–170 Å. In order to reject longer wave light C filter of 1200 Å is placed in front of MCP. The inflow of the ambient plasmas is rejected by a mesh at the entrance window. The normal incidence telescope for 170 Å is named NIT-1 and that for 130 Å is named NIT-2.

3.2. *Helium Line Monitor and UV Star Sensor*

Another multilayer telescope(HEM) is equipped to monitor He II lines(304 Å) from geocorona. One more telescope (UVT) with mono-layer coating is prepared as a attitude

Normal Incidence Telescope

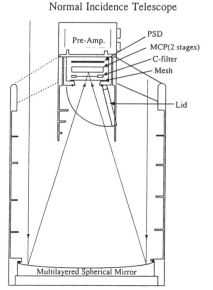

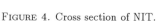

FIGURE 4. Cross section of NIT.   FIGURE 5. Picture of Hole Payload.

sensor with stars in UV band (1600–2800 Å). Four photomultiplyers are placed on the focal plane with rotating slits of three different patterns.

NIT-1 is placed at the top of the structure, while NIT-2, UVT and HEM are flooded inside the structure, which are to be deployed after the nose fairing is removed in the sky (Fig. 5). All these instruments are looking at the top direction of the structure.

4. Flight Operation

The telescopes were launched by a rocket S-520-19 at 16:00 (UT) on 1995 January 28 from Kagoshima Space Center of the Institute of Space and Astronautical Science. 40 s after the launch, nose fairing was blown off and the attitude control system, deploy of instruments, and high voltage operations were taken place in sequence. At X + 100 sec in the altitude of more than 150 km, all instruments were ready to observe the sky area around HZ 43. During 300 s pointing observation, payload achieved at the apogee of 350 km. Then the system was rotated 360 degree around the pointing direction for 50 sec. Telemetry data was successfully obtained by the end of our observation, with some short interruptions.

5. Results

5.1. *Light Curve*

Total counting rate of whole detector is examined for each detector after the start of operation at 100 s. It is almost constant at the level of 300 counts s^{-1} and 500 counts s^{-1} for NIT-1 and NIT-2, respectively, except for the three occasions at 130 s, 180 sec, and 400 s, when the telemetry was lost. Those counting rates could have altitude dependence, if there were some geophysical contaminations due to the air glow and the

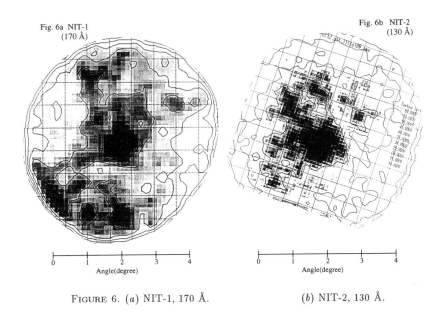

Fig. 6a NIT-1
(170 Å)

Fig. 6b NIT-2
(130 Å)

FIGURE 6. (a) NIT-1, 170 Å.   (b) NIT-2, 130 Å.

radiation belts. Electrons trapped by geomagnetic field should be sensitive to the angle between the line of sight and the magnetic field. We are able to put an upper limit of about 10% for the contamination from geophysical origins.

5.2. Images

Fig. 6a shows the NIT-1 image of 170 Å band integrated from X + 200 through X+ 350 s without attitude correction. The bright source at the center is the image of HZ 43. The image size of a point source is estimated to be 0.3 degree or so in this case. Other structure found right top, right bottom and a sort of "peninsula" to the left could be real feature.

Fig. 6b is the NIT-2 image of 130 Å band in the same period of time. The bright source is not so much out standing as in Fig. 6a, but there. Other structures seen in 170 Å image are also found.

6. Discussion

6.1. Detection Efficiency

The effective area and detection efficiency are roughly estimated as follows.

Eff. Area	Ref.	$d\lambda$	Detector	Filter	Mesh	
(300 cm^2) ×	(50%) ×	(5 Å) ×	(20%) ×	(30%) ×	(90%)	= 41 cm^2 Å (130 Å)
(300 cm^2) ×	(50%) ×	(10 Å) ×	(20%) ×	(20%) ×	(90%)	= 54 cm^2 Å (170 Å)

6.2. *Intensity of HZ 43*

The flux of HZ 43 in both bands is estimated to be 3 and 1.5 counts s^{-1} from the central enhancement from the diffuse emission. If one use the observe intensity of HZ 43 with EUVE, estimated counting rate is 7.6 counts s^{-1} and 3.5 counts s^{-1} for 170 and 130 Å, respectively. Though the absolute value is a few times lower, the flux ratio is consistent with the reported value.

6.3. *Diffuse EUV Emission*

The observed diffuse emission averaged over the field of view is 20 and 27 c/s/deg^2 for 170 and 130 Å. Using the effective area of 5.4 cm^2 and 8.2 cm^2 for a line, the flux of Fe IX at 170 Å is 3.7 c/s/cm^2/deg^2 and 3.3 c/s/cm^2/deg^2 for 130 Å. If the normal abundance is assumed, the temerature of gas is suggested to be a few time 10^5 based on the calculation in Fig. 1. However, it is necessary to evaluate the contamination to those wave bands very carefully.

7. Conclusion and Summary

This is the first observation of EUV diffuse emission with enough S/N ratio and with good spatial resolution of quarter degree, which were enabled by the multilayer normal incidence telescopes. It confirms the possibilities of multilayer telescopes in astrophysical observations. The data analysis is still going on and the results presented here are subjects to change.

REFERENCES

KATO, T. 1976, Radiation from Hot Thin Plasmas from 1 to 250 Å, ApJS, 30, 397

Active Galactic Nuclei in the Extreme Ultraviolet

ARIEH KÖNIGL

Department of Astronomy & Astrophysics, University of Chicago,
5640 S. Ellis Ave., Chicago, IL 60637, USA

EUV observations could provide valuable and unique information about the nature of active galactic nuclei. I discuss their potential usefulness and review some of the main results obtained so far with the *ROSAT* WFC and the *EUVE* instruments. About three dozen sources, mostly Seyfert galaxies and BL Lacertae objects, have already been identified, and spectroscopic measurements of several of them have yielded a number of surprises. However, more observations are clearly needed to confirm some of the reported spectral features and to validate their interpretation.

1. Introduction

The extreme ultraviolet (EUV) wavelength interval (commonly defined to constitute the range $100 - 912$ Å, or $13.6 - 124$ eV) lies between the intensively studied FUV ($912 - 3000$ Å) and X-ray ($1 - 100$ Å) spectral regimes, but until recently it has not been subjected to systematic investigations. This lack of information has been particularly acute in the case of active galactic nuclei (AGNs), where the very ability to carry out successful EUV observations has been seriously questioned. This has stemmed from the fact that, for a Galactic H I column density $N_G(HI) = 1 \times 10^{20}$ cm^{-2}, the interstellar transmission factor is 0.42 at 60 Å, but it is a factor of 10 smaller at 100 Å. Equivalently, each increment of 10^{20} in $N_G(HI)$ contributes an optical depth of 3.2 at 100 Å. However, there are no known regions of the interstellar medium (ISM) that have $[N_G(HI)/10^{20} \text{ cm}^{-2}] < 0.6$ (resp., 0.8) for positive (resp., negative) latitudes (e.g., Heiles 1991).

It has thus been gratifying to find out during the last few years that the "unobservable ultraviolet" could be measured after all, and that "clear windows to extragalactic space" do exist in the EUV wavelength regime. This opening up of the field of extragalactic EUV astronomy was made possible by the launch of the Wide Field Camera (WFC) on board *ROSAT* in 1990 and of the EUV-dedicated *Extreme Ultraviolet Explorer* with its three scanning telescopes and deep-survey spectrometer telescope in 1992 (see, e.g., Bowyer 1994 for a review). In this contribution, I give an overview of the results obtained so far by these observations and of their implications to the study of AGNs. After providing an interim tally of the identified objects, I discuss separately the two main classes of AGNs that have been found to exhibit significant EUV emission, namely Seyfert galaxies and BL Lacertae objects. I conclude with a brief discussion of future research prospects.

2. EUV Observations of AGNs

In accord with the anticipated effect of Galactic absorption, firm detections of AGNs have only been made in the shortest-wavelength bandpass of the *ROSAT* WFC ($60 - 140$ Å) and of the *EUVE* photometers ($58 - 174$ Å).[†] In the same vein, those objects that

---

[†] In the few cases where EUV spectra have been measured (see §§ 3 and 4 below), the sources were found to be detectable only up to $\sim 110 - 120$ Å.

S. Bowyer and R. F. Malina (eds.), Astrophysics in the Extreme Ultraviolet, 27–36.
© *1996 Kluwer Academic Publishers. Printed in the Netherlands.*

have been identified typically lie in the direction of Galactic neutral hydrogen "windows," along which the column density $N_G(HI)$ does not exceed $\sim 1 \times 10^{20}$ cm^{-2} and for which the He II/H I ratio may also be favorable. In fact, the very detectability of an AGN may be used to place tight constraints on the helium abundance along the line of sight to the source. For example, by using the best-fit EUV energy spectral index $\alpha_{euv} = 1.6$ obtained during the *EUVE* spectrometer calibration measurement of the BL Lac object PKS 2155-304 together with the value $N_G(HI) = 1.36 \times 10^{20}$ cm^{-2} that had been determined from 21 cm emission measurements along the line of sight to this source, Fruscione et al. (1994) derived an upper limit of 0.16 on the ratio of the He II to H I column densities (assuming a negligible column of He I in that direction).

The WFC Bright Source Catalog (Pounds et al. 1993) has identified 7 AGNs (4 Seyfert-like and 3 BL Lac objects [BLOs]), whereas *EUVE* observations have so far reported 7 AGNs (4 Seyferts [including Ton S 180, recently identified by Vennes et al. 1995] and 3 BL Lac objects) with a significance of 6 σ or better, and a total of 14 objects (8 Seyfert 1 galaxies, 5 BLOs, and 1 quasar [3C 273]) detected during the all-sky survey at 2.5 σ or better and lying within 60" of the AGN optical coordinates (Marshall et al. 1995b). Perhaps not surprisingly, the strongest EUV sources appear to be characterized by comparatively steep soft X-ray spectra. In fact, Marshall et al. (1995b) have argued that the relatively high number (compared to hard X-ray surveys) of detected BLOs is compatible with the steep X-ray spectra of the latter class of objects.† Although the *EUVE* all-sky survey lists only 6 AGNs, additional sources continue to be identified by various other means, including pointed observations, serendipitous detections (McDonald et al. 1994), and a comparison between EUV and X-ray samples (Fruscione 1995). According to the most recent tally (given in the Second *EUVE* Source Catalog; Bowyer et al. 1995), the total number of EUV-detected AGNs now stands at 36.

3. Radio-Quiet QSOs and Seyfert 1 Galaxies

3.1. *Origin of the EUV–Soft X-ray Component*

EUV observations of radio-quiet QSOs and of the spectroscopically similar Seyfert 1 galaxies could potentially provide valuable clues to the nature of these sources. The peak power of radio-quiet QSOs is known to be emitted in the EUV range (several hundred Å), and although Galactic absorption limits the useful data to $\lesssim 100$ Å (see § 1), one might still expect to obtain useful information about the dominant AGN emission mechanism from successful EUV measurements. A related issue is the origin of the distinct soft X-ray component (representing a "soft excess" over an extrapolation of the ISM-absorbed hard X-ray component) that is often measured in this class of AGNs. The exact properties of this component as well as its interpretation are still controversial. For instance, the first results of *ROSAT* Position Sensitive Proportional Counter (PSPC) observations of a complete sample of optically selected QSOs (Laor et al. 1994) have pointed to a spectral index $\alpha_x = 1.50 \pm 0.40$ in the $0.2 - 2$ keV range that flattens by $\Delta\alpha_x \approx 0.5$ above ~ 2 keV. This study has concluded that steep-α_x sources are characterized by a weak

---

† It is interesting to note in this connection that, at least for non-BLO AGNs, the work of Mittaz et al. reported at this conference indicates that steep soft X-ray spectra, which are determined from models that include only a cold gas component along the line of sight to the source, are often associated with a low measured Galactic H I column. This correlation, which has been interpreted by Mittaz et al. in terms of a partially ionized absorber at the source, suggests that the apparent steepness of the spectrum may itself be a consequence of the low value of $N_G(HI)$ and that the latter is, in fact, the primary factor that determines the EUV detectability of AGNs.

hard component. A similar distribution of α_x has been inferred also in other *ROSAT* observations of Seyfert 1 galaxies and radio-quiet QSOs (e.g., Turner et al. 1993; Walter & Fink 1993; Fiore et al. 1994). These measurements have yielded a mean value of α_x that is of the order of the mean value of α_{ox}, the (rest frame) optical to X-ray (3000 Å–2 keV) spectral index. In contrast, previous studies (enumerated in Laor et al. 1994) by the *Einstein* and *EXOSAT* satellites have deduced steeper soft X-ray spectral slopes ($\alpha_x \gtrsim 2$) and have in some cases attributed the appearance of steep-α_x sources to the presence of a strong soft X-ray component. The origin of the conflicting results could be either the better sensitivity of *ROSAT* below ~ 0.5 keV or a PSPC calibration problem. EUV measurements could in principle strongly constrain the values of α_x and of $\Delta\alpha_x$ and thereby help determine the nature of the underlying emission mechanisms.

Lieu et al. (1995) have addressed this question using the data already at hand. They identified 6 AGNs for which *EUVE* Lex/B count rates are available and for which *ROSAT* spectra that accurately define the soft X-ray spectral index have also been obtained. They found that the EUV count rates predicted by extrapolating the soft X-ray spectra are in fairly good agreement with the actual measurements over a significant dynamic range of count rates (~ 50) and intrinsic X-ray luminosities ($\sim 10^4$). Although this sample clearly needs to be increased, the successful consistency check seems to provide strong support for the *ROSAT* measurements. The catch in this argument is that the soft X-ray spectra of the AGN sample utilized by Lieu et al. are considerably steeper than those of the average radio-quiet QSOs and Seyfert 1 galaxies measured by *ROSAT*; in fact, the mean value of α_x for the 6 objects that they considered is 2.0, which is consistent with the average values inferred by *Einstein* and *EXOSAT* (see, however, the previous footnote).

One can summarize the current state of affairs as follows. The apparent consistency between the *ROSAT* and *EUVE* observations supports the soft X-ray measurements carried out by the *ROSAT* PSPC. The implications of these measurements to the average radio-quiet AGNs are still not entirely clear, and recent studies (e.g., Walter et al. 1994; Fiore et al. 1995) have indicated that the optical through soft X-ray spectra of these sources cannot be explained in terms of a single simple model. Furthermore, it appears that the AGNs detected in the EUV have a particularly steep soft X-ray spectrum whose origin is also not yet fully understood. The data analyzed so far point to optically thick emission by a hot ($\sim 10^6$ K) gas, possibly associated with the innermost region of a nuclear accretion disk, and at least in some cases appear to be *inconsistent* with reprocessing of the observed hard X-ray component (e.g., Gondhalekar et al. 1994; Brandt et al. 1995; Marshall et al. 1995a).

3.2. *Probing the Warm Absorber*

The presence of a warm (i.e., partially ionized) absorbing gas along the line of sight to the continuum source has been inferred in a number of Seyfert 1 galaxies and QSOs from X-ray observations. Recent studies of objects like 3C 212 (Mathur 1994), 3C 351 (Mathur et al. 1994), and NGC 5548 (Mathur et al. 1995) have indicated that the same material is also responsible for the measured UV absorption, and that it can be identified with an outflowing (at a speed $\gtrsim 10^3$ km s^{-1}) ionized gas of low density but high column density ($N_H \gtrsim 10^{22}$ cm^{-2}) that is situated outside the Broad Emission Line Region (BELR). An analysis of the properties of this gas has led to the prediction that it could produce potentially detectable X-ray and UV *emission* features (Netzer 1993), and, in fact, an *ASCA* measurement of a 0.57 keV feature (interpreted as an O VII emission line) that may be attributed to such a component has already been reported (George et al. 1995).

It is thus natural to expect that EUV observations could also probe this gas and help to further constrain its properties.

Very recent work discussed at this conference has provided tantalizing suggestions that EUV emission lines which could be associated with the warm absorber have been observed in the Seyfert 1 galaxy NGC 5548 (paper presented by Kaastra & Mewe; see also Kaastra et al. 1995) and possibly also Mrk 478 (paper presented by Liedahl et al.). The evidence for emission features in the *EUVE* spectra of these two sources has, however, been disputed by Marshall, who has interpreted the data in terms of pure continuum emission from an accretion disk (see also Marshall et al. 1995a). It is worth noting in this connection that both Kaastra et al. and Liedahl et al. have reached the rather unexpected conclusion that the emitting gas can be better described as being in collisional ionization equilibrium than in photoionization equilibrium (with an extra heat supply of the order of 10% of the bolometric luminosity indicated in the case of NGC 5548 in order to sustain the high inferred gas temperature). Thus, if the claims for the presence of EUV emission features in these objects and their explanation are corroborated, then the currently accepted interpretation of the warm absorber may need to be modified. In another presentation at this conference, Hwang et al. reported detecting emission features in the *EUVE* spectrum of the Seyfert 1 galaxy Mrk 279, but they found it difficult to obtain a good model fit to the data and, in any case, argued against an association of the features with a warm absorber. On the other hand, as was already alluded to in § 2, Mittaz et al. reported indirect evidence for a warm absorber in a number of radio-quiet AGNs (including, in particular, those detected by the *EUVE*) that are characterized by a low value of $N_G(\text{HI})$. This rather confusing state of affairs might possibly be just a reflection of the extreme youth of the field of EUV spectroscopy, but it clearly argues for exercising great caution in interpreting the data and for the need for further observations.

4. Blazars

4.1. *Emission Signatures of a Relativistic Jet Component*

In contrast to Seyfert 1 galaxies and similar radio-quiet AGNs, whose emission over most wavelengths is evidently dominated by a thermal component, BLOs and other members of the "blazar" class of AGNs (notably high-polarization and optically violently variable QSOs) are distinguished by the dominance of a nonthermal emission component. This component, which is represented by a highly variable and often strongly polarized featureless continuum that extends from radio to X-ray wavelengths, has been interpreted as synchrotron (and possibly also inverse-Compton) radiation originating in a relativistic jet (see, e.g., Königl 1989 for a review). In this picture, blazars are sources where the jet axis happens to be oriented at a small angle to the line of sight, so that its observed intensity is boosted by Doppler beaming. Certain AGNs, such as the bright nearby QSO 3C 273, show evidence for a blazar component that does not dominate the thermal emission, most likely because the jet is oriented at a comparatively large angle to the line of sight.

EUV observations of blazars could be useful for testing the relativistic jet interpretation and for helping to constrain the proposed models. They may be particularly helpful in the context of multiwavelength monitoring campaigns that attempt to determine the variability properties and the detailed spectra of these objects. Kafatos et al. have reported at this conference on preliminary results of such measurements for 3C 273 and for the BLO Mrk 421. Significant EUV variability that is consistent with the relativistic beaming scenario has been detected on time scales ranging from hrs to weeks to months in multi-epoch observations of the BLO PKS 2155-304 (Marshall et al. 1993; Fruscione et al. 1994; Königl et al. 1995). The latter source has been the subject of extensive multiwavelength observations in the past. In particular, a recent campaign in 1991 November has detected $\gtrsim 20\%$ soft X-ray flux variations occurring on a time scale of hrs and UV flux variations that followed the X-ray changes with a lag of $\lesssim 3$ hr, although a strict linkage was not maintained at all times (Urry et al. 1993; Brinkmann et al. 1994). These observations have been interpreted in terms of a shock wave propagating in the jet (Edelson et al. 1995). The soft X-ray measurements in the 1991 campaign have been carried out by *ROSAT* and included observations with the WFC that seemed to be consistent with the data acquired with the PSPC (Brinkmann et al. 1994). Further multiwavelength observations of this BLO that include the EUV spectral regime may help to clarify the UV–X-ray connection and test the shock interpretation (see also Marshall's contribution to these proceedings).

EUV measurements may also help to determine the precise contribution of the non-thermal component in radio-loud QSOs and the possible distinction between different types of BLOs. For example, Brunner et al. (1994) have recently presented *ROSAT* soft X-ray spectra for a complete sample of core-dominated, flat-radio-spectrum AGNs, which included 8 quasars and 5 BLOs. They found that the quasars and BLOs were characterized by a mean value of $(\alpha_{ox} - \alpha_x)$ of ~ 0.6 and 0, respectively, but that, in contrast to the quasars, the BLOs showed a relatively large dispersion ($\sigma \approx 0.5$) about the mean (similar to that in α_x). Furthermore, the BLOs appeared to separate into two groups, one with $\alpha_x < 1.0$ and the other with $\alpha_x > 1.7$. As in the case of radio-quiet AGNs discussed in § 3.1, EUV data could be used in conjunction with X-ray and optical/UV results to test the soft X-ray measurements and help isolate the various spectral components that may be present. Another promising possibility, discussed at this conference by Marshall, is to attempt to measure EUV polarization (with the *EUVE* Deep Survey detector in an off-axis observing mode or with an optimized future telescope). This would be useful both for separating the thermal and nonthermal emission components and, once the relevant polarization mechanism is identified, for determining such important attributes of the emission region as its geometry and the local magnetic field distribution.

4.2. *EUV Spectroscopy of BL Lac Objects*

BLOs are distinguished by a weakness or absence of BELR emission. This has been attributed to the dearth of dense gas clouds (e.g., Guilbert et al. 1983), although the detectability of optical and UV emission lines is evidently also influenced by the strength of the underlying nonthermal continuum. Any spectral features that can be attributed to a circumnuclear gas are therefore of particular interest in these objects. One such feature, a ~ 0.6 keV absorption trough or edge, was detected in PKS 2155-304 already in 1980 (Canizares & Kruper 1984) and identified as an O VIII Lyα line originating in a mildly relativistic ($\sim 0.1\,c$) outflow at the redshift ($z = 0.116$; Falomo et al. 1993) of the apparent host galaxy (Krolik et al. 1985). Tentative evidence for a similar feature has been reported also in other BLOs (Madejski et al. 1991). However, the 1980 *Einstein* Objective Grating Spectrometer (OGS) measurement of PKS 2155-304 appears to be formally inconsistent with a *BBXRT* detection of a $\gtrsim 0.5$ keV absorption feature carried out 10 years later (Madejski et al. 1994), which has made the original identification of this feature problematic. This issue is related to the controversy over BLO distances. While these objects often appear in association with low-redshift ($z \lesssim 0.1$) elliptical galaxies, it has been suggested that at least some of them might correspond to gravitationally lensed quasars located at higher redshifts (e.g., Ostriker & Vietri 1985). If PKS 2155-304 in fact has a much higher redshift than that of the associated galaxy, then the identification of the OGS-detected feature with an O VIII Lyα line could not be correct.

Recent spectroscopic observations of this object with the *EUVE* have shed important new light on these questions. The *EUVE* has a state-of-the art spectrometer employing variable line-spaced gratings and equipped with advanced detectors that contain 1680×1680 independent resolution elements, are linear to $\lesssim 0.5\%$, and have a quantum efficiency as high as 80% (e.g., Bowyer & Malina 1991). In the on-axis observing mode, the shortest-wavelength channel encompasses $\sim 72 - 190$ Å, and in the wavelength interval $72 - 100$ Å where most of the useful data have been obtained the spectral resolution ranges from ~ 160 to ~ 200. PKS 2155-304, which is one of the brightest X-ray selected BLOs, has turned out to be the brightest AGN in the EUV (0.29 *EUVE* cts s^{-1}), and has therefore been a prime candidate for *EUVE* spectroscopic measurements.

Königl et al. (1995) carried out two *EUVE* observations of PKS 2155-304 during 1993 June and July with useful exposure times of ~ 111 and ~ 157 ks, respectively. The source was detected in the $\sim 75 - 110$ Å range during both epochs, but the two spectra differed in detail, and the flux had increased by $\sim 50\%$ between the two measurements (see § 4.1). A power-law fit to the data has yielded an energy spectral index $\alpha_{\mathrm{euv}} \approx 3 - 4$ for the measured Galactic H I column density and likely choices of the He I and He II abundances. Such steep values are inconsistent with the soft X-ray spectral index of 1.65 measured by the *ROSAT* PSPC, which approximately corresponds also to the observed EUV to X-ray flux ratio.† Fitting a power law with $\alpha_{\mathrm{euv}} = 1.65$ to the EUV data implies strong absorption at the source between ~ 75 and ~ 85 Å (see Fig. 1). Königl et al. (1995) have argued that this absorption is not due to continuum opacity and demonstrated that it can be attributed, instead, to a superposition of Doppler-smeared absorption lines originating in high-velocity ($\lesssim 0.1\,c$), radially localized clouds of total column density $N_{\mathrm{tot}} \approx 5 \times 10^{20}$ cm^{-2} that are ionized by the beamed continuum of the associated relativistic jet. They identified the lines as mostly L-shell transitions of Mg and Ne and M-shell transitions of Fe. The inferred ionization parameters and densities of the absorbing clouds are comparable to those of BELR clouds, but their velocities are an order of magnitude higher and their total column density is a factor $\gtrsim 10^2$ lower than the column of a standard BELR cloud. These values are consistent with the apparent dearth of BELR gas in BLOs as well as with Marshall et al.'s (1995b) conservative estimate (based on EUV data) that BLOs do not contain more than 10^{20} cm^{-2} of neutral hydrogen (since the deduced column density refers to *ionized* gas). It is, however, interesting to note that the cloud velocities and column densities obtained from this model fit are similar to those attributed to the Broad Absorption Line Region clouds detected in certain radio-quiet QSOs (e.g., Hamann et al. 1993).

The model used to fit the EUV spectrum also implies a pronounced O VII Kα X-ray absorption feature at roughly the same energy as the feature detected in 1990 by *BBXRT* (see discussion above). The predicted feature is broadened and blueshifted (relative to the cosmologically redshifted rest energy) by the motion of the absorbing gas and it dominates the soft X-ray spectrum. The maximum optical depth of the model feature is about an order of magnitude greater than that measured by *BBXRT*, although in making the comparison one should bear in mind that the *BBXRT* and *EUVE* observations were not simultaneous. For a given combination of N_{tot} and of the Mg, Ne, and Fe abundances

---

† Given the various observational uncertainties, this value of α_{x} is also compatible with the EUV measurements by the *ROSAT* WFC (Brinkmann et al. 1994), and it thus appears that any break to the flatter ($\alpha < 1$) spectrum recorded by the *IUE* (Urry et al. 1993) does not occur at wavelengths shorter than ~ 124 Å.

PKS 2155-304

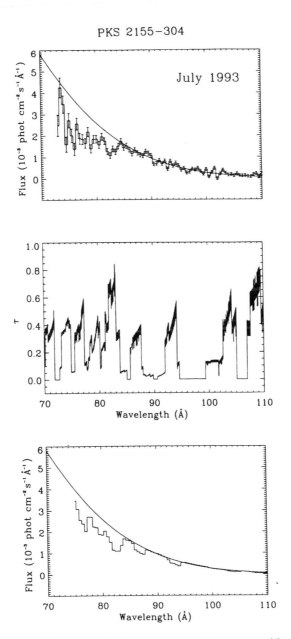

FIGURE 1. The *top* panel displays the background-subtracted spectrum (shown with 1 σ error bars) for the 1993 July *EUVE* observations of PKS 2155-304 (from Königl et al. 1995). The *solid* line represents a best-fit (in the range $85-110$ Å) power-law continuum of energy spectral index $\alpha_{euv} = 1.65$, modified by Galactic absorption. The *middle* panel presents the EUV line opacities calculated for a clumped outflow model, and the *bottom* panel shows the fit to the above data obtained by incorporating the effect of the outflow line and continuum opacities on the $\alpha_{euv} = 1.65$ power-law continuum (represented again by the *solid* curve).

that reproduces the EUV data, the predicted optical depth of the X-ray feature can be adjusted by varying the relative abundance of oxygen (which, in turn, does not affect the EUV spectrum below ~ 110 Å). It can be verified that the alternative interpretation of the *BBXRT* observations in terms of an oxygen K edge at a redshift in the range $0.18 - 0.42$ (Madejski et al. 1994) is inconsistent with the *EUVE* observations, basically because of the large values of N_{tot} ($\gtrsim 5 \times 10^{21}$ cm^{-2}) that it involves. The EUV and X-ray data together thus reinforce the identification of this BLO with a galaxy at $z = 0.116$.

The interpretation of the EUV and X-ray spectra of PKS 2155-304 in terms of high-velocity clouds that occupy a relatively narrow range of distances from the central object suggests a possible explanation of the apparent discrepancy between the X-ray spectra obtained by *Einstein* and *BBXRT*. In this picture, the difference between the two measurements could be attributed to the absorbing clouds observed by *Einstein* having crossed our line of sight closer to the continuum source than the clouds observed a decade later, so that their dominant oxygen ion was O VIII rather than O VII. These two groups of clouds might conceivably be associated with a single global event: for example, they could be related to a disk instability wherein a propagating front induces an outflow at progressively larger radii (a situation possibly resembling FU Orionis outbursts in pre–main-sequence stars; e.g., Calvet et al. 1993, Bell & Lin 1994). Coordinated EUV and X-ray observations could test whether the long-term X-ray spectral variability is correlated with changes in the EUV spectrum, as predicted by this scenario, and whether such variations indeed exhibit a systematic trend in the underlying ionization and velocity structures.

5. Conclusion

The field of EUV astronomy, particularly in regard to AGNs, is still in its infancy. These are exhilarating times, with new discoveries following each other in close succession, with cherished beliefs appearing to be in danger of imminent overhaul, and with conflicting claims and confusion clouding the scene. This brief review was merely meant to provide a synopsis of the recent developments in this field and of their relevance to AGN research rather than an evaluation of already well established results. The promising directions for future studies, predicated on the assumption that EUV telescopes will continue to be built and operated, can be summarized as follows.

- Radio-quiet QSOs and Seyfert 1 galaxies
  (*a*) Increase the data base for comparing EUV flux measurements with soft X-ray spectral extrapolations (§ 3.1).
  (*b*) Search for the signatures of the warm absorber (§ 3.2).
- Blazars
  (*a*) Probe for signatures of beamed nonthermal continuum (§ 4.1).
  (*b*) Attempt to measure the EUV spectra of other BLOs besides PKS 2155-304–e.g., Mrk 421.
  (*c*) Carry out off-axis *EUVE* observations of PKS 2155-304 (and other EUV-bright AGNs) with the aim of broadening the useful spectral range (down to ~ 64 Å).†
  (*d*) Monitor the spectral variability of bright BLOs for EUV–X-ray correlations and for systematic trends in the inferred ionization and velocity structures (§ 4.2).

I am grateful to my collaborators in the *EUVE* observations project, John Kartje, Stuart Bowyer, Steven Kahn, and Chorng-Yuan Hwang, for their valuable input into this

---

† Such observations have already been attempted; e.g., in the case of the Seyfert 1 galaxy Mrk 478 (Marshall et al. 1995a).

work. This research was supported in part by NASA grants NAG 5-2265 and NAGW-1636.

REFERENCES

BELL, K. R., & LIN, D. N. C. 1994, ApJ, 427, 987

BOWYER, S. 1994, Science, 263, 55

BOWYER, S., ET AL. 1995, ApJS, in press

BOWYER, S., & MALINA, R. F. 1991, in Extreme Ultraviolet Astronomy, ed. R. F. Malina & S. Bowyer, New York: Pergamon, 397

BRANDT, W. N., POUNDS, K. A., & FINK, H. 1995, MNRAS, 273, L47

BRINKMANN, W., ET AL. 1994, A&A, 288, 433

BRUNNER, H., LAMER, G., WORRALL, D. M., & STAUBERT, R. 1994, A&A, 287, 436

CALVET, N., HARTMANN, L., & KENYON, S. J. 1993, ApJ402, 623

CANIZARES, C. R., & KRUPER, J. 1984, ApJ, 278, L99

EDELSON, R., ET AL. 1995, ApJ, 438, 120

FALOMO, R., PESCE, J. E., & TREVES, A. 1993, ApJ, 411, L63

FIORE, F., ELVIS, M., McDOWELL, J., SIEMIGINOWSKA, A., & WILKES, B. J. 1994, ApJ, 431, 515

FIORE, F., ELVIS, M., SIEMIGINOWSKA, A., WILKES, B., McDOWELL, J., & MATHUR, S. 1995, ApJ, in press

FRUSCIONE, A. 1995, ApJ, submitted

FRUSCIONE, A., BOWYER, S., KÖNIGL, A., & KAHN, S. M. 1994, ApJ, 422, L55

GEORGE, I. M., TURNER, T. J., & NETZER, H. 1995, ApJ, 438, L67

GONDHALEKAR, P. M., KELLETT, B. J., POUNDS, K. A., MATTHEWS, L., & QUENBY, J. J. 1994, MNRAS, 268, 973

GUILBERT, P. W., FABIAN, A. C., & McCRAY, R. 1983, ApJ, 266, 466

HAMANN, F., KORISTA, K. T., & MORRIS, S. L. 1993, ApJ, 415, 541

HEILES, C. 1991, in Extreme Ultraviolet Astronomy, ed. R. F. Malina & S. Bowyer, New York: Pergamon, 318

KAASTRA, J. S., ROOS, N., & MEWE, R. 1995, A&A, in press

KÖNIGL, A. 1989, in BL Lac Objects, ed. L. Maraschi, T. Maccacaro, & M.-H. Ulrich, Springer, 321.

KÖNIGL, A., KARTJE, J. F., BOWYER, S., KAHN, S. M., & HWANG, C. -Y. 1995, ApJ, 446, 598

KROLIK, J. H., KALLMAN, T. R., FABIAN, A. C., & REES, M. J. 1985, ApJ, 295, 104

LAOR, A., FIORE, F., ELVIS, M., WILKES, B. J., & McDOWELL, J. C. 1994, ApJ, 435, 611

LIEU, R., MITTAZ, J., BOWYER, S., LEWIS, J., & HWANG, C. -Y. 1995, Adv. Space Res., 16(3), 81

MADEJSKI, G. M., ET AL. 1994, preprint

MADEJSKI, G. M., MUSHOTZKY, R. F., WEAVER, K. A., ARNAUD, K. A., & URRY, C. M. 1991, ApJ, 370, 198

MARSHALL, H. L., CARONE, T. E., & FRUSCIONE, A. 1993, ApJ, 414, L53

MARSHALL, H. L., CARONE, T. E., SHULL, J. M., MALKAN, M. A., & ELVIS, M. 1995a, ApJ, submitted

MARSHALL, H. L., FRUSCIONE, A., & CARONE, T. E. 1995b, ApJ, 439, 90

MATHUR, S. 1994, ApJ, 431, L75

MATHUR, S., ELVIS, M., & WILKES, B. 1995, ApJ, in press

MATHUR, S., WILKES, B., ELVIS, M., & FIORE, F. 1994, ApJ, 434, 493

McDONALD, K., ET AL. 1994, AJ, 108, 1843

NETZER, H. 1993, ApJ, 411, 594

OSTRIKER, J. P., & VIETRI, M. 1985, Nature, 318, 446

POUNDS, K. A., ET AL. 1993, MNRAS, 260, 77

TURNER, T. J., GEORGE, I. M., & MUSHOTZKY, R. F. 1993, ApJ, 412, 72

URRY, C. M., ET AL. 1993, ApJ, 414, 614

VENNES, S., POLOMSKY, E., BOWYER, S., & THORSTENSEN, J. R. 1995, ApJ, submitted

WALTER, R., ET AL. 1994, A&A, 285, 119

WALTER, R., & FINK, H. H. 1993, A&A, 274, 105

Discovery of Warm Gas in the Virgo Cluster

R. LIEU,[1] J. P. D. MITTAZ,[2] S. BOWYER,[1]
J. H. M. M. SCHMITT,[1,3] AND J. LEWIS[1]

[1] Center for EUV Astrophysics, 2150 Kittredge Street,
University of California, Berkeley, CA 94720-5030, USA

[2] Mullard Space Science Laboratory, Holmbury St. Mary, Dorking, Surrey RH5 6NT, UK

[3] Max Planck Institut für Extraterrestrische Physik, W-8046 Garching-bei-München, Germany

During the *EUVE* sky survey of the Virgo region, a central source positionally coincident with the X-ray emitting galaxy M87, and a surrounding halo of extended emission, were detected in the 0.065–0.248 keV band. A detailed comparison of these data with the *ROSAT* PSPC data of M87 revealed an excess flux at energies < 0.4 keV within the central 30' radius which cannot be associated with the well-known cluster gas at X-ray temperatures (kT ≥ a few keV). Instead, it is necessary to introduce a second gas component, of temperature $T \sim 5 \times 10^6 K$ (kT ~ 0.1 keV). The resulting two-component model (warm + hot) can account for all the data. The origin and stability of the warm component, with a temperature near the peak of the thermal plasma cooling curve, is unclear. Both the temperature and spatial extent argue against cooling flow as the primary process responsible for its production. Other mechanisms, such as a galactic wind and heating by galaxy motions, must be considered.

From soft X-ray ($\hbar\omega \sim 3$ keV) data gathered by the *EINSTEIN*, *EXOSAT* and *ROSAT* missions, it is well known that M87, the central galaxy of the Virgo cluster, has a halo of emission extending to a radius of 100' (Fabricant & Gorenstein 1983; Stewart et al. 1984; Forman, Jones & Binggeli 1985; Böhringer et al. 1994) and that a cooling flow exists in the innermost 20' region. We present new observational results of M87 obtained from *EUVE* and combine these results with data from the *ROSAT* mission for our analysis. The source was detected by the *EUVE* sky survey in the Lex/B (65–248 eV) filter passband, with an exposure of approximately 1,000 s. Figure 1 is a radial profile of the Lex/B count rates. The data are expressed in units of surface brightness of diffuse emission (i.e., counts ks^{-1} arcmin^{-2}); they are average values for concentric annuli moving outwards from the best-fit position of a central source, which is approximately 1' offset from the optical position of M87. The background at large radii (> 35') is flat, so that a sizable region is available for its determination with high accuracy. The brightness distribution within a radius of 5' is consistent with a point source detected at 4.2 σ significance. More surprisingly, an independent halo of extended emission was detected at 3.3 σ significance between 5' and 35' radii. These data represent the first reported detection of extended cluster emission in the EUV.

In order to compare the *EUVE* data with existing X-ray data of M87, we extracted from the public archive the results of a *ROSAT* PSPC pointed observation which took place in July 1992, with an exposure of nearly 10,000 s. The analysis was performed with the STARLINK ASTERIX X-ray data processing system. Periods of high background and poor attitude solutions were removed from the data, and the pulse-height spectra were computed for concentric annuli centered at the X-ray position of M87. We only considered energies $\hbar\omega > 0.18$ keV, and avoided the region between 19' and 22' to minimize the effects of the support structure on the analysis. The extracted spectra were also corrected for vignetting and dead times. An accurate model for the local soft

S. Bowyer and R. F. Malina (eds.), Astrophysics in the Extreme Ultraviolet, 37–43.

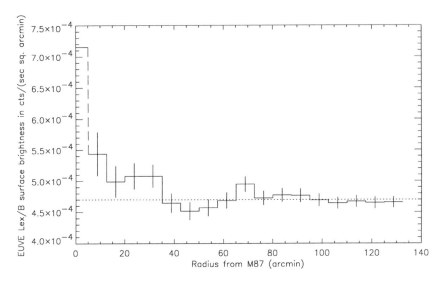

FIGURE 1. *EUVE* Lex/B (0.065–0.248 keV) filter count rates gathered during the sky survey are expressed in units of diffuse emission for concentric annuli centered at the best-fit position of a point source, which is offset from the optical position of M87 by $54.''2$. The innermost $5'$ region (i.e., first data point) corresponds to this point source, detected at 4.2 σ significance by the standard source search procedure of matched-filtering. The region between $5'$ and $35'$ corresponds to a diffuse excess which is 3.3 σ above a background (dotted line) determined by using all the data in the "flat" region beyond $35'$.

X-ray diffuse background was developed, essentially by fitting the 1/4 keV-band spectrum for the $45' - 57'$ region with the thin plasma emission code of Mewe et al. (1985, 1986), assuming a foreground absorption corresponding to a H I column density of 5.0×10^{18} cm^{-2} to account for the local solar cloud. This background model is a constant additive component used in the interpretation of data for all the inner annuli, after the model has been normalized to the appropriate sky areas of each annulus, taking into account obstruction by the detector wires and spokes. We also consulted the master-veto data to ensure that the particle background is negligible. Owing to the radial dependence of the emission properties, the PSPC spectra for concentric annuli extending out to a radius of $45'$ were individually fitted with the thin plasma emission code of Mewe et al., using the appropriate off-axis response matrix. Interstellar absorption corresponding to a Galactic H I column density of 2.5×10^{20} cm^{-2} (Stark et al. 1992) was assumed. A redshift of $z = 0.0043$ was also assumed.

We find that the *EUVE* and *ROSAT* data for the central ring of inner radius $5'$ and outer radius $35'$ (i.e., excluding the emission from M87) cannot simultaneously be fitted with a single temperature plasma model. While the X-ray temperature plasma can account for most of the PSPC count rates at energies $\hbar\omega > 0.4$ keV, it does not produce sufficient EUV radiation to match the observed Lex/B count rates. Instead, it is necessary to introduce a second gas component of considerably lower temperature (kT ~ 0.1 keV). Predictions of the resultant two-phase model agree well with the data in both passbands. They are listed in Table 1.

As we study the new emission component in more detail, we find that no single tem-

TABLE 1. M87 Halo between 5′ and 35′

	EUVE counts ks^{-1} (0.065 − 0.25 keV)	ROSAT X-ray counts s^{-1} (0.4 − 2.0 keV)
Observed	150.8 ± 45.2	18.97 ± 0.04
Warm Plasma Model $T = 9.6 \times 10^5$ K	35.0	0.2
Hot Plasma Model $T = 2.76 \times 10^7$ K	57.7	18.67
Total (two-phase model)	92.7	18.87

perature gas scenario can explain the PSPC spectrum of an inner annulus for the entire range of energies. Data at energies $\hbar\omega > 0.4$ keV are satisfactorily fitted by a single temperature plasma model of Mewe et al., with kT $\sim$ a few keV. At energies $\hbar\omega < 0.4$ keV, however, there exists a "soft excess" in the spectra. This is illustrated in Figure 2 where we show the spectrum and folded model for a typical inner annulus, of mean radius 12.′5 and width 2.′5. The dotted line represents the best-fit model for the > 0.4 keV band, which falls short of the data points at energies < 0.4 keV. It is unlikely for such a "soft excess" to be caused by systematic errors in the 1/4 keV-band calibration of the PSPC, because an extensive analysis of this band by Snowden et al. (1995) demonstrated that the published effective areas are *overestimated* by only 10%. Moreover, we find that the surface brightness of soft excess is peaked at the center of the cluster. This correlation strongly suggests that the excess is a real effect related to the M87 emission, rather than to any extraneous contamination not removed by the background subtraction procedure.

We find that the full PSPC spectrum for the 12.′5 ± 2.′5 annulus can be fitted by a two-temperature plasma with common metallicity and with a second component having kT $\sim$ 0.083 keV ($T \sim 9.6 \times 10^5$ K), which we call the "warm component." The solid line in Figure 2 shows the best-fit model. For all other annuli within a radius of 35′ equally good fits can be obtained by the use of two-temperatures. However, spectra taken outside 35′ can be fitted by a single temperature plasma, and moreover there is no detected Lex/B flux in this region. Although the PSPC does not have the spectral resolution to separate the different temperatures of the emitting material at various radii along the observational line-of-sight, it is the existence of much cooler plasmas at the inner region which characterizes the new result presented here.

The PSPC data have sufficient statistical quality to enable a comparison of the properties of the two gas components in the cluster. In Figures 3 and 4 we plot the radial temperature profiles of the hot and warm plasmas. It is clear that the hot component cools inside the central 20′ region. Within the accuracy of the data, there is no evidence of similar behavior in the warm plasma. In Figures 5 and 6 we plot the radial surface brightness profiles of the two plasmas, to demonstrate that the emission from the hot gas does not fall as rapidly w.r.t. radius. The hot component remains detectable beyond the radius of 35′ where the warm component can no longer be found. The best-fit common metallicity varies between 0.37 and 0.56 of the solar abundance, with a trend of higher

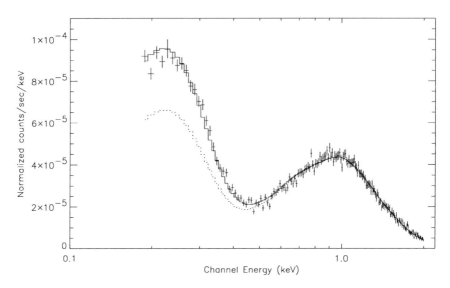

FIGURE 2. Data and folded model of the *ROSAT* PSPC spectrum for an annulus centered at M87, of mean radius 12′.5 and width 2′.5. Dotted line gives the best-fit single-temperature Mewe thin plasma emission model for the > 0.4 keV data, while solid line gives the best-fit two-temperature model for the entire spectral range. This model includes a "warm" plasma as second component.

metal content at smaller radii, consistent with results of the PSPC sky survey (Böhringer et al. 1994).

An important question concerning the newly detected gas component has to do with its maintenance, because an optically thin plasma at temperature $kT < 0.1$ keV cools rapidly by line emissions. Using the best-fit parameter values for the $31' \pm 4'$ (i.e., limiting) annulus, we estimated the plasma density to be $n_e \sim 4.73 \times 10^{-4}$ cm^{-3} and the cooling time $\tau \sim 3.16 \times 10^8$ years, assuming that the warm plasma fills the entire emission volume. Since the cooling time is much less than the age of the cluster, a mechanism of sustaining the gas is necessary, irrespective of its origin. The mass of warm gas within the same volume is estimated to be $7.92 \times 10^{10} M_\odot$.

We argue that the warm gas is unlikely to be associated with mass deposition throughout the cooling flow, because (a) given that the hot gas cools from $kT \sim 3$ keV at $20'$ radius to $kT \sim 1$ keV at the center, it is difficult to envisage the deposition of $kT \sim 0.1$ keV gas everywhere, unless the gas is extremely inhomogeneous; (b) there is no observational evidence of cooled material—the published result on the presence of cold absorbing matter in the center of M87 (White et al. 1991) is *inconsistent* with the data reported here, since our models of the PSPC spectra do not involve N_H values as high as 10^{21} cm^{-2}; (c) if the warm gas is continuously replenished by cooling flow as it rapidly leaves the peak of the cooling function to form cold material, we estimate the mass of such material to be $\sim 10^{13} M_\odot$ in the central region; this would provide very significant absorption to the EUV and X-ray photons, which is not observed.

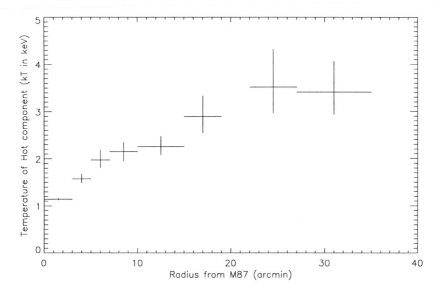

FIGURE 3. Radial profile of the best-fit hot component temperature

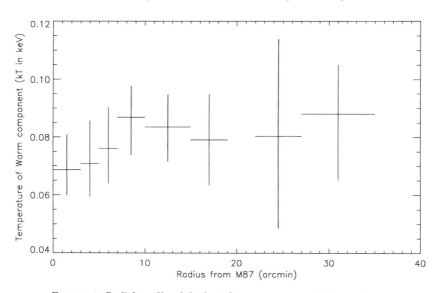

FIGURE 4. Radial profile of the best-fit warm component temperature

We conclude by suggesting an alternative mechanism, viz. a galactic wind of 10^6 K plasma (responsible for the soft X-ray background) driven out of the galaxies by cosmic ray pressure (Axford 1981), and which may further be heated as the bulk flow energy due to the motion of the galaxy w.r.t. the cluster is thermalized by collisions. The intracluster gas itself could also be heated through viscous forces between the gas and the moving galaxies (Sarazin 1986).

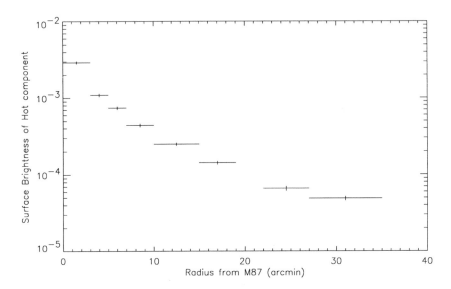

FIGURE 5. Radial profile of the best-fit surface brightness of the hot component. This surface brightness is related to the emission measure (EM, in cm^{-3}) by EM = 4.77 × 10^{66} × norm. × (area of annulus in arcmin^2)

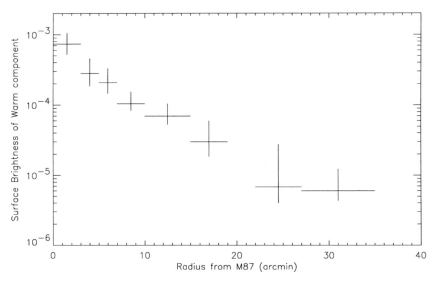

FIGURE 6. Radial profile of the best-fit surface brightness of the warm component, defined in the same way as in Fig. 5.

We thank R. Stern and D. Leahy for helpful discussions. R. Lieu, S. Bowyer and J. Lewis acknowledge the support of NASA contract NAS5-30180.

REFERENCES

AXFORD, W. I. 1981, 17th International Cosmic Ray Conference, 12, 155

BÖHRINGER, H., BRIEL, U. G., SCHWARZ, R. A., VOGES, W., HARTNER, G. & TRÜMPER, J. 1994, Nature, 368, 828

FABRICANT, D. & GORENSTEIN, P. 1983, ApJ, 267, 535

FORMAN, W. JONES, C. & DEFACCIO, M. 1985, ESO Workshop on the Virgo Cluster of Galaxies, ed. Richter, O.-G. & Binggeli, B., ESO Conf. Proc. No. 20, ESO, Garching, 323

MEWE, R., GRONENSCHILD, E. H. B. M. & VAN DEN OORD 1985, A&AS, 62, 197

MEWE, R., LEMEN, J. R. & VAN DEN OORD, G. H. J. 1986, A&AS, 65, 511

SARAZIN, C. L. 1986, Rev. Mod. Phys., 58, 1

SNOWDEN, S. ET AL. 1995, ApJ, submitted

STARK, A. A., GAMMIE, C. F., WILSON, R. W., BALLY, J., LINKE, R. A., HEILES, C. & HURWITZ, M. 1992, ApJS, 79, 77

STEWART, G. C., CANIZARES, C. R., FABIAN, A. C. & NULSEN, P. E. J. 1984, ApJ, 278, 536

WHITE, D. A., FABIAN, A. C., JOHNSTONE, R. M., MUSHOTZKY, R. F. & ARNAUD, K. A. 1991, MNRAS, 252, 72

EUV and Soft X-ray Evidence for Partially Ionized Gas in Active Galactic Nuclei

J. P. D. MITTAZ,[1] R. LIEU,[2] S. BOWYER,[2]
C. -Y. HWANG,[2] AND J. LEWIS[2]

[1]Mullard Space Science Laboratory, Holmbury St. Mary, Dorking, Surrey RH5 6NT, England

[2]Center for EUV Astrophysics, 2150 Kittredge Street, Berkeley,
University of California, CA 94720-5030, USA

We present a synoptic study of active galactic nuclei (AGN) detected by *EUVE*. We also present complementary *ROSAT* PSPC spectra for these sources and for other AGN in directions of low galactic absorption. It is found that the best-fit power-law photon indices of the X-ray spectra at 0.1–2.4 keV are anti-correlated with their galactic hydrogen columns. The indices for the 0.9–2.4 keV range do not show such a correlation, and are considerably smaller (i.e. flatter). We discuss a number of possible interpretations of this correlation but only one of these, the presence of a partially ionized absorbing gas in the AGN, explains the observations satisfactorily. The ubiquity of this effect suggests that this component be may very common in AGN.

Studies of the behavior of AGN in the EUV and soft X-rays have recently begun with the launch of *EUVE* and *ROSAT*. We report a collective effect which emerged from an examination of a sample of AGN observed by the two satellites. Our original sample was a set of 8 objects which are part of the much larger catalog of AGN detected in the Lex/B (65 − 190 eV) filter passband by *EUVE* during the survey and pointed phases of the mission. This subset is formed by selecting only objects for which we have well defined soft X-ray spectra from the *ROSAT* public archive. Moreover we have excluded BL Lac objects as they are special cases where a significant fraction of their emission comes from relativistic jets. Spectral models have been obtained by fitting the power-law parameters with galactic absorption to the PSPC data. Predicted Lex/B count rates using these best-fit parameters are found to be in reasonable agreement with the observed values (Lieu et al. 1995), indicating that the sources have not varied significantly and that we are not severely affected by any uncertainty in the soft response of the PSPC. In Table 1 we list the sources and their properties.

In this work we explore the relationship between intrinsic source absorption in the soft X-ray/EUV and absorption within our own galaxy, using the cross sections of Morrison and McCammon (1983). In Figure 1 we plot galactic H I column density (N_H, obtained from one of the following sources: Stark et al. 1987; Elvis et al. 1989; Dickey et al. 1990) against the best-fit photon indices. A striking anti-correlation exists between these two independently measured quantities. We do not believe that this is due to the method of fitting as in that case one would expect there to be a positive correlation between index and N_H. Further, in order to demonstrate that this is not an EUV selection effect, we have included in Figure 1 those non-BL Lac objects detected by the *ROSAT* sky survey (Walter & Fink 1993, Brinkmann & Siebert 1994) in directions of $N_H < 2.3 \times 10^{20}$ cm^{-2}, also detailed in Table 1. It is clear that these separately acquired data follow a very similar trend to that of the EUV data.

One possible explanation of the anti-correlation is that the spectra of AGN located in directions of high N_H are dominated by their > 1 keV photons since galactic absorption is removing the lower energy photons. A fit to the entire 0.1–2.4 keV band would then be biased by the spectral shape in the > 1 keV band. The spectrum from this higher energy band is known to be flatter and in Figure 2 we plotted the best-fit indices for

45

TABLE 1. AGN observed by EUVE and ROSAT

Name	RA			DEC			Type	Lex/B (c/ks)	Photon Index 0.11–2.4 keV	χ^2_{red}	Galactic N_H (10^{20} cm^{-2})	Fitted N_H (10^{20} cm^{-2})
PG 1415+451	14	17	0.6	+44	56	00	QSO	5.2±1.5	3.01±0.15	1.23	1.16	$1.61^{+0.30}_{-0.35}$
Mrk 478	14	42	6.5	+35	26	4	Sy1/QSO	62±10	3.38±0.15	1.24	1.11	$1.44^{+0.20}_{-0.20}$
NGC 4051	12	03	11.0	+44	32	24	Sy1	17±8	3.00±0.15	1.53	1.31	$1.95^{+0.28}_{-0.23}$
PG1116+215	11	19	8.0	+21	19	36	QSO	17±13	2.77±0.20	1.16	1.44	$1.62^{+0.20}_{-0.20}$
Ton 180	00	57	20	-22	22	36	Sy1.2	53±8	3.06±0.05	0.80	1.47	$1.46^{+0.10}_{-0.10}$
CG 0912	14	21	29.6	+47	47	00	AGN	3.4±1.0	2.12±0.30	1.02	1.74	$1.69^{+0.55}_{-0.55}$
3C 273	12	29	6.7	+02	03	00	QSO	79.2±0.8	2.18±0.30	1.09	1.8	$1.49^{+0.10}_{-0.10}$
NGC 7213	22	09	17.0	-47	09	36	Sy1	23±7	2.06±0.10	0.98	1.92	$1.68^{+0.32}_{-0.28}$
IR13349+2438	13	37	18.7	24	23	3	QSO	—	3.61±0.34	1.16	1.00	1.6 ± 0.6
CG 1043	12	42	10.5	33	17	4	Sy1	—	3.10±0.47	1.41	1.30	3.2 ± 1.7
3C279	12	56	11.2	-5	47	22	QSO	—	1.65±0.38	0.76	2.22	2.3 ± 1.1
NGC1566	04	19	00	-55	03	00	AGN	—	1.79 ± 0.41	0.35	1.55	0.7 ± 1.1
Mkn110	09	22	00	52	30	00	AGN	—	2.17 ± 0.27	0.25	1.57	1.3 ± 0.6
PG0953+451	09	54	41	30	30	00	AGN	—	2.93 ± 0.33	1.48	1.18	1.7 ± 0.8
Mkn142	10	22	00	51	55	00	AGN	—	3.12 ± 0.23	1.10	1.18	1.6 ± 0.5
3C263	11	37	00	66	04	00	AGN	—	2.66 ± 0.51	0.66	0.82	1.5 ± 1.2
Mkn766	12	16	00	30	06	00	AGN	—	2.75 ± 0.13	0.74	1.76	3.3 ± 0.5
NGC4593	12	37	00	-05	04	00	AGN	—	2.49 ± 0.42	1.07	1.97	2.0 ± 1.3
Mkn279	13	52	00	69	33	00	AGN	—	2.15 ± 0.17	1.27	1.64	1.8 ± 0.5
NGC5548	14	16	00	25	22	00	AGN	—	2.21 ± 0.15	1.00	1.93	1.7 ± 0.4
PG1444+407	14	45	00	40	48	00	AGN	—	3.41 ± 0.53	0.97	1.31	2.2 ± 1.5

the 0.9–2.4 keV band against N_H. For the fifteen objects where we have sufficient data to fit the 0.9–2.4 keV band it is clear that the high energy indices do not correlate with N_H and that the average 0.9–2.4 keV index is 2.20±0.10, considerably flatter than the average 0.1–2.4 keV index of 2.72±0.07, applicable to the data in Figure 1. However, in going from a column of 1×10^{20} cm^{-2} to 2×10^{20} cm^{-2} only half the photons below 0.5 keV are absorbed, and as an index of 2.2 is still quite steep it is difficult to see how a fit at 2×10^{20} cm^{-2} could be dominated solely by the higher energy photons.

We therefore must look for another explanation. In Figure 3 we display the best fit total absorption against the galactic absorption associated with a hydrogen column derived from 21 cm measurements. The best fit columns are larger than the galactic value for the low N_H sources, providing evidence for intrinsic absorption in the AGN. Recent evidence from *GINGA* showed that up to ~ 50% of bright Seyferts may have partially ionized absorbers (Nandra & Pounds 1994) and such absorbers can significantly alter the observed spectrum (Reynolds & Fabian 1995). Indeed, three of our sources (NGC4051, NGC7213, NGC4593) show evidence for a partially ionized absorber. Since we only have PSPC spectra it is not possible to exclude either the possibility of the AGN having a complex intrinsic spectrum giving rise to the observed shape in the PSPC spectrum, or the presence of a partial covering of the source by cold gas. Our correlation, however, is

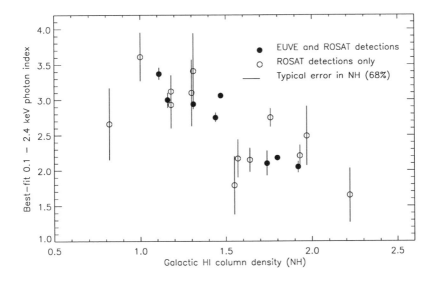

FIGURE 1. Galactic (21 cm) H I column density plotted against best-fit power-law indices for the PSPC spectra in the 0.11–2.4 keV range, for the 21 AGN shown in Table 1. The photon index is simultaneously fitted with a variable total absorption, so that the results are independent of galactic absorption from a measurement viewpoint. The errors in N_H are determined from the dispersion of results in the immediate vicinity of the sky location concerned; directional variations of the column within the radio beam size represents the dominant source of uncertainty.

unlikely to be caused by either of these two effects since in neither case can one introduce a distortion of the spectrum as a function of galactic N_H.

A more natural explanation would be the presence of a partially ionized intrinsic absorber in all of these objects. Such an absorber, with a temperature $\sim 50,000$ to $500,000$ K, transmits photons with lower energy than ~ 0.4 keV with very little absorption, but the O VII and O VIII edges give rise to a significant absorption at energies of 0.8 keV and above. Therefore, a partially ionized absorber intrinsic to the source will contribute a significant fraction of the total absorption above 0.8 keV but not below it. Any model based on a simple power-law spectrum combined with galactic absorption will be forced to steepen the fit to take into account the increased absorption of the hard X-ray photons relative to the < 0.8 keV photons. Indeed, by looking at the transmission curve of a typical warm partially ionized absorber (Reynolds & Fabian 1995), the ratio of absorption between the < 0.4 keV and > 0.8 keV photons can be as large as a factor of 10. This scenario then yields, at least qualitatively, a natural explanation for the anti-correlation between spectral index and N_H. At very low columns the effect of the partially ionized absorber will be significant, therefore steepening the observed spectrum. At higher column densities, however, the fraction of the absorption due to the partially ionized gas will decrease and therefore the spectrum will flatten. This therefore predicts an anti-correlation of observed spectral slope with N_H, exactly what we observe. The

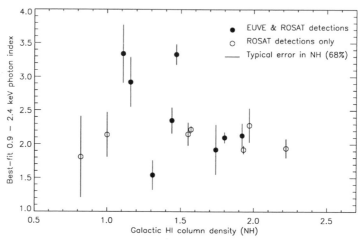

FIGURE 2. Galactic (21 cm) H I column density plotted against best-fit power-law indices for the *ROSAT* spectra in the 0.9–2.4 keV range, for 15 of the AGN shown in Table 1. No correlation between the power law indices and galactic absorption is seen.

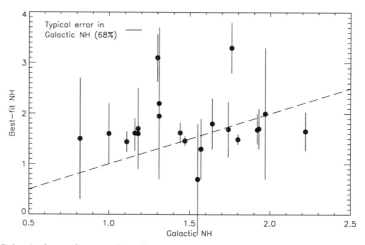

FIGURE 3. Galactic (21 cm) versus best-fit column densities for the AGNs listed in Table 1. The dashed line shows where the two sets of column densities agree. At low galactic hydrogen columns power-law models with absorption yield higher best-fit columns than the corresponding galactic values. This is indicative of intrinsic absorption which will be increasingly masked in AGNs observed through larger amounts of galactic absorption.

fact that we see such a good correlation for a sample of bright AGN selected purely on the value of the galactic absorption implies that partially ionized absorbers are commonplace, and indeed that such absorbers may exist in all AGN. The reason why they are not seen in all AGN is that the effect of this absorber is overwhelmed by the galactic absorption for many objects with large N_H.

R. Lieu acknowledges the support of NASA contract NAS5-30180.

REFERENCES

BRINKMANN, W. & SIEBERT, J. 1994, A&A, 285, 812

DICKEY, J. M. & LOCKMAN, F. J. 1990, Ann. Rev. Astr. Ap., 28, 215

ELVIS, M., LOCKMAN, F. J. & WILKES, B. J. 1989, AJ, 97, 777

LIEU, R., MITTAZ, J., BOWYER, S., LEWIS, J., & HWANG, C.-Y. 1995, Adv. Space Res., 16, 3,81

MORRISON, R. & MCCAMMON, D. 1983, ApJ, 270, 119

NANDRA, P. & POUNDS, K. A. 1994, MNRAS, 268, 405

REYNOLDS, C. S. & FABIAN, A. C. 1995, MNRAS, in press

STARK, A. A., ET AL. 1992, ApJS, 79, 77

WALTER, R. & FINK, H. H. 1993, A&A, 274, 105

Line Emission from Warm Material in NGC 5548

J. S. KAASTRA,[1] R. MEWE,[1] AND N. ROOS[2]

[1]SRON, Sorbonnelaan 2, 3584 CA Utrecht, The Netherlands

[2]Sterrewacht Leiden, Niels Bohrweg 2, 2300 RA Leiden, The Netherlands

EUVE observations of the Seyfert 1 galaxy NGC 5548 have shown the presence of line emission features identified as a Ne VII/Ne VIII blend at 88 Å and Si VII emission at 70 Å. The lines show significant broadening (FWHM 3800 km s^{-1}) placing the emitting region at the same distance as the inner broad-line region. A fit to a thermal plasma yields a temperature of 6×10^5 K. The line emission can be attributed to the warm absorbing material discovered before from oxygen and iron absorption features, which appears to be an optically thin (for the Lyman continuum), highly ionized phase of the broad-line region.

1. Introduction

NGC 5548 is one of the brightest Seyfert galaxies in the soft X-ray band (Branduardi-Raymont et al. 1984). The spectrum and variability of both soft and hard X-rays was studied extensively using *Exosat* observations by Kaastra & Barr (1989). The spectrum could be decomposed into a hard power-law component plus a modified-blackbody spectrum which was attributed to direct radiation from the accretion disk. The power-law was identified as disk radiation up-scattered by the inverse Compton process in a hot corona surrounding a part of the disk.

The spectrum of the soft component was highly variable, both in flux and shape, on a 0.5 day time scale. The spectral variations can be explained by correlated temperature - luminosity changes of the accretion disk (Kaastra 1991a,b), or alternatively by ionization of a warm absorber in the line of sight. Evidence for a warm absorber in NGC 5548 arises from *Ginga* observations of an Fe edge around 8 keV (Nandra et al. 1991) and from the discovery by *Rosat* of an O VII or O VIII edge near 0.8 keV (Nandra et al. 1993). The first *ASCA* observations (Fabian et al. 1994) confirmed the presence of the O VII and O VIII edge. The equivalent Hydrogen column densities implied are of the order of 10^{26} m^{-2}.

The low-energy spectrum of NGC 5548 is thus rather complicated, containing possible contributions from the accretion disk, the power-law, a warm absorber and galactic absorption. Observations with *EUVE* provide an excellent opportunity to study this complicated energy band.

2. *EUVE* Observations

Three *EUVE* observations of NGC 5548 were obtained in the spring of 1993. These observations have been described in detail elsewhere (Kaastra et al. 1995), but here we summarise them. The longest observation (exposure time 332000 s) was obtained between March 10–24, 1993; a second observation with exposure time 226000 s between April 26 and May 4, 1993; and a short third observation of 71000 s between May 12–14, 1993.

Especially during the first observation NGC 5548 showed significant line emission. At least two significant line features were visible. The strongest feature at ∼88 Å is identified

S. Bowyer and R. F. Malina (eds.), Astrophysics in the Extreme Ultraviolet, 51–55.
© 1996 *Kluwer Academic Publishers. Printed in the Netherlands.*

as a blend of Ne VII 88.13 Å (transition 2p $^3P_1$ - 4d $^3D_{2,3}$) and the Ne VIII 88.10 Å doublet (transition 2s $^2S_{1/2}$ (ground state) - 3p ^2P). The second line at 70.02 Å is a blend of Si VII (2p^4 ^3P - 2p^{3}3d $^3D,^3P$). The lines have intrinsic luminosities (corrected for galactic absorption) of 5.1×10^{35} W and 2.6×10^{35} W, respectively, similar to the average luminosity of e.g., the hydrogen Lyα and C IV 1549 Å lines.

The lines are significantly broadened, with a FWHM of 3800 ± 1700 km s^{-1}. If we interpret this line broadening as due to Doppler shifts corresponding to motion with typical Keplerian velocities around the central black hole, we can derive the distance of the emitting material. From optical/UV monitoring campaigns, the distance scale of the optical broad-line region is reasonably well constrained. The central mass derived from the correlation between the line width and the line lag of the optical/UV lines with respect to the UV continuum is 2.7×10^7 M$_\odot$ (Krolik et al. 1991) or 3.7×10^7 M$_\odot$ (Clavel et al. 1992). We adopt a central mass of 3×10^7 M$_\odot$. Using the FWHM of the Ne VII/VIII blend (3800 km s^{-1}), we derive a distance of 3×10^{14} m, with an uncertainty of a factor of 2. The plasma emitting the lines observed by *EUVE* then has a similar distance to the central source as the optical broad-line region.

3. Emission from a Warm Absorber

Recently it has been suggested by Netzer (1993) that the warm absorber that appears to be present in many AGN should manifest itself not only by absorption, but also by emission from the warm material.

The *EUVE* spectrum of NGC 5548 obtained during the first observation period was fitted by a thermal plasma in collisional ionization equilibrium (CIE), using the SPEX code (Kaastra & Mewe, 1993) which is an extended and updated version of the older code of Mewe et al. (1985). The best fit yields a temperature of (6.0 ± 1.4) 10^5 K, with an emission measure of 2.2×10^{72} m^{-3}. Of course the emitting region needs not to be in CIE, but could as well be in photo-ionization equilibrium (PIE). In the latter case, the temperature will be somewhat lower. Can this emission component be attributed to the warm absorber?

Estimates by Kaastra et al. (1995) (see also section 4) showed that the parameters of the emitting region are entirely consistent with the properties of the warm absorber as deduced from the Ginga, *Rosat* and *ASCA* observations.

Other evidence for line emission in Seyfert galaxies obtained from *EUVE* observations is found in Mkn 478 (Liedahl et al. 1995). In Mkn 478 there is evidence for a blend of Fe-M, Si and Mg emission, in particular near 85 Å; the derived emission measure, line broadening and temperature are within a factor of ~ 2 consistent with our findings for NGC 5548!

Also recently George et al. (1994) found evidence for O VII and O VIII line emission in the *ASCA* spectrum of NGC 3783, indicating emission from a plasma similar to the emitting regions in NGC 5548 and Mkn 478.

Emission from the warm absorber thus seems to be not an unusual feature in Seyfert galaxies.

4. The Warm Absorber As a Hot Phase of the Broad-Line Region

Recently Hamann et al. (1995a) studied a sample of $z \sim 1$ quasars with the HST. In several of these sources there was evidence for strong, broad-line emission of the Ne VIII 774 Å line. This line is unvisible in Seyfert galaxies due to the strong absorption by our galaxy, but in high redshift quasars it shifts beyond the Lyman edge and thus becomes visible. The line actually is the doublet of the 2s-2p transition in Li-like neon, while the 88.10 Å line detected by *EUVE* is the corresponding 2s-3p transition of the same ion. The luminosity deduced for the Ne VIII 774 Å line is of the same order of magnitude as the luminosity for the C IV 1549 Å line and Lyα. The same line is also seen in absorption in other $z \sim 1$ quasars (Hamann et al. 1995b). These authors also suggest that the Ne VIII line emitting material is the same as the warm absorber. Thus the warm absorber might be identified as a highly ionized phase of the broad-line clouds.

Shields et al. (1995) have elaborated this idea further. They conclude that the broad-line region consists of at least 3 phases. The low-ionization component, responsible for e.g., the Mg II emission and the high-ionization component responsible for e.g., Lyα and C IV 1549 Å are known for some time. A highly ionized component that is optically thin for the Lyman continuum and fully ionized in hydrogen is new in these models. This component can account for the warm absorber, as the column density and ionization stage are similar.

Shields et al. have modelled the BLR of NGC 5548 by such a multi-phase medium. They used clouds with a column density of 10^{25} m^{-2}, hydrogen density 10^{17} m^{-3}, and microturbulence velocity of the order of 100 km s^{-1}. They argue that the covering factor of the optically thin medium is probably high, and since the continuum in NGC 5548 is probably hard enough, emission by the warm absorber is likely. They also suspected that filtering of the intrinsic AGN continuum due to the warm absorber may be important for the radiation field as seen by the clouds at larger radii.

Let us now make some estimates for the Ne VIII 88 Å line of NGC 5548. In CIE this line contains some 60% of the total flux of the neon blend. In CIE at 6×10^5 K and for cosmic abundances, the optical depth in the continuum τ_c near the line is $0.46 N_{26}$, where N_{26} is the column density of a single cloud in units of 10^{26} m^{-2} (the warm absorber has typically $N_{26} \sim 1$). The optical depth at the line centre τ_ℓ is $560 N_{26}/\sqrt{1 + 20 v_{100}^2}$ where v_{100} is the microturbulence velocity of the clouds in units of 100 km s^{-1}. Under those PIE conditions where Ne VIII is the most abundant neon ion these optical depths are similar within a factor of 2–3.

In most cases the line is thus optically thick at its centre. Using random walk arguments for the line photons that are emitted and re-absorbed several times by line emission and absorption, before eventually escaping or being absorbed by the continuum, we find that only line photons produced in the outer fraction $f_e = (\tau_c \tau_\ell)^{-1/2}$ of the absorbing cloud can escape. Assuming a microturbulence velocity of the order of 100 km s^{-1}, we approximate $f_e \sim 0.13\sqrt{v_{100}}/N_{26}$. Thus the line becomes effectively optical thin for column densities below 10^{25} m^{-2}, and is optically thick otherwise. Contrary, the continuum absorption is only modest. As a result, for N_{26} of order unity, the lines will be suppressed considerably as compared to the continuum. Kaastra et al. (1995) showed that from the observed *EUVE* spectrum it follows that the line-to-continuum ratio can be suppressed by at most a factor of ~ 5. Thus the column density during the first *EUVE* observation was probably a few times 10^{25} m^{-2}. For smaller values of the microturbulence velocity the column density of the EUV-emitter needs to be much smaller than is consistent with the column density of the warm absorber. Note however that Shields et al. (1995) also required a typical micro-turbulence of 100 km s^{-1} in order to explain the optical/UV line ratio's in NGC 5548!

For the geometry of the source we assume a spherical shell with radius d, thickness Δ, filled with a fog of N small clouds with individual radius r, Hydrogen density n, temperature T and filling factor f. For $f = 1$ we have a homogeneously filled shell. The total emission measure is Y. The total column density of the layer is designated by $N_{\rm H}$. We use a ratio of electron to Hydrogen density of 1.2. For the distance we adopt the value of 3 10^{14} m, deduced from the line broadening (see section 2). The "observed" emission measure Y_{obs} as deduced e.g., from our CIE fit is the true emission measure Y multiplied by the fraction f_e. We use $Y = 1.2n^2 f 4\pi d^2 \Delta$, $N_H = nf\Delta$, and $f = \frac{4}{3}\pi r^3 N/4\pi d^2 \Delta$. Inserting the relevant numbers, we find that the typical density of the clouds must be 10^{17} m^{-3}, entirely consistent with the assumptions made for the optical thin high ionized broad-line region component used by Shields et al!

For the Ne VIII 774 Å line, τ_ℓ is a factor of 4.5 larger than for the 88.10 Å line; however the continuum optical depth is a factor of 30 smaller for the 774 Å line than for the 88.10 Å line. As a result, f_e is 2.5 times larger for the 774 Å line than for the 88.10 Å line. Since in CIE the photon emissivity of the 774 Å line is 190 times larger than for the 88.10 Å line, we expect the effective luminosity of the 774 Å line to be 54 times larger than for the 88.10 Å line (and hence than e.g., the C IV 1549 Å line). Of course in NGC 5548 the 774 Å line is completely blocked by the absorption in our own galaxy. In $z \sim 1$ quasars however, an intervening neutral hydrogen column of only $\sim 10^{21}$ m^{-2} would be sufficient to make the apparent flux of the 774 Å line similar to that of the UV lines. That material could be situated anywhere in the line of sight from the quasar host galaxy to the point where the photon energy passes the Lyman limit.

5. The DS Count Rate

The count rate obtained during our observations with the Deep Survey instrument with the Lexan/boron filter of $EUVE$ (DS) showed large amplitude variations on a typical time scale of half a day (Marshall 1995), similar to the behaviour as observed by $EXOSAT$ (Kaastra & Barr 1989). The average count rate during the first observation was about 0.06 c s^{-1}, and about 0.08 counts s^{-1} during the second observation period.

Our model for the warm emitter predicts a contribution to the DS count rate of only $\sim$0.02 c s^{-1} in the wavelength range 70–100 Å. The predicted contribution to the DS count rate below 70 Å is only 0.002 c s^{-1}. Since the deduced size of the line emitting region is $\sim$11 light days, the warm emitter is not expected to show rapid variability on a day time scale. It is thus evident that most of the DS flux must be produced by a different, strongly variable soft component, which we assume to be identical to the accretion disk spectrum (cf. Kaastra & Barr 1989).

It is interesting to note that the model of Kaastra & Barr for the spectrum of NGC 5548 of March 3, 1986 (a simple power law plus modified blackbody emission from the accretion disk) predicts a DS count rate of 0.065 c s^{-1}. Their model was based upon $EXOSAT$ observations. Since the transmission properties of the $EXOSAT$ Lexan filter and the DS Lexan filter are not too different, we may conclude that the average soft X-ray flux in March 1993 was similar to the average soft X-ray flux in March 1986. In NGC 5548, both the soft- and hard X-ray fluxes (e.g. Kaastra 1991b), and the UV continuum and the hard X-ray flux (Clavel et al. 1992) correlate well. Using these correlations we estimate that the 1450 Å continuum level must have been about 4×10^{-17} Wm^{-2}Å^{-1}. This appears to be only some 30% higher than the observed UV continuum at that time (cf., Peterson & Korista 1994).

Finally, our model also predicts strong lines to be present at 50.0, 51.8 and 60.6 Å due to S VII, which must be easily detectable by e.g., the LETGS aboard AXAF.

SRON, the Space Research Organization of the Netherlands is supported financially by NWO, the Netherlands Organization for Scientific Research.

REFERENCES

BRANDUARDI-RAYMONT, G., BELL-BURNELL, S. J., KELLETT, B., FINK, H., MOLTENI, D., & McHARDY, I. 1984, EXOSAT observations of active galactic nuclei. in X-ray and UV emission from active galactic nuclei, ed. W. Brinkmann & J. Trümper, MPE Report, 184., M.P.I., Garching, 88

CLAVEL, J., NANDRA, K., MAKINO, F., POUNDS, K. A., REICHERT, G. A., URRY, C. M., WAMSTEKER, W., PERACAULA-BOSCH, M., STEWART, G. C., & OTANI, C. 1992, Correlated hard X-ray and ultraviolet variability in NGC 5548, ApJ, 393, 113

FABIAN, A. C., NANDRA, K., BRANDT, W. N., HAYASHIDA, K., MAKINO, F., & YAMAUCHI, M. 1994, Simultaneous ASCA and Rosat observations of NGC 5548. in New Horizon of X-ray astronomy, ed. F. Makino & T. Ohashi, Tokyo: Universal Academy Press, 573

GEORGE, I. M., TURNER, T. J., & NETZER, H. 1995, The discovery of an O VII emission line in the ASCA spectrum of the Seyfert galaxy NGC 3783, Astroph. J.L, 438, L67

HAMANN, F., ZUO, L., & TYTLER, D. 1995a, Broad Ne VIIIλ774 in the HST-FOS snapshot survey ABSNAP, ApJL, in press

HAMANN, F., BARLOW, T., BEAVER, E. A., BURBIDGE, E. M., COHEN, R. D., JUNKKARINEN, V., & LYONS, R. 1995b, Ne VIIIλ774 and time-variable associated absorption in the QSO UM 675, ApJ, in press

KAASTRA, J. S. 1991a, The soft X-ray variability of NGC 5548, A&A, 249, 70

KAASTRA, J. S. 1991b, Soft X-ray variability and the accretion disk of NGC 5548. in Structure and emission properties of accretion disks, ed. C. Bertout, S. Collin, J-P. Lasota, J. Tran Than Van, IAU Colloq. 129, Gif sur Yvette: Editions Frontières, 449

KAASTRA, J. S. & BARR, P. 1989, Soft and hard X-ray variability from the accretion disk of NGC 5548, A&A, 226, 59

KAASTRA, J. S. & MEWE, R. 1993, The Mewe et al., plasma emission code, Legacy, 3, 16

KAASTRA, J. S., ROOS, N., & MEWE, R. 1995, EUVE observations of NGC 5548, A&A, in press

KROLIK, J. H., HORNE, K., KALLMAN, T. R., MALKAN M. A., & EDELSON, R. A. 1991, Ultraviolet variability of NGC 5548—dynamics of the continuum production region and geometry of the broad-line region, ApJ, 371, 541

LIEDAHL, D., PAERELS, F., HUR, M., FRUSCIONE, A., KAHN, S., & BOWYER, S. 1995, The EUV spectrum of the Seyfert 1 galaxy Mrk 478, these proceedings

MARSHALL, H. L. 1995, Variability and spectra of AGN in the EUV and the relation to other bands, these proceedings

MEWE, R. ., GRONENSCHILD, E. H. B. M., & VAN DEN, OORD, G. H. J. 1985, Calculated X-radiation from optically thin plasmas. V., A&AS, 62, 197

NANDRA, K., POUNDS, K. A., STEWART, G. C., GEORGE, I. M., & HAYASHIDA, K. 1991, Compton reflection and the variable X-ray spectrum of NGC 5548, MNRAS, 248, 760

NANDRA, K., FABIAN, A. C., GEORGE, I. M., BRANDUARDI-RAYMOND, G., LAWRENCE, A., MASON, K. O., McHARDY, I. M., POUNDS, K. A., STEWART, G. C., & WARD, M. J. 1993, A Rosat observation of NGC 5548, MNRAS, 260, 504

NETZER, H. 1993, Ionized absorbers, ionized emitters, and the X-ray spectrum of active galactic nuclei, ApJ, 411, 594

PETERSON, B. M., & KORISTA, K. T. 1994, Intensive spectroscopic monitoring of NGC 5548 with HST and IUE, in Multi-wavelength continuum emission of AGN, ed. T.J.-L. Courvoisier & A. Blecha, IAU Symp. 159, Dordrecht: Kluwer, 177

SHIELDS, J. C., FERLAND, G. J., & PETERSON, B. M. 1995, Optically thin broad-line clouds in active galactic nuclei, ApJ, 441, 507

Extreme Ultraviolet Spectroscopy of the Seyfert 1 Galaxy Markarian 478

D. A. LIEDAHL,[1] F. PAERELS,[2] M. Y. HUR,[2] S. M. KAHN,[2]
A. FRUSCIONE,[3] AND S. BOWYER[3]

[1] Lawrence Livermore National Laboratory, L-41, P.O. Box 808, Livermore, CA 94550, USA

[2] Physics Department and Space Sciences Laboratory,
University of California at Berkeley, Berkeley, CA 94720-7300, USA

[3] Center for EUV Astrophysics, 2150 Kittredge Street,
University of California, Berkeley, CA 94720-5030, USA

The Seyfert 1 galaxy Mrk 478, observed during the *EUVE* all-sky survey, is the brightest EUV source among its class. The SW spectrum of this object shows evidence of discrete emission, although this interpretation is tentative, since the source spectrum must be extracted against a bright background. If the EUV flux is, in fact, composed partly of line emission, we consider the implications if this is the result of emission from a collision-driven plasma at temperatures $\gtrsim 10^6$ K. In this context, we discuss some of the constraints imposed on the emission-line region by this observation.

1. Introduction

The opportunity to make high spectral resolution measurements of the spectra of active galactic nuclei (AGN) in the EUV has been made possible by the launch of the *Extreme Ultraviolet Explorer* (*EUVE*; Bowyer & Malina 1991). Those few AGN whose EUV fluxes are observable after attenuation by the interstellar medium (the neutral H column cannot greatly exceed 10^{20} cm^{-2}) are tempting targets for direct study of the blue bump, the broad UV/EUV peak in νL_ν.

Current models of the energy production mechanisms in AGN typically involve accretion onto supermassive black holes (Rees 1984), thereby requiring the presence of large quantities of fueling matter in a compact region. The distribution of this matter is an unsettled issue; it has long been expected that a disk, which then plays a key role in its capacity to transport angular momentum and to efficiently convert energy, describes the distribution near the black hole. The characteristic radiation temperature near the last stable orbit in the Schwarzschild potential is given by

$$T_C = 4.9 \times 10^5 \, M_7^{-1/4} \left(\frac{L}{L_E} \right)^{1/4} \left(\frac{\eta}{0.1} \right)^{-1/4} \text{K}, \qquad (1.1)$$

where L_E is the Eddington luminosity and M_7 is the central mass in units of 10^7 M$_\odot$. The accretion efficiency, η, is defined by $L = \eta \dot{M} c^2$, where L is the accretion-driven luminosity and $\dot{M}$ is the accretion rate. Because of the weak dependence of T_C on M, a wide range of central masses should produce a peak in the EUV band.

The existence of an EUV peak does not, however, unequivocally imply an accretion disk origin. Guilbert & Rees (1988) show that an EUV excess can arise even if the primary spectrum is purely non- thermal, produced, for example, by $e^+ - e^-$ annihilation (e.g., Zdziarski et al. 1990). Some fraction of the emitted radiation must be intercepted by matter with high Compton depth that subsequently re-emits this energy in the form of a blackbody spectrum. Assuming that such optically thick matter exists, we can estimate the resulting blackbody temperature. In terms of the Schwarzschild radius R_S, it is

S. Bowyer and R. F. Malina (eds.), Astrophysics in the Extreme Ultraviolet, 57–61.
© *1996 Kluwer Academic Publishers. Printed in the Netherlands.*

approximately

$$T_{BB} = 6.7 \times 10^5 \ \phi^{1/4} M_7^{-1/4} (1 - a)^{1/4} \left(\frac{L}{L_E}\right)^{1/4} \left(\frac{R}{R_S}\right)^{-1/2} \text{K}, \qquad (1.2)$$

for a cloud with albedo a at a distance R from the primary radiation source. The quantity ϕ is the ratio of the area of an individual cloud, projected along its line of sight to the central radiation source, to the total surface area. Therefore, if the clouds are in the vicinity of the primary radiation source, the blackbody spectrum can peak in the EUV range, provided that the accretion rate is not negligible compared to the Eddington limit.

Barvainis (1993) has suggested a model in which the blue bump arises from an optically thin thermal bremsstrahlung continuum. In the EUV band, the emission from a plasma in the temperature range $10^5 - 10^7$ K is entirely dominated by line emission (e.g., Mewe, Gronenschild, & van den Oord 1985), so that any quantitative arguments concerning the feasibility of this model for producing the EUV bump must consider discrete line emission, not just continuum emission. Evidence in the form of discrete emission (e.g., Ne VIII [Hamann, Zuo, & Tytler 1995; Kaastra, Roos, & Mewe 1995ab]) or continuum absorption (e.g., O VII [Nandra & Pounds 1992]) for the existence of a substantial circumnuclear plasma capable of supporting "warm" gas is beginning to accumulate. We note, however, that in none of the three cases discussed above does line emission enter into the arguments presented to date.

Among the AGN significantly detected during the *EUVE* all-sky survey, Mrk 478 is the second brightest Seyfert galaxy (Bowyer et al. 1995). Observations of Mrk 478 and a preliminary analysis of the spectral data are described in the next section. We find that the spectrum is consistent with discrete emission from a plasma with a temperature near 10^6 K. In § 3, we discuss some of the constraints and problems that accompany the existence of line emission in this band.

2. Observations and Data Analysis

Mrk 478 was observed with *EUVE* over the period 1993, April 9–18, for 303,000 s effective exposure time. Part of the observation (152,000 s exposure) was conducted with the source positioned off the optical axis of the Spectrometer Telescope, along the dispersion direction of the short-wavelength spectrometer, in an effort to extend the sensitivity of the instrument to shorter wavelengths.

The off-axis data were aspect corrected to the nominal on-axis position, and both datasets were reduced separately. For each image, we located the spectral image on the detector, and positioned a slightly curved mask of width 21 pixels on the spectrum along the dispersion direction; the mask curves with the spectrum. The count spectrum is obtained by integrating over the mask in the cross-dispersion direction. Next, we positioned background extraction regions above and below the spectral image. The width of these boxes was 84 pixels in the cross-dispersion direction; we verified that the background subtraction was insensitive to the actual width of the background regions. The total net source counts over the wavelength band 72–120 Å, after background subtraction, are 1093 ± 162, and 747 ± 138 for the on-axis and off-axis observations, respectively. In the final step, pixel numbers were converted to wavelength, and the observed count spectrum was divided by the effective area of the spectrometer to obtain the physical fluxes. This introduces an uncertainty, because of the unknown correction to the spectrometer effective area for off-axis position; but since statistical fluctuations dominate, this is probably a small effect. We summed both spectra in order to improve the signal-to-noise ratio.

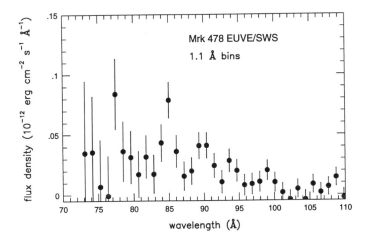

FIGURE 1. Average spectrum of Mrk478 observed with *EUVE* in April, 1993. The spectrum has been background subtracted, and is binned in 1.1 Å bins. Note the presence of discrete structure near 78, 85, and 90 Å, suggestive of emission line complexes.

The total summed spectrum, binned in 1.1 Å bins (16 pixels) is shown in Figure 1. The source is significantly detected over the 76–110 Å band and shows obvious structure.

We used a simple power law with interstellar absorption by cold gas, with slope and normalization determined with the *ROSAT* PSPC (Gondhalekar et al. 1994). This assumption is justified by the fact that the source has shown little X-ray spectral variability in the past. This assumed continuum reproduces the average flux densities as a function of wavelength reasonably well.

3. Interpretation of the Discrete Structure

The EUV spectrum in the spectral range corresponding to the SW bandpass is dominated by $2s - 2p$ intrashell emission from highly ionized Fe (more ionized than Fe^{16+}) if the electron temperature exceeds 3×10^6 K. We can easily rule out a high-temperature plasma as the source of EUV line emission, since, for example, we would not expect strong emission near 85 Å (79 Å in the rest frame; $z = 0.079$) for $T \gtrsim 10^{6.4}$ K. At temperatures nearer 10^6 K, the spectrum can be extremely complicated, consisting of L-shell emission from Ne, Mg, and Si, and Fe M-shell emission. These ions can also be important sources of line emission in a photoionized plasma, although the corresponding electron temperatures are roughly an order of magnitude lower. With lower temperatures, the ionic spectra can be markedly different because of the dominance of recombination in populating excited states. Theoretical spectra from these ions are poorly understood at this time. We are in the process of calculating these spectra in detail (Liedahl, Osterheld, & Goldstein 1995). Our analysis is still preliminary, but our calculated spectra of a photoionized plasma appear to be qualitatively incompatible with the observed spectra.

Therefore, we investigate the consequences of interpreting the spectra in terms of line emission from a plasma in collisional equilibrium.

The volume emission measure for a thin ($\Delta R << R$) spherical shell is given by $EM = 4\pi R^2 \Delta R n_e^2$. If we hypothesize that the complex near 85 Å is a blend of emission from Fe XII and Fe XIII plasma, then $EM \approx 10^{67}$ cm^{-3} for standard cosmic abundances (Allen 1973). We assume an electron temperature of $10^{6.2}$ K, the temperature at which the Fe charge state distribution is dominated by Fe$^{(11-13)+}$ (Arnaud & Raymond 1992).

Marshall et al. (1995) show that the SW flux can vary by a factor of two over approximately 0.5 d. If we assume that the varying flux includes the line emission, then the extent of the emission-line region, projected along our line of sight, D, is constrained by $D \lesssim ct_{\mathrm{var}}$, where t_{var} is the factor-of-two flux variation timescale. With spherical symmetry, we have $R \lesssim 1.3 \times 10^{15}$ cm.

If the lines are to originate in a plasma that is dominated by electron impact ionization, rather than by photoionization, then $n_e C > \beta$, where C is the collisional ionization rate coefficient, and β is the photoionization rate. For an estimate of the constraint imposed by this requirement, take $C = 6 \times 10^{11}$ cm^3 s^{-1} as a representative collisional ionization rate coefficient for an Fe M-shell ion in coronal equilibrium (Arnaud & Raymond 1992), and $\beta = 1.2 \times 10^3 R_{15}^{-1}$ s^{-1}, using the power-law spectrum given in Gondhalekar et al. (1994). The average radius of the shell is given in units of 10^{15} cm and is denoted by R_{15}. This provides a relationship between R and n_e. In fact, it provides the strictest constraint on the density: $n_e \gtrsim 1.2 \times 10^{13}$ cm^{-3}, where we have taken $R_{15} = 1.3$.

To check the consistency of the thin-shell assumption, use the relation $\Delta R/R = EM/4\pi R^3 n^2$. Using the most liberal constraints (maximum allowed radius and minimum allowed density), $\Delta R/R = 3 \times 10^{-6}$, which, though consistent, seems unlikely in the context of a shell distribution. Keep in mind, however, that this 'model' is simply a device to allow us to begin studying the plausibility of a relatively compact source of line emission in Mrk 478.

As noted by Elvis et al. (1991), the rapid time variations in AGN spectra often imply an optically thick emission region (cf., Barvainis 1993). We can estimate the possible effects of Compton scattering and resonant scattering. Let σ_T and τ_T denote the Thomson cross-section and the electron scattering optical depth, respectively. With $EM = 4\pi R^2 n_e \tau_T \sigma_T^{-1}$, $R = 1.3 \times 10^{15}$ cm, and $n_e = 1.2 \times 10^{13}$ cm^{-3}, we find a scattering depth of 3×10^{-2}, which is safely less than unity for a range of allowed parameters.

As an example of the effect of resonant scattering, consider an Fe ion, with elemental abundance A_{Fe}, ionic fraction F_{ion}, a fractional level population in the lower state of the transition Φ_{lev}, and an oscillator strength f. By eliminating ΔR in favor of EM, the line-center optical depth is given by

$$\tau_o = \frac{\pi^{1/2}e^2}{m_e c^2} A_{\mathrm{Fe}} F_{\mathrm{ion}} \Phi_{\mathrm{lev}} \left(\frac{n_H}{n_e}\right) \frac{EM}{4\pi R^2 n_e} \left(\frac{m_{\mathrm{Fe}}c^2}{2kT}\right)^{1/2} f\lambda, \qquad (3.3)$$

where the remaining quantities have their conventional meaning. Numerically, this is

$$\tau_o = 2.1 \times 10^3 \, F_{\mathrm{ion}} \Phi_{\mathrm{lev}} (EM)_{67} R_{15}^{-2} (n_e)_{13}^{-1} \left(\frac{\lambda}{100 \text{ Å}}\right) \left(\frac{f}{0.1}\right) \left(\frac{kT}{100 \text{ eV}}\right)^{-1/2}. \qquad (3.4)$$

Although this appears to be rather high, the escape of photons is still possible because the factor Φ_{lev} is less than 10^{-4} for the vast majority of transitions in the intermediate Fe M-shell ions (e.g., Fe XII, Fe XIII), which drops τ_o below unity. For example, this is the case for all levels above the 4th excited state in Fe^{12+}, which emits a complex of lines near 79 Å in its rest frame.

At a temperature of $10^{6.2}$ K, the dominant oxygen charge state is He-like, with a K photoionization edge at 739 eV. If we use the same emission measure as above, we find that the edge optical depth in O^{6+} is $\tau_{oxy} \approx 50 \ (EM)_{67} R_{15}^{-2} (n_e)_{13}^{-1}$. Although the O^{6+} zone is transparent to the EUV lines, it is opaque to the X-ray continuum above 0.74 keV. Therefore, in light of the $ROSAT$ spectrum, the EUV line-emitting material cannot entirely occult the X-ray continuum source. However, this does not seriously affect the above analysis, since modifications of the above equations with $4\pi \rightarrow \Omega < 4\pi$ can be tolerated.

The authors acknowledge useful discussions with Mike Lampton. Work at Lawrence Livermore National Laboratory was performed under the auspices of the US Department of Energy under contract No. W-7405-Eng-48, and at the Center for EUV Astrophysics by NASA contract NAS5-30180.

REFERENCES

ALLEN, C. W. 1973, Astrophysical Quantities, 3rd Ed., London, The Athlone Press

ARNAUD, M., & RAYMOND, J. C. 1992, ApJ, 398, 394

BARVAINIS, R. 1993, ApJ, 412, 513

BOWYER, S., & MALINA, R. F. 1991, in Extreme Ultraviolet Astronomy, ed. R. F. Malina & S. Bowyer Pergamon Press: New York, 397

ELVIS, M., GIOMMI, P., WILKES, B. J., & McDOWELL, J. 1991, ApJ, 378, 537

GONDHALEKAR, P. M., KELLETT, B. J., POUNDS, K. A., MATTHEWS, L., & QUENBY, J. J. 1994, MNRAS, 268, 973

GUILBERT, P. W., & REES, M. J. 1988, MNRAS, 233, 475

HAMANN, F., ZUO, L., & TYTLER, D. 1992, ApJ, 444, L69

KAASTRA, J. S., ROOS, N., & MEWE, R. 1995a, A&A, in press

KAASTRA, J. S., ROOS, N., & MEWE, R. 1995b, A&A, these proceedings

LIEDAHL, D. A., OSTERHELD, A. L., & GOLDSTEIN, W. H. 1995, in preparation

MARSHALL, H. L. 1995, these proceedings

MARSHALL, H. L., CARONE, T. E., SHULL, J. M, MALKAN, M. A., & ELVIS, M. 1995, ApJ, submitted

MARSHALL, H. L., FRUSCIONE, A., & CARONE, T. E. 1995, ApJ, 439, 90

MEWE, R., GRONENSCHILD, E. H. B. M., & VAN, DEN, OORD, G. H. J. 1985, A&AS, 62, 197

NANDRA, K., & POUNDS, K. A. 1992, Nature, 359, 215

POUNDS, K. A., ET AL. 1993, MNRAS, 260, 77

REES, M. J. 1984, ARA&A, 22, 471

ZDZIARSKI, A. A., GHISELLINI, G., GEORGE, I. M., SVENNSON, R., FABIAN, A. C., & DONE, C. 1990, ApJ, 363, L1

Variability and Spectra of AGN in the EUV and the Relation to Other Bands

HERMAN L. MARSHALL

Eureka Scientific, Inc., 5 Whipple Rd., Lexington, MA, 02173 USA

Data from several collaborations will be shown which demonstrate the utility of *EUVE* observations. For Mk 478, a Seyfert 1 galaxy, the rapidly variable EUV flux is shown to have a steep, featureless continuum. The EUV data are combined with UV and optical data to form an overall spectrum that is consistent with an accretion disk model; slight temperature variations in the innermost regions could cause the large EUV flux changes. *EUVE* data for other sources are presented: NGC 5548, which shows significant variations and has an EUV spectrum that shows no emission lines, contrary to a previous report; 3C 273, which did not vary much; and PKS 2155-304, which was observed simultaneously with *ASCA* and *IUE* when a hard X-ray flare was detected as an EUV polarization measurement was being attempted.

There were 13 AGN reported by Marshall et al. (1995a) that were detected in the *EUVE* all-sky survey. Of these, 6 have now been observed with the *EUVE* spectrometer and 4 of these observations will be discussed here.

1. Mrk 478

The observations of this source are reported in more detail by Marshall et al. (1995b). Figure 1 shows the light curve in the deep survey (DS) imaging detector, which obtains a direct image of the target during spectroscopic observations. The light curve shows variability of a factor of 2–10 over time scales of less than one day.

The spectrum shows no emission lines and is consistent with a very steep power law spectrum, with an energy index $\alpha = 4.70 \pm 0.65$ if the column density is fixed at the newly measured Galactic value, $\log N_H = 20.0$. Combining these data with (nonsimultaneous) UV and optical data indicate that a significant fraction of the total luminosity may be in the EUV band and appears to fit an accretion disk spectrum, such as modelled by Sun & Malkan (1989). The fit gave a central black hole mass of $1.3 \times 10^8 M_\odot$ and an accretion rate of 0.3 $M_\odot$/yr and requires a highly inclined disk so that the Eddington limit not be violated.

The smooth spectrum and rapid variability can be rationalized in a model where the EUV flux derives from the inner edge of an accretion disk where the peak temperature, T, is about 2.5×10^5 K. The EUV flux varies as $T^{8.2}$ because the bandpass is just over the black body peak and in the Wien tail of the spectrum. Thus a 10% change in the temperature would give a ×2 change in the EUV count rate but only a 10% change in the optical flux.

2. NGC 5548

Observations of this source will be reported in more detail by Marshall et al. (1995c). The light curve is shown in Figure 2 and shows variations of a factor of 2 in less than 2 days at TJD ($\equiv$ JD - 2440000) = 9062.5 and 9107.5. Even with simultaneous UV coverage, however, we could not discern the relationship to the UV band or the emission lines unambiguously. The data are consistent with zero lag or a 5 day lag (EUV leading) due to the sparse sampling of the UV data (only once per day).

S. Bowyer and R. F. Malina (eds.), Astrophysics in the Extreme Ultraviolet, 63–67.
© *1996 Kluwer Academic Publishers. Printed in the Netherlands.*

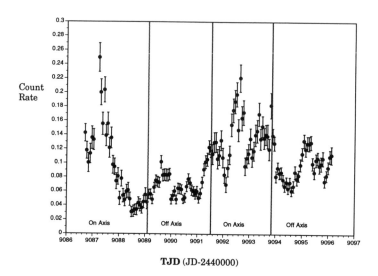

FIGURE 1. Time history of the Mrk 478 Deep Survey (DS) count rates (counts s^{-1}). Note the significant variations in less than 1 day. On-axis data have systematic uncertainties due to the deadspot correction, described by Marshall et al. (1995b).

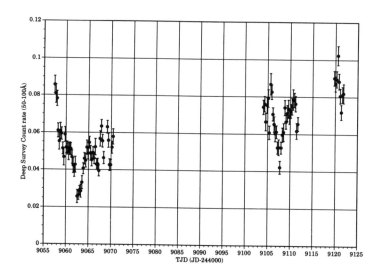

FIGURE 2. Time history of the NGC 5548 DS count rates (counts s^{-1}). Note the significant variations in less than 1 day. The dead spot was no factor in this analysis since the source was observed 0.3 degrees off-axis.

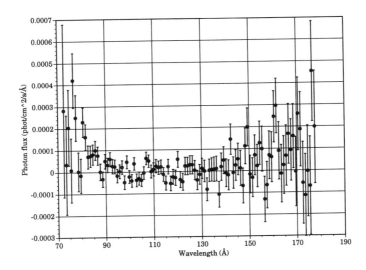

FIGURE 3. The EUV spectrum of 3C 273, taken in January, 1994. The EUV flux is lower than for Mrk 478, so the spectrum is not as easy to detect, although there is significant flux in the 75–90 Å region. Again, there are no apparent emission lines.

Contrary to the report by Kaastra et al. (1995), we find no emission lines in the EUV spectrum. Specifically, our data reduction shows no detectable emission in the 1 Å band centered at 90 Å where the strongest line reported by Kaastra et al. (1995) was. The spectrum appears to be strongly cut off by cold matter absorption so the DS data must be dominated by continuum or emission lines in the 50–70 Å region. Our reconciliation of these results involves the details of the background, which has a faint systematic ripple (probably due to microchannel plate fiber bundles) that is difficult to remove. An excess variance results from this background variation as well.

3. 3C 273

There is only weak evidence for variations in the DS count rate during the 6 day observation of 3C 273 in January, 1994. The count rate was nearly constant at 0.09 count/s but dipped for one day to 0.07 count/s. A preliminary spectrum is shown in Figure 3; the signal is very low but is detectable in the 75–90 Å region. Again, there are no obvious emission lines in the EUV spectrum, so the emission could originate in an accretion disk or from Comptonization of disk photons.

4. PKS 2155-304

The objective of the 9 day observation of PKS 2155-304 was to measure the EUV polarization by a new technique, first described by Marshall (1994). Briefly, a polarized source should produce a slight $cos2\theta$ variation of the azimuthal image intensity when observed off-axis by the DS telescope. This observation is very difficult to perform, since

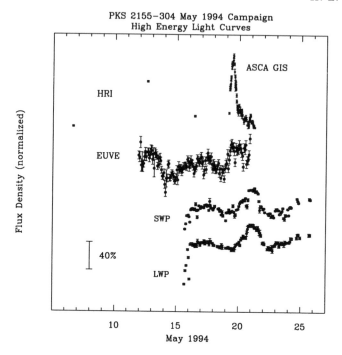

FIGURE 4. Light curves for PKS 2155-304 observed by the *ASCA* GIS and the *ROSAT* HRI (top), *EUVE* (middle) and the *IUE* SWP and LWP (bottom). The curves have been displaced for clarity. Note the UV feature on May 16 and the X-ray flare on May 19, neither of which are apparent in the EUV data.

the maximum modulation is expected to be about 4% when the source is 100% polarized. Results of analyzing these data are still in preparation.

Meanwhile, however, the light curve was generated very soon after the observation. It appeared to be uninteresting, showing a drop of 30% in 0.4 days at TJD 9486.3 and an increase of similar magnitude from TJD 9491.5 to 9492.0. These data became significantly more interesting when combined with the *ASCA* and *IUE* light curves, shown in Figure 4.

There are two remarkable features in this figure, one in each of the UV and X-ray light curves and neither of these show up in the *EUVE* band, which overlaps them both. At TJD 9488.4 (May 15.9, 1994), both the SWP and LWP detectors on *IUE* registered significant increases which were not accompanied by any significant change in the *EUVE* count rate. The X-ray coverage was too sparse to be of much help at this time. The second feature occurs between TJD 9491.6 and 9492.4 (on May 19, 1994), when the count rate in the *ASCA* GIS is seen to increase by a factor of two and then return. Due to the fine data sampling in both *IUE* and *EUVE*, we can firmly state that there is no corresponding flare in either of these two bands. At most, there may be a delayed response in the UV band lagging the X-ray band by 2 days. The case of the EUV flux is more curious; there is only the rate increase which is simultaneous with the *ASCA* rate increase but there is no corresponding drop of the EUV flux. These observations are being prepared for publication by Urry et al. (1995).

These observations would normally indicate that the X-ray and EUV bands are decoupled and that the UV band is only weakly coupled to either band. Previous observations (Edelson et al. 1995), however, showed a very tight correlation between these bands, so the behavior of the overall spectrum of PKS 2155-304 (and possibly BL Lac objects in general) is now in doubt. We have no explanation for the apparently disparate behaviors.

5. Summary

We have observed many AGN and the EUV count rates of all of them show at least 30% variability over time scales of a day. Mrk 478 shows the most dramatic variability, sometimes reaching a factor of 10. None of them show emission lines (see Fruscione et al. 1994 for a spectrum of PKS 2155-304). Except for PKS 2155-304, the EUV flux from AGN may be due to the high energy tail of an accretion disk spectrum and the best evidence has been obtained for Mrk 478. There is still a poorly understood mismatch of the fluxes in different wavebands from PKS 2155-304—longer coverage may help resolve the discrepancy.

This work has been supported under NASA contracts NAS5-32485 and NAS5-32678 to Eureka Scientific, Inc. a private corporation created to promote the active pursuit of scientific research.

REFERENCES

EDELSON, R., ET AL. 1995, Multiwavelength Monitoring of the BL Lacertae Object PKS 2155. IV. Multiwavelength Analysis, ApJ, 438, 120

FRUSCIONE, A., BOWYER, S. KONIGL, A. & KAHN, S. M. 1994, Extreme Ultraviolet Spectroscopy of PKS 2155, ApJ, 422, 55

KAASTRA, J. S., ROOS, N. & MEWE, R. 1995, EUVE Observations of NGC 5548, Astr. Ap., submitted

MARSHALL, H. L., CARONE, T. E., SHULL, J. M., MALKAN, M. A., & ELVIS, M. 1994, Conceptual Design of a Fast Soft X-ray Polarimeter, Proc. SPIE, 2283, 75

MARSHALL, H. L., FRUSCIONE, A., & CARONE, T. E. 1995, Active Galaxies Observed during the Extreme Ultraviolet Explorer All Sky Survey., Ap. J., 439, 90

MARSHALL, H. L., CARONE, T. E., SHULL, J. M., MALKAN, M. A., & ELVIS, M. 1995, The Steep Soft X-ray Spectrum of the Highly Variable Active Nucleus in Mk 478, ApJ, submitted

MARSHALL, H. L., ET AL. 1995, in preparation

SUN, H-S., & MALKAN, M. A. 1989, Fitting improved accretion disk models to the multiwavelength continua of quasars and active galactic nuclei, ApJ, 346, 68

URRY, C. M., ET AL. 1995, in preparation

X-Ray Selected EUV Galaxies: A Quest for the Faintest Extragalactic EUV Sources

ANTONELLA FRUSCIONE

Center for EUV Astrophysics, 2150 Kittredge Street, University of California,
Berkeley, CA 94720–5030, USA

Using data from the public archive of the *Extreme Ultraviolet Explorer* (*EUVE*) all-sky survey, we have systematically searched for extreme ultraviolet (EUV) emission (58–174 Å, 0.07–0.21 keV) around approximately 2500 distinct positions in the sky corresponding to known X-ray emitting extragalactic sources. We find that 20 X-ray galaxies are EUV bright and were detected with significance above 4 σ during the *EUVE* survey: 8 are reported here for the first time (MS 0037.7−0156, Mrk 142, M 65, EXO 1128.1+6908, M 87, Mrk 507, PKS 2005−489 and 1H 2351−315.A). 68 additional galaxies are detected with a lower significance (3 < σ < 4), but the list is affected by a high percentage of spurious sources.

1. Introduction

Observations at the shortest ($\lambda \lesssim 175$ Å) EUV wavelengths with the *ROSAT* Wide Field Camera (WFC) and the *EUVE* photometers, have established Active Galactic Nuclei (AGN) (Seyfert galaxies, BL Lacertae objects and quasars) as a distinct class of EUV source. The brightest members of this class were discovered during the WFC and the *EUVE* all-sky surveys (Pounds et al. 1993; Malina et al. 1994; Bowyer et al. 1994; Marshall et al. 1995, hereafter MFC; Vennes et al. 1995). Although small in number, and until very recently thought to be invisible at EUV wavelengths, galaxies detected in the EUV represent a crucial sample toward the understanding of the overall spectral energy distribution in AGN. In Seyfert galaxies, for example, two spectral components dominate the emission in the optical to X-ray range: the so called "big blue bump" (extending from the optical-UV to the EUV or soft X-ray range) and an hard X-ray power law component. Soft X-ray excesses have been detected above the hard X-ray component (e.g., Wilkes & Elvis 1987; Walter & Fink 1993) and they have been usually interpreted as a high-frequency tail of accretion disk spectra. However the connection between the soft X-ray excess and the optical/UV bump is still under debate (e.g., Czerny & Życky 1994) and EUV observations may provide the missing link.

All the EUV AGN discovered to date share the common property of having been detected in X-rays. This is in part explained by the fact that the spectra of AGN are generally steep toward soft X-ray energies therefore they can compete with the catastrophic effect, at EUV wavelengths, of the absorption by the interstellar gas. Indeed, all the EUV galaxies lie in directions of low Galactic neutral hydrogen column density ($N(\text{H I}) \lesssim 3 \times 10^{20}$ cm^{-2}). The brightest EUV extragalactic sources included in the published EUV catalogs were discovered by automatic source detection and analysis routines which are designed to extract the largest possible number of new sources while keeping a conservative significance level in order to avoid excessive contamination by spurious sources. In this process the high significance threshold used (e.g., 6 σ in Malina et al. 1994) eliminates many potentially interesting fainter sources. MFC analyzed the *EUVE* all-sky survey data immediately after the end of the survey and reported the detection of 13 AGN at 2.5 σ or better. Additional data were collected after the first 6 months of the survey and the analysis software and calibration data have improved since the

69

S. Bowyer and R. F. Malina (eds.), Astrophysics in the Extreme Ultraviolet, 69–73.

MFC analysis. A more in depth and systematic look at the *EUVE* all-sky survey data to search for new and fainter EUV galaxies is therefore highly motivated.

2. *EUVE* Observations and Data Analysis

We have analyzed the entire Lex/B (58–174 Å) data set (interstellar absorption prevents the observation of any extragalactic source longward of ≈ 100 Å) collected during the *EUVE* all-sky survey and recently made available through the *EUVE* public archive. The analysis was done independently from the previous all-sky survey results, taking advantage of all the improvements in the analysis software and in the calibration data made since the time of the earlier studies. We also included in our analysis "gap filling" data not previously available.

Unbiased automatic source detection and analysis routines applied to low signal-to-noise ratio data, such as the *EUVE* sky maps, are not the best method for finding new faint extragalactic sources. This is because such methods need conservative significance thresholds to limit the number of spurious detections. MFC introduced a "biased" search to find EUV AGN by searching around a sample of about 200 "interesting" positions in the sky, i.e., where a X-ray AGN is present and its predicted EUV count rate is > 0.002 count s^{-1}. In this work we adopted a similar, but less restrictive, approach and did not eliminate any X-ray source a priori, but look for EUV emission at *all* positions in the sky corresponding to a cataloged X-ray extragalactic source (AGN and normal galaxies).

2.1. *Sample Selection*

Our sample of X-ray selected galaxies to test for EUV emission was constructed by collecting all the published lists, known to the author, of extragalactic sources classified as AGN or galaxies detected during surveys or pointed observations performed by the *HEAO 1, Einstein, EXOSAT* and *ROSAT* X-ray satellites. The total sample includes 5305 X-ray sources, 2430 of which are unique.

2.2. *Count Rate Determination*

We determined EUV position, source and background count rates, and detection significance, simultaneously at a given position in the sky, from the *EUVE* sky maps, with the publicly accessible *EUVE* software. The detection significance is expressed in terms of a χ^2 value with 1 degree of freedom ($\approx \sigma^2$ Gaussian) and corresponds to the probability that a detection is spurious, and it is caused by a random background fluctuation: we chose as our first detection threshold $\chi^2 = 9$ ($\approx 3\,\sigma$). Statistically this value corresponds to a formal spurious rate probability of 2.7×10^{-3}. Out of the 2430 X-ray sources analyzed, 92 were detected with significance above $3\,\sigma$, 24 of which were found to be above $4\,\sigma$.

An hidden complication when searching the *EUVE* sky maps at low significance levels, is that spurious detections can be generated by the wing of the point spread function of bright sources. In order to eliminate this possibility with a high degree of confidence, we constructed an empirical model of the effect of the point spread function of bright sources on the surrounding pixels and compared the 92 significant detections in our sample with the First *EUVE* Catalog (Bowyer et al. 1994) setting conservative thresholds around all the sources in the catalog. We rejected as spurious three sources (all with $\chi^2 \geq 16$) among the 92 with $\chi^2 \geq 9$, because the source positions are possibly affected by bright white dwarfs.

TABLE 1. X-ray Selected EUV Extragalactic Sources

X-ray Source Name	Other Name	Type	Count rate (count s^{-1})	Signif σ^2	Log $N(H\ I)$	Flux $10^{-12}\frac{erg}{cm^2s}$
MS0037.7−0156		Sy	0.035±0.012	16.6	20.5	4.33
1H1023+513C	Mrk142	Sy1	0.026±0.008	16.6	20.1	1.36
1WGA118.8+1306	M65	LI	0.049±0.015	23.7	20.3	3.84
EXO1128.1+6908		Sy1	0.022±0.006	17.4	20.1	1.15
1H1226+128	M87	Sy	0.031±0.009	17.8	20.4	3.06
2E1748.8+6842	Mrk507	Sy2	0.005±0.001	18.3	20.6	0.76
RXJ20094−4849	PKS2005−489	BL	0.027±0.009	16.3	20.6	4.08
1H2351−315.A	H2356−309	BL	0.034±0.010	19.4	20.1	1.78
WGA0057−2222	TONS180	Sy1.2	0.054±0.012	37.6	20.2	3.42
1H0419−577	1ES0425−573	Sy	0.036±0.006	72.0	20.4	3.55
REJ1034+393	X12325	Sy	0.023±0.007	19.4	20.2	1.46
1H1104+382	Mrk421	BL	0.052±0.010	50.2	20.2	3.29
H1226+023	3C273	QSO	0.050±0.015	20.0	20.3	3.92
1H1350+696	Mrk279	Sy1	0.024±0.005	44.4	20.2	1.52
1H1415+255	NGC5548	Sy1.5	0.029±0.008	24.1	20.2	1.83
1H1430+423	WGAJ1428+4240	BL	0.029±0.007	29.2	20.1	1.51
1H1429+370	Mrk478	Sy1	0.061±0.010	83.6	20.0	2.71
1H1651+398	Mrk501	BL	0.038±0.007	56.6	20.2	2.40
1H2156−304	PKS2155−304	BL	0.269±0.021	571.7	20.1	14.1
1H2209−470	NGC7213	Sy1	0.032±0.009	20.0	20.4	3.16

2.3. Monte Carlo Random Sample Test

On a purely statistical basis we expect 60 spurious detections at $9 < \chi^2 < 16$ in our sample of 2430 sources. In order to check the reliability of the significance threshold values calculated by the public analysis software, we have performed a Monte Carlo simulation and constructed a control sample of pseudo-random positions as large as the real sample. We generated a pseudo-random sample similar to the real source distribution by converting the coordinates of the sources from celestial to galactic, inverting the sign of the galactic latitude and finally transforming the coordinates back to celestial. We analyzed the random sample the same way as the real sample: 54 random positions yielded $\chi^2 > 9$, 3 of which had $\chi^2 > 16$. However all three 4 σ detections were eliminated because they coincide with EUV bright cataloged objects. In conclusion, 51 random positions (a priori not corresponding with any known or suspected EUV source) were found with significance $9 < \chi^2 < 16$ ($3 < \sigma < 4$). This number is in reasonable agreement with the value of 60 obtained on purely statistical grounds. Given that 68 X-ray AGN and galaxies were found with $9 < \chi^2 < 16$, only about 10 to 20 of them are likely to be real. Therefore we will concentrate from now on, on the highly reliable 4 σ sources.

3. Discussion

Our systematic search found that 21 X-ray selected extragalactic objects have been detected as likely sources of EUV radiation ($\sigma > 4$). To exclude any other possible candidate, we performed a catalog search in a 3 arcmin circle around the 21 positions using the NED and SIMBAD databases: in all but one case (EXO 1429.9+371 which we excluded) the X-ray galaxy is the only cataloged object.

Table 1 gives the final list of the 20 X-ray selected AGN detected during the *EUVE*

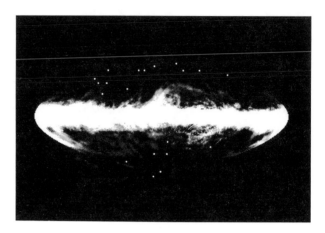

FIGURE 1. Galactic distribution of the X-ray selected AGN detected during the *EUVE* all-sky survey, superimposed on the N(H I) sky map (Dickey & Lockman 1990). The map is centered on the galactic center with longitude increasing to the left.

all-sky survey. Eight of them are reported here for the first time, and are listed in the top section of the table. Twelve were previously known EUV sources (MFC, Vennes et al. 1995). Two EUV galaxies reported in MFC (NGC 4051 and 1H 1013+489) are not detected with $\chi^2 > 16$: the first was detected at $\sigma = 3.6$ in agreement with MFC results, the second showed a low significance of 2.7 σ. We want to emphasize the fact that our independent analysis recovered all the previously discovered EUV AGN.

In the table we list the source type, as Sy (Seyfert), BL (BL Lac object), LI (LINER) or QSO: all the 4 σ sources are AGN. The EUV position of all the sources is within 2 arcmin of the X-ray position, and even closer to the object optical coordinates. The Lex/B count rate and the corresponding standard deviation are in column 4. The significance (in column 5) is expressed as a χ^2 value and it is approximately equal to the Gaussian σ^2. Column 6 lists the N(H I) values from Stark et al. (1992) or Dickey & Lockman (1990) unless a more accurate narrow beam pointed 21 cm measurement was found in the literature (e.g., Elvis et al. 1989)–interstellar absorption (both in the Galaxy and intrinsic to the source) is the single most relevant limiting factor in the detection of EUV galaxies. As expected, the column densities in the directions of the detected EUV AGN are below $\approx 3 \times 10^{20}$ cm^{-2} in all cases but Mrk 507 and PKS 2005−489; however Mrk 507 was observed for more than 35000 s due to its high ecliptic latitude which could explain the detection despite the relatively high N(H I). A conspicuous demonstration of the anticorrelation between AGN detection in the EUV and N(H I) distribution is given in Figure 1 where the galactic distribution of the 20 X-ray selected EUV AGN is superimposed on the N(H I) map (Dickey & Lockman 1990).

MFC discussed the problems related to a proper determination of the EUV flux from a single broad-band EUV count rate, namely the strong dependence on N(H I) and the fact that the EUV spectral shape is poorly known unless EUV and soft X-ray observations are taken simultaneously. We used Table 3 from MFC and the count rate from column 4 to estimate the approximate source flux reported in column 7. We assumed the Galactic absorption in column 6 and a "sample" energy index $\alpha = 2$ for all the sources (varying α in the 0.5–3 interval results in a maximum of 30% change in the total flux).

Most of the brightest EUV extragalactic sources have been studied in detail and EUV spectroscopy has been performed on 6 of them. We refer to the individual papers for comments on PKS 2155−304 (Marshall et al. 1993, Fruscione et al. 1994, Königl et al. 1995), 3C 273 (Kafatos et al. this volume), Mrk 421 (Bruhweiler et al. 1995), Mrk 478 (Marshall et al. 1995b), NGC 5548 (Kaastra et al. 1994), TON S180 (Vennes et al. 1995) and Mrk 279 (Hwang, Bowyer, & Lampton this volume). Detailed analysis on each individual new EUV AGN will be presented in a forthcoming paper (Fruscione 1995).

4. Conclusion

We have systematically searched the *EUVE* all-sky survey data for EUV emission from a large sample of X-ray selected AGN and galaxies. 20 AGN emit detectable EUV radiation, 8 of which (6 Seyfert galaxies and 2 BL Lac objects) are reported here for the first time. This sample includes the brightest EUV extragalactic sources and it is likely to represent nearly all the known AGN that were detected during the *EUVE* all-sky survey. However, many unidentified sources are still listed in the EUV catalogs and could hide many more new extragalactic counterparts.

This research has made extensive use of the SIMBAD database, operated at CDS, Strasbourg (France), of the HEASARC service, provided by NASA/GSFC, and of the NED database operated by the JPL-NASA, Caltech. This work has been supported by NASA contract NAS5-29298 to UCB.

REFERENCES

BOWYER, S., LIEU, R., LAMPTON, M., LEWIS, J., WU, X., DRAKE, J. J., & MALINA, R. F. 1994, ApJS, 93, 569

BRUHWEILER, C., ET AL. 1995, in prereparation

CZERNY, B., & ŻICKY, P. T. 1994, ApJ, 431, L5

DICKEY, J. M., & LOCKMAN, F. J. 1990, ARA&A, 28, 215

ELVIS, M., LOCKMAN, F. J., & WILKES, B. J. 1989, AJ, 97, 777

FRUSCIONE, A., BOWYER, S., KÖNIGL, A., & KAHN, S. M. 1994, 422, L55

FRUSCIONE, A. 1995, ApJ, submitted

KAASTRA, J. S., ROOS, N., MEWE, R. 1994, A&A, in press

KÖNIGL, A., KARTJE, J. F., BOWYER, S., KAHN, S. M., & HWANG, C. -Y. 1995, ApJ, in press

MALINA, R. F. ET AL. 1994, AJ, 107, 751

MARSHALL, H. L., CARONE, T. E., & FRUSCIONE, A. 1993, ApJ, 414, L53

MARSHALL, H. L., FRUSCIONE, A., & CARONE, T. E. 1995, ApJ, 439, 90 MFC

MARSHALL, H. L., CARONE, T. E., SHULL, J. M., MALKAN, M. A., ELVIS, M. 1995b, ApJ, submitted

POUNDS, K. A. ET AL. 1993, MNRAS, 260, 77

STARK, A. A., GAMMIE, C. F., WILSON, R. W., BALLY, J., LINKE, R. A., HEILES, C., & HURWITZ, M. 1992, ApJS, 79, 77

VENNES, S., POLOMSKI, E., BOWYER, S., & THORSTENSEN, J. R. 1995, ApJ, in press

WALTER, R., & FINK, H. H. 1993 A&A, 274, 105

WILKES, B. J., & ELVIS, M. 1987 ApJ, 323, 243

EUVE Observations of the Seyfert Galaxy MRK 279

C.-Y. HWANG, S. BOWYER, AND M. LAMPTON

Center for EUV Astrophysics, 2150 Kittredge Street,
University of California, Berkeley, CA 94720–5030, USA

We report *EUVE* spectral and photometric data of the Seyfert 1 galaxy MRK 279. The photo-metric data show large amplitude variations over time scales less than 10,000 s. The spectrum is characterized by several features between 80 and 100 Å. We compare the observed data with several models. We can rule out the possibility that the EUV emission is from a diffuse corona or intercloud medium. Models that assume the soft X-ray/EUV emission results from reprocessing in an optical BLR region are also inconsistent with the data. A collisional excitation model is consistent with the observations but requires a cloud density $\geq 10^{11}$ cm^{-3}.

1. Introduction

The relation between the optical-/ultraviolet (OUV) blue bump and soft X-ray excess is an important topic in AGN studies. It has been conjectured that the accretion disk around the black hole of an AGN will manifest itself as a strong EUV emitter (e.g., Ross et al. 1992). On the other hand, reprocessing of X-ray photons due to cold (neutral) and warm (partially ionized) material in the AGN can significantly alter the appearance of the emergent spectrum in the EUV and soft–X-ray range (e.g., Netzer 1993). Observations of AGNs in the EUV range have the potential to clarify the physical mechanisms that are operative in these systems.

The Seyfert 1 galaxy MRK 279 is one of the brightest Seyfert galaxies detected during the *EUVE* all-sky survey (Bowyer et al. 1995) where it is designated J1352+692. The Galactic hydrogen column density toward this object determined by high angular-resolution H I measurements (Elvis et al. 1989) is $N_{HI} = 1.64 \times 10^{20}$ cm^2, which suggests that this object may be detected well into the EUV range. We report here an EUV observation of MRK 279 obtained with *EUVE*.

2. Observations and Data Reduction

The Seyfert 1 galaxy MRK 279 was observed simultaneously with the short wavelength spectrometer (70−190 Å) and the Deep Survey telescope on the *Extreme Ultraviolet Explorer* (*EUVE*) over a period of 7 days from 1994 April 22 to April 29. We first plotted the photometer and spectrometer counts in time sequence and rejected time intervals with obvious excess count rates. The remaining data were further processed by eliminating points that deviated more than 2 σ from the median. This left 199,191 s of exposure time for analysis.

Deep Survey data of MRK 279 were binned with bin sizes of 5400 s and 600 s. The light curve with bin sizes of 5400 s is shown in Figure 1a; this shows amplitude variations by a factor of 2 on time scales of one day and a trend of increasing fluxes toward the end of the observation. The light curve with bin sizes of 600 s shown in Figure 1b. This is a magnified portion of the Figure 1a data, revealing for the first time the substantial short term EUV variability of the object.

S. Bowyer and R. F. Malina (eds.), Astrophysics in the Extreme Ultraviolet, 75–80.

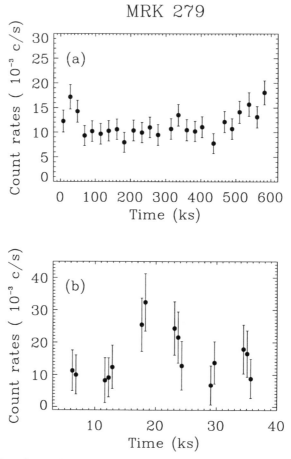

FIGURE 1. *EUVE* Deep Survey light curves of MRK 279. (*a*) The light curve with a bin size of 5400 s illustrates the EUV flux variation over the 7-day observation. The data begin at JD = 2449465.53. Amplitude variations by a factor of 2 are detected on time scales of 1 day. (*b*) The light curve with a bin size of 600 s. This subset of the data begins at the same starting point as (*a*) and illustrates the rapid EUV flux variation. Amplitude variations by a factor of 3 are detected on time scales less than 10,000 s.

The spectrum of MRK 279 was obtained from the spectrometer by subtracting backgrounds evaluated from both sides of the spectrum. The spectrum with $\lambda < 80$ Å has been excluded because of uncertain background estimations. The spectrum is shown in Figure 2. It is characterized by several line features. Some of these features exceed the extrapolation of the *ROSAT* power-law fit (Walter & Fink 1993) by 2-3 σ. The error bar shown on Figure 2 provides an estimate of the standard deviation of the data. Comparing the error bar with these features, we conclude that it is very unlikely that these features are all due to statistical deviations of a smooth photon distribution.

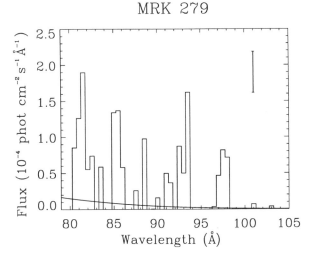

FIGURE 2. *EUVE* spectrum of MRK 279. The histogram is the background-substracted EUV flux. The solid curve is an extrapolation of the *ROSAT* power-law fit.

3. Models

Considerable quantities of warm and/or cold gas have been found near active galactic nuclei with optical and X-ray observations (e.g., Antonucci & Miller 1985; Turner et al. 1991). This gas can affect the emergent spectra in the EUV and soft X-rays by absorbing and reflecting the intrinsic spectra. In addition, warm gas can also produce soft X-ray and EUV emission. A complete model should take into account all these processes in the gas. In this section we discuss various models suggested for Seyfert galaxies and compare these models with the observed data.

3.1. *Power Law Model*

A simple power-law extrapolation of X-ray can not fit the observed EUV flux. The comparison of the observed EUV spectrum with the extrapolation of the *ROSAT* power-law fit (Walter & Fink 1994) shows a difference with 98% significance. The power-law extrapolations from other X-ray observations (Malaguti et al. 1994) have even larger deviations from the EUV spectrum. The best power-law fit differs from the *EUVE* spectrum at the 96.7% significance level, and is inconsistent with the observed features in the EUV spectra discussed below.

3.2. *Photoionization Model*

Line and line-like features can be produced by reflection from a gas photoionized by a continuum flux. However, a photoionization model that can take into account all necessary line emission mechanisms in the EUV region is not available. For example, line emission from photoionized gas in this wavelength band comes mostly from dielectronic recombination, which requires a detailed many-body treatment and is only treated approximately in most models. We must therefore be careful regarding conclusions from a comparison of current models with the *EUVE* data.

We have calculated the emission from a photoionized gas using the photoionization code XSTAR (Kallman & Krolik 1993) for several representative ionization parameters and column densities. The model consists of a spherical gas cloud photoionized with a point source of continuum radiation at the center. A power law of $\alpha = 0.84$, which is the best-fit power law of X-ray observations (Malaguti et al. 1994), has been used for the continuum radiation at the center. The ionization parameters explored range from 1 to 1000 and the column densities from 5.0×10^{20} to 1.0×10^{22}. Despite the wide range of parameters explored, we were unable to find suitable physical parameters to explain the *EUVE* spectrum. It is unclear whether this is due to limitations in the theoretical models or if the model is fundamentally inappropriate.

3.3. *Collisional Excitation Model*

The EUV features of MRK 279 may be emission lines from a collisionally ionized thermal plasma around the central source of the AGN. We have calculated the theoretical EUV spectra at different temperatures using the thermal plasma code of Landini and Monsignori Fossi (Landini & Monsignori Fossi 1990; Monsignori Fossi & Landini 1994) and compared the results with our data. The best fit temperatures occur in a relatively narrow range from 3.0×10^5 to 6.0×10^5 K where the χ^2 has an improvement $\geq 90\%$ significance over the other temperature fits. We have modeled the emission using the Landini-Monsignori Fossi model plus the *ROSAT* power law $\alpha = 1.15$ (Walter & Fink 1994). The results are shown in Figure 3 where they are compared with the data. The theoretical spectrum has been averaged over 0.5 Å to simulate the data binning. Most of the lines are emission lines of the ionized heavy elements Mg, Si, Ne, and Fe. The total luminosity between 70 and 100 Å is 5.6×10^{42} erg s^{-1} assuming $H_0 = 50$ km s^{-1} Mpc^{-1} and a red shift z = 0.029 (Bowyer et al. 1984, Helou et al. 1991). The volume emission measure required is $\gtrsim 1.0 \times 10^{66}$ cm^{-3}. The time scale of the EUV variation (10,000 s) sets an upper bound to the size of the emitting region of 3×10^{14} cm by the argument of light travel time. To obtain the line emission that we have observed with *EUVE*, the density of the plasma must exceed 1×10^{11} *cm*$^{-3}$. If the emitting gas is in a form of homogeneous corona, the Thomson optical depth would be ~ 20 and would wash out the lines completely. Thus the *EUVE* results are inconsistent with models of diffuse corona with a homogeneous intercloud medium (Emmering et al. 1992).

One way to circumvent this problem is to assume that the gas is clumped into clouds and the total gas density is $> 10^{11}$ cm^{-3}. For example, let us assume each individual cloud has density of 10^{14} cm^{-3} (e.g. Ferland & Rees 1988) and column density of 10^{22} cm^{-2}. The Thomson optical depth is ~ 0.007 for every cloud. The filling factor of these clouds is only 10^{-6} and a typical line of sight encounters only $\sim$ three clouds. Such clouds would be similar to the optical BLR clouds except that the EUV emitting clouds are closer to central source of the AGN and thus have higher densities and ionization parameters than their optical counterparts.

The total luminosity required to heat the gas to the inferred temperature in this model is 1.6×10^{44}, which is comparable to the luminosity 2.6×10^{44} derived from dynamical models (Wandel 1991).

4. Conclusions

We have obtained *EUVE* spectral and photometric observations of MRK 279. The light curve of MRK 279, which is among the first EUV light curves of Seyfert galaxies obtained, sets a strong constraint of 3.0×10^{14} cm on the size of the EUV emitting region assuming that the emission is correlated. This size is much smaller than the optical BLR region, which is order of 10 light days (Stripe et al. 1994). We can thus rule out models assuming the EUV emission arising from the BLR region. Models that assume a

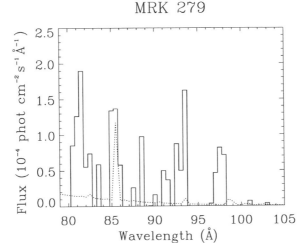

FIGURE 3. Comparison of line emission from a collisionally excited plasma with the EUV data. The histogram is the EUV data. The dotted-line emission features are red-shifted lines of Mg, Si, Ne, and Fe from a thermal plasma at temperature 4.5×10^5 K as predicted by the Landini & Monsignori Fossi model. A power law of $\alpha = 1.15$ has been added. The line at 85.5 Å is from Si VI.

diffuse corona or intercloud medium for the origin of EUV excess in MRK 279 are also inconsistent with the observations. A model with collisionally excited gas clouds is more consistent with the line features of the spectrum. The photometric data in combination with the spectrometer data suggest emission from clouds with densities $\geq 10^{11}$ cm^{-3} near the center of the AGN.

We thank Arieh Königl and Steven Kahn for valuable comments. This work has been supported by NASA contract NAS5-30180.

REFERENCES

Antonucci, R. R., & Miller, J. S. 1985, ApJ, 297, 621

Bowyer, S., Lampton, M., Lewis, J., Wu, X., Jelinsky, P., Lieu, R., & Malina, R. F. 1995, ApJ, submitted

Bowyer, S., Brodie, J. P., Clarke, J. T., & Henry, J. P. 1984, ApJ, 278, L103

Elvis, M., Lockman, F. J., & Wilkes, B. J. 1989, AJ, 97, 777

Emmering, R. T., Blandford, R. D., & Shlosman, I. 1992, ApJ, 385, 460

Ferland, G. J., & Rees, M. J. 1988, ApJ, 332, 141

Helou, G., Madore, B. F., Schmitz, M., Bicay, M. D., Wu, X., & Bennett, J. 1991, in Databases and On-Line Data in Astronomy, ed. M.A. Albretch & D. Egret, Dordrecht: Kluwer, 89

Kallman, T. R., & Krolik, J. H. 1993, NASA Internal Report

Landini, M., & Monsignori, Fossi, B. C. 1990, A&AS, 82, 229

Malaguti, G., Bassani, L., & Caroli, E. 1994, ApJS, 94, 517

Monsignori Fossi, B. C., & Landini, M. 1994, Solar Physics, 152, 81

Nandra, K. & Pounds, K. A. 1992, Nature, 359, 215

Netzer, H. 1993, ApJ, 411, 594

Ross, R. R., Fabian, A. C., & Mineshige, S. 1992, MNRAS, 258, 189

Stripe, G. M., et al. 1994, A&A, 285, 857

Turner, T. J., Weaver, K. A., Mushotzky, R. F., Holt, S. S., & Madejski, G. M. 1991, ApJ, 381, 85

Walter, R., & Fink, H. H. 1993, A&A, 274, 105

Wandel, A. 1991, A&A, 241, 5

EUVE Spectra of Coronae and Flares

CAROLE JORDAN

Department of Physics (Theoretical Physics), University of Oxford,
1 Keble Road, Oxford, OX1 3NP, UK

Following a summary of early solar EUV spectroscopy the spectra of some late-type stars obtained with the *Extreme Ultraviolet Explorer* (*EUVE*) are briefly surveyed. Some transitions which are not included in current emissivity codes but could lead to numerous weak lines, and an apparent continuum in the *EUVE* short wavelength region, are discussed. The importance of the geometry adopted when interpreting the emission measure distribution is stressed, since radial factors can lead to an *apparent* emission measure distribution gradient that is steeper than the value of 3/2 expected in plane parallel geometry.

1. Introduction

The launch of the *Extreme Ultraviolet Explorer* (*EUVE*) began a new era in the study of stellar coronae and stellar flares. The detailed spectra of stars that are being obtained are essential in understanding the physics of stellar coronae and flares, and will aid the interpretation of data with low spectral resolution.

Our present understanding of stellar EUV spectra builds on over 30 years of research in solar spectroscopy, concerning line identifications, plasma diagnostic techniques and methods of modelling from emission line fluxes (see, e.g., Feldman, Doschek, & Seely 1988; Mason & Monsignori Fossi 1995; and Jordan & Brown 1981; respectively).

2. Early Solar EUV Spectroscopy

Tousey (1967) reviewed early observations of the solar spectrum in the wavelength range 170 Å to 370 Å, which date from 1960. The group of strong lines between 170 Å and 220 Å were shown to be transitions of the type $3p^n - 3p^{n-1}3d$ in Fe VIII to Fe XIV (Gabriel, Fawcett & Jordan 1966; see also Fawcett 1974, 1981). Flare spectra obtained from instruments on the *Skylab* Apollo Telescope Mount (ATM) allowed Sandlin et al. (1976) to identify lines of highly ionized iron and of less abundant elements, in the *EUVE* medium wavelength (MW) range. A more complete list was given by Dere (1978). The lines of Fe XVII ($2p^53s - 2p^53p$) are of particular interest since this ion is not represented by strong lines in the *EUVE* short wavelength (SW) region.

Transitions of the type $3s^23p^n - 3s3p^{n+1}$ in iron lie in both the *EUVE* MW and long wavelength (LW) ranges. Early spectra contained many second order lines, as do *EUVE* spectra when the coronal temperature is about $1 - 2 \times 10^6$ K. The identifications are discussed by Bromage, Cowan & Fawcett (1978) and in references therein.

In the *EUVE* SW range the solar spectra analyzed by Malinovsky & Heroux (1973) still provide a useful comparison for stars which have coronal temperatures around $1 - 2 \times 10^6$ K. In addition to lines of Ne, Mg and Si, this region contains many lines originating from transitions in Fe IX to Fe XII of the type $3p^n - 3p^{n-1}4s, 4d$ (e.g., Fawcett et al. 1972), and the $3p^n - 3p^{n-1}3d$ transitions in Ni IX $-$ Ni XIV. The first solar flare spectrum was obtained from OSO-5 by Kastner, Neupert, & Swartz (1974), but see Fawcett & Cowan (1975) for some revised identifications. The flare spectra are dominated by transitions of the type $2s^22p^n - 2s2p^{n+1}$ in Fe XVIII to Fe XXIII and provide a useful reference for sources where the temperature is around 6×10^6 K to 2×10^7 K.

S. Bowyer and R. F. Malina (eds.), Astrophysics in the Extreme Ultraviolet, 81–88.

3. Survey of *EUVE* Spectra of Coronae and Flares

3.1. *Procyon: α CMi, F5 IV–V*

The spectrum of Procyon is very similar to that of the Sun. The strong group of lines of Fe IX to Fe XI around 171 Å to 180 Å was observed previously at lower resolution from *EXOSAT* (Schrijver 1985). Observations from the *Einstein* Observatory were analyzed by Schmitt et al. (1985) and Jordan et al. (1986). The emission measure distribution found by Drake, Laming & Widing (1995) from the *EUVE* spectra is consistent with the earlier work, giving a maximum emission measure at about 1.6×10^6 K. Although the identifications of the stronger lines of iron are secure at the resolution of *EUVE* lines of elements such as magnesium, silicon and sulphur usually occur in blends. Drake et al. (1995) have deconvolved such blends through spectral modelling. Longer exposures are required to confirm the existence of the weakest lines. Since these provide at least upper limits to line fluxes the conclusion by Drake et al. (1995), that the spectra are best fitted with solar photospheric abundances is unlikely to change. Lines of oxygen and neon are particularly important since they provide the link to the region observed with the *International Ultraviolet Explorer (IUE)*, while silicon is the only element with observable lines formed below and above 2×10^5 K.

3.2. *ξ Boo A (+B); G8 V (+ K4 V)*

Since observations with *IUE* show that ξ Boo A dominates the total flux from the system in the UV region, we assume here that the *EUVE* flux arises mainly from ξ Boo A.

Figure 1 shows segments of the spectra obtained in our Guest Investigator observations. The SW spectrum shows the stronger of the Fe XVIII lines, and probably lines of Fe XIX and Fe XX. A line near 98 Å may not be entirely due to Ne VIII. The MW spectrum shows strong lines of Fe XV and Fe XVI (the latter appear also in the LW spectrum), but only very weak emission in the region of the Fe IX to Fe XI lines. The group of Fe XIII lines appears around 202 Å, and lines Fe XIV at 211 Å, 219+220 Å and 264 Å may be detected. Longer exposures are required to confirm the existence of weak features.

3.3. *Capella (G1 III + G8 III) and AU Mic (dM0e)*

These two stars are discussed together since it is useful to make direct comparisons of the calibration spectra. The SW spectra of these stars all show lines of Fe XVIII to Fe XXIII. Fe XIV appears at 192 Å, although Ca XVII may also be present. Fe XV is observed in the MW spectrum and Fe XVI in the MW and LW spectra. The line identifications in Capella have been discussed by Dupree et al. (1993). The higher temperature of AU Mic compared to Capella is apparent through the presence of lines of Fe XXII at 114.41 Å and Fe XXI at 102.22 Å, which are absent or very weak in Capella. In AU Mic the ratio of the Fe XXII lines to those of Fe XX suggest that the blend at 132.8 Å is dominated by Fe XXIII, rather than Fe XX. Conversely, the line of Fe XIX at 101.55 Å is significantly weaker in AU Mic than Capella. If one examines the precise positions of features near the Fe XXI lines at 97.88 Å and 102.22 Å (whose flux ratio is not sensitive to the electron density), in AU Mic and Capella, it appears that the contribution of Fe XXI to the feature at 98 Å has been overestimated by Dupree et al. (1993). They give F(97.88)/F(102.22) = 1.7, whereas Brickhouse et al. (1995) calculate a ratio of 0.44, closer to the ratio observed in AU Mic. Monsignori Fossi & Landini (1994) have discussed the spectra of AU Mic.

Regarding electron densities, the ratio of the fluxes of either of the Fe XXI lines at 97.88 Å or 102.22 Å to that of Fe XXI at 128.73 Å is sensitive to N_e (Mason et al. 1979). The clear *change* in the ratios between Capella and AU Mic shows that in AU Mic, at least, the ratio is in the density sensitive regime. Dupree et al. (1993) find a ratio of

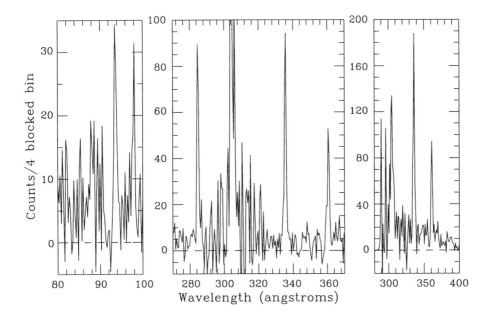

FIGURE 1. EUVE spectra of ξ Boo A, showing lines of Fe XVIII (93.9 Å), Fe XV (284 Å) and Fe XVI (335 Å and 360 Å).

$F(128.73)/F(102.22) = 4.76$, which is close to the low density limit value of 4.6 from the calculations by Brickhouse et al. (1995). The ratio of these lines in AU Mic leads to a density of about $2 - 4 \times 10^{12}$ cm^{-3} (see also Monsignori Fossi & Landini 1994). A higher density in AU Mic is also suggested by a doublet transition in Fe XX at 110.63 Å, which is absent in Capella. The line of Fe XXI at 146.65 Å, whose relative flux increases with N_e, may just be detectable in AU Mic, but a longer exposure is required to confirm its presence in Capella. The density in Capella seems to be $\leq 10^{10}$ cm^{-3}, lower than proposed by Dupree et al. (1993).

4. Sources of Weak Lines

It has been proposed that the continuum observed in the *EUVE* SW region below 150 Å, and the line to continuum ratio, requires an abundance of iron that is lower than in the solar photosphere (e.g., Stern et al. 1995). The present emissivity codes include the stronger lines, but many weaker transitions that could contribute to the continuum are absent. The transitions discussed below need to be further investigated.

4.1. Weak Lines Formed at $T_e \simeq 1 - 3 \times 10^6$ K

Although transitions of the type $3p^n - 3p^{n-1}4s, 4d$ in the ions Fe IX to Fe XIV were identified in early solar spectra some lines of moderate strength remained unidentified. Jordan (1968) suggested that these could be $3p^{n-1}3d - 3p^{n-1}4p, 4f$ transitions, excited via the non-dipole transitions $3p^n - 3p^{n-1}4p, 4f$ which can have collision strengths larger than or comparable to dipole transitions. (See reviews of atomic data in Lang 1994). In

Fe IX to Fe XIV the $3d - 4p, 4f$ transitions lie in groups between 117 Å to 73 Å, and 170 Å to 91 Å, respectively, and some do correspond to lines in the solar spectrum, but not all of the 4p and 4f levels are known. When $T_{corona} \simeq 1 - 3 \times 10^6$ K these lines will contribute to the apparent continuum in the $EUVE$ SW spectra. Mewe, Gronenschild & van den Oord (1985) do include some of the $3p^n - 3p^{n-1}4s, 4d$ transitions and a few $3d - 4f$ transitions in Fe XIV to Fe XVI, but the majority of the possible transitions are absent from the current emissivity codes. Much of the atomic data required to predict individual line fluxes has not yet been calculated.

4.2. Weak Lines Formed at 5×10^6 to 2×10^7 K

At these temperatures the *strong* lines observed in the $EUVE$ SW spectra are of the type $2s^2 2p^n - 2s2p^{n+1}$ in Fe XVIII to Fe XXIII. The $2p^n - 2p^{n-1}3s, 3d$ transitions in Fe XVII to Fe XXII occur between 11.8 Å and 17.1 Å. Mewe et al. (1985) do include the above transitions, but only some of the non-dipole $2p^n - 2p^{n-1}3p$ transitions.

Transitions of the type $2s^2 2p^n - 2s2p^n 3s$, 3p and 3d can also occur, with alternative decay routes in transitions of the type $2s^2 2p^{n-1}3l$ - $2s2p^n 3l$. These $2s - 2p$ transitions in the presence of a spectator $3l$ electron lie in broadly the same wavelength region as the strong $2s^2 2p^n - 2s2p^{n+1}$ transitions, as will all transitions with a spectator nl ($n \geq 4$) electron. Apart from Fe XVII, for which Loulergue & Nussbaumer (1975) made a detailed theoretical study, the energy levels involved and emissivities are mostly unknown. Together, the above transitions in several stages of ionization could appear as a weak continuum in the $EUVE$ SW region.

Solar flare spectra do show transitions with $\Delta n = 2$, between about 8 Å and 13 Å, e.g., $2p^n - 2p^{n-1}4s, 4d$, but many were identified only recently (Fawcett et al. 1987), and are not systematically included in the emissivity codes. Drake et al. (1994) find that the "MEKA" code (Mewe et al. 1985; Kaastra 1992) underestimates the flux observed in the $ASCA$ spectrum of β Cet (K0 III) around 9.5 Å, and they increase the abundance of magnesium to achieve a fit. However, the iron lines with $\Delta n = 2$ lie around the Mg XI (He I-like) lines, and could provide the additional flux without increasing the abundance of magnesium. Decays of the type $3l - 4(l \pm 1)$ are known only in Fe XVII, but in Fe XVII to Fe XXII will lie between about 35 Å to 75 Å.

In view of the large number of weak transitions not yet included in the emissivity codes it seems premature to attribute discrepancies between the observations and the predictions of the codes to non-photospheric abundances.

5. Emission Measures and Their Interpretation

5.1. Apparent and True Emission Measures

The *volume* emission measure derived from an X-ray flux is usually defined through

$$F_{\oplus} = \int \epsilon_\lambda N_e^2 dV/(4\pi d^2) \tag{5.1}$$

where d is the distance to the star and ϵ_λ is the line emissivity. This expression assumes that *all* the photons escape from the emitting region without interception by the star and gives an *apparent* volume emission measure

$$Em(V)_{app} = 4\pi d^2 F_{\oplus}/\epsilon_\lambda \tag{5.2}$$

However, for a spherically symmetric atmosphere, the fraction of photons *not* intercepted by the star is given by

$$G(r) = 0.5[1 + (1 - (R_*/r)^2)^{1/2}] \tag{5.3}$$

In interpreting transition region fluxes one usually assumes a plane parallel atmosphere with $r = R_*$, so $G(r) = 0.5$.

The *apparent* emission measure over height is then

$$Em(h)_{app} = (F_\oplus d^2)/(\epsilon_\lambda R_*^2) = Em(V)_{app}/4\pi R_*^2 \tag{5.4}$$

However, if the *true* volume emission measure is

$$Em(V) = \int N_e^2 4\pi r^2 dr \tag{5.5}$$

then what is observed is

$$Em(V)_{app} = \int N_e^2 4\pi r^2 G(r) dr \tag{5.6}$$

Thus the relation between the *true* emission measure over radial extent, $Em(r)$, and the apparent volume emission measure is

$$Em(r) = Em(V)_{app}(R_*/r)^2/4\pi R_*^2 G(r) \tag{5.7}$$

This spherically symmetric formulation has been used by Harper (1992) and Pan & Jordan (1995).

The important point is that the *apparent* emission measure distribution will not have the same gradient (with temperature) as the true emission measure distribution, because of the additional geometrical factors.

5.2. *What Determines the Shape of the Emission Measure Distribution?*

Jordan et al. (1987) have shown that between $T_e = 2 \times 10^4$ K and 10^5 K the shape of the emission measure distribution is determined by the inverse of the radiative power loss function $P_{rad}(T_e)$.

Above about $T_e = 2 \times 10^5$ K, one can make two alternative simple approximations to show that the *true* emission measure distribution depends on T_e to a power that is close to 3/2. Consider the energy balance

$$dF_M/dr = -dF_C/dr - dF_R/dr \tag{5.8}$$

where F_M, F_C and F_R are the non-thermal, conductive and radiative fluxes, respectively.

Now

$$F_C = -\kappa T_e^{5/2} dT_e/dr \tag{5.9}$$

Writing the emission measure over a fixed interval of $\Delta \log T_e = 0.3$ as

$$Em(0.3) = P_e^2 (dr/dT_e)/\sqrt{2} k^2 T_e \tag{5.10}$$

and substituting for dr/dT_e in terms of F_C gives

$$Em(0.3) = \kappa P_e^2 T_e^{3/2}/\sqrt{2} k^2 F_C \tag{5.11}$$

When F_C and P_e are constant, one obtains the well-known result that $Em(r)$ scales as $T_e^{3/2}$.

Alternatively one can assume that over the first pressure-squared scale-height, from which the emission is mainly observed, conduction balances radiation so that in plane parallel geometry (see Pan & Jordan 1995 for the spherically symmetric case).

$$dF_C/dT_e = -dF_R/dT_e \tag{5.12}$$

Using

$$dF_R/dT_e = 0.8 N_e^2 P_{rad}(T_e) dr/dT_e \tag{5.13}$$

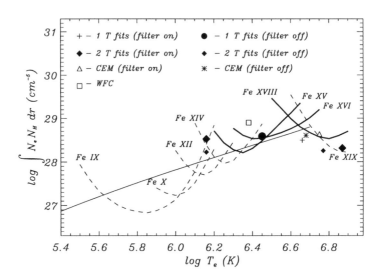

FIGURE 2. The apparent emission measure distribution for ξ Boo A from *EUVE* line fluxes. Short dashed lines indicate upper limits. The full line is the energy balance model using spherical symmetry. See key to symbols and text for results from ROSAT.

leads to

$$d \log Em(0.3)/d \log T_e = 3/2 + 2d \log P_e/d \log T_e - 1.6k^2 Em(0.3)^2 P_{rad}(T_e)/\kappa P_e^2 T_e^{3/2} \tag{5.14}$$

The last term is essentially the coronal value of F_R/F_C, and substituting typical numbers shows that this ratio is usually much less than 1.

The emission measure distribution can then be computed as a function of temperature using the apparent coronal emission measure and temperature as the initial conditions, where the initial pressure is found by assuming that the coronal emission is formed over a pressure-squared scale height. If only the gas pressure is included in hydrostatic equilibrium then the radial extent is not large even in coronal giants. But if a wave-pressure term is included, consistent with the widths of UV lines (see Harper 1992), then the radial extent becomes quite large. The *apparent* emission measure distribution can then have a gradient steeper than 3/2, while the *true* emission measure distribution remains close to a 3/2 gradient. The same approach can be used with other geometries, e.g., those appropriate to loop structures, when the loop area is constrained by the fact that equation (5.14) has no solution if the starting emission measure is larger than a certain value.

5.3. *Example Results*

Figure 2 shows the apparent emission measure loci found from lines of Fe XV, XVI and XVIII observed in the spectra of ξ Boo A obtained with EUVE. The results from *alternative* fits to our *ROSAT* PSPC spectra and the point obtained from the WFC all-sky survey (Pye et al. 1995) are also shown. The emission measure from *IUE* spectra (Jordan et al. 1987) is about 10^{27} at 10^5 K. The WFC filter ratio assigns all the emission measure

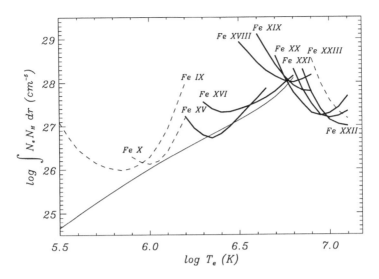

FIGURE 3. The apparent emission measure distribution for Capella from *EUVE* line fluxes. The full line is the result of an energy balance model using spherical symmetry and including wave pressure.

to the average temperature of the lines in the WFC wavebands and therefore tends to overestimate the emission measure. With two temperature fits the higher temperature tends to be larger than that of the peak coronal emission measure. The emission measure at the lower temperature is significantly higher than the upper limit found from the *EUVE* line fluxes. A continuous emission measure fit gives the best agreement with the peak emission measure from the *EUVE* line fluxes. In the fits illustrated the 1994 version of the Raymond-Smith code has been used. For ξ Boo A there is little difference between the solutions using plane parallel and spherically symmetric geometry. The difference between the observed distribution and that computed at temperatures between 3×10^5 K and 10^6 K may be due to the area of the emitting region varying with temperature.

Figure 3 shows the apparent emission measure distribution for Capella, derived from the fluxes given by Dupree et al. (1993) and the emissivities of Brickhouse et al. (1995) at $N_e = 10^{10}$ cm^{-3}. Between $T_e = 4 \times 10^5$ K and 6×10^6 K the *apparent* emission measure gradient is clearly steeper than 3/2. This gradient can be reproduced by an energy balance model using spherical symmetry and *including wave pressure* in the equation of hydrostatic equilibrium. Without the wave pressure the predicted gradient is around 1.7. In either case the *true* emission measure distribution has a gradient close to 3/2.

I am grateful to Dr Hongchao Pan for extracting the *EUVE* spectra used above, and to Miss D. Philippides for allowing me to use the spectral fits she has made to our spectra of ξ Boo A obtained with ROSAT.

REFERENCES

BRICKHOUSE, N. S., RAYMOND, J. C. & SMITH, B. W. 1995, ApJS, 97, 551

BROMAGE, G. E., COWAN, R. D. & FAWCETT, B. C. 1978, MNRAS, 183, 19

DERE, K. P. 1978, ApJ, 221, 1062

DUPREE, A. K., BRICKHOUSE, N. S. DOSCHEK, G. A., GREEN, J. C., & RAYMOND, J. C. 1993, ApJL, 418, L41

DRAKE, J. J., LAMING, J. M. & WIDING, K. G. 1995, ApJ, 443, 393

DRAKE, S. A., SINGH, K. P., WHITE, N. E. & SIMON, T. 1994, ApJL, 436, L87

FAWCETT, B. C. 1974, Adv. At. Mol. Phys., 10, 223

FAWCETT, B. C. 1981, Phys. Scripta, 24, 663

FAWCETT, B. C. & COWAN, R. D. 1975, MNRAS, 171, 1

FAWCETT, B. C., COWAN, R. D., KONONOV, E. Y. & HAYES, R. W. 1972, J. Phys. B, 5, 1255

FAWCETT, B. C., JORDAN, C., LEMEN, J. & PHILLIPS, K. 1987, MNRAS, 225, 1013

FELDMAN, U., DOSCHEK, G. A. & SEELY, J. F. 1988, J. Opt. Soc. Am. B, 5, 2237

GABRIEL, A. H., FAWCETT, B. C. & JORDAN, C. 1966, Proc. Phys. Soc., 87, 825

HARPER, G. M. 1992, MNRAS, 256, 37

JORDAN, C. 1968, J. Phys. B, 1, 1004

JORDAN, C. & BROWN, A. 1981, In Solar Phenomena in Stars and Stellar Systems, ed. R. M. Bonnet & A. K. Dupree, Dordrecht: Reidel, 199.

JORDAN, C., BROWN, A., WALTER, F. M. & LINSKY, J. L. 1986, MNRAS, 218, 465

JORDAN, C., AYRES, T. R., BROWN, A., LINSKY, J. L. & SIMON, T. 1987, MNRAS, 225, 903

KAASTRA, J. E. 1992, An X-ray Spectral Code for Optically Thin Plasmas, Internal SRON-Leiden Report V2

KASTNER, S. O., NEUPERT, W. M. & SWARTZ, M. 1974, ApJ, 191, 261

KELLY, R. L. 1987, J. Phys. Chem. Ref. Data, 16, S, No. 1, Part II

LANG, J. 1994, Special Ed. Atomic Data Nucl. Data, 57

LOULERGUE, M. & NUSSBAUMER, H. 1975, A&A, 45, 125L

MALINOVSKY, M. & HEROUX, L. 1973, ApJ, 181, 1009

MASON, H. E. & MONSIGNORI, FOSSI, B. C. 1995, A&AR, in press

MASON, H. E., DOSCHEK, G. A., FELDMAN, U. & BHATIA, A. K. 1979, A&A, 73, 74

MEWE, R., GRONENSCHILD, E. H. B. M. & VAN, DEN, OORD, G. H. J. 1985, A&AS, 62, 197

MONSIGNORI, FOSSI, B. C. & LANDINI, M. 1994, A&A, 284, 900

PAN, H. C. & JORDAN, C. 1995, Mon. Not. R. astr. Soc., 272, 11

PYE, J. P. ET AL. 1995, MNRAS, in press

SANDLIN, G. D., BRUECKNER, G. E., SCHERRER, V. E. & TOUSEY, R. 1976, ApJL, 205, L47

SCHMITT, J. H. M. M., HARNDEN, F. R., PERES, G., ROSNER, R. & SERIO, S. 1995, ApJ, 288, 751

SCHRIJVER, C. J. 1985, Space Sci. Rev., 40, 3

STERN, R. A., LEMEN, J. R., SCHMITT, J. H. M. M. & PYE, J. P. 1995, ApJ, in press

TOUSEY, R. 1967, ApJ, 149, 239

Cool Stars in the EUV: Spectral and Structural Variability

ALEXANDER BROWN

Center for Astrophysics and Space Astronomy,
University of Colorado, Campus Box 389, Boulder, CO 80309-0389, USA

EUV observations of RS CVn binaries and M dwarf flare stars are used to illustrate the variability of coronal emission seen in *EUVE* spectra and photometry. The *EUVE* emission line spectra show that the quiescent coronae of active stars are extremely hot (≥ 10 million K) and flares, which are even hotter, are common. Spectral sequences showing the evolution of flare temperature and density are presented. Very high coronal electron densities ($\log N_e \sim 13$) have been detected during M dwarf flares. *EUVE* data for the RS CVn binary HR 1099 are compared with those from simultaneous observations in other spectral regions, including X-ray (*ASCA*), ultraviolet (*IUE*), and radio (VLA, AT) data. These combined data show that the peak coronal temperature is above the sensitivity of *EUVE* and that the time of maximum flare brightness is spectral region dependent.

1. Introduction

The study of stellar coronae has been hindered greatly by a lack of detailed information on the physical properties, such as the ranges of temperature and density, present in the outer atmospheres of stars other than the Sun. This situation has changed radically with the availability of coronal EUV emission line spectra from the *Extreme Ultraviolet Explorer* (*EUVE*) satellite (Bowyer & Malina 1991). Earlier X-ray telescopes, such as *ROSAT* and *Einstein*, provided data of such low spectral resolution that, while the coronal fluxes could be estimated, attempts to characterise coronal temperature distributions were crude if not altogether misleading. Pointed *EUVE* observations consist of data from four coaligned instruments; three spectrometers and the Deep Survey (D/S) photometer. These spectrometers provide coverage over the following wavelength ranges: 70–190 Å ("short wavelength," SW); 140–380 Å ("medium wavelength," MW); and 280–760 Å ("long wavelength," LW). The Deep Survey photometer observes through a broadband Lexan/B filter with a passband almost identical to the SW spectrometer and can thus provide higher time resolution information on coronal flux variations than possible from the measurements of individual spectral features.

The spectral region observable with *EUVE* (70–700 Å) contains many important emission lines formed at temperatures between a few 10^4 and a few 10^7 K. Thus stellar EUV spectra are very sensitive probes of the dominant conditions within stellar outer atmospheres (See e.g. Brown 1994). However, much of this spectral interval is difficult to observe due to interstellar absorption. Consequently, for all but a few of the nearest stars, very few emission lines are seen above 400 Å. For solar-like stars EUV emission is dominated by plasma at temperatures of a few 10^6 K and strong emission lines of ions such as Fe IX–XII are observed. Very active stars, including the RS CVn binaries and flare stars discussed in this paper, show strikingly different spectra dominated by lines of Fe XVIII–XXIV formed at temperatures near or above 10^7 K, as shown in Figure 1.

S. Bowyer and R. F. Malina (eds.), Astrophysics in the Extreme Ultraviolet, 89–96.

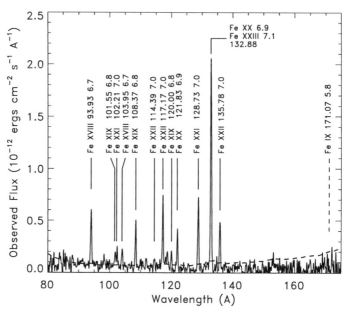

FIGURE 1. The integrated *EUVE* short wavelength spectrum of σ^2 CrB obtained on 1994 February 16–21 during a 144 ks exposure. The dashed line shows the 3 σ detection threshold. Prominent emission line are identified by ion, wavelength, and temperature (log T) of formation. The expected position of the strongest Fe IX line is noted.

2. Coronal Variability of RS CVn Binaries

RS CVn binaries are close binaries with typical orbital periods of a few days. Due to tidal action that acts to enforce orbital and rotational synchronism many stars in these systems rotate extremely rapidly and thus show very high levels of coronal activity.

As an illustration of RS CVn EUV coronal variability I present results from an observation of σ^2 CrB obtained over a 4.5 day period (Brown, Linsky, & Dempsey 1994). The stars in this system have spectral types of F6 V and a G0 V, and an orbital period of 1.14 days (= 98.5 ks). Being a dwarf system the rotational periods of the stars are 2%–3% longer than the orbital period. σ^2 CrB is at a distance of 21 pc and thus has a low interstellar hydrogen column density, allowing good transmission of EUV photons. The *EUVE* exposure times were ~140 ks and spanned over 4 binary orbital periods; therefore, flaring and rotation modulation are easily distinguished. That direct analysis of the integrated mean *EUVE* emission line spectra (illustrated in Fig. 1) would be injudicious can be seen from the D/S light curve presented in Figure 2. Two intervals of flaring occur separated by two binary orbital periods, with peak increases of factors of 5 and 3 relative to quiescence. In both cases the outbursts last ~ 11 hrs and show evidence for the occurrence of multiple flares. Figure 3 shows the spectra from the two flare events and the three surrounding intervals of quiescent emission. Note that the emission line fluxes increase significantly during the flare events, showing that these are not artifacts within the D/S data. However, while the five spectra vary in strength, there is no dramatic change in the appearance of the spectra between flares and quiescence. This behaviour is typical of RS CVn flares and indicates that the quiescent temperature is close to or above the maximum temperature sensitivity of *EUVE* spectra. *EUVE* MW

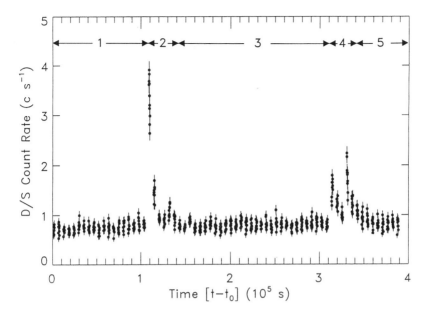

FIGURE 2. The *EUVE* Deep Survey light curve of σ^2 CrB obtained over the period 1994 February 16–21. The five time intervals used for temporal analysis are shown at the top.

spectra are dominated by He II 304 Å but also contain the hottest lines visible in *EUVE* spectra, i.e., Fe XXIV (192 Å and 255 Å) formed at $\log T_e = 7.2$ or higher. The MW spectra of most stars show a similar behaviour during flares with fluxes rising but no appearance of hotter lines. However, the MW spectra of σ^2 CrB are unique to this point in showing dramatic (a factor of 10 for the first flare) increases in Fe XXIV line fluxes during the two flare events. This implies that the quiescent coronal temperature of σ^2 CrB peaks at about $\log T_e = 7.0$, and higher temperatures present in flare plasma are seen as large increases in Fe XXIV emission.

Large EUV flares are common in the coronae of RS CVn binaries. During *EUVE* observations of HR1099 (G5 IV + K1 IV, orbital period 2.84 days) obtained over 9 days in 1993 September and 1994 August at least **five** flare outbursts occurred, with four long outbursts lasting between 12 and 24 hours. In contrast a 3 day observation in 1992 October showed only weak flaring with one definite hour-long flare occurring at phase $\phi=0.1$. Drake et al. (1994) found possible evidence for rotational modulation due to a long-lived coronal structure by comparing both *EUVE* pointed and all-sky survey data taken approximately 2 months apart. However, flare-like brightenings could not be discounted as the cause, reinforcing the need for long, continuous observations in variability analyses.

In all three years the onset of flares occur in two narrow phase intervals, $\phi = 0.1$–0.3 and 0.75–0.85. It is as yet unclear whether this correlation is significant or how flaring might be related to orbital quadratures in HR1099. During this same time interval the location of starspots and active regions on the surface of the K1 star derived from optical data have changed significantly.

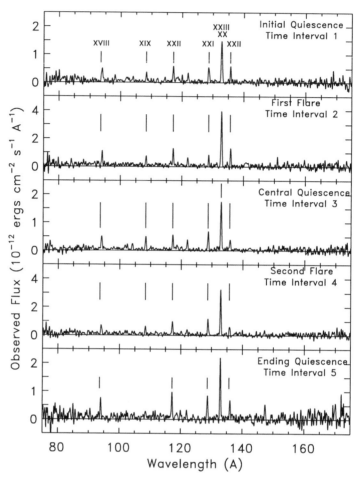

FIGURE 3. Time-resolved *EUVE* SW spectra of σ^2 CrB showing the response of coronal emission lines due to stellar flares. Roman numerals indentify the ionisation stage of different Fe emission lines.

3. EUV Flaring on dMe Flare Stars

Flares on M dwarf stars have been known for almost 50 years due to the detection of optical U and B band enhancements associated with the impulsive phase of flares that last typically a few minutes. Similarly X-ray observations have shown numerous flares with typical timescales of minutes to hrs (see e.g. Pallavicini, Tagliaferri, & Stella 1990). Consequently *EUVE* observations of dMe stars were expected to provide significant new information on the coronal conditions during flares. This has proved correct with the detection of numerous EUV flares, e.g. flares on AU Mic (Cully et al. 1993), AD Leo (Hawley et al. 1995), EQ Peg (Monsignori-Fossi et al. 1995), EV Lac (Ambruster et al. 1994), and AT Mic (Monsignori-Fossi et al. 1994). Typically these flares have durations of hrs to more than a day. In Figure 4 the D/S light curve of a relatively short flare on EV Lac is shown. This flare has a simpler form than most EUV flares with a smooth

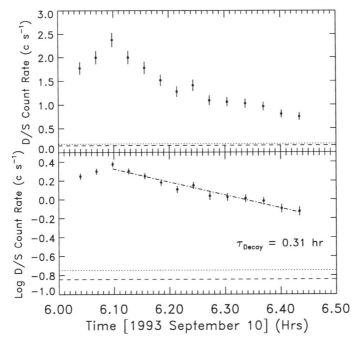

FIGURE 4. *EUVE* Deep Survey light curve of a flare on the M dwarf EV Lac. The dot-dashed line shows an exponential decay fit to the flare decay. The dotted line shows the mean D/S count rate for the complete observation, while the dashed line shows the mean count rate excluding this large flare.

decline that is almost exponential in form, with only slight evidence for additional flaring. However, many of the long duration flares, such as those of AU Mic and AD Leo, show the presence of strong flares superimposed on the decay of a larger flare. From the photometry alone it is not possible to determine the interrelationship between such flares. In the case of the 1992 July 15 AU Mic flare it is possible to determine the coronal (10^7 K) electron density during the course of the flare. Figure 5 shows the variation of the Fe XXII 114 Å to 117 Å ratio during different time intervals of the D/S variations (see Fig. 1 of Cully et al. 1993). High electron densities are found during the initial decay of the large flare and in the second shorter flare. These high densities indicate the presence of compact sources of intense magnetic heating. Interestingly, the peak of the large flare does not show high density, contrary to the results of Monsignori-Fossi & Landini (1994). Such large densities during the flare decay suggest that continued heating of the plasma must be occurring contrary to the coronal mass ejection model of Cully et al. (1994).

4. The Importance of Multiwavelength Observations for Studying Flare Phenomena

While *EUVE* has provided extremely important opportunities for the detailed study of stellar coronae over long time intervals, *EUVE* spectra and photometry can still only provide a limited insight into the physical processes operating in these magnetically controlled and highly dynamic plasmas. Observations involving data from a range of

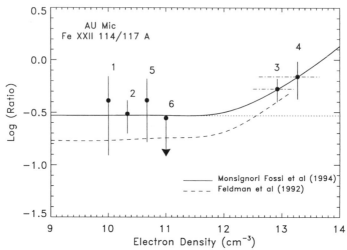

FIGURE 5. Estimates of the log electron density in the corona of AU Mic based on the ratio of Fe XXII 114.4 Å to 117.2 Å for different intervals in the 1992 *EUVE* observation. The solid and dashed curves show two calibrations of this ratio; the Monsignori-Fossi et al. calibration is much more consistent with the data.

spectral regions can lead to an even more detailed investigation of coronal phenomena. Routinely we try to obtain observations simultaneous with *EUVE* in spectral regions such as the radio, ultraviolet, and X-ray to the extent that observatory schedules allow. As an example I show a portion of our 1994 August observations of HR1099 in Figure 6. This time interval covers a single day from a 5 day observation. The unique aspect of this time interval is the first simultaneous observation of a stellar corona by *ASCA* and *EUVE*. *ASCA* observes in the soft X-ray region (0.5–10 KeV). Figure 6 shows the presence of a small flare with a duration of $\sim$ 4 hrs and a peak luminosity in the 1–300 Å region of 2×10^{31} ergs s^{-1}. Analysis of the *ASCA* spectra show that the hottest coronal component ranges from a quiescent temperature of $\log T_e = 7.4$ to 7.5 at the peak of the flare, confirming that the peak temperature in many RS CVn flares is beyond the temperature sensitivity of *EUVE*. The relative timing of the flare in different spectral regions is particularly interesting. The flare timing and shape is almost identical in the *EUVE* D/S and *ASCA* SIS data, and this implies that both instruments are seeing the flare decay rather than the initial impulsive phase of the flare. The UV flare response is seen at the very onset of the flare rise in X-rays/EUV; although the 8 hour break in *IUE* coverage eliminates UV coverage of most of the flare. The 3 cm radio continuum emission, due to gyrosynchrotron emission from nonthermal coronal electrons, shows a flare starting roughly an hour earlier than in the other spectral regions.

5. Summary

• Most active cool stars (RS CVn binaries and flare stars) observed by *EUVE* have shown variability, usually in the form of readily identifiable stellar flares.

• RS CVn EUV flares typically have longer time scales than those on the Sun. EUV flares on dMe stars can also last up to a day but are typically somewhat shorter than RS CVn flares.

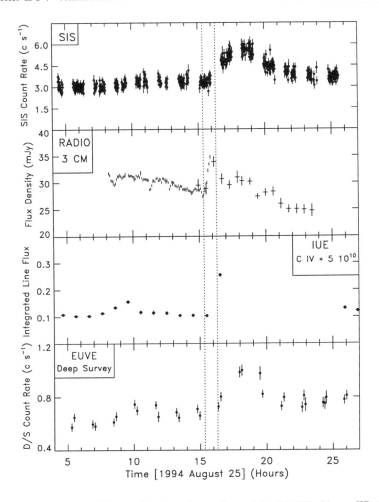

FIGURE 6. Simultaneous multispectral-region observations of the RS CVn binary HR1099 obtained on 1994 August 24. The data shown (from top to bottom) are soft X-ray count rate from the *ASCA* SIS detector, 3 cm radio continuum flux density from the VLA (vertical lines) and AT (crosses), ultraviolet CIV line flux (formed at 10^5 K) from *IUE* SWP-LO spectra, and the *EUVE* D/S count rate. Dotted lines mark times of flare onset.

- RS CVn flares seem to occur preferentially at particular orbital phases, although this result is based on as yet limited statistics.
- Coronal temperatures of active stars are close to the upper limit of the *EUVE* temperature sensitivity; usually flares increase the flux in all emission lines but no new hotter lines appear.
- *EUVE* does not see the flare impulsive phase.
- High (log $N_e \sim 13$) electron densities have been seen during short rapid flares. Rapid spiky" flares are often seen superimposed on the decays of large outbursts.
- *EUVE* is an excellent instrument for studying coronal flares, particularly due to the long (*days*) time intervals that can be studied.

I am extremely grateful to my colleagues Carol Ambruster, Grodon Bromage, Robert Dempsey, Jeremy Drake, Marc Gagne, Jeffrey Linsky, and David Wonnacott with whom I have collaborated on many of the investigations described in this paper. I thank the staff of the Center for EUV Astrophysics at UC Berkeley for their assistance in the analysis and understanding of *EUVE* data. This work is supported by NASA grants NAG5-2259 and NAG5-2530 to the University of Colorado.

REFERENCES

AMBRUSTER, C. W., BROWN, A., PETTERSEN, B., & GERSHBERG, R. E. 1994, EUVE observations of a flare on EV Lac, BAAS, 26, 866

BOWYER, S. & MALINA, R. F. 1991, The EUVE, mission, in Extreme Ultraviolet Astronomy, ed. R. F. Malina & S. Bowyer, New York: Pergamon Press, 397

BROWN, A. 1994, Coronal and transition region spectroscopy of cool stars using EUVE, in Cool Stars, Stellar Systems, & The Sun: Eighth Cambridge Workshop, ed. J.-P. Caillault, ASP Conf. Ser., 64, 23

BROWN, A., LINSKY, J. L., & DEMPSEY, R. C. 1994, EUVE coronal spectroscopy of the RS CVn binaries σ^2 CrB and II Peg, BAAS, 26, 865

CULLY, S. L., SIEGMUND, O. H. W., VEDDER, P. W., & VALLERGA, J. V. 1993, Extreme Ultraviolet Explorer Deep Survey observations of a large flare on AU Microscopii, ApJL, 414, L49

CULLY, S. L., FISHER, G. H., ABBOTT, M. J., & SIEGMUND, O. H. W. 1994, A coronal mass ejection model for the 1992, July 15 flare on AU Microscopii observed by the Extreme Ultraviolet Explorer, ApJ, 435, L449

DRAKE, J. J., BROWN, A., PATTERER, R. J., VEDDER, P. W., BOWYER, S., & GUINAN, E. F. 1994, Detection of rotational modulation in the coronal extreme-ultraviolet emission from V711 Tauri?, ApJL, 421, L43

FELDMAN, U., MANDELBAUM, P., SEELY, J. F., DOSCHEK, G. A., & GURSKY, H. 1992, The potential for plasma diagnostics from stellar extreme-ultraviolet observations, ApJS, 81, 387

HAWLEY, S. L., ET AL. 1995, Simultaneous EUVE and Optical Observations of AD Leonis: Evidence for large Coronal Loops and the Neupert Effect in Stellar Flares, ApJ, in press

MONSIGNORI-FOSSI, B. C., LANDINI, M. 1994, The EUV Spectrum of AU Microscopii: Temperature and density diagnostics from EUVE spectrometers observations, A&A, 284, 900

MONSIGNORI-FOSSI, B. C., LANDINI, M., DEL ZANNA, G., & DRAKE, J. J. 1994, The EUV Spectrum of AT Mic. in Cool Stars, Stellar Systems, & The Sun: Eighth Cambridge Workshop, ed. J.-P. Caillault,. ASP Conf. Ser., 64, 44

MONSIGNORI-FOSSI, B. C., LANDINI, M., FRUSCIONE, A., & DUPUIS, J. 1995, Extreme Ultraviolet Spectroscopy and Photometry of EQ Pegasi, A&A, in press

PALLAVICINI, R., TAGLIAFERRI, G., & STELLA, L. 1990, X-ray emission from solar neighbourhood flare stars: A comprehensive survey of EXOSAT results, A&A, 228, 403

The FIP Effect and Abundance Anomalies in Late-Type Stellar Coronae

J. J. DRAKE,[1] J. M. LAMING,[2] AND K. G. WIDING[3]

[1] Center for EUV Astrophysics, 2150 Kittredge Street,
University of California, Berkeley, CA 94720-5030, USA

[2] SFA Inc., Landover, MD 20785, and Naval Research Laboratory,
Code 7674L, Washington, DC 20375, USA

[3] Naval Research Laboratory, Code 7674W, Washington, DC 20375, USA

"Yes, it will be a long time before people know what I know. How much of iron and other metal there is in the sun and the stars is easy to find out, but anything which exposes our swinishness is difficult, terribly difficult" (Tolstoy 1889).
In the solar corona, the abundances of elements appear to differ from the photospheric values in a manner related to the element first ionization potential (FIP): species with FIP ≤ 10 eV are observed to be enhanced relative to the photosphere by factors of $\sim$3–4. The first studies of *stellar* coronal composition with *EUVE* suggest that some stars exhibit a solar-like FIP effect, whereas others do not. We briefly review the latest results, and we argue that element abundance anomalies, such as the FIP effect, can provide potentially powerful new coronal diagnostics. Moreover, knowledge of the composition of a stellar corona is crucial for interpreting its spectrum—for understanding its structure and energy balance, and for testing its possible heating mechanisms: We must begin to understand coronal abundance anomalies and the compositions of active stars in order to begin to understand their coronal physics.

1. Introduction

As hinted by Tolstoy (1889), swinish suggestions that the abundances of elements in the solar corona might differ significantly from the corresponding values in the underlying photosphere are not entirely recent. The first hint in modern times came from the pioneering work of Pottasch (1963), who obtained significantly different abundances for some elements, most notably Mg, Si and Fe, than the generally accepted photospheric abundances of the day (those of Goldberg, Müller & Aller 1960). Regarding Mg, Si and Fe, Pottasch stated that the abundance differences could not be explained by experimental uncertainties—conclusions which are not compromised by today's revised solar photospheric abundances. Perhaps more interestingly, Pottasch also remarked that the agreement between his abundances and those observed in cosmic rays was much closer than with the photospheric values.

The story of the study of abundances in the solar corona and wind during the intervening 30 years or so has been comprehensively reviewed by Meyer (1985a,b; 1993) and Feldman (1992),† and will also be reviewed briefly elsewhere in this volume (Haisch, Saba & Meyer 1995; see also Saba 1995 for a review of the *Solar Maximum Mission* results). In summary, the last three decades have seen the early hints of abundance anomalies raised by the Pottasch analysis grow into a substantial body of evidence that points to a significant abundance anomaly in the solar corona. This anomaly appears to have a common underlying trend related to the element first ionization potential (FIP), similar to that

† Note that the coronal abundances adopted by Feldman (1992) differ from those of Meyer (1985) in the normalization with respect to H: Meyer (1985) suggested that the *high* FIP species are *depleted* with respect to H, whereas Feldman argues that the *low* FIP species are *enhanced* relative to H; Meyer (1993) tends to side with Feldman's (1992) assessment.

S. Bowyer and R. F. Malina (eds.), Astrophysics in the Extreme Ultraviolet, 97–104.
© *1996 Kluwer Academic Publishers. Printed in the Netherlands.*

uncovered for the low to medium energy cosmic rays by Cassé & Goret (1978): elements
with a low FIP ($\lesssim$ 10 eV—e.g., Mg, Si, Fe) appear to be enhanced relative to elements
with high FIP ($\gtrsim$ 10 eV—e.g., O, Ne, S) by average factors of 3–4. The phenomenon
is now known as the "FIP Effect," but the sites and mechanisms responsible for this
elemental fractionation according to FIP are currently not known or understood. The
fractionation energy of 10 eV points to chromospheric temperatures, but a photospheric
fractionation site has not been ruled out.

In this short paper, we point out the importance of coronal abundances and review
recent results. Studies of abundances in *stellar* coronae might be extremely valuable for
providing key insights into the FIP Effect which solar observations alone cannot. Three
important questions of wide astrophysical importance are raised: (1) Do other stars
exhibit coronal abundance anomalies? (2) If so, are the anomalies related to FIP, as in
the solar case? (3) Is the solar FIP Effect connected to that observed in cosmic rays?

2. The Importance of Stellar Coronal Abundances

The flux, F_{ji}, emitted through a spectral line $j \rightarrow i$ by an optically thin stellar corona is
essentially proportional to the abundance of the element in question. The expression for
the observed line flux can be closely approximated by the integral over the temperature
interval, ΔT_{ji}, over which the line flux is non-negligible (usually, $\Delta \log T_{ji} \sim 0.3$),

$$F_{ji} = A K_{ji} \int_{\Delta T_{ji}} G_{ji}(T) \overline{n_e^2}(T) \frac{dV(T)}{dT} \, dT \text{ erg cm}^{-2} \text{ s}^{-1} \qquad (2.1)$$

where A is the abundance of the element in question, K_{ji} is a known constant which
includes the frequency of the transition and the stellar distance, $n_e(T)$ is the number
density of electrons at temperature T within the plasma volume $V(T)$, and $G_{ji}(T)$ is
the "contribution" function of the line. This last parameter defines the temperature
interval, ΔT_{ji}, over which the line is formed, and is dependent on the atomic physics of
the particular transition. For most transitions of interest, $G_{ji}(T)$ can be fairly reliably
estimated theoretically. The integrand $\overline{n_e^2}(T) \frac{dV(T)}{dT}$ is the "differential emission measure".

The abundance pattern in the emitting plasma, then, influences both the spectral
shape and intensity: deriving basic quantities such as the emission measure cannot be
done without prior knowledge of the composition. In short, the plasma composition is
fundamental to the basic interpretation and analysis of spectroscopic *and photometric*
observations of stellar outer atmospheres.

Since stellar coronae cannot be spatially resolved, the global emission measure distribution as a function of temperature, hopefully with some clues as to the density for at
least one temperature, will form the extent of the tangible data from which we must
begin to understand stellar coronal structure. Moreover, radiative loss, together with
the emission measure form the key ingredients for estimating the outer atmosphere energy budget, which is a necessary step towards pinpointing and testing possible heating
mechanisms. Similarly, the energetics of transient phenomena such as flares and coronal
mass ejections cannot be deduced without knowledge of the plasma element abundances,
since line emission from species such as Fe, Mg, Si and O control radiative cooling times
and the onset of thermal instabilities (see, e.g., Cook et al. 1989).

A situation in which the abundances change from one coronal structure to another,
and vary with time, such as appears to be the case for the solar corona, and in which the
underlying patterns of abundance variations might differ widely from one spectral type
and activity level to the next, such as is suggested by recent *EUVE* and *ASCA* analyses,
will pose somewhat of a challenge to those hoping to understand the observations.

It should also be pointed out that the *photospheric* compositions are either unknown or extremely uncertain for nearly all of the stars whose coronae are bright enough for detailed spectroscopic study. In particular, the RS CVn's, the algols and the late K and M dwarf single and binary stars generally do not have reliable abundances and often lack even rough metallicity estimates—measurements are very difficult because of rapid rotation, binarity, and, in the case of the late K and M dwarfs, severe line blanketing. We can probably not expect to know the photospheric compositions of these types of stars to much better than a factor of 2. Thus, even in the absence of any coronal abundance anomalies, uncertainties in composition still pose significant difficulties for coronal physics.

2.1. *Clues to Coronal Structure and Underlying Physical Processes?*

While compositional uncertainties complicate the interpretation of coronal observations, we speculate that, if the abundance anomalies can be understood, then coronal abundances might provide powerful diagnostics of the physics and structure of stellar coronae. This is prompted by observations of the solar corona which suggest that there are systematic variations of abundances in different types of coronal structure. These observations have been discussed by Feldman (1992) and by Meyer (1993). We can summarize briefly the main findings as follows:

(*a*) Above sunspots, coronal loops and material have been observed to have photospheric compositions.

(*b*) In active regions, *newly emerged* low-lying bipolar loops have *photospheric* compositions.

(*c*) In active regions, *evolved loops* more than 1 day old have enhancements of low FIP species by factors of $\sim 4-5$; no enhancements larger than these factors are observed in regions of closed magnetic fields.

(*d*) Features associated with closed magnetic fields and compact loop structures, such as young active regions and flares, show photospheric abundance ratios.

(*e*) Features associated with open unipolar magnetic fields such as polar plumes show extreme coronal type composition. On the other hand the diffuse emission over polar coronal holes, and the fast solar wind, appear to have composition between photospheric and coronal.

To these we add:

(*f*) The supergranulation network emission has a photospheric composition; the FIP effect appears also to be a function of *altitude*.

2.2. *Do Cosmic Rays Originate from Late-Type Stellar Coronae?*

The problem of the origin of cosmic rays remains an outstanding question in high energy astrophysics. A possible clue to the galactic cosmic ray sources (GCRS; $\lesssim 100$ GeV) was uncovered a decade or so ago by Cassé and Goret (1978), whose analysis of their elemental composition revealed the underlying correlation of heavy element abundances with FIP. Subsequently, Meyer (1985b) pointed out that this composition was very similar to that of solar energetic particles and suggested that most GCRs could originate in unevolved late-type star surface material (with some small fraction coming from massive WC stars).

Direct *in situ* acceleration of stellar coronal or wind material meets with difficulties on energetic grounds (Mullan 1979; Gorenstein 1981). A two-step process was then suggested, whereby some fraction of the $\sim$MeV particles "injected" by stellar coronae are boosted to relativistic energies by stochastic or diffusive shock acceleration provided mainly by supernova remnants (e.g., Axford et al. 1977). This model has been successful in producing quite naturally a CR energy spectrum which is consistent with the CR source spectrum inferred from observations (e.g. van Bloemen 1987). Moreover, it is particularly appealing since other non-stellar injection models suffer the disadvantage of both having to explain the fractionation through other means, and facing the prospect that the remarkable similarity between the CR source composition and the composition of solar energetic particles is purely accidental (van Bloemen 1987). Stellar population considerations then suggest that the K and M dwarfs would be the major injectors of suprathermal CR seed particles (Gorenstein 1981). However, this hypothesis is based *entirely* on extrapolation of the solar coronal abundance case, since, before the launch of *EUVE*, it was not possible to determine the abundances of elements in the coronae of other stars.

3. Determining Abundances in Stellar Coronae with *EUVE*

There are three ways in which *EUVE* spectra might be used for gleaning compositional information in coronal plasmas, all of which have been previously applied to the Sun.

(*a*) Comparing the EM predicted by a line of one element with a baseline EM distribution (essentially the "differential emission measure" $n_e^2(T)\frac{dV(T)}{dT}\,dT$ of Eqn. 2.1) derived using lines of another element (in practice, Fe). The ratios of the EM's predicted by elements X and Y then yield their abundance ratios relative to the composition assumed for the plasma in the EM calculations: for an assumed solar composition, $\log(EM_X/EM_Y)$ yields the abundance ratio [X/Y], using the conventional spectroscopic bracket notation.

(*b*) Temperature insensitive line ratios—comparing the fluxes of lines which have approximately the same temperature dependence in their $G_{ij}(T)$ functions: in the expression describing the *ratio* of the two line fluxes, from Eqn. 2.1, all terms cancel except for the element abundance and the relevant atomic physics parameters. This method does not then require detailed knowledge of the EM distribution.

(*c*) Line-to-continuum ratios. If the EM distribution is known, the abundance of an element relative to H can be derived by measuring the strength of an emission line relative to the thermal bremsstralung continuum, which is formed primarily by electron free-free interactions with protons. In order for such measurements to be possible, the continuum needs to be firmly detected. In the case of *EUVE* observations, this is only likely for the very hot ($T \sim 10^7$ K) coronae found in rapidly rotating stars, or for long observations of less active stars (however, see § 4.3).

4. Recent Stellar Coronal Abundance Studies in the EUV

4.1. *The FIP Effect in the Full Solar Disk*

We have recently investigated the FIP effect in the *full disk* ("Sun as a star") solar corona (Laming, Drake & Widing 1995; Figure 1a). At $\log T \sim 6.3$, the low FIP elements are enhanced by about a factor of 3. However, this enhancement *varies with temperature*, and at temperatures cooler than $\log T \sim 5.8$—the temperatures dominated by supergranulation emission—there is no discernible FIP effect in the full disk average corona.

Laming, Drake & Widing (1995) suggested that in full disk spectra, the dominant emission at temperatures of a few 10^5 K comes from low lying structures associated with the supergranulation network with essentially no FIP effect, while true coronal emissions comes from active loops and much higher altitudes where the FIP effect is strong. Such an idea of the "discontinuity" between the corona and the transition region has also been discussed by Feldman & Laming (1994) on the basis of observations of the different

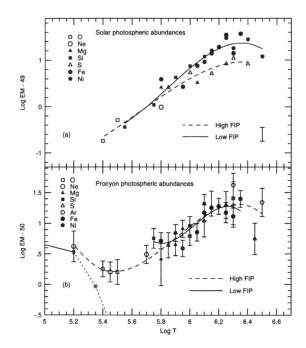

FIGURE 1. EM distributions derived for low and high FIP elements in (a) the solar full disk corona, and (b) the corona of Procyon. The curves illustrate best-fit splines to the high and low FIP points. There appears to be no FIP effect in Procyon, while in the full disk solar corona there is no significant FIP effect for $\log T < 5.8$.

in morphology between lines formed at temperatures above and below $\sim 10^6$ K, and the apparent absence of footpoints emitting transition region radiation at the base of otherwise coronal loops.

4.2. Evidence for No FIP Effect in Procyon—Acoustic Heating or a Photospheric Fractionation Site?

In a recent analysis of *EUVE* spectra we have found evidence for *photospheric* abundances in the corona of the F5 subgiant Procyon (Drake et al. 1995; Drake, Laming & Widing 1995a; Figure 1b). This result might be an indication that Procyon's corona is mainly heated by acoustic means, as has been suggested by some workers (e.g. Mullan & Cheng 1994; Simon & Drake 1989). For diffusion models of element fractionation to work, small spatial scales are required, otherwise, neutral species have sufficient time to become ionized during the separation process (von Steiger & Geiss 1989). Drake et al. (1995a) pointed out that such small spatial scales can occur quite naturally with magnetic fields, but *not* with acoustic waves.

Alternatively, the only high FIP lines available in the Procyon spectrum at *coronal* temperatures are due various charge states of S, and two rather unsatisfactory blended lines from Ar XII (224.23) and Ar XV (221.20). If the FIP fractionation site is photospheric, then it is possible that S (IP $\sim$ 10 eV) could be acting as a low FIP species in the hotter (6500 K) photosphere of Procyon. This hypothesis can be also tested by *EUVE* observations of stars cooler than the Sun—see below.

4.3. *Depleted Fe in the Coronal Emission from Algol and CF Tuc?*

Based on line-to-continuum ratios of Fe lines observed in the quiescent Algol *EUVE* spectrum, Stern et al. (1994) have demonstrated evidence that Fe must be underabundant (relative to the solar photospheric abundances of Anders and Grevesse 1989) by a factor of ∼ 4. For an object as young as Algol, such an Fe deficiency is not likely; the only obvious alternative is that the corona is deficient in Fe. The case for a low coronal Fe abundance was first suggested by the *ASCA* observations of Antunes et al. (1994), who derived an abundance for Fe of 32% of the solar value (see also § 4.5).

More extreme than the Algol case is that of the RS CVn system CF Tuc, whose *EUVE* spectrum exhibits almost no lines! Schmitt et al. (1995) have shown that most of the EUV flux from CF Tuc is in the form of continuum. No plausible EM distribution can explain the absence of the strong Fe lines in charge states XVIII–XXIV expected from such an object unless the Fe abundance is severely (by a factor of ∼ 10) depleted relative to solar, prompting Schmitt et al. to dub the phenomenon the "MAD" (metal abundance deficient) syndrome. However, the coronal composition might simply reflect the underlying *photospheric* composition as expected: Randich, Gratton & Pallavicini (1993) have derived Fe abundances for the two components of [Fe/H]=−0.5 and −0.9. Randich et al. caution that their Fe abundances might be systematically too low due to line filling from plage or similar emission; also, since one would expect the Fe abundances to be the same in both stars, the composition of CF Tuc warrants further investigation.

4.4. *Evidence for a FIP Effect in G–M Dwarfs—The Sources of Cosmic Rays?*

We have very recently derived abundance ratios for low and high FIP elements observed in the *EUVE* coronal spectrum of the α Cen AB (G0 V + K1 V) system using temperature insensitive line ratios (Drake, Laming & Widing 1995b). The α Cen AB system was not spatially resolved by *EUVE*, and it is currently not known which component was brightest in the EUV at the time of the observation.† It is possible that the solar analogue and cooler K dwarf show *different* coronal abundance anomalies.

The active G8 dwarf ξ Boo A also appears to exhibit a FIP effect based on an EM distribution analysis similar to that carried out for Procyon (Figure 2). Though still preliminary, this result would be extremely important because it would represent the first detection of the FIP effect in a star which is significantly cooler than the Sun. It argues strongly against a *photospheric* site for the FIP fractionation, since ions such as Fe II, Si II and Mg II will not be such major species in the photosphere of this star as compared to the Sun.

The same type of analysis applied to the intermediate activity K2 dwarf ε Eri yielded similar results, though not without some equivocation (Laming, Drake & Widing 1995). However, Laming et al. were also able to place limits on the magnitude of the FIP effect in ε Eri, arguing that it could not be substantially larger than that found in the sun. This suggests that the FIP effect is not "linearly" related to stellar activity, as one might naively expect from the fact that the fractionation according to FIP must be driven by electromagnetic forces—forces directly related to stellar activity.

Preliminary results for the dMe binary FK Aqr also indicate the presence of a FIP effect. Unfortunately, the exposure was too short to detect the high FIP lines needed to quantify the result. Nevertheless, upper limits to the fluxes from these lines do indicate

---

† Golub et al. (1982) found B to be brightest in X-rays from an observation in 1979. In view of the large cyclic and stochastic variability in the solar X-ray flux—by more than a factor of 10—extrapolation of this result to the time of the *EUVE* observation is perilous.

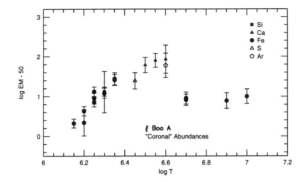

FIGURE 2. The EM distribution for low and high FIP species in the *EUVE* spectrum of
ξ Boo A for Feldman (1992) "coronal" abundances.

an enhancement of low FIP species relative to the high FIPs by at least the same amount
as seen on the Sun—factors of 4 or more.

The results for these stars are of central importance for the origin of cosmic rays: we
find that the types of stars proposed by Meyer (1985b) to supply the cosmic ray seed
particles appear to exhibit a solar-like FIP effect. Our results therefore support the
hypothesis that the galactic cosmic rays originate predominantly from the surfaces of
late-type stars.

4.5. Abundance Studies with ASCA

In addition to the work of Antunes et al. (1994), mentioned above, Drake et al. (1994)
have derived abundances using two temperature coronal models from *ASCA* observations
of π^1 UMa (G1.5 V) and β Ceti (K0 III). Interestingly, they obtained a bias similar to
that of a FIP-effect, *but with some deviations*. Most notably, Fe was observed to be
depleted rather than enhanced, relative to the solar mixture. White et al. (1994) also
found evidence for metal deficiency in the corona of the RS CVn system AR Lac, and
pointed to similar hints from *GINGA* observations; they ventured that these abundances
possibly just reflected the photospheric compositions of the RS CVn secondaries, which
some optical observations had also suggested were metal poor. Despite the possibility
that the underabundances might just reflect the stellar compositions, and also that the
real uncertainties in the *ASCA* abundances remain to be quantified, these results, and
those for Algol and CF Tuc, raise some very interesting questions: is FIP not the universally common factor in coronal abundances? The depletion of Fe would be particularly
mysterious and does not fit with *any* existing fractionation models.

5. Summary

Recent studies with *EUVE* and *ASCA* indicate that the Sun is not alone in exhibiting
coronal abundance anomalies. G–M dwarfs appear to exhibit solar-like FIP effects. This
is consistent with Meyer's (1985b) hypothesis that cosmic rays originate from suprathermal seed particles injected into the ISM by late-type stellar coronae.

Are the coronae of the very active stars really metal deficient ("MAD"!) relative to
their photospheric compositions? If so, FIP might not be the universal underlying factor
controlling coronal abundance anomalies. It is vital to know the photospheric and coronal

compositions of the stars whose coronae we wish to study and attempt to understand. Based on the solar analogy, coronal abundance anomalies, if understood, promise exciting new diagnostics for processes occurring in stellar outer atmospheres. Future abundance studies with *ASCA* should concentrate on stars with EM distributions determined from *EUVE* observations.

Much longer exposures with *EUVE* and *ASCA* than have hitherto been the norm are essential for making the next advancements.

REFERENCES

ANDERS, E. & GREVESSE, N. 1989, Geochim. Cosmochim. Acta., 53, 197

ANTUNES, A., NAGASE, F., & WHITE, N. E. 1995, ApJ, 436, L83

AXFORD, W. I., LEER, E., & SKANDRON, K. G. 1977, Proc. 15th Int. Cosmic Ray Conf., 11, 32

BLOEMEN, H. 1987, in "Interstellar Processes," ed. D. J. Hollenbach & H. A. Thronson, Jr., Reidel, 143

CASSÉ, M., & GORET, P. 1978, ApJ, 221, 703

COOK, J. W., CHENG, C. -C., JACOBS, V. L., & ANTIOCHOS, S. K. 1989, ApJ, 338, 1176

DRAKE, J. J., LAMING, J. M., & WIDING, K. G. 1995a, ApJ, 443, 393

DRAKE, J. J., LAMING, J. M., & WIDING, K. G. 1995b, ApJ, submitted

DRAKE, J. J., LAMING, J. M., WIDING, K. G., SCHMITT, J. H. M. M., HAISCH, B., & BOWYER, S. 1995, Science, 267, 1470

DRAKE, S. A., SINGH, K. P., WHITE, N. E., & SIMON, T. 1994, ApJ, 436, L87

FELDMAN, U. 1992, Phys. Scripta, 46, 202

FELDMAN, U. & LAMING, J. M. 1994, ApJ, 434, 370

GOLDBERG, L., MÜLLER, E., & ALLER, L. H. 1960, ApJS, 5, 45

GOLUB, L., HARNDEN, F. R., PALLAVICINI, R., ROSNER, R., & VAIANA, G. S. 1982, ApJ, 253, 242

GORENSTEIN, P. 1981, in Proc. 17th Int. Cosmic Ray Conf., Paris, 12, 99

HAISCH, B., SABA, J. L. R., & MEYER, J. -P. 1995, this proceedings

LAMING, J. M., DRAKE, J. J., & WIDING, K. G. 1995a, ApJ, 443, 416

LAMING, J. M., DRAKE, J. J., & WIDING, K. G. 1995b, ApJ, submitted

MEYER, J. -P. 1985a, ApJS, 57, 151

MEYER, J. -P. 1985b, ApJS, 57, 172

MEYER, J. -P. 1993, in Origin and Evolution of the Elements, ed. N. Prantzos, E. Vangioni-Flam, & M. Cassé, Cambridge: Cambridge University Press

MULLAN, D. J. 1979, ApJ, 234, 588

MULLAN, D. J. & CHENG, Q. Q. 1994, ApJ, 435, 435

POTTASCH, S. R. 1963, ApJ, 137, 945

RANDICH, S., GRATTON, R., & PALLAVICINI, R. 1993, A&A, 273, 194

SABA, J. L. R. 1995, Adv. Sp. Res., 15(7), 13

SCHMITT, J. H. M. M., STERN, R. A., DRAKE, J. J., & KÜRSTER, M. 1995, ApJ, submitted

SIMON, T. & DRAKE, S. A. 1989, ApJ, 346, 303

STERN, R. A., LEMEN, J. R., SCHMITT, J. H. M. M., & PYE, J. P. 1995, ApJ, 444, L45

TOLSTOY, L. N. 1889, The Kreutzer Sonata

VON STEIGER, R., & GEISS, J. 1989, A&A, 225, 222

WHITE, N. E., ET AL. 1994, PASJ, 46, L97

Dissecting the EUV Spectrum of Capella

NANCY S. BRICKHOUSE

Harvard-Smithsonian Center for Astrophysics, 60 Garden St., Cambridge, MA 02138 USA

Extreme ultraviolet spectra of Capella, obtained at various orbital phases over the past two years by the *EUVE* satellite, show strong emission lines from a continuous distribution of temperatures ($\sim 10^5 - 10^{7.3}$ K). In addition to the strong He II $\lambda303.8$, the spectra are dominated by emission lines of highly ionized iron. Strong lines of Fe IX, XV, XVI, and XVIII–XXIV are used to construct emission measure distributions for the individual pointings, which show several striking features, including a minimum near 10^6 K and a local maximum at $10^{6.8}$ K. Furthermore, intensities of the highest temperature lines ($T_e > 10^7$ K) show variations (factors of 2–3) at different orbital phases, while the lower temperature Fe lines show variations of about 30% or less. The low variability of most of the strong low temperature features motivates a detailed analysis of the summed spectrum. With $\sim$ 280 ks of total exposure time, we have measured over 200 emission features with S/N ≥ 3.0 in the summed spectrum. We report here initial results from the analysis of this spectrum. We can now identify lines of Fe VIII and X–XIV, as well as a number of electron density and abundance diagnostic lines.

We also report here the first *direct* measurement of the continuum flux around ~ 100 Å in a cool star atmosphere with *EUVE*. The continuum flux can be predicted from the emission measure model based on Fe line emission, and demonstrates that the Fe/H abundance ratio is close to the solar photospheric value.

1. Introduction

Capella (Alpha Aurigae; HD 34029) is a bright nearby multiple star system that has been the target of numerous observations over the past two decades in the UV and X-ray regions. Under the Guest Observer Program of the *EUVE* satellite, we have now obtained multiple observations of this source. All of the individual pointings confirm the initial results of the calibration data analysis, reported by Dupree et al. (1993): (a) a continuous distribution of temperatures from the transition region to the hot corona; (b) a minimum in the emission measure distribution (EMD) near 10^6 K; (c) a local maximum in the EMD at $10^{6.8}$; and, (d) a number of transition region lines that are weak relative to their ultraviolet counterparts. It is the goal of this work to explore the predictive capability of the EMD derived from strong Fe lines. Using the summed spectrum from all the *EUVE* observations, we have now measured over 200 emission features with signal to noise (S/N) ≥ 3 in the three spectrometers. A more detailed analysis can be found in Brickhouse et al. (1995b).

The analysis of strong EUV Fe lines reveals complex structure in the Capella atmosphere not previously discerned by low resolution X-ray spectroscopy. For example, Swank et al. (1981) characterize a number of cool binary coronae as two-temperature distributions using observations taken with the *Einstein* Solid State Spectrometer. Even the moderate resolution *EXOSAT* Transmission Grating Spectrometer was unable to resolve line blends, and thus the multithermal emission models of Lemen et al. (1989) are simple parameterizations reliant on global spectral fitting. Among the early results from *EUVE* spectroscopy are clean, strong emission lines of Fe IX, XV, XVI, and XVIII–XXIV in a number of bright active binaries (Dupree et al. 1993; Landini & Monsignori Fossi 1993; Stern et al. 1995). For example, Dupree et al. (1993) report about twenty Fe lines from which an EMD is constructed. While the Capella distribution is indeed reminiscent

S. Bowyer and R. F. Malina (eds.), Astrophysics in the Extreme Ultraviolet, 105–112.

of two dominant temperatures (unlike certain other sources), a continuous distribution is required to fit the *EUVE* lines.

Given the sharpness of features in Capella's atmosphere and the rich diversity of structures found in other sources (Dupree et al. 1995), a number of questions arise as to the physical conditions at various temperatures in the transition region and corona. Dupree et al. (1993) find evidence for high densities (10^{12} to 10^{13} cm^{-3}) from N_e-sensitive lines of Fe XXI. Brickhouse et al. (1995a) discuss apparent inconsistencies in the densities derived from different line ratios, suggesting that some combination of weak lines, blends, and different source regions for Fe XXI might be responsible. Since these densities at $T_e \sim 10^7$ K imply small emitting volumes and confining magnetic fields of several hundred gauss, the confirmation of such high densities is a strong motivation for longer observations.

At lower temperatures, the EMD has been much less well constrained, relying solely on Fe IX below $T_e \sim 2 \times 10^6$ K to connect the transition region to the corona. Since the emission measure minimum is at a significantly higher temperature ($\sim 10^6$ K) than that of the sun ($\sim 10^5$ K), the role of various components in the energy balance of Capella must be quite different from the sun. The summed spectrum now fills in much of the gap in coverage of Fe ionization stages, adding Fe VIII and X–XIV. Many of these lines are also N_e-sensitive.

Drake et al. (1995) discuss the importance of elemental abundance measurements in stellar coronae for understanding coronal heating, and as Cook et al. (1989) point out, significant variations from solar photospheric abundances can also change the shape of the radiative loss function. The summed spectrum of Capella includes lines from many of the abundant elements; we report here the analysis of emission formed at high temperature from elements other than Fe. Discussion of EUV lines formed in the transition region and near the emission measure minimum is forthcoming (Brickhouse et al. 1995b).

2. Observations and Data Reduction

The individual spectra were acquired over five separate pointings. The extraction techniques, described elsewhere (Dupree et al. 1995a), have been applied to the individual observations separately. The spectra are co-added to produce the summed spectra for a total of 2.82, 2.77, and 2.83 ks for the SW, MW, and LW spectrometers, respectively. Figure 1 shows these spectra, with the S/N curve overplotted. The S/N of individual lines is calculated over the entire line profile, and thus is somewhat better than the peak value shown. Since the individual spectra were accumulated under different conditions, the simple summation leads to a minor degradation of the effective spectral resolution. The strong, isolated spectral lines maintain Gaussian profiles, with fits that are generally improved more by higher S/N than they are degraded by decreased spectral resolution; they are consistent with the performance characteristics described by Boyd et al. (1994). For individual lines, the centroid and flux determinations, the continuum flux subtraction and the correction for interstellar absorption are done as in Dupree et al. (1993). For blended lines, we require that the line widths, deviations from laboratory wavelengths, and wavelength separations of the individual lines be consistent with the spectrometer characteristics before accepting line identification and fluxes.

Variations during the individual pointings are at the 20% level. Dupree et al. (1995b) find that the intensities of the highest temperature lines ($T_e > 10^7$ K) show modest variations (factors of 2 to 3) with different pointings, while the lower temperature lines ($T_e < 10^7$ K) show variations of about 30% or less. Ayres (1988) shows that the UV line fluxes are remarkably constant with time as well. While variability in the high

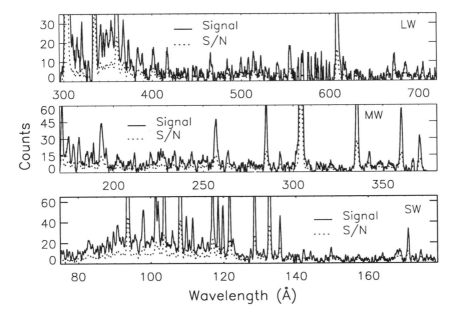

FIGURE 1. Capella summed spectrum in the three *EUVE* spectrometers (*solid line*); signal to noise (*dotted line*). Curves are smoothed by 5 and 7 pixels, respectively.

temperature lines is useful for line identification and deblending (such as in the blend of Fe XXIV, Ca XVII, Fe XII, and O V around 192 Å), the lack of variability over most of the temperature range gives confidence that this analysis is appropriate for typical atmospheric conditions.

3. The EMD for the Summed Spectrum

The analysis of a coronal source through its EMD provides key insights into heating, cooling, and confinement. While these issues are central to much of the EUV work on stellar coronae, the more immediate goal here is to use the predictive capabilities of an EMD, determined by strong lines from a single element, to examine the rich spectrum for other diagnostics. This goal sets the criterion for the "best fit" as the agreement of strong lines with the model. Figure 2 shows our model for the Capella spectrum, and the agreement of the lines used in constructing it.

For $T_e \geq 5 \times 10^5$ K the EMD depends on the the EUV lines of highly ionized Fe. Newly identified lines are discussed by Brickhouse et al. (1995b). Although the high temperature extent is not well constrained using EUV lines, this model is now consistent with the *ASCA* continuum above 2 keV (Singh 1994), as well as with the carefully deblended doublet lines of Fe XXIV. We include the X-ray fluxes observed by Vedder & Canizares (1983) in Figure 2, although we do not use these lines in the fitting. It is interesting to note that the two lines formed predominantly at $T_e = 10^{6.8}$ K (O VIII $\lambda 18.97$ and Fe XVII $\lambda 15.01$) are in good agreement with the model, while the higher temperature line (Fe XX $\lambda 12.83$) exceeds the prediction of the summed spectrum, but is consistent within the range of variability.

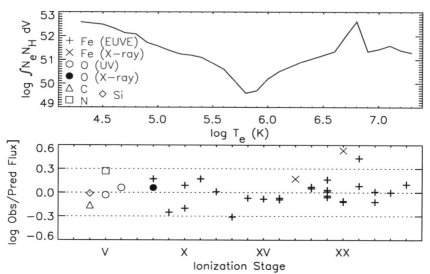

FIGURE 2. Capella model. *Upper:* log Emission Measure (cm^{-3}) vs. log T_e (K). *Lower:* comparison of observed to predicted fluxes for the strong lines. See text for sources of X-ray and UV lines from different elements.

Below this temperature we use the total UV line fluxes measured by Linsky et al. (1995) with *HST* and Hurwitz et al. (1995) with *ORFEUS*. As Linsky et al. (1995) discuss, the N V resonance line flux may reflect N enrichment in the evolved star, and thus we have forced our fit to the weaker O V line. Many EUV transition region lines are extremely weak relative to the prediction based on the UV emission lines, as initially reported in Dupree et al. (1993) and discussed in more detail in Brickhouse et al. (1995b).

We cannot emphasize too strongly that the features in the EMD are required by the fluxes of strong lines. For example, Capella's local maximum at $10^{6.8}$ K is determined by eight strong lines of Fe XVIII and XIX, for a total of 7600 counts in the summed spectrum. Continuous distribution models which smooth over this feature will underpredict the emission from other lines formed predominantly at this temperature. On the other hand, a two-temperature model has poor predictive powers for lines formed away from the two dominant temperatures. While the single emission measure at $10^{6.8}$ K (i.e. the integral over the interval $\Delta T_e = \pm 0.05$ dex) in our model accounts for about 80% of Fe XVIII and XIX line fluxes (with most of the rest of the flux coming from $10^{6.7}$ K), it only accounts for 43%, 13%, and 1% of the Fe XVI, XV, and XIV line fluxes, respectively. Thus many temperatures are required to produce the range of Fe ionization stages.

With high S/N spectra and excellent flux calibration, errors in the atomic models begin to dominate the uncertainties. For the more highly ionized Fe species, some collision strengths are as accurate as 10%; for the lower ionization stages (Fe VIII to XIV), collision strengths may be inaccurate at the 30% level or worse. Dielectronic recombination rates are the largest source of uncertainty in the ionization balance, as discussed by Brickhouse et al. (1995a). In addition to these uncertainties, many of the lower temperature Fe lines are N_e-sensitive, and may complicate the emission measure analysis. We find from a synthesis of the Fe XII–XIV spectra that $N_e \sim 10^9$ cm^{-3} is more consistent with observations than 10^{10} or 10^{11} cm^{-3}. While we are not able to measure unblended line ratios from these ions, we are able to isolate several line groups.

At high temperature, line ratio diagnostics give $\log N_e$ (cm^{-3}) = 11.1 (+0.3, −no constraint) from Fe XXI $\lambda128.7/(\lambda97.9 + \lambda102.2)$; 12.5 (+0.2, −0.1) from Fe XXI $\lambda128.7/\lambda113.3$; 12.35 (+0.2, −0.3) from Fe XXI $\lambda128.7/ \lambda142.2$; 11.73 (+0.2, −0.3) from Fe XXI $\lambda128.7/\lambda145.6$; and, 12.05 (+0.3, −no constraint) from Fe XXII $\lambda117.1/\lambda114.4$. Uncertainties given correspond to 1 σ errors in the line ratios. Although the summed spectra provide a great improvement in S/N over the calibration data (Dupree et al. 1993), and allow reasonable separation of the blended lines, a large spread remains, which seems to require multiple densities at the source. As noted in Brickhouse et al. (1995a), Fe XXI in the model is formed about equally at $T_e = 10^{6.8}$ K and at $T_e > 10^7$ K.

4. Measurement of the Continuum and the Fe/H Abundance Ratio

Stern et al. (1995) measure the continuum emission for *Algol* by summing the flux in 4 bins to achieve sufficient S/N. We are able to measure *directly* the minimum flux level in the Capella summed spectrum, since S/N > 3.0 in every bin throughout the wavelength range 81–123 Å. Figure 3 is a detail of this spectral region. We attribute the minimum flux to real continuum emission (mostly bremsstrahlung), and compare it to model predictions to derive an Fe/H abundance ratio. We find that the Fe/H abundance ratio is 0.88 ± 0.13 solar, assuming the solar values of Anders & Grevesse (1989).

The flux in the pixels which contain the lowest flux values determines the continuum, and the quandary is to select the appropriate number of pixels in the flux averaging. We select the lowest points in the spectrum that agree in value with each other to within their statistical error limits. With smoothing by 3 pixels, this condition produces 14 bins to define the continuum, and the derived error is the standard deviation. The value of the continuum defined in this way is consistent with the measurement determined (from fewer points) with smoothing by 5 or 7 pixels. Furthermore, inspection of line lists gives additional confidence that emission lines are avoided. For all but 3 bins, there are no close wavelength matches with lines listed in the Doschek & Cowan (1984) line list, which includes lines from flaring as well as other solar conditions. We can estimate the contribution from weak, highly ionized Ni by scaling from lines we observe relative to their theoretical Fe counterparts, for which the tokamak line lists of Davé et al. (1987) and Stratton et al. (1985) are invaluable. The existence of pseudo-continua produced by unknown weak lines is difficult to rule out without models for these weak lines, but their contribution should be well within the errors on the continuum flux measurement.

Figure 3 also shows the temperature-dependence of bremsstrahlung emission. In our model the emission comes predominantly from the $10^{6.8}$ K local maximum. The model predicts that the continuum emission decreases by about 20% over the range from 85 to 140 Å; we are not able to determine a slope observationally.

5. Elemental Abundances at High Temperature

The lines used to determine abundances are listed in Table 1. Line emissivities are from Raymond (1988), modified by the atomic data of Zhang et al. (1990) for Li-like ions. Figure 4 shows the dependences of model line emissivities and integrated intensities. For the high-Z (above Ne) Li-like lines, the emission in our model comes primarily from $10^{6.8}$ K. The doublet lines are particularly useful, since their line ratio can give additional confidence in the flux measurement. The O and Ni lines in the Table similarly come from this temperature in the model. The errors listed are observational, but there are systematic errors as well, particularly errors in the correction for interstellar absorption. The effect of assuming neutral He is comparable to the errors in $N_{\rm H}$ given

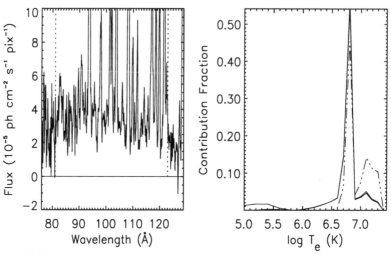

FIGURE 3. *Left*: continuum emission in flux units, where 1 pix = 0.067 Å. Dotted vertical lines indicate the wavelength range for which S/N > 3. *Right*: model of continuum flux at 100 Å (0.12 keV) (*solid*) and at 6 Å (2.0 keV) (*dash-dotted*) vs. temperature. The contribution fraction is the integral of the continuum emissivity over the EMD relative to the total.

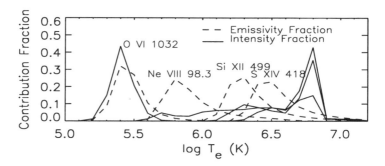

FIGURE 4. The fractional emissivity (*dashed*) and intensity (*solid* contributions as functions of temperature for selected lines from Li-like ions.

by Linsky et al. (1993). Vennes et al. (1993) show that the fraction of ionized He might be as high as 25%. The abundances in Table 1 derived from lines between 227 and 504 Å would then be somewhat reduced. For Si XII the relative abundance would decrease from 1.99 to 1.65, and would bring the doublet ratio into better agreement with theory.

6. Conclusions

High quality *EUVE* spectra of cool stars, such as the summed spectrum of Capella, are rich with information on the physical conditions in the atmosphere. Long observations confirm the presence in Capella of high densities ($\geq 10^{12}$ cm^{-3}) at high temperature, but lower densities are present as well at other temperatures. We have now measured the relative abundances of a number of elements which emit radiation at the local maximum

TABLE 1. Elemental Abundances at $T_e \sim 10^{6.8}$ K[†]

Element	Emission Lines	Capella[‡]	Capella[¶]	Solar Corona[‖]
O	O VIII λ102.45[††]	0.42 ± 0.18	0.13(0.03,0.27)	0.9
	O VIII λ18.97[‡‡]	1.02 ± 0.28		
Si	Si XII λ499.40	1.99 ± 0.38	0.85(0.67,1.07)	3.6
	Si XII λ520.67			
S	S XIV λ417.61	1.15 ± 0.18	0.73(0.39,1.14)	1.1
	S XIV λ445.77			
Ar	Ar XVI λ353.92	2.41 ± 0.31		1.0
	Ar XVI λ389.14			
Fe	(all)	0.88 ± 0.13	0.46(0.36,0.61)	2.7
Ni	Ni XVII λ249.18	1.81 ± 0.20		3.9
	Ni XVIII λ291.97			
	Ni XVIII λ320.54			

[†] Abundances are relative to solar photospheric abundances of Anders & Grevesse (1989).

[‡] This work. Errors are 1 σ estimates of measurement uncertainties, including statistical errors, deblending of weak lines and high order flux subtraction.

[¶] From Drake et al. (1994). Allowed ranges reflect 90% confidence levels from X-ray spectral fitting with two temperatures.

[‖] Feldman (1992)

[††] This line was deblended from Fe XXI λ102.2.

[‡‡] Observed by Vedder & Canizares (1983). Prediction uses the EMD from this work.

at $\sim 10^{6.8}$ K, and find that they are more consistent with solar photospheric than coronal values. Density and abundance measurements may provide keys to understanding the nature of this enhanced emission region. Variability in the N_e-sensitive lines at the highest temperatures may provide a tool for distinguishing different sources of emission.

The author appreciates the contributions of Andrea Dupree, John Raymond and Greg Hanson. This work has been supported by NAGW-528 and NAG5-2330 to the Smithsonian Institution.

REFERENCES

ANDERS, E., & GREVESSE, N. 1989, Abundances of the Elements: Meteoritic and Solar, Geochim. Cosmochim. Acta, 53, 197

AYRES, T. R. 1988, A Spectral Dissection of the Ultraviolet Emissions of Capella, ApJ, 331, 467

BOYD, W., JELINSKY, P., FINLEY, D. S., DUPUIS, J., ABBOTT, M., CHRISTIAN, C., & MALINA, R. F. 1994, In-Orbit Performance of the Spectrometers of the Extreme Ultraviolet Explorer, Proc. SPIE, 2280, 280

BRICKHOUSE, N. S., RAYMOND, J. C., & SMITH, B. W. 1995a, New Model of Iron Spectra in the Extreme Ultraviolet and Application to SERTS, and EUVE, Observations: A Solar Active Region and Capella, ApJS, 97, 551

BRICKHOUSE, N. S., DUPREE, A. K., & RAYMOND, J. C. 1995b, Analysis of the EUVE, Spectrum of Capella from the Transition Region to the Hot Corona, in progress

Cook, J. W., Cheng, C. -C., Jacobs, V. L., & Antiochos, S. K. 1989, Effect of Coronal Elemental Abundances on the Radiative Loss Function, ApJ, 338, 1176

Davé et al. 1987, Time-resolved spectra in the 80 Å Wavelength Region from Princeton Large Torus Tokamak Plasmas, J. Opt. Soc. Am., B4, 635

Doschek, G. A., & Cowan, R. D. 1984, A Solar Spectral Line List Between 10 and 200 Å Modified for Application to High Spectral Resolution X-ray Astronomy, ApJS, 56, 67

Drake, J. J., Laming, J. M., & Widing, K. G. 1995, Stellar Coronal Abundances. II. The First Ionization Potential Effect and Its Absence in the Corona of Procyon, ApJ, 443, 393

Drake, S. A., Singh, K. P., White, N. E., & Simon, T. 1994, ASCA, X-ray Spectra of the Active Single Stars β Ceti and Π^1 Ursae Majoris, ApJ, 436, L87

Dupree, A. K., Brickhouse, N. S., Doschek, G. A., Green, J. C., & Raymond, J. C. 1993, The Extreme Ultraviolet Spectrum of Alpha Aurigae Capella, ApJ, 418, L41

Dupree, A. K., Brickhouse, N. S., & Hanson, G. J. 1995a, High Temperature Structure in Cool Binary Stars, these proceedings

Dupree, A. K., Brickhouse, N. S., Doschek, G. A., Hanson, G. J., & Raymond, J. C. 1995b, EUVE Spectra of Alpha Aurigae Capella, at Different Phases, in preparation

Feldman, U. 1992, Elemental Abundances in the Upper Solar Atmosphere, Phys. Scripta, 46, 202

Hurwitz, M., Bowyer, S., & Dupree, A. K. 1995, ORFEUS, Observations of Capella, in preparation

Landini, M., & Monsignori, Fossi, B. C. 1993, Extreme Ultra Violet Plasma Diagnostic: a Test Using EUVE Calibration Data, A&A, 275, L17

Lemen, J. R., Mewe, R., Schrijver, C. J., & Fludra, A. 1989, Coronal Activity in F-, G-, and K-Type Stars. III. The Coronal Differential Emission Measure Distribution of Capella, σ^2 CrB, and Procyon, ApJ, 341, 471

Linsky, J. L et al. 1993, Goddard High-Resolution Spectrograph Observations of the Local Interstellar Medium and the Deuterium/Hydrogen Ratio along the Line of Sight toward Capella, ApJ, 402, 694

Linsky, J. L., Wood, B. E., Judge, P., Brown, A., Andrulis, C., & Ayres, T. R. 1995, The Transition Regions of Capella, ApJ, 442, 381

Raymond, J. C. 1988, Radiation from Hot, Thin Plasmas, in Hot Thin Plasmas in Astrophysics, ed. R. Pallavicini, Dordrecht: Kluwer, 3

Singh, KP 1994, private communication

Stern, R. A., Lemen, J. R., Schmitt, J. H. M. M., & Pye, J. P. 1995, EUVE Observations of Algol: Detection of a Continuum and Implications for the Coronal [Fe/H] Abundance, ApJ, 444, L45

Stratton, et al. 1985, Relative Intensities of $2s^2 2p^k$ s $2p^{k+1}$ transitions in FI- to BI-like Ti, Cr, Fe, Ni, and Ge in a Tokamak Plasma: A Comparison of Experiment and Theory, Phys. Rev. A, 31, 2534

Swank, J. H., White, N. E., Holt, S. S., & Becker, R. H. 1981, Two-Component X-ray Emission from RS Canum Venaticorum Binaries, ApJ, 246, 208

Vedder, P. W., & Canizares, C. R. 1983, Measurement of Coronal X-ray Emission Lines from Capella, ApJ, 270, 666

Vennes, S. et al. 1993, The First Detection of Ionized Helium in the Local ISM: EUVE, and IUE, Spectroscopy of the Hot DA White Dwarf GD 246, ApJ, 410, L119

Zhang, H. L., Sampson, D. H., & Fontes, C. J. 1990, Relativistic Distorted-Wave Collision Strengths and Oscillator Strengths for the 85 Li-like Ions with $8 \leq Z \leq 92$, At. Data Nucl. Data Tables, 44, 31

Hot Times in the Hertzsprung Gap

THOMAS R. AYRES

Center for Astrophysics and Space Astronomy, Campus Box 389,
University of Colorado, Boulder, CO 80309, USA

Moderate-mass giants represent a touchstone for probing the mechanisms of magnetic activity among fast-rotating convective stars. Extreme ultraviolet and soft X-ray observations of such stars detect generally hot coronae: the Hertzsprung-gap giants (F5–G2), in particular, have remarkable high-excitation peaks (10^7 K, or hotter) in their emission-measure distributions. While the high-temperature coronal plasmas are reminiscent of violent solar *flares*, the high-energy spectra of the Hertzsprung-gap giants appear to be quite steady over time; in contrast to other hot-corona objects whose optical light curves carry the stamps of *starspots*, and whose high-excitation emissions are sporadically—and dramatically—variable. The constancy of the Hertzsprung-gap stars is particularly puzzling in light of high-dispersion FUV spectroscopy that reveals supersonic flows in their 10^5 K subcoronal "transition zones."

1. Introduction

The exploration of high-excitation *coronae* among Main Sequence late-type stars is relatively advanced; not surprising given that we have a prime example—the Sun—close at hand (e.g., Cox, Livingston & Matthews 1991). As one moves away from the MS, however, the situation becomes more complex; in part because of the dramatic structural changes imposed by the rapid post-MS evolution of moderate-mass ($\approx 2 - 4M_\odot$) stars. One key group are the G8–K0 giants in the "Clump" at the base of the red giant branch. Most of such stars are in the post-flash core-helium burning phase. A second key group are the F5–G2 giants in the Hertzsprung gap, blueward of the Clump. They are the predecessors of the Clump stars, and are relatively rare thanks to the brief first-crossing time for moderate-mass giants. The denizens of the gap are the fresh descendants of A- and B-type dwarfs; consequently one finds many fast-rotators among them. Because the gap giants also are convective, one might anticipate enhanced spin-catalyzed magnetic activity. However, just redward of G0 is a sharp break in the rotational speeds of the giants (Gray 1989). The braking mechanism is controversial. Internal redistribution of angular momentum certainly plays a role, but magnetospheric mass loss (crucial for dwarf stars) might be important as well (e.g., Schrijver & Pols 1993: hereafter SP).

A pivotal aspect of the Hertzsprung-gap stars, originally noted by Simon and Drake (1989: hereafter SD), is that they appear to be deficient in their X-ray luminosities compared with the coronal proxy C IV $\lambda1549$. At the same time, the active Clump giants are more solar-like in their X-ray/C IV ratios. Exploring the dichotomy with the powerful tools provided by the *EUVE* and *HST*/GHRS has provided new clues to the nature of the underlying magnetic "engine" that powers stellar activity.

2. *ROSAT* and *IUE* Observations of Coronal Giants

Although the first recognition of the abnormal coronae of the Hertzsprung-gap giants came from *Einstein* pointings, the recent *ROSAT* all-sky survey has provided a clear, and minimally-biased, view of the dichotomy. Figure 1 illustrates an X-ray H–R diagram based on material from the "RIASS" campaign (which coordinated UV spectroscopy by the *IUE* satellite with the *ROSAT* survey passes: see Ayres et al. 1995).

Figure 2 compares the normalized fluxes (f/f_{bol}) of X-rays and C IV $\lambda1549$ for the

113

S. Bowyer and R. F. Malina (eds.), Astrophysics in the Extreme Ultraviolet, 113–120.

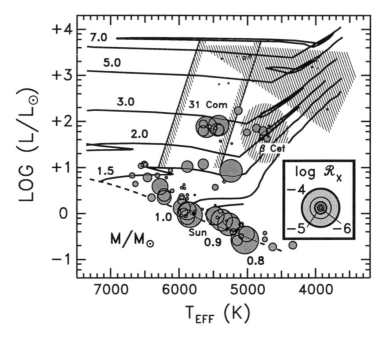

FIGURE 1. X-ray H–R diagram. Size of each "bubble" is proportional to $\mathcal{R}_X \equiv f_X/f_{bol}$ according to key. Slanted hatched lines delimit the Hertzsprung gap; hatched oval, the "Clump;" shaded wedge, the "hybrid-chromosphere" stars (not discussed here). Solid curves depict evolutionary trajectories; dashed curve is ZAMS. Dot at base of $1 M_\odot$ track indicates the Sun.

evolved stars of the RIASS sample. The thick dashed line depicts the power-law correlation obeyed by dwarfs of solar-type and cooler. The Clump giants generally follow the MS trend, but the Hertzsprung-gap stars (as well as supergiants of all types) fall below it. The behavior is ascribed to an X-ray *deficiency*, because C IV/Mg II diagrams are perfectly normal for the F5–G2 giants (and G/K supergiants: Ayres et al. 1995).

SD offered an explanation for the X-ray deficiency of the warm giants based on classical acoustic-wave heating of their coronae, rather than the magnetic agency believed to operate in solar-type stars and active late-type binaries. The latter all show evidence for *starspots, flares,* and other signatures of solar-like magnetic phenomena. In the SD model, the acoustic waves strongly heat the intermediate temperature layers (up to, say, a few$\times 10^5$ K), but the lack of closed magnetic loops permits the heated gas to flow away from the star in a pervasive but tenuous wind. The low densities of the expanding exospheric flow depress the emission measure ($EM \equiv \int_V n_e^2\, dV$) relative to a normal magnetically-bottled (closed-field) corona, yielding a much reduced X-ray luminosity. At the same time, the wind—particularly if slightly magnetized—sheds considerable angular momentum from the initially fast-rotating star, promoting the dramatic spindown seen at the "G0 break." The acoustic heating model also explains the apparent constancy of the FUV emissions of the Hertzsprung-gap giants (e.g., Ayres et al. 1995), noted originally in the Capella system (α Aur = HD 34029: G8 III + G1 III; see, e.g., Ayres (1988), and references therein). Because the acoustic heating is a byproduct of the global convective

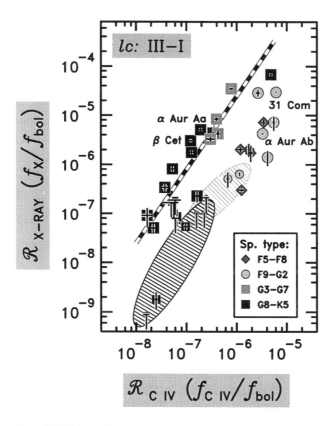

FIGURE 2. X-ray/C IV "correlogram" from RIASS campaign. Luminosity classes III–I are depicted explicitly; thick dashed line is mean power-law trend for F9–K5 dwarfs. Larger symbols indicate class III; intermediate-size symbols, II–I; small symbols with downward "tails," upper limits. Middle (light) oval encompasses the G supergiants; lower (dark) oval, the K-type hybrids. Symbol for G8 primary of Capella (α Aur Aa) lies on dashed curve, partially obscured.

envelope, it should be comparably uniform in its surface effects. In contrast, solar magnetic fields erupt in localized *active regions* that ultimately can produce hemispherical asymmetries in X-ray or UV irradiances.

3. New Insights from the Extreme Ultraviolet

A recent breakthrough in the exploration of the X-ray deficiency "syndrome" came with the *EUVE* satellite (Bowyer & Malina 1991). The key binary Capella was a calibration target for the spectroscopy modes, and an early 80 ks pointing was reported by Dupree et al. (1993). The Capella SW and LW spectra are dominated by hot iron lines, with an apparent strong peak in the *EM* distribution at 6×10^6 K. Unfortunately, however, the Capella EUV spectra are an unknown blend of the coronal emissions of the two companions (which might be more nearly equal in the EUV: the large FUV advantage of the G1 secondary might be offset by its X-ray deficiency, e.g., Ayres 1988). Nevertheless,

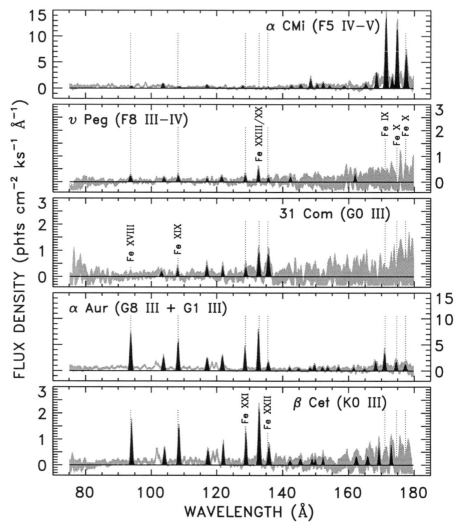

FIGURE 3. *EUVE* SW spectra of Hertzsprung-gap and Clump giants. Light shaded envelopes are traced spectra ±1 σ; solid peaks represent detections of selected "features." The instrumental sensitivity has been corrected, but there was no compensation for interstellar absorption.

the Capella spectra show no evidence for the very soft ($T < 10^6$ K) coronal component predicted by the acoustic-heating scenario (cf., SD).

Another of the *EUVE* targets during the early Guest Observer phase was 31 Comae (HD 111812: G0 III), an archetype gap star. Remarkably, its emissions (in an 80 ks pointing) are completely dominated by hyper-ionized iron, with little material cooler than about 10^7 K. Figure 3 compares the *EUVE* SW spectra of Capella and 31 Com with those of Procyon (F5 IV–V; a cool-corona dwarf like the Sun), β Ceti (K0 III; a Clump

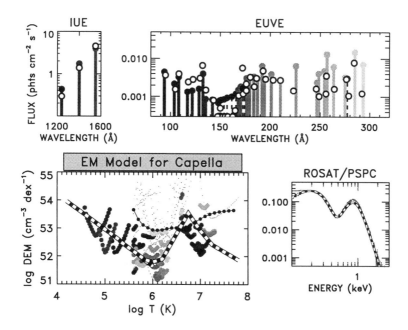

FIGURE 4. Emission-measure modeling of Capella based on *EUVE* SW and MW spectra, constrained by *IUE* and *ROSAT* fluxes. In the upper panels, FUV and EUV detections (corrected for ISM absorption) are marked by vertical solid bars; upper limits by dashed bars. *ROSAT* PSPC pulse-height spectrum in lower right panel is a light-shaded band (barely visible here): observed counts (per 0.010 keV bin) ±1 σ. Open circles in *IUE* and *EUVE* frames, and dashed curve in *ROSAT* frame, are predictions based on *EM* model in lower left panel. In it, heavy dashed curve is model; larger dots are "monothermal" curves for individual spectral features (connected dots for PSPC); small dots depict upper limit curves for nondetections.

giant like the G8 primary of Capella), and v Pegasi (F8 III–IV; another Hertzsprung-gap giant, observed recently in AO-2). It is clear that while the Hertzsprung-gap stars might be X-ray deficient, they are no slouches coronal-*temperature*-wise.

4. Emission-Measure Modeling

Figures 4 and 5 illustrate the inferred *EM* distributions for Capella and 31 Comae, respectively, based on fitting *IUE* fluxes of key FUV species at the low-temperature end; *EUVE* lines at the high-temperature end; and *ROSAT* PSPC pulse-height spectra as a secondary constraint. The EUV modeling was based on prominent "features" (e.g., Fig. 3); that is, a specific set of wavelengths where there appears, over one or more temperature intervals, a strong emission line (not necessarily from the same species over the full temperature range). Emissivity curves were developed for each "feature" by a width-constrained measurement of temperature-resolved spectra simulated with the line list of Mewe, Gronenschild & van den Oord (1985), assuming cosmic abundances. The philosophy differs somewhat from the approach of assigning a specific identification (and thus emitting species) to an apparent detected peak in the *EUVE* spectra, but

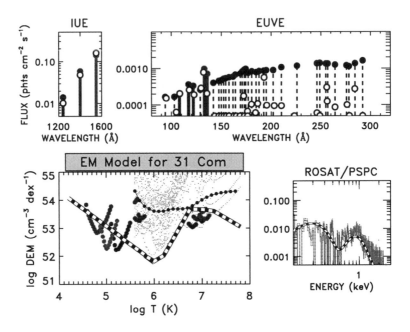

FIGURE 5. Same as previous figure, for 31 Comae. PSPC observation was taken with Boron filter, whose response has been corrected.

neatly avoids the quandary of blended lines and/or ambiguous assignments. The iterative χ^2 minimization procedure—with a weighting scheme to accommodate upper limits— empirically describes the coronal *EM* distribution in sufficient detail to compare with theoretical expectations, given the uncertainties in the underlying atomic physics (cf., discussions at this meeting), and the further ambiguities posed by possible abundance anomalies (cf., the "FIP" effect) and gas dynamics.

5. Discussion and Conclusions

On the one hand, the very hot coronae of the Hertzsprung-gap giants argues forcefully against the acoustic heating scenario. 1 keV gas implies very large post-shock velocity contrasts, thousands of km s^{-1}. But, as illustrated in figure 6, the attendant line Doppler broadening is not seen in high-dispersion *HST*/GHRS spectra. On the other hand, the presence of solar-flare temperature material and supersonic flows, but without the overt signatures of flare activity, presents a dilemma for the magnetic heating scenario.

One way out is to appeal to the *nanoflare* hypothesis (e.g., Parker 1988). Namely, the magnetosphere of a typical Hertzsprung-gap giant is in a continual state of flaring, but the demographics of the flares is dominated by numerous, relatively compact, widely dispersed events. Such a population would be equivalent to that proposed for heating the solar corona (Parker 1988), but without the tail towards larger events evident in the Sun's case. The truncated population might be a natural consequence of the fast rotation

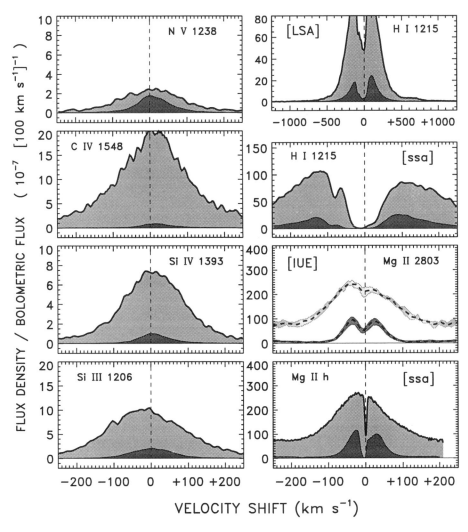

FIGURE 6. High-dispersion UV spectra of 31 Com (light shaded) and β Cet (dark shaded). Data are taken from *HST*/GHRS, except for Mg II λ2803 panel labeled "IUE." There, shaded envelopes represent ±1 σ (temporal) deviations of available Mg II profiles over 15-year span of *IUE* mission. In upper left panels, the anomalous strength of N V, and weakness of C IV, in β Cet probably are due to "evolved" abundances. Note *redshifts* of N V, C IV, and Si IV in both stars, but apparent *blueshifts* of Mg II *h* and H I Lyα in 31 Com. Despite partial blurring of 31 Com lineshapes by fast rotation ($v \sin i \approx 70$ km s^{-1}), the large Doppler widths indicate supersonic motions in $T \approx 10^5$ K regime. The profiles show no evidence, however, for 1000's of km s^{-1} broadening expected for acoustic waves of sufficient amplitude to produce 1 keV coronal material detected by *EUVE*.

of the Hertzsprung-gap giants coupled with their thin convective envelopes: a limiting variant of Dynamo action.

If the *nanoflare-ona* mechanism applies to the Hertzsprung-gap giants, what about the Clump stars? After all, they too show a high-T peak in their coronal *EM* distribution, even if their overall multi-spectral properties (e.g., $\mathcal{R}_X/\mathcal{R}_{CIV}$) are more solar-like. The author was reminded gently at the meeting that it is obvious that giant stars *must* have hot coronae, owing to their low surface gravities. Yet, most modern theories of coronal structures (i.e., constant pressure loops) dismiss the influence of surface gravity precisely because the plasma is *field-dominated* (see, e.g., Antiochos 1994; gravity is important only for $T \approx 10^5$ K "cool loop" solutions). Furthermore, the full range of possible "coronal" temperatures (i.e., $10^5 - 10^7$ K) is seen in discrete structures on the Sun. Thus gravity, *per se*, must play a subordinate role to the heating mechanism.

In short, the puzzle of the structure and heating of the Hertzsprung-gap coronae is mirrored by their counterparts in the Clump. Indeed, it is curious that the Clump giants exhibit *any* high-excitation activity at all, given the ample opportunities during their post-MS evolution to lose virtually all of their Dynamo-catalyzing spin. Even weak magnetospheric winds in giant stars shed angular momentum prodigiously (SP), and there is the disrupting influence of helium flash as well.

The *EUVE* has offered tantalizing clues concerning the nature of coronae among the moderate-mass evolved stars. The full resolution of the open issues undoubtedly lies in future observations; particularly high-dispersion, high-sensitivity measurements in the 100–400 Å spectral range, where *EUVE* has taken the pioneering first steps.

I thank my collaborators—A. Brown, S. Drake, T. Simon, and R. Stern—for their help with the project. My participation was supported by NASA grants NAG5-2274 (*EUVE*), GO-5323.01-93A (*HST*), and NAG5-2451 (*ROSAT*).

REFERENCES

Antiochos, S. K. 1994, The physics of coronal closed-field structures, Adv. Space Res., 14(4), 139

Ayres, T. R. 1988, A spectral dissection of the ultraviolet emissions of Capella, ApJ, 331, 467

Ayres, T. R., et al. 1995, The RIASS Coronathon: Joint X-ray and ultraviolet observations of normal F–K stars, ApJS, 96, 223

Bowyer, S. & Malina, R. F. 1991, The EUVE Mission, in Extreme Ultraviolet Astronomy, ed. R. F. Malina & S. Bowyer, New York: Pergamon, 397

Cox, A. N., Livingston, W. C. & Matthews, M. S. (editors) 1991, Solar Interior and Atmosphere, Univ. of Arizona Press

Dupree, A. K., Brickhouse, N. S., Doschek, G. A., Green, J. C., & Raymond, J. C. 1993, The extreme ultraviolet spectrum of Alpha Aurigae Capella, ApJL, 418, L41

Gray, D. F. 1989, The rotational break for G giants, ApJ, 347, 1021

Mewe, R., Gronenschild, E. H. B. M. & van den Oord, G. H. J. 1985, Calculated X-radiation from optically-thin plasmas. V., A&AS, 62, 197

Parker, E. N. 1988, Nanoflares and the solar X-ray corona, ApJ, 330, 474

Schrijver, C. J. & Pols, O. R. 1993, Rotation, magnetic braking, and dynamos in cool giants and subgiants, A&A, 278, 51

Simon, T. & Drake, S. A. 1989, The evolution of chromospheric activity of cool giants and subgiant stars, ApJ, 346, 303

Are Some Stellar Coronae Optically Thick?

C. J. SCHRIJVER,[1] G. H. J. VAN DEN OORD,[2]
R. MEWE,[3] AND J. S. KAASTRA[3]

[1] Lockheed Palo Alto Research Laboratories, Palo Alto, CA 94304, USA

[2] Sterrekundig Instituut, Utrecht, The Netherlands

[3] Space Research Organization of The Netherlands, Utrecht, The Netherlands

We discuss the coronal spectra of a sample of cool stars observed with the spectrometers of the *Extreme Ultraviolet Explorer* (*EUVE*). The emission measure distributions show (a) a relatively weak component between 0.1 MK and 1 MK, (b) a dominant component somewhere between 2 MK and 10 MK, and (c) in all cases but one a component in the formal solution at temperatures exceeding ≈ 20 MK. Where this hot tail is not associated with a real hot component, it is a spurious result reflecting a lowered line-to-continuum ratio, which, for instance, may be the result of a low abundance of heavy elements or of resonant scattering in some of the strongest coronal lines. We suggest that in Procyon's corona photons in the strongest lines formed around a few million Kelvin undergo resonant scattering in a circumstellar medium, possibly a stellar wind. The flare spectrum of AU Mic suggests that resonant scattering may also occur in dense, hot flare plasmas. The electron densities of the 5–15 MK component are some three orders of magnitude higher than typical of the solar-like component around 2 MK; the volume filling factors of the hot components are therefore expected to be relatively small.

1. Introduction

The soft X-ray emission from stellar coronae has been measured by satellites such as *EINSTEIN*, *EXOSAT*, and, most recently, *ROSAT*. Most of these observations made use of either broad-band filters or of imaging proportional counters with a spectral resolution of order $\lambda/\Delta\lambda \approx 1$. Broad-band observations, however, constrain the temperature structure of stellar coronae only weakly. Observations with moderate to high spectral resolution are required to determine the "differential emission measure" (DEM), i.e., the weighting function $D(T) \equiv n_e n_H dV/d\log(T)$ which, together with the volume emissivity, measures the contribution of plasma of a given temperature to the emitted spectrum. Recently, *EUVE* has observed EUV spectra with a resolution of 0.5 to 2 Å, observing lines characteristic of temperatures in the range from $\sim 10^5$ to $\sim 10^7$ K, and a few lines outside that interval. We analyzed the spectra of eight cool stars, presented in detail in papers by Mewe et al. (1995) and Schrijver et al. (1995). This paper discusses the main conclusions for all sources but AY Cet. The DEM inversion method is described in these proceedings by Mewe et al.

2. The Temperature Structure of Coronae

Fig. 1 shows the $D(T)$ curves for our sample (see also Table 1). In order to facilitate the comparison of these results for plasma around temperatures typical of the quiescent solar corona, the curves are normalized to a comparable level for the (peak) emission measure between 2 to 4 MK.

Procyon and α Cen both have a relatively strong tail extending to well below 1 MK. Their $D(T)$ are alike between 10^5 K and 10^7 K. A scaled $D(T)$ for a solar active region is comparable to that of α Cen over a substantial temperature range.

Capella and ξ UMa form another pair with comparable $D(T)$: both have a component

S. Bowyer and R. F. Malina (eds.), Astrophysics in the Extreme Ultraviolet, 121–128.

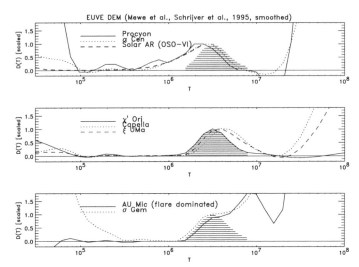

FIGURE 1. Comparison of $D(T)$ (as defined in Section 1; taken from Mewe et al. (1995), and Schrijver et al. (1995) —listed in Table 2 in the latter paper, which also gives the uncertainties). The curves are scaled to approximately unity around 2–4 MK, and are smoothed over two bins of $\log(T) = 0.1$. The dashed region repeated in the three panels is the $D(T)$ of χ^1 Ori between 1.5 MK and 8 MK. The top panel also shows a differential emission measure curve for solar active region data observed with OSO-VI (from Dere (1982))

below 10^7 K that extends to higher temperatures than those of χ^1 Ori, Procyon, and α Cen, while they show little evidence of material with temperatures below about 1.5 MK.

The emission measure of σ Gem and AU Mic continues to increase above about 2 MK. Note that the spectrum of AU Mic is dominated by a major flare followed by a few smaller ones (see Brown & Gagné et al. in these proceedings).

The $D(T)$ curves below 10^5 K are dominated by the strong He II lines 256 Å and 304 Å. Their presence is indicative of upper chromospheric and transition-region emission, but as these lines are effectively thick, the $D(T)$ serves only as a formal solution.

All sources in our sample show a strong high-temperature tail, with the exception of the solar-like star χ^1 Ori (G0V). The strong increase in $D(T)$ at very high temperatures, cut off in Fig. 1, corresponds to a total emission measure that exceeds the total emission measure of cooler plasma in the range from ≈ 20 MK down to several hundred thousand Kelvin by factors of 2 up to 15 (see Table 1). This hot component is associated with a "continuum" in the SW range between 80 Å and 140 Å (that is observed in other cool stars too, see, for instance, the review by Brickhouse in these proceedings for another analysis of the Capella-spectrum, and the contributed paper by Walter on AR Lac). Even when assuming pure thermal equilibrium there are several plausible reasons for such a "continuum" to exist (also discussed by Schrijver et al. 1994 and 1995, Mewe et al. (1995), and in these proceedings by, e.g., Schmitt et al.): (1) an improper correction for the instrumental background; (2) a substantial number of closely spaced lines that are not (yet) included in the spectral code(s); (3) an optically thick continuum source; (4) a relatively low abundance for the elements producing most of the strong lines (predominantly Fe); (5) an asymmetric plasma that is optically thick in the strongest lines; or (6) a real hot component. We rule out an improper background correction, because the shape of the apparent continuum does not conform with the gra-

TABLE 1. Summary of results of emission-measure analyses by Mewe et al. (1995) and Schrijver et al. (1995)

Source	T_{max} (MK)	T 2nd comp. (MK)	Hot comp. (rel.)	Origin of hot comp.	Electron density (cm^{-3})	(Char. temp.) Fe lines
Procyon (F5IV–V)	2	<1	15	thick?	10^9–10^{10}	(1–2MK) X–XIV
α Cen (G2V+K2V)	3		9	thick?	$2\ 10^8$–$2\ 10^9$	(1–2MK) X–XIV
AU Mic (M0Ve)	9	3	15	real& thick?	2–$5\ 10^{12}$	(6–11MK)* XXI–XXII
Capella (G5III+G0III)	4		3	real or thick?	10^{12}–$2\ 10^{13}$	(6–11MK) XIX–XXII
σ Gem (K1III+...)	15	5	2	real or abund.?	10^{12}	(10–20MK) XXI–XXII
ξ UMa B (G0V+...)	4		1.7	real or abund.?	$5\ 10^{12}$	(10–20MK) XXI–XXII

* Strongly weighted towards the initial phases of a strong flare.
NOTES: The component(s) of binaries expected to dominate the observed spectrum have been underlined. Note that the ratio of the emission measures in the tail above 28 MK to that in the rest of the distribution above 0.9 MK (above 0.1 MK for Procyon), should be reduced by a factor of about four to six for a comparison in terms of counts if the continuum is interpreted as corresponding to the main component around about 3 MK rather than to a component at over 10 MK (compare Schrijver et al. (1995)). The inferred densities are consistent with independent analyses of, e.g., Brickhouse, Gagné et al., Hanson et al., and Keenan et al. in these proceedings; see also further references in Mewe et al. (1995) and Schrijver et al. (1995).

dient in the instrumental background. Based on the *EUVE* spectra alone one cannot determine uniquely which (combination of) the remaining causes is responsible for the high-temperature tail in the $D(T)$ so that presently we rely on indirect arguments and tests of consistency. Additional information from instruments with other energy pass bands, such as *ASCA*, and *EINSTEIN*, however, can be used to exclude certain causes. Schmitt, Drake & Stern (these proceedings) argue that for Procyon the $D(T)$-tail is incompatible with the *ROSAT* spectrum and can consequently not be real. A comparison with *EXOSAT*-TGS observations by Schrijver et al. (1995) resulted in the same conclusion. Schmitt et al. suggest that a number of lines missing from the spectral code are responsible for the "quasi-continuum." Although a number of weak lines are not included in the available spectral codes, it is doubtful that the missing lines are (1) sufficient in number and (2) have the appropriate strengths to mimic a flat continuum. The ensemble of missing lines should have the same effect in spectra of coronae as distinct as Procyon and AU Mic (flare dominated), but also have no effect in the spectrum of χ^1 Ori. We therefore decided to investigate alternative options. We do not consider an optically thick continuum source a likely candidate. The last three options in the above list appear to us as the most viable: solar coronal abundances have been observed to differ from the photospheric abundances (see papers by Haisch and by J. J. Drake in these proceedings), while ever since the *EINSTEIN* mission a real hot component is known to be present in very active stars (see, e.g., Ayres, Brickhouse, Brown, Dupree et al., Gagné et al., and Walter in these proceedings).

Lowering the abundance of an element reduces the line-to-continuum ratio. Schrijver et al. (1995) tested the case of a reduced metal abundance for the stars in their sample. In each case the high-temperature tail is indeed reduced in strength by lowering the abundance (cf., Mewe et al. in these proceedings; see their Fig. 3), while the emission measure for the cooler plasma is increased as required in order to fit the emission lines now reduced in strength in the model. The quality of the fit does not, however, improve significantly. In the cases of σ Gem and ξ UMa, the quality of the fit with and without reduced iron abundances is about the same, with a reduced χ^2-value below unity. Hence the possibility is left open in Table 1 that in some systems the tail may in fact be due to low iron abundances. The hot tail for AU Mic and Capella is likely to reflect a real hot component (see Schrijver et al. (1995) for details). For α Cen and Procyon, however, we suggest that resonant scattering in an asymmetric corona with line photon destruction upon impact on the stellar surface causes the spurious hot tail.

The usual assumption that stellar coronae are so tenuous that emitted photons escape without interaction with the stellar outer atmosphere or circumstellar plasma has been questioned for a number of strong lines based on solar observations (see below). A crude estimate of the optical depth can be made as follows: in the case of thermal Doppler broadening (which is likely to dominate in coronae) the optical depth at line centre is given by (e.g., Mariska 1992):

$$\tau_0 = 1.2 \, 10^{-17} (\frac{n_i}{n_{el}}) A_Z (\frac{n_H}{n_e}) \lambda f \sqrt{\frac{M}{T}} n_e \ell \equiv 10^{-19} C_d \left(\frac{A_Z}{A_{Z,\odot}} \right) \left(\frac{n_e \ell}{\sqrt{T_6}} \right) \qquad (2.1)$$

where (n_i/n_{el}) is the ion fraction, $A_Z = n_{el}/n_H$ the abundance, $n_H/n_e \approx 0.85$ the ratio of hydrogen to electron density (in cm^{-3}), λ the wavelength in Å, f the oscillator strength, M the atomic weight, T the temperature, and ℓ a characteristic dimension (in cm). The product $n_e \ell$, the column density, depends on the details of the coronal geometry. We tested two cases: a corona comprised of an ensemble of loops and a corona dominated by a spherically symmetric, hydrostatically settled envelope (see Table 2). The effective optical depth is strongly dependent on the geometry and temperature structure. In an elongated coronal loop photons suffer more scattering along the loop than perpendicular to it. In an actual stellar corona other loops overlay the originally emitting one, so that a photon can be scattered even if it escaped the loop in which it was originally emitted. The values for the loop-dominated atmosphere in Table 2 formally hold for radiation propagating along the loop, but in a more or less spherical volume of nested loops with the lowest most brightly emitting, the estimate approximates the net optical depth of coronal condensations over active regions.

Table 2 shows that some line-centre optical depths approach unity in loop-dominated geometries, particularly if sufficient loops overlie each other. But even for the values of the density and temperature for the base of the solar wind, optical depths can approach unity, while an increase in the base density or in the density scale height (as would be the case in a stellar wind) can raise these values. Hence, effects of resonant scattering should be expected in strong coronal emission lines.

3. Resonant Scattering in the Solar Atmosphere

Resonance scattering has been suggested many years ago for the solar corona. Based on observational data by Acton & Catura (1976), Acton (1978) already investigated the problem of resonance scattering of X-ray emission lines in the solar corona. More recently, Rugge & McKenzie (1985), for instance, discuss the effects of resonance scattering for the Fe XVII line ratios. Schmelz et al. (1992) use the resonance scattering of the 15.01 Å Fe XVII line as a density diagnostic for solar active regions; they find that for a "typical" active region, over 50% of the photons for this resonance line could be scattered out of

TABLE 2. Fe Resonance Lines

Ion	$\lambda(\text{Å})$	T_6	f	b	C_d	τ_{loop}	τ_{env}
Fe VIII	167.49	0.7	0.11	1.0	2.00	0.2	0.5
Fe IX	171.08	0.9	3.05	1.0	4.00	0.4	1.1
Fe X	174.53	1.1	1.28	1.0	1.33	0.2	0.4
Fe X	177.24	1.1	0.8	0.7	1.22	0.2	0.4
Fe XI	180.41	1.3	0.93	1.0	1.50	0.3	0.5
Fe XI	188.22	1.3	0.59	0.8	0.99	0.2	0.3
Fe XII	193.51	1.4	0.51	1.0	0.71	0.2	0.3
Fe XII	195.12	1.4	0.76	1.0	1.07	0.2	0.4
Fe XIII	197.43	1.6	0.73	0.3	1.49	0.4	0.6
Fe XIII	202.04	1.6	0.7	0.6	1.46	0.4	0.6
Fe XV	284.15	2.0	0.82	1.0	1.20	0.4	0.5
Fe XXI	128.73	8.9	0.09	0.9	.091	0.3	0.1
Fe XXIII	132.85	12.	0.16	1.0	.170	0.9	0.2
Fe XXIV	192.02	15.	0.05	0.7	.095	0.7	0.1
Fe XXIV	255.10	15.	0.02	0.3	.047	0.4	0.1

NOTES: Data for the strongest Fe resonance lines in the α Cen and AU Mic *EUVE* spectra (modified after Mewe et al. (1995), and Schrijver et al. (1995)), with line-centre optical depths τ_{loop} for a loop-dominated atmosphere and τ_{env} for a spherically symmetric gravitationally settled envelope of at least 0.3. The value of τ_{loop} is computed using the scaling law for quasi-static coronal loops derived by Rosner et al. (1978). The value of τ_{env} is computed using a wind base density $n_e = 6\,10^8\,\text{cm}^{-3}$, a temperature T_6 (in MK) of the maximum line formation temperature, and a density scale height equal to the pressure scale height in a static, gravitationally stratified atmosphere, and the solar surface gravity. The envelope optical depth is proportional to the density scale height and to the base density, and inversely proportional to surface gravity. The loop optical depth depends on the topology and structure of the surrounding loops. Both quantities are therefore crude estimates. Also listed are the absorption oscillator strength, f, and the branching ratio, b. Solar photospheric abundances are used.

the line of sight, while optical depths range from 1.7 to 3 in their sample of four active regions. Waljeski et al. (1994) also study these lines as well as a few others. They show that the effects of resonance scattering for a pair of Fe XVII lines at 15.01 Å and 15.26 Å with substantially different oscillator strengths can be used as an abundance-independent density diagnostic. They also argue that O VIII at 18.97 Å and Ne IX at 13.44 Å also have optical depths exceeding unity for disk-center active regions. Note that the Fe XVII and O VIII lines strongly contribute to the spectra of stars as observed by the *ROSAT* PSPC, so that even in those data optical thickness effects need to be tested for, particularly in active stars where the emission is likely to be dominated by active-region sources.

Other data sets also support radiative-transfer effects in the solar corona. The simple examples in Fig. 2 show that as lines become stronger, the emission measure that is to be assigned to them to explain the observed total emission tends to decrease as the optical depth, compatible with resonant scattering as explained in the next section. We selected lines of Fe in a limited temperature range, thus eliminating abundance effects and excluding effects of strongly changing intrinsic emission measures with temperature. M. Laming pointed out that scattering in the quiet solar corona or base of the solar wind cannot be very strong, because for optical depths significantly larger than unity, off-limb structures would appear much more diffuse than actually observed (see, for instance, the images in Cheng (1980)). Hence, for the Sun, scattering inside the coronal condensations themselves seems to be the most important mechanism.

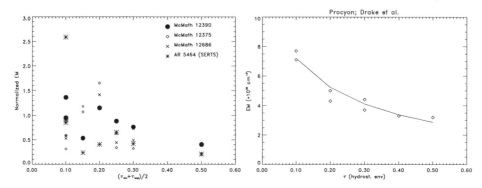

FIGURE 2. **Left:** Evidence for resonant scattering in the solar corona. Data for three active regions, observed at the solar limb with the NRL EUV slitless spectroheliograph, are taken from Dere (1982), and SERTS data for a fourth region from Thomas & Neupert (1994). Plotted are normalized inferred emission measures, i.e., ratios of the observed line strengths for the strongest Fe XII through XV lines for which optical depths are listed by Mewe et al. (1995) to the emissivity from the Mewe et al. (1985) plasma code. The optical depths are taken from Table 2. Note that these are expected to differ from region to region and should be regarded as a relative measure only. The comparison of lines formed at different temperatures is made easier by a correction for the temperature dependence of the emission measure using the solar $D(T)$ plotted in Fig. 1. In order to allow the comparison of data for different regions, the EM values are scaled to unity for lines with optical depths (average of estimates for envelope and loop-dominated atmosphere) less then 0.26. McM 12375, McM 12686 and the SERTS region contained flaring or post-flare loops when observed, which probably accounts for part of the scatter. **Right:** Comparison of the estimated emission measure needed to explain observed line strengths for Procyon (from Drake et al. (1995)) compared to the expected optical depth in a hydrostatic coronal envelope with a temperature of roughly 2 MK and a base density of $6\,10^{8}\,\mathrm{cm}^{-3}$ (as in the solar wind), as listed by Mewe et al. (1995). The lines from Fe XI up to Fe XVI range in wavelength from 192 Å up to 359 Å, and are characteristic of temperatures from 1.3 MK up to 2.5 MK. The optical depth should be multiplied by a constant which depends on the actual base density and on the density scale height. (The curve shows the relationship $EM \propto 1/(1 + 6\tau)$, which Schrijver et al. (1995) argue to be indicative of scattering in a wind, although scattering in coronal condensations cannot be ruled out.)

These few references are not exhaustive, but they provide the reader with entry points into the solar literature, which does not appear to be substantially larger than the papers quoted here and the references therein; although the phenomenon of resonant scattering is known to solar coronal physicists, the complexity of the problem and the relative scarcity of extensive spectral data sets hampered developments in the field. Our discussion also shows, however, that existing data should be subjected to a careful analysis looking for evidence of a non-negligible optical thickness.

4. Some Comments on Radiative Transport in Stellar Coronae

Under coronal conditions, even repeated scattering rarely leads to photon destruction in the corona itself, because the probability of collisional de-excitation is very small. The scattered photons do, however, have a substantial chance of being destroyed upon impact on the stellar surface. If the scattering and emitting media were homogeneous and identical, any downward scattered photon would on average be compensated for by an upward scattering of an initially downward-emitted photon, so that no net effect would be observed. If, however, the emitting plasma lies, on average, below the scattering medium, an asymmetry is created in which more line photons are scattered down than up (see Schrijver et al. (1994)). As a result, the lines that weigh most strongly in the inversion process appear to be too weak compared to the continuum (not necessarily observable at the same wavelengths!) introducing a spurious hot component to $D(T)$. This model predicts that the strongest lines are affected most, as is indeed the case in Fig. 2. Note that the reduced scatter for Procyon as compared to the solar data in that Figure may be the result of (a) disk-averaging of the stellar emission, and (b) a reduced importance of individual flares on the mean stellar $D(T)$.

If resonant scattering is indeed important in stellar coronae, the implicitly assumed linearity of the $D(T)$-inversion problem breaks down, and detailed radiative transfer calculations will be required if all spectral information is to be used. All the usual problems complicating the recovery of the atmospheric structure from the observed emission that are encountered in the studies of stellar photospheres and chromospheres will have to be faced in coronal studies as well.

Frequency redistribution: The simple model discussed by Schrijver et al. (1994) is formulated for a single frequency, and therefore implicitly assumes that the resonant scattering is coherent. In reality significant frequency redistribution will occur. Thomas (1957), for instance, already argued that within the Doppler core of the line, the assumption of complete redistribution (CRD) over the thermal Doppler profile provides a very good approximation (see, for instance, discussions in Athay (1972), and Mihalas (1978)). With CRD and multiple scattering, line photons predominantly escape by "diffusion" in wavelength to a value where the optical depth becomes less than unity. One can define the probability P_e for photon escape as the fraction of the photons that have an optical depth less than unity as computed from the thermal absorption profile (e.g., Athay (1972)): $P_e = (\tau_0 \sqrt{\pi \ln \tau_0})^{-1}$. For a line-center optical depth of $\tau_0 = 2$ to 5, some 34% to 9%, respectively, of the photons escape without suffering a single scattering simply because of the lower optical depth in the line wings. This limits the strength of the effects on the line-to-continuum ratio.

Anisotropy and geometry: The optical depth τ_ν depends on the plasma which photons encounter in the outer atmosphere, and is therefore dependent on the detailed topology and the photon's path. In a spherical stellar atmosphere photons have a higher probability to escape in the radial direction resulting in an anisotropic distribution that is peaked in the direction of the normal (see Haisch & Claflin (1985) for a discussion of these effects). If the distribution of emitting regions, possibly embedded in a scattering envelope, is not spherically symmetric, the number of photons escaping into the direction of the observer can be significantly affected.

5. Electron Densities in Stellar Coronae

The inferred electron densities listed in Table 1 and reported by others during the meeting (see caption to Table 1) suggest that the electron densities associated with the solar-like component at approximately 3 MK are typically about three orders of magnitude lower than found for hotter components around 5-15 MK. Brickhouse (these proceedings) discusses evidence from the Capella-spectrum that this contrast is present within the same system. It is at present unknown whether the hot 5-15MK component is associated with hot loops or with flaring events, but regardless of this the high volume emissivity of the hot component implies that the volume filling factors of the hot components are much smaller than those associated with the 3 MK component.

6. Concluding Remarks

The line-to-continuum ratio in the *EUVE* spectra of Procyon, AU Mic and α Cen and possibly Capella is suggestive of the occurrence of scattering of photons in the strongest spectral lines. In the case of Procyon this could be due to a stellar envelope or wind with a temperature of approximately 1–3 MK. For AU Mic and Capella we suggest that the compact hot structures are embedded in an equally hot but less dense environment, in the case of Capella possibly consisting of larger and therefore less dense quasi-static magnetic loops, while in the case of AU Mic the (predominantly) flaring plasma could be embedded in a hot, more tenuous plasma that itself may be either flare-related or part of a hot, quiescent coronal component. The shallow, dynamic convective envelope of Procyon and the flaring corona of AU Mic may result in coronae that are quite different from those of the Sun, so that a significant increase of the optical depth as compared to the solar corona may not come as a surprise. The fact that we also found evidence for line-photon scattering in the spectrum of α Cen is surprising, but solar observations suggest that upon careful examination the solar disk-integrated spectrum may also show effects of resonant scattering. Deeper *EUVE* exposures and instruments like the *SUMER* and *CDS* spectrographs on board *SOHO*, the *ORFEUS* instrument, the spectrometers planned for *AXAF* and *XMM*, but even full-disk imaging as planned with *TRACE* can be used to test the role of resonant scattering in stellar coronae further. For such studies the study of lines and continuum together is crucial (see Mewe et al. in these proceedings). Whenever resonant scattering occurs, the $D(T)$ results for the corresponding stars should be interpreted as lower limits in the temperature range in which scattering is important.

We thank T. Ayres, J. Bruls, J. J. Drake, C. Jordan, M. Laming, D. Liedahl, R. Rutten, J. Schmitt, R. Stern, and K. Strong for insightful comments.

REFERENCES

Acton, L. 1978, ApJ, 225, 1069

Acton, L. & Catura, R. 1976, Phil. Trans. Roy. Soc. London, A, 281, 383

Athay, R. 1972, Radiation Transport in Spectral Lines, Dordrecht: Reidel

Cheng, C. 1980, SPh, 65, 347

Dere, K. 1982, SPh, 77, 77

Drake, J. J., Laming, J., & Widing, K. 1995, ApJ, in press

Haisch, B. & Claflin, E. 1985, SPh, 99, 101

Mariska, J. T. 1992, The Solar Transition Region, Cambridge University Press

Mewe, R., Kaastra, J., Schrijver, C., van den Oord, G., & Alkemade, F. 1995, AA, 296, 477

Mihalas, D. 1978, Stellar Atmospheres, San Francisco: Freeman

Rosner, R., Tucker, W., & Vaiana, G. 1978, ApJ, 220, 643

Rugge, H. & McKenzie, D. 1985, ApJ, 279, 338

Schmelz, J., Saba, J., & Strong, K. 1992, ApJL, 398, 115

Schrijver, C., Mewe, R., van den Oord, G., & Kaastra, J. 1995, AA, in press

Schrijver, C., van den Oord, G., & Mewe, R. 1994, AA, 289, L23

Schrijver, C. J. 1993, AA, 269, 446

Thomas, R. 1957, ApJ, 125, 260

Thomas, R. J., Neupert, W. M. 1994, ApJS, 91, 461

Waljeski, K., Moses, D., Dere, K., Saba, J., Strong, K., Webb, D., & Zarro, D. 1994, ApJ, 429, 909

EUVE Observations of AR Lacertae: The Differential Emission Measure and Evidence for Extended Prominences

FREDERICK M. WALTER

Department of Earth and Space Sciences, SUNY, Stony Brook, NY 11794–2100, USA

EUVE observed the eclipsing RS CVn system AR Lac in October 1993. The differential emission measure shows a double-peaked structure similar to the two-temperature distribution seen in X-rays. The best-fit Fe abundance is 0.4 solar. The DS light curve shows a deep primary eclipse. The slow egress from eclipse may be due the presence of spatially extended, optically-thick "prominences" in the equatorial plane of the K star.

1. Introduction and Observations

All late-type stars, so far as we know, possess magnetic fields and the consequent phenomena termed "solar-like activity." This activity results in non-radiative heating of the outer atmosphere, and the outwardly increasing temperatures that manifest themselves as chromospheres, transition regions, and coronae. We know far less about this non-radiative heating mechanism than we would like, even in the case of the Sun. Studies of how the emissions vary with stellar parameters (mass, rotation, age) give some clues to the coupling of the heating to the magnetic fields. Studies of the the differential emission measure (DEM) tell us of how the energy is deposited in the stellar atmosphere.

I observed the well-studied active star, AR Lacertae, with the *EUVE* to determine the DEM for comparison with the X-ray and UV measurements. AR Lacertae (HD210334) is the brightest eclipsing RS CVn system known, with a 1.983 day orbital period.

The *EUVE* observed AR Lac on 12–15 October 1993. The first half of the observation was made with the target located at the nominal boresight (which by then was known to have suffered reduced sensitivity). For the latter half of the observation, the target was offset to two other positions [$(-0.5', -10.8')$, $(+1.1', -5.4')$] for boresight calibrations. Data were obtained over 48 spacecraft orbits. The total time in the observation is 104 ks.

The spectra were extracted at CEA, using IRAF/APEXTRACT, with the variance-weighting method. Lines Fe XVI through Fe XXIV are present (see Table 1).

The interstellar column n_H has generally been assumed to be 10^{19} cm^{-2} for X-ray spectral fits. I estimated n_H from the $\lambda\lambda335,360$ Å Fe XVI line ratio. The ratio of emissivities $\frac{\lambda335}{\lambda360}$ is 2.14, independent of temperature or density. The observed ratio of 2.72 ± 0.84 implies $n_H=1.8^{+1.9}_{-1.8}\times10^{18}$cm^{-2}.

2. The Emission Measure Distribution

I used the EUVE_FIT software, written by R.A. Stern, to fit the spectrum. The code simultaneously fits the emission line fluxes and the continuum intensity, yielding a DEM and the iron abundances. The ratio of the emission lines to the continuum is proportional to the metallicity [$\frac{Fe}{H}$]. The best fit DEM is shown in Figure 1.

The two-component thermal fits to the X-ray spectra, overplotted in Figure 1, agree remarkably well with the DEM from *EUVE*. The Ottmann et al. (1993) fit used a Raymond-Smith (1977) solar-abundance plasma; the *ASCA* and my PSPC fits use the MEKA code (Kaastra 1992) with the abundances as free parameters.

S. Bowyer and R. F. Malina (eds.), Astrophysics in the Extreme Ultraviolet, 129–133.
© *1996 Kluwer Academic Publishers. Printed in the Netherlands.*

TABLE 1. Measured line fluxes

λ_{obs}	Line ID	λ_{pred} [†]	Flux [‡]	λ_{obs}	Line ID	λ_{pred}	Flux
90.80	Fe XIX	91.02	3.2 ±0.2	121.73	Fe XX	121.83	4.1 ±0.4
93.80	Fe XVIII	93.92	9.4 ±0.4	128.53	Fe XXI	128.73	7.8 ±0.6
97.73	Fe XXI	97.88	0.9 ±0.2	132.69	Fe XXIII	132.85	30.1 ±2.0
101.55	Fe XIX	101.55	2.5 ±0.3	135.65	Fe XXII	135.78	5.7 ±0.6
102.49	Fe XXI	102.22	1.7 ±0.3	142.08	Fe XXI	142.27	3.4 ±0.9
103.98	Fe XVIII	103.94	1.7 ±0.2	192.32	Fe XXIV	192.04	18.5 ±1.5
108.26	Fe XIX	108.37	5.8 ±0.5	256.41	He II	256.32	3.5 ±0.5
109.72	Fe XIX	109.97	0.6 ±0.2	284.89	Fe XV	284.15	2.5 ±0.5
117.03	Fe XXII	117.17	14.7 ±1.0	303.89	He II	303.91	63.9 ±2.6
118.60	Fe XX	118.66	2.5 ±0.3	335.30	Fe XVI	335.41	7.9 ±1.5
119.81	Fe XIX	120.00	1.5 ±0.3	359.93	Fe XVI	360.80	2.9 ±0.7

[†] Line list from Brickhouse et al. 1995.
[‡] Fluxes in units of 10^{-14} erg cm^{-2} s^{-1}.

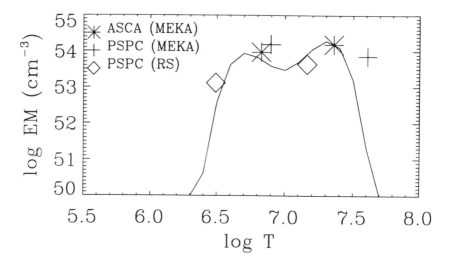

FIGURE 1. The differential emission measure (DEM) from the $EUVE$ (continuous curve). The emission measures and temperatures from two component thermal fits to X-ray spectra are overplotted.

2.1. The Iron Abundance

The best estimate of $\frac{Fe}{H}$ is 0.39 solar. Sub-solar abundances are routinely observed in $ASCA$ X-ray spectra (e.g., White et al. 1994), and even $ROSAT$ PSPC spectral fits for active stars are greatly improved if the abundances are permitted to fall below solar. These low abundances have been explained away as an artifact of incomplete line lists in the models (the Fe L transitions, which contribute significantly around 1 keV, are missing): the missing lines are modeled as continuum, leading to low line-to-continuum

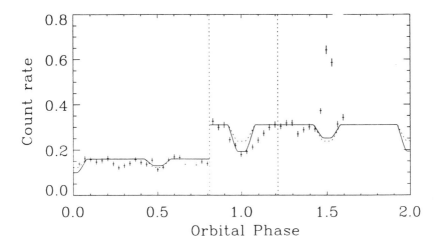

FIGURE 2. The DS light curve, plotted versus binary phase. The dotted vertical lines separate intervals with different boresight offsets. The low count rate in the first interval is due to the reduced sensitivity at the original boresight location. The solid line is the optical light curve, scaled to the mean out-of-eclipse level. Purely geometrical eclipses (the dotted line) are shallower. Primary eclipse has a depth of 42%, and is clearly asymmetric on egress. A flare coincides with secondary minimum.

ratios. *EUVE* spectra should not be subject to this problem, so the low abundance is somewhat of a mystery.

As the DEM fitting procedure weights all points by their errors, and the continuum points greatly outnumber the lines, it is possible that the criterion used to signal convergence of the fit is relatively insensitive to the lines. I checked this by artificially enhancing the S/N for the lines, to no avail. I then ran models at a range of n_H. Since the mean wavelength of the continuum is shorter than that of the lines, the line-to-continuum ratio depends on the assumed n_H. For $n_H = 7 \times 10^{18}$ cm^{-2} the best fit abundance does reach solar. However, examination of the fits shows that fits with $n_H \geq 4 \times 10^{18}$ cm^{-2} greatly underestimate the fluxes of the lines longward of $\sim$250 Å.

Could this metal abundance deficiency be real? Naftilan & Drake (1977) claimed that the K star was metal-deficient relative to the Sun by about 1 dex, while the G star showed solar abundances. They note that is difficult to understand how the two stars in a close binary can have different abundances. Randich et al. (1993) show that many RS CVn systems appear underabundant in $[\frac{Fe}{H}]$, but suggest that the weak Fe lines are more likely a consequence of chromospheric emission which fills in the lines than of real underabundances. Both components of AR Lac are chromospherically active.

3. The Light Curve

The DS light curve is shown in Figure 2. The dotted vertical lines divide the regions with different boresight offsets. The loss in sensitivity the DS suffered in February 1993 is evident from the low count rate seen in the first half of the observation. The steep spatial gradient of the sensitivity loss could account for most, if not all, of the apparent variations during that time. The light curve during the latter half of the observation clearly shows primary minimum; secondary minimum coincides exactly with a flare.

The residual intensity at primary minimum is 58%±4% of that outside eclipse. A purely geometrical eclipse of emitting regions with small scale heights would yield an eclipse depth of 33%. The mean surface flux of the emission from the G star must be 2.5× that of the K star. Note that the EUV light curve is asymmetric with respect to primary minimum. The emission returned to the pre-eclipse level only slowly.

The depth of primary eclipse is similar to that seen with *Einstein* (Walter et al. 1983), *EXOSAT* (White et al. 1990), *ROSAT* (Ottmann et al. 1993), and *ASCA* (White et al. 1994). This is not surprising, as all instruments are probing similar coronal temperatures.

3.1. The Asymmetric Egress from Primary Minimum

There is deficiency of emission from the G star upon egress from primary eclipse. Such asymmetric light curves have been seen before in the *EXOSAT* LE light curve (White et al. 1990), the *ASCA* light curve (White et al. 1994), and the spectrally-resolved Mg II light curve (Neff et al. 1989).

The attenuation is highest close to the star, and falls slowly, reaching zero by phase 0.2 (projected separation 5.8 R_K). If the attenuation is caused by absorption or scattering, the responsible material must extend a significant fraction of the distance between the stars. Four possible explanations for the slow egress from primary minimum are:

Enhanced photoelectric absorption on the line of sight: The maximum n_H necessary to produce the observed attenuation is just over 10^{19} cm^{-2}. The neutral hydrogen can exist either *(a)* within an extended and presumably hot corona, or *(b)* in a cool extended region well above the K star. Possibility *a* conflicts with the inferred coronal emission measure $n_e^2 V$. For a neutral fraction of 10^{-6}, the total column is of order 10^{24-25} cm^{-2}. Over a length of order $2R_K$, this requires coronal densities $n_e \sim 10^{12}$ cm^{-3} and $n_e^2 V \sim 10^{60}$ cm^{-3}, which exceeds all coronal emission measure estimates by some 7 orders of magnitude. For possibility *b*, I assume $n_e = 10^{10}$ cm^{-3} (Solar prominences have densities up to 10^{12}). This requires a path length of order 10^8 to 10^9 cm, which is appropriate for a prominence. However, the scale height of the prominence exceeds 10^{11} cm, since it must cover the entire surface area of the G star. The absorbing structure must then be a thin sheet parallel to our line of sight, which seems unlikely.

Enhanced electron scattering on the line of sight: This possibility is attractive inasmuch as the X-ray, EUV, and Mg II light curves appear similar, despite the different sensitivities to interstellar absorption. $\tau = 1$ for electron scattering requires $n_H \sim 10^{24}$ cm^{-2}. Over a path length of order $2R_K$, n_e must exceed 10^{12} cm^{-3}, which leads to the same absurdly large volume emission measures discussed above.

Reduced emission on the visible hemisphere of the G star: The attenuated flux is observed over an interval of only 0.2 in phase. There is no evidence that the emission from the G star varies at any other phase. It is not possible to construct a surface intensity distribution that is constant for 80% of a cycle and dips for the other 20%. While is possible to contrive such a light curve using patchy flux distributions on both stars (e.g., White et al. 1990), I consider this an unlikely and ad-hoc explanation.

Geometric obscuration by optically thick material: If the obscuring material is optically thick (say, $n_H > 10^{22}$), but only obscures a portion of the G star, then one would see similar attenuations in *EUVE* and ASCA. Cool, dense material confined within 15° of equator of the K star, and extending outward by $\sim R_K$, could account for the observed attenuation by geometrical obscuration alone (Figure 3). The inferred n_e

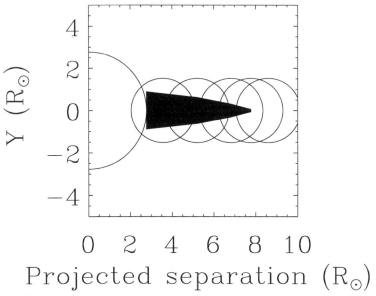

FIGURE 3. Simple geometrical model of the obscuration needed to account for the attenuation seen on egress from primary eclipse. The G star is plotted at the five phases observed following primary minimum. The shaded area is optically thick.

is from 10^{11} to 10^{12} cm^{-3}, which is similar to that seen in some solar prominences. Such cool structures would not contribute to the coronal emission measure. There is growing evidence for such large structures in active stars (Linsky 1994).

I am indebted to Bob Stern for donating his *EUVE* line + continuum fitting code, and to Jürgen Schmitt for useful discussions about MADs. I thank the staff at CEA for assisting me with the preliminary data reductions. Lastly, we must acknowledge Stu Bowyer for his persistence, which ensured the success of the *EUVE*.

REFERENCES

BRICKHOUSE, N. S., RAYMOND, J. C., & SMITH, B. W. 1995, ApJS, in press

KAASTRA, J. S. 1992, An X-Ray Spectral Code for Optically Thin Plasmas SRON-Leiden Report

LINSKY, J. L. 1994, in Solar Coronal Structures, ed. V. Ruslin, P. Heinzel, & J.-C. Vial, 641

NAFTILAN, S. A. & DRAKE, S. A. 1977, ApJ, 216, 508

NEFF, J. E., ET AL. 1989, A&A, 215, 79

OTTMANN, R., SCHMITT, J. H. M. M., & KÜRSTER, M. 1993, ApJ, 413, 710

RANDICH, S., GRATTON, R., & PALLAVICINI, R. 1993, A&A, 273, 194

RAYMOND, J. C. & SMITH, B. W. 1977, ApJS, 35, 419

WALTER, F. M., GIBSON, D. M., & BASRI, G. S. 1983, ApJ, 267, 665

WHITE, N. E, SHAFER, R. A., HORNE, K., PARMAR, A. N., & CULHANE, J. L. 1990, ApJ, 350, 776

WHITE, N. E., ARNAUD, K., DAY, C. S. R., EBISAWA, K., GOTTHELF, E. V., MUKAI, K., SOONG, Y., YAQOOB, T., & ANTUNES, A. 1994, PASJ, 46, L97

EUVE Observations of BY Dra Systems

R. A. STERN[1,3] AND J. J. DRAKE[2]

[1] Solar and Astrophysics Laboratory, Lockheed Martin Palo Alto Research Lab,
O/91-30 Bldg 252, 3251 Hanover Street, Palo Alto, CA, 94304, USA

[2] Center for EUV Astrophysics, 2150 Kittredge Street,
University of California, Berkeley, CA 94720-5030, USA

[3] Visiting Investigator, Center for EUV Astrophysics

We have observed 3 nearby BY Dra systems, FK Aqr, BF Lyn, and DH Leo, with the *EUVE* spectrometers. All 3 show evidence of high-temperature ($\sim 10^7$ K) plasma; FK Aqr and DH Leo show significant variability in their Deep Survey lightcurves. In FK Aqr, spectral differences between its "quiescent" and "active" states suggest possible differences in the plasma density. In DH Leo, the Deep Survey lightcurve, taken over nearly 8 days, shows a distinct period of ~ 1.05 days, consistent with the photometric period. The emission measure distributions of all three systems are rather similar in shape, and can be well-represented by a power law with slope ~ 1.5 from 6.2–7.0 in $\log T$.

1. Introduction

The BY Draconis stars are a group of red (dKe–dMe) low-amplitude variables with photometric periods of typically a few days, Ca II and often Hα emission, and unusually bright X-ray emission for their spectral type, with a typical $L_x \approx 5 \times 10^{29}$ erg s^{-1} (Bopp & Fekel 1977; Caillault 1982). Most, but not all, are binaries in short period (few days) orbits, with rapid rotation the crucial factor in their high level of activity. A few are dMe flare stars (Bopp & Fekel 1977; Strassmeier et al. 1988). From the results of EUV sky surveys (Pounds et al. 1993; Bowyer et al. 1994), it now appears that many of the BY Dra systems are strong EUV emitters as well: these include KZ And, FK Aqr, V1396 Cyg, V775 Her, DH Leo, BF Lyn, YY Gem, V833 Tau, CC Eri, and BY Dra, the prototypical system. All of these systems are within ≈ 30 pc of the Sun. Because they are relatively young main sequence stars, with X-ray activity levels typically 10^2 or more times solar, they represent some of our most accessible main sequence stellar laboratories in the EUV with which to examine the effect of strong magnetic heating on coronal temperature. We have performed a detailed study of 3 representative, nearby, EUV-bright BY Dra systems: FK Aqr, DH Leo, and BF Lyn. The characteristics of these systems are listed in Table 1. The column densities ($N_{\rm H}$) to these systems are uncertain, but are probably $\lesssim 1$–2×10^{18} cm^{-2}.

2. The FK Aqr Lightcurves

The Deep Survey (DS) telescope feeds both an imaging detector and the 3 *EUVE* spectrometers (Bowyer & Malina 1991), thus providing simultaneous photometry and spectrophotometry. Sources near telescope boresight are imaged in the DS/Lexan band (70–180 Å). In Figure 1 we show the lightcurves obtained, in DS/Lexan and by summing the 80–140 Åregion in the SW spectrometer. The data are binned over ≈ 1 satellite orbit, with ~ 2 ks effective exposure time in each bin. The light curves are quite similar (the DS/Lexan has smaller errors due to the higher count rate), even to the extent of recording a brief flare at ~ 135 ks into the observation. The most striking aspect of the lightcurves is the dramatic rise in EUV flux at ~ 220 ks into the observation. Since FK Aqr is not an eclipsing system, we may attribute this rise either to rapid growth of a

S. Bowyer and R. F. Malina (eds.), Astrophysics in the Extreme Ultraviolet, 135–140.

TABLE 1. BY Dra Systems Observed (Lex/B and Al/C are *EUVE* scanner count rates ks^{-1})

Star	Sp. Type	P_{orb}	d (pc)	Lex/B	Al/C
FK Aqr	dM2e/dM2e	4.08	7.7	150	50
BF Lyn	K2 V/dK	3.8	29	90	50
DH Leo	K0/K7/K5	1.07	33	50	40

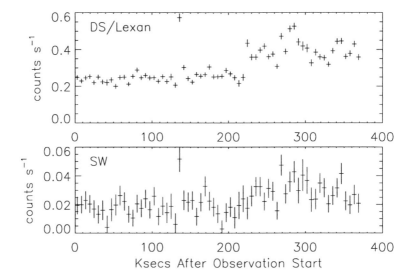

FIGURE 1. Deep Survey/Lexan (40–190Å) lightcurve for FK Aqr (*top*); Summed SW spectrometer (80–140 Å) lightcurve (*bottom*). Each data point covers about 1 spacecraft orbit (96 mins.)

large active region or complex of active regions, or perhaps to the rotation of an already existing active region complex onto the visible disk of one of the stars in the system. Since the length of the observation ($\approx$ 4.28 d) is slightly longer than the orbital period (4.08 d), but slightly shorter than the photometric period (4.39 d), we cannot rule out either explanation or a combination of both.

3. The FK Aqr Spectra

Given the appearance of "quiescent" and "active" states in the FK Aqr lightcurve, we have extracted *EUVE* short wavelength and medium wavelength spectra in two parts: one including the period before the lightcurve "transition" (but excluding the flare at 135 ks), and a second part including both the flare and the time period after the transition. These are shown in Figure 2. The most noticeable differences between the spectra are the relative strengths of the Fe XXI 128.7 Å line compared to the other Fe XXI line and lines from nearby ionization stages. This could be an indicator of higher ($\sim 10^{13}$ cm^{-3}) densities; however, an independent confirmation through the Fe XXII density-sensitive 117/114 line ratio is problematic because of the lower statistical significance of these line fluxes.

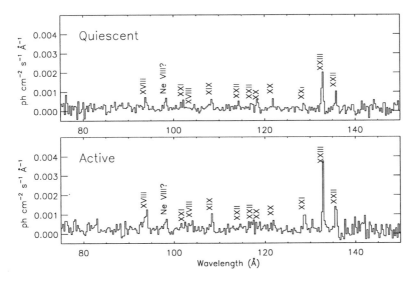

FIGURE 2. Quiescent (*top*) and active (*bottom*) SW spectra for FK Aqr. The data have been rebinned by a factor of 4 to about 0.26 A. The effective exposure times are 74 and 54 ks, respectively. In each plot, the ionization stage of Fe for each line is indicated.

4. DH Leo

4.1. *The Lightcurve: Rotational Modulation?*

Our observation of DH Leo covered a time span of ≈ 670 ks, or over 7 binary or photometric periods. The DS/Lexan lightcurve for this observation is shown in Figure 3, along with a periodogram analysis (using the "fast" algorithm of Press & Rybicki 1989). The most significant peak occurs at a period of ≈ 1.05 days. Because this period is close to ~ 1 day as well as the photometric period of 1.067 days (Barden et al. 1986), we also performed a periodogram analysis of the detector background, but failed to find a significant peak at either of these values. Thus, for the moment we conclude that the periodicity is real, and is close enough (within an estimated error of ± 0.05 d) that we may associate it with the photometric period of DH Leo. A phased lightcurve at this period is shown in Figure 4, indicating a definite "ripple" in the EUV flux. No other obvious periodicity in the *EUVE* satellite at ~ 1 day is known to produce such a variation in the source count rate, thus we tentatively conclude that our lightcurve is an indication of rotational modulation in the DH Leo system.

4.2. *Spectrum*

Dividing up the DH Leo spectrum by phase into "low level" and "high level" count rate did not produce significant spectral differences, unlike in the case of FK Aqr. The relative strengths of the Fe XXII 117/114 line ratio are consistent with the low-density limit (i.e. $\lesssim 10^{12}$ cm^{-3}).

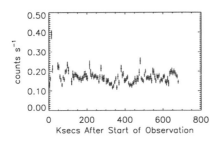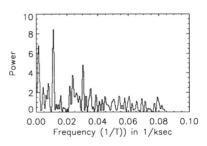

FIGURE 3. DS/Lexan lightcurve for DH Leo (left) and periodogram analysis (right)

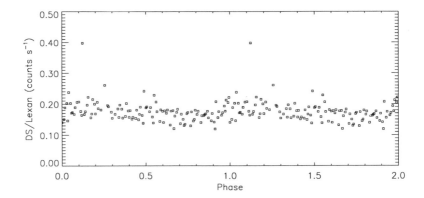

FIGURE 4. Phased lightcurve for DH Leo (phase = 0. has been arbitrarily chosen at the start of the observation)

5. The BF Lyn Lightcurve and Spectrum

The BF Lyn DS/Lexan lightcurve exhibited variability of up to 50% during the course of the observation ($\approx$ 2.5 d). Since the total observation coverage extends for less than a single orbital or photometric period of the system (3.8 days), the remaining variability is probably due to a combination of flares, rotational modulation, or growth and decay of active regions. BF Lyn is a relatively weak EUV source, and the 70 ks effective exposure time reveals fewer lines (many at low significance) than the other 2 BY Dra systems. However, the Fe XXIII/XX feature is prominent, and other ionization stages from XV–XXII are also present, thus indicating a relatively hot corona.

6. Emission Measure Distributions

In Figure 5 we show preliminary emission measure distributions for the BY Dra systems. These have been computed using the Fe emissivities of Brickhouse et al. (1995), and the "Pottasch" method of using the peak temperature of the emissivity function. As such they represent a quick estimate of the form of the EM distribution; further analysis using spectral synthesis methods may yield a better estimate of the EM. Note that overall shapes of the EM distributions are quite similar for all the systems, and in the range $\log T = 6.2 - 7.0$, can be well approximated by a power law of slope ~ 1.5. DH Leo, the shortest period system, also appears to have the greatest quantity of the highest

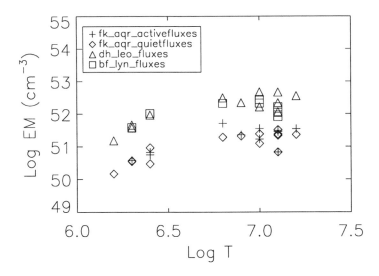

FIGURE 5. Emission measure distributions for the BY Dra systems (see text)

temperature material. Even allowing for the scatter in the plot due to a combination of statistical uncertainties (up to a factor of 2 for the weaker lines), and atomic physics uncertainties, the enhancement in the EM at log T $\sim$ 6.8 as seen in the Fe XVIII 93.9 Åline for the "active" FK Aqr spectrum is still quite significant, and somewhat reminiscent of the "bump" seen in the Capella EM distribution in Dupree et al. (1993) and Brickhouse et al. (1995). We note that deriving the EM for the quiescent spectrum assuming $n_e = 10^{13}$ cm^{-3} made only a slight difference in the EM distribution.

7. Summary

EUVE observations provide one of the best ways to study both the variability and the temperature structure of the coronae of BY Dra systems. Both periodic modulations and stochastic variability are amply evident in the short wavelength high temperature lines seen by EUVE: the DH Leo study for over 7 binary orbits is probably the best example to date of rotational modulation of stellar coronal emission. In addition, there is a hint of a high density structure in the *quiescent* corona of FK Aqr as compared to the "active" or flaring corona.

R.A.S. wishes to thank the scientists and staff of the Center for EUV Astrophysics for making his stay there enjoyable and productive. R.A.S. was supported in part by NASA contract NAS5-32640 and by the Lockheed Independent Research Program. J.J.D. was supported by NASA Grant AST91-15090.

REFERENCES

BARDEN, S. C., RAMSEY, L. W., FRIED, R. E, GUINAN, E. F., & WACKER, S. W. 1986, in Cool Stars, Stellar Systems, and The Sun, ed. M. Zeilik & D.M. Gibson, Berlin: Springer, 291

BOPP, B. W. & FEKEL, F. C., JR. 1977, AJ, 82, 490

BOWYER, S., LIEU, R., LAMPTON, M, LEWIS, J., WU, X., DRAKE, J. J., & MALINA, R. F. 1994, ApJS, 93, 569

BOWYER, S. & MALINA, R. F. 1991, in Extreme Ultraviolet Astronomy, ed. R.F. Malina & S. Bowyer, New York: Pergamon Press, 397

BRICKHOUSE, N. S., RAYMOND, J. C., & SMITH, B. W. 1995, ApJS, in press

CAILLAULT, J. P. 1982, AJ, 87, 558

DUPREE, A. K., BRICKHOUSE, N. S., DOSCHEK, G. A., GREEN, J. C., & RAYMOND, J. C. 1993, ApJ, 418, L41

POUNDS, K. A., ET AL. 1993, MNRAS, 260, 77

PRESS, W. H. & RYBICKI, G.B. 1989, ApJ, 338, 277

STRASSMEIER, K. G., ET AL. 1988, A&AS, 72, 291

High Temperature Structure in Cool Binary Stars

A. K. DUPREE, N. S. BRICKHOUSE, AND G. J. HANSON

Harvard-Smithsonian Center for Astrophysics, 60 Garden Street, Cambridge, MA 02138 USA

Strong high temperature emission lines in the *EUVE* spectra of binary stars containing cool components (Alpha Aur [Capella], 44ι Boo, Lambda And, and VY Ari) provide the basis to define reliably the differential emission measure of hot plasma. The emission measure distributions for the short-period (P$\leq$ 13 d) binary systems show a high temperature enhancement over a relatively narrow temperature region similar to that originally found in Capella (Dupree et al. 1993). The emission measure distributions of rapidly rotating single stars 31 Com and AB Dor also contain a local enhancement of the emission measure although at different temperatures and width from Capella, suggesting that the enhancement in these objects may be characteristic of rapid rotation of a stellar corona. This feature might be identified with a (polar) active region, although its density and absolute size are unknown; in the binaries Capella and VY Ari, the feature is narrow and it may arise from an interaction region between the components.

1. Introduction

The emission measure defined by the well-exposed EUV spectra of the binary system Capella (G8 III + G0 III) surprisingly revealed a *narrow* high temperature "bump" (Dupree et al. 1993) occurring at $T = 6.3 \times 10^6$ K. The feature is continuously present irrespective of the orbital phase of the system (Dupree et al. 1994; 1995), and may arise from a continuously visible hot region on one component, or from an interacting region between the stars. The distribution of the high temperature emission measure in the Capella system differs from the broad emission measure distribution found in a solar active region (see Brickhouse et al. 1995). We investigate whether this feature is characteristic of binary systems and whether rotation controls its presence.

2. Spectra and the Emission Measure Distribution

EUVE spectra from our Guest Observer program and from the *EUVE* Science Data Archives were obtained for four binary systems (VY Ari, Capella, 44ι Boo, λ And) and two rapidly rotating single stars (AB Dor and 31 Com). EUV spectra were extracted from the images and calibrated in a uniform manner. The reduced spectra are shown in Fig. 1. In spite of the varying physical characteristics of the stars, there are similarities among the spectra. The resonance line of He II (λ304) dominates the MW spectra; strong resonance lines of iron dominate the SW spectra. However, the relative strengths of the lines differ among these objects, and so require detailed evaluation of the emission measure. It is the strength of the Fe XVIII and Fe XIX transitions relative to Fe XX, XXIII that controls the presence of the locally enhanced emission measure (the "bump") in Capella.

Signal to noise ratios were evaluated at the *EUVE* spectral resolution, and only the strongest features (S/N $\approx$ 9 $-$ 15) were selected to determine the emission measure. Correction of the observed fluxes for interstellar absorption was made with N_H from Table 1. The H/He abundance ratio was set to 11.6 (Kimble et al. 1993) and helium was assumed to be neutral. Assumption of higher values for N_H would strengthen the

141

S. Bowyer and R. F. Malina (eds.), Astrophysics in the Extreme Ultraviolet, 141–145.
© *1996 Kluwer Academic Publishers. Printed in the Netherlands.*

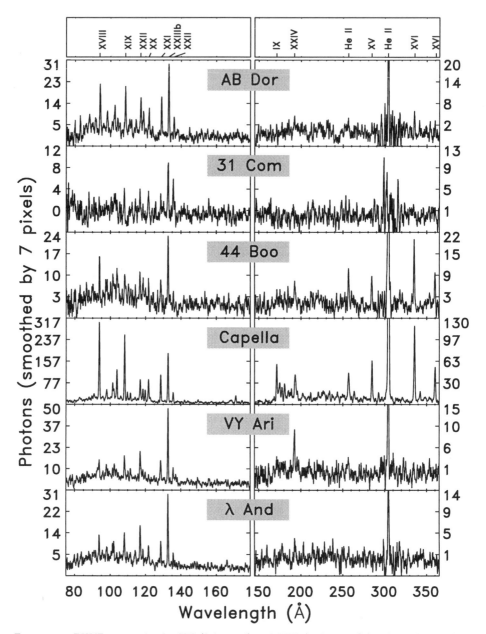

FIGURE 1. *EUVE* spectra in the SW (*left panel*) and MW (*right panel*) band for the rapidly rotating single stars AB Dor and 31 Com and the binaries 44 Boo, Capella, VY Ari, and λ And. Species of Fe are marked by ionization stage. Fe XXIII (λ132.8) is blended with Fe XX.

TABLE 1. Parameters of Target Stars

HD	Star	Sp. Type/Lum.	P_{orbital} (days)	$P_{\text{phtm.}}$ (days)	N_H (cm^{-2})	SW Exp (s)
17433	VY Ari†	K3–4V–IV	13.208	16.64	1.0×10^{18}	168557
34029	α Aur‡	G0 III/G8 III	104.0	8/80	1.8×10^{18}	281571
36705	AB Dor¶	K1 IIIp	$\cdots$	0.514	1.0×10^{18}	166460
111812	31 Com‖	G0 III	< 6	$\cdots$	1.0×10^{18}	84503
133640	44ι Boo††	G0 V	0.268	$\cdots$	1.0×10^{18}	104145
222107	λ And†	G8IV–III	20.52	53.95	4.0×10^{18}	105701

† RS CVn-type binary.

‡ RS CVn-type binary; *EUVE* spectra of Capella taken at several phases were averaged together.

¶ Rapidly rotating single star.

‖ 31 Com is a rapidly rotating single star thought to be a Hertzsprung Gap giant. The "orbital" period was evaluated from the $v \sin i$ value of 77 km s^{-1}, assuming a radius of 9 $R_\odot$.

†† W UMa-type contact binary.

Fe XVI emission (λ336) relative to the Fe XV emission (λ284) where observed, but leave the fluxes in the short wavelength region (λ85–λ170) essentially unchanged.

The emission measure distribution is determined iteratively, by fully integrating the line emissivities through a trial emission measure distribution, and comparing the predicted and observed line fluxes. There is no restriction on the "smoothness" of the emission measure distribution such as occurs with spectral fitting procedures; moreover the fit is not degraded by bins containing noise or weak signals. Only strong lines are considered in evaluating the emission measure. We require that these fluxes agree within acceptable uncertainties which in this case is better than a factor of two. Detailed current models for all iron ions (Brickhouse et al. 1995) were used. Results for λ And are taken from another publication (Hanson et al. 1995) where fluxes of ultraviolet lines measured with *IUE* were included so that the emission measure below 2×10^5 K is well-defined.

3. Conclusions

All of the target stars exhibit a range of high ion stages indicating a generally continuous distribution of temperature. The spectral resolution of *EUVE* demonstrates that two-temperature fits which rely on energy band measures do not represent these stellar atmospheres. For the weaker spectra, the strongest observable lines arise from high stages of ionization (Fe XVIII ... XXIII), the Fe XV (λ284) and Fe XVI (λ336 and λ361) emission is not prominent, and the emission measure can only be reliably defined in a restricted temperature range.

Inspection of the emission measure distributions contained in Fig. 2 and 3 shows that the Capella "bump" is not unique. The binary systems (VY Ari, 44ι Boo) clearly show the enhanced emission measure distribution similar to Capella. The appearance of the emission measure enhancement is most pronounced in binary systems with short orbital period ($\leq$13 days) and/or high rotational velocity. The orbital period of Capella is 104 days, however the hot component of Capella is rotating rapidly with a period of $\approx$ 8 days (Fekel et al. 1986) so that the system may be classified as a rapid rotator for these purposes. The strong lines of Fe XVIII (λ93.9) and Fe XIX (λ108.4) demand the presence

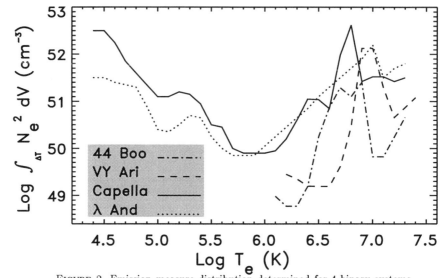

FIGURE 2. Emission measure distribution determined for 4 binary systems.

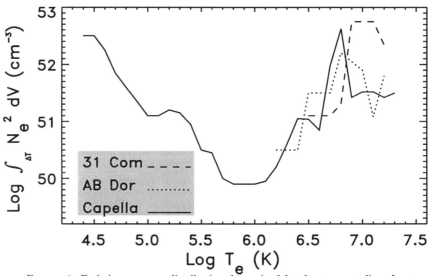

FIGURE 3. Emission measure distribution determined by the strongest lines for two rapidly-rotating single stars (31 Com and AB Dor) and the binary system Capella.

of an increased emission measure in Capella for their production. That 44ι Boo would have a similar feature is apparent from inspection of its SW spectrum; the enhancement is broader in temperature than the Capella bump. The SW spectrum of VY Ari shows an exceptionally strong feature of Fe XX, XXIII (λ132.8) relative to the Fe XVIII and XIX transitions; thus the local enhancement of the emission measure occurs at a temperature higher by 0.2 dex, namely $T(K) = 7.0$ dex. The long period system, λ And has a much less well defined emission measure enhancement (Hanson et al. 1995). The slopes of the emission measures with temperature are consistent with models of magnetic loops or arcades.

Two single stars can address the question of whether the "bump" is unique to binaries. The rapidly-rotating single giant 31 Com is of particular interest because it lies in the Hertzsprung gap (Wallerstein et al. 1994). Since the hot rapidly-rotating secondary in the Capella system is believed (Pilachowski & Sowell 1992) to be passing through the Hertzsprung gap too, 31 Com might prove helpful in disentangling the composite EUV spectrum of Capella. The star 31 Com has a broad enhancement required by the dominant Fe XXII and Fe XXIII transitions, at a temperature higher (7.0 dex) than the Capella feature. Such a broad hot component is not found in the Capella emission measure. AB Dor, another rapidly-rotating single giant shows a broad enhancement, also different from Capella. Thus, rapidly-rotating single stars have "bumps" but they appear different from those in Capella and VY Ari.

Rotation appears to be a significant physical parameter in producing an enhanced emission measure feature in cool star atmospheres. This feature is generally present in those systems with periods $\leq$ 13 days. However the temperature of the maximum and the width and strength of the enhancement differ from star to star. It is plausible to associate this feature with magnetic structures in a rapidly-rotating corona, in which case it could be dense and of small scale; however, in the binary systems Capella and VY Ari, a *narrow* enhancement (of unknown density) is found. It may be that such a feature results from interaction between the components. Binaries of longer period, represented by λ And and σ Gem (Hanson et al. 1995) do not show such well-defined enhancements in the emission measure, although they possess equally hot plasma. However longer EUV exposures are needed on many of these stars to detect weaker species and to define the atmospheric structure over a wider temperature range.

This work is partly supported by NASA NAG5-2330 to the Smithsonian Institution.

REFERENCES

BRICKHOUSE, N. S., RAYMOND, J. C., & SMITH, B. W. 1995, New Model of Iron Spectra in the Extreme Ultraviolet and Application to SERTS and EUVE Observations of a Solar Active Region and Capella, ApJS, 97, 551

DUPREE, A. K., BRICKHOUSE, N. S., DOSCHEK, G. A., GREEN, J. C., & RAYMOND, J. C. 1993, The Extreme Ultraviolet Spectrum of Alpha Aurigae Capella, ApJ, 418, L41-L44

DUPREE, A. K., BRICKHOUSE, N. S., DOSCHEK, G. A., HANSON, G. J., & RAYMOND, J. C. 1994, EUV Spectra of Alpha Aurigae Capella, at Different Phases, BAAS, 26, 864

DUPREE, A. K., BRICKHOUSE, N. S., & HANSON, G. J. 1995, Inhomogeneity of Coronal Structures in Capella, in prereparation

FEKEL, F. C., MOFFETT, T. J., & HENRY, G. W. 1986, A Survey of Chromospherically Active Stars, ApJS, 60, 551

HANSON, G. J., BRICKHOUSE, N. S., & DUPREE, A. K. 1995, EUVE Spectra of Lambda Andromedae and Sigma Geminorum, ApJ, submitted

KIMBLE, R. PLUS, 13 AUTHORS 1993, Extreme Ultraviolet Observations of G191-B2B and the Local Interstellar Medium with the Hopkins Ultraviolet Telescope, ApJ, 404, 663

PILACHOWSKI, C. A., & SOWELL, J. R. 1992, The Lithium Abundance of the Capella Giants, AJ, 103, 1668

WALLERSTEIN, G., BOHM-VITENSE, E., VANTURE, A. D., & GONZALEZ, G. 1994, The Lithium Content and Other Properties of F2-G5 Giants in the Hertzsprung Gap, AJ, 107, 2211

A Re-Analysis of the X-ray Spectrum of HR 1099 and Its Implications for Coronal Abundances

STEPHEN A. DRAKE,[1,2] KULINDER P. SINGH,[1,3]
AND NICHOLAS E. WHITE[1]

[1]HEASARC, Code 668, NASA/GSFC, Greenbelt, MD 20771, USA

[2]USRA, Code 610.3, NASA/GSFC, Greenbelt, MD 20771, USA

[3]TIFR, Homi Bhabha Road, Colaba, Bombay, 400 005, India

We have analyzed archival X-ray spectra of the RS CVn binary system HR 1099 in an attempt to see if we can obtain a consistent picture of the state of the X-ray emitting plasma. We have modeled six spectra obtained by the *Exosat* ME and LE telescopes, two spectra obtained by the *Einstein* SSS and MPC instruments, as well as a more recent *ROSAT* PSPC spectrum. We find that these spectra are in general poorly fitted by solar-abundance Raymond and Smith or Mewe and Kaastra thermal plasma models, and that no simple combination of these models significantly improves these fits: the observed continuum is too strong relative to the line features. We find acceptable fits for thermal models with two or more components in which the heavy elements are depleted by a factor of 2 to 4 relative to their solar photospheric values. These results are consistent with those obtained from the analysis of higher-resolution *EUVE* and *ASCA* spectra of active binary and single stars. We discuss the implications of these findings on ongoing analyses of the EUV spectra of HR 1099 and other RS CVn binaries.

1. Introduction

Recent results from moderate-resolution X-ray spectroscopy with *ASCA* and high-resolution EUV spectroscopy with *EUVE* of the coronae of active, late-type stars have yielded surprising results concerning the elemental abundances of these high-temperature plasmas. Most of the stars for which detailed analyses have been obtained appear to show a general underabundance of metals, relative to their solar photospheric values of factors from $\sim 2 - 6$ (e.g., White et al. 1994; Drake et al. 1994; Singh et al. 1995a; Stern et al. 1995; Rucinski et al. 1995). This result is quite unexpected, since most of these solar neighborhood stars are thought to have photospheric abundances similar to the sun. The solar corona does indeed exhibit a different pattern of abundances than the solar photosphere, but in a completely opposite sense to that inferred for stellar coronae: elements like Fe with first-ionization potentials (FIPs) less than ~ 10 eV are **enhanced** by factors from $\sim 1 - 10$ compared to their photospheric values (e.g., Anders & Grevesse 1989; Feldman 1992).

Before attempting to explain the inferred peculiar abundances found in stellar coronae, it is essential that their reality be firmly established. In particular, why was evidence for this metal-deficiency effect not inferred from previous (admittedly generally either lower spectral-resolution and/or signal-to-noise) X-ray observations of stellar coronae? To answer this question, we examine *Einstein*, *EXOSAT*, and *ROSAT* spectra of the prototypical RS CVn binary star HR 1099 (V711 Tau). As secondary goals of the present study we want to (i) refute or verify the existence of a high-temperature ($T_e \gg 10^{7.5}$K) plasma component of its corona, such as has been inferred might be present from analyses of *EUVE* spectra of similar active stars, by searching for a high-energy tail in the 3 to 20 keV region of its X-ray spectrum; and (ii) derive best-fit models from earlier epoch

147

S. Bowyer and R. F. Malina (eds.), Astrophysics in the Extreme Ultraviolet, 147–151.
© 1996 *Kluwer Academic Publishers. Printed in the Netherlands.*

observations of HR 1099 that can be compared with more recent *EUVE* and *ASCA* observations of this system.

2. Observations and Analysis

The following archival X-ray spectra of HR 1099 have been analyzed:

(i) Two observations by the *Einstein* Observatory Solid State Spectrometer (SSS) (spectral resolution ΔE of ~ 160 eV covering 0.5 to 4.5 keV) and the Monitor Proportional Counter (MPC) (8 logarithmically spaced energy channels covering the 1 to 20 keV band) made in 1979, with a total exposure of ~ 10 kiloseconds (ks).

(ii) Six observations made by the *EXOSAT* Observatory Low Energy (LE) telescope and Medium Energy (ME) proportional counter array in the period 1983 to 1986, with a total exposure of ~ 200 ks. The LE essentially measures the X-ray flux in the 0.1 to 2.0 keV band, while the ME covers the 1 to 20 keV band with an energy resolution increasing from 0.5 keV at low energies to 2.0 keV at high energies.

(iii) An observation made by the *ROSAT* Observatory Position Sensitive Proportional Counter (PSPC) in 1992 with an exposure time of 2.7 ks.

We have used the XSPEC (Version 8.50) spectral analysis package to fit the data with the Mewe-Kaastra, or "meka" spectral models (Mewe et al. 1985; Kaastra 1992) for thermal-equilibrium plasmas. (Similar results are obtained using the Raymond & Smith model (Raymond 1990; Raymond & Smith 1977), except for a small difference in the derived temperatures). We used the "meka" model for ease of comparison with our earlier work on active, stars where we also found somewhat better fits using "meka" model (Drake et al. 1994).

We tried (a) single-temperature or isothermal plasma models, (b) a model consisting of two or more discrete plasma components at different temperatures, and (c) a plasma model with continuous emission measure (CEM) distribution which is a power-law function of the temperature of the type, EM(T) $\propto (T/T_{max})^{\alpha}$, where T_{max} is the maximum temperature of the plasma and α is the slope of the emission measure (EM) distribution (Schmitt et al. 1990). The last of these is a good approximation to models that have been used to describe the EM distribution of individual coronal loops on the sun (e.g., Antiochos & Noci 1986). It has also been used to fit the X-ray emission from stellar coronae which would normally involve an ensemble of such loops (e.g., Stern et al. 1986; Schmitt et al. 1990). We have kept α and T_{max} as free parameters. The elemental abundances in the plasma for all of these different models, were varied with respect to the solar photospheric values taken from Anders & Grevesse (1989).

3. Results

A subset of the results of this analysis are given in Table 1, where we list the parameters of the best-fit 2-component models and their corresponding reduced chi-squared fit statistics, S. Some examples of the best-fit models together with the observed spectra are shown in figures 1 and 2.

The *EXOSAT* LE + ME, *Einstein* SSS + MPC, and *ROSAT* PSPC spectra are all best fitted by two-component plasma emission models with temperatures of 0.6–0.7 keV and 2–3 keV, and having sub-solar abundances. There is no evidence for significant amounts of hotter ($T_e \sim 10^8$K) plasma being present. We find that discrete (two) component plasma emission models can usually better fit the data than the power-law CEM models based on models of single solar loops; the latter models are even found to be unacceptable in several cases. Given the low-resolution nature of the spectra discussed here, we have

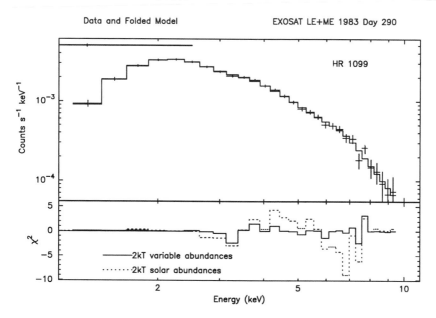

FIGURE 1. The observed spectrum of HR 1099 obtained with the *EXOSAT* LE and ME (the crosses) on 1983, day 290, compared to the best-fit 2-T variable abundance model (the histogram). The lower panel shows the χ^2 residuals both for this model and, for comparison, the best-fit 2-T solar abundance model.

TABLE 1. Summary of Results of Two-Component Models

Data Set	Date	Best solar abund. model			Best variable abund. model			
		T(low)	T(high)	S	T(low)	T(high)	f	S
SSS + MPC	1979.222	0.66	3.6	1.42	0.68	3.4	0.54	1.31
SSS + MPC	1979.223	0.67	4.0	1.37	0.68	3.9	0.49	1.23
LE + ME	1983.290	0.41	2.7	1.97	1.34	4.2	0.24	0.60
LE + ME	1985.267	0.92	2.3	1.45	1.26	5.0	< 0.88	1.35
LE + ME	1985.267	1.06	2.5	0.94	1.25	3.3	< 0.87	0.85
LE + ME	1986.033	0.54	2.2	1.97	0.73	2.3	< 0.67	1.70
LE + ME	1986.034	0.45	1.9	2.60	1.17	2.7	< 0.19	1.02
LE + ME	1986.035	0.57	2.0	1.43	0.96	2.2	< 0.63	1.21
PSPC	1992.026	0.19	1.0	6.70	0.57	1.3	0.29	1.21

limited our consideration to models with either solar (photospheric) abundances or those having all the non-hydrogenic atomic abundances varying from solar by a common factor f. Not surprisingly, the introduction of the latter additional free parameter does improve the fits to the observed spectra: some spectra, however, can simply not be fit by solar-abundance plasmas with the admittedly simple thermal structures that we have tested. The abundance results can be summarized thus: (i) In all cases the inferred values of the abundance factor f are subsolar, with best-fit values in the range of $0.2- \leq 0.9$; (ii) The abundance factor f appears to be variable: some spectra appear to be only mildly metal-deficient, while others (e.g., the 1986 Day 33/34 *EXOSAT* spectrum) appear to be

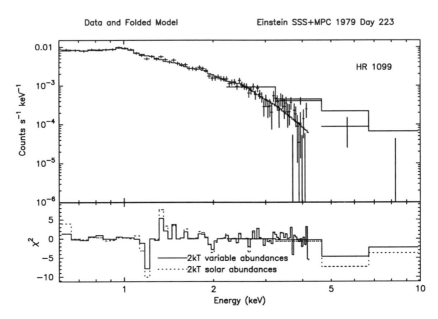

FIGURE 2. Similar to Fig. 1, but showing the observed and model spectra for the *Einstein* SSS and MPC data obtained on 1979, day 223.

much more depleted; and (iii) The abundance results are broadly compatible with those that we and others have inferred from *ASCA* spectra of active, late-type stars.

4. Discussion

Thus, the analysis of the archival X-ray spectra of HR 1099 has confirmed the metal-deficiency effect seen in *ASCA* and *EUVE* spectra of active stars. The reason that these abundance peculiarities were not inferred 15 years ago is apparently that most previous studies of coronal spectra have been made without allowing the abundances to vary from their solar photospheric values. The next question to address is whether these non-solar abundances are intrinsic (i.e., real) or are an artifact somehow resulting from our analysis. We believe that we have demonstrated here that they are not due to an instrumental artifact, since similar results are obtained from several different X-ray instruments of widely differing properties such as spectral resolution.

If there is a problem then, it must be associated with the coronal plasma models used in the fits. Indeed, it is known that the 'meka' and Raymond and Smith codes incorporated into the present version of XSPEC poorly represent the the Fe L shell region of 0.7 to 1.4 keV, due to a large number of errors and omissions, and that strong lines in this complex can be over- or underestimated by as much as factors of 2 or 3. Until we have the next-generation 'meka' code available for comparison, it is impossible for us to say how the problems with the current-generation codes will affect the inferred Fe abundance. For example, in the worst-case scenario, if the emission from the whole Fe L complex has been overestimated by a factor of 3, then the underabundance of Fe that has been derived herein and in previous papers discussing *ASCA* spectra of late-type stars would be spurious. In fact, according to Liedahl et al. (1995), the picture is not as simple as that: some lines are underestimated, while some lines are over-estimated, so that, based on this preliminary report, it seems difficult to attribute the factor of 3 deficiency in Fe (relative to solar photospheric) found here to this known problem. Furthermore, similar Fe underabundances are inferred from studies of the Fe K 6.7 keV feature for which the atomic physics is known to be fairly accurate.

The other major result of this study is that there is no upturn in the emission measure distribution at temperatures $T_e \geq 10^8$ K. X-ray spectra would easily detect this material if it were present: it is not. As is discussed in more detail in Singh et al. (1995b), there is a relatively small amount of information that can be derived about the detailed temperature structure of coronae from low or even moderate spectral resolution X-ray spectra: $EUVE$ spectra, if of high enough signal-to-noise, are clearly superior in this regard. The best fits to X-ray spectra of cool stars are simple 2-temperature models, although poorer, but still acceptable fits can usually be obtained by more complex continuous emission measure models. However, these same analyses also show that the elemental abundances inferred from X-ray spectra are in general very insensitive to the assumed temperature structure. Thus, it is much easier to derive information on abundances from $ASCA$ spectra than $EUVE$ spectra, both because of this reason, as well as the presence of strong features due to Si, Mg, S, Ar, and Ca in the $ASCA$ spectral range.

This research has made use of data obtained through the High Energy Astrophysics Science Archive Research Center Online Service, provided by the NASA-Goddard Space Flight Center.

REFERENCES

ANDERS, E. & GREVESSE, N. 1989, Geochimica et Cosmochimica Acta, 53, 197

ANTIOCHOS, S. K. & NOCI, G. 1986, ApJ, 301, 440

DRAKE, S. A. ET AL. 1994, ApJ, 436, L87–L90

FELDMAN, U. 1992, Physica Scripta, 46, 202

KAASTRA, J. 1992, An X-ray Spectral Code for Optically thin plasmas, SRON-Leiden

LIEDAHL, D. A. ET AL. 1995, ApJ, 438, L115–L118

MEWE, R. ET AL. 1985, A&ApS, 62, 197

RAYMOND, J. C. 1990, private communication

RAYMOND, J. C. & SMITH, B. W. 1977, ApJS, 35, 419

RUCINSKI, S. M. ET AL. 1995, this volume

SCHMITT, J. H. M. M. ET AL. 1990, ApJ, 365, 704

SINGH, K. P. ET AL. 1995a, ApJ, in press

SINGH, K. P. ET AL. 1995b, ApJ, submitted

STERN, R. A. ET AL. 1986, ApJ, 305, 417

STERN, R. A. ET AL. 1995, ApJ, 444, L45–L48

WHITE, N. E. ET AL. 1994, PASJ, 46, L97–L100

Continued Analysis of *EUVE* and Optical Observations of a Flare on AD Leonis

S. L. CULLY[1,2] G. H. FISHER,[1] S. L. HAWLEY,[3] AND T. SIMON[4]

[1] Experimental Astrophysics Group, Space Sciences Laboratory,
University of California, Berkeley CA 94720 USA

[2] Center for Extreme Ultraviolet Astrophysics, 2150 Kittredge Street,
University of California, Berkeley, CA 94720-5030, USA

[3] Department of Physics and Astronomy,
Michigan State University, East Lansing, MI 48824, USA

[4] Institute for Astronomy, University of Hawaii, 2680 Woodlawn Dr., Honolulu, HI 96822, USA

The flare star AD Leo (dM3.5e, 4.9 pc) was observed by *EUVE* from 1993 March 1–March 3 UT. A flare was detected by the *EUVE* DS/S and seen in optical photometry on 1993 March 2 UT. We summarize an analysis of the flare's physical parameters, and present differential emission measure (DEM) curves calculated for the quiescent, flare peak and flare decay phases of the observation.

The flare star AD Leo (dM3.5e, 4.9 pc) was observed by the *EUVE* DS/S from 1993 March 1–March 3 UT. Two flares were clearly visible in the lightcurve of the Lexan/boron (40–190 Å) band of the Deep Survey Instrument (Fig. 1) occurring March 2 and 3 UT. The impulsive phase at the beginning of the larger flare (Flare 1 in Fig. 1) was also observed with optical photometry at Lick Observatory (Fig. 2). The 0.6 magnitude optical U band (3000–4300 Å) flare had a peak DS Lex/B count rate of 1.0 cps and was visible for 7 hours. Hawley et al. (1995) estimated the total EUV energy released during the 7 hour peak of the larger flare to be $\sim 2 \times 10^{33}$ ergs. Using the stellar flare loop model developed by Fisher & Hawley (1990), Hawley et al. (1995) found that the flare could be modeled as a constant area loop with a half length of 4×10^{10} cm ($\sim R_*$), a coronal cross sectional area of 9×10^{19} cm^2 and an average electron density of 3×10^{10} cm^{-3}. They also used the contemporaneous optical data taken in the Johnson U, B and V bands during the flare to compute a photospheric fractional stellar area coverage of 0.01% (1×10^{18} cm^2) and a blackbody temperature of 9000 K. ($T_{\text{eff}} \sim 3000$ K during quiescence).

In Fig. 3, we show that the *EUVE* DS Lex/B and optical U Band (3000–4300 Å) luminousities roughly follow the relation, $L_{\text{EUV}} \propto \int_0^t L_U \, dt'$ during the impulsive phase of the EUV flare Hawley et al. (1995). This is reminiscent of the Solar "Neupert" effect where the EUV and Soft X Ray emission from hot, evaporated, chromospheric material in the flare loop(s) is seen to be proportional to the time integral of the Hard X-ray and white light emission (Neupert (1968), Dennis & Zarro (1993)).

Figure 4 shows the optimally extracted, fluxed *EUVE* (60–600 Å) spectra taken during quiescence, the peak of the large flare and the "tail" between the two flares (see Fig. 1). Figure 5 and Table 1 show the results of two different methods of estimating the volume differential emission measure (DEM) ($n_e^2 dV/d\ln(T)$) of the emitting plasma during each of the time periods mentioned above. Both methods assumed a column density of 2×10^{18} cm^{-2} and used the updated Landini and Mosignori-Fossi plasma line list (Monsignori Fossi & Landini (1994)) in the low density limit. Abundances 0.4 dex lower than Solar coronal abundances (Naftilan, Sandmann & Pettersen (1992)) were also assumed.

153

S. Bowyer and R. F. Malina (eds.), Astrophysics in the Extreme Ultraviolet, 153–158.

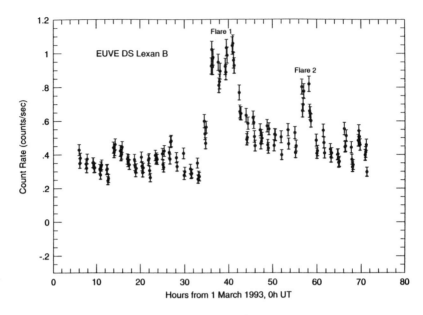

FIGURE 1. EUVE DS Lex/B Lightcurve

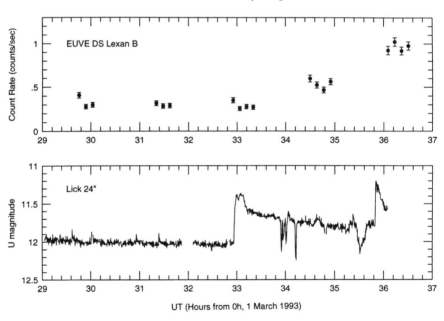

FIGURE 2. EUVE DS Lex/B and optical U band lightcurves at the onset of the large flare shown in Fig. 1.

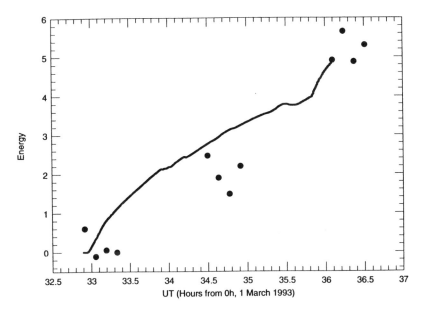

FIGURE 3. Illustration of the Neupert effect calculated from the data shown in Fig. 2;
— U-band ergs, $* \, 10^{-32}$, • $EUVE$ DS, - ergs–s^{-1}, $* \, (3 \times 10^{-27})$

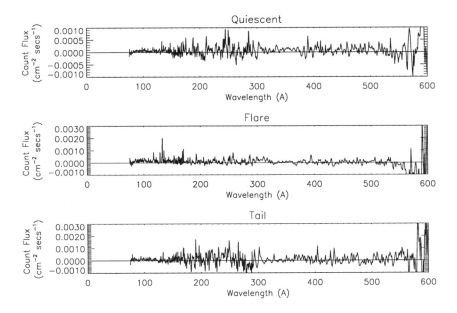

FIGURE 4. Fluxed Spectra taken during the peak of the large flare, the "tail" between the two flares and Quiescence

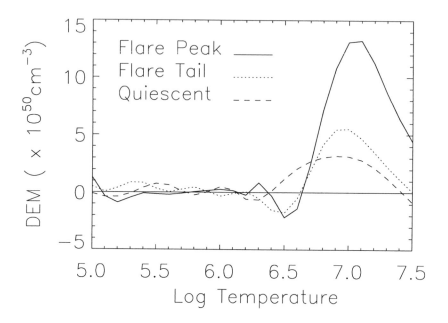

FIGURE 5. DEM analysis for the spectra shown in Fig. 4

Table 1 shows the results of a differential emission measure (DEM) analysis using unblended single lines. In the cases, where a positive detection is shown the line was detected with a certainty of at least 2 σ. Each DEM value was calculated by using the formula

$$DEM(T_{\mathrm{Max}}) = \frac{4\pi d^2 f_{\mathrm{obs}}}{\int \epsilon(T') \, dT'}$$

where $\epsilon(T)$ is the line power output as a function of temperature at our assumed elemental abundance, d is the distance to the source, f_{obs} is the observed line flux and T_{Max} is the peak line power temperature.

Figure 5 shows the continuous DEMs found using an inversion procedure similar to that described in Craig & Brown (1986) and Mewe et al. (1995). In this method, the entire spectrum is used, as opposed to just a few lines. This increases the S/N and allows analysis of heavily blended spectra where there are no obvious single unblended lines available for separate analysis. It also finds a shape for the DEM consistent with each of the single line points when the lines are strong (e.g. around Log $T \sim 7.0$ in Table 1). However, when a spectrum is dominated by noise a single strong line can give an improved upper limit (e.g. Fe IX/O V 171.1 Å, Log $T \sim 5.8$). The inversion method can also oscillate or produce spurious high temperature tails when there is insufficient data to constrain the solution. Therefore it is desirable to use both methods to gain a more accurate picture.

Figure 5 and Table 1 show that the spectra in all three cases are dominated by the high temperature ($\sim 10^7$ K) Fe lines. The curves show progressively cooler peak temperatures

TABLE 1. Summary of Differential Emission Measures Calculated from Individual
Spectral Lines

Lambda (Å)	Ion	Log T (K)	Log Quiescent DEM (cm^{-3})	Log Flare Peak DEM (cm^{-3})	Log Flare Tail DEM (cm^{-3})
192.0	Fe XXIV	7.2	< 50.9	< 51.1	—
255.1	Fe XXIV	7.2	< 51.3	< 51.6	< 51.8
132.9	Fe XXIII/XX	7.2	50.6	51.4	50.8
135.8	Fe XXII	7.1	51.0	51.5	< 51.2
116.3	Fe XXII	7.1	< 51.1	< 51.5	< 51.4
117.2	Fe XXII	7.1	51.0	51.6	< 51.2
118.2	Fe XX/XXI	7.0	< 51.1	51.4	< 51.4
121.6	Fe XX	7.0	< 50.8	51.3	< 51.1
128.7	Fe XXI	7.0	50.6	51.4	50.9
108.4	Fe XIX	6.9	50.9	51.3	51.0
93.7	Fe XVIII	6.8	50.7	51.2	51.2
334.2	Fe XVI	6.4	50.4	< 50.4	< 50.5
284.1	Fe XV	6.3	50.1	< 50.1	< 50.4
202.0	Fe XIII	6.2	< 49.7	< 49.9	< 50.2
181.2	Fe XI	6.1	< 49.8	< 50.6	< 50.6
196.6	Fe XII	6.1	< 49.9	< 50.0	< 50.5
174.5	Fe X	6.0	< 50.0	< 49.8	< 50.0
177.2	Fe X	6.0	< 49.9	< 50.7	< 50.6
171.1	Fe IX/O V	5.9	49.2	< 49.1	< 49.5
401.7	Ne VI	5.7	< 50.3	< 50.4	< 50.6
358.9	Ne V	5.5	< 50.4	< 50.5	< 50.6
238.5	O IV	5.3	< 49.9	< 50.0	< 50.4
554.4	O IV	5.3	< 49.5	< 49.7	< 49.9
303.7	He II	4.9	< 49.7	< 50.0	< 50.0

and smaller emission measures as the flare evolves from its peak, through the decay tail and back to quiescence. This behavior is suggestive of a general cooling of the flare loops but may also be due to the superposition of progressively cooler flaring events.

REFERENCES

CRAIG, I. J. D. & BROWN, J. C., EDS. 1986, Inverse problems in astronomy: A guide to inversion strategies for remotely sensed data, Bristol: Adam Hilger Ltd.

CULLY, S. L. ET AL. 1995, EUVE spectral observations of a flare on AD Leonis, in preparation

DENNIS, B. R. & ZARRO, D. M. 1993, The Neupert effect—What it can tell us about the impulsive and gradual phases of solar flares, Sol. Phys., 146, 177

FISHER, G. H. & HAWLEY, S. L. 1990, An equation for the evolution of solar and stellar Flare Loops, ApJ, 357, 243

HAWLEY, S. L., ET AL. 1995, Simultaneous EUVE and optical observations of AD Leonis: Evidence for large coronal loops and the Neupert effect in stellar flares, ApJ, in press

MONSIGNORI, FOSSI, B. C. & LANDINI, M. 1994, The X-Ray–EUV Spectrum of Optically Thin Plasmas, Solar Phys., 152, 81

MEWE, R., ET AL. 1995, EUV spectroscopy of cool stars: 1. The corona of Alpha Centauri observed with EUVE, A&A, 296, 477

NEUPERT, W. M. 1968, Comparison of Solar X-ray line emission with microwave emission during flares, ApJ, 153, L59

Naftilan, S. A., Sandmann, W. S. & Pettersen, B. R. 1992, The spectrum of the dwarf Me flare star AD Leonis, PASP, 104, 1045

Spectroscopic *EUVE* Observations of the Active Star AB Doradus

SLAVEK M. RUCINSKI,[1] ROLF MEWE,[2] JELLE S. KAASTRA,[2]
OSMI VILHU,[3] AND STEPHEN M. WHITE[4]

[1]Institute for Space and Terrestrial Science and York University,
4850 Keele St., Toronto, Ontario, M3J 3K1, Canada

[2]Space Research Organization Netherlands (SRON),
Sorbonnelaan 2, NL 3584 CA Utrecht, The Netherlands

[3]Observatory and Astrophysics Laboratory,
Box 14, FIN-00014 University of Helsinki, Finland

[4]Department of Astronomy, University of Maryland, College Park, MD 20742, USA

We present observations of the pre-Main Sequence, rapidly-rotating (0.515 day) late-type star, AB Doradus (HD 36705), made by the *Extreme Ultraviolet Explorer* (*EUVE*) satellite. A high-quality spectrum was accumulated between November 4–11, 1993, with an effective exposure time of 40 hours. The data constrain the coronal temperature structure between several 10^4 K up to roughly 2×10^7 K through a differential emission measure analysis using an optically-thin plasma model. The resulting differential emission measure (DEM) distribution shows: a) dominant emission from plasma between about 2×10^6 K and 2×10^7 K, b) very little emission from plasma between 10^5 K and 2×10^6 K, and c) emission from plasma below about 10^5 K. If solar photospheric abundances are assumed, then the *formal* DEM solution also requires the presence of a strong high-temperature component (above about 3×10^7 K) in order to fit the strong continuum emission below about 150 Å; however, we believe that this component of the solution is not physical. The DEM analysis gives a best-fit value for the interstellar hydrogen column density of $N_H = (2.4 \pm 0.5) \times 10^{18}$ cm^{-2}.

1. Introduction

Since its identification as a flaring *EINSTEIN* X-ray source (Pakull 1981), AB Doradus (HD 36705) has been one of the most frequently observed active late-type stars. Its particular importance is due both to its very short rotation period (0.515 day) for its spectral type of K0-K2 IV-V, and to its proximity to the Sun (distance 20–30 pc; Rucinski 1985, Innis, Thompson & Coates 1986). The discoveries of a strong lithium absorption line in its spectrum (Rucinski 1982, 1985) and of kinematic properties characteristic of the Pleiades group (Innis et al. 1986) have established the star as one of the nearest pre-Main Sequence objects in the solar neighborhood, with an inferred age of 10^{6-7} years.

AB Doradus is easily detected in all spectral bands from the X-rays to radio. Following the *EINSTEIN* observations, Collier et al. (1988) using the *EXOSAT* ME detector found a hot thermal corona. The spectral information in this observation was scant but the whole emission in the low (0.05–2 keV) and medium (1–10 keV) energy ranges of *EXOSAT* could be explained by a one-component plasma at about 2×10^7 K with an emission measure of about 2.3×10^{53} cm^{-3} (for an assumed distance of 25 pc). X-ray flares were observed with a rate of about one per 0.5-day stellar rotation. The same rate was observed during an extensive multifrequency campaign of coordinated observations extending from X-rays (GINGA satellite) to radio (3 cm, Parkes radio-telescope) conducted by Vilhu et al. (1993). The *ROSAT* X-ray all–sky survey observations of AB Dor (Kürster et al. 1992)

S. Bowyer and R. F. Malina (eds.), Astrophysics in the Extreme Ultraviolet, 159–164.
© *1996 Kluwer Academic Publishers. Printed in the Netherlands.*

obtained over a period of more than a month showed erratic variability with very weak rotational modulation.

The corona of AB Dor is also practically always visible in the radio as non-thermal emission at frequencies as low as 843 MHz (Beasley & Cram 1993), and as high as 8.6 GHz (Vilhu et al. 1993). At 6 cm, the star has often shown a clear and regular rotational modulation (Lim et al. 1992, 1994). Spectroscopic optical studies of Collier (1982) and Vilhu, Gustafsson & Edvardson (1987) gave $v \sin i = 100 \pm 5$ km s^{-1} and a radial velocity constant to ± 2 km/s.

The chromospheric and transition-region emission of AB Doradus was analyzed with the *IUE* satellite by Rucinski (1985). The overall level of emission was found to be very high and the star seemed to be rather uniformly covered by active regions at that time. This result, together with the hot coronal component derived by Collier et al. (1988) from the *EXOSAT* ME spectra suggested an obvious follow-up with the *EUVE* satellite whose spectral range probes the region of temperatures between 10^5 and 10^7 K.

AB Dor forms a wide binary with the young rapidly-rotating M-type dwarf Rst 137B. Since this secondary is bolometrically about 60 times fainter than AB Dor, and the "saturation" phenomenon (Vilhu & Walter 1987) should apply to Rst 137B as it does to AB Dor, we do not expect Rst 137B to contaminate significantly the EUV spectrum of AB Dor.

2. Observations

The *Extreme Ultraviolet Explorer* observations of AB Dor started on November 4 and continued for 14 rotations of the star until November 11, 1993. The data were collected in three spectroscopic bands, SW: 80–180 Å, MW: 150–350 Å, and LW: 300–700 Å, as well as with the Deep Sky Survey (DSS) imager (60–150 Å). Results of a time variability analysis to study rotational modulation in the EUV spectra, in the DSS channel and from extensive optical and radio ground-based supporting observations will be presented by White et al. (1995a).

An observation of AB Dor with the *ASCA* X-ray satellite obtained during our *EUVE* observations (White et al. 1995b) helped to constrain our interpretations. AB Dor did not show any major flares in the EUV, and was in a low-radio-flux state. Optical monitoring (White et al. 1995a) showed no flaring except during a brief period amounting to no more than 5% of the *EUVE* observation, so that time-averaging of our spectra should result in data representative of the quiescent state of AB Dor with no significant contamination from flare emission.

The effective exposure times for our spectra were about 40 hours. For the final spectral analysis, we rebinned the spectra into wavelength intervals of 0.25 Å, 0.5 Å, and 1.0 Å for the SW, MW, and LW, respectively. This binning corresponds to about 4 original pixels and ensures that the number of counts (source plus background) is large enough in each bin so that the assumption of Gaussian statistics, as implicitly assumed in the χ^2 statistics, is warranted. Typically all bins in the raw data contain 50–100 background counts. Such rebinned spectra over-sample the resolution of the *EUVE* spectrographs by a factor of 2.

The spectra expressed in total accumulated counts, corrected for the background are shown for the SW and MW bands in Figures 1 and 2. For the stronger lines we indicate the ions from which they originate. In cursory examinations of the spectra, we note that, in addition to the prominent lines in the SW band, we see a relatively strong continuum between about 80 Å and 150 Å. The noise level in the vicinity of the helium lines (e.g. He II at 304 Å line, see Figure 2) is relatively high, even after background subtraction,

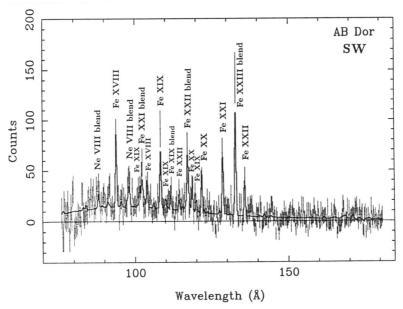

FIGURE 1. The *EUVE* background–corrected spectrum of AB Dor in the SW passband. The best–fit spectrum, of which the differential emission measure distribution **D**(T) is plotted in Figure 3.

due to strong geocoronal contributions in these lines. This is because AB Dor is situated at a high ecliptic latitude $(-87°)$ and therefore on average the angle between the pointing direction of the telescope and the bright Earth is smaller than for most other sources, leading to relatively more scattered geocoronal light entering the detectors.

3. Spectral Analysis

To analyze the *EUVE* spectrum, we calculated isothermal equilibrium spectra utilizing a spectral code which has evolved from the well–known and widely used optically–thin plasma code developed by Mewe and co–workers (e.g. Mewe et al. 1985, 1986), called SPEX (cf. Kaastra and Mewe 1993, Mewe and Kaastra 1994).

Each of the calculated spectra is modified by interstellar absorption using the absorption cross sections of Rumph et al. (1994), and convolved with the instrument response. In the analysis we assume a source distance $d = 25$ pc, and initially we adopt solar photospheric abundances from Anders and Grevesse (1989). When calculating interstellar absorption we use an interstellar hydrogen column density of $N_H = 2.4 \; 10^{18}$ cm^{-2} which followed from a best-fit $(N_H = (2.4 \pm 0.5) \times 10^{18}$ cm$^{-2})$ to the data, with adopted abundance ratios He I/H I = 0.1 and He II/H I = 0.01. However, the final results are not very sensitive to these ratios.

In our analysis, we follow the procedure described by Mewe et al. (1995a and these proceedings). The DEM function, defined to be $\mathbf{D}(T) = n_e n_H dV/d \log T$, is the weighting function which measures how strongly any particular temperature contributes to the observed spectrum (T is the electron temperature, n_e the electron density, n_H the hydrogen density, and V the plasma volume). The observed spectrum is interpreted as a

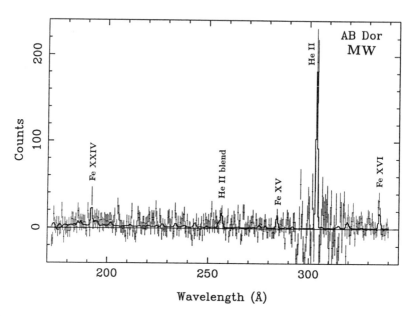

FIGURE 2. Same as in Figure 1, for the MW passband. We do not show the LW spectrum
which is weak because of the interstellar absorption.

statistical realization of a linear combination of isothermal spectra. The shape of the
DEM distribution is derived from the observed emission line *and* continuum fluxes. We
stress here that it is important that we include the continuum in our analysis since it is
the line-to-continuum ratio that can provide important information on the abundances
(in particular the [Fe/H] ratio) and line scattering effects.

The *EUVE* spectra contain information about plasma emission for the range between
about $5 - 7 \times 10^4$ K and 2×10^7 K. The maximum in the continuum spectrum moves
shortward of the *EUVE* wavelength range for temperatures exceeding roughly 3×10^6 K,
while the associated long-wavelength component in the *EUVE* range is rather insensitive
to changes in the temperature. Therefore, the continuum from plasma with temperatures
exceeding roughly 3×10^6 K can only yield evidence that plasma at these temperatures
is present, but cannot yield more detailed information on the actual temperature distribution.

The resulting **D** is plotted in Figure 3 as a solid line with error bars and the corresponding best-fit spectrum is shown in Figures 1–2 by the thick solid line. The **D** curve
is based on 41 logarithmically spaced temperatures between 0.001–10 keV. In Figure 3,
the emission measure is plotted as $\mathbf{D}(T)\Delta \log T$, with $\Delta \log T = 0.1$.

We note that, because our method is in principle a linear decomposition method, it
does not prevent **D** from taking negative values. The DEM inversion procedure produces
an essentially perfect fit to the observed spectrum, for the assumed errors as given by
Poisson counting statistics. The fit residuals are nowhere larger than 3 σ and we found no
evidence for systematic tendencies, in the sense of the fit residuals being systematically
above or below zero in certain wavelength ranges.

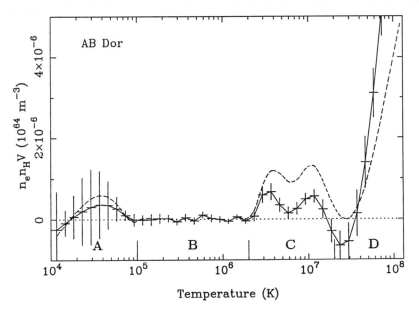

Temperature (K)

FIGURE 3. Differential emission measure (DEM) curve $\mathbf{D}(T)\Delta\log T$ for AB Dor corresponding to the best–fit theoretical spectrum shown in Figures 1 and 2, based on standard solar abundances (solid line with error bars) and the DEM derived for the case with the iron abundance reduced by a factor of 3.3 (dashed line). Further reduction of the iron abundance to about 1/7 solar entirely removes the high-temperature component.

4. Conclusions

The main results of our determination of DEM are as follows:

• At temperatures below about 10^5 K (region "A" in Figure 3) a "chromospheric" component is predominantly determined by the strong He II line at 304 Å and (upper limits of) a few other, weaker lines (e.g. He II 256 Å and O III 507 Å). The strong He II 304 Å line is certainly optically thick, so that the $\mathbf{D}(T)$ at these temperatures should be regarded only as a formal, certainly not unique solution.

• There is very little emission from plasma between 10^5 K and 2×10^6 K (region "B"). The striking absence of emission in this region is also found for other active cool stars such as α Aur and ξ UMa (Schrijver et al. 1995 and these proceedings).

• Most of the lines in the spectrum of AB Dor are due to various ions of iron, from Fe XV up to Fe XXIV, forming at temperatures from about 2.5×10^6 K up to $\sim 1.6 \times 10^7$ K. This pronounced emission corresponds to region "C" between 2×10^6 K and 2×10^7 K.

• In region "D" the formal $\mathbf{D}(T)$ solution shows a strong component above about 2.0×10^7 K as a high-temperature "tail", which reflects the presence of a featureless continuum in, foremost, the SW spectrum (cf. Figure 1).

The high-temperature component appears to be a common feature in the formal DEM solutions for many other cool stars, while for the quiet Sun itself such a hot component is never detected (Mewe et al. 1995a, Schrijver et al. 1995). Simultaneous *ASCA* observations of AB Dor do not show any extraordinary hot component (cf. White et al. 1995b, Mewe et al. 1995b). One of possible interpretations of the SW continuum is that its strength relative to the (predominantly Fe) line emission is due to a lower–than–assumed

Fe abundance, rather than the presence of very hot plasma (cf. Mewe et al., these proceedings). Through extensive modelling experiments, we found that a reduction of the iron abundance by a factor of seven leads to disappearance of the hot component, and that the *EUVE* data can be reconciled with the *ASCA* data. An alternative possibility is that the emitting region is an asymmetric, optically thick but effectively thin plasma in which the photons of the stronger lines are resonantly scattered and destroyed upon impact on the lower chromosphere (cf. Schrijver et al., these proceedings).

The full analysis of the *EUVE* spectra of AB Dor will appear in *Astrophysical Journal*, August 20, 1995.

REFERENCES

ANDERS, E. & GREVESSE, N. 1989, Geochimica et Cosmochimica Acta, 53, 197

BEASLEY, A. J. & CRAM, L. E. 1993, MNRAS, 264, 570

COLLIER, A. C. 1982, MNRAS, 200, 489

COLLIER, CAMERON, A., BEDFORD, D. K., RUCINSKI, S. M., VILHU, O. & WHITE, N. E. 1988, MNRAS, 231, 131

INNIS, J. L., THOMPSON, K. & COATES, D. W. 1986, MNRAS, 223, 183

KAASTRA, J. S. & MEWE, R. 1993, Legacy, 3, 16

KÜRSTER, M., SCHMITT, J. H. M. M. & FLEMING, T. A. 1992, Cool Stars, Stellar Systems, and the Sun, ed. M. S. Giampapa & J. A. Bookbinder, ASP Conf. Ser. 26, 109

LIM, J., NELSON, G. J., CASTRO, C., KILKENNY, D. & VAN WYK, F. 1992, ApJ, 388, L27

LIM, J., WHITE, S. M., NELSON, G. J., & BENZ, A. O. 1994, ApJ, 430, 332

MEWE, R., GRONENSCHILD, E. H. B. M., VAN DEN OORD, G. H. J. 1985, A&AS, 62, 197

MEWE, R., LEMEN, J. R., VAN DEN OORD, G. H. J. 1986, A&AS, 65, 511

MEWE, R., KAASTRA, J. S. 1994, European Astronomical Society Newsletter, 8, 3

MEWE, R., KAASTRA, J. S., SCHRIJVER, C. J., VAN DEN OORD, G. H. J. & ALKEMADE, F. J. M. 1995a, A&A, 296, 477

MEWE, R., KAASTRA, J. S., PALLAVICINI, R., WHITE, S. M. 1995b, in prepreparation

PAKULL, M. W. 1981, A&A, 104, 33

RUCINSKI, S. M. 1982, Inf. Bull. Var. Stars, 2203

RUCINSKI, S. M. 1985, MNRAS, 215, 591

RUMPH, T., BOWYER, S. & VENNES, S. 1994, AJ, 107, 2108

SCHRIJVER, C. J., MEWE, R., VAN DEN OORD, G. H. J., KAASTRA, J. S. 1995, A&A, in press

VILHU, O., GUSTAFSSON, B. & EDVARDSON, B. 1987, ApJ, 320, 850

VILHU, O., TSURU, T., COLLIER, CAMERON, A., BUDDING, E., BANKS, T., SLEE, O. B., EHRENFREUND, P. & FOING, B. H. 1993, A&A, 278, 467

VILHU, O. & WALTER, F. M. 1987, ApJ, 321, 958

WHITE, S. M., LIM, J., RUCINSKI, S., ROBERTS, G., RYAN, S., PRADO, P., KILKENNY, D., & KUNDU, M. R. 1985a, ApJ, in prepreparation

WHITE, S. M., PALLAVICINI, R., & LIM, J. 1995b, ApJ, in prepreparation

A Search for Rotational Modulation in the EUV Emission from AB Doradus

S. M. WHITE,[1] J. LIM,[2] S. M. RUCINSKI,[3] G. ROBERTS,[4]
D. KILKENNY,[4] S. G. RYAN,[5] P. PRADO,[6] AND M. R. KUNDU[1]

[1] Dept. of Astronomy, Univ. of Maryland, College Park MD 20742, USA

[2] IAA, Academica Sinica, PO Box 1-87, Nankang, Taipei, Taiwan

[3] Inst. for Space and Terr. Science, and York Univ., 4850 Keele St., Toronto,
Ontario M3J 3K1, Canada

[4] South African Astronomical Observatory, PO Box 9, Observatory 7935, South Africa

[5] Anglo-Australian Observatory, PO Box 296, Epping NSW 2121, UK

[6] Las Campanas Observatory Casilla 601, La Serena, Chile

One of the goals of the *EUVE* observation of AB Doradus was to search for rotational modulation of the EUV emissions. In support of this goal we carried out optical photometry in Chile, Australia and South Africa and radio observations in Australia. In addition, an *ASCA* observation of AB Dor was scheduled to occur during the campaign. Several spectacular X-ray and optical flares with accompanying brightening in the EUV were seen, but no rotational modulation of the EUV emission was evident, except for a dip in the He II 304 Å line coincident with the optical minimum. An X-ray flare was seen with no accompanying EUV flare.

1. Introduction

AB Doradus is a very active nearby southern K dwarf star. This object has attracted attention for a number of reasons. It is one of the most rapidly-rotating single stars known, with a period of only 12.4 hours, and it is relatively close, at 25 pc. Its rotation axis is inclined at 60° to our line of sight. Further, it is a young star: it has a strong Li absorption line (Rucinski 1985), and kinematic properties characteristic of the Pleiades group (Innis, Thompson, & Coates 1986). This makes it probably the nearest pre–main-sequence star known, with an age of approximately 50 million years; its closeness makes it an easily-studied analogue of the numerous rapidly-rotating dwarf stars in the Pleiades open cluster. Prior X-ray observations by *EINSTEIN* (Pakull 1981), *EXOSAT* (Cameron et al. 1988), *GINGA* (Vilhu et al. 1993) and *ROSAT* (Kürster, Schmitt, & Fleming 1992) revealed AB Dor to be a strong and interesting coronal source. In addition, radio observations have shown a remarkable modulation of the light curve at the rotation period (Lim et al. 1992) which can be used to analyze the distribution of radio-emitting material in the stellar atmosphere (Lim et al. 1994).

EUVE observed AB Dor from 1993 Nov 4.4–11.2 (total exposure of 140 ks). Rucinski et al. (1995 and this volume) present a detailed analysis of the EUV spectrum of AB Dor; in this paper we focus on a search for rotational modulation in the photometric data and on the results of a coordinated campaign of supporting observations. This included U-band photometry at the South African Astronomical Observatory SAAO), with 2 s time resolution, on Nov 4–8; U-band CCD photometry at the U. Toronto Las Campanas facility on Nov 4, 5, & 7; and U-band CCD photometry at Mount Stromlo Siding Spring Observatory (MSSSO) on Nov 5–7. CCD photometry was used at the latter two sites to establish the relevance of AB Dor's M dwarf companion, Rst 137B, which is 9″ away and therefore is not resolved by EUVE. Rst 137B showed several very impulsive optical flares but should not be contributing in any of the data shown. Both optically and in

S. Bowyer and R. F. Malina (eds.), Astrophysics in the Extreme Ultraviolet, 165–169.
© 1996 Kluwer Academic Publishers. Printed in the Netherlands.

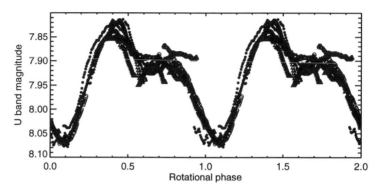

FIGURE 1. Optical U-band data taken during the campaign plotted versus rotational phase using the ephemeris of Innis et al. (1988). The largest optical flare has beem removed from the data.

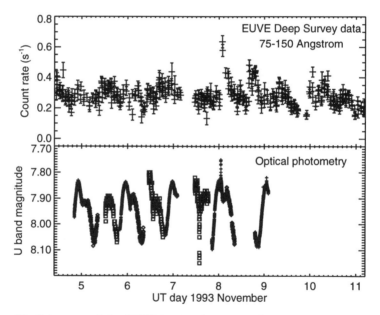

FIGURE 2. The light curve of the *EUVE* DS data (upper panel) and the U-band optical light curve (lower panel) for the duration of the *EUVE* observation.

X-rays it is a factor of 60 fainter than AB Dor. We obtained radio observations at 1.4, 2.4, 5.0, & 8.0 GHz with the Australia Telescope on November 5 and 6. In addition the Japanese X-ray satellite *ASCA* observed AB Dor from Nov 7.55–8.25 (total exposure of 40 ks).

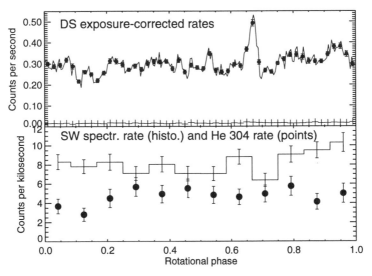

FIGURE 3. The *EUVE* DS data plotted against rotational phase (upper panel). The points with error bars correspond to 50 bins per orbit, while the solid line corresponds to 200 bins per orbit. The crosses at the bottom of the upper panel show the background count rate. In the lower panel we plot the count rate in the SW spectrometer (histogram with 1 σ error bars), which is a diagnostic of the hot plasma in the corona similar to the DS data, and the counts in the strong He II 304 Å line (points), which is a transition-region diagnostic.

2. Observational Data

Figure 1 shows all the photometry data plotted against rotational phase using the ephemeris of Innis et al. (1986). In Figure 1, filled circles represent SAAO data, binned to 2 minute resolution; open circles represent Las Campanas data; and open triangles represent MSSSO data. While the intrinsic scatter in the data is less than 0.01 magnitudes, the absolute level is not well determined at the three observatories, and relative shifts of up to 0.1 magnitudes were used to line up the light curves. The data show a clear modulation of 0.2 magnitudes in U. The star is brightest at a phase of 0.4, and faintest at 0.1. AB Dor's light curve has been monitored for over a decade and this is the largest level of variability seen in the light curve in that time.

In Figure 2 we show the EUV light curve using data from the Deep Survey (DS) imager, which is the most sensitive detector for EUV photometry. Its passband corresponds to that of the SW spectrometer. In this figure we have binned the photons into 400 s bins. Due to AB Dor's high ecliptic latitude the observing efficiency was only 24%. Several weak flares may be seen. In Figure 3 we have divided the rotation period into a number of bins and allocated the DS data according to rotational phase. The top panel shows the count rate as a function of phase, while the bottom panel shows two diagnostics from spectrometer data: total counts in the 80–170 Å range in the SW detector; and counts in the He II 304 Å line (MW and LW combined). Unlike the optical data, the DS data show no striking modulation with rotational period. The narrow feature in the phase-binned DS data at phase 0.67 is due to the combination of a low exposure time in that particular bin, and the coincidence of having several flares contribute to the bin.

Figure 4 shows the *ASCA* count rate, the *EUVE* DS data and the optical photometry

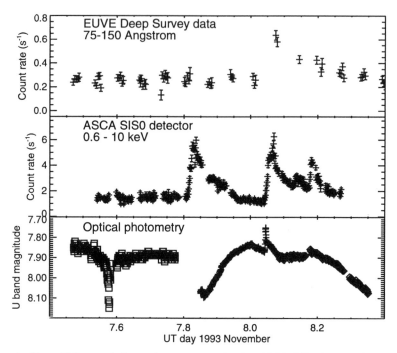

FIGURE 4. The middle panel shows the count rate in the *ASCA* SIS0 detector (0.6–10 keV) during the *ASCA* observation of AB Dor on Nov 7–8. The upper panel shows the light curve of the *EUVE* DS data and the lower panel the optical light curve during the same period.

during the period of the *ASCA* observation. The *ASCA* data show a very steady level of just under 2 counts s^{-1} for the first 6 hrs of the observation, but the remaining period is dominated by a succession of flares reaching up to 6 counts s^{-1}. Two of the X-ray flares (at Nov 8.05 and 8.17) are coincident with optical flares. The *EUVE* light curve is unchanged following the first major flare at Nov 7.8, but at least doubles following the second major X-ray flare and shows a gradual decay over the following 6 hours. Since the two large flares occur about 6 hrs apart, then they either come from the always-visible region near the pole, or else they come from opposite hemispheres of the star. The second large flare is coincident with a large optical flare, as is one of the small later events.

3. Discussion

The main conclusion of the data collected during the *EUVE* AB Dor observation is that there was little evidence for rotational modulation of the coronal diagnostics (EUV, radio and X-ray). This result can be interpreted as either an approximately uniform distribution of material with longitude on the visible hemisphere, or else a very extended corona in which most of the coronal material is visible at all times. Unlike the radio data (Lim et al. 1992, Lim et al. 1994), the previous EUV (Pagano et al. 1993) and X-ray observations (Cameron et al. 1988, Vilhu et al. 1993, Kürster, Schmitt, & Fleming 1992) have also found little compelling evidence for rotational modulation.

However, there is evidence for modulation of the transition-region temperature diagnostic, the He II 304 Å line, which appears to have a minimum at the minimum in the optical light curve. This suggests weak transition-region emission above a large spot at the minimum of the optical light curve.

We believe that AB Dor was in a low-activity state until about Nov 7.8. Three pieces of evidence suggest this: although about two-thirds of the optical photometry took place prior to this time, no optical flares were detected prior to this time, but four optical flares were detected following it; no EUV flares were detected prior to this time, whereas at least 2 were detected subsequently; and the X-ray light curve is very steady until this time, but shows a succession of flares thereafter. The radio data (not shown) were taken prior to this time and show clear variability, but no unambiguous rotational modulation. The relatively low level of radio emission (a few mJy at all frequencies) is consistent with the discussion by Lim et al. (1994), who noted that the radio modulation is most pronounced when the radio flux is high (tens of mJy).

It is a puzzle to understand why one large X-ray flare should produce an obvious response in SW EUV emission when another of similar size did not. The EUV emission in the DS detector is dominated by lines of highly-ionized Fe, and we expect it to respond to the same (in this case, 10–30 MK) plasma which produces the X-rays, as it did in the second flare.

SMW gratefully acknowledges support from NASA's *EUVE* and *ASCA* Guest Investigator programs (grants NAG-5-2364 and NAG-5-2531, respectively), and from NSF grant AST 92-17891. SMR's research has been supported by a grant from the Natural Sciences and Engineering Research Council of Canada.

REFERENCES

Cameron, A. C., Bedford, D. K., Rucinski, S. M., Vilhu, O., & White, N. E. 1988, MNRAS, 231, 131

Innis, J. L., Thompson, K., & Coates, D. W. 1986, MNRAS, 223, 183

Kürster, M., Schmitt, J. H. M. M., & Fleming, T. A. 1992, in Cool Stars, Stellar Systems and the Sun, 7th Cambridge Workshop, ed. M. S. Giampapa & J. A. Bookbinder, San Francisco: ASP Conf. Ser. 26, 109

Lim, J., Nelson, G., Castro, C., Kilkenny, D., & van Wyk, F. 1992, ApJL, 388, 27

Lim, J., White, S. M., Nelson, G. J., & Benz, A. O. 1994, ApJ, 430, 332

Pagano, I., et al. 1993, in Physics of Solar and Stellar Coronae, ed. J. F. Linsky & S. Serio Dordrecht: Kluwer Academic Press, 457

Pakull, M. W. 1981, A&A, 104, 33

Rucinski, S. M. 1985, MNRAS, 215, 591

Rucinski, S. M., Mewe, R., Kaastra, J. S., Vilhu, O., & White, S. M. 1995, ApJ, in press

Vilhu, O., Tsuru, T., Collier, Cameron, A., Budding, E., Banks, T., Slee, B., Ehrenfreund, P., & Foing, B. H. 1993, A&A, 278, 467

EUV Emission Sources in Gas-Dynamic Models of Stellar Flares

M. A. LIVSHITS[1] AND M. M. KATSOVA[2]

[1]Institute of Terrestrial Magnetism, Ionosphere, and Radio Wave Propagation, Russian Academy of Sciences, 142092 Troitsk, Moscow Region, Russia

[2]Sternberg State Astronomical Institute, Moscow State University, 119899 Moscow, Russia

A stellar flare model in which the main energy release is located above the chromosphere based on a set of elementary acts of the electron acceleration or impulsive heating of plasma is discussed. The response of the chromosphere to impulsive heating for both a single burst and the simultaneous effect of a set of the bursts is considered.

The results of numerical modeling of the process of explosive evaporation of the stellar chromosphere allow us to select 3 classes of EUV emission source: (1) the bursts of the EUV emission in the temperature range of $3 \cdot 10^4 - 3 \cdot 10^5$ K at the beginning of each of elementary act of the energy release; (2) the bursts of the EUV emission at temperatures $T \approx 10^6$ K, accompanied by outflow of heated plasma; (3) the EUV emission, caused by new features of the process, namely, when the maxima of the distribution of the pressure are forming in the region of the downward-moving thermal front.

The properties of the EUV sources, velocities of the plasma motions therein, and possible behaviour of the light curve for an elementary burst are discussed.

1. Introduction

In previous papers we concentrated on understanding the dynamics of plasma in stellar flaring processes. The acceleration of particles takes place often enough, especially in small-scale regions, and apparently there are different mechanisms that accelerate electrons up to $20 - 100$ keV energies. The response of the chromosphere to impulsive heating by nonthermal electrons should lead to a development of specific plasma motions: the low-temperature condensation moves downward, and the hot plasma evaporates upward. The first one emits in the Balmer lines and, in some cases, in the optical continuum; the evaporated plasma turns out to be the source of the soft X-ray emission. The modelling of gas-dynamical processes helps us to interpret the flare light curves in the optical and X-ray ranges, and to estimate the area of the source of the optical continuum and other characteristics of the emission sources (see details in Katsova & Livshits 1995 and references therein).

Here we would like to discuss the properties of the sources of EUV emission formed during the gas-dynamical process referred to above. We will mainly discuss the process of explosive evaporation which is realized during impulsive flares on red dwarf stars. Gentle evaporation can appear to develop on late-type subgiants and giants, and it will be accompanied by the EUV emission of stellar wind and long-duration flares. However, we are at the beginning of understanding these problems.

2. The Source of EUV Emission at the Start of the Gas-Dynamical Process

Recently we carried out new modelling of this gas-dynamical process using modern numerical methods (Boiko & Livshits 1995; Katsova et al. 1995). The system of equa-

S. Bowyer and R. F. Malina (eds.), Astrophysics in the Extreme Ultraviolet, 171–174.
© *1996 Kluwer Academic Publishers. Printed in the Netherlands.*

tions of the gas-dynamics for 2T- and one-fluid plasmas includes heating flux, radiative energy losses, thermal conduction flux, and the function of the energy change between the electron and ion plasma components. In the first tenth of a second the upper chromosphere is heated suddenly without a remarkable change in the total density, because bulk motions are still practically absent. The temperature turns out to be constant in these layers of the upper chromosphere because the heating has been compensated for by radiative losses, and this value of T varies from $\approx 10^4$ to $\leq 10^5$ K. These layers (the source I in Figure 1) emit in the UV-lines with $\lambda = 1000$–2000 Å, and in the resonance line of He II with $\lambda = 304$ Å. This emission appears after each elementary act of an acceleration, and lasts less than one second. The same type of emission should arise when the thermal wave arrives at the second footpoint of the large-scale coronal loop.

The UV and EUV burst should be shorter than one second (see Fisher 1987 for the case of solar flares, and Katsova & Livshits 1989 for the flares on red dwarfs). The burst in the He II line can be longer than 1 s because the ionization of He I by the EUV and soft X-ray emission can be remarkable in some cases, and this effect is able to prolong the emission at 304 Å.

Unfortunately, there is no direct evidence for an identification of such very short UV and EUV spikes with the source arising in the beginning of the process of explosive evaporation. Our interpretation (Katsova & Livshits 1989) of observations of a very short burst in the C IV line, detected on EV Lac by the ASTRON (Burnasheva et al. 1989) showed that this is possible, but it supposes that the response of the chromosphere to the heating extends over a large area. This enlargement of the flare area is understandable, if we consider the more complicated scenario of an impulsive stellar flare like that proposed by Katsova & Livshits (1992).

3. The EUV Emission of the Evaporated Plasma

The computations give $5 \cdot 10^{19}$ for the number of protons that evaporate per cm^2 during one elementary burst lasting 10 s. The main part of the evaporated plasma has a temperature of $T \geq 10^7$ K and a velocity close to 1000 km/s. However, the temperature of evaporated plasma reaches only $(1-4) \cdot 10^6$ K in the beginning of this process, especially in the lower part of the upward flow. This is shown in Figure 1 for the time $t = 0.4$ s. Note that in this moment 30% of the hot plasma moves downward with velocities of $100 - 180$ km/s, and the remaining 70% of the gas moves upward with velocities from a few km s^{-1} up to 1000 km/s.

Thus, the plasma with temperature $(1 - 4) \cdot 10^6$ K in the lower part of the evaporated flow is the source of the EUV-lines for the few seconds after the beginning of the process. These lines have to be broaded by the Doppler effect, but they shouldn't have remarkable blueshift. The blueshifted components of high temperature lines observable during flares are those emitted by ions such as Fe XXIV and Ca XIX. The intensities of the EUV lines depend on the area of a flare, which is close to the area of the optical continuum source. The rise of the intensity of these lines is 1–2 s ahead of the appearance of the optical continuum, and is about 30 s ahead of the soft X-ray radiation during powerful events with duration not longer than a few minutes; such events can be considered a response to a single act of primary energy release.

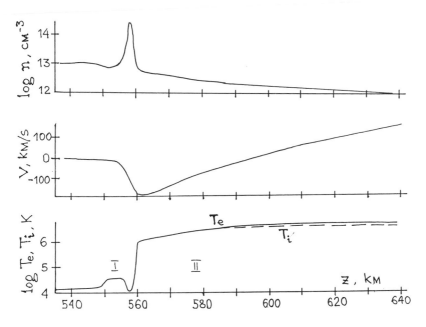

FIGURE 1. The density, velocity and temperature distributions versus height in the AD Leo atmosphere through 0.4 s after the beginning of heating, when the radiative shock wave has just been formed. The zero-point for the height is the photospheric level with $\tau(5000) = 1$. The origin of the EUV source (I) is caused by the initial heating of the upper chromosphere. The source (II), i.e., the low layers of evaporated plasma, exists for a few seconds after the beginning of heating.

4. The Features of Explosive Evaporation and the UV and EUV Radiation

In addition to the previously discussed possibilities for the appearance of the UV and EUV emission, there is another possibility caused by the character of the propagation of the downward thermal wave front. A quasi-stationary regime of the flow is almost always set, if powerful impulsive heating exists. It takes place $0.5 - 1$ s after the beginning of the process. The character of the flow, i.e., the downward motion of the thermal wave front with a radiative shock wave ahead of it, holds until the heating operates.

In real processes going on in stellar atmospheres the temperature jump will break down, if this region is exposed to the external flow of the plasma or to another disturbance. The downward thermal front will be strongly changed when heating is varied significantly or when this thermal wave encounters some inhomogeneities of density. These events should occur with additional UV and EUV radiation.

Furthermore, the gas-dynamical modelling showed that even with no external disturbance of the quasi-steady flow, the propagation of the temperature jump downward takes place not continuously, but like a set of weak jumps (Boiko & Livshits 1995). As a rule, the pressure is kept constant on the front of the thermal wave, i.e., the density is inversely proportional to the temperature. To simplify the gas-dynamical computations, one adopts a constant pressure for the thermal wave front. In the calculations by Boiko & Livshits (1995) no assumptions were made in this region. These results were obtained for continual heating. In this version of gas-dynamical simulation, the small maximum of the pressure distribution is forming in the region of the downward-moving thermal wave front from time to time. Thereafter this maximum begins to break down, and two weak disturbances start to propagate upward and downward. This brings the gas-dynamical process back to the original regime with a smooth shape for the pressure ($p =$const on the thermal wave front). This event is repeated time and again, and some kind of periodic process arises.

The process of the restoration of this quasi-stationary regime of the flow is accompanied by enhancement of the UV and EUV emission. The flux at $1 - 2000$ Å is close to 10^6 ergs cm^{-2} s when explosive evaporation takes place. As a result of this effect the flare luminosities in the X-ray, UV, and optical ranges should approach each other.

The authors would like to thank the International Science Foundation for the financial support of their participation in this Colloquium. The work of M.A.L. is partly supported by the grant 4-129 of the State Program "Astronomy" of the Russian Ministry of Science.

REFERENCES

BOIKO, A. YA., & LIVSHITS, M. A. 1995, Astronomy Reports of the Russian Academy of Sciences, 39(3)

BURNASHEVA, B. A., GERSHBERG, R. E., ZVEREVA, A. M., ET AL. 1989, Sov. Astron., 33, 165

FISHER, G. 1987, ApJ, 317, 502

KATSOVA, M. M., BOIKO, A. YA., & LIVSHITS, M. A. 1995, in preparation

KATSOVA, M. M. & LIVSHITS, M. A. 1989, Sov. Astron., 33, 155

KATSOVA, M. M. & LIVSHITS, M. A. 1992, Astron. Astrophis. Trans., 3, 67

KATSOVA, M. M. & LIVSHITS, M. A. 1995, in Flares and Flashes. IAU Colloq. 151, ed. J. Greiner, H. M. Duerbeck, R. E. Gershberg, Lect. Notes in Physics, Springer, 454, 177

Post-Eruptive Flare Energy Release as Detected on AU Mic by *EUVE*

M. M. KATSOVA,[1] J. J. DRAKE,[2] AND M. A. LIVSHITS[3]

[1]Sternberg State Astronomical Institute, Moscow State University, 119899 Moscow, Russia

[2]Center for EUV Astrophysics, 2150 Kittredge Street,
University of California, Berkeley CA 94720-5030, USA

[3]Institute of Terrestrial Magnetism, Ionosphere and Radio Wave Propagation, Russian
Academy of Sciences, 142092 Troitsk, Moscow Region, Russia

The long-duration emission arising after the impulsive rise and decay in a flaring event observed by the *Extreme Ultraviolet Explorer* on the red dwarf star AU Mic is discussed. The decay of the intensity in the Deep Survey 65–190 Å band and in the Fe XVIII line during this prolonged event is 10 times slower than the time of radiative cooling of coronal loops with the typical for the flare plasma density. The temporal behavior of the emission measure is determined for both the 65–190 Å band and the Fe XVIII line fluxes. The total energy emitted in the 1–2000 Å region over nearly 12 hrs is $3 \cdot 10^{35}$ ergs. We first point out some difficulties with earlier explanations proposed for this event; we then propose the following physical model: the source of the prolonged emission is a system of high coronal loops, the size of which is more than the active region scale, but less than the stellar radius. Such systems are observed in soft X-rays during large solar flares after coronal mass ejections. Some additional post-flare energy input into this high coronal loop system can be caused by reconnection in a vertical current sheet, and this post-eruptive energy release provides prolonged and intensive EUV emission.

Apparently, we are faced here with new kind of the surface activity on late-type stars which is intermediate between impulsive flares on red dwarfs and long-duration, powerful events the subgiants components of the RS CVn binaries.

1. Observations

AU Mic was observed by *EUVE* from 12:28 UT 14 July 1992 to 8:09 UT 18 July 1992. The Deep Survey (DS) data have been discussed in previous papers (Cully et al. 1994; Drake et al. 1994; Landini & Monsignori Fossi 1994). We illustrate the DS light curve from Cully et al. in Figure 1. The description of the spectrometer data reduction will appear in a forthcoming paper (Brown et al. 1995); a very brief description is provided in Drake et al. (1994).

In this study, we represent the flare decay using an isothermal model with a time dependent temperature $T(t)$. The physical conditions were determined from the DS photometer data and the intensity of lines of Fe seen in the SW and MW spectrometers. The estimation of the temperature and its temporal behaviour was based on the behaviour of Fe lines from charge states XVIII, XXI, XXII, XXIII and XXIV. These lines indicate that, following the impulsive phase, the temperature decreases gradually from the value of $\sim 1 \cdot 10^7$ K. The estimated EM, based on the DS and Fe XVIII data, and T as a function of time are presented in Figure 2. The intensity of the Fe XVIII line emission decreases faster than the DS count rate independently from the chosen temporal behaviour of the temperature. The maximum EM of the flare decay estimated from both from the DS and spectral data is $1.6 \cdot 10^{53}$ cm^{-3}—similar to the values derived by Cully et al. (1994) from the DS photometer data.

Three possible explanations of this long duration EUV radiation are discussed:

(1) the flare X-ray loops;

S. Bowyer and R. F. Malina (eds.), Astrophysics in the Extreme Ultraviolet, 175–180.

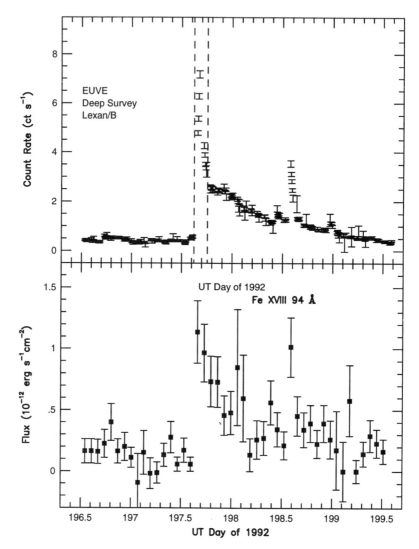

FIGURE 1. Temporal behaviour of the EUV emission observed from the July 14–17 1992 flare on AU Mic. Top panel: the 65–190 Å Deep Survey count rate (from Cully et al. 1994). Bottom panel: the Fe XVIII 93.9 Å line (Drake et al. 1994).

(2) very high coronal loops;
(3) coronal mass ejection (CME), cf. Cully et al., 1994.

The soft X-ray emission of stellar flares does not usually exceed $2.3t_{rad}$ or about of 1 hour; in these cases the soft X-ray intensity decay is caused by radiative cooling. The plasma radiative cooling time is given by

$$t_{rad} = \frac{3kT}{nL(T)},$$

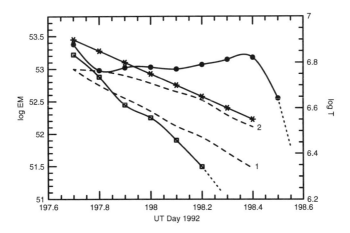

FIGURE 2. Emission measure vs. time calculated for the flare decay for the 65–190 Å DS data (solid dots) and the Fe XVIII line (hollow squares). The curve delineated by stars shows the chosen behaviour of the temperature as a function of time, $T(t)$. Theoretical calculations (dashed lines) are given for the case of 30 % compensation of the radiative losses by additional heating and the initial density $n_o \sim 5 \cdot 10^9$ cm^{-3}: (1) for the chosen temporal temperature dependence $T(t)$; (2) for the isothermal case.

	Volume cm^{-3}	Density cm^{-3}	t_{rad} s	t_{obs}/t_{rad}
1	$2 \cdot 10^{29}$	10^{12}	140	320
2	$2 \cdot 10^{31}$	$> 10^{10}$	1700	26
3	10^{35}	$1 \cdot 10^9$	$> 10^5$	< 1

where $n^2 L(T)$ is the radiative loss function (1–2000 Å) in erg cm^{-3} s^{-1}. We compare t_{rad} for typical parameters of the three scenarios mentioned above in Table 1.

As seen in Table 1, the observed duration of the EUV emission decay is much longer than the radiative cooling time in the model (1) and (2). Indeed, the total duration of the decay of the DS 65–190 Å band count rate and of the Fe XVIII line during the prolonged event on AU Mic is 10 times longer than the radiative cooling time of coronal loops with typical flare densities.

The long-duration event emits as a whole

$$W = \int n^2 L(T) dt dV = \int EM_{DS} L(T) dt$$

i.e., about of $W = 3 \cdot 10^{35}$ ergs. This exceeds the energetics of typical flares on red dwarf stars, and even more so the energetics of solar flares.

2. How Does the Solar Analogy Help to Explain the Stellar Phenomenon?

The similarity in the general appearance of the AU Mic event and large solar flares was pointed out by Drake et al. (1994), who noted that such an analogy would require substantial post-flare heating in order to compensate for radiative losses.

Several long-duration soft X-ray events were observed on the Sun during solar cycle XXII. The 1–8 Å radiation lasts from about 0.5 to 1 day. Different aspects of such solar events with soft X-ray emission lasting longer than 0.5 day are discussed by Cliver et al. (1986), Harrison (1986), Smith et al. (1994) and Akimov et al. (1993). Based on the combination of these results, Chertok (1993) have suggested that the post-eruptive energy release for such long duration solar flares can be explained within the framework of the theory of Martens (1988): a coronal mass ejection radially distends the magnetic loop force lines which then form a large-scale vertical current sheet. Subsequent plasma instabilities and reconnection in this vertical current sheet are able, apparently, to accelerate the particles up to very high energies like 10^{10} eV. Note that this phenomenon is observed as a system of cool H_α post-flare loops lying lower than the soft X-ray loops. Such systems were observed during powerful solar flares such as those of September 29 1989, and June 15 1991 within 12 hrs of the onset of the impulsive phase. If additional heating does compensate for the radiative losses, then the hot dense loops can exist as long as this heating lasts. These solar observations are now considered as some of the main evidence for this post-eruptive phase of solar flares.

The solar results also support the point of view that a CME alone is not able to provide the observed fluxes of EUV and soft X-ray emission over many hours. Therefore, despite the fact that the energetics of CME and the type of event considered above are rather similar, the third model (CME) is probably not the best explanation of long-duration EUV emission both on the Sun and on red dwarfs. In particular, the interpretation of the AU Mic event in terms of a CME alone by Cully et al. (1994) might be questioned, because the theoretical temporal behaviour of the density in the CME does not agree with optical data for solar CME's. A possible test of this model is the observation or not of the expected blue-shifted or red-shifted lines indicative of velocities of $\sim 1000\,kms^{-1}$ or so.

Thus, we have to explain the long timescale for the decay of the observed post-flare EUV flux from AU Mic. In principle, it could be due to flaring loops, the size of which does not exceed the typical scale of active regions (the Model 1 in Table 1). In such a case, emission should be observed not only in the EUV but also in soft X-rays such as was detected by *EXOSAT* and EINSTEIN. We note that there are no cases of such prolonged events observed with these satellites (Pallavicini et al., 1990). These dense ($n_e \sim 10^{13}$ cm^{-3} in the second pulse; Brown, 1994) flaring loops should cool down rapidly due to radiative losses in the radiative cooling timescale t_{rad}. In order for such an event to last for $100t_{rad}$ or so, total compensation for the radiative losses over this period is required. Small-scale flaring processes are not able, in principle, to support additional heating with total energy $W = 3 \cdot 10^{35}$ ergs in the $1 - 2000$ Å range. Therefore, there we are lead naturally to consider processes occurring on a larger scale (comparable to, or larger than the stellar radius). This, then, is the Model 2 in the Table).

Thus, the following model can be proposed for the explanation of these EUV observations:

If, as on the Sun, we consider the first pulse of the EUV light curve to be caused by an impulsive flare, so the gradual decay after this impulse can be presented as the radiation of the system of coronal loops which are forming after the CME. Figure 3 shows the

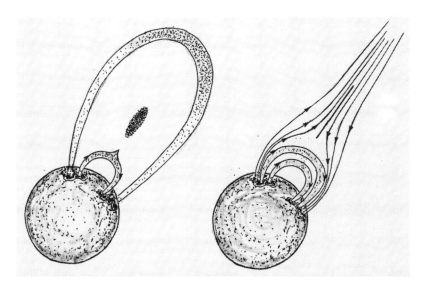

FIGURE 3. A schematic of the post-eruptive energy release. Left: CME as upper part of the highest loop; active prominence or plasmoid; hot coronal flare loop. Right: the phase after the coronal mass ejection (post-eruptive phase); magnetic force lines, vertical current sheet and long-lived hot, high coronal loops emitting in the soft X-ray and EUV range (see also Martens 1988).

CME as the upper part of an extended coronal loop and it is observed often in white light. The surge (active prominence) follows it, and the energy release in this casp (X-) point heats the plasma inside the X-ray coronal loop. The plasma motions extend the force lines of quasi-dipole magnetic field, forming the large-scale vertical current sheet (Martens et al. 1986), which make available additional heating of plasma in the coronal loops. This prolongs the life time of the coronal loops, which emit in the EUV spectral range. This process is similar to the solar case, but the energy input during the long-duration EUV event should be $(2-10) \cdot 10^{30}$ ergs s^{-1}—a factor of 30 more than the relevant solar values.

We consider the balance of the energy in the EUV source without taking into account the detailed magnetohydrodynamic phenomena. If, for simplicity, we assume that the heating is proportional to the pressure, then we can write for 1 cm^3 (or for all the volume which stays constant during this process):

where $E = 3nkT$ is inner thermal energy, the time independent value β erg cm^{-3}s is the initial heating, p_o is the initial pressure.

We have also derived the analytical solution of the equation of the energy balance when radiative losses are partially compensated by additional heating. Calculated values of emission measure for the isothermal case and the temperature dependence, $T(t)$, chosen above are illustrated in Figure 2 for the case of additional post-flare heating amounting to 30% of the radiative losses.

To explain the long duration of the post-flare EUV emission we need a large value of t_{rad}, which in turn implies a lower density during this phase as compared to the impulsive phase. Available spectral data and this theoretical consideration lead us to conclude that Model 2 in Table 1 seems to be acceptable.

3. Conclusions

The event observed on AU Mic on July 15–17, 1992 can be considered as a stellar analog of a typical solar flare with post-eruptive energy release. We argue that the long timescale for this event requires additional post-flare energy input. This energy is provided by reconnection in a vertical current sheet in a system of high coronal loops formed after a CME. Apparently, we are faced here with a new kind of surface activity on late-type stars which is intermediate between impulsive flares commonly seen on red dwarfs and long-duration, powerful events on RS CVn binaries.

MMK and MAL would like to thank the International Science Foundation for financial support of their participation in this Colloquium. JJD was supported by NASA grant AST91-15090 administered by the Center for EUV Astrophysics, University of California.

REFERENCES

AKIMOV, V. V. ET AL. 1993, in Proc. 23rd International Cosmic Rays Conference, Calgary

BROWN, A. 1994, in Cool Stars, Stellar Systems and the Sun, The 8th Cambridge Workshop, ed. J.-P. Caillaut, ASP Conf. Ser., 64, 23

CHERTOK, I. M. 1993, Astronomy Reports of the Russian Academy of Sciences, 37(1), 87

CLIVER, E. W., DENNIS, B. R., & KIPLINGER, A. L., et al., 1986, ApJ, 305, 920

CULLY, S. L., FISHER, G., ABBOTT, M. J., & SIEGMUND, O. H. W., 1994, ApJ, 435, 449

DRAKE, J. J., BROWN, A., BOWYER, S., JELINSKY, P., MALINA, R. F., & WU, X. Y. 1994, in Cool Stars, Stellar Systems and the Sun. The 8th Cambridge Workshop, ed. J.-P. Caillaut, ASP Conf. Ser., 64, 35

HARRISON, R. H. 1986, A&A, 162, 283

MONSIGNORI FOSSI, B. C. & LANDINI, M. 1994, A&A, 284, 900

MARTENS, P. C. H. 1988, ApJ, 330, L131

PALLAVICINI, R., TAGLIAFERRI, G., & STELLA, L. 1990, A&A, 228, 403

SMITH, K. L., SVESTKA, Z., STRONG, K. T., & McCABE, M. K. 1994, Sol. Phys., 149, 363

The EUV Flux of Chromospherically Active Binary Stars

ALVARO GIMÉNEZ[1] AND CONSTANZE LA DOUS[2]

[1]LAEFF, INTA, Villafranca del Castillo, Apartado 50727, 28080 Madrid, Spain

[2]IUE Observatory, ESA, Villafranca del Castillo, Apartado 50727, 28080 Madrid, Spain

We have compared *EUVE* and *ROSAT*/WFC photometric measurements of chromospherically active binary systems at around 100 Å. Long term variations have found for some 20% of the systems as well as a tight linear correlation between the EUV and soft X-ray fluxes.

The recent all-sky survey of the *EUVE* satellite has permitted a study of the emission flux in this spectral range for chromospherically active late-type stars. These binaries are among the most numerous objects observed by *EUVE* due to the fact that they show the brightest coronae among late-type, relatively nearby, stars.

The EUV emission from active binaries originates in a hot plasma in their coronae. They are known to have two temperature components: one around 20 million K (similar to solar flares plasma) and a cooler one of some 5 million K. The bulk of the emission is concentrated in lines of highly ionized species of elements like iron and silicon. In principle, by measuring the relative strengths of the coronal emission lines, the densities and temperatures of the plasma could be inferred. But a detailed plasma diagnostic is only possible through high resolution spectra which are difficult to obtain except for the very brightest systems.

We have performed a comparison of photometric data in the EUV range of chromospherically active binaries, looking for long term variability. For this purpose, we have taken the *EUVE* Source Catalog published by Bowyer et al. (1994), which contains a total of 410 objects and the *ROSAT* WFC Bright Source Catalogue by Pounds et al. (1993), which includes 383 sources. Only 46 of the identified objects by *EUVE* are also included in the Catalog of Chromospherically Active Binary Stars by Strassmeier et al. (1993), which contains 206 systems in total. As expected, only those binaries with a high level of coronal activity and at short distances were detected. Four sources were not detected with the Wide Field Camera (WFC) of *ROSAT* (AG Dor, DK Dra, 42 Cap and HD 57853) while *EUVE* did not observe 6 WFC sources included in Strassmeier et al. (1993) catalogue (ζ And, V1285 Aql, OU Gem, DQ Leo, TW Lep and AR Psc). The Second *EUVE* Source Catalog (Bowyer et al. 1996) became available to us too late to be fully used in this comparison. Nevertheless, a preliminary evaluation led us to the same overall conclusions, with a number of detected active binaries increased to 53 and only 2 systems from the WFC catalogue not detected by *EUVE*.

Long term variability in the EUV range was checked by comparing the obtained fluxes with the *EUVE* satellite during 1992 and by the WFC during 1990. Short term variations, like flares, are not considered in the tabulated input data. 42 systems were observed by the two satellites and 6 of them did show relatively large differences in flux: mainly, UX Ari and α Aur, but also VY Ari, BH CVn, II Peg and AR Lac. All of them are well known for their conspicuous levels of activity and variability in other wavelength ranges. Differences are not so obvious in less active systems and, when instrumental corrections are taken into account, only 4 more binaries were found to have flux differences larger than 50% of the average value (UX For, V815 Her, BY Dra and ER Vul). This indicates that some 15 to 25% of the sample show large variations, in good agreement with similar

181

S. Bowyer and R. F. Malina (eds.), Astrophysics in the Extreme Ultraviolet, 181–182.
© *1996 Kluwer Academic Publishers. Printed in the Netherlands.*

results in the far UV and soft X-ray ranges. Figure 1 shows the comparison of *EUVE* and WFC fluxes with indication of the variable sources.

Taking out the variable systems, a good linear relation is found between the *EUVE* (Lexan/B) and WFC(S1) fluxes, or instrumental correction, that can be expressed in terms of counts s^{-1} as,

$$EUVE = -0.012(\pm 0.007) + 2.14(\pm 0.07)\text{WFC}$$

To check the soft X-ray variability, we have compared the fluxes measured by the PSPC on board *ROSAT* (Dempsey et al. 1993) and the IPC of the *Einstein* observatory (Schmitt et al. 1990). Only 14 binaries of our sample of active binaries were observed by the two instruments and 4 of them were found to be variable, actually the same systems detected in the EUV (UX Ari, BH CVn, II Peg, and VY Ari) within the subset. Finally, we have related the EUV flux in the *EUVE* Lexan/B band with the PSPC measurements for the non-variable binaries in common to both catalogues (24 systems) and, again, a good linear relation was found though the position of σ CrB may show an anomalous behaviour with a too high EUV flux.

REFERENCES

BOWYER, S. ET AL. 1994, ApJS, 93, 569
BOWYER, S. ET AL. 1996, ApJS, in press
DEMPSEY, R. C. ET AL. 1993, ApJS, 86, 599
POUNDS, K. A. ET AL. 1993, MNRAS, 260, 77
SCHMITT, J. H. M. M. ET AL. 1990, ApJ, 365, 704
STRASSMEIER, K. G. ET AL. 1993, A&AS, 100, 173

A Catalogue of Ultraviolet Observations of Chromospherically Active Binary Stars

C. LA DOUS[1] AND ALVARO GIMÉNEZ[2]

[1]IUE Observatory, ESA, Villafranca del Castillo, Apartado 50727, 28080 Madrid, Spain

[2]LAEFF, INTA, Villafranca del Castillo, Apartado 50727, 28080 Madrid, Spain

We present the recently-published IUE-ULDA Access Guide on Chromospherically Active Binary Stars which should be of interest for the interpretation and analysis of EUV data obtained for this type of objects. We provide background information on both high and low resolution *IUE* spectra of chromospherically active late-type binary stars that have been taken until the end of 1992. Physical information on all systems, arranged by variable star name, together with characteristics of the individual exposures, the position of the observation in the orbital light curves, and an average low resolution ultraviolet spectrum are given.

Many of the sources of EUV radiation are cool late-type stars closer than 100 pc. Many of them (those having spectral types later than F5) show indicators of solar-like magnetic activity, such as photospheric spots, chromospheric emission, coronal X-ray and radio emission, as well as flare activity (see e.g., Guinan & Giménez 1993). Apart from very young stellar objects, such as T Tauri stars and related pre-main sequence stars, those with the highest activity levels are close binary systems with cool, G- to M-type, components. Otherwise, the majority of solar-like stars in the Sun's neighborhood are relatively old and have activity levels comparable to that of our Sun.

The currently accepted general picture of the nature of active late-type binaries is based on the emergence of magnetic flux tubes producing active regions in analogy to those in the Sun with spots, active chromospheres and coronae. The combination of deep convective envelopes with magnetic fields and forced stellar rotation lies at the origin of this kind of stellar activity. The strength of the indicators is expected to increase as a function of stellar rotation and of the depth of the convective envelope.

The most prominent groups among the late-type chromospherically active binary stars are those known under the names of RS CVn stars (Hall 1976)—where the stellar components are dwarfs subgiants and giants of spectral type F to K (Hall 1981)—and BY Dra stars (Bopp & Feckel 1977)—where always at least one component is of type dK or dM. Observationally they are characterized by strong strong CaII H and K emission. Following the solar analog, these emissions may be identified with enhanced chromospheric emission from plage-like regions and the chromospheric network. Therefore, emission lines which originate in the same stellar region as the Ca II resonance lines but lie at ultraviolet wavelengths (as a large number of them do) should be observable with a better contrast with respect to the continuum flux. The most interesting lines in this context are the chromospheric emission features of MgII h and k at 2800 Å(see Budding and Giménez 1982; Smith et al. 1991) and the transition region line emissions of NV at 1240 Å, SiII at 1400 Å, and CIV at 1550 Å.

Active late-type binaries are key sources of information on the physical mechanisms responsible for coronal and chromospheric activity. The *IUE* on *Chromospherically Active Binary Stars* (ESA SP-1181, 1994) was originally compiled in order to facilitate usage of the information now contained in the *IUE* Archive; we call attention to it in the context of *EUVE* astronomy for the interpretation and analysis of *EUVE* data obtained for these stars. In addition to details on the available observations, it provides background

S. Bowyer and R. F. Malina (eds.), Astrophysics in the Extreme Ultraviolet, 183–184.
© *1996 Kluwer Academic Publishers. Printed in the Netherlands.*

information on each system. The object list used is the one provided by Strassmeier et al. (1993). For each object we provide plots of the average low-resolution *IUE* spectrum and, whenever available, the high-resolution profile of the Mg II line. Various tables provide details on each *IUE* image (camera, image number, spectral resolution, date and time of observation, FES magnitude, orbital phase, and image quality) and physical parameters (alternative stars names, coordinates, the linear ephemeris, system parameters, activity parameters and photometric parameters from different catalogues.

REFERENCES

Bopp, B. W. & Feckel, F. C.   1977, Binary incidence among the BY Draconis variables, A&A, 82, 490

Budding, E. & Giménez, A.   1982, Stellar activity in the short period subgroup of the RS CVn systems, Third European IUE Conference, ed. E. Rolf et al., ESA SP-176, 169

Guinan, E. F. & Giménez, A.   1993, in The Realm of Interacting Binary Stars, ed. J. Sahade et al., Dordrecht: Kluwer, 51

Hall, D. S.   1976, The RS CVn binaries and binaries with similar properties, in Multiple Periodic Phenomena in Variable Stars, IAU Colloq. 29, ed. W. S. Fitch, Dordrecht: Reidel, 287

Hall, D. S.   1981, The RS Canum Venaticorum binaries, in Solar Phenomena in Stars and Stellar Systems, ed. R. M. Bonnet & A. K. Dupree, Reidel, Dordrecht/Holland, 431

Smith, G. H., Burstein, D., Fanelli, M. N., O'Connell, R. W. & Wu, C. C 1991, On the utility of low resolution IUE spectroscopy of the 2800 A MgII lines as a stellar chromosphere indicator, AJ, 101, 655

Strassmeier, K. G., Hall, D. S., Feckel, F. C. & Scheck, M 1993, A catalogue of chromospherically active binary stars second edition, A&AS, 100, 173

New Developments in Hot White Dwarf Models

DETLEV KOESTER

Institut für Astronomie und Astrophysik, Universität Kiel, D-24098 Kiel, Germany

A short history of EUV observations of white dwarfs is presented, and how they have forced us to use increasingly sophisticated theoretical models—starting from blackbody spectra up to the present generation of NLTE models which include the blanketing effect of millions of atomic lines.

1. Introduction

Observations of white dwarfs mark the beginning of extrasolar EUV astronomy and these objects continue to play a very important role as targets until today. About one third of all EUV sources found in the *ROSAT*/WFC (Pounds et al. 1993) as well as the *EUVE* (Bowyer et al. 1994) surveys are white dwarfs. This is very easy to understand: to be detectable in the EUV through the strongly absorbing interstellar hydrogen and helium the sources should not be too far away; to have enough photospheric flux at these wavelengths they should be hot. Both criteria are met by white dwarfs. Very simple estimates using the luminosity function of white dwarfs give the expected number of white dwarfs within 100 pc and hotter than 25,000 K as about 300; about 60 of these should be hotter than 60,000 K. This corresponds well with the typically 120 white dwarfs detected in the surveys. On the other hand, the nearest O-type star is at a distance of about 200 pc.

A second reason for the prominence of white dwarfs among EUV sources is provided by their peculiar chemical abundances. Most white dwarfs belong to the spectral type DA, which means that they show only the Balmer lines of hydrogen in the visible part of the spectrum. And hydrogen seemed indeed to be the only element present in these atmospheres, an assumption shown to be incorrect for the hottest of them only fairly recently through EUV observations. In a pure hydrogen photosphere the absorption coefficient decreases as λ^3 going from the Lyman absorption edge to the soft X-ray region, and, conversely, the radiation leaving the star comes from progressively deeper and hotter layers. As a consequence, the photospheric radiation of a pure hydrogen DA in the range 100–300 Å can be orders of magnitude larger than that of a blackbody of the same temperature. This effect is demonstrated in Fig. 1 for an effective temperature of 50,000 K.

The addition of helium, even at the level of 1/10 the solar value, changes the spectrum completely (Fig. 2). Now all flux below the He II groundstate absorption edge at 228 Å is completely suppressed, but the flux between 250 and 400 Å is still much higher than the corresponding blackbody flux. These two figures demonstrate that

- DA white dwarfs should be very powerful emitters of EUV radiation
- very small admixtures of helium, which would be completely invisible in the optical and UV, should be easily detectable in the EUV.

2. Early Observations and Interpretations

The first extrasolar EUV source detected was the DA white dwarf HZ43. It was observed by the Apollo-Soyuz extreme-ultraviolet telescope (Lampton et al. 1976). Between

185

S. Bowyer and R. F. Malina (eds.), Astrophysics in the Extreme Ultraviolet, 185–192.
© 1996 *Kluwer Academic Publishers. Printed in the Netherlands.*

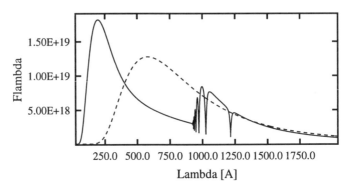

FIGURE 1. Comparison of a pure hydrogen white dwarf model atmosphere for $T_{\rm eff} = 50,000$ K, $\log g = 8$ with a blackbody spectrum of the same temperature

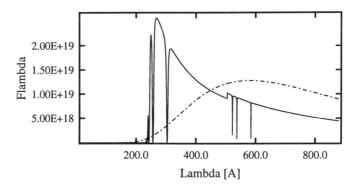

FIGURE 2. A similar comparison as in Fig. 1, except for a He/H ratio of 0.01 in the model atmosphere.

very simple theoretical spectra, like power law and blackbody, a blackbody of 110,000 K was found to give the best fit to the observations. This is much higher than the photospheric temperature of HZ43—a modern value is 49,000 K (Napiwotzki et al. 1993)—but the discrepancy is easily understood in light of the discussion in the preceding section.

White dwarf model atmosphere calculations were first used to interpret the same observations by Auer & Shipman (1977). The effective temperature in their analysis depended on the *assumption* about the helium abundance, which could not be determined independently from the data, and ranged from 55,000 K (no photospheric helium) to 70,000 K for a helium abundance of 1/100 the solar value (the upper limit compatible with the absence of the He II 4686 Å line.

The EUV radiation observed from a white dwarf photosphere depends on a large number of parameters: effective temperature, surface gravity, solid angle of the star, photospheric He/H ratio or abundances of other elements, interstellar column densities of H I, He I, and He II. Observations in the EUV have never been detailed enough to determine all of these parameters, and it is therefore necessary to include in the analysis all available information from other spectral ranges. Even then, the information gained from the EUV up to the *EXOSAT* experiment has been rather limited.

3. Observations and Status Up to 1988—Pre *ROSAT* and *EUVE*

About 20 DA white dwarfs were observed by *EINSTEIN* and *EXOSAT*, in part through different filter bands giving a limited spectral information. While some of the objects could be explained with simple, pure hydrogen model atmospheres, many others, especially among the hotter DA showed clear indications of additional absorbers in the photosphere. The most natural assumption at that time seemed to be helium, and the observations have been used by a number of groups to determine He abundances in these DA (Kahn et al. 1984; Petre et al. 1986; Jordan et al. 1987; Paerels & Heise 1989). It was already known that theoretically He was not expected in the atmospheres, because it should diffuse rapidly towards the deeper layers and radiative levitation is not effective enough to support it (e.g., Vennes et al. 1988).

A possible solution to this problem was the assumption that the pure hydrogen layer is extremely thin. Because of the decrease of hydrogen absorption in the EUV the underlying He could become visible in the EUV, whereas in the optical and UV the star would appear as a pure hydrogen object. The next step in the refinement of the theoretical atmosphere models and the first step beyond simple, classical LTE atmospheres, was therefore to include this diffusion equilibrium H/He stratification consistently in the computations (e.g. Jordan and Koester 1986). These theoretical models were equally able to interpret the *EINSTEIN* and *EXOSAT* observations—it was not possible to distinguish between homogeneous He/H mixtures and a stratified atmosphere with a thin hydrogen layer on top of the helium and a transition zone as calculated from the diffusion equations (Koester 1989). The signature of helium on the spectrum is quite distinct in the two cases; nevertheless, with only two very broad bandpass measurements available, it was always possible to fit the observations and determine either He/H, or the thickness of the hydrogen layer as parameter of the model.

The only exception, where an interpretation of the absorption by helium was not possible, was the *EXOSAT spectrum* of Feige24. In this case Vennes et al. (1989) showed that the spectrum could be reproduced with a rather ad hoc mixture of heavy elements ("metal soup"). Vennes (1992) also demonstrated that similar mixtures of heavy elements could explain the observations of a few other hot DA.

4. Accumulating Evidence Against Helium

High resolution *IUE* spectra had since a long time shown the presence of weak metal lines (C IV, Si IV, N V, Fe V, Ni V) in some hot DA white dwarfs, which was also confirmed later by HST/FOS and HST/GHRS spectra (Bruhweiler & Kondo 1981,1983; Tweedy 1991; Sion et al. 1992; Vennes et al. 1992; Vidal-Madjar et al. 1994; Holberg et al. 1993; Werner and Dreizler 1994). It was not always easy to distinguish between interstellar, circumstellar, and photospheric lines, but in several cases a photospheric origin was clearly established. The real breakthrough, however, came only with EUV observations. A rocket observation of G191-B2B (Wilkinson et al. 1992) showed a number of metal features in the EUV, but no helium. And, finally, the analysis of a large number of DA white dwarfs observed with the Wide Field Camera on *ROSAT* (Barstow et al. 1993) demonstrated convincingly that for several objects the absorption in the EUV could not be caused by helium, neither in a homogeneous mixture nor in a stratified atmosphere. Another very significant result of this study was that all DA below a certain temperature threshold of about 40,000 K could be explained by pure hydrogen atmospheres, whereas almost all hotter objects showed additional absorption. This result was confirmed with larger samples and in much more detail by Jordan et al.(1994), Wolff et al. (1995a,b), and Finley (1995).

It thus became obvious that the next step in the development of white dwarf model atmospheres would have to be the inclusion of EUV bound-free and line absorption by heavy elements up to iron and nickel, typically in the ionization stages of III–VIII, a task which would have been next to impossible a few years ago.

5. LTE Model Atmospheres and Atomic Data

The situation concerning the availability of necessary atomic data has been greatly improved by the calculations of the Opacity Project (Seaton et al. 1992). Photoionization cross sections for all important elements up to iron are available in the TOPBASE database at the CDS. Unfortunately, at present the lowest ionization state for iron is III, and other iron peak elements (especially Ni) are completely missing, so that for these hydrogen-like estimates have to be used.

An invaluable database for the bound-bound transitions are the line lists compiled by Bob Kurucz over many years and distributed on CDROM (Kurucz 1991, 1992). They contain more than 40 million atomic lines, mostly of the iron group elements. The absorption by these lines is not only important for the calculation of a detailed synthetic spectrum, but also for the structure of the model atmosphere, that is the blanketing effect of all these lines has to be taken into account. The large amount of data, and the complexity of handling it poses challenges to the calculation of white dwarf atmosphere models, which we have not had to face in this field before. The two general methods discussed in the literature to solve this problem in the context of LTE models are Opacity Distribution Functions (ODF) and Opacity Sampling (OS). Rather than repeating this discussion here, I will present a short description of my own atmosphere codes. These codes, or models calculated with it, have been widely used by our own group in Kiel as well as by others (Finley, Barstow, Gemmo, O'Donoghue, Saffer, Kepler, Robinson, and others).

The calculations proceed in three steps

• Calculation of an opacity table for a given chemical composition. This is by far the most expensive calculation in the case of LTE atmospheres, and it is therefore economically reasonable to separate this step from the calculation of the model structure. The typical size of these tables is 600 temperature/pressure pairs and 35,000–75,000 wavelengths. This large number of wavelength values, which in the EUV corresponds to spacings of 0.005 to 0.015 Å, is necessary because we use Opacity Sampling. The code steps through the complete lists of photoionization cross sections (from the Opacity Project) and then through the complete Kurucz line lists, at each wavelength adding the contribution of the lines, as long as they are stronger than a predefined threshold. Calculation of such a table takes about 24 hrs on a typical workstation, and we have therefore developed a number of shortcuts, e.g., calculating tables for different mixtures from precalculated tables for individual elements, etc. The typical size of such a table is 30 MB.

• Calculation of the atmospheric structure. This uses the standard assumption of LTE atmospheres and includes convection, if the stratification is unstable. The numerical method is the Feautrier method with variable Eddington factors, and the only unusual feature is the large number of wavelength points (up to 75,000) necessary to model the blanketing effect as accurately as possible.

• The final step is the calculation of a detailed synthetic spectrum for a fixed model structure. In the case considered here the spectrum provided by the previous step is

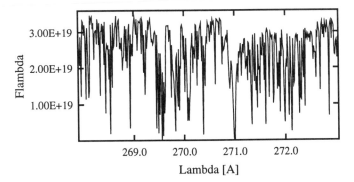

FIGURE 3. Arbitrary 5 Å section of a DA model with metals, $T_{\text{eff}} = 52{,}500$ K, $\log g = 7.5$, using about 5,000 bound-bound absorption edges and 9,500,000 atomic lines calculated individually with detailed profiles.

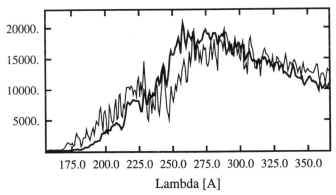

FIGURE 4. Comparison of the *EUVE* spectrum of G191-B2B (thick line) with an LTE model (thin line), using only abundances derived from UV observations as described in the text.

already detailed enough for a comparison with most observations, so the effect of this step is only to add a few more wavelength points, use more sophisticated broadening theories for the H and He lines, and change output formats.

Fig. 3 shows the result of such a model calculation for $T_{\text{eff}} = 52{,}500$ K and $\log g = 7.5$, which has included about 5,000 bound-free edges and 9.5 million lines.

The aim of this paper is not a spectral analysis of individual objects, so I will present only one comparison of models with the best studied of all white dwarfs, G191-B2B. If we take the metal abundances as determined from the UV (e.g. Vidal-Madjar et al. 1994), the interstellar column densities from the *HUT* observations (Kimble et al. 1993), and the solid angle from fitting the models at the V magnitude, we arrive at the comparison of Fig. 4. This is not a perfect fit, and not the final result, but considering the fact that the only free parameter we are left with is the effective temperature, the good agreement is very encouraging. The effective temperature for this model is 52,500 K, which is significantly lower than T_{eff} determined from optical spectra with pure H atmospheres (57,000–60,000 K). This result should not be a surprise: compared to a pure H atmosphere the energy flux in the EUV of G191-B2B is decreased by orders of magnitude, leading to a very significant backwarming effect in the visible. And indeed, our calculations show that the Balmer line spectrum for the metal-rich 52,500 K model in Fig.3 is practically indistinguishable from that of a pure H model at $T_{\rm eff} = 57,500$ K. In other words, the blanketing effect in this temperature range, and for the rather extreme metal abundances as in G191-B2B, can amount to 5,000 K or more.

6. NLTE Versus LTE Models for White Dwarfs

NLTE effects are known to be very important for the very hot PG1159 and similar objects above 100,000 K (e.g. Werner et al. 1991). Almost all EUV sources, however, are DA white dwarfs at much lower temperature, and for these NLTE effects had been considered to be unimportant, with the exception of the $H\alpha$ line core. This view has been challenged only recently with the necessity of considering metal absorbers in the EUV, because first NLTE calculations have predicted large effects on the ionization balance of the heavy elements.

As described in the previous section, even in LTE models the inclusion of so many metal lines is a challenging problem. To calculate occupation numbers and the blanketing effect consistently in NLTE is a formidable task, which would have been impossible just a few years ago. New, very powerful numerical techniques and ingenious ways of handling huge sets of atomic data had to be developed to make this possible. There are currently only two groups capable of providing such calculations for white dwarfs: Klaus Werner and coworkers in Kiel, and Ivan Hubeny and coworkers at NASA/Goddard. Both groups have very recently given detailed descriptions of their codes (Werner & Dreizler 1995; Hubeny & Lanz 1995). Although the implementations are different in the two codes, the two major ideas which have made these calculations feasible are common in both

• The use of "Approximate Λ-Operators". This idea goes back to Cannon (1973) and Scharmer (1981), and has been further developed by the Kiel group (Hamann 1985, Werner & Husfeld 1985, Werner 1986).

• The introduction of "superlevels" and "superlines" goes back to Anderson (1985). The enormous number of levels and transitions of the iron group elements is reduced by orders of magnitude, and thus made tractable, by grouping many individual levels into superlevels with combined energies, statistical weights, and transition probabilities.

Both groups have calculated grids of metal line blanketed NLTE white dwarf atmospheres and published first applications. Both groups also claim significant NLTE effects on the ionization balance of the metals, leading to substantial differences between NLTE and LTE model predictions for the EUV continuum flux as well as line strengths (see e.g., Dreizler & Werner 1993; Lanz & Hubeny 1995).

There is no doubt that the ultimate goal must be to apply the correct physical description, that is the NLTE calculations. Only with real NLTE calculations will it be possible to determine and justify the ranges of parameters, where the much simpler LTE calculations will be sufficient, and the groups taking on this immense task deserve our strongest encouragements. We are, however, still a long way from this goal; at present, in spite of all the improvements of the past few years, there are still many compromises that have to be made in NLTE calculations; there are also still many uncertainties in the atomic data, which may influence the final results. It is my personal view, that at the moment it is not at all clear, which compromises are more severe: neglecting NLTE effects, or the limited detail in the treatment of levels and transitions in the NLTE models. I. Hubeny, for example, showed at this meeting, that by including 300,000 atomic lines instead of 30,000 a model came much closer to the observations of G191-B2B (and also to the LTE model!).

The present generation of NLTE models has not yet been very successful in the interpretation of *EUVE* spectroscopic observations; in my experience the LTE models seem to be closer to the truth at the moment.

7. Outlook: What Is the Next Step?

Gravitational separation would lead to a very fast depletion of heavy elements in the atmospheres of hot DA—the timescales are of the order of a few years only. If they are still present, there must be forces at work counteracting the influence of gravitation, and the most likely effect is selective radiation pressure on ions, which have strong absorption lines in regions of high stellar radiation flux. Recent calculations of this effect have been performed for stellar *envelopes* (Chayer et al. 1994, 1995), leading to the prediction of a single value for an element abundance at $\tau_{Rosseland} = 2/3$. However, the conditions (e.g. ionization stage of elements) vary considerably within the stellar atmosphere, and we have to expect not a homogeneously mixed atmosphere, but rather one in which some elements accumulate in certain layers. This distribution of elements in turn will influence the distribution of the photons over wavelength at that depths, and the complete problems can only be solved in a self-consistent way, considering at the same time the equilibrium solution of the diffusion equations including radiation pressure *and* the condition of radiative equilibrium for the atmosphere.

Such calculations do not yet exist, but will ultimately be necessary—probably even for the NLTE case—if we really want to understand the EUV observations. Work in this direction is currently underway in Kiel, and probably at other institutions as well.

REFERENCES

ANDERSON, L. S. 1985, ApJ, 298, 848

AUER, L. H. & SHIPMAN, H. L. 1977, ApJ, 211, L103

BARSTOW, M. A., FLEMING, T. A., FINLEY, D. S. ET AL. 1993, MNRAS, 260, 631

BOWYER, S., LIEU, R., LAMPTON, M. ET AL. 1994, ApJS, 93, 569

BRUHWEILER, F. & KONDO, Y. 1981, ApJ, 248, L123

BRUHWEILER, F. & KONDO, Y. 1983, ApJ, 269, 657

CANNON, C. J. 1973, JQSRT, 13, 627

CHAYER, P., LEBLANC, F., FONTAINE, G., ET AL. 1994, ApJ, 436, L161

CHAYER, P., FONTAINE, G. & WESEMAEL, F. 1995, ApJS, in press

DREIZLER, S. & WERNER, K. 1993, A&A, 278, 199

FINLEY, D. S. 1995, these proceedings

HAMANN, W. -R. 1985, A&A, 148, 364

HOLBERG, J. B., BARSTOW, M. A., BUCKLEY, D. A. H., ET AL. 1993, ApJ, 416, 806

HUBENY, I. & LANZ, T. 1995, ApJ, 439, 875

JORDAN, S., KOESTER, D., WULF-MATHIES, C. & BRUNNER, H. 1987, A&A, 185, 253

JORDAN, S. & KOESTER, D. 1985, A&AS, 65, 367

JORDAN, S., WOLFF, B., KOESTER, D. & NAPIWOTZKI, R. 1994, A&A, 290, 834

KAHN, S. M., WESEMAEL, F., LIEBERT, J., ET AL. 1984, ApJ, 278, 255

KIMBLE, R. A., DAVIDSEN, A. F., LONG, K. S., & FELDMAN, P. D. 1993, ApJ, 408, L41

KOESTER, D. 1989, ApJ, 342, 999

KURUCZ, R. L. 1991, in Stellar Atmospheres: Beyond Classical Methods, ed. L. Crivellari, I. Hubeny & D. G. Hummer, NATO ASI Ser. 341, 441

KURUCZ, R. L. 1992, Rev. Mexicana Astron. Af., 23, 45

LAMPTON, M., MARGON, B., PARESCE, F. ET AL. 1976, ApJ, 203, L71

LANZ, T. & HUBENY, I. 1995, ApJ, 439, 905

NAPIWOTZKI, R., BARSTOW, M. A., FLEMING, T. ET AL. 1993, A&A, 278, 478

PAERELS, F. B. S. & HEISE, J. 1989, ApJ, 339, 1000

PETRE, R., SHIPMAN, H. L. & CANIZARES, C. R. 1986, ApJ, 304, 356

POUNDS, K. A., ALLAN, D. J., BARBER, C. ET AL. 1993, MNRAS, 260, 77

SCHARMER, G. 1981, ApJ, 249, 720

SEATON, M. J., ZEIPPEN, C. J., TULLY, J. A. ET AL. 1992, Rev. Mexicana Astron. Af., 23, 19

SION, E. M., BOHLIN, R. C., TWEEDY, R. W. & VAUCLAIR, G. 1992, ApJ, 391, L29

TWEEDY, R. W. 1991, Ph.D. Thesis, Univ. Leicester

VENNES, S. 1992, ApJ, 390, 590

VENNES, S., PELLETIER, C., FONTAINE, G. & WESEMAEL, F. 1988, ApJ, 331, 876

VENNES, S., CHAYER, P., FONTAINE, G. & WESEMAEL, F. 1989, ApJ, 336, L25

VENNES, S., CHAYER, P., THORSTENSEN, J. R. ET AL. 1992, ApJ, 392, L27

VIDAL-MADJAR, A., ALLARD, N. F., KOESTER, D. ET AL. 1994, A&A, 287, 175

WERNER, K., HEBER, U. & HUNGER, K. 1991, A&A, 244, 437

WERNER, K. 1986, A&A, 161, 177

WERNER, K. & HUSFELD, D. 1985, A&A, 148, 417

WERNER, K. & DREIZLER, S. 1994, A&A, 286, L31

WERNER, K. & DREIZLER, S. 1995, in Computational Astrophysics, II Stellar Physics, ed. R. P. Kudritzki, D. Mihalas, K. Nomoto, & F.-K. Thielemann, submitted

WILKINSON, E., GREEN, J. C. & CASH, W. 1992, ApJ, 397, L51

WOLFF, B., JORDAN, S., BADE, N. & REIMERS, D. 1995a, A&A, 294, 183

WOLFF, B., JORDAN, S. & KOESTER, D. 1995b, A&A, submitted

The Hot White Dwarf Population in the *EUVE* Survey

STÉPHANE VENNES

Center for EUV Astrophysics, 2150 Kittredge Street,
University of California at Berkeley, Berkeley, CA 94720-5030, USA

The processes leading to the formation of white dwarf stars are known only in their most general principles; post-asymptotic giant branch evolution, leading to the formation of C-O degenerate cores, is possibly the main formation channel of white dwarf stars. In contrast, observations of hot white dwarf stars and studies of their main population characteristics offer detailed insights into the origin and evolution of these objects. We examine some new facts uncovered in the study of the survey of hot white dwarf stars at extreme ultraviolet (EUV) wavelengths. We describe model atmosphere techniques required to interpret these observations and discuss some implications of our findings for stellar evolution theory.

1. The White Dwarf Population

The white dwarf population has been surveyed using various means and methodologies, often using proper motion surveys (e.g., Luyten catalogs) or color sensitive surveys (e.g., Palomar-Green & Montréal-Cambridge-Tololo catalogs). Bergeron, Saffer, & Liebert (1992; BSL) completed an analysis of the sample of hydrogen-rich white dwarfs compiled by McCook & Sion (1987) and established some of their properties; they noted in particular the extreme narrowness of the mass distribution. On the other hand, the temperature distribution, or alternatively the luminosity function, of hot white dwarf stars ($T_{\rm eff} \geq 25,000$ K) is not well established (see Fleming, Liebert, & Green 1986). The atmospheric composition of hot white dwarf stars has been the subject of a number of investigations and theoretical studies. Apart from the existence of two parallel channels, helium- and hydrogen-rich (see MacDonald & Vennes 1991), it has also been established that heavy elements are found in larger concentrations in the hottest white dwarfs (see Vennes & Fontaine 1992). Although diffusion in white dwarf atmospheric layers is extremely efficient, agreement between theory and observation has not yet been achieved (see Chayer et al. 1995).

The *Extreme Ultraviolet Explorer* (*EUVE*) survey of hot white dwarf stars is the most comprehensive collection of such objects (Bowyer et al. 1994). A variety of phenomena are represented in the sample and such occurrences are useful in characterizing the white dwarf population, its origin and destiny. I will describe in the following sections a comprehensive research program aimed at a complete description of the white dwarf stars detected at EUV wavelengths. To achieve this goal, we have obtained optical spectroscopy for most objects in the sample; these data are primarily used to describe the mass distribution and space density of the white dwarfs in the survey for comparison with evolutionary scenarios (§ 2). I also describe far ultraviolet (FUV) and EUV spectroscopy of representative objects. These data help determine the chemical structure of the atmospheric layers of hot white dwarf stars (§ 3). Additional EUV spectroscopic data are also presented and analyzed by Dupuis & Vennes (1995a,b). Many white dwarfs in the EUV sample are found in binaries, some in short-period pairs, and we briefly examine the properties of these systems and possible implications for evolutionary scenarios (§ 4). We summarize our findings in § 5.

S. Bowyer and R. F. Malina (eds.), Astrophysics in the Extreme Ultraviolet, 193–202.
© *1996 Kluwer Academic Publishers. Printed in the Netherlands.*

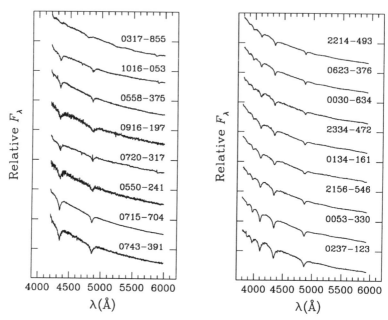

FIGURE 1. Optical spectra of southern-hemisphere white dwarfs obtained at CTIO (Vennes et al. 1995d). The white dwarfs are ordered, from top to bottom, with decreasing temperatures. Note the composite DAO+dM spectra of EUVE J0720-317 and EUVE J1016-053. The star EUVE J0317-855 is a new magnetic white dwarf (DAp).

2. Surface Gravity and Temperature

Over a hundred white dwarf stars have been detected in the *EUVE* all-sky survey. We have obtained follow-up optical spectroscopy for a sample subset at the Michigan-MIT-Dartmouth (MDM) observatory, the Lick observatory, and the Cerro-Tololo Inter-American (CTIO) observatory. I present in Figure 1 some remarkable spectra obtained at CTIO. The measured effective temperatures range approximately from 60×10^3 K (*top*) down to $30 - 40 \times 10^3$ K (*bottom*). In particular, both EUVE J0743-391 and EUVE J0916-197 have high surface gravities and masses in excess of $1 M_\odot$. We also noted the DAO+dM composite spectra of EUVE J0720-317 and EUVE J1016-053; further investigations revealed short orbital periods (Vennes & Thorstensen 1994; Thorstensen, Vennes, & Bowyer 1996). These objects are commonly viewed as recent survivors of common-envelope evolution (see de Kool & Ritter 1993). The peculiar spectrum of EUVE J0317-855 shows evidence of a large magnetic field. Other magnetic white dwarfs have been detected in the *EUVE* survey, most notably PG 1658+441 (=EUVE J1659+440), and population statistics can be derived from the complete sample.

We have determined the effective temperature and surface gravity for 57 white dwarfs from the *EUVE* sample by modeling Balmer line profiles (Vennes et al. 1995d). We used techniques similar to BSL. I present in Figure 2 (*top panel*) the surface gravity distribution of our sample and compare it to BSL's sample. Two main differences are apparent: the BSL distribution is significantly sharper than the *EUVE* distribution and a larger number of high-gravity stars are identified in the *EUVE* sample. The former is

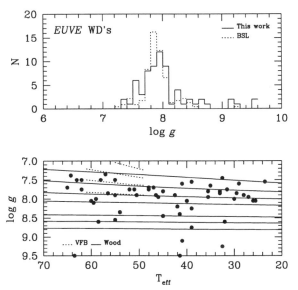

FIGURE 2. (*top panel*) Surface gravity distribution of a sample of ~ 60 hot DA white dwarfs discovered in the *EUVE* survey. The wide temperature range explored with *EUVE* explains the broader distribution relative to the "cooler" sample studied by Bergeron, Saffer, & Liebert (1992; BSL). (*lower panel*) Surface gravity as a function of effective temperature compared to theoretical mass radius relations at 0.4, 0.45, 0.5 and 0.6 $M_\odot$ (VFB; Vennes, Fontaine, & Brassard 1995), and 0.4, 0.5, 0.6, 0.7, 0.9, 1.0 and 1.1 $M_\odot$ (Wood; Wood 1995). Note the low-mass ($M \leq 0.4 M_\odot$) and high-mass ($M \geq 1.0 M_\odot$) white dwarfs identified in the survey.

easily explained by the relative homogeneity of the BSL sample which is composed of cooler objects, while the *EUVE* sample extends up to $\sim 65 \times 10^3$ K where a significant radius increase is predicted. The second difference is more fundamental and is one of the main surprises in the *EUVE* survey: several new high-mass white dwarfs have been discovered, four of them exceeding $1.2 M_\odot$. Such a large turn-out may offer support to a merger model, in which degenerate cores in a close binary merge into a single massive white dwarf. The measured temperatures and surface gravities are compared to theoretical mass-radius relations in the lower panel. The mass distribution, using Wood's mass-radius relations, is presented in Figure 3 and compared with BSL's distribution. The *EUVE* distribution is significantly broader which is attributed to a lesser accuracy in gravity determinations of extremely hot white dwarfs. To summarize, we find that the average masses of the *EUVE* and BSL samples are identical ($< M > \sim 0.55 M_\odot$). We find that eight objects in the *EUVE* sample exceed $1.0 M_\odot$ while two of them are below $0.4 M_\odot$ and are possibly the results of binary evolution.

3. Chemical Composition

The extreme diversity of white dwarf chemical compositions is already evident in the classification nomenclature: DA (by number, $X \sim 1$ and $Y, Z \leq 10^{-5}$), DO ($X \leq 0.5$, $Y \geq 0.5$), DB ($X \leq 10^{-5}$, $Y \sim 1$), etc. (see MacDonald & Vennes 1991). What became apparent only in the *EUVE* survey is the extreme diversity *within* the hydrogen-rich DA class itself. Following the *EUVE* photometric survey (completed in 1993 January, see

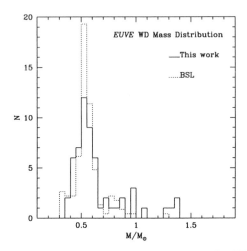

FIGURE 3. Mass distribution of hydrogen-rich white dwarfs in the *EUVE* survey. The *EUVE* distribution is compared to a renormalized BSL's distribution; several massive white dwarfs were discovered in the *EUVE* survey.

Finley 1995), a systematic program of EUV spectroscopic observations demonstrated for the first time the extent of abundance variations in white dwarf atmospheres. I illustrate this phenomenon in Figure 4 (*left panel*) using two extreme cases with metallicity ranging between upper limits of $Z \leq 10^{-8}$ (HZ 43) and actual measurements near $Z = 0.00002$ (MCT 0455-2812). Both objects are separated by less than 10^4 K (i.e., separated in age by less than 10^6 yrs) and yet show metallicity variations of a factor of more than a thousand. The distinction between helium- and hydrogen-rich spectra is illustrated by a comparison with the helium-rich star MCT 0501-2858 (Vennes et al. 1994).

The effect of metallicity on the measured energy distribution of hydrogen-rich stars is also predicted using model spectra (*right panel*). We computed model atmospheres in radiative equilibrium including detailed opacities of C, N, O, Si and Fe based on Opacity Project radiative cross-sections (see Pradhan 1995). The abundance ratios amount to a metallicity of $Z = 0.00002$ with iron accounting for 50% of the heavy element concentration. This high-metallicity model displays a steep downturn near 260 Å relative to the pure-hydrogen model. We interpret the helium-rich white dwarf MCT 0455-2858 with a model at $T_{\text{eff}} = 70 \times 10^3$ K, $\log g = 7$ and abundances of C/He= 0.8%, N/He= 0.01%, and O/He= 0.02%. Our knowledge of the true atmospheric composition of these objects (MCT 0455-2812 and MCT 0501-2858) and similar cases (e.g., G191 B2B) is, at best, fragmentary.

An exhaustive description of the atmospheric composition of a hot white dwarf is not available yet; high-dispersion FUV spectra currently offer the most detailed abundance estimates in hot white dwarf stars. The case of G191 B2B is quite exemplary: using *International Ultraviolet Explorer* and *ORFEUS* data (Fig. 5) we have assembled the most complete description of a hot white dwarf atmosphere yet. I summarize in Table 1 the abundances measured in G191 B2B. The emergence of Fe as the most abundant trace element in hot white dwarf atmospheres only reinforces the need for accurate atomic data for this element (e.g., Iron Project, see Pradhan 1995). Considerable discrepancy is observed between diffusion model predictions and abundance measurements. It is now

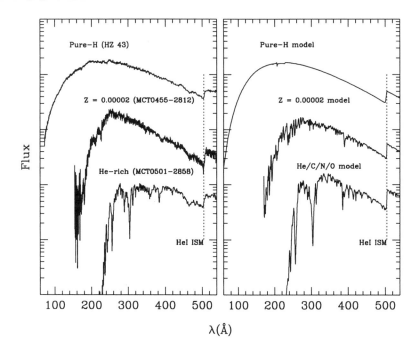

FIGURE 4. A comparison between actual *EUVE* spectroscopy (*left panel*) and model spectra using *Opacity Project* radiative cross-sections (see Pradhan 1995) (*right panel*). We show three spectra, each representative of a class of white dwarfs: HZ 43 for pure-hydrogen objects, MCT 0455-2812 for high-metallicity white dwarfs, and MCT 0501-2858 for helium-rich objects.

acknowledged that the observed abundance pattern may be dictated by yet unidentified fundamental mechanisms that operate in these atmospheres. Some suggest the presence of a weak mass loss (see Chayer et al. 1995).

4. Binaries

The white dwarf population in the EUV spectral range shows a substantial fraction of white dwarfs in binaries, some with very short orbital periods—evidence for past interaction—and most in wide pairs. Vennes & Thorstensen (1994, 1995) noted the presence in EUV all-sky surveys of a class of short-period binaries that emerge from a common envelope phase (de Kool & Ritter 1993). Detailed studies of BD+08°102 (K V+DA; Kellett et al. 1995), HR 1608 and HR 8210 (K0 IV+DA and A8+DA, respectively; Landsman, Simon, & Bergeron 1993; Wonnacott, Kellett, & Stickland 1993), HD 33959C (F+DA; Hodgkin et al. 1993), and β Crt (A+DA; Fleming et al. 1991) all demonstrate the importance of a multiwavelength approach to optical identification of EUV sources. Both β Crt and HD 33959C are candidate close binaries. The study of binaries with white dwarf companions may help constrain models of close-binary evolution as well as models of binary star formation; a well-defined sample of these objects may help resolve the issue of mass-correlation in binaries. We also examine interaction mechanisms between close binary components: in particular, the peculiar helium enrichment

TABLE 1. Photospheric composition of the hot DA white dwarf G191 B2B.

Element	Z/H	References
C	2×10^{-6}	Vennes et al. (1991)
N	3×10^{-6}	Vennes et al. (1991)
O	1×10^{-6}	Vennes et al. (1995a)
Si	3×10^{-7}	Vennes et al. (1991)
P	1×10^{-8}	Vennes et al. (1995a)
S	1×10^{-7}	Vennes et al. (1995a)
Fe	3×10^{-6}	Vennes et al. (1992), Holberg et al. (1994), Werner & Dreizler (1994)
Ni	1×10^{-6}	Holberg et al. (1994), Werner & Dreizler (1994)
He	$\leq 2 \times 10^{-4}$	Vennes et al. (1995a)
Cl	$\leq 3 \times 10^{-9}$	Vennes et al. (1995a)

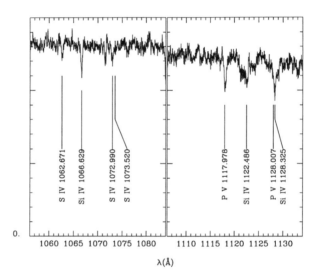

FIGURE 5. Silicon, sulfur and phosphorus in *ORFEUS* (see Hurwitz & Bowyer 1995) spectra of G191 B2B. Similar detections have been made in the atmosphere of MCT 0455-2812. The silicon abundance is consistent with previsous measurements obtained with *IUE* (Vennes, Thejll, & Shipman 1991). New abundances of $P/H = 1 \times 10^{-8}$ and $S/H = 1 \times 10^{-7}$ have been secured with the *ORFEUS* data (Vennes et al. 1995a).

of the white dwarf is possibly the spectroscopic signature of mass exchange between a mass-losing red dwarf and an accreting white dwarf.

As noted above, the optical spectrum of WD+MS binaries is often dominated by luminous main sequence stars. New ultraviolet (UV) observations of late-type stars detected in the *EUVE* all-sky survey revealed an unsuspected white dwarf companion to the K0 star HD18131 (Fig. 6). The *International Ultraviolet Explorer* (*IUE*) spectrum shows a composite of a white dwarf and a late-type star. The white dwarf dominates the emission below 2000 Å while the K0 star prevails at longer wavelengths. A model atmosphere analysis of the new ultraviolet spectrophotometry and of the EUV photometry reveals a

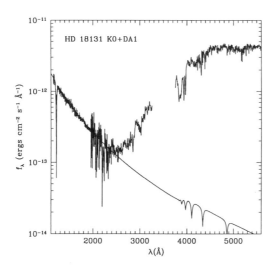

FIGURE 6. Ultraviolet and optical energy distribution of the K0 IV+DA1 binary HD 18131. The hot white dwarf binary is identified with the EUV source EUVE J0254-053 (Vennes et al. 1995c).

hot, hydrogen-rich (DA) white dwarf ($T_{\text{eff}} \sim 30,000$ K) that is the most likely source of the EUV emission (EUVE J0254-053). Optical spectroscopy revealed that the K0 star is a subgiant (K0 IV) and most likely constitutes with the white dwarf a wide pair at a distance of ~ 70 pc. This discovery has important implications for the EUV white dwarf population survey and, in particular, for the binary frequency.

The binary HD 18131 shares some properties with the binary HR 1608 (Landsman et al. 1993) with the distinction that HR 1608's K0 star has a measured orbital velocity of 5 km s^{-1}. Both components of these systems are evolved, i.e., a young white dwarf and a subgiant, and because the components in each system were presumably formed together, they share a similar evolutionary age, and therefore a similar initial mass. The sample of binaries comprised of one or two evolved components offers important insights to the formation and evolution of stars. In particular we find evidence that, in cases like HR 1608 and HD 18131, the mass distribution in binaries tends to favor components with similar mass. This mass correlation between components allows us, in turn, to identify cases that establish *initial mass–final mass* relations for white dwarf stars (Weidemann & Koester 1984). The case of HD 18131 requires further investigation, but we note that the adopted atmospheric parameters for the white dwarf are consistent with a mass of $0.4 - 0.6$ $M_{\odot}$ and that the progenitor was possibly a G star with a mass of ~ 1.3 $M_{\odot}$.

The hot white dwarf in the close binary EUVE J1016-053 has been classified a DAO white dwarf; this mixed H/He composition has been attributed to steady accretion from its close red dwarf companion (as in EUVE J2013+400; see Thorstensen, Vennes, & Shambrook 1994). Spectroscopic observations obtained with *EUVE* also show the effect of heavy element opacities on the white dwarf EUV energy distribution (Fig. 7). Spectral synthesis including trace opacities (see Pradhan 1995) of helium and a group of heavy elements (C, N, O, Si, S, Fe) in the otherwise hydrogen-rich atmosphere constrain their abundance to $Y/Y_{\odot} = Z/Z_{\odot} = 2 \times 10^{-3}$, in support of a simple accretion model. The low

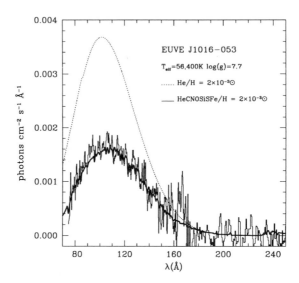

FIGURE 7. EUV spectrum of the white dwarf in the close binary EUVE J1016-053. The respective effects of helium and heavy element opacities are compared in two model simulations. The white dwarf is probably accreting from its companion at a low rate of $10^{-18} M_\odot \, \mathrm{yr}^{-1}$ (Vennes et al. 1995b).

surface-averaged helium abundance measured in the white dwarf atmosphere limits the accretion rate to $10^{-19} - 10^{-18} \, M_\odot \, \mathrm{yr}^{-1}$, i.e., much lower than the Bondi-Hoyle accretion rate which is of the same order as the red dwarf mass loss rate ($\geq 10^{-14} \, M_\odot \, \mathrm{yr}^{-1}$), therefore invalidating a simple wind-accretion model. We speculate that interaction with a weak white dwarf mass loss or a magnetosphere may inhibit accretion onto the white dwarf. Accretion of heavy elements may also be restricted to smaller areas possibly associated with magnetic poles.

EUV spectroscopy of the hot, hydrogen-rich white dwarf GD 50 reveals an unusual photospheric mixture of hydrogen and helium (Fig. 8). This hot DA white dwarf is also remarkable for its mass ($\approx 1.3 \, M_\odot$) near the Chandrasekhar limit. The spectra obtained with *EUVE* show a prominent He II resonance line series and constrain its atmospheric parameters to $T_{\mathrm{eff}} = 40,300 \pm 100$ K and $n(\mathrm{He})/n(\mathrm{H}) = 2.4 \pm 0.1 \times 10^{-4}$ (assuming $\log g = 9.0$). The optical photometry excludes the presence of a companion earlier than dM7-8. The presence of helium in an isolated, massive DA white dwarf is paradoxical; radiative levitation of helium is clearly undone by the high surface gravity; we exclude the possibility that helium is accreted from the ISM at the Bondi-Hoyle rate. Because accretion of ISM material onto GD 50 would constitute a unique occurrence and because we find no traces of heavier elements, we examine an alternative. If massive white dwarfs do result from binary evolution with a stellar merger, large orbital angular momentum may be transferred to rotation angular momentum which may induce large meridional circulation current, possibly dredging-up helium from the envelope.

5. Conclusions

The *EUVE* survey of the white dwarf population and the follow-up multi-wavelength spectroscopic observations, revealed unsuspected classes of objects and phenomena. Fol-

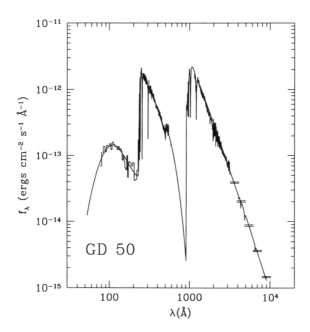

FIGURE 8. Complete energy distribution (*EUVE*, *IUE*, and Landolt's *UBVRI* photometry) of the massive DA white dwarf GD 50. Note the depression below 911 Å associated with neutral hydrogen absorption in the local ISM. The dramatic effect of the helium concentration (He/H= 2.4×10^{-4}) is evidenced by strong He II ground state transitions. We also constrain the local interstellar medium column densities of neutral helium and hydrogen to $n_{\mathrm{He\,I}} = 6 \times 10^{16}$ and $n_{\mathrm{H\,I}} = 9 \times 10^{17}$ cm^{-2}. (Vennes, Dupuis, & Bowyer 1995).

lowing optical spectroscopic observations we have discovered new massive ($M \geq 1.0 M_\odot$) white dwarfs, some of which, like GD 50, may be the outcome of binary evolution and stellar mergers. The class of EUV-emitting white dwarfs displays large abundance variations with a metallicity index ranging from $Z \leq 10^{-8}$ up to $Z = 0.00002$. Dupuis & Vennes (1995a,b) analyzed EUV spectroscopic observations of a large sample of hot DA white dwarfs and presented evidence of an abrupt decline in the metallicity index near a temperature of 50,000 K. This effect may be the spectroscopic signature of a weak mass loss. The *EUVE* sample of white dwarfs also displays the effects of binary evolution with the discovery of several new binaries and some post-common-envelope systems. Vennes & Thorstensen (1994, 1995) summarize the properties of these systems. We found evidence of interaction between components of close binaries: the EUV spectrum of the white dwarf in EUVE J1016-053 displays traces of heavy elements, possibly accreted from the red dwarf companion. This work is supported by NASA contract NAS5-30180 and NASA grant NAG5-2405.

REFERENCES

BERGERON, P., SAFFER, R., & LIEBERT, J. 1992, ApJ, 394, 228 BSL

BOWYER, S., LIEU, R., LAMPTON, M., LEWIS, J., WU, X., DRAKE, J. J., & MALINA, R. F.

1994, ApJS, 93, 569

CHAYER, P., VENNES, S., PRADHAN, A. K., THEJLL, P., BEAUCHAMP, A., FONTAINE, G., & WESEMAEL, F. 1995, these proceedings

DE, KOOL, M., & RITTER, H. 1993, A&A, 267, 397

DUPUIS, J., & VENNES, S. 1995a, in Lecture Notes in Physics 443, ed. D. Koester & K. Werner Berlin: Springer, 323

DUPUIS, J., & VENNES, S. 1995b, these proceedings

FINLEY, D. 1995, these proceedings

FLEMING, T. A., LIEBERT, J., & GREEN, R. F. 1986, ApJ, 308, 176

FLEMING, T. A., SCHMITT, J. H. M. M., BARSTOW, M. A., & MITTAZ, J. P. D. 1991, A&A, 246, L47

HODGKIN, S. T., BARSTOW, M. A., FLEMING, T. A., MONIER, R., & PYE, J. P. 1993, MNRAS, 263, 229

HOLBERG, J. B., HUBENY, I., BARSTOW, M. A., LANZ, T., SION, E. M., & TWEEDY, R. W. 1994, ApJ, 425, L105

HURWITZ, M., & BOWYER, S. 1995, these proceedings

KELLETT, B. J., ET AL. 1995, ApJ, 438, 364

LANDOLT, A. U. 1992, AJ, 104, 340

LANDSMAN, W., SIMON, T., & BERGERON, P. 1993, PASP, 105, 841

MACDONALD, J., & VENNES, S. 1991, ApJ, 371, 719

McCOOK, G. P., & SION, E. M. 1987, ApJS, 65, 603

PRADHAN, A. K. 1995, these proceedings

THORSTENSEN, J. R., VENNES, S., & BOWYER, S. 1996, ApJ, 457, in press

THORSTENSEN, J. R., VENNES, S., & SHAMBROOK, A. 1994, AJ, 108, 1924

VENNES, S., CHAYER, P., THORSTENSEN, J. R., BOWYER, S., & SHIPMAN, H. L. 1992, ApJ, 392, L27

VENNES, S., CHAYER, P., HURWITZ, M., & BOWYER, S. 1995a, ApJ, submitted

VENNES, S., DUPUIS, J., & BOWYER, S. 1995, ApJ, submitted

VENNES, S., DUPUIS, J., BOWYER, S., FONTAINE, G., WIERCIGROCH, A., JELINSKY, P., WESEMAEL, F., & MALINA, R. F. 1994, ApJ, 421, L35

VENNES, S., DUPUIS, J., BOWYER, S., & PRADHAN, A. K. 1995b, ApJ, submitted

VENNES, S., & FONTAINE, G. 1992, ApJ, 401, 288

VENNES, S., FONTAINE, G., & BRASSARD, P. 1995, A&A, 296, 117

VENNES, S., MATHIOUDAKIS, M., DOYLE, J. G., THORSTENSEN, J. R., & BYRNE, P. B. 1995c, A&A, 299, L29

VENNES, S., THEJLL, P., DUPUIS, J., GÉNOVA, R., THORSTENSEN, J. R., FONTAINE, G., WESEMAEL, F., & LAMONTAGNE, R. 1995d, ApJ, in preparation

VENNES, S., THEJLL, P., & SHIPMAN, H. L. 1991, in White Dwarfs, ed. G. Vauclair & E.M. Sion, Dordrecht: Kluwer, 235

VENNES, S., & THORSTENSEN, J. R. 1994, ApJ, 433, L29

VENNES, S., & THORSTENSEN, J. R. 1995, in Lecture Notes in Physics 443, ed. D. Koester & K. Werner, Berlin: Springer, 313

WEIDEMANN, V., KOESTER, D. 1984, A&A, 132, 195

WERNER, K., & DREIZLER, S. 1994, A&A, 286, L31

WONNACOTT, D., KELLETT, B. J., & STICKLAND, D. J. 1993, MNRAS, 262, 277

WOOD, M. A. 1995, in Lecture Notes in Physics 443, ed. D. Koester & K. Werner, Berlin: Springer, in press

The Composition and Structure of White Dwarf Atmospheres Revealed by Extreme Ultraviolet Spectroscopy

MARTIN A. BARSTOW,[1] IVAN HUBENY,[2] THIERRY LANZ,[2]
JAY B. HOLBERG,[3] AND EDWARD M. SION[4]

[1]Department of Physics and Astronomy, University of Leicester,
University Road, Leicester LE1 7RH, UK

[2] Universities Space Research Association NASA/GSFC, Greenbelt, MA 20711, USA

[3] Lunar and Planetary Laboratory, University of Arizona, Tucson, AZ 85721, USA

[4] Department of Astronomy and Astrophysics, Villanova University, Villanova, PA 19085, USA

The *ROSAT* and *EUVE* all-sky surveys have resulted in an important change in our understanding of the general composition of hydrogen-rich DA white dwarf atmospheres, with the photospheric opacity dominated by heavy elements rather than helium in the hottest stars ($T > 40,000$ K). Most stars cooler than 40,000 K have more or less pure H atmospheres. However, one question, which has not been resolved, concerned the specific nature of the heavy elements and the role of helium in the hottest white dwarfs. One view of white dwarf evolution requires that H-rich DA stars form by gravitational settling of He from either DAO or He-rich central stars of planetary nebulae. In this case, the youngest (hottest) DA white dwarfs may still contain visible traces of He. Spectroscopic observations now available with *EUVE* provide a crucial test of these ideas. Analysis of data from the *EUVE* Guest Observer programme and *EUVE* public archive allows quantitative consideration of the sources of EUV opacity and places limits on the abundance of He which may be present.

1. Introduction—The View from EUV Surveys

Among the most significant problems in the study of hot white dwarf evolution has been the existence of two distinct groups having either H or He dominated atmospheres and the possible relationships between them and their proposed progenitors, the diverse types of central stars of planetary nebulae (CPN). While the very hottest H-rich DA white dwarfs outnumber He-rich DOs by a factor 7 (Fleming, Liebert & Green 1986), the relative number of H- and He-rich CPN is only about 3:1. In addition, there is an apparent absence of He-rich stars in the temperature range 30,000–45,000 K, the so-called DB gap, suggesting that H- and He-dominated groups are not entirely distinct. Several competing processes can affect the composition of a white dwarf atmosphere. He and heavier elements tend to sink out of the photosphere under the influence of gravity but this can be counteracted by radiation pressure. Convective mixing, accretion or mass loss via a weak wind may also play a significant role.

The first soft X-ray and EUV studies of hot DA white dwarfs made the simplifying assumption that He was the sole photospheric opacity source, although the He in these stars was never identified spectroscopically. However, current evidence indicates that heavier elements have an important role to play. The *ROSAT* EUV/X-ray sky survey data demonstrated that the hottest stars ($T > 40,000$ K) must have heavy elements in their atmospheres to explain the observed fluxes, but stars below 40,000 K seem to have little opacity other than that provided by hydrogen (Barstow et al. 1993). For the hottest stars, it is clear that He alone is insufficient to account for all the opacity and even heavier elements must be involved. Unfortunately, none of the available data permit

S. Bowyer and R. F. Malina (eds.), Astrophysics in the Extreme Ultraviolet, 203–210.

us to decide whether or not He does play a role in DA atmospheres. Optical and *IUE* data typically only allow upper limits of a few $\times 10^{-4}$ to be placed on the He content and the broadband photometric *ROSAT* data do not allow separation of the possible opacity contributions from He and the heavier elements.

2. The Role of Helium in DA White Dwarfs

The availability of *EUVE* spectroscopy has provided us with an opportunity of resolving some of these issues, in particular searching for He with considerably greater sensitivity than has been possible in the past. *EUVE* allows spectra to be obtained throughout most of the EUV band, from $\approx$ 60–700 Å, but the crucial region is that of the HeII lyman absorption series between 304 Å and the limit at 228 Å.

From an analysis point of view, the white dwarfs studied (drawn from both GO programme and *EUVE* archive) divide into two distinct groups—those with and those without significant heavy element opacity. EUV spectra of the latter category can be described completely by a synthetic spectrum calculated for an H+He model atmosphere, together with a suitable model of the interstellar medium (Rumph, Bowyer, & Vennes 1994). If heavy elements are present, the stars can only be properly studied using complex line-blanketed models with a comprehensive selection of opacity sources. However, it is possible to study just the region of the spectrum near the HeII lines, reproducing the general shape of the spectrum with an H+He model plus an exponential roll-off towards short wavelengths. A complete description of our data reduction and analysis techniques has been published elsewhere (Barstow, Holberg, & Koester 1994a, 1994b, 1995).

During the first year of *EUVE* spectroscopic observations the signal-to-noise was limited to $\approx \pm 20\%$ by the detector fixed pattern efficiency variation. Subsequently, this problem has been largely diminished by implementation of the 'dither mode' to average out the fixed pattern and leave a residual efficiency variation of $\pm 5\%$. Figure 1 shows the complete *EUVE* spectrum of the EUV brightest DA white dwarf HZ43. Although it has a temperature above 40,000 K, the star is known to have a remarkably pure H atmosphere. It is well fitted by a pure H model, within the range of temperature and gravity determined by Napiwotzki et al. (1993) and the systematic uncertainties in the models and instrument calibration, although the best fit does require a finite He abundance of 1.3×10^{-6}. However, since the shape of the continuum flux is quite sensitive to possible systematic effects in the models and instrument calibration, we are cautious about interpreting this as a real detection of He in HZ43. Concentrating on the region spanned by the HeII Lyman series (Figure 2) allows a 90% upper limit of 3×10^{-7} to be placed on the He abundance (Barstow, Holberg, & Koester 1995).

Table 1 summarises the He abundance limits determined for several stars, with their effective temperatures, gravities and noting whether or not the observation was dithered. The value obtained for HZ43 is substantially lower than any of the others. However, more importantly it is well below the He abundance predicted by diffusion theory (see Vennes et al. 1988), which should be $\approx 3 \times 10^{-6}$ for a star with the temperature and mass of HZ43. This has important consequences for our understanding of the physical mechanisms that determine the photospheric composition of white dwarfs.

3. Heavy Elements in DA White Dwarfs

Recently, Marsh et al. (1995) have extended the analysis of *ROSAT* observations begun by Barstow et al. (1993) to a much larger sample of stars, including most of the new white dwarfs discovered in the survey. While the general conclusions that may be drawn

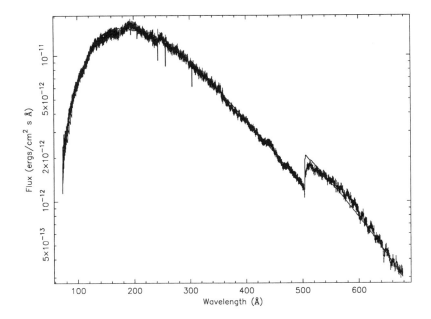

FIGURE 1. Complete *EUVE* spectrum of the DA white dwarf HZ43 after deconvolution with the instrument response (error bars). The model histogram corresponds to the best fit homogeneous H+He model with T=50,560 K, log g=7.9, and He/H= 1.3 × 10^{-6}.

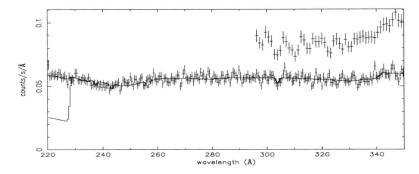

FIGURE 2. Count spectrum of HZ43 between 220 and 310 Å, showing a model with 3 × 10^{-7} of photospheric He and demonstrating the size of the interstellar 228 Å edge which would be present if the weak 304 Å "feature" was completely accounted for by an interstellar component.

Table 1. He abundance 90% upper limits in DA white dwarfs

Star	$T(\pm 1\ \sigma)$	$\log g\ (\pm 1\ \sigma)$	He/H	Dithered?
GD659	35,300(500)	8.00(0.20)	1.3×10^{-6}	Y
Feige 24	59,800(3,400)	7.50(0.51)	3.0×10^{-6}	N *
GD 71	32,045(148)	7.72(0.05)	1.1×10^{-6}	N
HZ43	49,000(2,000)	7.70(0.20)	3.0×10^{-7}	Y **
RE J2156−546	49,800(3,300)	7.75(0.18)	2.0×10^{-6}	N *
GD246	54,987(931)	7.71(0.09)	5.0×10^{-6}	N *

** This upper limit is *below* the level of the He abundance predicted by diffusion theory for a star of this T and $\log g$.
* Close to the predicted value.

Table 2. Limits on heavy element abundances in "pure H" DA white dwarfs

Element Abundance	HZ43		GD659	
	IUE high res	*EUVE*	*IUE* high res	*EUVE*
log C/H	−8.5	−5.5	−7.7	−4.6
log N/H	−5.0	−6.0	−4.2	−5.4
log O/H	−6.0	−6.5	—	−6.0
log Si/H	−8.0	−6.0	−9.0	−7.0
log Fe/H	−4.5	−4.5	—	—

remain unaltered, more detail is seen in the 40,000–60,000 K region. The increase in photospheric opacity is gradual in the 40,000–50,000 K range but becomes dramatic above 50,000 K. However, it is also apparent that the dispersion in opacities is considerable within a selected narrow temperature range and any theory which seeks to explain the composition of DA white dwarf atmospheres must be able to account for this. EUV spectroscopy can contribute to the solution of this problem through direct study of the sources of EUV opacity in those stars having significant heavy element abundances and measurement of upper limits in stars with near pure H envelopes.

As an example, upper limits to the abundances of C, N, O, Si, and Fe have been estimated from the *EUVE* spectra of HZ43 (T=49,000 K) and GD659 (T=35,300 K; see also Holberg et al. 1995). The strengths of EUV absorption lines were calculated for a range of abundances, using the TLUSTY and SYNSPEC non-LTE codes (Hubeny 1988; Hubeny & Lanz 1992, 1995), and compared with the data. Any lines with a depth approximately twice the amplitude of the signal-to-noise of the spectra should be readily detected and the upper limits are defined as the abundance at which the strongest features just reach this threshold. These results are summarised in Table 2 and compared with limits, derived in a similar way, for coadded *IUE* high dispersion spectra. While *IUE* provides the best constraints on C and Si, the EUV data are much more sensitive to N and O. At the effective temperatures in question, NIV and OIV are the most populated ionisation states and most of their bound-bound transitions are found in the EUV rather than the far-UV.

Currently, analysis of *EUVE* spectra of white dwarfs containing large quantities of

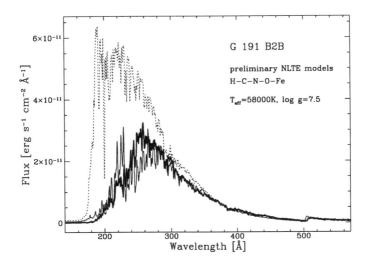

FIGURE 3. *EUVE* spectrum of G191–B2B (thick solid line) compared with two non-LTE line blanketed models for T=58,000 K, log g=7.5, C/H= 2×10^{-6}, N/H= 1.75×10^{-7}, O/H=1.0×10^{-6}, and Fe/H= 6.5×10^{-5}. Dashed line—6,000 lines; thin solid line—300,000 lines. Interstellar absorption is included with H I= 2.5×10^{18} cm^{-2}, He I/H I=0.09 and HeII/HI=0.01

heavy elements poses some difficulty. Some progress has been made for objects with less extreme abundances, such as PG1234+482 (see Jordan et al. these proceedings), using LTE models. However, with considerable blanketing from Fe group lines contributing to the overall opacity, non-LTE effects are expected to be important. Unfortunately, non-LTE models of stars like G191–B2B have been completely unable to match the observations whereas LTE models can yield a reasonable representation (see Koester, these proceedings). The problem seems to stem from the difficulties of dealing with a large enough number of heavy element lines in non-LTE. Recent state-of-the-art calculations (Lanz & Hubeny 1995) have typically used the Kurucz line list, considering only the 36,000 lines between observed levels with some 6,000 contributing to the EUV opacity in the range from 25 to 600 Å, but cannot match the steep fall in the spectrum below 250 Å (thin dashed line, Figure 3). Taking all the lines predicted by Kurucz for the higher levels of Fe IV, Fe V, and Fe VI increases the total number of lines to about 730,000. About 300,000 of these contribute significantly to the opacity and the new calculation yields a large improvement in the level of agreement between model (thin solid line) and data (thick solid line). Reworking of the codes to deal more easily with such large numbers of lines should allow progress to be made in this area in the near future.

4. The Composition of RE J0503−289

The hot helium-rich DO white dwarf RE J0503−289 lies in a region of extremely low interstellar hydrogen density (Barstow et al. 1994). As the only DO white dwarf detected

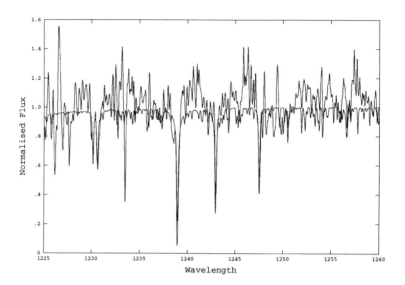

FIGURE 4. Region of the *IUE* high dispersion spectrum of RE J0503−289 showing N V, C IV and C III features compared to a 63,000 K model with C:He=0.1%, N:He=0.01%, O:He=0.01%, Si:He=0.01% and Ni:He=0.001%.

in both *ROSAT* WFC and *EUVE* sky surveys, it is a very important target for further study with the *EUVE* spectrometer. The star is also bright enough to be observed at high dispersion with *IUE* and we have acquired four spectra during the past two years. Detection of weak P Cygni profiles in one of these is evidence for the presence of an episodic weak wind in this object (Barstow & Sion 1994).

The *IUE* and *EUVE* data show that, in addition to the carbon already discovered in our earlier optical and low dispersion *IUE* observations, the atmosphere of the star contains significant quantities of nitrogen, oxygen and silicon. There is no sign of any Fe, found in DA white dwarfs of similar temperature, but many of the weaker features in the spectrum coincide with Ni V transitions. More detailed analysis of the photospheric composition of RE J0503−289 has proved difficult. It is possible to estimate photospheric abundances by comparing non-LTE model spectra with the measured UV line strengths. However, the EUV fluxes predicted by the same models which match the UV data are up to an order of magnitude greater than observed. In an attempt to reconcile the EUV and far-UV data, we constructed a grid of non-LTE spectra for a range of temperature and composition and fit both spectral ranges simultaneously. A consistent solution is achieved with a temperature of 63000 K (cf. 70000 from earlier work) and a C abundance of 0.1%, a factor 10 below that originally estimated. Abundances of N, O, and Si are all about 0.01% and that of Ni 0.001%, with respect to helium. Figure 4 shows a section of the *IUE* spectrum spanning the wavelength range from 1225 to 1260 Å, including N V, C IV and C III features compared to a non-LTE model computed for these abundances. The EUV model spectrum is now a very good match to the data (Figure 5), although the predicted He II line profiles are rather broader than observed. This probably arises from the absence of a satisfactory broadening theory for the He II Lyman series. We hope to address this problem with theorists working on the atomic data.

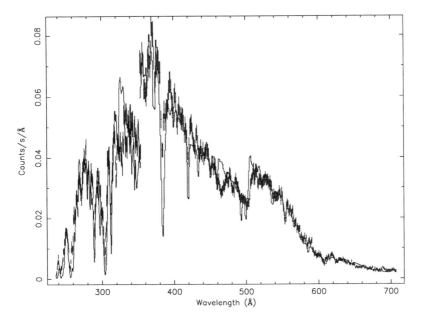

FIGURE 5. *EUVE* spectrum of RE J0503−289 compared to the best fit model discussed in the text.

5. Conclusion

It is clear that EUV spectroscopy is already making significant contributions to our understanding of white dwarf atmospheres. For example, important progress has been made in understanding the role of helium in DA stars. Limits placed on the He abundance of a number of typical DAs, spanning a temperature range from 30,000 to 60,000 K are 2 to 3 orders of magnitude below the values typical of DAO white dwarfs. Furthermore, the allowed abundances of He are below a level where the Balmer line temperature measurement becomes problematic. In those DA stars which have low levels of heavy element abundances in their atmospheres, the *EUVE* spectra can also be used to place limit on their abundances. The higher spectral resolution of the *IUE* echelle gives the best limits on carbon and silicon but the most populated ions of nitrogen and oxygen, N IV and O IV, have their strongest transitions in the EUV. Hence, *EUVE* data give more stringent limits on these elements.

Study of the very hottest DA white dwarfs such as G191−B2B, which contain significant quantities of heavy elements, has proved difficult because it has not been possible to calculate detailed model spectra which can match even the general shape of the spectral data. However, it seems that this is a result of not including sufficient numbers of detailed line transitions, mainly for iron ions, in the non-LTE calculations. We hope that imminent improvements in the computer programmes will solve this problem. In contrast, we have been able to make significant progress in understanding the atmosphere of the DO white dwarf RE J0503−289, which contains significant quantities of carbon, nitrogen, oxygen, and silicon but shows no evidence for iron.

MAB acknowledges the support of PPARC, UK. The work of IH and TL was supported in part by NASA grant NAGW-3834 and that of EMS by NASA LTSA grant NAGW-5716 to Villanova University.

REFERENCES

BARSTOW, M. A. ET AL. 1993, MNRAS, 264, 16

BARSTOW, M. A. ET AL. 1994, MNRAS, 267, 653

BARSTOW, M. A., HOLBERG, J. B. & KOESTER, D. 1994a, MNRAS, 268, L35

BARSTOW, M. A., HOLBERG, J. B. & KOESTER, D. 1994b, MNRAS, 270, 516

BARSTOW, M. A., HOLBERG, J. B. & KOESTER, D. 1995, MNRAS, in press

BARSTOW, M. A. & SION, E. M. 1994, MNRAS, 271, L52

BERGERON, P., ET AL. 1994, ApJ, 432, 305

FLEMING, T. A., LIEBERT, J. & GREEN, R. F. 1986, ApJ, 308, 176

HOLBERG, J. B. ET AL. 1995, ApJ, in press

HUBENY, I. 1988, Comp. Phys. Comm., 52, 103

HUBENY, I. & LANZ, T. 1992, A& A, 262, 501

HUBENY, I. & LANZ, T. 1995, ApJ, 439, 875

LANZ, T. & HUBENY, T. 1995, ApJ, 439, 905

MARSH, M. C. ET AL. 1995, in Proceedings of the 9th European Worshop on White Dwarfs, ed. D. Koester & K. Werner, Lecture Notes in Physics, Springer, in press

NAPIWOTZKI, R. 1993, A& A, 278, 478

RUMPH, T., BOWYER, S. & VENNES, S. 1994, AJ, 107, 2108

VENNES, S., PELLETIER, C., FONTAINE, G., & WESEMAEL, F. 1988, ApJ, 331, 876

Equilibrium Abundances of Heavy Elements Supported by Radiative Levitation in the Atmospheres of Hot DA White Dwarfs

P. CHAYER,[1] S. VENNES,[1] A. K. PRADHAN,[2] P. THEJLL,[3]
A. BEAUCHAMP,[4] G. FONTAINE,[4] AND F. WESEMAEL[4]

[1] Center for EUV Astrophysics, 2150 Kittredge St., University of California at Berkeley, Berkeley, CA 94720–5030, USA

[2] Department of Astronomy, Ohio State University, Columbus, OH 43210–1106, USA

[3] Niels Bohr Institute, Blegdamsvej 17, DK-2100, København Ø, Denmark

[4] Département de Physique, Université de Montréal, C.P. 6128, Succ. Centre-Ville, Montréal, Québec, Canada, H3C 3J7

We present revised estimates of the equilibrium abundances of heavy elements supported by radiative levitation in the atmospheres of hot DA white dwarfs. We emphasize, in particular, the role of trace pollutants that may be present in the background plasma, an effect which has been heretofore neglected. We take advantage of the availability of a table of detailed monochromatic opacities calculated for a plasma made of H containing small amounts of C, N, O, and Fe to illustrate how the equilibrium abundances of levitating elements react to the flux redistribution caused by the addition of these small traces of opaque material. We also consider two other improvements: a more sophisticated treatment of the momentum redistribution process and ion experiences following a photoexcitation, and use of an upgraded value for the line profile width associated with pressure broadening.

1. Introduction

The presence of pollutants in the high-gravity atmospheres of hot white dwarfs has received considerable attention in the last several years, thanks to important discoveries made in the UV, far ultraviolet (FUV), and extreme ultraviolet (EUV) spectral ranges. On the basis of such observations, quantitative abundance analyses have revealed, so far, the presence of C, N, O, Si, Fe, and Ni in a handful of bright hot white dwarfs. To understand better how radiative levitation could potentially account for these and other possible atmospheric contaminants, Chayer, Fontaine, & Wesemael (1995; hereafter referred to as CFW) recently presented the results of detailed calculations of radiative accelerations and equilibrium abundances for several elements in the envelopes of hot DA and DO/DB white dwarf models. This set of computations uses equilibrium surface abundances derived under the assumption of a strict balance between the radiative acceleration on a given element and the effective gravity at the Rosseland photosphere. This *equilibrium approach*, which ignores the potential influence of other competing mechanisms (e.g., stellar winds or accretion), constitutes but the first step toward understanding the observed patterns of trace heavy elements in hot white dwarfs.

CFW (see also Chayer et al. 1994) have pointed out that the most significant aspect of this equilibrium theory, which remains to be addressed in the white dwarf context, is the problem of flux blocking (or, more properly, flux redistribution) resulting from the simultaneous presence of traces of several different heavy elements in the atmospheres of hot stars. In order to assess the importance of this effect, we have carried out exploratory calculations in DA model atmospheres based on a *representative* mixture of C, N, O, and Fe that, together with the dominant element (H here), may provide a more realistic

211

S. Bowyer and R. F. Malina (eds.), Astrophysics in the Extreme Ultraviolet, 211–215.

background monochromatic opacity than that provided by H alone. We further upgrade our approach to equilibrium radiative levitation theory in white dwarfs by combining the model atmosphere description with an implementation of the momentum redistribution process recently discussed by Gonzalez et al. (1995). As mentioned by CFW (their § 4), these considerations are the last remaining major improvements that need to be tackled in the framework of equilibrium radiative levitation theory.

2. A Model Atmosphere Approach

2.1. Pure Hydrogen Models

The computation of the total radiative acceleration on a given trace element is still based on equation (1) of CFW, except that the background monochromatic Eddington flux, $H_\nu(b)$, is evaluated through the full solution of the radiative transfer problem in the atmospheric layers instead of relying on approximations only formally valid at large optical depths. This approach is similar to the one followed by Bergeron et al. (1988) in the context of hot B subdwarfs. We compute the equilibrium abundances of levitating elements at the Rosseland photosphere ($\tau_R = {}^2/_3$) of our model atmospheres, as these abundances are *representative* of the observable values. In order to provide a comparison point with the results of CFW as well as with the results of more complicated models below, we first consider the case of *pure* H atmosphere models.

The result is shown in Figure 1 where the expected equilibrium surface abundances of levitating elements are shown as functions of the effective temperature for pure H DA models which have $\log g = 7.5$ and $T_{\rm eff} = 52{,}000, 56{,}000, 60{,}000$, and $64{,}000$ K. The solid curves show the results of CFW, while the dashed curves correspond to the improved results obtained with pure hydrogen model atmospheres. With one particularly outstanding exception, the predicted abundances of the two different approaches generally agree well within 0.3 dex. The outstanding exception occurs when the physical conditions are such that a closed-shell electronic configuration dominates the ionization balance.

2.2. Hydrogen-Rich Models with Trace Heavy Elements

The representative chemical composition of the background plasma we consider consists of a H-dominated mixture with small traces (by number) of carbon (C/H $= 4 \times 10^{-7}$), nitrogen (N/H $= 5 \times 10^{-6}$), oxygen (O/H $= 10^{-6}$), and iron (Fe/H $= 10^{-5}$). The C and N abundances have been determined by Wesemael, Henry, & Shipman (1984), the O abundance by Chayer & Vennes (1995), and the Fe abundance by Vennes et al. (1992) and Werner & Dreizler (1994). These abundances are representative of the DA Feige 24.

The opacity tables were originally constructed to compute a small grid of model atmospheres suitable for the analysis of the hot DA star Feige 24 and its siblings (G191–B2B, RE 2214–492, and RE 0623–377), all of which have comparable atmospheric parameters. They are LTE models, and include a detailed treatment of blanketing effects through the use of some 30,000 frequency points. This choice was made on purpose in order to oversample the monochromatic opacity given in the tables. The models were computed by Vennes, Pradhan, & Thejll (1995). They will be referred to as the HZ models in what follows.

We compare, in Figure 1, the predicted atmospheric abundances of our twelve standard heavy elements as functions of the effective temperature of the HZ (long dashed curves) and H (dashed curves) atmosphere models. The expected equilibrium abundances of S, Ar, and Ca are larger, while those of Na, Mg, Al, and Fe are smaller in presence of a contaminated background plasma compared to the pure H case. Likewise, in both

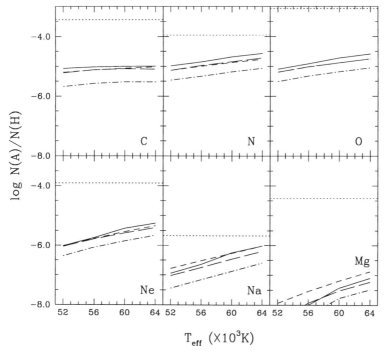

$$T_{eff} \ (\times 10^3 K)$$

FIGURE 1. Expected equilibrium abundances of heavy elements levitating at the surface of hot DA white dwarfs. The equilibrium abundances are given with respect to that of hydrogen by number. The models have $\log g = 7.5$. The solid curves show the results of CFW; the dashed curves show the results obtained with pure hydrogen model atmospheres; the long dashed curves correspond to the HZ models; the dot-dashed curves correspond to the HZ models but incorporate the improvements suggested by Gonzalez et al. (1995). The horizontal dotted line in each panel gives the cosmic abundance of the element of interest.

approaches, the abundances of Ne are comparable for the two kinds of background. For the levitating elements that do not belong to the background mixture (i.e., those other than C, N, O, and Fe), and except for possible accidental occurrences, the narrow absorption lines generally fall in the continuum of the energy distribution of an HZ model. For the needs of the present discussion, this "continuum" also includes the relatively broad hydrogen lines present in both the HZ and H models. Because the continuum flux of an HZ model is larger than that of a comparable H model for $\lambda \gtrsim 200$ Å, and the converse is true for shorter wavelengths, a heavy element will receive more (less) radiative support in an HZ model if its line opacity is more (less) important in the spectral range $\lambda \gtrsim 200$ Å than in the spectral region $\lambda \lesssim 200$ Å.

2.3. Other Improvements

In an important paper, Gonzalez et al. (1995) recently discussed a number of improvements in the computations of radiative accelerations in stellar envelopes. It turns out that, along with the questions considered in § 2.1 and § 2.2 above, the last significant problem addressed in the context of equilibrium radiative levitation theory in white dwarfs is how to describe the momentum redistribution process following the absorption

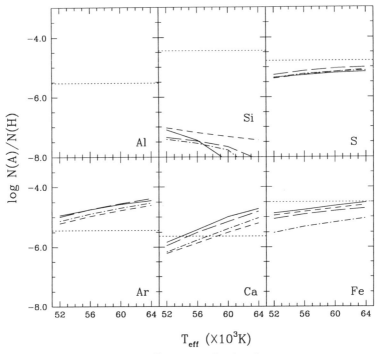

$$T_{eff} \ (\times 10^3 K)$$

FIGURE 1. *Continued*

of a photon by an ion. In the method used by CFW, the probabilities that the newly gained momentum of the excited ion is spent either in its initial ionization state i or in the higher state $i + 1$ are assigned to the ion as *a whole*. In contrast, in the improved discussion of Gonzalez et al. (1995), such probabilities are assigned to *each and every* individual radiative transition considered in the calculation of the radiative acceleration. Moreover, Gonzalez et al. (1995) show that all previous studies (including the work of CFW) have underestimated the effects of electronic collisions on the ionization rates. As a result, the improved treatment of Gonzalez et al. (1995) leads to a momentum redistribution process that tends to favor the higher ionization state $i + 1$. In contrast, most of the newly gained momentum is spent in state i in the method used by CFW. Because the higher ionization state $i + 1$ is less mobile than state i (its higher charge leads to a larger interaction with the background plasma), the prescription of Gonzalez et al. (1995) leads to *smaller* radiative accelerations than the method used by CFW and many others before.

To follow up on the work of CFW, we present the results of additional calculations that incorporate the detailed recipe proposed by Gonzalez et al. (1995). Along with this improved prescription, our calculations also incorporate an upgraded value for the line profile width associated with pressure broadening. This new line profile width is also suggested by the work of Gonzalez et al. (1995). This change leads to a relatively small but systematic *increase* of the radiative accelerations and equilibrium abundances. Figure 1 compares the predicted surface equilibrium abundances of our twelve selected heavy elements calculated according to the above procedures (solid, dashed, and long-dashed curves) to those calculated according to the improved procedure just described (dot-dashed curves). In the latter case, the Eddington flux used in the radiative accelerations calculations was taken from our grid of HZ model atmospheres. We observe that the improved estimates of the equilibrium abundances are generally systematically smaller than before, indicating that the momentum redistribution process of Gonzalez et al. (1995) is the dominant new mechanism. A few exceptions arise for Si, S, Ar, and Ca. Taking into account the contributing effects of increasing the line width (which leads to an increase of the equilibrium abundances), we find that the estimate of CFW, which suggests an overall decrease of the abundances by a factor ~ 3 because of the momentum redistribution process, is quite realistic.

3. Conclusion

In their concluding remarks, CFW suggest a number of avenues for improving the calculations of radiative accelerations in the context of equilibrium radiative levitation theory in hot white dwarfs. We have addressed these issues here, and the present effort can be considered to complement the work of CFW. Our "best" results are given by the dot-dashed curves in Figure 1. We again emphasize that these results can only be considered as *illustrative* because of the inconsistencies between the fixed chemical composition of the background plasma and the derived equilibrium abundances of the various heavy elements. They show, however, that the relative abundances of levitating elements can be significantly altered, particularly in the present examples, by the presence of an opaque element such as Fe.

This work is supported by NASA contract NAS5-30180.

REFERENCES

BERGERON, P., WESEMAEL, F., MICHAUD, G., & FONTAINE, G. 1988, ApJ, 332, 964

CHAYER, P., & VENNES, S. 1995, in preparation

CHAYER, P., FONTAINE, G., & WESEMAEL, F. 1995, ApJ, in press

CHAYER, P., LeBLANC, F., FONTAINE, G., WESEMAEL, F., MICHAUD, G., & VENNES, S. 1994, ApJ, 436, L161

GONZALEZ, J. -F., LeBLANC, F., ARTRU, M. -C., MICHAUD, G. 1995, A&A, in press

VENNES, S., PRADHAN, A. K., THEJLL, P. 1995, in preparation

VENNES, S., CHAYER, P., THORSTENSEN, J. R., BOWYER, S., & SHIPMAN, H. L. 1992, ApJ, 392, L27

WERNER, K., & DREIZLER, S. 1994, A&A, 286, L31

WESEMAEL, F., HENRY, R. B. C., & SHIPMAN, H. L. 1984, ApJ, 287, 868

A Spectroscopic Survey in the EUV of the "Coolest" Hot DA Stars

JEAN DUPUIS AND STÉPHANE VENNES

Center for EUV Astrophysics, University of California,
2150 Kittredge Street, Berkeley, CA 94720-5030, USA

We present an analysis of the extreme ultraviolet (EUV) spectroscopy of a sample of 10 DA white dwarfs observed by the *Extreme Ultraviolet Explorer* (*EUVE*). We have selected white dwarfs cooler than about 50,000 K and with presumably low heavy element abundances. The goal of this study is to determine the fundamental atmospheric parameters, namely the effective temperature and chemical composition, of these stars by fitting their continua with synthetic spectra computed from pure hydrogen LTE/line-blanketed model atmospheres. The question of the presence (or absence) of trace elements is explored by comparing EUV-determined effective temperatures to the one obtained from a fit of hydrogen balmer lines. It is found that the majority of the DA in the sample are consistent with having a pure hydrogen atmosphere. One of the star, MCT0027-634, is another possible example of a HZ 43-type white dwarf, having an effective temperature above 50000 K and a low heavy element abundance, i.e., much lower than predicted by diffusion theory.

1. Introduction

The spectra of white dwarfs in the extreme ultraviolet (EUV) are generally more complex than their optical and ultraviolet spectra traditionally used in the determination of their fundamental atmospheric parameters. EUV spectra of hot DA ($T_{\text{eff}} \geq 50000$ K) often show evidences for the presence of heavy elements as indicated by strong absorption features and a notable flux depression at wavelengths shorter than about 200 Å. It was initially recognized by Vauclair, Vauclair, & Greenstein (1979) that selective radiative pressure could defeat gravitational settling in hot white dwarfs atmospheres and they predicted that small abundances of C, N, and O should be present. In general, the metal rich spectra cannot not explained by simple mixtures of He, C, N, O, and Si, but other trace elements such as Fe are likely to be dominant source of photospheric opacity in the EUV (Vennes et al. 1989, 1991). Chayer, Fontaine, & Wesemael (1995), in their comprehensive study of radiative pressure in hot white dwarfs, are indeed predicting substantial abundances for many more heavy elements. The interpretation of hot DA spectra is further complicated by continuum absorption by hydrogen and helium in the interstellar medium. However, it is possible to separate the effect of interstellar absorption when fitting the EUV white dwarfs spectra. In fact, it turns out that white dwarfs EUV continua are excellent background sources against which to measure column densities of H I, He I, and He II in the local interstellar medium (see Bowyer 1995) permitting then to gain information on its morphology and ionization state (Vennes et al. 1993, 1994; Dupuis et al. 1995).

We have selected a sample of 10 DA with effective temperatures ranging from 25000 K to 50000 K and therefore more likely to have lower metal abundances according to Chayer, Fontaine, & Wesemael (1995). Even then, the predicted abundances are in some case sufficiently high to be detected in *EUVE* spectra. It is an important test of the diffusion to search for trace of heavy elements in the somewhat cooler white dwarfs considered here. There is indeed a class of hot DA characterized by a low metallicity with HZ 43

S. Bowyer and R. F. Malina (eds.), Astrophysics in the Extreme Ultraviolet, 217–222.

being a prototypical case. Until now, HZ 43 has been a rather unique object, one of our objective is to determine how unique it is. Most of the stars we have selected are cooler than HZ 43 but are hot enough for theory to predict detectable heavy element abundance in their EUV spectra. Instead of trying to identify individual spectral features from heavy elements, we consider the effect of the metallic opacity on the shape of the continua in the EUV. Metals cause deviations from the pure hydrogen ideal case and attempt to fit the EUV spectra will result in effective temperature and gravity that disagree with optical results. The effect is that a consistent fit is obtained for a cooler temperature. We have therefore determined effective temperature obtained by fitting pure hydrogen spectra for all the stars in the sample and made a comparison with temperature based on optical spectroscopy (balmer lines).

2. Analysis

The group of stars analyzed in this paper were either observed as part of our own guest observers program with *EUVE* or obtained from the *EUVE* public archive. All observations were reprocessed with the most recent calibration data and the spectra were re-extracted, corrected for overlapping spectral orders, and converted to flux units using effective areas in the first order only. The spectra were binned by 4 pixels or more to improve the signal-to-noise ratio.

Our analysis is based on fitting pure hydrogen model, which is equivalent to assuming no contribution from heavy elements to the photospheric opacity. The synthetic spectra are computed from pure hydrogen LTE-blanketed white dwarf models described in Vennes (1992). In this paper, we have not attempted to fit the gravity, instead values obtained from fitting optical spectra were used. In several cases, determinations are available in the literature as indicated in Table 1. In other cases, we have made determinations based on our database of optical spectra (Vennes et al. 1995). Therefore, we are left with 4 parameters, the effective temperature and the H I, He I, and He II interstellar column densities, to be determined using *EUVE* spectroscopy. Fortunately, in many cases, the He I column density is directly measurable from either the 504 Å photoionization edge or the 206 Å auto-ionization transition (Vennes et al. 1993; Rumph, Bowyer, & Vennes 1994). Similarly, the He II column density can be determined from the photoionization edge at 228 Å. In the few cases for which that cannot be done, we set He I/H I to 0.07 and He II/H I to 0.03. The effective temperature and the H I column density is determined by performing a χ^2 minimization. The model spectra are normalized to the flux at 5500 Å (V magnitude) and convolved with the spectrometer response. The best fit effective temperatures are given in Table 1 along with the gravity and the effective temperatures obtained from optical spectroscopy. Table 2 is listing the interstellar column densities we have determined. Figure 1 shows a mosaic of the spectra for the stars in the sample along with the best fit models.

3. Results

Deviations between optical and EUV spectroscopy determinations of the effective temperatures are good indicators of additional heavy element opacities in atmospheres of DA white dwarfs. We find, in Table 1, that for most of the stars in the sample, the two effective temperature determinations agree within the assigned uncertainties meaning these stars are consistent with having pure hydrogen photospheres. Several stars (EUVE0715-704, EUVE1032+534, EUVE2009-604, and EUVE2156-546) have effective temperature in excess of 40000 K and appear not to be contaminated by a significant heavy element

TABLE 1. Effective temperatures and gravities

EUVE J	Name	$T_{\rm eff}$ (EUV) (K)	$T_{\rm eff}$ (Optical) (K)	$\log g$ (cm s^{-2})
0029−634	MCT	49000	55000	7.90
0053−330	GD 659	34550	35500	7.95
0552+158	GD 71	32560	33000	7.85
0715−704	EUVE,RE	42358	42500	7.90
1032+534	EUVE,RE	44200	45000	7.80
1257+220	GD 153	39125	39000	7.55
1316+290	HZ 43	51015	49000	7.70
1623−392	CD−38°10980	24387	25000	8.10
2009−604	EUVE,RE	42100	...	8.00
2156−546	MCT	44320	45753	7.98

TABLE 2. Interstellar column densities of hydrogen and helium

EUVE J	$N_{\rm H\,I}$ (10^{18} cm^{-2})	$N_{\rm He\,I}$ (10^{17} cm^{-2})	$N_{\rm He\,II}$ (10^{17} cm^{-2})
0029−634	20.0	14.0	6.0[†]
0053−330	2.9	2.6	1.9
0552+158	0.56	0.52	≤ 5.0
0715−704	21.0	15.0	6.4†
1032+534	5.8	4.0	1.7†
1257+220	0.80	.66	≤ 0.30
1316+290	0.82	0.55	≤ 0.50
1623−392	3.2	2.2	0.96†
2009−604	18.0	12.0	3.2
2156−546	9.6	2.3	4.1

[†] He I/H I and He II/H I assumed equal to 0.07 and 0.03

abundance. Previous analysis of EUV/soft X-ray photometry of hot DA indicated that below 35000–40000 K, no EUV flux deficiency is observed (Vennes & Fontaine 1992; Barstow et al. 1993; Finley 1995; Wolff et al. 1995). In addition to HZ 43, we find several other hot white dwarfs that, contrary to diffusion theory predictions, have very low metallicity in spite of their high temperature and normal gravity. As pointed out by Chayer, Fontaine, and Wesemael (1995), this probably indicates that other physical processes, such as weak mass loss, are determining the heavy elements pattern in the atmospheres of hot white dwarfs. It might also be that the metallicity of hot white dwarfs photospheres depend on whether or not the atmosphere of the white dwarfs progenitors were already depleted of heavy elements.

The remaining stars in the sample are cooler than 40000 K (GD 659, GD 71, GD 153, and CD-38) and it less a surprise that their atmospheres are of pure hydrogen composition. However, the excellent agreement obtained between EUV effective temperature determinations and optical determinations is remarkable. The EUV continuum is very sensitive to the effective temperature and gravity, especially at short wavelengths where

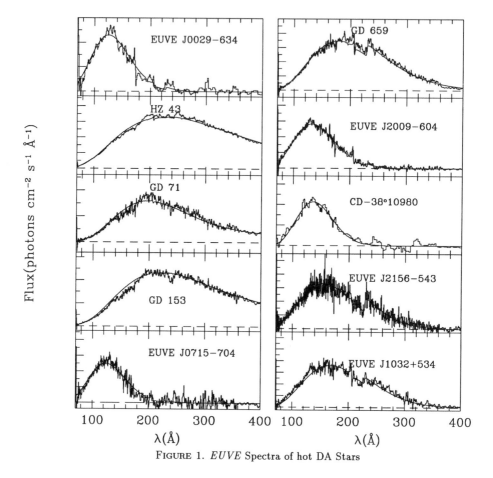

FIGURE 1. *EUVE* Spectra of hot DA Stars

the effect of interstellar absorption is less important. This explain the relatively small uncertainties of our temperature determinations. EUV spectroscopy therefore provides an independent determination of the temperature scale for hot DA stars.

EUVE 0029−634 is the only star for which its EUV effective temperature is significantly cooler than its optical temperature by about 5000 K. It is by far the hottest star in our sample with an effective temperature approximatively equal to 55000 K. At such a high temperature, radiative pressure is certainly able to support relatively high metal abundances. In this case, we infer that their must be a certain level of contamination by heavy elements, but not as high as in hot DA like Feige 24 and G191-B2B. In fact, if it would have been the case, considering EUVE 0029−634 has a much higher column density of interstellar hydrogen, it would not have been detected by *EUVE*. Therefore, this object also shares some resemblance with HZ 43.

HZ 43 is, as expected, well reproduced by a pure hydrogen model. In addition to the usual interstellar features (possibly the 228 Å photoionization edge of He II), the HZ 43 spectrum show several absorption features that could either possibly due to a low level of metal abundances or detector features. Dupuis, Vennes, & Pradhan (1995) presented a preliminary analysis showing that these features could not be attributed to C, N, or O.

Column densities of H I, He I, and He II are derived for all the stars in the sample. It is very interesting that several white dwarfs exhibit the interstellar 228 Å photoionization edge as first observed by *EUVE* in the spectrum of GD 246 (Vennes et al. 1993). These measurements provide estimates of the ionization fractions of hydrogen and helium in the nearby interstellar medium. At this point, we can rule out high ionization fraction of hydrogen in most of the directions. These stars also cover a wide range of column densities, and unlike the low columns stars like HZ 43 (Dupuis et al. 1995), they are probing other clouds in addition to the so-called local cloud. It is not quite clear whether these ionization fractions are representative of the physical conditions of the local interstellar medium or of clumps of partially ionized matter distributed along the line of sight.

4. Conclusions

We have determined the effective temperatures and the interstellar column densities of H I, He I, and He II for 10 hot DA stars with effective temperatures ranging from 25000 K to 55000 K. By comparing the effective temperatures derived from a fit of pure hydrogen model to the extreme-ultraviolet spectra with temperatures derived from optical spectroscopy, we conclude that most of these stars are consistent with having pure hydrogen photospheres. A fraction of these stars have temperatures in excess of 40000 K and in which detectable level of contamination by heavy elements are expected theoretically. The lack of heavy elements in these stars stress the need to consider competing physical processes to selective radiative pressure. Several of these stars are useful probes of the local interstellar medium since they provide a direct measurements of He I and He II column densities.

We thank Dr. Pierre Chayer for interesting discussions about diffusion theory in hot white dwarfs. We thank S. Bowyer for his advice. This research is supported by NASA contract NAS5-30180 and NASA grant NAG5-2405.

REFERENCES

Barstow, M., Fleming, T. A., Diamond, C. J. et al. 1993, MNRAS, 264, 16

Chayer, P., Fontaine, G., & Wesemael, F. 1995, ApJ, in press

Dupuis, J., Vennes, S., & Pradhan, A. K. 1995, BAAS, Late Abstract of January 1995 meeting

Dupuis, J., Vennes, S., Bowyer, S., Pradhan, A. K., & Thejll, P. 1995, ApJ, submitted

Finley, D. S. 1995, these proceedings

Rumph, T., Bowyer, S., & Vennes, S. 1994, AJ, 107, 2108

Vauclair, G., Vauclair, S., & Greenstein, J. L. 1979, A&A, 80, 79

Vennes, S., Chayer, P., Fontaine, G., & Wesemael, F. 1989, ApJ, 336, L25

Vennes, S., Chayer, P., Thorstensen, J. R., Bowyer, S., Shipman, H. L. 1992, ApJ, 392, L27

Vennes, S. 1992, ApJ, 390, 590

Vennes, S. & Fontaine, G. 1992, ApJ, 401, 288

Vennes, S., Dupuis, J., Rumph, T., Drake, J. J., Bowyer, S., Chayer, P., & Fontaine, G. 1993, ApJ, 410, L119

Vennes, S., Dupuis, J., Bowyer, S., Fontaine, G., Wiercigroch, A., Jelinsky, P. Wesemael, F., & Malina, R. F. 1994, ApJ, 421, L35

Vennes, S., Thejll, P., Dupuis, J., et al. 1995, in preparation

Wolff, B., Jordan, S., Bade, N., & Reimers, D. 1995, A&A, 294, 183

The Metallicity of Hot DA White Dwarfs as Inferred from EUV Photometry

DAVID S. FINLEY

Center for EUV Astrophysics, University of California,
2150 Kittredge Street, Berkeley, CA 94720-5030

Photometric EUV observations have shown that the hotter DA white dwarfs tend to show significant excess opacity relative to hydrogen. EUV and high-resolution FUV spectroscopy have conclusively demonstrated that the excess opacity is due to the presence of trace heavy elements in the white dwarf photospheres. In the past, the general abundance distribution in hot DA has been studied as a function of temperature, assuming that He was the trace absorber. We present here the first determination of the variation of relative total heavy element abundances in DA as a function of both temperature and gravity using realistic models that include metals. We compare the observational results with theoretical calculations of equilibrium abundances due to radiative acceleration.

1. Introduction

The DA spectral class is defined by the lack of any observable spectral features in the optical due to elements other than hydrogen. However, FUV, EUV and soft X-ray observations have now made it abundantly clear that the hotter DA can have significant amounts of trace heavy elements in their photospheres. (See Koester (1995), Jordan et al. (1995), and Vennes (1995) and citations therein for more details regarding trace element abundances in DA white dwarfs). Detailed studies of the abundances of individual trace elements in hot DA white dwarfs are currently being pursued using EUV and FUV spectroscopy. Suitable data, though, are available for only a modest number of stars. Therefore, the overall distribution of trace element abundances in DA can best be determined using the full set of EUV photometric observations, which includes over 100 white dwarfs.

2. Analysis Method

EUV fluxes of white dwarfs are determined by the effective temperatures, abundances, and ISM columns. In order to derive abundances and columns from the photometric EUV data, it is therefore necessary to constrain effective temperatures by other means. This has been accomplished for nearly all the DA white dwarfs that have been detected in the EUV via detailed fitting of the Balmer line profiles obtained as part of a recent optical spectroscopic survey of hot DA white dwarfs (Finley, Koester, & Basri (1995)). The model spectra used in this analysis were calculated using D. Koester's model atmosphere code (Koester, Schulz, & Weidemann (1979), Koester (1995)), including only bound-free opacities of C through Ni. (The additional complexity involved in including millions of lines was not justifiable for the analysis of photometric data). Cross sections were taken from the Opacity Project database where available (Seaton et al. 1992), otherwise hydrogenic cross sections were calculated. He abundances were set at a negligible value given the non-detection of photospheric He in any EUV spectra of DA other than the extremely massive white dwarf GD 50 (Vennes 1995), which appears to have only H and He in its photosphere. The other hot DA that are not pure H that have been observed spectroscopically in the EUV (11 of which are known to the author) are clearly dominated

S. Bowyer and R. F. Malina (eds.), Astrophysics in the Extreme Ultraviolet, 223–228.
© 1996 Kluwer Academic Publishers. Printed in the Netherlands.

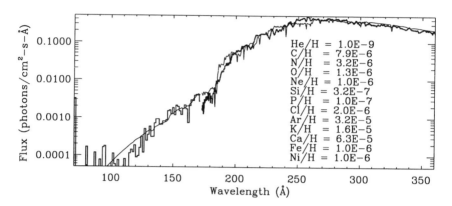

FIGURE 1. Model fit to the observed EUV spectrum of G191-B2B, including only bound-free heavy element opacities, for $T_{eff} = 55,000$ K, $\log g = 7.5$, with abundances as listed. Model is thin curve, observed spectrum is thick histogram.

by metal opacity. The only other DA that are likely to be dominated by He opacity in the EUV are the two EUV-detected DAO, RE1016-052 and RE2013+400, and they have been excluded from this analysis.

Given the goal of obtaining the variation of overall metal abundances as a function of T_{eff} and $\log g$, absolute abundances of individual elements are unimportant if the models used in the analysis accurately reproduce the observed EUV fluxes. The latter criterion was explicitly satisfied by determining an initial set of trace element abundances that matched the *EUVE* spectrum of G191-B2B (see Fig. 1). Models of different "metallicity" were calculated by varying all heavy element abundances relative to H by the same factor. As seen in Fig. 1, the match at high metallicities was quite good. At sufficiently low abundances (between -2.5 and -3 dex relative to the G191-B2B values) the models became indistinguishable from pure H. The models also worked well for the intermediate metallicity DA GD 246. Independent fits were made to the EUV photometry and to the *EUVE* spectra for that object. Identical abundances were obtained, requiring a difference in T_{eff} of only 2,000 K and a difference in the HI column of only 0.2 dex.

Relative abundances and ISM columns were obtained by comparing the observed count rates with theoretical count rates for those DA for which optical temperatures and gravities were available from the Finley, Koester, & Basri (1995) sample. (T_{eff} and $\log g$ for WD 2331-375 (MCT2331-4731) were kindly provided by S. Vennes). Theoretical fluxes were scaled using published V magnitudes or V magnitudes derived from the optical spectra; the latter were generally accurate to about 0.1 magnitude. Interstellar absorption was calculated assuming an HI/HeI ratio of 10:1, using the cross-sections of Rumph, Bowyer, & Vennes (1994). Observed count rates were taken from the *EUVE* catalogs (Bowyer et al. (1994), Malina et al. (1994)), the *ROSAT* Wide Field Camera catalog (Pounds et al. (1993)), and pointed *EUVE* observations. The error determinations included the addition of an assumed 15% calibration error in quadrature with the counting statistics errors.

Some typical confidence contours obtained for representative objects are shown in Fig. 2. WD 2111+498 (GD394) and WD 2211-495 (RE2214-491) were detected in multiple EUV bandpasses. ISM opacity reduces the flux more strongly toward longer wave-

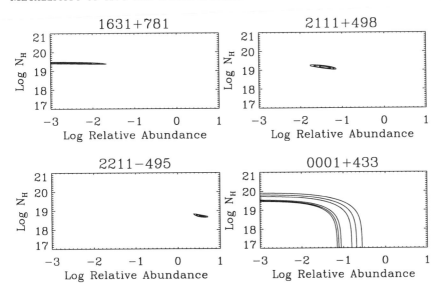

FIGURE 2. Fits for some representative objects, labeled at top with WD numbers. Contours plotted are 90%, 99%, and 99.9% confidence intervals.

lengths, while trace metals preferentially attenuate the flux toward shorter wavelengths; hence unique solutions requiring significant trace element abundances were obtainable in both these cases. WD 1631+781 (RE1629+780) also has well-determined fluxes in multiple EUV bandpasses, but is consistent with a pure H atmosphere. WD 0001+433 (RE0003+433) was detected in only one EUV bandpass, thereby precluding a unique solution for both the trace element abundances and the ISM column.

3. Results

The abundance determinations for the DA hotter than 33,000 K and with $\log g < 8.3$ are displayed in Fig. 3 as a function of temperature and gravity. Both the measurements and the 90% confidence upper limits such as that obtained for 1631+781 are included. However, limits for objects similar to WD 0001+433, for which both the columns and abundances were effectively unconstrained, were excluded. The figure shows that abundances increase strongly as gravity decreases and increase less strongly with T_{eff}, as is expected for radiatively supported heavy elements. The absence of detectable trace elements at lower temperatures and higher gravities is strengthened by additional EUV spectroscopic results presented in this volume by Dupuis & Vennes (1995). In addition to the DA shown in Fig. 3 as being spectroscopically pure H (GD 153, (PG)1057+719, HZ43, GD 2, WD 1029+537 (RE1032+532)) based on our analyses of *EUVE* spectra, Dupuis and Vennes present spectroscopic confirmation of the photometric determination that trace metals are also absent in GD 659, WD 0715-703 (RE0715-702), and WD 2152-548 (RE2156-543).

The dependences of the observed abundances on different parameters are shown in Fig. 4. The only correlation with T_{eff} alone was the apparent lower temperature threshold. A much better correlation was obtained with gravity alone. A good correlation was also

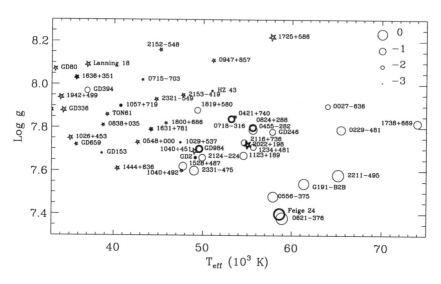

FIGURE 3. Inferred trace element abundances relative to G191-B2B. The sizes of the symbols are proportional to the logs of the relative abundances (see key). Circles are measurements, stars are upper limits, and solid circles are spectroscopic determinations that are consistent with pure H atmospheres. Error bars are omitted for clarity, but typical errors in T_{eff} range from 150 K at the cool end to 1,000 K at the hot end, while the average error in $\log g$ is 0.05 dex. Symbols drawn with thick lines represent WD in binaries. Numerical labels are WD numbers.

obtained with age, but that could due to a fortuitous correlation between the cooling isochrones and the isocontours of observed abundances. The dependence of derived abundances on the interstellar HI column is displayed to show that there is no obvious bias present except for the absence of DA with both very high abundances and HI columns, which were undetected in the EUV or were detected in only one bandpass.

We also compared the relative abundance determinations with recent extensive calculations of radiative support of heavy elements presented in Chayer, Fontaine & Wesemael (1995, hereafter CFW). CFW calculated equilibrium photospheric abundances for individual heavy elements in an otherwise pure H atmosphere as a function of temperature and gravity. Using their results (kindly provided by P. Chayer), we interpolated in T_{eff} and $\log g$ to find predicted abundances (normalized to 60,000 K, $\log g = 7.5$) for each individual object. The comparisons for Fe+H and C+H are presented in Fig. 5. Linear regressions were performed on the logs of the predicted and measured abundances for the 20 DA in which metals were definitely present, resulting in correlation coefficients near 0.9 and residuals of about 0.3 dex for both Fe and C. However, the slopes of the fits were 2.5 for Fe and 3.6 for C, and about 2 for N, O and Ar. A slope near unity was obtained for Ca, but our analysis of the G191-B2B spectrum with models including the Ca lines from Kurucz (1991, 1992) requires Ca/H $< 10^{-8}$, hence Ca cannot contribute significantly to the observed opacities. We conclude that the observed variation of trace element abundances with T_{eff} and $\log g$ is much greater than predicted by current theory.

In the case of the predicted abundances for carbon, applying a linear scale factor of 3.6 to the logs of the predicted abundances (and a shift of 0.2) produced excellent agreement with the measured abundances. All the abundance measurements were consistent with a

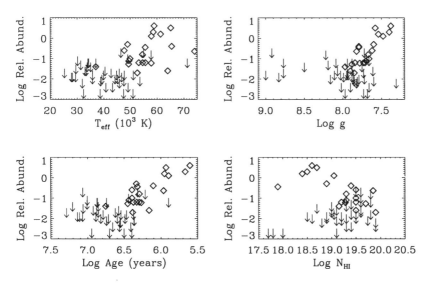

FIGURE 4. Abundance measurements (diamonds) and upper limits (arrows) vs. stellar parameters and ISM columns. Ages are taken from Wood's latest evolutionary models (Wood 1990, 1994).

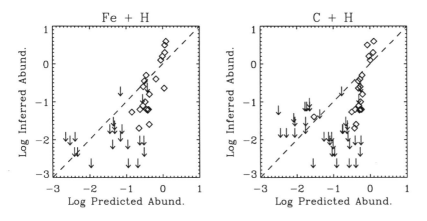

FIGURE 5. Relative abundance measurements (diamonds) and upper limits (arrows) vs. single-element relative abundances predicted by CFW. Dashed line has slope of unity.

linear fit (except GD394) and the 1 σ dispersion was only 0.34 dex. The only significant discrepancies found were for HZ 43, GD 2, WD 1040+492 (RE1043+492), and WD 1029+537 (RE1032+532), all of which had observational upper limits of about an order of magnitude below the adjusted predicted abundances.

4. Summary

Many previous studies have quoted a 40,000 K lower threshold for the appearance of detectable trace elements in hot DA. This analysis shows that there are no DA cooler than 48,000 K or with $\log g > 7.9$ that *require* the presence of trace heavy elements, with the exception of GD 394. The theoretical abundances calculated by CFW correlate well with the observed abundances, but vary much more slowly with $T_{\rm eff}$ and $\log g$. Further investigations of the joint effects of a mix of trace heavy elements may modify the predicted abundances; some initial results are presented in this volume (Chayer et al. 1995). The current discrepancy between theory and observation may also support the suggestion by CFW that radiatively driven winds may deplete the photospheric trace elements over time. However, if mass loss is occurring, differential effects must be small as long as significant amounts of photospheric trace elements are present, given the tight correlation obtained between the equilibrium abundance predictions and the observed abundances for the DA with definite metal detections. It is notable that the four objects that depart significantly from the adjusted predictions are underabundant by at least 1 dex and lie near the "dividing line" (see Fig. 3) between effectively pure H and metal-rich DA. The challenge to theorists is to find a way to accelerate the reduction in abundances as the DA cool (by factors of 2 to 3.5 in the logarithm relative to current equilibrium calculations) without altering the observed regular dependence on $T_{\rm eff}$ and $\log g$ and yet allow some DA to have effectively eliminated their trace metals when abundances as high as 10% of G191-B2B are expected.

The model atmosphere codes used for this analysis were kindly provided by D. Koester. This work was supported by NASA grant NAGW-2478.

REFERENCES

Bowyer, S. et al. 1994, ApJS, 93, 569

Chayer, P., Fontaine, G., & Wesemael, F. 1995, ApJ, in press

Chayer, P., Vennes, S., Pradhan, A. K., Thejll, P., Beauchamp, A., Fontaine, G., & Wesemael, F. 1995, these proceedings

Dupuis, J. & Vennes, S. 1995, these proceedings

Finley, D. S., Koester, D., & Basri, G. 1995, in prereparation

Jordan, S., Koester, D., & Finley, D. S. 1995, these proceedings

Koester, D., Schulz, H., & Weidemann, V. 1979, A&A, 76, 262

Koester, D. 1995, these proceedings

Kurucz, R. L. 1991, in Stellar Atmospheres: Beyond Classical Methods, ed. L. Crivellari, I. Hubeny & D. G. Hummer, NATO ASI Ser. 341, 441

Kurucz, R. L. 1992, Rev. Mexicana Astron. Af., 23, 45

Malina, R. F. et al. 1994, AJ, 107, 751

Pounds, K. et al. 1993, MNRAS, 260, 77

Rumph, T., Bowyer, S., & Vennes, S. 1994, AJ, 107, 2108

Seaton, M. J., Zeippen, C. J., Tully, J. A. et al. 1992, Rev. Mexicana Astron. Af., 23, 19

Vennes, S. 1995, these proceedings

Wood, M. A. 1990, Ph.D. thesis, University of Texas at Austin

Wood, M. A. 1994, private communication

Confining the Edges of the GW Vir Instability Strip

KLAUS WERNER,[1] STEFAN DREIZLER,[1]
ULRICH HEBER,[2] AND THOMAS RAUCH[1]

[1]Institut für Astronomie und Astrophysik der Universität, 24098 Kiel, Germany

[2]Dr. Remeis Sternwarte, Universität Erlangen-Nürnberg, 96049 Bamberg, Germany

We report on our NLTE model atmosphere analyses of PG 1159 stars. The results enable us to confine the location of the GW Vir instability region in the Hertzsprung-Russell diagram. The analysis of a spectrum of the non-pulsator PG 1520+525 taken with the *EUVE* satellite in comparison with HST data of the pulsating protoype PG 1159-035 (=GW Vir) locates the blue edge of the instability strip near $T_{\rm eff}=140\,000$ K for stars in the respective luminosity range.

1. The GW Vir Instability Strip

PG 1159 stars are hydrogen-deficient and extremely hot pre-White Dwarfs. According to our current understanding they have lost their hydrogen envelope either by extraordinarily strong mass loss events or by ingestion and burning of hydrogen. This might be a consequence of a late thermal pulse during previous post-AGB evolution. Spectroscopically the PG 1159 stars are characterized by strong C IV and occasionally O VI lines, together with He II, but H Balmer lines are absent. Within the last years the number of known PG 1159 stars was brought up to 26. Eight stars of this group were identified as multimode nonradial g-mode pulsators (and four of the eight pulsators are Planetary Nebula central stars). They allow probing their interior structure by means of asteroseismology and thus play a key role in our understanding of post-AGB stellar evolution.

The GW Vir instability strip is defined by the pulsating PG 1159 stars. The pulsations are driven by cyclic ionization of C and O (Starrfield et al. 1984). The stellar pulsational model calculations can be subject to a stringent test if we compare the predicted instability region in the HRD with the observationally determined position of the PG 1159 stars.

2. Spectroscopic Analyses

Non-LTE model atmosphere analyses are in progress in order to find the photospheric parameters of all known PG 1159 stars. Figure 1 shows the positions of most of these objects in the $T_{\rm eff}$–surface gravity plane. The models include H, He, C, O and are computed with a NLTE code developed in Kiel and Bamberg. Most recent models include line blanketing by millions of lines from iron group elements. As an example Figure 2 shows the EUV flux of such an iron line blanketed model for a H-rich central star. More details on work done so far can be found in a recent review given by Dreizler, Werner & Heber (1995).

Among the first PG 1159 stars analysed were the spectroscopic twins PG 1159-035 and PG 1520+525. This pulsator/non-pulsator pair holds the key for the determination of the blue edge of the instability strip. Our optical analysis gave an equal $T_{\rm eff}$ (140 kK) for both of them, however, within a large error range ($\pm$ 15 kK). The determination of more precise atmospheric parameters is hampered by the fact that temperature indicators in

S. Bowyer and R. F. Malina (eds.), Astrophysics in the Extreme Ultraviolet, 229–234.

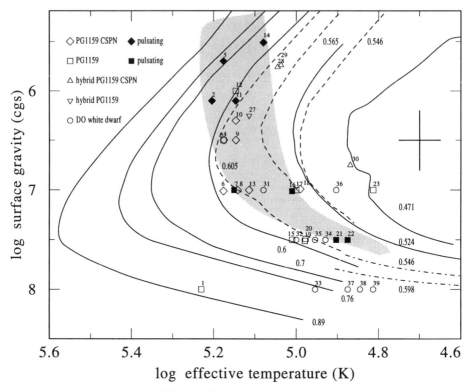

FIGURE 1. Location of the GW Vir instability strip (shaded region), comprising the pulsating PG 1159 stars (filled symbols). Evolutionary tracks are labeled with the respective stellar mass. A typical error bar is shown. For object identification see Table 1 (PG 1159 stars, labels 1–26) and Dreizler et al. (1995) for other objects

the optical spectral region are relatively insensitive. In the UV region, however, we can exploit the O V/O VI ionisation equilibrium by line profile fitting. Alternatively, the EUV region can be used for line and continuum fitting. The latter is particularly useful because the PG 1159 stars have their flux maximum in the EUV range and the continuum flux and shape are very temperature sensitive. The oxygen ionization balance method has been used to derive a rather precise value for $T_{\rm eff}$ of PG 1159-035 from a HST-FOS spectrum (Werner & Heber 1993). The result was again 140kK but now with a much smaller uncertainty ($\pm$ 5 kK). No usable UV spectrum of the other twin was recorded up to now, so we decided to observe both twins with *EUVE*.

3. *EUVE* Observation of PG 1520+525

Model atmospheres predict a very strong EUV flux, however, interstellar column densities are high. Therefore weak but still detectable fluxes at earth were expected in the region around 100Å. Unfortunately the extinction towards PG 1159-035 turned out to be higher than expected: A 38ks exposure taken during Apr 06–07 (1993) is underexposed. On the other hand a 155ks exposure of PG 1520+525 taken on Feb 09–15 (1994) shows a noisy but still usable spectrum (Fig. 3). Compared to a 140kK model spectrum we

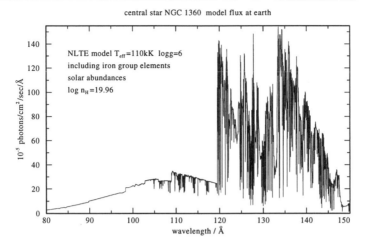

central star NGC 1360 model flux at earth

NLTE model T_{eff}=110kK logg=6
including iron group elements
solar abundances
log n_H =19.96

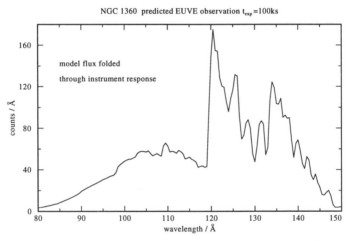

NGC 1360 predicted EUVE observation t_{exp}=100ks

model flux folded
through instrument response

FIGURE 2. Top: Non-LTE model flux for a hot planetary nebula central star. Line blanketing by iron group elements is included. Bottom: Model folded through the *EUVE* SW instrument response. Such models are currently under construction in order to analyze an *EUVE* spectrum of NGC 1360

would expect to see a strong O v absorption edge at 120Å, but it is not detectable. We must have a higher T_{eff} in order to depopulate O v and to weaken this edge: It can be seen that the 150kK model has a much weaker O v edge. We conclude that T_{eff} of PG 1520+525 (150kK) is slightly higher than that of the pulsating prototype (140kK). Hence the blue edge of the GW Vir instability strip runs right between the loci of these two stars in the T_{eff}–$\log g$ plane.

Another pulsator/non-pulsator pair comprises PG 1707+427 and PG 1424+535. These define the red edge of the instability strip, which is found near T_{eff}=100kK. Analogously, this is what we learned from optical spectra. Their temperatures are much too low to produce enough EUV photons that could penetrate the ISM, so they cannot be detected

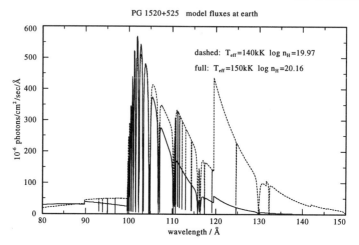

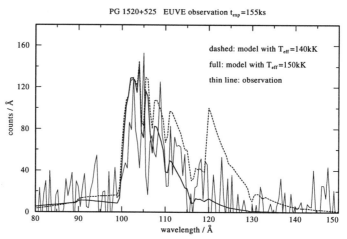

FIGURE 3. Top: Non-LTE model fluxes (for two different effective temperatures) attenuated by the ISM. Note the strong O V absorption edge at 120Å in the cooler model. It almost disappears when T_{eff} is increased by only 10kK. Another strong edge (at 100Å) is due to O VI (2p). The absorption line spectrum is dominated by the respective O VI series as well as C IV lines. Other model parameters, taken from Werner et al. (1991): $\log g = 7$, He/C/O=0.33/0.50/0.17 (mass fractions).
Bottom: Both models folded through the *EUVE* SW instrument response and compared to the observation. The non-detection of the 120Å edge suggests that T_{eff} exceeds 140 000 K

by *EUVE*. Instead, HST-GHRS observations are approved for Cycle 5 to find subtle differences in their atmospheric parameters.

After all, the picture as we have described it up to now is not as simple. The location of the instability region is not only dependent on surface gravity (and hence luminosity), but also on C and O abundances in the pulsation driving regions of the stars. Although these are very close below the photosphere, their composition may differ from the photospheric abundances. This fact might explain the presence of non-pulsators in the instability

TABLE 1. The complete list of known PG 1159 stars (ordered by decreasing effective temperature), their spectroscopic subtype and atmospheric parameters (: =uncertain). Columns 3 and 4 denote if the star is variable or has a planetary nebula. Abundances are given in % mass fraction, T_{eff} in kK. The consecutive numbers in the last column refer to the labels in Fig. 1. For more details see Dreizler et al. (1995)

Star	Type	Var.	PN	T_{eff}	log g	H	He	C	N	O	Nr.
H 1504+65	Ep	no	no	170	8.0			50		50	1
RX J 2117+3412	lgE	yes	yes	160	6.1		38	56		6	2
PG 1144+005	Ep	no	no	150	6.5		39	58	1.5	1.6	3
Jn 1	E	no	yes	150:	6.5		30	46		24	4
NGC 246	lgE	yes	yes	150:	5.7		38	56		6	5
PG 1520+525	E	no	yes	150	7.0	<8	30	46	<0.3	16	6
PG 1159-035	E	yes	no	140	7.0	<8	30	46	<0.3	16	7
NGC 650	E		yes	140:	7.0						8
Abell 21=Ym29	E		yes	140:	6.5		35	51		14	9
Longmore 3	lgE	no	yes	140:	6.3		38	56		6	10
K 1-16	lgE	yes	yes	140:	6.1		38	56		6	11
PG 1151-029	lgE	no	no	140:	6.0		35	51		14	12
VV 47	E	no	yes	130:	7.0		35	51		14	13
Longmore 4	lgEp	yes	yes	120	5.5		46	43		11	14
HS 0444+0453	A		no	100:	7.5						15
PG 1707+427	A	yes	no	100	7.0	<8	30	46	<0.3	16	16
PG 1424+535	A	no	no	100	7.0	<8	30	46	<0.3	16	17
IW 1	A	no	yes	100:	7.0						18
MCT 0130-1937	A	no	no	95	7.5		50:	30		20	19
HS 1517+7403	A		no	95	7.5		61	37		2	20
PG 2131+066	A	yes	no	80	7.5		50:	30		20	21
PG 0122+200	A	yes	no	75	7.5		50	30		20	22
HS 0704+6153	A		no	65	7.0	<10	44	26		20	23
NGC 6852	lgE:		yes								24
NGC 6765	lgE:		yes								25
Sh 2-78	A		yes								26

region. One example is HS 2324+3944, a so-called hybrid-PG 1159 star (nr. 27 in Fig. 1). It does have hydrogen in the photosphere. If present in deeper layers, too, hydrogen would "poison" the pulsations.

K.W. acknowledges IAU and DFG (We 1312/7-1) travel grants and thanks Anne Miller (CEA) for support during *EUVE* data reduction in Berkeley. This research is sponsored by the DFG (He 1356/16-2, We 1312/6-1) and the DARA (50 OR 9409 1).

REFERENCES

DREIZLER, S., WERNER, K. & HEBER, U. 1995, Analysis of PG 1159 stars and related objects, n White Dwarfs, ed. D. Koester & K. Werner, Lect. Notes Phys., Springer, 443, 160

STARRFIELD, S., COX, A. N., KIDMAN, R. B., & PESNELL, W. D. 1984, Nonradial instability strips based on carbon and oxygen partial ionization in hot, evolved stars, ApJ, 281, 800

WERNER, K. & HEBER, U. 1993, UV spectroscopy of PG 1159 with HST and a prospective view of future EUVE observations, in White Dwarfs: Advances in Observation and Theory, ed. M. A. Barstow, Dordrecht: Kluwer, NATO ASI Series C, 403, 303

WERNER, K., HEBER, U., HUNGER, K. 1991, Non-LTE analysis of four PG1159 stars, A&A, 244, 437

Detection of Heavy Elements in the *EUVE* Spectrum of a Hot White Dwarf

S. JORDAN,[1] D. KOESTER,[1] AND D. FINLEY[2]

[1]Institut für Astronomie und Astrophysik, Universität Kiel, D-24098 Kiel, Germany

[2]Center for EUV Astrophysics, 2150 Kittredge Street,
University of California, Berkeley, CA 94720-5030, USA

Observations with the *ROSAT* satellite have already indicated that metal absorbers must be present in the atmosphere of the hot DA white dwarf PG 1234+482. This is now confirmed by strong absorption features found in the short and medium wavelength *EUVE* spectrum of the star. With fully blanked model atmospheres, taking into account several million lines of heavy elements, we could attribute the strongest features to absorption by FeVI and FeVII. Since the spectrum has not been dithered during the observation other elements could not be identified with the same level of confidence, but upper limits could be determined. These are in general lower than predicted by models, which attempt to explain the presence of the metals by theoretical calculations of radiative forces in hot DA white dwarf atmospheres.

1. Introduction

Observations in the EUV and soft X-ray region of the electromagnetic spectrum have revealed that hot white dwarfs of spectral type DA, showing only the Balmer series of hydrogen in the optical, can possess significant amounts of heavier elements. This conclusion has been reached since in many objects the energy flux measured in the EUV and soft X-ray by the *EINSTEIN* and *EXOSAT* satellites turned out to be smaller than predicted by pure hydrogen atmospheres (Kahn et al. 1984, Petre et al. 1986, Jordan et al. 1987, Paerels & Heise 1989) It is believed that radiative levitation is responsible for the presence of heavy elements in the atmospheres of white dwarfs, since these ions would otherwise sink down into deeper layers due to the strong gravitational acceleration. The recent results from the *ROSAT* observations (Barstow et al. 1993; Jordan et al. 1994; Wolff et al. 1994, 1995) have shown that the ensemble of hot DA stars ($\gtrsim 25,000$ K) can be divided into two groups: all stars with $T_{\text{eff}} \lesssim 38,000$ K are compatible with pure hydrogen atmospheres while at higher temperatures most objects contain additional opacity.

With X-ray and EUV photometry only, the presence of these absorbers could be established, but not their nature. Vennes et al. (1989) analyzed one of the very few EUV spectra obtained by *EXOSAT* and concluded that a mixture of several heavy elements is necessary to explain the observed energy distribution of the hot DA Feige 24. In some case metal absorbers have also been found in *IUE* high resolution and HST GHRS spectra of the bright DA star G191-B2B (Bruhweiler & Kondo 1991; Vennes et al. 1992; Holberg et al. 1993, 1994; Werner & Dreizler 1994; Vidal-Madjar et al. 1994). Now the *EUVE* satellite provides a unique tool to obtain high resolution spectra in a spectral region where strong absorption edges and lines of several elements can directly be detected.

2. Observations and Model Atmosphere Analysis

PG 1234+482 was identified as a hot DA by Jordan et al. (1991). From the optical and UV (*IUE*) spectrum we determined an effective temperature of $55,600 \pm 1,500$ K with a pure H model atmosphere. However, the soft X-ray flux measured by *ROSAT* is about a factor of 12 lower than predicted by such a model. Moreover, the PSPC pulse height

S. Bowyer and R. F. Malina (eds.), Astrophysics in the Extreme Ultraviolet, 235–240.

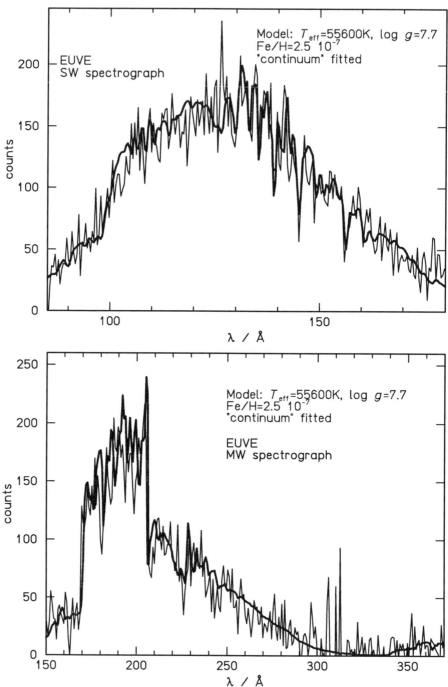

FIGURE 1. The SW (top) and MW (bottom) spectrum of PG 1234+482 is compared to a synthetic spectrum for $T_{\text{eff}} = 55,600$ K, $\log g = 7.7$, and Fe/H=$2.5 \cdot 10^{-7}$

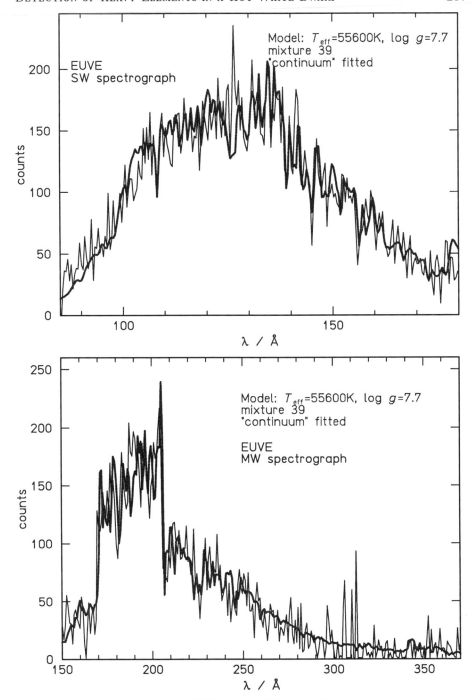

FIGURE 2. Comparison of the *EUVE* spectra of PG 1234+482 to a model spectrum for $T_{\text{eff}} = 55,600$ K, $\log g = 7.7$, and a chemical composition listed in the table

distribution could not be reproduced by any model atmosphere containing H and He only, so that the opacity in the short wavelength region must be due to metals (Jordan 1993). This is now confirmed by the high resolution *EUVE* spectrum (exposure time: 99 ksec). No strong HeII photoionization edge at 228 Å but many strong absorption features were detected with the short (SW) and medium (MW) wavelength spectrographs.

The interpretation of such a spectrum with model atmospheres has become possible just recently (Koester, these proceedings). For our analysis we used the Koester model atmosphere code which now takes into account blanketing by typically 10^7 metal lines listed in the Kurucz (1991) tables. For the photoionization cross sections Opacity Project data (Seaton et al. 1992) were used with the exception of nickel, for which the hydrogenic approximation had to be applied. The model flux is folded with the detector response matrix of the spectrograph and and normalized at the *V* magnitude of 14^m38; the interstellar absorption is calculated according to Rumph et al. (1994) and Morrison & MacCammon(1983).

In order to find out which absorbers may be responsible for the observed features we calculated model atmospheres and synthetic spectra for a mixture of hydrogen and at each time one of the following elements: Fe, Ni, Ca, C, N, O, and Si. The flux level predicted by these bielemental mixtures cannot be expected to agree with the observed spectra, so that we had to artificially reduce the theoretical *EUVE* spectrum with respect to the observed one (typically by a factor of 2 or 3 in the SW, and some 10% in the MW). The overall shape was also adjusted by multiplication with a quadratic function.

Fig. 1 shows that many of the dominant absorption features (especially in the SW) can be reasonably fitted with a theoretical spectrum for an atmosphere consisting of H and Fe with Fe/H = $2.5 \cdot 10^{-7}$ by numbers. The conclusion that iron ions (FeVI, FeVII) are the most important absorbers in this spectral region could already be reached by comparing the *EUVE* spectrum to opacities for a temperature and pressure representative for the line forming region in the atmosphere (see Fig. 1 in Jordan et al. 1995).

None of the other bielemental compositions led to equally convincing results, although there are indications that nickel and calcium are the next strongest absorbers. However, observational uncertainties may result from the fact that the spectrum of PG 1234+482 has not been dithered during the observation so that part of the structure may be due to fixed pattern noise.

Mixture used for Fig. 2:

number ratio	predictions
Fe/H = $2 \cdot 10^{-7}$	$1.5 \cdot 10^{-5}$
Ni/H = $2 \cdot 10^{-8}$	$2.0 \cdot 10^{-6}$
Ca/H = $2 \cdot 10^{-8}$	$2.0 \cdot 10^{-6}$
C/H = $1 \cdot 10^{-5}$	$1.8 \cdot 10^{-4}$
N/H = $2 \cdot 10^{-6}$	$2.5 \cdot 10^{-4}$
O/H = $3 \cdot 10^{-8}$	$1.3 \cdot 10^{-4}$
Si/H = $1 \cdot 10^{-8}$	$< 10^{-8}$

Nevertheless, we tried to calculated upper limits by increasing the abundance of the metals until the predicted features became stronger than observed. After this procedure was performed for each element individually, we calculated a synthetic spectrum for the full mixture of metals (listed in the Table). The result is shown in Fig. 2. Compared to the calculation with H and Fe only, there is a slight improvement of the overall fit, but there is still a significant flux discrepancy in the SW region of the order of 50%. Therefore we have to conclude that other elements, not yet included in our calculations, are probably present.

However, the abundances can be regarded as upper limits; to be more careful the values may be multiplied by a factor of 2. Even then the abundances are in general lower than predicted by Chayer et al. (1995, for $\log g = 7.5$), who performed detailed calculations of the radiative forces in hot DA white dwarf atmospheres. Currently, these models neglect the influence of the metals on the temperature and pressure structure of the atmosphere, and do not include the effects on each trace element of the flux blocking by the other elements that are present.

The flux distribution in the MW spectrum is relatively well reproduced by our mixture of heavy elements under the assumption of an interstellar hydrogen column density of $N_H = 10^{19}\,\text{cm}^{-2}$. Modeling of the HeI autoionization transition, first detected in the spectrum of GD 246 by Vennes et al. (1993), we estimated $N_{He} = 10^{18}\,\text{cm}^{-2}$, meaning that the ISM is mostly neutral. A careful determination of the helium ionization fraction will be performed when a final model for PG 1234+482 is found.

3. Future Prospects

PG 1234+482 will be reobserved with *EUVE* in the dithered mode and with an exposure time of 200 ksec. This will result in a high quality spectrum so that we may be able to find a unique chemical composition reproducing both the strength of the absorption features and the overall flux distribution. We may then arrive at a somewhat lower effective temperature since up to now we have neglected the blanketing effect of the metals on the optical temperature determination. Finally, since NLTE effects cannot be excluded at temperatures above 50,000 K, we plan to repeat our analysis with NLTE atmospheres which also account for the large number of metal lines (Dreizler & Werner 1993).

Work on *ROSAT* data in Kiel was supported by the DARA (grant 50 OR 9302 4).

REFERENCES

BARSTOW, M. A., FLEMING, T. A., DIAMOND, C. J., ET AL. 1993, MNRAS, 264, 16

BRUHWEILER, F. C., & KONDO, Y. 1981, ApJ, 248, L123

CHAYER, P., FONTAINE, G., & WESEMAEL, F. 1995, ApJ, in press

DREIZLER, S., & WERNER, K. 1993, A&A, 278, 199

HOLBERG, J. B., BARSTOW, M. A., BUCKLEY, D. A. H. ET AL. 1993, ApJ, 416, 806

HOLBERG, J. B., HUBENY, I., BARSTOW, M. A., ET AL. 1994, ApJ, 425, L105

JORDAN, S. 1993, in Advances in Space Research, 13, no. 12, Pergamon Press, 319

JORDAN, S., KOESTER, D., WULF-MATHIES, C., & BRUNNER, H. 1987, A&A, 185, 253

JORDAN, S., HEBER, U., & WEIDEMANN, V. 1991, in White Dwarfs, ed. Vauclair & Sion, 121

JORDAN, S., WOLFF, B., KOESTER, D., & NAPIWOTZKI, R. 1994, A&A, 290, 834

KOESTER, D., WERNER, K., FINLEY, D. S., & DREIZLER, S. 1995, Lecture Notes in Physics, 443, 332

KAHN, S. M., WESEMAEL, F., LIEBERT, J., ET AL. 1984, ApJ, 278, 255

KURUCZ, R. L. 1991, in, NATO ASI Series, 341, 441

MORRISON, R., & MACCAMMON, D. 1983, ApJ, 270, 119

PAERELS, F. B. S., & HEISE, J. 1989, ApJ, 339, 1000

PETRE, R., SHIPMAN, H. L., & CANIZARES, C. R. 1986, ApJ, 304, 356

RUMPH, T., BOWYER, S., & VENNES, S. 1994, AJ, 107, 2108

SEATON, M. J., ZEIPPEN, C. J., TULLY, J. A., ET AL. 1992, Rev. Mexicana Astron. Af., 23, 19

VENNES, S., CHAYER, P., FONTAINE, G., & WESEMAEL, F. 1989, ApJ, 336, L25

VENNES, S., CHAYER, P., THORSTENSEN, J. R., BOWYER, S., & SHIPMAN, H. L. 1992, ApJ, 392, L27

VENNES, S., DUPUIS, J., RUMPH, T. ET AL. 1993, ApJ, 410, L119

VIDAL-MADJAR, A., ALLARD, N. F., KOESTER, D. ET AL. 1994, A&A, 287, 175

WERNER, K., & DREIZLER, S. 1994, A&A, 286, L31

WOLFF, B., JORDAN, S., BADE, N., & REIMERS, D. 1994, A&A, 294, 794

WOLFF, B., JORDAN, S., & KOESTER, D. 1995, A&A, submitted

EUVE and *ORFEUS* Observations of the Cool DO White Dwarf HD 149499 B

R. NAPIWOTZKI,[1] S. JORDAN,[2] S. BOWYER,[3] M. HURWITZ,[3] D. KOESTER,[2] T. RAUCH,[2] AND V. WEIDEMANN[2]

[1]Dr. Remeis-Sternwarte, Sternwartstr. 7, 96049 Bamberg, Germany

[2]Institut für Astronomie und Astrophysik der Universität, 24098 Kiel, Germany

[3]Center for EUV Astrophysics, 2150 Kittredge Street, University of California, Berkeley, CA 94720-5030, USA

We present the results of a recent spectroscopic investigation of the cool DO white dwarf HD 149499 B in the EUV and FUV ranges. Observations were performed with the spectrograph of the *EUVE* satellite and the Berkeley EUV/FUV spectrometer of the *ORFEUS* space experiment.† The analysis of the *ORFEUS* spectrum, performed with a grid of LTE model atmospheres, yielded the basic parameters $T_{\text{eff}} = 49500 \pm 500$ K and $\log g = 7.97 \pm 0.08$. This result is confirmed by the *EUVE* spectra. The photospheric hydrogen Lyman lines in the FUV spectrum indicate the presence of hydrogen: $\log n_{\text{H}}/n_{\text{He}} = -0.65 \pm 0.12$. The implications of this finding for the spectral evolution of white dwarfs are discussed. A check of the LTE assumption was performed by a comparison with NLTE atmospheres calculated for appropriate parameters. The interstellar hydrogen column towards the HD 149499 system amounts to $N_{\text{H}} = (7 \pm 2) \cdot 10^{18}$ cm^{-2}.

---

1. Introduction

The class of hot helium-rich white dwarfs (spectral type DO) is divided into three subgroups: cool DOs, hot DOs, and PG 1159 stars (Wesemael et al. 1985). In the last years spectral analyses of PG 1159 stars and hot DO white dwarfs have been carried out by the Kiel/Bamberg NLTE group (see Dreizler et al. 1995 for a review). The PG 1159 stars are very carbon- and oxygen-rich: He:C:O = 0.61:0.31:0.08 (by number). Analyses of hot DO stars resulted in C/He ranging from below the observational limit $< 10^{-3}$ to $\approx 10^{-2}$.

For an understanding of the probable evolutionary sequence PG 1159 → hot DO → cool DO, analyses of cool DO white dwarfs are needed. The primary candidate for the search for weak metal features in the FUV and EUV range is the DO HD 149499 B because it is by far the brightest star of this class ($V \approx 11^{\text{m}}7$). It is the secondary star of a binary system with a K0V primary (separated by only 1″5; Holden 1977), dominating in the optical range ($3^{\text{m}}5$ brighter in V; Holden 1977). Therefore optical observations of the white dwarf are nearly impossible. However, in the FUV region the white dwarf flux is nearly uncontaminated, but The temperature determination remains controversial. From *IUE* data Wray et al. (1979) derived a temperature range 70000 K$\leq T_{\text{eff}} \leq 100000$ K, while Sion et al. (1982) determined $T_{\text{eff}} = 55000^{+15000}_{-5000}$ K. The probably most reliable T_{eff} result was derived by Poulin et al. (1989) from combined *IUE* and *Voyager* spectra: $T_{\text{eff}} = 54000$ K. However, more accurate data are highly desirable. In this paper we present the results of new observations of HD 149499 B in the EUV and FUV with the *EUVE* satellite and the *ORFEUS* telescope, respectively.

† Based on the development and utilization of *ORFEUS* (*Orbiting and Retrievable Far and Extreme Ultraviolet Spectrometers*), a collaboration of the Astronomical Institute of the University of Tuebingen, the Space Astrophysics Group, University of California, Berkeley, and the Landessternwarte Heidelberg.

S. Bowyer and R. F. Malina (eds.), Astrophysics in the Extreme Ultraviolet, 241–246.

2. Observations and Data Reduction

EUV spectrograms of HD 149499 B were obtained with the *EUVE* satellite in pointed mode. *EUVE* provides three spectrometers covering the ranges 70 ... 190 Å (short wavelength; SW), 140 ... 380 Å (medium wavelength; MW), and 280 ... 760 Å (long wavelength; LW). A spectral resolution of $\lambda/\Delta\lambda \approx 300$ is achieved. Due to the strong He II 228 Å edge, HD 149499 B was not detected in the SW range. The MW and LW spectra are displayed in Fig. 2. The flux peak in the 500 ... 600 Å region is contaminated by second order flux.

HD 149499 B was observed for 1675 s with the Berkeley spectrometer of the *ORFEUS* (*Orbiting Retrievable Far and Extreme Ultraviolet Spectrograph*) experiment (Hurwitz & Bowyer 1995). *ORFEUS* was mounted aboard the free flying Astro-SPAS platform which was deployed from and later recovered by the space shuttle Discovery in September 1993. The Berkeley spectrometer covers the 390 ... 1170 Å band simultaneously, with an achieved spectral resolution of $\lambda/3000$. Because the column density of interstellar hydrogen is relatively high (see below), only the spectral range longward of the Lyman edge at 912 Å is useful for this target (for further details see Napiwotzki et al. 1995 and references therein).

3. Analysis

The LTE model atmosphere codes of Koester were used to calculate an extensive grid of H/He models around the expected parameters of HD 149499 B. The model code uses the classical assumptions of LTE models: plane-parallel stratification, hydrostatic equilibrium, radiative equilibrium. Convection is included, if the stratification is unstable, using the ML1 version of the mixing length approximation with $l/H_p =1$. Line blanketing by hydrogen and helium (neutral and ionized) lines is fully included.

Since it was shown by Napiwotzki (1995a,b) from an investigation of optical lines that NLTE effects can be important even in cool DO white dwarfs, it is necessary to check the validity of the LTE assumption for our analysis. For this purpose we calculated a H/He NLTE model atmosphere for HD 149499 B with the ALI code developed by Werner (1986). It turned out that LTE and NLTE spectra are in very good agreement for the relevant parameter range and differences can be neglected for practical purposes. Only at $\lambda < 228$ Å moderate NLTE deviations occur. However, in the case of HD 149499 B the flux is essentially zero below the He II absorption edge.

Analysis of spectral lines in the spectral range covered by our *ORFEUS* spectrogram is hampered by a lack of up-to-date line broadening tables. New data based on recent theoretical developments are highly desirable. For our analysis we applied the line broadening tables for hydrogen of Edmonds et al. (1967; ESW) for the He II $2 \to n$ series.

The flux calibrated FUV spectrum of HD 149499 B was used for the determination of temperature, gravity, and hydrogen content. We scaled the model flux according to the observed flux at 1150 Å and performed a χ^2 fitting with the synthetic spectra. The final results are: $T_{\rm eff} = 49500 \pm 500$ K, $\log g = 7.97 \pm 0.08$, $\log n_{\rm H}/n_{\rm He} = -0.65 \pm 0.12$. The best fit is displayed in Fig. 1.

The error ranges are formal 1 σ errors provided by the fitting procedure. However, one has to be aware of possible systematic errors, especially those caused by the applied line broadening theory. The detection of hydrogen is of high statistical significance. On the other hand all hydrogen lines coincide with He II lines and our result depends on a correct description of the He II line broadening. Thus it is possible that the quantitative result might change, if an improved line broadening is applied. Independent of the exact

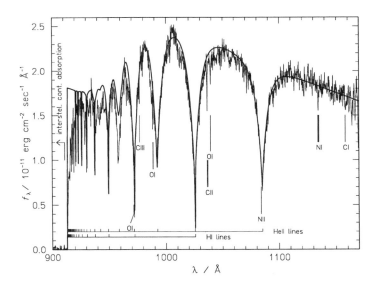

FIGURE 1. FUV spectrogram obtained with the Berkeley EUV/FUV spectrograph of *ORFEUS* compared to our best fitting model. Important photospheric and interstellar lines are indicated.

abundance, a comparison with a pure He model shows that the pure He II lines can be fitted exactly, while those possibly containing a hydrogen contribution are discrepant. We are therefore confident that the detection of hydrogen is real.

An independent check of the results can be performed with the *EUVE* spectra. We kept g and n_H/n_{He} fixed at the values determined from the *ORFEUS* spectrum and varied T_{eff}. The optimum value of the interstellar hydrogen column density N_H was calculated for each model. The results are shown in Fig. 2. The derived limits $49000 \leq T_{eff} \leq 51000$ are in excellent agreement with our *ORFEUS* results. The corresponding column density amounts to $(7 \pm 2) \cdot 10^{18}\,\mathrm{cm}^{-2}$.

The knowledge of the interstellar hydrogen column density is crucial for the temperature determination. In our analysis of the *EUVE* data it was determined simultaneously. However, the interstellar Lyman lines of hydrogen in the *ORFEUS* spectrum provide us with a further test. These lines are best fitted with $N_H = 1 \cdot 10^{19}\,\mathrm{cm}^{-2}$. This result is accurate within a factor of two. It is in good agreement with the determination from our *EUVE* spectra.

4. Discussion

We have presented an analysis of FUV and EUV spectra of the cool DO HD 149499 B. The He II $2 \rightarrow n$ series and the coinciding Lyman lines are used for the determination of the basic parameters T_{eff}, g, and n_H/n_{He} from the *ORFEUS* spectrum. The LTE analysis yields $T_{eff} = 49500\,\mathrm{K}$, $\log g = 7.97$, and $\log n_H/n_{He} = -0.65$. An independent check of our T_{eff} determination is performed by means of the *EUVE* observations. NLTE effects

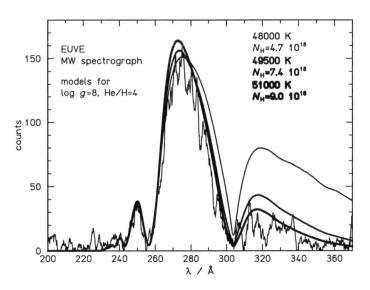

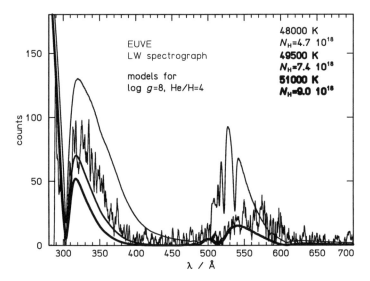

FIGURE 2. *EUVE* spectra of HD 149499 B and model spectra demonstrating the strong temperature dependence.

on the investigated lines can be neglected for the parameters of HD 149499 B. This is the first accurate determination of the atmospheric parameters of this bright DO star. Only the FUV observations presented here allow to overcome the spectroscopic contamination by the main sequence companion.

The temperature of 49500 K places HD 149499 B at the cool end of the DO sequence, close to the so-called DB gap (28000 K $\leq$ T$_{\rm eff}$ $\leq$ 45000 K) in which no helium-rich white dwarf is known. This makes HD 149499 B an important object for an understanding of the processes responsible for this gap. The float-up hypothesis of Fontaine & Wesemael (1987; cf. MacDonald & Vennes 1991) predicts that hydrogen, which remained highly diluted in the helium envelope, floats up when the DO stars cool down and transform them into DA white dwarfs with a very thin hydrogen layer on top. Thus the detection of hydrogen in the atmosphere of HD 149499 B is of extreme interest.

The only other DO white dwarf with a hydrogen detection is HZ 21 (Koester et al. 1979; Wesemael et al. 1985) with a temperature similar to HD 149499 B ($\approx$50000 K). However, these analyses used much poorer observational material and for a quantitative confirmation a repeated analysis with new observations and up-to-date model atmospheres is needed. For all other DO stars only upper limits on the hydrogen abundance exist. In most cases, especially for the hot DO and PG 1159 stars, these are not very conclusive, because the limits are higher than or comparable to the hydrogen abundance detected in HD 149499 B, even with the use of good spectra and modern model atmospheres. The detection of hydrogen in cool DO white dwarfs supports the float-up hypothesis, but improved analyses of more DO stars are necessary before we can draw firm conclusions.

Work on *ORFEUS* observations in Kiel was supported by DARA grants 50 OR 90073, 50 OR 93024, and 50 OR 94091. *ORFEUS* is funded by NASA grant NAG5-696 and DARA grant WE 3 50 OB 8501 3. This research has made use of the Simbad database, operated at CDS, Strasbourg, France.

REFERENCES

DREIZLER, S., WERNER, K., & HEBER, U. 1995, in White dwarfs, ed. D. Koester & K. Werner, Berlin: Springer Verlag, in press

EDMONDS, F. N., SCHLÜTER, H., & WELLS, III, D. C. 1976, MmRAS, 71, 271

FONTAINE, G., & WESEMAEL, F. 1987, in: IAU Colloq. 95, The second conference on faint blue stars, ed. A.G.D. Phillip, D.S. Hayes, & J. Liebert, L. David Press, 319

HOLDEN, F. 1977, PASP, 89, 588

HURWITZ, M. & BOWYER, S. 1995, these proceedings

IANNA, P. A., ROHDE, J. R., & NEWELL, E. B. 1982, ApJ, 259, L71

KOESTER, D., LIEBERT, J., & HEGE, E. K. 1979, A&A, 71, 163

MACDONALD, J., & VENNES, S. 1991, ApJ, 371, 719

NAPIWOTZKI, R. 1995a, in White dwarfs, ed. D. Koester & K. Werner, Berlin: Springer Verlag, in press

NAPIWOTZKI, R. 1995b, A&A, in prep

NAPIWOTZKI, R., HURWITZ, M., JORDAN, S., BOWYER, S., KOESTER, D., WEIDEMANN, V., LAMPTON, M., & EDELSTEIN, J. 1995, A&A, submitted

POULIN, E., WESEMAEL, F., HOLBERG, J. B., & FONTAINE, G. 1989, in White Dwarfs, ed. G. Wegner, Berlin: Springer-Verlag, 144

SION, E. M., GUINAN, F., & WESEMAEL, F. 1982, ApJ, 255, 232

WERNER, K. 1986, A&A, 161, 177

WESEMAEL, F., GREEN, R. F., & LIEBERT, J. 1985, ApJ, 58, 379

Wray, J. D., Parsons, S. B., & Henize, K. G. 1979, ApJ, 234, L187

Guide to Modeling the Interstellar Medium

DONALD P. COX

University of Wisconsin-Madison, Dept. of Physics,
1150 University Avenue, Madison, WI 53706 USA

We are in a period that is rich with data and ideas, but the collection of ideas is fundamentally incomplete, leaving us with no useful model for the general characteristics of the interstellar medium (Cox 1995). At the same time, we wrestle with tentative models at various levels, and concepts that such synthesis modeling must involve. In the list and commentary that comprise this paper I offer to future explorers what limited wisdom I can.

Heating-Cooling Balance: It is not clear that the heating of interstellar clouds is sufficiently well understood, but what is clear is that at densities above 10 cm^{-3}, thermal equilibration can be regarded as rapid. The temperature is effectively just a function of the density, as is the thermal pressure. At densities of order 0.1 cm^{-3}, impulsively heated gas will fairly rapidly drop to 10^4K, but below that there is a long slow transient. Even so, the slogan "the gas cools before it recombines" from the days of stochastic heating models (from assumed SNe radiations) is still apropos. Warm intercloud gas is likely often out of thermal balance, and even more often out of ionization equilibrium.

Interphase Pressure Balance: It has often been suggested that contiguous regions of very different density should have about equal thermal pressures. Unless we have been badly deceived, however, the Local Bubble surrounding the Local Cloud (or wisp, if you prefer) has more than an order of magnitude higher thermal pressure than the cloud (e.g., Bowyer et al. 1995). For those comfortable with magnetic fields, this poses no great problem, one merely supposes that the typical magnetic pressure of the ISM is absent from the very low density hot bubble. There is still rough total pressure balance, the thermal pressure of the hot gas against the magnetic pressure in the wisp. A few microgauss is sufficient, tangent to the wisp surface. This tangent field also helps keep the wisp from evaporating into the hot gas—without such help it should long ago have disappeared (Cox & Reynolds 1987).

Interphase pressure balance is still a useful concept, but it must be employed somewhat cautiously—including magnetic field pressure and tension (more on this later). By and large, thermal pressure appears to be relatively small except in hot low density regions. In cooler regions, one can perhaps think of the surrounding magnetic field pressure as providing an upper limit to the contained thermal pressure. For short distances along flux tubes, thermal pressure should be relatively uniform, but needn't equal that of the surroundings. In addition, the absence of a significant bulk modulus associated with the thermal pressure at typical cloud parameters could lead to some very interesting effects reminiscent of critical point opalescence (Cox 1988).

Hydrostatics: The vertical structure of the Galaxy is far from static, but it is close enough to stationary that the midplane pressure is essentially the weight of the interstellar matter. One has only to be a little careful in defining pressure, to be certain that it contains all relevant forms. Boulares & Cox (1990) found the total weight of the interstellar matter to have p/k~25,000 cm^{-3} K, divided about equally between dynamic, cosmic ray, and magnetic forms. Unless there is an extensive hot phase at high thermal pressure (see below), volume averaged thermal pressure is negligible.

Hydrostatics in the ISM is a bit more complicated than on Earth, however; the weight

S. Bowyer and R. F. Malina (eds.), Astrophysics in the Extreme Ultraviolet, 247–254.

depends not only on the mass per unit area, but also on the vertical distribution, because gravity does.

As a consequence, vertical models of the ISM must look either for reasons why the mass has the vertical distribution it does—what specifies the scale heights of the various components, or for reasons why the structure has the midplane pressure it does. In the two phase model of Field, Goldsmith, & Habing (1969), only intercloud gas was stable at low pressure and formed an atmosphere. Because there was more than enough matter, the lower reaches of that atmosphere had sufficient thermal pressure to confine stable clouds. These low lying clouds contained much of the mass but had little weight. Thus the thermal pressure adjusted itself to make the clouds stable. In the McKee & Ostriker (1977) model, supernova remnants grew to very large sizes in a low intercloud density environment, the average pressure of the medium being roughly the time and volume averaged energy density. Those two different models found two totally different reasons for the ISM to have about the pressure it does.

My sense is that the pressure problem is badly overdetermined and that we are missing some vital links. Although the FGH model is not valid in detail, there are similar two phase models based on photoelectric rather than cosmic ray heating (e.g., Wolfire, Hollenbach, McKee, Tielens & Bakes 1995) that lead to predictions of the thermal pressure required within stable clouds. One might suppose that the medium, via a mechanism like that sketched in FGH, conspires to stabilize clouds to avoid having too dense an intercloud medium. But the situation is precarious. For example, if the atmosphere of the Earth were removed, and if the oceans were pure water, water would evaporate to create an atmosphere sufficient to prevent the oceans' further boiling. Only the top few meters of ocean would be required. But the ISM is very different. Roughly half the mass is in molecular clouds, the other half split fairly evenly between the diffuse clouds and the warm intercloud medium—there is not a huge reservoir of evaporable material in the clouds.

But that's only half the problem. I can't shake the feeling that McKee & Ostriker were correct in their notion that the SNR pressure is important. Some of the energy is distributed very widely and is not fully dissipated for a long time. The lower the pressure of the surroundings, the larger the region perturbed and the longer it lasts, raising the remnant contribution to the pressure. Let us suppose we could formulate a theory that would determine the SNR contribution to the pressure as a function of the intercloud density, the SN rate per unit volume, and the total pressure. Plug in the observed values and bingo, we find out what the remnants can accomplish. I have done this more than once over the years, essentially always with the same result, $p/k \sim 10^4$ cm^{-3} K. Maybe I got that result because I wanted it, or maybe because it is unavoidable. In any case it is very strange. It is a sufficient pressure to bind the clouds. But in today's models, the pressure required to bind diffuse clouds depends on gas metallicity, depletion onto grains, details of the properties of very small grains, and the abundance of starlight. Why do both pressure estimates, depending on totally different quantities and physics, concur to within better than a factor of 3? Which is in control? Are dust grains modified in the intercloud environment until they have properties that will admit stable clouds in the pressure defined by the supernovae? Is the supernova rate controlled by the interstellar pressure, so that cloud existence is accomplished via adjustment of the SNR pressure, and only secondarily through the associated adjustment of the weight of the intercloud gas?

Or is something altogether different "in charge"? For example, somewhere I heard that cold disks are unstable, buckling and heating themselves spontaneously to achieve an acceptable scale height. I have wondered whether this might happen to the interstellar component as well as to the stars. If it does, then one presumes that a certain minimum interstellar pressure is required by the disk itself. Much of this pressure could be ballistic, but my sense is that the ISM cannot consist of ballistic bits as the stars do—more on that below. Maybe in many regions, this disk stabilization pressure is all there is. If there is too little surface density of interstellar material, it will all be in intercloud form, with no star formation, and no supernovae. With more material, some will be able to condense, leaving clouds, stars, supernovae, dust, and starlight in the right relative proportions to make things behave the way they do in our vicinity. Perhaps with even more interstellar matter, the supernovae become important contributors to the pressure and the disk inflates beyond the disk-required thickness. Perhaps star formation enters a runaway mode, a starburst with its own controls, until the gaseous surface density drops to a more quiescent level. In short, perhaps the Galaxy has the gaseous surface density it does because if it had more it would rapidly dump it into stars. So we sit just below a critical surface density with all sorts of odd coincidences baffling us. At a lower surface density yet, one supposes we wouldn't be here to wonder why.

Dynamic Pressure: Spitzer told us that the mean free path between clouds is 160 pc and that at a mean speed of $\sim$10 km s^{-1} $\sim$10 pc Myr^{-1} they would collide every 20 Myr or so. That was long ago and I don't know whether he would say it now, but it probably isn't a very accurate picture. Clouds are not isolated entities, they are probably structures within structures, interacting over considerable distances via their connecting magnetic fields. You really want a model? How about the modern (and American) version of plum pudding? String some beads of lead shot on fishing line and suspend many such strings in rapidly cooling Jell-o. When jelling is complete, shake it. Notice that the lead shot moves around but there are never contact collisions. The motion of the Jell-o and shot is a form of dynamic pressure.

Turbulence: Turbulence doesn't mean what I think it should, not anymore. I don't like the word, but it has become a fixture and there's not much I can do about it; mathematically inclined plasma physicists got there first. Turb, turbine, twist, torque, eddy, swirl, twirl, turn. Apparently turbulence hasn't to do exclusively with these familiar forms. For example, I wouldn't be inclined to say that Jell-o could contain turbulence. Not unless I put it in a blender anyway. But its chief characteristic is a healthy wave field, and if through nonlinearities that wave field can cascade to shorter scales, then nowadays it's turbulence. So, my only words of wisdom on this subject are that when others speak of interstellar turbulence, cascades, power spectra and the like, do not immediately assume they are speaking of an eddy field.

Radiation Pressure: Radiation pressure is not always negligible. You'd be surprised how large it can be sometimes—take a look at articles on levitation of dust grains and interstellar material from Ferrini, Franco, & Ferrara, for example, or work it out for yourself. Do you really think it's an accident that the galactic disk has roughly optical depth unity and 1 eV cm^{-3} of starlight?

Cosmic Ray Pressure: There are people who believe that cosmic rays are accelerated by shock waves in supernova remnants. The initial surge of excitement came from discovery of the fact that a shock with the unique adiabatic compression factor of 4 would accelerate particles to the observed power law spectral index. Pre-radiative supernovae are a copious supplier of such shocks; because supernovae (with a reasonable available power) had long been suspected as the source, the leap was made. Last year, Don Ellison and Steve Reynolds made an extremely responsible attempt to bring together people who study SNRs and those who model cosmic ray acceleration. A very fine review appears in PASP (1994). My reaction (not widely shared among the participants) is that there is no more evidence that SNRs are the primary source of cosmic rays than there was say 25 years ago. (The synchrotron emission at the outer shocks of older remnants is insufficient evidence before correction for the Van der Laan mechanism.)

Nevertheless, cosmic ray acceleration, propagation, pressure, and escape from the Galaxy are all vital components of interstellar activity. You'd best not leave them out of your model. Does the rate of star formation depend on the population of cosmic rays that penetrate dense clouds and drive the ion chemistry there?

Magnetic Field Pressure, Tension, and Flux Ropes: It is easy to be uncomfortable with magnetic fields, or with the uses they are put to by modelers who seem to need the field to save their theories. But it is folly to pursue a model very far without inquiring whether the magnetic field expected to be present will not totally change the basic features. Even the most rudimentary inclusion is better than neglect. In the ISM, the magnetic field is one of the major pressure components, is difficult to compress but highly elastic, and has a major effect on transport processes—of the wave field, thermal conduction, and cosmic ray propagation in particular.

A striking feature of the field is that it pushes in two directions and pulls in the third. In force free configurations (e.g., the field around a current carrying wire or of a dipole), these two features are balanced. Tension provides a negative pressure, $-B^2/8\pi$, acting to straighten flux tubes, propagate transverse Alfven waves, drive the interchange instability, support clouds against the galactic gravity (Cox 1988; Boulares & Cox 1990), etc. The gradient of the tension provides net force even when the field lines are straight. The tension in twisted flux tubes supports torsional Alfven waves, and therefore transmission of torque.

Twisted flux tubes also pinch, and can conceivably be a major player in the mass transfer from low to high density. Because the field is strong and difficult to compress, intercloud material must normally be gathered along field lines or collect in reconnection regions (to rid itself of excess flux) to move to higher density. But the field can be used to overwhelm itself if it can be twisted. With sufficient torque, material can be squeezed in a flux rope much like water in a damp twisted towel. If ambipolar diffusion and/or reconnection is faster in this denser environment, a net transfer to higher density is achieved, even with the relaxation of the torque. Perhaps clouds drip out of the twisted towel.

I expect flux ropes to be common features in MHD models of spiral density waves, essentially because rolling motions will be unavoidable when the vertical structure of the waves is fully explored (Martos 1993). The waves are likely to resemble tidal bores more than simple shocks. (Tidal bores are hydraulic jumps; look for one in the floor of your kitchen sink next time you turn on the water.)

Flux ropes also provide one hope for understanding the presence of extremely dense interstellar material as a common feature (see e.g., Frail et al. 1994). We need to know, however, whether their formation requires extraordinary conditions or is a natural consequence of instabilities in already well accepted interstellar behaviors.

Note: I think it's an established fact that mass in the ISM tends to be concentrated at the largest scales—but if you start working on hierarchical models, you'd best check up on this.

Sheets and Shells: They exist, apparently, and are a major organizing feature of the mass distribution in the ISM (e.g., Bregman & Ashe 1991). Spectrally identifiable clouds are likely subunits within these larger structures. Presumably the structures arise via SN and superbubble activity.

Superbubbles, Chimneys, and Worms: These are easily confused concepts. Superbubbles are two things, large roundish structures observed in space and velocity, and the theoretical structures produced in models of the activity of OB associations. Chimneys are the nearly vertical walls of model superbubbles large enough to experience the density gradient normal to the galactic plane. Worms, on the other hand, are observed vertical structures containing an unknown fraction of the high z neutral hydrogen, that may or may not be related to the theoretical concept of chimneys. Opinion varies on whether it takes one or two worms to make a chimney. In one instance (Maciejewski et al. 1995) two worms appear to open into a V which is topped by an apparent superbubble $\sim$400 pc in diameter, extending about 600 pc off the plane.

In the past it has been fairly common to suppose that OB associations spewed effluent hot gas out into the galactic halo through their chimneys. That, however was prior to full appreciation of the vertical extent of the interstellar mass and field distribution (e.g., Cox 1989). At present it seems to be more fashionable to include the mass and pressure distribution as a uniform layer in which the chimneys of the largest OB associations blow large bubbles (e.g., Ferrière 1995), rather like the observed bubble described above.

Porosity: Porosity is a theoretical measure of the degree to which supernovae (or generalized to OB association bubbles) will disrupt a hypothetical ambient interstellar medium. It is a characteristic of a model. Specifically, for a given ISM model and SNR evolution within it, one calculates the volume fraction occupied by the population of remnants under the assumption that they do not overlap or interact.

If the result is not small compared to 1, one concludes that the remnants would disrupt the assumed medium and force it into an entirely different state in which hot gas and low density play a prominent part. McKee & Ostriker (1977) found that the porosity of a medium with warm intercloud density less than 0.3 cm^{-3} was at least 3. The calculation seemed simple, the logic unassailable. This result, which I took to calling the porosity imperative, was the one of three major motivations behind a broad acceptance of the idea that most of interstellar space was hot.

Slavin & Cox (1993), however, have shown that with current parameters and inclusion of nonthermal pressure, the porosity of a warm intercloud model could be less that 0.1. Our results do not guarantee that the supernovae would not disrupt the medium, but the porosity imperative has lost its oomph. Things are no longer so clear, though they certainly would be if Slavin's remnants were observed with their predicted characteristics in the FUV.

Hot Gas in the ISM: We have discussed this elsewhere at length (e.g., Cox 1990, 1991, 1993; Shelton & Cox 1994; Slavin & Cox 1993). The bottom line is that with the porosity imperative gone, there is very little to suggest that hot gas might be common in the ISM. The remaining evidence involves only O VI and other high stage ions, the soft x-ray background, and the assumption of interphase thermal pressure balance. A reanalysis of the Copernicus O VI data (Shelton & Cox 1994) has shown that the observed ions are probably located within discrete disturbances (SNRs and superbubbles) and should not be attributed to "interfaces" of clouds immersed in a pervasive hot interstellar component. Thus, the details of the existing O VI data speak against the picture they have long been supposed to support. The jury is still out on the soft x-ray background, but much of it arises locally; that which does not is patchy, probably more consistent with an origin in discrete disturbances than in a pervasive phase. In addition, along some very low density sightlines, there is no appreciable x-ray excess as might be expected from a pervasive hot phase. And finally, we come to the evidence from interphase pressure balance—what confines the high latitude clouds? The answer is unclear, but from my earlier remarks on this general topic, it is clearly not appropriate to assume that it has to be a comparable thermal pressure from hot gas.

Diffuse Ionized Gas: Miller & Cox (1993) showed that the high latitude ionized gas studied by Ron Reynolds could be due to O star radiation, but that one would then expect

H alpha to be brighter above regions with concentrations of O stars. (This appears to be true in Perseus.) Another feature of this picture is that much of the apparently diffuse H alpha should arise in cloud boundaries illuminated by the ionizing radiation. The correlation between H alpha and 21 cm has been explored for one field (Reynolds, Tufte, Heiles, Kung, & McCullough 1995), with curious results. It will be explored much more generally when results are in from Reynolds's WHAM (Wisconsin H-Alpha Mapper) telescope.

Meanwhile there is a fly in this ointment also. Helium appears to be significantly less ionized in the diffuse ISM than hydrogen is (Reynolds & Tufte 1995; Heiles et al., in preparation). Dennis Sciama is probably quite happy about this (e.g., Sciama 1994), but the rest of us are wondering whether radiative transfer of diffuse radiation off cloud boundaries or some other effect could be the culprit. Perhaps the overluminosity of ionizing radiation from some B stars, as found with *EUVE* (Cassinelli et al. 1995) will turn out to alter the expected ratio of ionized helium to hydrogen sufficiently and save the day.

Warm Intercloud Gas in the ISM: Following the demise of the porosity imperative and the clarification of the O VI evidence, my personal view was that the warm inter-cloud component would soon return to fashion. (If Slavin's SNR bubbles are eventually observed, fashion will be much too weak a word.) But I am presently uneasy because there are GHRS spectra that appear to show that there are very long lines of sight with virtually nothing filling the space along them (e.g., Spitzer & Fitzpatrick 1993). I am uncertain whether a quasi-homogeneous warm intercloud medium could have been seen, and whether velocity crowding could have created apparent clouds where there are none. But I am steeling myself for the possibility that much of space is actually very close to empty.

Empty Spaces: In 1986, Priscilla Frisch asked me why I didn't consider the possibility that much of interstellar space could be effectively empty. At the time, I couldn't conceive of it, but I've been trying.

In a model of the z-dependent porosity due to SNRs and superbubbles, Ferrière (1995) found a very high porosity well off the plane of the galaxy, due almost entirely to defunct and dying superbubbles. They take a long time to dissipate. It is quite possible that their interiors could have cooled to essentially negligible thermal pressure well before the bubbles disappear. In such a situation, the bubble walls would be seen as ensembles of high latitude clouds, while the cavity could be a transient emptiness.

If flux ropes turn out to be a major feature of the interstellar mass and field configuration, then individual filaments of material could have appreciable cohesion, the space between them could perhaps be largely empty.

One might suppose that if near emptiness were common, it would soon become very hot, there are after all supernovae still, and even cosmic ray heating will be important at sufficiently low density. McKee & Ostriker used thermal evaporation of clouds to stabilize their model against thermal runaway, radiative cooling being less important at higher temperatures and lower density. But at sufficiently low density, things change. If there is too little material to thermalize a supernova ejecta's kinetic energy, a remnant will not "heat"; the ejecta will sweep through the emptiness until it finds matter to splat against. If the void is sufficiently large, the mass density within it remains negligible, the shock into the surrounding medium is immediately radiative and soon unobservable (S.J. Arthur 1995, private communication). Do you recall that the Crab Nebula appears to have hit only nothingness so far? Was that nothingness of the star's own making or was it a characteristic of the ambient medium?

Thermal Conduction and Thermal Evaporation: Thermal conduction is commonly neglected in hot gas studies, but it probably rarely should be (Slavin & Cox 1992). It can be suppressed by a tangled magnetic field, but it is often such a powerful transport phenomenon that a considerable amount of suppression would be required to curtail its tendency to flatten temperature profiles in hot gas (Tao 1995), and to prevent excavation of ridiculously low density cavities such as one finds, for example, in the Sedov solution. In situations in which the electron and ion temperatures are not equilibrated, one should even not ignore the effects of ion conduction (Cui & Cox 1992). Our group is presently constructing a model of the SNR W44 from which it clear that the centrally brightened thermal x-ray emission has just the characteristics expected of a remnant with thermal conduction included (Shelton, Smith, & Cox 1995). It would probably be instructive to consult students of the solar wind to learn what is known about the effective radial thermal conductivity in that context, across the spiraling magnetic field.

But saying that thermal conduction is active in regions of hot gas is not an invitation to suppose its effects are never suppressed. The structure of the strong gradient boundary between hot and cold gas is certainly altered from the field free case, and can have an enormous impact on one's conclusions regarding thermal evaporation of clouds (e.g., Slavin 1989; Borkowski, Balbus, & Fristrom 1990). One hopes that in time, observations of the ionic and kinetic structure in the boundary between the Local Cloud and the surrounding hot gas will clarify matters somewhat. At present, it is fairly safe to say that thermal evaporation of clouds is a theoretical possibility which has never found confirmation in the ISM.

Galactic Fountains: This is a thoroughly intriguing concept which has seen a fair amount of redefinition over the years. It was conceived in the days when hot gas was widely believed to pervade the galactic disk and burble up out of it into a fountain. Later it was the returning chimney effluent. Next it will probably be redefined to be the cooling hot gas in the high latitude extensions of superbubbles. It's chief purpose has been to explain the existence of high stage ions well off the plane of the Galaxy. There could well be no identifiable galactic fountain, but calculations of one's properties are useful in setting limits on the rate at which supernovae generate hot gas, and on the conditions in which that gas cools and recombines.

Galactic Wind: Cosmic rays leave the Galaxy; does anything else? One is sometimes left with the impression from studies of hot gas in clusters that as much mass leaves a galaxy in a wind as is condensed into stars. But this seems to involve mainly elliptical galaxies. Spirals with starbursts might also have appreciable winds, as might regions around some galactic nuclei, but when it comes to the Solar Neighborhood of the Milky Way, there is no evidence I know of to support the idea that there is a wind, other than that of the cosmic rays. But I'm not sure that the door is fully closed on interesting possibilities. Radiation pressure ejection of dust grains could change our sense of chemical evolution; Lyman alpha pressure on neutral hydrogen might be exciting. It could be important to know whether cosmic rays diffuse and escape the Galaxy individually, or leave more collectively when their local pressure overwhelms magnetic tension and they flare out of the Galaxy. (The later provides a natural way to understand the magnitude of the trapped cosmic ray pressure.)

Galactic Flares and Microflares: If there are galactic flares, do they suddenly hoist great quantities of material to high z, after which it rains back into the disk? Is such material heated to the point that it contains high stage ions? Do microflares (Raymond 1992) occur in the galactic halo, and if so, what are their characteristics? Are they significant actors or merely a source of noise in our observables, the lithium like ions of carbon, nitrogen, and oxygen?

REFERENCES

BREGMAN, J. N. & ASHE, G. A. 1991, in The Interstellar Disk-Halo Connection in Galaxies, ed. H. Bloemen, 387. Dordrecht: Kluwer

BORKOWSKI, K. J., BALBUS, S. A., & FRISTROM, C. C. 1990, ApJ, 355, 501

BOULARES, A. & COX, D. P. 1988, ApJ, 365, 544

BOWYER, S,, LIEU, R., SIDHER, S. D., LAMPTON, M., & KNUDE, J. 1995, Nature, 375, 212

COX, D. P. 1995, Nature, 375, 185

CASSINELLI, J. P. ET AL. 1995, ApJ, 438, 932; and Cassinelli these proceedings

COX, D. P., 1988, in Supernova Remnants and the Interstellar Medium, ed. R. S. Roger & T. L. Landecker, Cambridge Univ. Press, 73

COX, D. P. 1989, in Structure and Dynamics of the Interstellar Medium, ed. Tenorio-Tagle et al., Springer-Verlag, 500

COX, D. P. 1990, in The ISM in External Galaxies, ed. Thronson & Shull, Dordrecht: Kluwer, 181

COX, D. P. 1991, in The Interstellar Disk-Halo Connection in Galaxies, ed. H. Bloemen, Dordrecht: Kluwer, 143

COX, D. P. 1993, in Massive Stars: Their Lives In The Interstellar Medium, 1993, ed. J. P. Cassinelli & E. B. Churchwell, ASP Conf. Series, 35, 402

COX, D. P. & REYNOLDS, R. J. 1987, ARA&AA, 25, 303

CUI, W. & COX, D. P. 1992, ApJ, 401, 206

ELLISON, D. C., REYNOLDS, S. P., ET AL. 1994, PASP, 106, 780

FERRARA, A. 1993, ApJ, 407, 157

FERRIÈRE, K. M. 1995, ApJ, 441, 281

FRANCO, J. FERRINI, F., & FERRARA, A., & BARSELLA, B. 1991, ApJ, 366, 443

FIELD, G. B., GOLDSMITH, D. W., & HABING, H. J. 1969, ApJ, 155, L149 (FGH)

FRAIL, D. A. WEISBERG, J. M. CORDES, J. M., & MATHERS, C. 1994, ApJ, 436, 144

MACIEJEWSKI, W., MURPHY, E., LOCKMAN, F. J., & SAVAGE, B. D., 1995, ApJ, in preparation

MARTOS, M. 1993, Ph.D. thesis, UW-Madison

McKEE, C. F. & OSTRIKER, J. P. 1977, ApJ, 218, 148 MO

MILLER, W. W. III, & COX, D. P. 1993, ApJ, 417, 579

RAYMOND, J. R. 1992, 384, 502

REYNOLDS, R. J., TUFTE, S., HEILES, C., KUNG, D., & McCULLOUGH, P. 1995, ApJ, in press

REYNOLDS, R. J. & TUFTE, S. 1995, ApJ, 439, L17

SCIAMA, D. 1994, ApJ, 426, 65

SHELTON, R. L. & COX, D. P. 1994, ApJ, 434, 599

SHELTON, R. L., SMITH, R. K., &, COX, D. P. 1995, BAAS, 186th AAS, late abstracts

SLAVIN, J. D. 1989, ApJ, 346, 718

SLAVIN, J. D. & COX, D. P. 1992, ApJ, 392, 131

SLAVIN, J. D. & COX, D. P. 1993, ApJ, 417, 187

SPITZER, L. & FITZPATRICK, E. L. 1993, ApJ, 409, 299

TAO, L. 1995, MNRAS, in press

WOLFIRE, M. G., HOLLENBACH, D., McKEE, C. F., TIELENS, A. G. G. M., & BAKES, E. L. O. 1995, ApJ, 443, 152

New Insights on the Interstellar Medium from EUV Observations

STUART BOWYER

Center for EUV Astrophysics, 2150 Kittredge Street,
University of California, Berkeley, CA 94720-5030, USA

Observations in the EUV band have provided new insights into the interstellar medium. In the following I discuss two areas in which EUV observations are providing unique information: the ionization state of the ISM, and the pressure of the hot phase of the local interstellar medium and the bearing of this work on the McKee-Ostriker model of the ISM.

1. The Pressure of the Hot Phase of the Local Interstellar Medium and Implications Regarding the Mckee-Ostriker Model

Numerous estimates of the state of the hot ISM have been made based on theoretical models and meager observational data. In Table 1, I list some of the estimates of this pressure. They are highly disparate.

In principle, the pressure can be obtained directly from the relation:

$$P = 1.92kT\sqrt{EM/L} \tag{1.1}$$

where T is the temperature, EM is the emission measure, L is the line-of-sight length of the emitting region, k is Boltzmann's constant and the factor of 1.92 is appropriate for the total number of particles in a near fully-ionized plasma that is 90% hydrogen and 10% helium.

A number of shadows have been detected in the X-ray background; I summarize these in Table 2. I also list an EUV shadow detected with *EUVE*. Attempts have been made to use shadows in the X-ray background to obtain the pressure of the hot phase of the ISM using Equation 1.1. Unfortunately, because of the relatively high transmission of the ISM to X-rays, the X-ray flux observed is a mixture of nearby emission convolved with partially absorbed emission from more distant regions. Nonetheless, estimates of the pressure have been made from the X-ray measurements using various assumptions and extensive modeling; I list the result of one recent analysis using X-ray shadows in Table 1. In Table 2 I also list the wavelengths of the bandpasses of the X-ray observations of cloud shadows at 10% of peak transmission. I also show the mean effective wavelength determined by a weighted integral over a nominal plasma spectrum having $T = 7 \times 10^5$ K and a foreground H I column of $N_H = 5 \times 10^{18}$ cm^{-2} to account for absorption by the local cloud around the sun. Finally, I show the column density of hydrogen associated with a non-depleted ISM that would produce one optical depth of absorption for the mean effective wavelength of the observation. The substantial column needed to absorb X-rays highlights the difficulties of using cloud shadows in the X-ray band to define a definitive emission path length.

During the Extreme Ultraviolet Explorer's all-sky survey, a long exposure along the ecliptic plane was obtained by the deep survey telescope. Only part of this data has been fully analyzed but in the analyzed data an EUV shadow was detected in the Lexan/boron (Lx/B) filter (65–190 Å) (Lieu et al. 1993). A statistical analysis using a sliding rectangular box of 0.3° width revealed that the core of this shadow is 3.9 σ below the average background. I show the relevant parameters of this observation in Table 2.

In contrast with X-ray shadow data, EUV shadows have great potential for determining

255

TABLE 1. Estimates of the Hot ISM Pressure

Method	Reference	P/k, K/cm^3
Static snr model	Cox & Reynolds 1987	10000
Halo equilibrium model	Spitzer 1990	11000 (thermal plus magnetic)
C IV & O III] emission data	Martin & Bowyer 1990 (corrected)	3000
CIV modeled to midplane	Shull & Slavin 1994	4400–7400
Pressure equilibrium with cool diffuse clouds	Data from Jenkins, Jura & Loewenstein 1983	280–50,000 4000 (typical)
X-ray shadows	Snowden, McCammon, & Verter 1993	12500

TABLE 2. Shadows in the Diffuse Background

Instrument	Reference	Band (keV)	10% Wavelengths (Å)	Mean Wavelength (Å)	$N_{\mathrm{H\,I}}$ at $1/e$ (cm^{-2})
ROSAT PSPC	Burrows & Mendenhall 1991; Snowden et al. 1991	0.25	44–85	69	1×10^{20}
	Snowden, McCammon, & Verter 1993	0.75	11–25	18	2×10^{21}
	Snowden, McCammon, & Verter 1993	1.5	7–10	8	1×10^{22}
EUVE DS	Bowyer et al. 1995	Lx/B	65–190	116	2×10^{19}

the pressure of the local ISM because EUV emission is completely absorbed by even thin clouds, thus defining the path length of the region observed. Bowyer et al. (1995) have compared the EUV cloud shadow data with *IRAS* 100 μm flux averaged over the deep survey telescope field of view. In the *IRAS* data, compact molecular clouds and isolated cirrus appear as enhanced features above the background levels established by zodiacal light and diffuse cirrus. This background was fitted by a cubic spline function and was subtracted to isolate discrete clouds. An enhancement was found in the *IRAS* data which is aligned with the EUV absorption dip.

The mean intensity of the *IRAS* feature averaged over the 3 σ EUV absorption region is 0.518 MJy sr^{-1} above the baseline *IRAS* flux. Using a standard conversion ratio, the associated hydrogen column is 6×10^{19} cm^{-2}. This is about three optical depths at the mean wavelength of the *EUVE* Lx/B filter. Hence EUV flux from behind the cloud is completely attenuated, and the count rate in the direction of the cloud in this band is due entirely to foreground emission.

Bowyer et al. determined the distance to the EUV-absorbing cloud using Strömgren photometry; this technique can be used to derive the absolute magnitude and reddening of most classes of main sequence stars (Strömgren 1966). High accuracy is achievable provided a large number of pertinent standards are observed. Thirty-six of 100 stars we studied met all criteria required for the determination of both distance and color excess; the majority of the stars rejected were metal-poor reddened giants and stars that had evolved somewhat off the main sequence. Fifteen of the acceptable stars were at distances greater than 200 pc and provide no information of relevance to the cloud distance. The color excess vs. distance for the remaining 21 stars in the 16 square degrees centered on the cloud show the presence of a cloud in the direction of the EUV shadow that is at a distance of ≤ 40 pc.

We derived the emission measure for the plasma in the line of sight to the cloud by folding the *EUVE* deep survey count rate in the direction of the cloud, 0.437 ± 0.02 counts s^{-1}, with the Landini & Fossi code for a temperature of 7×10^5 K. The emission measure in the direction of the cloud is found to be 0.0077 cm^{-6} pc.

Combining the results in Equation 1.1, we immediately obtain the pressure of the hot phase of the ISM in the direction of the absorbing feature. Using a distance to the cloud of 40 pc, we find the pressure is $p/k = 19000$ K cm^{-3}. Because we do not have unreddened stars in the direction of the cloud at distances less than 40 pc, the cloud could, in principle, be closer. In that case the pressure would be even higher.

The McKee-Ostriker model (1977) assumes eventual pressure equilibrium between cool denser clouds and the hot phase of the ISM. The pressure at any particular location is established by the effects of past supernovas. The pressure of the local cloud surrounding the Sun has been well established by a variety of techniques, the most direct is through the analysis of solar HeI 584 Å radiation resonantly scattered by helium in the inflowing cloud. The pressure obtained for the local cloud is 730 ± 30 cm^{-3}K. (Frisch 1994)

There has been a general belief that the pressure of the local bubble is larger than the local cloud, but because of the penetrating power of X-rays it is not clear where this higher temperature/pressure gas originates. This radiation may be produced nearby, or further away, or from a boundary wall, or from all these locations. Wherever its origin, the flux would be modified by absorption by unknown amounts of intervening cooler material.

However, we can now, for the first time, compare two direct measurements of the pressure of nearby ISM material that has experienced the same supernovae shock conditions: (1) the pressure of the parcel of gas in the path between the Sun and the cloud reported here, and (2) the pressure of the local interstellar cloud flowing through the solar system. It is almost superfluous to note that the shock conditions are the same for these two local components. There has been no recent supernovae in this region and Frisch (1994) and Bertin et al. (1993) have provided independent evidence that this region has not been violently heated within the past million years. Hence material in this region has had more than sufficient time to establish pressure equilibrium as per the McKee-Ostriker hypothesis.

The wide difference between the pressure of the local solar cloud and the pressure of the hot quiescent EUV emitting material discussed here directly challenges the underlying concepts of the McKee-Ostriker model.

2. The Ionization State of the ISM

The ionization state of the ISM depends on only a few key mechanisms which produce this ionization, hence a measurement of this parameter will provide crucial information on the current status and evolution of the ISM. This topic has been the subject of over a dozen studies; the results before EUV observations were essentially a scatter plot in a χ_H, χ_{He} diagram.

Hot white dwarfs ($T_{eff} \geq 25,000$ K) are copious sources of EUV continuum radiation, making them excellent objects to use in studying the ISM. Interstellar column densities of hydrogen and both neutral and singly ionized helium can be directly obtained from their EUV absorption signature in hot white dwarf spectra given the known wavelength dependence of the photoelectric absorption and the photoionization edges of HeI at 504

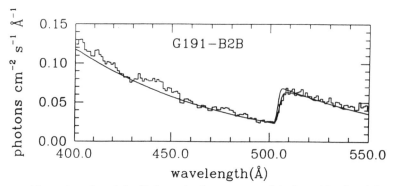

FIGURE 1. Absorption edge of the He I seen in the spectrum of the hot white dwarf G191-B2B.

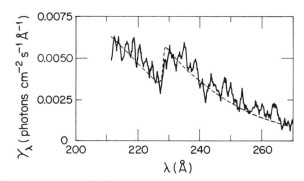

FIGURE 2. Absorption edge of He II seen in the spectrum of the hot white dwarf GD 246.

Å and of He II at 228 Å. This technique offers a direct and model independent way of measuring densities for the main constituents of the ISM.

A measurement of the He I/H I made at longer EUV wavelengths from a sounding rocket observation of the hot white dwarf G191-B2B (Green et al. 1990) resulted in a ratio greater than 10 suggesting that hydrogen is not preferentially ionized in the ISM. Further observations at longer EUV wavelengths by the Hopkins Ultraviolet Telescope of the hot white dwarfs G191-B2B and HZ 43 (Kimble et al. 1993a; 1993b) led to similar conclusions but also opened the prospect of preferential ionization of helium.

The comprehensive wavelength coverage of the *EUVE* spectrographs offers the capability of direct, model independent measurements of both neutral and ionized helium in the ISM. In Figure 1 I show an example of absorption due to interstellar He I. In Figure 2 I show an example of absorption due to interstellar He II. The power of these observations for ISM studies is immediately obvious. *EUVE* has observed several white dwarfs for which a determination of the interstellar column densities of H I, He I, and He II is possible. Vennes et al. (1993) and Dupuis et al. (1995) have used these spectra to explore this topic. In Table 3 I show the result of these analyses and Figure 3 summarizes the most important features. The Vennes et al. (1993) result provides direct spectral data of the He II edge in the hot white dwarf GD 246 that shows helium is ionized at 25% in this view direction, while hydrogen is certainly less than 25% ionized and may be much less; and the work by Dupuis et al. (1995) shows, using similar data on the He I edge, that H I/He I = 14 and that the hydrogen is predominantly neutral.

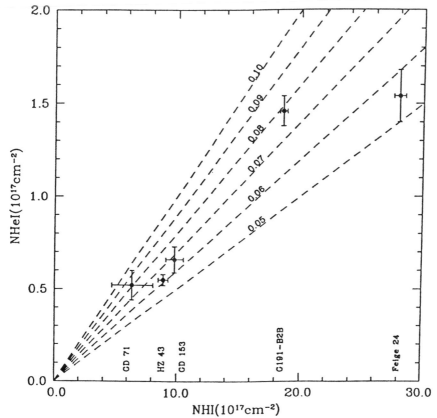

FIGURE 3. The neutral helium and hydrogen columns toward a number of hot white dwarfs.

This work was supported by NASA contract NAS5-30180.

REFERENCES

BERTIN, P., LALLEMENT, R., FERLET, R., & VIDAL-MADJAR, A. 1993, A&A, 278, 549

BOWYER, S., LIEU, R., SIDHER, S. D., LAMPTON, M., & KNUDE, J. 1995, Nature, 375, 212

BURROWS, D. N., & MENDENHALL, J. A. 1991, Nature, 351, 629

COX, D. P., & REYNOLDS, R. J. 1987, ARA&A, 25, 303

DUPUIS, J., ET AL. 1995, ApJ, submitted

FRISCH, P. C., & YORK, D. G. 1983, ApJ, 272, L59

FRISCH, P. C. 1994, Science, 265, 1423

GREEN, J., JELINSKY, P., & BOWYER, S. 1990, ApJ, 359, 499

JAKOBSEN, P., & KAHN, S. M. 1986, ApJ, 309, 682

JENKINS, E. B., JURA, M., & LOEWENSTEIN, M. 1983, ApJ, 270, 88

KIMBLE, R. A., ET AL. 1993a, ApJ, 404, 663

KIMBLE, R. A., DAVIDSEN, A. F., LONG, K. S., & FELDMAN, P. D. 1993b, ApJ, 408, L41

LIEU, R., SUMNER, T. J., BOWYER, S., & SIDHER, S. D. 1993, BAAS, 25 #2, 860

MARTIN, C. & BOWYER, S. 1990, ApJ, 350, 242

McKee, C., & Ostriker, J. 1977, ApJ, 218, 148

Paresce, F. 1984, AJ, 89, 1022

Shull, J. M., & Slavin, J. D. 1994, ApJ, 427, 784

Spitzer, L., Jr. 1990, ARA&A, 28, 71

Snowden, S. L., Mebold, U., Hirth, W., Herbstmeier, U., & Schmitt, J. H. M. M. 1991, Science, 252, 1529

Snowden, S. L., McCammon, D., & Verter, F. 1993, ApJ, 409, L21

Strömgren, B. 1966, ARA&A, 4, 433

Vennes, S., Dupuis, J., Rumph, T., Drake, J., Bowyer, S., Chayer, P., & Fontaine, G. 1993, ApJ, 410, L119

The Morphology and Physics of the Local Interstellar Medium

FREDRICK C. BRUHWEILER

Department of Physics, The Catholic Univ. of America, Washington, DC 20064, USA

We are finally on the threshold of obtaining a coherent morphological and physical picture for the local interstellar medium (LISM), especially the region within 300 pc of the Sun. The *EUVE* is playing a special role in revealing this picture. This instrument can provide direct measurements of the the radiation field that photoionizes both hydrogen and helium. It also can yield direct measurements of the column densities of hydrogen, but especially He I and He II toward nearby white dwarfs. These observations suggest that the ionization in the Local Cloud, the cloud in which the Sun is embedded, is not in equilibrium, but in a recombination phase. Heuristic calculations imply that the the present ionization is due to the passage of shocks, at times greater than 3×10^6 years ago. The origin of these shocks are probably linked to the supernova which was responsible for the expanding nebular complex of clouds know as the Loop I supernova remnant, of which the Local Cloud is a part. extreme-UV radiation field, that which ionizes both hydrogen and helium in the LISM. Of the ISM within 300 pc, the volume appears to be predominantly filled by hot (10^6 K) coronal gas. This gas is laced with six largescale shell structures with diameters $\sim$100–150 pc including the long-recognized radio loops, Loop I–IV, as well as the Orion-Eridanus and Gum Nebulae are identified. An idea that has evolved in the literature for over two decades is that the kinematically-linked OB associations representing Gould's Belt, plus the gas and dust of Lindblad's Ring, require that previous supernova activity and stellar winds carved out a 400–600 pc diameter cavity some 3 to 6×10^7 yr ago. This activity produced a pre-existing low density region, into which the present young loop structures have expanded. The outer boundaries of the identified expanding loop structures, inside this pre-existing cavity, delineate the periphery of the the mis-named "local interstellar bubble." Thus, this picture naturally explains some of the problems often associated with the presence of this low density region exterior to Loop I.

---

1. The Emerging Picture Within 50 pc

The importance of the local interstellar medium is sometimes not fully appreciated by many astronomers, even though the study of this region is crucial to understanding the basic physical processes that arise in the interstellar gas at greater distances. For example, absorption studies for lines-of-sight toward nearby stars, where the distances are relatively certain, can we unambiguously determine the gas morphology and physical conditions in this gas. By studying long lines-of-sight, one usually does not know at what distance or in what cloud absorption lines arise. As we shall see, absorption studies with the *EUVE* of He I and He II photoionization edges for objects within 50 pc are now enabling us to place strong constraints on the basic ionization mechanisms for this gas.

Because stars within 50 pc generally exhibit no detectable extinction and only negligible optical absorption, efforts to probe this region had to await for the advent of UV astronomy. Only then could the strong resonance lines of the cosmically abundant ionic species be seen. Consequently, practically all of our knowledge of the LISM within 50 pc has come from spacecraft.

The faint glow of the interstellar gas, immediately in the vicinity of the Sun (within 0.03 pc) can be seen through backscattering of radiation initially produced in the solar chromosphere by the strong emission lines of H I (λ1216) and He I (λ586). Analysis of this backscattered radiation indicates an interstellar hydrogen particle density near the

S. Bowyer and R. F. Malina (eds.), Astrophysics in the Extreme Ultraviolet, 261–268.

Sun of $n(\mathrm{H\ I}) \sim 0.1\ \mathrm{cm}^{-3}$, with a temperature of $T \approx 8 - 15,000$ K (Bertaux et al. 1985; Chassefiere et al. 1988).

At greater distances, interstellar H I–Lα absorption superposed upon the corresponding chromospheric emission of nearby late-type stars has provided a means to sample the gas within ~ 5 pc of the Sun. Besides yielding estimates for H I column densities, these studies have also obtained estimates for the cosmically important D/H ratios for these lines-of-sight. Early investigations using the *Copernicus* satellite (McClintock et al. 1978, Dupree et al. 1978), and more recent studies using the *IUE* (Landsmann et al. 1985; Murthy et al. 1987) and HST (Linsky et al. 1993), are all consistent and yield $n(\mathrm{H\ I}) \sim 0.1\ \mathrm{cm}^{-3}$ and $T \sim 10,000$ K. An analysis of the collected dataset also suggests that the density increases toward $l^{\mathrm{II}} \sim 0°$ in the general direction of Sco-Cen.

Within ~ 50 pc, *IUE* observations of nearby, hot white dwarfs (Bruhweiler & Kondo 1982; also see Bruhweiler & Kondo 1981 and Dupree & Raymond 1982) have clearly indicated that neutral hydrogen is concentrated within a few parsecs of the Sun (the Local Cloud) with no other clouds detected within 50 pc. The largest detected H I column density for the white dwarfs studied was $N(\mathrm{H\ I}) \approx 2 \times 10^{18}\ \mathrm{cm}^{-2}$.

Almost every line-of-sight observed through the Local cloud shows multiple velocity components to interstellar lines as seen in high resolution spectral data acquired from the ground (Lallement et al. 1986, 1992) and from the Goddard High Resolution Spectrograph aboard the HST as any perusal of the archival data will show. Although this might be interpreted as evidence for multiple distinct clouds, given the implied history of this gas and the nature of the surrounding substate in which this gas is embedded, the multiple velocity components are more likely due to conduction fronts at the cloud interface (McKee & Cowie 1977) and shocks recently passing through this cloud (also see Bruhweiler, Smith, and Lyu 1995).

Soft energy X-ray data reveal a uniform soft X-ray background with no evidence of cloud shadows (Fried et al. 1980; McCammon et al. 1983). The collected ultraviolet and soft X-ray data indicate the Local Cloud is surrounded by a hot, coronal gas at $T \approx 10^{6}$ K and $n \approx 10^{-2.5}\ \mathrm{cm}^{-3}$.

Evidence for a conductive interface between the Local Cloud and the surrounding 10^{6} K gas is seen in GHRS data for the line-of-sight toward the B star, α Gru ($d = 26$ pc). Weak interstellar features of C IV at levels of 3.5 and 1.7 mÅ are measured for the members of the resonance doublet, while a much stronger feature is seen in Si III (Bruhweiler, Smith, & Lyu 1995). Since both of these ions have high charge-exchange rates with neutral hydrogen, they must arise in a pure H II regions. The ratios of the implied column densities are compatible with simple conductive interface models.

A further comparison of the very low level optical polarization for nearby stars (Tinbergen 1981) and ultraviolet absorption line studies has revealed that the Sun is embedded in, near the edge of, a rather diffuse cloud, with a total extent less than 15 pc (Bruhweiler 1982). For directions toward the hemisphere in the anti-galactic center direction, away from the main body of the Local Cloud, low H I column densities (i.e. $N(\mathrm{H\ I}) = 1 - 2 \times 10^{18}$ cm^{-2}) are intercepted through the outer cloud skin. Directions toward $l \approx 0°$, intercepts the main body of the cloud and typically yield $N(\mathrm{H\ I}) \approx 1 - 2 \times 10^{19}\ \mathrm{cm}^{-2}$. The core of the Local Cloud lies near $l = 5°$, $b = -20°$ (Also see Fig. 3 of Bruhweiler & Vidal-Madjar 1987). The direction of the Local Cloud core also coincides with the region showing an absence of soft X-ray sources as seen in *Rosat* WPC data (Warwick et al. 1993). The absence of these sources is due to attenuation from gas corresponding to $N(\mathrm{H}$ I)$= 1 - 2 \times 10^{19}\ \mathrm{cm}^{-2}$.

Comparisons with radio data (discussed more thoroughly below) also show that the Local Cloud lies at the periphery and shares motion of the expanding filaments of the

Loop I supernova remnant. This association with Loop I would make the Local Cloud a shell fragment, or perhaps better described as a wisp in the filaments of this supernova remnant.

2. The Ionization of Hydrogen & Helium in the Local Cloud

Before the launch of the *EUVE*, Bruhweiler & Cheng (1988) and Cheng & Bruhweiler (1990) had calculated photoionization models to predict the ionization equilibrium of H, He and heavy elements in the Local Cloud. Although we now have better estimates for the ionizing radiation fields, thanks for observations from the *EUVE*, these works are illustrative of what physical processes are important for the ionization of the two most abundant elements, H and He.

In the calculations of Cheng & Bruhweiler, contributions due to photoionization by the Extreme-UV/X-ray radiation field from the nearby hot stars and surrounding coronal substrate as well as collisional ionization, especially in the conductive interface of the Local Cloud skin are included. Effects of charge-exchange and Auger ionization are also considered. The high attenuation toward the hot stars of Sco-Cen precludes any ionizing flux shortward of the H I Lyman edge reaching the Sun from that direction. Thus, all the ionizing radiation is predicted to come from the direction of low column density, toward $l \sim 180°$.

With the then "best guess" to the EUV radiation field, these calculations showed that the ionization fraction was higher for He than for H at the Sun($X_{H+} = 0.17$ and $X_{He+} = 0.3$). Specifically, it was found that the H-ionization was determined by the ambient stellar Extreme-UV radiation field, while the He-ionization was fixed by the diffuse radiation field from the surrounding hot coronal substrate. Specifically, it is the line emission in the hot substrate, principally from the Fe-complex near 190Å that was predicted to be the most significant source of ionization.

Recent Extreme-UV observations with *HUT* of the white dwarf, G191-B2B (Kimble et al. 1993), and both WD 2309+105 (Vennes et al. 1993) and GD659 (Holberg et al. 1995) are in striking agreement with these predictions. Yet, other results indicate that this agreement may be fortuitous.

EUVE observations are now providing direct measurements for both the stellar and diffuse components to the EUV radiation field in the LISM. Perhaps the biggest surprise has been the discovery that the B star, ϵ CMa is the dominant photoionizing source for hydrogen, much stronger than the numerous, hot, hydrogen-atmosphere DA white dwarfs. An estimate of the photoionization rate for H by summing the observed EUV fluxes from these stellar sources gives $\Gamma_{H I} = 1.4 \times 10^{15}$ s^{-1} at the Sun (Vallerga & Welsh 1995). this result is not too different from that predicted by Bruhweiler & Cheng (1988). However, the contribution from the individual sources are definitely different than predicted, and the large contribution from ϵ CMa was not foreseen.

The *EUVE* is also giving us some observational limits on the diffuse EUV radiation field capable of ionizing He in the LISM. The recent non-detection of Jelinsky et al. (1995) is a factor of 12 below the level predicted by Cheng & Bruhweiler 1990). However, Jelinsky et al. did not recognize that the direction scanned in their observations overlap extensively with that of the Local Cloud core where high attenuation should be expected due to the $1 - 2 \times 10^{19}$ cm^{-2} column density for the main body of the cloud. For the emission line complex near 190Å, this corresponds to an optical depth of $\tau_{190A} \sim 1 - 2$. Notwithstanding, these observations still imply that the EUV flux is well below the Cheng & Bruhweiler predictions. Conversations with J. Vallerga also indicate that other unpublished *EUVE* observations of directions pointing away from the Local Cloud core show no diffuse EUV background radiation. Without further in depth analysis of these data, we can only estimate that the upper limit for the He ionization rate at the Sun is at least a factor of four below the Cheng & Bruhweiler prediction. *If so, this upper limit would make it highly unlikely that the observed He ionization can be due to ionization equilibrium.*

If ionization equilibrium is ruled out, then one must assume that the observed He ionization reflects gas that is undergoing a time-dependent recombination. The initial high ionization would have been produced by a supernova explosion, presumably the one that gave rise to the observed Loop I remnant (also see Cox & Anderson 1982). This ionizing event can be either a UV-flash or the passage of shocks through the gas now recognized as the Local Cloud.

We have recently performed time-dependent ionization calculations appropriate for the LISM (Lyu & Bruhweiler 1995). in which we explored the ionization and recombination from both a photoionization pulse and shocks produced by a typical supernova in the Sco-Cen association.

Our calculations indicate that unless a supernova occurs at distances less than 20 pc of a cloud, the cloud will not be significantly ionized. The high velocity mass loss from stars in Sco-Cen should have carved out a pre-existing hot cavity before the supernova explosion and the subsequent momentum of the ejecta drove the Sco-Cen shell to much larger radius. Conservative calculations indicate that the precursor cavity was much larger than 20 pc. Thus, the effects of any UV-flash should have been minimal.

The high velocity SN-ejecta after traversing the evacuated cavity would hit the surrounding clouds and compress them to high temperatures and high ionized levels. The peak temperatures in these high velocity shocks, as they pass through these clouds, should be well above 10^5 K. The temperature versus time profiles for gas at constant density above 10^5 K are very similar for a wide range of conditions (cf., Shapiro & Moore 1977). Our models for the recombining H and He below 10^5 K, with the effects of the EUV radiation field included, can easily reproduce the deduced H and He ionization fractions deduced from *EUVE* observations of sightlines toward hot white dwarfs (see above). For both the observed ionization fractions and the limits for the ionizing fluxes, acceptable fits are obtained for 2 to 3 million years since the beginning of the recombination phase (defined to be the time when T dropped below 10^5 K). Other assumptions place the time when the supernova event occurred at roughly 3.8 million years ago. These calculations indicate that time-dependent ionizations can give a reasonable explanation to the observed H and He ionization fractions deduced from *EUVE* data.

The *EUVE* is basically a Helium machine. The wavelength coverage of its spectrographs spans both the ionization edges of He I and He II. For hot white dwarfs with significant fluxes down to 228Å, there is the possibility of measuring both the absorption of He I and He II toward the same object. For objects with large H I column densities such that the He I edge at 504Å experiences large IS attenuation, the He I autoionization feature at 204Å can be used (cf., Vennes et al. 1993; Holberg et al. 1995).

All the information about the He ionization fractions are derived from observations of hot hydrogen atmosphere (DA) white dwarfs. However, one major complication in using these objects is the observed opacity from trace metals below 300Å. This opacity is mainly due to photospheric absorption from resonance lines and low-lying metastable levels of C, N, O, and Fe-peak elements. However many of the contributors have not been identified. Until proper line-blanketed non-LTE models becomes available, it will be difficult to obtain reliable He ionization ratios for the interstellar gas toward these objects. Thus, the current reliable estimates have come from white dwarfs which show little hint of metals in their spectra; such objects as GD659. (Also see contributions by I. Hubeny and M. Barstow, these proceedings.)

3. The Local Supershell Structures & the Origin of the Local Bubble

3.1. *Determining Gas Morphology Out to ∼300 pc*

Delineating the gas morphology beyond 50 pc requires a synthesis of observations spanning the electromagnetic spectrum. Perhaps the best clues to this structure of the LISM comes from radio wavelengths. The radio loops, Loop I through Loop IV, are all contained within 250–300 pc (Berkhuijsen 1971; Spoelstra 1972). The most notable of the radio loops is Loop I, which spans more than 100° of the sky. The interior of Loop I is the most prominent diffuse soft X-ray source ($\approx$ 90° in diameter) seen in the M band (440–930 eV) of the full-sky Wisconsin sounding rocket data (McCammon et al. 1983). These combined results demonstrate Loop I is filled with hot gas, hotter than that seen in all directions at lower X-ray energies. Loop I is thought to have been produced by a recent supernova that occurred in the Sco-Cen stellar association within the last few 10^6 yr. As previously mentioned, the position of the Local Cloud and its implied motion suggest that it is a shell fragment of Loop I. We emphasize that the low measured column densities, $N(\mathrm{H\ I}) \leq 2 \times 10^{18}$ cm^{-2}, are measured in the direction away from Loop I in the region often labeled the "Local Bubble" by many. This suggests that the shell complex of Loop I is actually expanding into a hot (10^6 K), pre-existing low density region.

Just where the observed diffuse X-ray emission originates is not necessarily straightforward. However, IRAS, Radio, UV absorption line, and extinction results all point to significant concentrations of cool gas at distances beyond 100 pc in the direction of the Sco-Cen, especially near $l \sim 30°$ and the clouds where the stars ζ Oph and ρ Oph are found (see de Geus 1991). Since the emission measure is proportional to $(n_e)^2$, the observed X-ray emission in the M band probably arises in front of, on or near, the dense clouds close to the back wall of the Loop I cavity.

The shells representing Loop I and the other loops are not like the surfaces of a balloon, but are more like a complex of expanding cloud fragments. Thus, the hotter gas of Loop I's interior is probably freely mixing with the hot gas exterior to Loop I.

Determining temperatures of the X-ray emitting gas is not a simple matter. Most of the X-ray flux at temperatures of interest here ($10^6 \leq T \leq 10^7$ K) is emitted as line emission. But the ion recombination times for this low density plasma are quite long. Thus, the ionic abundances, as for H and He, are not represented by electron temperature, but depend upon the ionization history of the gas. The actual emission line intensities should be far different from those predicted from ionization equilibrium. Surely as the spectral resolution improves in new X-ray instrumentation, much that we have inferred about the characteristics of this gas may well change (see Edgar et al. 1993 and also this conference).

Besides the four radio loops, Loop I–IV, two additional shell structures are also found within 300 pc. The stellar winds and supernovae from stars in the Orion OB1 association ($d \sim 400$ pc) have produced a large expanding shell complex (Cowie et al. 1979) spanning the Orion and Eridanus constellations. This complex is defined by Barnard's Loop on the East and a system of nebular filaments traced out in [O III] and H-α extending to the West in the sky (Reynolds & Ogden 1979). Burrows et al. (1993) in a re-interpretation of these data, incorporating X-ray and *IRAS* observations, suggest that this complex may extend to within 100 pc (near $l \sim 180°$) of the Sun.

Another nearby shell complex is the well-known Gum Nebula (Brandt et al. 1977;

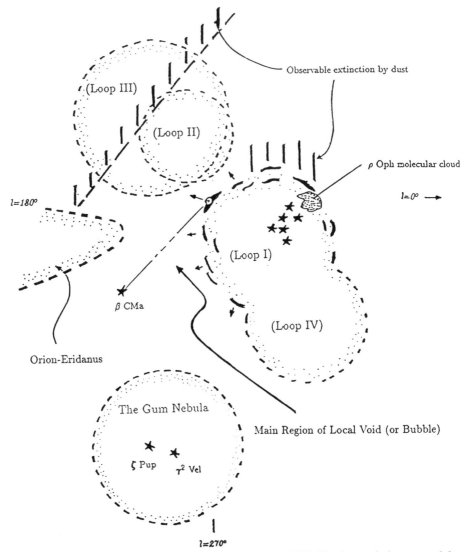

FIGURE 1. Distribution of largescale shell structures in the LISM. The sizes and placement of the radio loops, Loops I–IV were determined from Spoeltra (1973). A portion of the Orion-Eridanus Complex, as well as the Gum Nebula, are shown. The line-of-sight between the Sun and β CMa is depicted. This denotes the principal region of the Local Void (or Bubble). See text for discussion.

Reynolds 1977). This structure, seen near $l = 270°$, is centered about 400 pc away and has a radius of 120 pc as defined by the H-α filaments. This nebula and its cavity contain the stars ζ Pup (O4) and γ^2 Vel (WR + O9 Ia). The stellar winds from these stars appear to be responsible for carving out the observed cavity (cf., Weaver et al. 1977). The Vela supernova remnant lies either in, or more likely behind, the Gum Nebula.

It appears that Loop I is expanding into a pre-existing hot, coronal region, which extends at least 200 pc in the direction of β and ϵ CMa. (This is the region often referred to as the Local Bubble.) As illustrated in Figure 2, the term, Local Bubble, is clearly a misnomer, for the its boundary actually seems to denote the peripheries of the expanding neighboring shell complexes. Thus, the Local Bubble, itself, is not a bubble.

3.2. What Is the Origin of the Pre-Existing Cavity?

If the Local Bubble (or Void) is part of a pre-existing low density region filled with 10^6 K gas, then what produced this region? One possibility is that the entire region within 400-600 pc of the Sun was previously the site of a much older, much more energetic shell complex. This shell may have triggered the star formation now represented by the kinematically expanding Gould's Belt. Gould's Belt is defined by the stellar associations of Orion, Perseus, Lacertae, and Sco-Cen, some of which lie outside the plane (Blaauw 1956; Hughes & Routledge 1972; Lesh 1972). Also, an expanding ring of gas and dust, called Lindblad's Ring has been linked with these stellar associations (Lindblad 1973; FitzGerald 1968; Olano 1982). More recently, Taylor, Dickman, Scoville (1987) has found that the molecular clouds, too, show this same kinematic expansion.

If it does, the inferred age of this supershell would be in 3 to 6×10^7 yr (Elmegreen 1982). The original stellar associations responsible for carving out this large shell complex may have been part of the Taurus group and the α Per cluster. Unfortunately, most of the "smoking guns," namely the stars giving rise to the stellar winds and supernovae driving the expansion of this yet putative shell, are long gone.

Both Gould's Belt and Lindblad's Ring have been topics in the literature for years, and may present evidence that the region now within 400-600 pc was previously evacuated and filled with hot coronal gas. An old pre-evacuated cavity would allow any shells, carved out by the stellar winds and supernovae from the young nearby stellar associations, a natural setting for expansion.

4. Summary

Multi-frequency observations, radio through X-ray wavelengths, have been essential in revealing the gas and dust morphology within 300 pc. Observations with the *EUVE* are revealing much about the ionization and physical conditions of the two most fundamental elements, H and He, in the LISM. Within 300–450 pc, there are six identified large shell nebulae, which appear to be expanding into a pre-existing hot, low density cavity. This older cavity may owe its origin to stars of the α Per cluster, members of the Taurus moving group. This scenario provides a natural explanation for the misnamed "Local Bubble," which has been inferred to explain the soft X-ray background and the low H I column densities seen in the LISM in the galactic anti-center direction.

Thanks are extended to my fellow colleagues, K.-P. Cheng, J. Holberg, C.-H. Lyu, Y. Kondo, & A. Smith, who I have had the pleasure of working with is aspects of this work.

REFERENCES

BERKHUIJSEN, E. 1971, A&A, 14, 359
BERTAUX, J. ET AL. 1985, A&A, 150, 82
BLAAUW, A. 1956, ApJ, 123, 406
BOHLIN, R. ET AL. 1976, ApJ, 224, 132
BRANDT, J. ET AL. 1977, ApJ, 208, 109

Bruhweiler, F. 1982, in Advances in Ultraviolet Astronomy, Based on Six Years of IUE
    Research, ed. Y. Kondo, R. Chapman, & J. Mead, NASA, 125
Bruhweiler, F., & Cheng, K. -P. 1988, ApJ, 335, 188
Bruhweiler, F., & Kondo, Y. 1981, ApJL, 248, L123
Bruhweiler, F., & Kondo, Y. 1982, ApJ, 259, 232
Burrows, D. 1993, ApJ, 406, 97
Cheng, K. -P., & Bruhweiler, F. 1990, ApJ, 364, 573
Cox, D., & Anderson, P. 1982, ApJ, 252, 268
Chassefiere, E. et al. 1988, A&A, 174, 239
de Geus, E. 1988, Ph.D. Dissertation
Dupree, A., & Raymond, J. 1982, ApJL, 263, L63
Edgar, R. J. et al. 1993, BAAS, 25, 805
Elmegreen, B. 1982, in Submillimetre Astronomy, ed. J. Beckman,& J. Philips, Cambridge
    Press, 3
Fried, P. M. et al. 1980, ApJ, 242, 987
Gry, C. et al. 1985, ApJ, 296, 593
Hughes, V., & Routledge, D. 1972, AJ, 77, 210
Kimble, R. et al. 1993, ApJ, 404, 663
Lallement, R., Bertaux, J. L., Sandel, B. R., & Chassefiere, E. 1992, in Solar Wind
    Seven, Proc. of the 3rd COSPAR Colloq., Goslar, Germany, (A93-33554 13-92), 209
Lallement, R. et al. 1986, A&A, 168, 225
Lesh, J. 1972, ApJS, 17, 371
Lindblad, P. 1973, A&A, 24, 309
Linsky, J. et al. 1993, ApJ, 402, 694
McCammon, D. et al. 1983, ApJ, 269, 107
Murthy, J. et al. 1987, ApJ, 315, 675
Olano, R. 1982, A&A, 112, 195
Perry, C. et al. 1982, AJ, 87, 1751
Reynolds, R. 1976, ApJ, 203, 151
Reynolds, R. 1989, ApJL, 339, L29
Reynolds, R., & Ogden, J. 1979, ApJ, 229, 942
Spoelstra, J. 1973, A&A, 24, 149
Taylor, J. et al. 1987, ApJ, 324, 149
Tinbergen, J. 1982, A&A, 105, 53
Vennes, et al. 1993, ApJ, 400,400
Warwick, et al. 1993, MNRAS, 100, 100
Weaver, R. et al. 1977, ApJ, 218, 377

Observations of Diffuse Emission from the Hot ISM

W. T. SANDERS[1,2] AND R. J. EDGAR[1,2]

[1]Department of Physics, University of Wisconsin-Madison,
1150 University Avenue, Madison, WI 53706, USA

[2]Space Science & Engineering Center, University of Wisconsin-Madison,
1225 W. Dayton Street, Madison, WI 53706, USA

The Diffuse X-ray Spectrometer (DXS) is a Bragg-crystal spectrometer designed to obtain spectra of the diffuse x-ray background in the 83–44 Å (150–284 eV) range, with ~ 3 Å spectral resolution (10–25 eV), and $\sim 15°$ angular resolution. It was flown successfully as an attached Shuttle payload on the STS 54 mission of NASA's Space Shuttle Endeavour in January 1993, and spectra were obtained from the diffuse background along an arc extending roughly along the galactic plane from longitude 150° to longitude 300°. The primary conclusions so far from the analysis of the DXS data are: (1) The spectra of the diffuse background in the 83–44 Å range show emission lines or emission-line blends, indicating that the emission is thermal. Although most models of this emission have assumed that it is of thermal origin, this is the first detection of lines in the diffuse background in this wavelength range. (2) The detected spectra do not resemble the model spectra of cosmic abundance equilibrium plasmas at any temperature in the $10^5 - 10^7$ K range. This is independent of the particular plasma model used, Raymond & Smith, Mewe & Kaastra, or Monsignori-Fossi & Landini. (3) The detected spectra do not resemble the model spectra of depleted abundance equilibrium plasmas at any temperature in the $10^5 - 10^7$ K range, for a variety of assumed elemental depletions and the same emission models. This aspect of the analysis is not completed. (4) Tentative line identifications can be made, but other lines predicted to arise from the same ions must be of consistent strength in both the DXS and *EUVE* data sets.

1. Introduction

In this talk, we discuss the soft x-ray observations of the hot interstellar medium (ISM). The first part summarizes the spatial structure of the ISM as revealed by sounding rocket and satellite experiments. The second part summarizes spectral results, in particular, the data from the Diffuse X-ray Spectrometer (DXS) experiment that our group built and that NASA flew as an attached Shuttle payload in 1993.

A basic assumption in all this is that observations of the low energy x-ray diffuse background are in fact observations of x-rays emitted by the hot ISM. The implied picture then is that the solar system is surrounded by a 10^6 K interstellar plasma. We will address the temperature more later, but it is approximately correct. Further, we know that the *local* ISM is relatively deficient in neutral material, which would easily absorb low energy x-rays, out to distances of 50–100 pc in all directions. So the natural home for this surrounding x-ray emitting plasma is within this "local cavity."

All-sky surveys of the diffuse background have been done in the low energy x-ray band from 160–284 eV (44–80 Å, also known as the carbon band, the C band, and the $\frac{1}{4}$ keV band) by a series of sounding rockets (McCammon et al. 1983), by the *SAS 3* satellite (Marshall & Clark 1984), by the *HEAO 1* satellite (Garmire et al. 1992), and by the *ROSAT* satellite (Snowden et al. 1995). One all-sky survey has also been done in the 120–188 eV band (66–100 Å, also known as the boron band, or the B band) by a series of sounding rockets (McCammon et al. 1983), and there are ~ 20 pointings around the

269

sky in a band from 70–111 eV (112–177 Å, also known as the beryllium band or the Be band) from sounding rockets (Bloch et al. 1986) (Juda et al. 1991) (Edwards 1990).

When the count rates from all four $\frac{1}{4}$ keV band experiments are compared, with allowances made for the different instrumental response functions, they agree with each other to better than 10% (Snowden et al. 1995). This tells us that the features seen on the maps are real, and that they have persisted over a twenty-year time interval. The first observations of the soft x-ray background (Bowyer, Field, & Mack 1968) have features that also are apparent on today's maps, as can be seen if Bowyer's map is compared to the $ROSAT$ $\frac{1}{4}$ keV band map (Snowden 1993).

2. Spatial Structure of the Local Hot ISM

When one compares the $\frac{1}{4}$ keV $ROSAT$ map (Snowden 1993) to maps of 21-cm data (Dickey & Lockman 1990), the most striking overall effect is the general "anticorrelation" of the two maps. The x-ray intensity is lowest in the Galactic plane and increases by a factor of four or five in some high latitude directions.

For x-rays in the $\frac{1}{4}$ keV band, one absorption optical depth is $\sim 1 \times 10^{20}$ cm^{-2}. Thus the interstellar galactic gas is opaque at low Galactic latitudes, $|b| \lesssim 30°$, while its transmission rises to $\sim 50\%$ at high Galactic latitudes in the directions of lowest $N_{\rm H}$. The finite intensity of the x-ray background seen at low latitudes must be Galactic in origin and originate relatively near to the Sun, within the closest $\sim 1 - 2 \times 10^{20}$ cm^{-2}. This corresponds to a distance of roughly 100 pc, plus or minus a factor of 2. At the high Galactic latitudes there is a significant fraction of the sky, 15%, where the total $N_{\rm H} \lesssim 2 \times 10^{20}$ cm^{-2}, and contributions from extragalactic or halo sources are possible in addition to the local emission.

Prior to the launch of $ROSAT$, the data did not $require$ that the hot interstellar medium have more than one component, the emission from the local hot plasma located in the local cavity. This component is required by the finite intensity in the Galactic plane. It had been found that simple halo models, models in which a Galactic halo exhibited symmetries about the Galactic center and Galactic plane and whose soft x-ray emission was smoothly distributed, did not work very well, so many thought that the halo emission was negligible. The higher measured count rate of the diffuse background at high latitudes was taken to be due to a greater physical extent of the emitting region in those directions, plus some small contribution from extragalactic sources (Snowden et al. 1990). One advantage of this picture was that of simplicity: no more components were introduced than were absolutely necessary. Another advantage was that it naturally explained why the B band map looked so similar to the C band maps, and why the Be band count rates tracked the B band count rates so well even though the column density for unit optical depth in the Be band was less than 10% of that in the C band. The implication was that observed intensity differences were not due to absorption.

$ROSAT$ demonstrated that this view was wrong. The arcminute angular resolution of $ROSAT$ and its high sensitivity to diffuse emission features made possible for the first time the detection in the x-ray diffuse background of "shadows"–the absorption of distant soft x-ray emission by foreground clouds of neutral material. Soft x-ray emission from the Galactic halo was demonstrated by the detection of absorption of the soft x-ray background in Draco (Snowden et al. 1991) (Burrows & Mendenhall 1991) by a cloud at roughly ($l \sim 90°, b \sim 40°$) whose distance was found to be more than 300 pc, well outside the accepted boundary of the local cavity. Half of the x-ray background count rate in this direction is from the halo or beyond, and half is local emission, originating closer than the Draco cloud. This local emission is still twice the value typically found in the galactic plane, so the local emission region may extend twice as far in this direction as it does in the plane. Shadows were also seen in Ursa Major (Snowden et al. 1994), both toward an intermediate negative velocity cloud complex at a distance of 300–400 pc (Benjamin et al. 1995), as well as towards low velocity clouds whose distances are not yet known. About 70% of the count rate in this direction originates closer than the clouds, with an intensity almost three times that typically found in the Galactic plane, while 30% comes from the halo or beyond. In this low N_H direction, the extragalactic component accounts for 20% of the total rate, leaving $\sim$ 10% as halo emission (Snowden et al. 1994). The intensity of the halo emission in this directions appears to change by a factor of two on angular scales as small as 2°.

In summary, the count rate due to the halo emission can be as large as 10–50% of the total count rate in some high latitude directions where (a) the N_H is very low (Ursa Major), or (b) the halo is very bright (Draco), but the local component is still required and its intensity at high latitudes is 2–3 times greater than its intensity in the plane.

Evidence that the local component is truly local comes from the *ROSAT* observation of MBM 12 (Snowden, McCammon, & Verter 1993), a nearby (d $\lesssim$ 65 pc) molecular cloud at intermediate latitude ($l \sim 159°$, $b \sim -34°$). Toward this cloud, practically no shadow was seen, implying that this cloud is at the edge of the local emission region and that almost all the emission in this direction originates closer than 65 pc. These authors use this observation to determine a pressure for the hot interstellar medium of 1.25×10^4 K cm^{-3}. They assume that the plasma generating the local emission is described by a Raymond & Smith equilibrium model at a temperature near 10^6 K. Data from the DXS and *EUVE* spectrometers do not support this assumption, so while the pressure is likely near 10^4 K cm^{-3}, the exact value is not determined by this method.

Figure 1 is a schematic diagram of some of the major features of the local interstellar medium relevant to the emission of the soft X-ray background. The view is a cut taken along longitudes 160° - 340° and passing through the north and south Galactic poles. It illustrates some of the count rates and spatial geometries discussed above. Count rates are indicated in units of 10^{-6} *ROSAT* counts s^{-1} arcmin^{-2} (SXU). Count rates after absorption are indicated in parentheses.

3. Spectroscopic Observations of the Hot Interstellar Medium

Until very recently, spectroscopic observations of the soft x-ray diffuse background, which was presumed to originate in the ISM, had the limited spectral resolution of proportional counters for the energy range below 284 eV. From proportional counters, spectral information comes from the shape of the distribution of pulse heights within a given band, and from the ratio of count rates from one band to another. Over the past 20 years, it has been frequently noted that the observed pulse height distributions and band count rate ratios of the diffuse background are consistent with those predicted from a plasma in ionization equilibrium at a temperature near one million degrees, independent of the particular model used. Since there were no viable non-thermal mechanisms, and there were reasons to think that the diffuse background should have a thermal origin (e.g., supernova remnants), the standard interpretation of the diffuse background spectra became that it originated from a million degree plasma in the ISM, the hot ISM.

In January 1993, NASA flew the Diffuse X-ray Spectrometer (DXS), built by our group at the University of Wisconsin, as an attached shuttle payload. DXS is a Bragg-crystal spectrometer designed to obtain spectra of the diffuse x-ray background in the 83–44 Å (150–284 eV) range, with $\sim$ 3 Å spectral resolution (10–25 eV), and $\sim$ 15° angular resolution. The primary objective of this experiment was to find out if there were lines

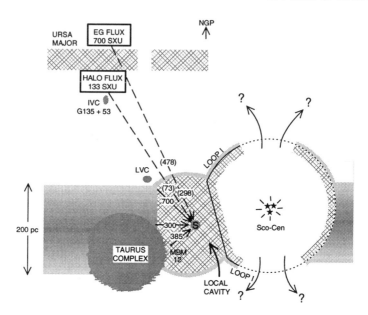

FIGURE 1. Schematic diagram of some major features of the local interstellar medium relevant
to the emission of the soft X-ray background. The view is a cut taken through longitudes 160°
- 340°. Numbers shown are count rates in different directions in units of SXU, defined in the
text.

in the spectrum of the diffuse background. If lines were present, then the emission has
a thermal origin and the DXS experiment could answer additional questions about the
emission:

- Is the plasma in coronal equilibrium?
- If so, what is the temperature?
- If not, which non-equilibrium model(s) fit the data?
- What are the wavelengths and fluxes of the dominant lines?
- What are the likely x-ray emitting elements and their ionization states?
- Are spectral differences apparent from one part of the sky to another?

DXS uses a curved crystal panel to Bragg-reflect incident soft x-rays towards a colli-
mated position-sensitive proportional counter. The collimator allows each position along
the counter to "see" only a restricted piece of the curved crystal panel, and each position
sees its piece of the crystal panel at a different angle. Thus, each position along the
counter sees a different wavelength x-ray from a different direction on the sky than does
its neighboring position. By rotating the counter about an axis perpendicular to the
dispersion direction of the crystals, full wavelength coverage over the 83–44 Å range is
obtained from the arc of the sky along which the counter scans. The DXS detector is
described more fully elsewhere (Sanders et al. 1992).

The calibration of the DXS detectors has determined that the effective area-solid angle
product rises from ~ 0.01 cm^2 sr at 151 eV, to ~ 0.02 cm^2 sr at 183 eV, and falls again
to ~ 0.01 cm^2 sr at 277 eV. The spectral resolution (FWHM) is $\lesssim 3$ Å over the full
spectral range of the detectors. During the data collection times of the shuttle mission,
the Orbiter was oriented such that the DXS detectors repeatedly scanned the same arc

on the sky–an arc running almost in the Galactic plane from longitudes $150° - 300°$. Although the charged particle induced background in the detectors produced a total count rate of 500–1000 counts s^{-1}, the instrumental background rejection efficiency was such that the net count rate was ~ 0.1 s^{-1} when the detectors were stowed, and ~ 0.6 s^{-1} when they were scanning the sky. Pre-flight modeling had shown that a net count rate of ~ 0.5 s^{-1} was expected from this region of the sky so we have confidence that the DXS instrument was working as expected during its data collection times.

Figure 2 shows the data collected while DXS scanned the Galactic longitude range $213°–252°$, mostly in the constellation Puppis. The wavelength scale is still preliminary, as is the flux axis. The data do show lines, indicating that the source of the emission is thermal–we are indeed observing a hot interstellar medium. The solid line is the best-fit Raymond-Smith model folded through the detector response function. Both the temperature of the plasma and the amount of intervening absorbing material were allowed to float. Although the best-fit parameters are similar to those found by earlier experiments, $kT = 0.094$ $(1.09 \times 10^6$ K$)$ and $N_H = 6 \times 10^{19}$ cm^{-2}, it is easy to see that the model does not fit the data. No single-temperature equilibrium plasma model (Raymond-Smith, Kaastra-Mewe, Landini-Fossi) in the temperature range $10^5 - 10^6$ K fits the data any better. We examined two-temperature models, and found that by combining a million-degree plasma and a $\sim 2.5 \times 10^6$ K plasma we could do reasonably well in the DXS bandpass, but exceeded by factors of 3–4 the measured intensity in the 0.5–1 keV range. Models with depleted elemental abundances were also tried, but without success. We conclude that either the models are seriously deficient, or that the emitting plasma is not in ionization equilibrium. We tried to fit the data with several non-equilibrium models, specifically blast wave models (Edgar & Cox 1993) and cooling plasmas, both with solar elemental abundances and with depleted abundances. None of these fit any better than the single temperature equilibrium model. We may need to develop more appropriate models, incorporating new atomic physics data as they become available, and new astrophysical ideas.

We have compared the spectra from different regions of the sky with one another and our tentative conclusion is that we see small but probably significant differences in different directions. More analysis is needed to strengthen and quantify this conclusion.

The current analysis is focusing on completing the reduction and incorporation into the analysis models of all of the calibration data to better determine the wavelength scale, to better understand the detectors' efficiency at all wavelengths, and better remove any remaining instrumental artifacts in the data. Another facet of the current analysis is to place upper limits on the flux that could be present from individual ions as a function of temperature. The latter requires better models of the individual ionic spectra than are currently available from the standard models.

To summarize the state of spectroscopic observations of the hot interstellar medium:

• The low energy diffuse background shows evidence for lines, implying the emission is thermal.

• The detected spectra do not match model spectra of cosmic abundance equilibrium plasmas for $10^5 \lesssim T \lesssim 10^7$ K.

• We have not yet found a model with depleted abundances or a model of a non-equilibrium plasma that fits the data. We may need improved models.

• The data show weak evidence for different spectra in different directions, but further analysis is needed to confirm this conclusion.

• Additional data is needed, particularly with higher spectral resolution, higher angular resolution, and a wider range of directions on the sky. High latitude data would be very enlightening.

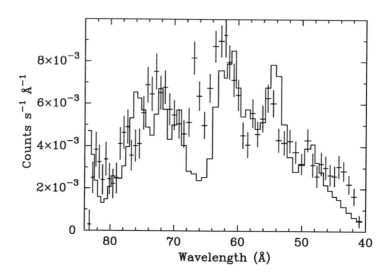

Figure 2. DXS data from the Galactic longitude range 213°- 252°. The solid line is the best-fit Raymond & Smith model, temperature of 1.09×10^6 K, N_H of 6×10^{19} cm^{-2}. The scales of both axes are preliminary.

The authors would like to acknowledge helpful discussions with Steve Snowden, Bill Kraushaar, and Dan McCammon in preparing this talk. This work was supported in part by NASA contract NAS 5-26078.

REFERENCES

Benjamin, R. A., Venn, K. A., Hiltgen, D. D., & Sneden, C. 1995, The Distance to an X-Ray Shadowing Molecular Cloud in Ursa Major. ApJ, submitted

Bloch, J. J., Jahoda, K., Juda, M., McCammon, D., Sanders, W. T., & Snowden, S. L. 1986, Observations of the Soft X-Ray Background at 0.1 keV, ApJ, 308, L59–L62

Bowyer, C. S., Field, G. B., & Mack, J. E. 1968, Detection of an Anisotropic Soft X-ray Background Flux, Nature, 217, 32

Burrows, D. N., & Mendenhall, J. A. 1991, Soft X-ray Shadowing by the Draco Cloud, Nature, 351, 629

Dickey, J. & Lockman, F. J. 1990, H I in the Galaxy, Ann. Rev. Astr. Ap., 28, 215

Edgar, R. J. & Cox, D. P. 1993, Hot Bubbles in a Magnetic Interstellar Medium: Another Look at the Soft X-ray Background, ApJ, 413, 190

Edwards, B. C. 1990, Observations of the Soft X-ray Background, PhD Thesis, University of Wisconsin, Madison

Garmire, G. P., Apparao, K. M. V., Burrows, D. N., Fink, R. L., & Kraft, R. P. 1992, The Soft X-ray Diffuse Background Observed with the HEAO 1 Low-Energy Detectors, ApJ, 399, 694

Juda, M., Bloch, J. J., Edwards, B. C., McCammon, D., Sanders, W. T., Snowden, S. L., & Zhang, J. 1991, Limits on the Density of Neutral Gas within 100 Parsecs from

Observations of the Soft X-Ray Background, ApJ, 367, 182

MARSHALL, F. J. & CLARK, G. W. 1984, *SAS 3* Survey of the Soft X-ray Background, ApJ, 287, 633

McCAMMON, D., BURROWS, D. N., SANDERS, W. T., & KRAUSHAAR, W. L. 1983, The Soft X-Ray Background, ApJ, 269, 107

SANDERS, W. T., EDGAR, R. J., JUDA, M., KRAUSHAAR, W. L., McCAMMON, D., SNOWDEN, S. L., ZHANG, J., & SKINNER, M. A. 1992, The Diffuse X-ray Spectrometer Experiment, Proceeding of the SPIE: EUV, X-Ray, and Gamma-Ray Instrumentation for Astronomy III, 1743, 60

SNOWDEN, S. L., COX, D. P., McCAMMON, D., & SANDERS, W. T. 1990, A Model for the Distribution of Material Generating the Soft X-Ray Background, ApJ, 354, 211

SNOWDEN, S. L., MEBOLD, U., HIRTH, W., HERBSTMEIER, U., & SCHMITT, J. H. M. M. 1991, *ROSAT* Detection of an X-ray Shadow in the $\frac{1}{4}$ keV Diffuse Background in the Draco Nebula, Science, 252, 1529

SNOWDEN, S. L. 1993, Implications of *ROSAT* Observations for the Local Hot Bubble, Adv. Space Res., 13(12), 103–111

SNOWDEN, S. L., McCAMMON, D., & VERTER, F. 1993, The X-ray Shadow of the High-Latitude Molecular Cloud MBM 12, ApJ, 409, L21–L24

SNOWDEN, S. L., HASINGER, G., JAHODA, K., LOCKMAN, F. J., McCAMMON, D., & SANDERS, W. T. 1994, Soft X-Ray and H I Surveys of the Low N_H Region in Ursa Major, ApJ, 430, 601

SNOWDEN, S. L., FREYBERG, M. J., PLUCINSKY, P. P., SCHMITT, J. H. M. M., TRÜMPER, J., VOGES, W., EDGAR, R. J., McCAMMON, D., & SANDERS, W. T. 1995, First Maps of the Soft X-ray Diffuse Background from the *ROSAT* XRT/PSPC All-Sky Survey, ApJ, submitted

ε Canis Majoris and the Ionization of the Local Cloud

J. V. VALLERGA AND B. Y. WELSH

Eureka Scientific, 2452 Delmer St., Oakland, CA 94602, USA

Using the EUV (70 − 730 Å) spectrum of the brightest EUV source, ε CMa taken with the *Extreme Ultraviolet Explorer Satellite* (*EUVE*) and simple models that extrapolate this spectrum to the Lyman edge at 912 Å, we have determined the local interstellar hydrogen photoionization parameter, Γ, solely from this B2 II star to be 1.1×10^{-15} s^{-1}. This figure is a factor of 7 greater than previous estimates of Γ calculated for *all* nearby stars combined (Bruhweiler & Cheng 1988). Using measured values of the density and temperature of neutral interstellar hydrogen gas in the Local Cloud, we derive a particle density of ionized hydrogen, $n(H^+)$, and electrons, n_e, of 0.015 − 0.019 cm^{-3} assuming ionization equilibrium and a helium ionization fraction of less than 20%. These values correspond to a hydrogen ionization fraction, X_H from 19% to 15%, respectively.

1. Introduction

One of the surprising results of the all-sky photometric survey carried out by *EUVE* was the discovery that the B2 II star ε CMa was the brightest source in the EUV sky (Vallerga, Vedder, & Welsh 1993). Most of the EUV flux from ε CMa was detected in the "tin" filter bandpass (500 − 700 Å), which is very susceptible to absorbing interstellar hydrogen. At 100 counts per second in the tin band, ε CMa was over an order of magnitude brighter than β CMa, HZ 43, and Sirius B. It was already known that in the direction of β CMa (B1 II–III) an "interstellar tunnel" of very low interstellar neutral hydrogen column extends as far as ∼ 300 pc and includes the star ε CMa at 187 pc (Welsh 1991). It was totally unexpected that the slightly later type star, ε CMa, would have a brighter Lyman continuum flux at Earth than the earlier type β CMa, since small temperature differences have such a large effect on the Wien tail of the Planck function. It was pointed out by Vallerga, Vedder, & Welsh (1993) that such a large flux in the EUV would affect the ionization state of the Local Cloud, but the single flux measurement in the tin bandpass could not constrain the stellar models and interstellar medium (ISM) absorption well enough to derive a spectrum that could be used to calculate the local ionization rate. However, the EUV spectrum of ε CMa taken using the spectrometers aboard *EUVE* (Cassinelli et al. 1994) effectively represents the local stellar EUV radiation field above 500 Å. From this EUV spectrum a photoionization rate of the local ISM can then be derived. A more detailed approach to this problem can be found in Vallerga & Welsh (1995).

2. Observations and Data Analysis

Observations of ε CMa with the three spectrometers aboard *EUVE* took place in January, 1993, and covered the entire EUV range from 70 to 730 Å. Details of the observation and reduction to the one-dimensional spectrum are given in Cassinelli et al. (1994). The long wavelength fluxed spectrum of ε CMa at Earth is shown in Figure 1. Because the source is so bright, the errors in the intensity for each bin are dominated by systematic rather than statistical effects, and the relative errors are estimated to be

S. Bowyer and R. F. Malina (eds.), Astrophysics in the Extreme Ultraviolet, 277–282.
© 1996 *Kluwer Academic Publishers. Printed in the Netherlands.*

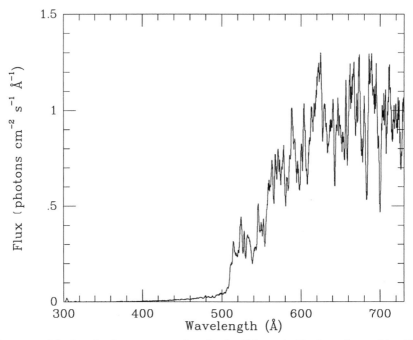

FIGURE 1. The flux density versus wavelength of ϵ CMa at the Earth as observed by *EUVE* showing the intense emission above the neutral helium edge at 504 Å.

less than 10%. Errors in the derived absolute flux density depend on the effective area calibration of *EUVE* and await a complete analysis of in-orbit calibration data from many more stars. Assuming the instrument response is unchanged from prelaunch values, the absolute effective areas will be accurate to 25%.

The hydrogen photoionization parameter,

$$\Gamma = \int_0^{912\,\text{Å}} N_\lambda \ \sigma_H d\lambda \tag{2.1}$$

is the radiation field flux, N_λ (photons cm^{-2} s^{-1} Å^{-1}), weighted by the hydrogen cross section, σ_H (cm^2).

The total ionizing flux from ϵ CMa, as well as the corresponding hydrogen photoionization parameter, are given in Table 1. This includes both the measured values derived by integrating the flux using only the measured spectrum from 400 to 730 Å as well as the unobserved values derived by integrating the extrapolated spectrum of an absorbed stellar blackbody from 730 to 912 Å ($T = 17300$ K, $N_H = 9 \times 10^{17}$ cm^{-2}). The errors in the extrapolated numbers were estimated by using the extreme curves plotted in Figure 2 and amount to no more than 6% of the total ionizing photon flux and 9% of the photoionization parameter. Also shown in Table 1 is the flux and ionization parameter at the edge of the Local Cloud assuming all of the observed absorption toward ϵ CMa is due to the cloud itself. Not shown in Table 1 is the absolute error of 25% because of the uncertainty of the *EUVE* effective area.

TABLE 1. Ionizing flux and hydrogen photoionization rate from ϵ CMa

	At Earth	Local Cloud Surface[†]
Integrated Flux (photons cm^{-2} s^{-1})[‡]		
Measured (504–730 Å)	169	2150
Extrapolated (730–912 Å)	152 ± 20	11000 ± 3000
Total (504–912 Å)	320 ± 20	13150 ± 3000
Photoionization Rate, Γ_{H^0} (s^{-1})[‡]		
Measured (504–730 Å)	4.1×10^{-16}	5.9×10^{-15}
Extrapolated (730–912 Å)	$6.8(\pm 1.1) \times 10^{-16}$	$5.6(\pm 1.6) \times 10^{-14}$
Total (504–912 Å)	$1.1(\pm 0.1) \times 10^{-15}$	$6.2(\pm 1.6) \times 10^{-14}$

[†] Assumed that all absorption is due to a Local Cloud and the total neutral hydrogen column density is 9×10^{17} cm^{-2}.

[‡] Errors shown are only those resulting from the extrapolation of the spectrum to the Lyman edge. The absoute flux is only known to 25% because of the calibration of the effective area of *EUVE*. Statistical errors are insignificant.

The value of Γ_H in the Local Cloud resulting from ϵ CMa is a factor of 6.9 greater than that predicted for the sum of all the bright EUV emitting stars known or predicted by Bruhweiler & Cheng (1988). Previous authors, although cognizant of the low interstellar column to ϵ CMa, did not realize that ϵ CMa is brighter than β CMa in the Lyman continuum by a factor of ~ 16 even though the effective temperature of β CMa (B1 II–III) is 4000 K hotter than ϵ CMa (B2 II). With the *EUVE* all-sky photometric survey now complete from 70 to 760 Å, we can now rule out the possibility of another unknown, constant *stellar* source of Lyman continuum radiation that would dominate the EUV radiation field.

The *diffuse* contribution to the hydrogen ionization parameter of emission from the 10^6 K gas surrounding the Local Cloud was calculated to be small when compared to stellar sources (Cheng & Bruhweiler 1990). The possibility exists, however, that the gas in the conductive interface at a temperature of $\sim 10^5$ K between the Local Cloud and the surrounding hot, ionized gas, will dominate the EUV radiation field because of strong EUV emission lines and the large solid angle of this postulated source. Slavin (1989) developed a model of this interface and predicted a diffuse EUV spectrum at the Sun assuming a hydrogen column density of 10^{18} cm^{-2} in all directions between the Sun and this interface. The spectrum was dominated by bound-bound emission lines resulting in a local Γ_H ranging from 1.6×10^{-15} s^{-1} to 5.1×10^{-15} s^{-1} depending on the angle of the magnetic field with respect to the cloud's radial direction. Therefore this (as yet) unobserved radiation is comparable to, or a factor of 5 greater than, that produced by ϵ CMa (Table 2). Recent results using *EUVE* to measure diffuse radiation have been able to place 3 σ upper limits less than Slavin's predictions for the shorter wavelength emission lines (Jelinsky, Edelstein, & Vallerga 1994). However, better limits at the longer wavelengths must await further analysis because of the difficulty of removing the local 584 Å background radiation from the diffuse spectra. We will proceed in the following section assuming that ϵ CMa is the dominant source hydrogen ionization with the understanding

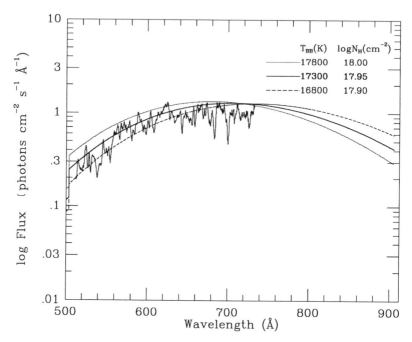

FIGURE 2. The same spectrum of ϵ CMa as in Fig. 1 but plotted logarithmically. The three lines overplotted represent simple blackbody curves attenuated by neutral hydrogen absorption. These models were used to extrapolate the ϵ CMa spectrum from 730 to 912 Å which could not be measured with *EUVE*. Solid line: $T = 17300$ K, $\log N_{H^\circ} = 17.95$. Dotted line: $T = 17800$ K, $\log N_{H^\circ} = 18.00$. Dashed line: $T = 17300$ K, $\log N_{H^\circ} = 17.90$.

that a lower limit will result and hydrogen ionization fraction will increase if the diffuse emission from the conductive interface is eventually determined to be significant.

3. Ionization State of the Gas Near the Sun

The relevant equations describing ionization equilibrium and charge balance are given in Bruhweiler & Cheng 1988, and the choice of parameters required to determine Γ are fully explained in Vallerga & Welsh (1995). We assume that the temperature of the Local Cloud near the Sun is 7000 ± 1000 K. We base this assumption on recent high-resolution absorption measurements of interstellar Lyα observed toward the nearby stars of Cappella (Linsky et al. 1993) and Procyon (Linsky et al. 1994). This value is consistent with the slightly higher temperature of 8000 ± 1000 K derived by Bertaux et al. (1985) using hydrogen gas cell Lyα line widths of the local interstellar wind. At 7000 K, the hydrogen recombination rate, $\alpha(H^\circ)$, is 3.4×10^{-13} cm^3 s^{-1} (Halpern & Grindlay 1980). The collisional ionization rates are insubstantial compared to the photoionization rates at this temperature (Lotz 1967).

If we assume a neutral hydrogen density of 0.065 cm^{-3} and a neutral helium density of 0.001 cm^{-3} from the solar backscatter measurements (Chassefiere et al. 1986), and that $n(He^\circ) > n(He^+) > n(He^{++})$, then the photoionization parameter dominates the charge

TABLE 2. Electron density and hydrogen ionization fraction near the sun resulting from ϵ CMa

Assumed $n(H^\circ)$ (cm^{-3})	n_e(cm^{-3})	X_H [†]
.065 (Chassifiere et al. 1986)	.015	.19
.110 (Linsky et al. 1993)	.019	.15

[†] $X_H \equiv n(H^+)/[n(H^+) + n(H^\circ)]$

exchange terms in the ionization balance equations (Vallerga & Welsh 1995). By then equating n_e with $n(H^+)$, assuming that helium is mostly neutral, results in the simplified formula:

$$n_e^2 \cong n(H^+) \cdot n_e = \frac{\Gamma(H^+) \cdot n(H^\circ)}{\alpha(H^\circ)}. \tag{3.2}$$

This equation shows the square root dependence of the electron density on the ionization rate and local neutral hydrogen density. Table 2 gives the electron densities and ionization fraction X_H ($\equiv n(H^+)/[n(H^+) + n(H^\circ)]$) assuming the neutral density measured in the local interstellar wind as well as the higher density determined from average line-of-sight measurements to nearby stars. To check our assumption about the charge exchange rates and helium densities a posteriori, we can take our derived ionized hydrogen density together with the measured values of the neutral densities (Table 2) to determine the ionization fraction of helium. Using this technique, a helium ionization fraction, X_{He}, can be constrained to be between 0% and 33% assuming the two extremes of hydrogen density used in Table 2. In both density cases the charge exchange contribution to the hydrogen ionization balance is still less than 1% of the photoionization caused by ϵ CMa. We can compare this range of helium ionization fraction to the only directly measured ionization fraction of helium observed toward the hot white dwarf GD246 (Vennes et al. 1993). Using the EUVE derived spectrum, Vennes et al. were able to detect the interstellar He II edge at 228 Å as well as the autoionization feature of He I at 206 Å giving both the neutral and singly ionized column densities of helium. They derived a helium ionization fraction of 25%. However, it is not clear whether the high hydrogen column in this direction is an extension of the Local Cloud or a separate cloud nearer to the white dwarf ($d = 65$ pc). Note that the ion densities derived in Table 2 assume no helium ionization. If the ionization fraction of helium was as high as 33% then the electron density would increase by 25% with a corresponding decrease of $n(H^+)$.

3.1. Comparison with Previous Work

The present empirical derivation of the ionization fraction of hydrogen from ϵ CMa alone gives a lower limit to the ionization state of the local hydrogen gas. Therefore, this result cannot rule out other sources of ionization (such as conductive boundary EUV flux or recent supernova shocks) that might explain the high ionization fraction implied by the ratios of neutral densities found inside the heliosphere. A compilation of neutral hydrogen and helium density measurements inside the heliosphere taken by various instruments gives typical densities $n(H^\circ) \sim 0.05$ cm^{-3} and $n(He^\circ) \sim 0.01$ cm^{-3} (Chassefiere et al. 1986). The fact that the ratio of these densities is not the canonical cosmic ratio of 10 implies that hydrogen must be partially ionized in the solar system with an ionization fraction $X_H = 0.3$–0.7 if helium is completely neutral and greater if helium is partially ionized.

Our derived electron densities can be compared to those determined by the Mg absorption line ratio technique (Cox & Reynolds 1987). The local electron density is derived by measuring the interstellar column densities of Mg I and Mg II toward nearby stars and using the standard ionization equilibrium equations. This technique is fraught with uncertainties, not only in the measurement of the ratio since the Mg II lines are usually highly saturated, but also in the assumed ionization and recombination constants, Γ and α, because of uncertainties in the radiation field and temperature. Results for the derived local electron density range from 0.026 to 0.10 cm^{-3} averaged over the line of sight, though there are certain directions where this value is much smaller (Bruhweiler et al. 1984; Cox & Reynolds 1987). These densities equal or exceed those found from ϵ CMa alone, though they come from measurements of stars with distances 8, 20, and 40 pc whose lines of sight certainly sample the hot, ionized gas and the conductive interface with the Local Cloud.

We wish to thank David Cohen, Joe MacFarland, Janet Drew, Peter Vedder, and other members of the *EUVE* B star collaboration for their help in deriving the stellar parameters of ϵ CMa. We also wish to thank Joe Cassinelli for leading this effort to understand B stars, which unexpectedly resulted in a better understanding of the LISM. This work was supported by NASA grant NAGW 5-2282, administered through the University of Wisconsin.

REFERENCES

Aldrovandi, S. M. V. & Pequignot, D. 1973, A&A, 25, 137

Bertaux, J. L., Lallement, R., Kurt, V. G., & Mironova, E. N. 1985, A&A, 150, 1

Bruhweiler, F. C. & Cheng, K. P. 1988, ApJ, 355, 188

Bruhweiler, F. C., Oegerle, W., Weiler, E., Stencel, R., & Kondo, Y. 1984, In IAU Colloq. 81, The Local Interstellar Medium, ed. Y. Kondo, F. C. Bruhweiler, & B. D. Savage, NASA CP-2345, 64

Cassinelli, J. P., Cohen, D. H., MacFarlane, J. J., Drew, J. E., Hoare, M. G., Vallerga, J. V., Vedder, P. W., Welsh, B. Y. & Hubeny, I. 1994, ApJ, in press

Chassefiere, E., Bertaux, J. L., Lallement, R., & Kurt, V. G. 1986, A&A, 160, 229

Cheng, K. P. & Bruhweiler, F. C. 1990, ApJ, 364, 573

Cox, D. P. & Reynolds, R. J. 1987, ARA&A, 25, 303

Halpern, J. P. & Grindlay, J. E. 1980, ApJ, 242, 1041

Jelinsky, P., Edelstein, J., & Vallerga, J. 1994, ApJ, submitted

Linsky, J. L., Brown, A., Gayley, K., Diplas, A., Savage, B. D., Ayres, T. R., Landsman, W., Shore, S. N., & Heap, S. R. 1993, ApJ, 402, 694

Linsky, J. L., Diplas, A., Andrulis, C., Brown, A., Savage, B. D., & Ebbets, D. 1994, In "Frontiers of Space and Ground-Based Astronomy," ed. W. Wamsteker, M. S. Longair, & Y Kondo, Dordrecht: Kluwer, 102

Lotz, W. 1967, ApJS, 14, 207

Slavin, J. D. 1989, ApJ, 346, 718

Vallerga, J. & Welsh, B. Y. 1995, ApJ, 444, 702

Vallerga, J., Vedder, P., & Welsh, B. 1993, ApJ, 414, L65

Vennes, S., Dupuis, J., Rumph, T., Drake, J. J., Bowyer, S., Chayer, P., & Fontaine, G. 1993, ApJ, 410, L119

Welsh, B. Y. 1991, ApJ, 373, 556

ALEXIS Observations of the Diffuse Cosmic Background in the Extreme Ultraviolet

BARHAM W. SMITH, T. E. PFAFMAN,
J. J. BLOCH, AND B. C. EDWARDS

Astrophysics and Radiation Measurements, NIS-2, Mail Stop D436,
Los Alamos National Laboratory, Los Alamos, NM 87545, USA

We present preliminary results of *ALEXIS* satellite observations of the extreme ultraviolet diffuse sky. *ALEXIS* was designed to be a half-sky monitor in three narrow wavelength bands between 13.0 and 19.0 nm. In our band centered at 172 Å we find a clear signal from the diffuse sky that is about 20 counts per second above the signal when looking at the dark earth from the same part of an orbit. This difference corresponds to an upper limit on the true cosmic diffuse background signal in this narrow band. When estimates of the geocoronal contributions, both in and out of band, are removed, our upper limit is reduced.

1. Introduction

The *ALEXIS* satellite was designed to make robust measurements of the diffuse cosmic background in the extreme ultraviolet. For a description of the instrument and its performance, see Bloch et al. 1990, or Bloch, this volume. We have selected a small number of time intervals to examine during which telescope pair 1, which looks out approximately along the satellite spin equator, alternately looks at the dark earth and the sky on each spin. If the earth is truly dark, i.e., no significant airglow or aurora, the difference between sky-looking and earth-looking fluxes should be an absolute upper limit to the diffuse cosmic background flux. As can be seen from Figure 1, our strict upper limit is at the limit of detectability of previous instruments. When all our analysis is done, we should be able to decrease this limit. Here we confine ourselves to telescope 1B which has a bandpass from approximately 171 to 186 Å, with peak sensitivity at 176 Å.

2. Analysis

Data is selected by a nested set of criteria. This was done in order to eliminate data that was contaminated by the anomalous background that is observed when the angle between the ram of the spacecraft orbital motion and the look direction of the telescope falls below 90° to 100° (Bloch et al. 1994). The method used excluded data with a ram angle below 100 or a count rate above 100 counts s^{-1}. This removed the unwanted ram background and also excluded data when the spacecraft passed through the auroral zones or the South Atlantic Anomaly. The ram background can be seen in Figure 2, which shows the smoothed count rate scalers and the horizon angle for a number of *ALEXIS* spins as a function of time.

Figure 3 shows the count rate as a function of angle from the telescope optical axis to the earth's horizon in 5° bins. At 0° on this axis, half the 33° field of view is on the earth and half is on the sky. The data is shown for telescope 1B (176 Å) where a difference of more than 10 counts s^{-1} is observed between the earth-looking and sky-looking count rates. Telescope 1A, with a bandpass at about 133 Å, shows no difference.

We are in the process of compiling a number of long data sets to look for correlations in the sky-ground count rates at different times of the year. These sets of data are being compiled on a spin-by-spin subtraction of the average count rate at 30° above the horizon

S. Bowyer and R. F. Malina (eds.), Astrophysics in the Extreme Ultraviolet, 283–287.
© *1996 Kluwer Academic Publishers. Printed in the Netherlands.*

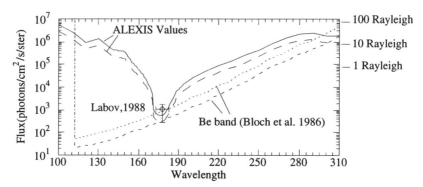

FIGURE 1. A comparison of upper limits on the diffuse cosmic background in the extreme ultraviolet. The sounding rocket results of the Wisconsin group (Bloch et al. 1986) are shown for the "beryllium band." The result of Labov (1988) for a single line is shown. The *ALEXIS* upper limit is shown for excess sky rates of 10 (lower curve) and 20 counts s^{-1}.

from the average count rate at 30° below the horizon. Differences range from −10 to 50 counts s^{-1} as can be seen in the histogram of a data set from January shown in Figure 4. This histogram peaks around a sky-to-ground difference of 10 to 20 counts s^{-1}. The variation might be due to variations in geocoronal or interplanetary contributions to the sky flux, or to true variations in the diffuse soft X-ray cosmic background, which is known to vary by factors of three across the sky.

To derive Figure 1, we assumed that the net sky counts could be converted into flux using an area-solid angle product of 4.22 cm^2-sr and an effective system efficiency of 0.004 counts per incident photon, as measured directly during pre-flight calibration. Our measured flux for HZ43 leads us to believe that our in-flight efficiency is close to the pre-flight calibration.

3. Discussion

The observed flux contains contributions from geocoronal He II 304 Å (0.1 to 1.6 R; e.g., Chakrabarti et al. 1982) and He I 584 Å (0.4 R; e.g., Chakrabarti et al. 1984), as well as interplanetary 584 Å backscatter, which has a seasonal variation when looking-antisun, being largest in December, up to 10 R, and smallest in June, 1 R (R. R. Meier, private communication). The sensitivity of telescope 1B to 304 Å is about 1 count per Rayleigh, while the sensitivity to 584 Å is about 25 times smaller. Therefore, it appears that no more that a few counts per second can be easily attributed to helium line fluxes. Other potential sources of sky flux include scattered solar X-rays and EUV, which we estimate to be smaller than 304 Å contributions. Local particle flux is not variable in a repeatable way over time scales of less than 1 minute, and we believe our subtraction technique should eliminate them as an explanation for the observed sky-earth difference. Auroras in the field of view, when we are earth-pointed, would decrease the observed net flux, and indeed one can see spins then this seems to be happening.

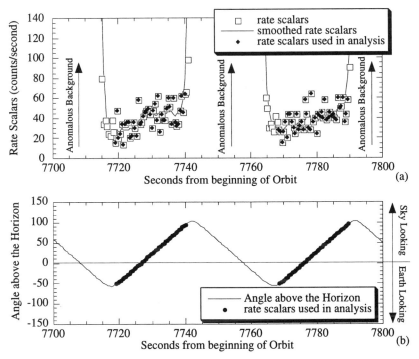

FIGURE 2. The count rate scalers (a) and the horizon angle (b) for two rotations of telescope 1B. Times dominated by the anomolous background (a) and our selected data intervals (b) are shown.

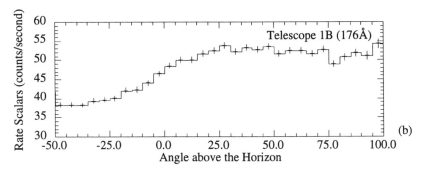

FIGURE 3. Count rates for telescope 1B as a function of the angle between the telescope's optical axis and the earth's horizon, in 5° bins.

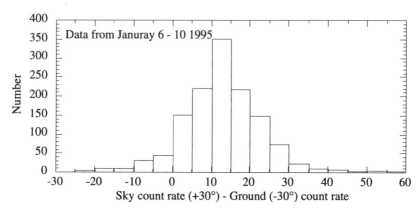

FIGURE 4. Histogram of the count rate differences between 30° above and 30° below the horizon for a data set in January.

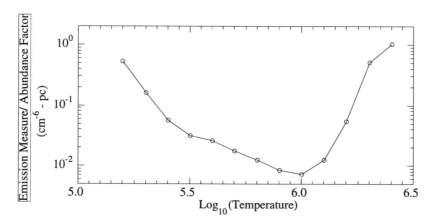

FIGURE 5. Upper limit on the emission measure of a hypothetical hot interstellar plasma with cosmic abundances and no absorption, as a function of the logarithm of the plasma temperature in kelvin. The spectrum is from Brickhouse et al. 1995.

In Figure 5 we use the latest plasma spectrum model of Brickhouse, Raymond, & Smith (1995) with the response curve of *ALEXIS* telescope 1B, to derive upper limits on the emission measure required to make up the observed net sky flux without interstellar absorption. When absorption is added, these limits will rise. Therefore these are preliminary upper limits to the possible emission measure assuming normal cosmic abundances.

This work was supported by the United States Department of Energy.

REFERENCES

Bloch, J. J., et al. 1986, ApJL, 308, L59
Bloch, J. J., et al. 1990, Proceedings SPIE, 1344, 154
Bloch, J. J., et al. 1994, Proceedings SPIE, 2280, 297
Brickhouse, N. S., Raymond, J. C., & Smith, B. W. 1995, ApJS, in press
Chakrabarti, S., et al. 1982, Geophys. Res. L., 9, 151
Chakrabarti, S. et al. 1984, J. G. R., 89, 5660
Edwards, B. C. 1990, Ph. D. thesis, U. Wisconsin, Madison
Labov, S. 1988, Ph. D. thesis, U. California, Berkeley

A Search for the Signature of the Diffuse Soft X-ray Background in the *ROSAT* Wide-Field Camera All-Sky Survey

R. G. WEST,[1] R. WILLINGALE,[1] J. P. PYE,[1] AND T. J. SUMNER[2]

[1] Department of Physics & Astronomy, University of Leicester, University Road, Leicester LE1 7RH, UK

[2] Department of Physics, Imperial College, London, UK

We present the results of an attempt to locate the signature of the diffuse soft X-ray background in the *ROSAT* Wide-Field Camera (WFC) all-sky survey. After removal of non-cosmic background sources (eg. energetic charged particles), the field-of-view integrated count rate in the WFC S1a filter (90–185 eV) shows no consistent variation with Galactic latitude or longitude. We place limits on the signal from the soft X-ray background (SXRB) in the WFC, and show that these limits conflict with the observations of the Wisconsin Sky Survey if the SXRB in this energy range is assumed to be produced by a thermal plasma of cosmic abundance and a temperature $T \sim 10^6$ K within $d \sim 100$ pc of the Sun.

1. Introduction

The *ROSAT* X-ray satellite was launched in June 1990 in to a circular orbit at an altitude of 575km, and an inclination of 53°. It carries two co-aligned telescopes, the German X-Ray Telescope (Trümper 1984) working in the 0.1–2keV energy band, and the UK Wide-Field Camera (Sims et al. 1990) which is an imaging telescope with a field-of-view of 2.5° operating in the extreme ultraviolet (90–200 eV).

2. The Wide-Field Camera All-Sky Survey

The dataset produced by the *ROSAT* WFC all-sky survey, carried out between August 1990 and February 1991, has been extensively searched for point sources with the final cataloguing of 479 EUV sources (Pounds et al. 1993, Pye et al. 1995). In addition to locating point sources, the WFC also presented an excellent opportunity to map with unprecedented sensitivity and spatial and spectral resolution the hot X-ray emitting gas thought to fill the local cavity in the interstellar medium. The bandpasses of the WFC survey filters ("S1" and "S2", Figure 1) are comparable to the "B"- and "Be"-bands of the sky surveys performed by the Wisconsin group (McCammon et al. 1983, Bloch et al. 1986), and although the throughput of the WFC is much smaller than the instruments used in those rocket flights, the average sky exposure in the WFC survey ranges from $\sim$ 2 ks in the plane of the Ecliptic to $\sim$ 80 ks near the poles yielding potentially much greater sensitivity. The WFC survey also confers the advantages of a six-month baseline over which to measure slowly varying extraneous background components, and the imaging capabilities of the WFC allow unequivocal exclusion of the contribution of point-like and extended sources (SNRs) from the detected diffuse emission.

S. Bowyer and R. F. Malina (eds.), Astrophysics in the Extreme Ultraviolet, 289–293.
© *1996 Kluwer Academic Publishers. Printed in the Netherlands.*

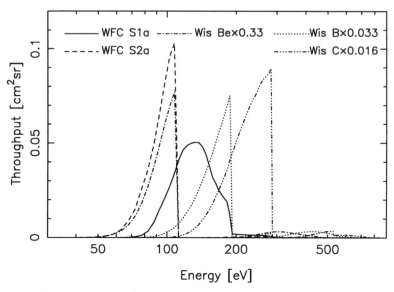

FIGURE 1. Throughput of the WFC survey filters and the Wisconsin "Be"-, "B"- and "C"-bands

TABLE 1. Typical count rates from various background components in the WFC

Component	Intensity $(ct\,s^{-1})$
Spacecraft glow	0–1000
He II geocorona	$\ll 1$ (S1)
	10–40 (S2)
Auroral electrons	0–1000
Cosmic rays	4–25
MCP intrinsic noise	1.97 ± 0.17

3. Extraneous Background Components

Five sources of non-cosmic background have been identified in the WFC survey and are listed below. Typical count rate ranges experienced under normal operating conditions during the survey are given in Table 1.

(a) **Spacecraft glow** A persistent background phenomenon which shows a clear correlation with the orientation of the spacecraft with respect to its velocity vector, and is thought to be due to an interaction between those surfaces and the impinging atmosphere producing a "spacecraft glow" in the far ultraviolet, possibly via a chemiluminscent reaction on the exposed spacecraft surfaces (West et al. 1994). This background component has been well characterised in terms of orbital parameters and the affected time periods can be reliably removed from the dataset.

(b) **He II 304 Å geocoronal radiation** Solar helium line emission resonantly scattered from the He II in the plasmasphere. The WFC filters are designed to suppress geocoronal radiation and the S1 survey filter is highly opaque at the wavelengths (transmission $\lesssim 10^{-9}$). This component has been positively identified in the survey data taken in the S2 filter, which has allowed us to show that this background source is not present in the S1 survey data.

(c) **Auroral electrons** Soft ($E \sim 50\,\text{keV}$) electrons precipitated at high geographic latitudes ($\sim 50\text{-}70°$). Excluding data taken while the spacecraft is in the auroral zones is highly effective at excluding this background component.

(d) **Cosmic rays** are modelled by correlating the WFC count rate (after removal of the spacecraft glow and auroral components) with the contemporaneous Master Veto Rate (MVR) from the *ROSAT* PSPC . A convincing linear correlation is found at low geomagnetic latitudes ($|\Lambda_B| \lesssim 40°$) which allows the cosmic ray induced background in the WFC to be modelled to better than $\sim 10\%$.

(e) **MCP intrinsic noise** caused by β-decay of ^{40}K in the substrate of the microchannel plate detector. The half-life of the decay process is extremely long ($t_{1/2} \simeq 1.28 \times 10^9\,\text{yr}$) and this background component is essentially constant throughout the survey.

4. Limits on the Diffuse Background in the WFC

After removal (by discrimination) of the spacecraft glow and auroral background, and subtraction of the cosmic ray contribution, cleaned data are available covering $\sim 40\%$ of the sky, with a total exposure of $\sim 130\,\text{ks}$. The residual field-of-view averaged count rate shows no statistically significant variation with Galactic longitude or latitude, and the sky averaged count rate ($2.00 \pm 0.12\,\text{counts s}^{-1}$) is consistent with the intrinsic noise in the WFC microchannel plate detector ($1.97 \pm 0.17\,\text{counts s}^{-1}$). Clearly this result cannot be regarded as a detection of a signal from the diffuse background in the WFC, rather we interpret this as a conservative upper-limit on the field-of-view averaged count rate in the S1 band of around $0.5\,\text{counts s}^{-1}$.

To compare the cleaned WFC survey data with the results from the Wisconsin survey we have prepared "simulated" WFC skymaps from the data of McCammon et al. (1983). Proceeding under the assumption that the measured B-band count rate represents emission from an optically thin thermal plasma (Raymond & Smith 1977) with Solar abundance with intervening neutral hydrogen column $N_H = 2 \times 10^{18}\,\text{cm}^{-2}$, we can predict S1/B band ratios for a range of plasma temperature. By scaling the Wisconsin B-band map by the relevant factor, adding a constant $1.97\,\text{counts s}^{-1}$ to represent the MCP intrinsic noise and masking areas of sky for which clean WFC data are not available, we can produce synthetic WFC skymaps for comparison with the data. These maps (Figure 2) show that large-scale structure should be visible in the WFC survey data and is demonstrably not present. Typically the error in the WFC skymap is dominated by uncertainties in the MCP intrinsic noise ($\sim 10\%$) and in the correlation with the MVR, rather than by counting statistics.

Jelinsky et al. (1995) have recently presented the results of an analysis of around 15% of the data taken with the *EUVE* Deep Survey Spectrometers during the *EUVE* sky survey, and have also derived limits on the plasma emission measure which are a factor around 5 to 10 below those implied from the Wisconsin C-band measurements. In Figure 3 we plot the emission measure derived from the Wisconsin Be, B and C-bands compared to the upper-limits from the WFC and *EUVE* data. The latter are taken directly from Jelinsky et al. and are calculated using a different plasma code (Landini & Monsignori-Fossi 1990). Jelinsky suggests that these values may differ from equivalent limits calculated using the Raymond & Smith code by 40% in the temperature

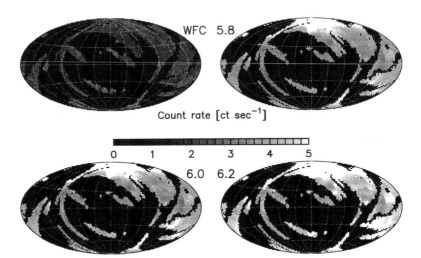

FIGURE 2. Comparsion of WFC survey data with maps synthesised from the Wisconsin B-band assuming optically thin thermal plasma ($\log T = 5.8, 6.0$ and 6.2) with Solar abundance and intervening column $N_H = 2 \times 10^{18}\,\mathrm{cm}^{-2}$. Errors in the WFC count rate are typically less than 20%. Areas for which no clean data are available are displayed as zero count rate, i.e., filled black.

range $T = 10^5$–10^6 K and by a factor 2 in the range $T = 10^6$–10^7 K. This uncertainty aside, however, it is evident from the plot that the WFC and *EUVE* results are in conflict with the higher energy Wisconsin measurements over a broad temperature range $5.5 \lesssim \log T < 6.3$.

An additional complication is that probably not all the B-band count rate is from a local cavity as evidenced by the Draco shadow (Burrows & Mendenhall 1991). In fact Sidher et al. (1995) have shown that there seems to be a global non-local component which accounts for about 50% of the B-band count rate and which they associate with the Galactic halo.

5. Conclusion

We have presented an analysis of the WFC all-sky survey database tuned to locate the signature of the diffuse soft X-ray background. This work, unlike previous efforts (Lieu et al. 1992), utilises the entire survey dataset and incorporates a more detailed understanding of the non-cosmic background sources to which the WFC is subjected in-orbit.

Our conclusions are that no identifiable signal from the diffuse background is visible in the WFC S1 dataset at a level $\gtrsim 0.5\,\mathrm{counts\ s^{-1}}$. Our results are inconsistent with predicted WFC count rates made using the Wisconsin Be, B and C-band intensities and assuming that the spectrum of the SXRB is that of a hot ($T \sim 10^6$ K), optically thin gas in thermal equilibrium with normal metallic abundances, and foreground column of absorbing neutral material is that solely of the "local fluff" ($N_H \lesssim 10^{19}\,\mathrm{cm}^{-2}$). To reconcile our results with the Wisconsin surveys may require depletion of heavy elements in the hot gas or non-equilibrium ionization.

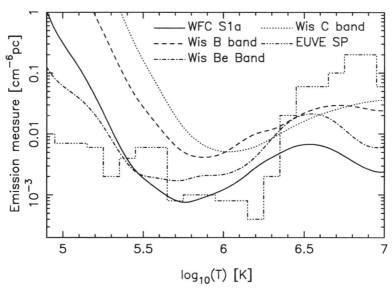

FIGURE 3.   Emission measure calculated from Wisconsin survey results compared to the upper-limits derived from the WFC and *EUVE* surveys.   Assumed count rates are $B = 49$ counts s^{-1}, $C = 133$ counts s^{-1} and $Be = 5.4$ counts s^{-1} (calculated from the B-band assuming $Be/B = 0.11$). The assumed plasma model is that of Raymond & Smith for the Wisconsin and WFC curves, and Landini & Monsignori-Fossi for the *EUVE* limit. Solar abundance is assumed, and an intervening neutral column of $N_H = 2 \times 10^{18}$

REFERENCES

BLOCH, J. J., JAHODA, K., JUDA, M., McCAMMON, D., SANDERS, W. T., & SNOWDEN, S. L. 1986, ApJ, 308, Sept., L59

BURROWS, D. N. & MENDENHALL, J. A. 1991, Soft X-ray shadowing by the Draco cloud, Nature, 351, 629

JELINSKY, P., VALLERGA, J. V., & EDELSTEIN, J. 1995, ApJ

LANDINI, M., & MONSIGNORI-FOSSI, B. C. 1990, A&AS, 82, Feb., 229

LIEU, R., QUENBY, J. J., SIDHER, S. D., SUMNER, T. J., WILLINGALE, R., WEST, R., HARRIS, A. W., BICKERT, & S. L. SNOWDEN K. 1992, ApJ, 397, Sept., 158

McCAMMON, D., BURROWS, D. N., SANDERS, W. T., & KRAUSHAAR, W. L. 1983, ApJ, 269, June, 107

POUNDS, K. A., ET AL. 1993, MNRAS, 260, 1, 77

PYE, J. P., ET AL. 1995, MNRAS, in press

RAYMOND, J. C. & SMITH, B. W. 1977, ApJS, 35, 419

SIDHER, S. D, SUMNER, T. J, QUENBY, J. J. & GAMBHIR, M. 1995, A&A, in press

SIMS, M. R., ET AL. 1990, Optical Engineering, 26(9), 649

TRÜMPER, J. 1984, Physica Scripta, T7, 216

WEST, R. G., SIMS, M. R. & WILLINGALE, R. 1994, Planet Space Sci., 42(1), 71

Features of the Soft X-ray Background and Implications for the EUV Background

C. R. BARBER AND R. S. WARWICK

Department of Physics and Astronomy, University of Leicester,
University Road, Leicester, LE17RH, UK

A series of eight overlapping *ROSAT* PSPC observations situated in the Lockman hole region have been analysed in the $0.1 - 0.4$ keV band. The total exposure of ~ 137 ks and the ultra-low interstellar column density make the data extremely suitable for studies of the diffuse signal. After excluding the contribution of bright discrete sources the remaining signal was examined on a scale of $20'$. As expected the $0.1 - 0.4$ keV signal exhibits a strong anticorrelation with column density. However, there is also evidence for significant excess fluctuations superimposed on this anticorrelation which may be attributable to some form of cloud structures within the hot ISM.

The diffuse signal expected in the *ROSAT* WFC all-sky survey for this region is predicted and shown to be far in excess of that observed. The origin of the fluctuations in the soft X-ray background and their possible relevance to the Extreme-Ultraviolet background are discussed.

1. Introduction

In the 30 years since the discovery of the X-ray background (XRB) considerable advances have been made in our understanding of this intriguing source of diffuse radiation. However, even with the enormous observational progress that has been made a comprehensive description of its origin remains a far distant prospect.

At energies > 2 keV the diffuse emission of the XRB is without doubt of extragalactic origin. However, at lower energies, at least one additional component emerges and dominates the emission. In the $0.1 - 0.4$ keV band, where this work concentrates, the diffuse intensity is a factor ≥ 3 above an extrapolation of the high energy power law and anticorrelates with galactic N_H. All-sky maps of the diffuse emission, in well separated energy channels within the $0.1-0.4$ keV band, exhibit strong spatial intensity correlations (e.g. McCammon & Sanders, 1990). As the absorbing cross section changes considerably across the $0.1 - 0.4$ keV range this is strong evidence for the emission originating in front of the majority of absorbing material.

Studies of the interstellar medium (ISM) using observations of the hydrogen Lyman series, interstellar absorption lines and Extreme-Ultraviolet (EUV) absorption have shown that the solar system is located near the center of a low-density cavity. A model has therefore been developed to account for the $0.1 - 0.4$ keV emission in which the local cavity is filled with an X-ray emitting plasma. The observed anticorrelation between X-ray intensity and N_H is then reproduced by a displacement mechanism; as the plasma extends further into the HI layer so the N_H reduces, the plasma emitting path length increases, and with it the observed X-ray intensity. The emissivity inferred for the cavity from this model is consistent with a 10^6 K gas at an acceptable interstellar pressure.

The aim of the work presented here is to test over a small area of sky the anticorrelation between N_H and $0.1 - 0.4$ keV X-rays, to decompose the soft X-ray signal into its constituents and to predict the contribution to the EUV background. To this end a series of eight overlapping *ROSAT* PSPC observations are employed (see Figure 1). They subtend ~ 14 degree^2 and have a cumulative exposure of 137 ks. High quality HI data for the region (kindly provided by F.J. Lockman) show that the north-south strip of

S. Bowyer and R. F. Malina (eds.), Astrophysics in the Extreme Ultraviolet, 295–298.

observations, the "UK Medium Sensitivity Survey", span a region of relatively uniform HI ($N_H = 5.6 - 9.3 \times 10^{19}$ cm^{-2}). In contrast, the two observations which extend to the north-east, the "Lockman Spur" pointings, follow a clear HI enhancement (maximum $N_H = 14.0 \times 10^{19}$ cm^{-2}).

2. *ROSAT* PSPC Data Reduction

Isolation of the diffuse soft XRB signal in a *ROSAT* PSPC dataset requires the careful removal of non-cosmic contaminants. Thus data has been rejected if taken at a time when the charged-particle detection rate was high. Similarly, only data accumulated during periods when the satellite was in earth shadow has been accepted. The light curves of the remaining data have been examined and all appear highly stable. This stringent process reduced the cumulative exposure of all 8 observations to 63.5 ks.

The $0.1 - 0.4$ keV band corresponds to *ROSAT* PSPC pulse invariant channels 11-40 and images have been cast in this range for use in the remainder of this analysis. Exposure and vignetting corrections have been performed using the flat-fielding technique described by Snowden et al. (1993). As the long and short term enhancement contaminants would not necessarily appear in the light curves there remains the possibility of an offset intensity between adjacent observations. This has been corrected by applying relative field-to-field offsets which reduce the overall baseline to the minimum recorded in the whole set of observations. The resulting images have been spatially "mosaiced" together for further analysis.

In order to reduce the discrete source contribution to the diffuse signal the mosaiced image has been searched for sources. The detection algorithm takes into account the varying background signal and detects sources brighter than 10^{-2} count s^{-1}. Assuming a power law of energy index $\alpha = 1$ and $N_H = 5.7 - 12.5 \times 10^{19}$ cm^{-2} this count rate corresponds to $0.5 - 2$ keV fluxes of $6.4 - 10.4 \times 10^{-14}$ erg cm^{-2} s^{-1}. A mask is created which when applied to the raw mosaiced image masks out at least 95% of the flux from each detected source (the remaining surface brightness is insignificant). To produce a map of the residual diffuse emission (Figure 1) the raw mosaiced image is convolved with a $20' \times 20'$ top hat (the resolution of the pixels used in the remainder of this analysis), ignoring the masked regions. For statistical analysis a sample of 97 independent pixels are extracted from this map excluding at this stage those coincident with Abell clusters. The mean count rate in a pixel is $947 \pm 65.6 \times 10^{-6}$ count s^{-1} arcmin^{-2}. There is clearly considerable variation in the observed count rate, a significant proportion of which is caused by a general anticorrelation with N_H.

3. Examination of the Diffuse $0.1 - 0.4$ keV Band Signal

Considering the diffuse signal in each individual field there are two potential sources of intrinsic scatter namely counting statistics and confusion noise. For a typical exposure and count rate the former induces a scatter of $\sim 2\%$ in the diffuse intensity measured in a pixel. The latter can be estimated as the distribution in flux per unit solid angle of discrete sources is known (Hasinger et al. 1993); simulations show that the typical scatter caused by this effect is $\sim 2.8\%$ of the total diffuse intensity for the pixel size used here. Monte Carlo simulations including both of these effects yield a predicted rms scatter between the pixels in any individual field of 37×10^{-6} count s^{-1} arcmin^{-2}. Table 1 shows the mean count rate and scatter observed in each field. It is clear that for seven of the eight observations the scatter is considerably more than expected from just counting statistics and confusion noise.

Table 1. Mean, scatter and standard error ($\times 10^{-6}$ count s^{-1} arcmin^{-2}) for each field.

Field	Mean	Standard error
UKMS 1	998 ± 34.8	9.3
UKMS 2	917 ± 44.3	12.3
UKMS 3	939 ± 43.6	10.6
UKMS 4	982 ± 49.5	11.4
UKMS 5	983 ± 64.3	15.6
UKMS 6	905 ± 51.5	21.0
SPUR 1	949 ± 51.1	11.4
SPUR 2	875 ± 64.5	16.1

Probing field-to-field diffuse signal fluctuations requires consideration of the scatter induced by the field-to-field correction process. This is typically 15×10^{-6} count s^{-1} arcmin^{-2}. Combining this scatter with the standard errors given in Table 1 does not account for the observed field-to-field variation implying that there are also significant fluctuations present on this scale.

Comparing the HI map for the region and the $0.1 - 0.4$ keV diffuse signal map (Figure 2) shows that a significant part of the non-statistical scatter highlighted above arises from the general anticorrelation between $N_{\rm H}$ and X-ray intensity. To explore this further the data have been fit with an absorption model which assumes spatially uniform foreground and background intensities, the latter being subject to varying attenuation dictated by the line-of-sight $N_{\rm H}$. (The absorption model is chosen rather than a displacement model as Snowden et al. (1991) and Burrows & Mendenhall (1991) have shown that in some directions a large fraction of the soft X-ray background originates well beyond the local cavity. Further the quality of the fit is relatively independent of the assumed model.) In this case all eight fields are considered together and the error bars are as measured from the pixels and include the scatter caused by the field-to-field correction process. The best fit has $\chi_v^2 = 1.5$, a local component of $646 \pm 70 \times 10^{-6}$ count s^{-1} arcmin^{-2} and a distant component of $675 \pm 159 \times 10^{-6}$ count s^{-1} arcmin^{-2} (where the errors are for 90% confidence). The rms scatter about the predicted model is 52×10^{-6} count s^{-1} arcmin^{-2}; considerably greater than the 40×10^{-6} count s^{-1} arcmin^{-2} predicted by our simulations. An image of the observed diffuse count rate with the best fit absorption model subtracted is shown if Figure 2 and is characterised by many deviations to $\pm 100 \times 10^{-6}$ count s^{-1} arcmin^{-2}. Further, these deviations extend over scales $>> 20'$ and appear to form coherent structures.

4. Implications for the Diffuse EUV Background Signal

Having isolated the $0.1 - 0.4$ keV diffuse signal the EUV diffuse signal can be predicted. For the purpose of this comparison the results from the *ROSAT* WFC S1a filter are utilised (see West et al., these proceedings). In the energy range of the S1a filter an $N_{\rm H}$ of 8×10^{19} cm^{-2} presents seven optical depths to a 10^6 K thermal plasma spectrum. Therefore the distant component derived in fitting the absorption model to the $0.1 - 0.4$ keV band data above can be ignored. Folding the derived foreground count rate through the relevant responses using a Raymond & Smith 10^6 K plasma model spectrum (Raymond & Smith 1977, Raymond 1991 privately communicated computer code update) yields a predicted S1a count rate of 2.7 count s^{-1} across the WFC field-of-view. West et al. however, derive a 99% upper limit of 0.5 count s^{-1}.

Data from the Diffuse X-ray Spectrometer (Sanders, these proceedings) has shown that current thermal plasma models, such as that used here, fail to fit high resolution $150 - 284$ eV spectra (even with a variety of assumed elemental depletions). The analysis presented here illustrates a similar failure of the models but over a considerably wider energy range. It is clear that there are fundamental flaws in our assumptions about the state of the plasma that is responsible for the soft X-ray background. Detailed fitting of high quality spectra and a comprehensive understanding of the physics in the plasma are required to resolve this conundrum.

This analysis has revealed an important feature of the morphology of the soft X-ray background—the existence of small scale fluctuations superimposed on the general anticorrelation with $N_{\rm H}$. The $ROSAT$ observations imply that a typical fluctuation has a magnitude of $\sim 30 \times 10^{-6}$ count s^{-1} arcmin^{-2}. It is likely that these fluctuations are caused by clumping of the hot ISM along the line-of-sight. To illustrate the physical conditions of this plasma assume that it is contained within a spherical cloud. Given the count rate of the fluctuation and the temperature of the plasma the emission measure, $\int n_e^2 \, dl$, of the plasma in the cloud can be computed. The emitting path length, dl, is limited by the angle subtended by the cloud, (taken to be $20'$), and therefore the electron density, n_e and thermal pressure, $n_e T$, can be derived. Cox & Reynolds (1987) have estimated that hot X-ray emitting plasma in the ISM has a pressure budget $\sim 10^4$ cm^{-3} K. (In this analysis an acceptable range of $0.5 - 1.5 \times 10^4$ cm^{-3} K is assumed.) Therefore acceptable distances to the cloud can be determined. If the cloud is placed in front of galactic $N_{\rm H}$ then its distance is constrained to $150 - 500$ pc. Alternatively if the cloud is beyond the galactic $N_{\rm H}$ then it is constrained to a distance $500 - 3000$ pc. A further cloud diagnostic is their predicted lifetime which can be estimated using the adiabatic sound crossing time. For the cloud in front of the galactic $N_{\rm H}$ the lifetime is $2 \times 10^4 - 7 \times 10^4$ years and for the cloud behind the galactic $N_{\rm H}$ it is $3 \times 10^4 - 1 \times 10^6$ years.

Currently it is unclear as to whether the fluctuations arise within the local cavity or beyond. Further work is being undertaken to characterise the angular scales of the fluctuations. Also a large sample of $ROSAT$ PSPC pointings are being examined to determine the amplitude of the fluctuations as a function of galactic $N_{\rm H}$. It is possible that if the fluctuations are generated locally then the local plasma arises not from the cooling of an elderly supernova remnant, but rather from a more vigorous phase of its existence. If this is the case then the fluctuations represent a morphological constraint that will have to be satisfied by models which produce spectra of the sort observed by the DXS. If, however, the fluctuations have a distant origin then they may represent significant evidence for circulatory models of the ISM.

REFERENCES

BURROWS, D. & MENDENHALL, J. 1991, Nature, 351, 629

COX, D. & REYNOLDS, R. 1987, ARA&A, 25, 303

HASINGER, G., BURG, R., GIACCONI, R., HARTNER, G., SCHMIDT, M., TRUMPER, J. & ZAMORANI, G. 1993, A&A, 275, 1

McCAMMON, D. & SANDERS, W. 1990, ARA&A, 28, 657

RAYMOND, J. & SMITH, B. 1977, ApJS, 35, 419

SNOWDEN, S., McCAMMON, D., BURROWS, D. & MENDENHALL, J. 1993, ApJ, 28, 657

SNOWDEN, S. ET AL. 1991, Science, 252, 1529

The Interstellar Gas in the Line of Sight to ε Canis Majoris

C. GRY,[1] L. LEMONON,[1] A. VIDAL-MADJAR,[2] M. LEMOINE,[2] AND R. FERLET[2]

[1]Laboratoire d'astronomie Spatiale, CNRS, B.P.8, F-13376 Marseille cedex 12, France

[2]Institut d'Astrophysique, CNRS, 98 bis Boulevard Arago, F-75014 Paris, France

We analyse Hubble Space Telescope GHRS observations of the interstellar medium in the direction to ε CMa, the strongest EUV source in the sky located 200 pc away in a region deficient in neutral gas. We show that the neutral gas density is the lowest yet measured in a galactic sight-line. The line of sight contains three main components among which the Local Cloud, and we derive their column densities, their velocity their temperature and their turbulence velocity. We discuss the ionization of the Local Cloud and we show that we detect the conductive interface between diffuse local cloud and the hot local bubble.

We analyse HST GHRS observations of the interstellar medium in the direction to ε CMa, the strongest EUV source in the sky, located in a region deficient in neutral gas in the high resolution Ech B observations. Three weak components are also detected. We derive their heliocentric velocities, their column densities in Fe II, Mg II, Mg I, Si III and C IV, upper limits for N I and H I, as well as the temperature and turbulence velocity of the three main components.

The analysis leads to the following conclusions :

(1) The neutral gas column density is very low indeed : less than 5×10^{17} cm^{-2}. This value is considerably lower than what had been measured before and it puts constraints on the EUV ionizing flux from ε CMa.

(2) Two of the main clouds are identified as the two components observed in the nearby Sirius line of sight, which shows that they lie less than 3 pc away from the Sun. Beyond this distance, the remaining 180 pc line of sight is thus an interstellar region with the lowest neutral mean density yet measured, with less than 2×10^{-4} cm^{-3}.

This confirms that the EUV flux is essentially unattenuated until the nearby clouds surface, which must therefore be much more ionized than the solar environment.

(3) One of the main components is the local interstellar cloud (LIC) in which the Sun is embedded. Its column density and its temperature (~ 7000 K) are in agreement with the lines of sight to Sirius and Capella. We derive a high electron density $n_e = 0.09^{+.23}_{-.07}$ cm^{-3} implying a high ionization fraction which could reach 75%.

(4) High ionization state elements like Si III and C IV are detected for the LIC and the third main component, showing that a hot phase is associated with these clouds. An extra absorption well described by a hot H I counterpart for the LIC is also required by the Lyman α profile fitting. We suggest that we are observing the conduction fronts between the clouds and the hot bubble in which they are embedded.

Note: A more complete description of this work is in press in A&A.

S. Bowyer and R. F. Malina (eds.), Astrophysics in the Extreme Ultraviolet, 299.

The *EUVE* Observations of Dwarf Novae

KNOX S. LONG

Space Telescope Science Institute, 3700 San Martin Drive, Baltimore, MD 21218, USA

In the standard theory of dwarf novae in outburst, the boundary layer region between the inner edge of the accretion disk and the white dwarf surface radiates primarily in the extreme ultraviolet. Using *EUVE*, observers have been able to obtain spectra with sufficient spectral resolution to characterize accurately the emission from several dwarf novae in outburst, including U Geminorum and SS Cygni. I present an overview of the observations and early analyses of the dwarf nova observations. The spectra obtained of dwarf novae are complex compared to the EUV spectra of magnetic cataclysmic variables and single white dwarfs. Detailed spectral modeling of an expanding atmosphere will most like be required to fully understand the spectra. Nevertheless, we already know there were significant differences in the effective temperatures and other properties of the EUV emissions. If we assume the EUV emission arises primarily from the boundary layer and parameterize the EUV spectrum in terms of a blackbody, then for U Gem the derived boundary luminosity is comparable to the disk luminosity, consistent with the standard theory, and the minimum size of the emitting region is about that of the white dwarf surface. The count rates from U Gem were modulated strongly with the orbital period; differences in the shape of the spectrum in eclipse and out of eclipse suggest that while the bulk of the emission arose from the vicinity of the white dwarf, there was an extended source of emission as well. For SS Cyg, however, the derived boundary layer luminosity was a small fraction of the disk luminosity. In U Gem, the effective temperature dropped during the decline from outburst. In contrast, in SS Cyg, the effective temperature remained constant as the count rate rose by a factor of 100 and the effective size increased. Thus while the observations of U Gem seem broadly consistent with the standard theory for the boundary layer emission from dwarf novae, SS Cyg appears to present fundamental challenges to that theory.

1. Introduction

Dwarf novae are mass-exchanging binary star systems containing a white dwarf (WD) and a relatively normal companion star. In dwarf novae, the magnetic field is low and accretion occurs by means of a viscous disk. Dwarf novae are characterized by quasiperiodic outbursts of 3–5 magnitudes at visible wavelengths. Although alternatives have been proposed, the outbursts are probably due to a thermal instability associated with a dramatic change in the opacity of disk material near 10,000 K, in essence, the temperature for hydrogen and helium ionization (Osaki 1974; Meyer & Meyer-Hofmeister 1982). In outburst in the "standard" theory, the disk is optically thick and the effective temperature of the disk is typically 20,000 K–40,000 K, depending on the accretion rate. In outburst, most of the luminosity of the disk emerges in the UV. In contrast, in quiescence, the disk should be optically thin and cool (2,000 K to 7,000 K) and very little UV emission should emerge from the disk (Cannizzo & Wheeler 1984).

In a steady-state accretion disk, half the energy of accretion (GMM/r) is radiated as matter traverses the disk, but the other half reaches the inner edge of the accretion disk in the form of kinetic energy and is radiated away in a transition region which is known as the boundary layer. Because the boundary layer region is smaller than the accretion disk, the effective temperature of the boundary layer is higher. In outburst the boundary layer region should be optically thick, have an effective temperature of 100,000 - 300,000 K, and radiate primarily in the EUV (Pringle 1977). In contrast, in quiescence, primarily due to the density dependence of the volume emissivity, the boundary layer should be be optically thin and relatively hot (10^7 K).

301

S. Bowyer and R. F. Malina (eds.), Astrophysics in the Extreme Ultraviolet, 301–308.
© 1996 *Kluwer Academic Publishers. Printed in the Netherlands.*

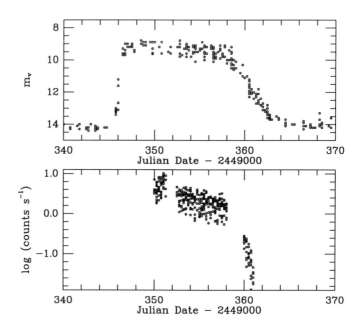

FIGURE 1. The AAVSO lightcurve (upper panel) and the count rate observed in the SW spectrometer (lower panel) for U Gem. The short term variations in the *EUVE* lightcurve are largely due to changes in the flux as a function of orbital phase.

EUVE is an important research tool for research on dwarf novae because it is the first experiment with sufficient sensitivity and spectral resolution to be able to characterize the spectrum of the boundary layer region. A fundamental question, which one hopes to address with *EUVE*, is whether the boundary layer actually behaves in the manner that the conventional theory suggests. Is the boundary layer luminosity equal to that of the disk and what is the size of the emitting region? Three dwarf novae have been observed with the spectrometers aboard *EUVE*—U Geminorum, SS Cygni, and VW Hydri. Analysis of the observations of VW Hyi is not yet complete. Here, I will attempt to describe what has been learned thus far from the observations of U Gem and SS Cyg.

2. U Geminorum

U Geminorum, which lies a a distance of about 90 pc (Marsh et al. 1990), is the prototypical dwarf nova, undergoing quasiperiodic outbursts ($\Delta m_v \sim 5$) lasting typically 7-14 days about once every 118 days (Ritter 1990). In the far UV, the quiescent spectrum is dominated by the WD in the system (Panek & Holm 1984). The average temperature of the WD surface drops from $\sim 38,000$ K shortly after outburst to $\sim 30,000$ K far from outburst (Long et al. 1993; 1994). The $\sim 1 M_\odot$ WD is rotating very slowly, if at all ($v \sin(i) \leq 100 \, \mathrm{km \, s^{-1}}$; Sion et al. 1994). In outburst, the far UV flux rises by a factor of 100 and is dominated by the optically-thick accretion disk (Panek & Holm 1984).

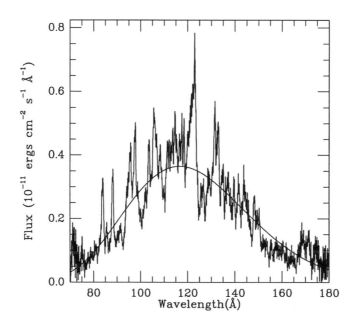

FIGURE 2. The fluxed spectrum of U Gem as observed with the SW spectrometer during the second observing interval. The best fit blackbody model assuming $N_H = 3 \times 10^{19}$ cm^{-1} is shown as the solid curve. The best fit temperature is 125,000 K.

In outburst, U Gem is a bright soft X-ray source, as evidenced by observations made with the soft X-ray proportional counters on HEAO-A by Córdova et al. (1984) and with the *Einstein* IPC by Córdova & Mason (1984). Though the actual spectral shape was very uncertain given the very limited spectral resolution of the proportional counters being used, model fits to spectra were consistent with that expected from a blackbody having a temperature kT=10–30 eV. The luminosity of U Gem was not accurately determined by these observations; Córdova et al. (1984) quote values between 2×10^{33} ergs s^{-1} and 1×10^{35} ergs s^{-1}. If the highest luminosity is correct and if the mass accretion rate was 8×10^{-9} M$_\odot$ yr^{-1} as was suggested by Panek & Holm, then the luminosity of the disk is 8×10^{34} (D/90pc)^2 ergs s^{-1} and it is quite possible that in U Gem the boundary layer and the accretion disk luminosities are similar. As predicted by the simple model, the X-ray spectrum of U Gem hardens in quiescence.

Following reports from the AAVSO that an outburst of U Gem had begun on the evening of 24 December 1993 and consultation with the *EUVE* project scientist, three sets of observations of U Gem were carried out with the spectrometers on *EUVE*, beginning on 28 December and extending through 8 January 1994. The outburst was a typical "wide" outburst of the U Gem system, rising from quiescence to optical maximum in ~2 days, remaining at maximum for ~14 days, and then declining to quiescence in ~5 days. The observations cover the peak and decline of the optical outburst. At peak, U Gem was one of the brightest sources in the EUV sky. As shown in Figure 1, long term (1 day) averages of the count rate in the SW spectrometer show a steady decline through the plateau phase of the optical outburst. During the first observing interval, the count rate in the SW spectrometer was about 5 cts s^{-1}. A preliminary analysis of the data was presented by Long et al. (1995). The basic results are as follows:

The time-averaged spectra are quite complex, showing numerous emission and/or absorption lines, quite unlike that of most other types of objects—WDs, polars, or late-type stars—observed with high signal to noise with *EUVE*. The spectra do not resemble those of a thin plasma or a simple WD atmosphere. Assuming the emission peaks are emission lines then a substantial number of them can be identified with lines which would be present in a thin plasma ionization temperature of about 500,000 K. Detailed model calculations will be required to properly interpret the spectra.

Mauche (1991) has measured the column density N_H of cool interstellar material along the line of sight to U Gem to be 3×10^{19} cm^{-2}. Unit optical depth for this column density occurs at 95 Å, which accounts for the decrease in flux at long wavelengths. In order to make a rough estimate of the luminosity and size of the emitting region, the data were fitted to a blackbody spectrum assuming the absorption is given by the interstellar value. During the first observing interval, the apparent temperature U Gem was $\sim 134,000$ K and the luminosity was $\sim 6 \times 10^{34}$ ergs s^{-1}. Given this luminosity and temperature, the minimum size of the emitting region is $\sim 4.5 \times 10^8$ cm, similar to that of a WD. If the EUV emission arises primarily from the boundary layer, then the boundary layer luminosity in U Gem is comparable to the disk luminosity, consistent with the standard theory of dwarf nova outbursts. The temperature declines as the outburst proceeds; fits to the spectra obtained in the second and third intervals yield $\sim 125,000$ and $\sim 100,000$ K respectively. The characteristic size of the emitting region declines, but only slightly, to $\sim 3.0 \times 10^8$ cm.

The count rates observed in both the SW and MW spectrometers are modulated on the orbital period, with broad a absorption dip near phase 0.7, and additional structure near phase 0 (given an ephemeris in which secondary conjunction occurs at phase 0.0). These absorption dips had been observed with *EXOSAT* during a very anomalous 45 day long outburst (Mason et al. 1988). The new observation shows that the absorption dips are a common phenomenon in U Gem.

As shown in Figure 3, the lines in the spectrum appear to be less eclipsed than the continuum. This supports the hypothesis that the emission peaks are really lines and argues that the lines arise from an extended region while the continuum arises from the region of the WD.

The dips are surprising because of their phase and because the orbital inclination of U Gem is relatively low ($\sim 65°$). The primary dip is not directly associated with the hot spot where the mass stream from the primary encounters the accretion disk which occurs near phase 0.9. Mason et al. (1988), on the basis of one very rapid, nearly complete eclipse, suggested that the absorbing material must lie near the Roche lobe radius of the WD, considerably outside the corotation radius for the incoming mass stream, and a distance of $\sim 3 \times 10^{10}$ cm above the disk plane. Hirose, Osaki & Mineshige's (1991) 3-dimensional disk simulations do show that the outer regions of the disk have scale heights $H \sim 0.1 r$, an order of magnitude greater than estimates based on on hydrostatic balance. These models suggest the maximum scale height of the disk is $\sim 0.15 r$ at phase 0.8 close to the phase at which the absorption dip is centered. In U Gem, however, the absorbing material has to lie approximately 0.42 r from the disk mid-plane. The amount of material required to occult the source is quite small ($N_H < 3 \times 10^{20}$ cm^{-2}), very much less than the surface density of the disk ($N_H > 10^{24}$ cm^{-2}).

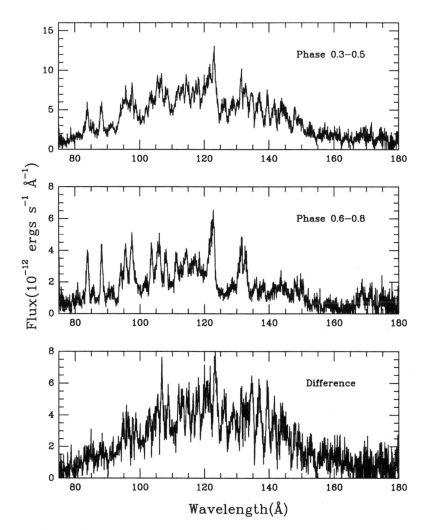

FIGURE 3. Fluxed SW spectra of U Gem from intervals 1 and 2 as a function of orbital phase. The *EUVE* emission is partially eclipsed near phase 0.7. The upper, middle and lower panels show the uneclipsed spectrum, the eclipsed spectrum, and the difference between the uneclipsed and eclipsed spectrum. The eclipse is mainly an eclipse of the continuum source.

3. SS Cygni

Like U Gem, SS Cyg is one of the brightest and best studied of dwarf novae. The system, as summarized by Ritter (1990), consists of a $\sim 1.2\,M_\odot$ WD and a $\sim 0.7\,M_\odot$ K5 V star orbiting one another every 6.60 hour seen at an orbital inclination of $\sim 37°$. During outbursts, which occur on average once every 49 days (Cannizzo & Mattei 1992), the visual magnitude rises from 12 to values between 8 and 9. Although Holm & Polidan

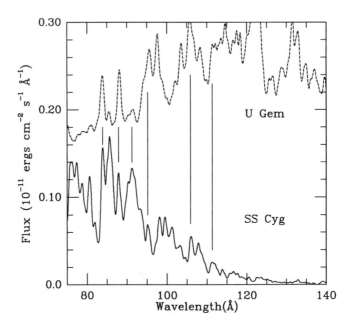

FIGURE 4. A comparison between the fluxed SW spectrum of SS Cyg and U Gem. The spectra have been convolved with a gaussian with a FWHM of 0.5 Å to facilitate comparison of the two spectra.

(1988) have interpreted *IUE* spectra of SS Cygni in quiescence in terms of a WD with a temperature of $34,000 \pm 5000$, FUV (850-1850 Å) spectra obtained with the *Hopkins Ultraviolet Telescope* show none of the absorption lines seen in spectra of U Gem in quiescence and indicate that disk dominates the FUV emission in quiescence.

SS Cyg was the first dwarf nova to be detected at X-ray wavelengths (Rappaport et al. 1974). In quiescence, SS Cyg is a bright soft X-ray source; hard X-ray emission is generally suppressed, in basic agreement with the standard model (Jones & Watson 1992), though not necessarily absent (Nousek et al. 1994).

SS Cyg has been observed twice with *EUVE*, first in 1993 August and then in 1994 June/July, Mauche and his collaborators; only the first observation has been reported in detail (Mauche, Raymond & Mattei 1995), although a preliminary analysis of *EUVE* photometry of both outbursts is described in the volume (Mauche 1995). The discussion here is confined to the first outburst.

Though most outbursts in SS Cyg are classified as narrow or wide, depending primarily on the length of the outburst, the 1993 August outburst of SS Cyg was a a rare "symmetric" or "anomalous" outburst, in which the rise to outburst was exceptionally long ~ 5 days, comparable to the decay time. The observations began about 1 day after the beginning of the optical outburst and lasted ~ 6.5 days, covering the rise and the beginning of the plateau phase of the optical outburst. During the observation, the count rate in the SW spectrometer rose by about 2 orders of magnitude to a maximum rate of about 0.75 cts s^{-1}. Surprisingly, the ratio of the count rate in the SW spectrometer in the wavelength range 90-130 Å to the count rate in the range 72-90 Å remained constant at a value of 1.39 ± 0.05 throughout the rise and during the peak portion of the outburst, suggesting a large change in the size of, but not the physical conditions in, the emitting region.

A comparison between the average spectrum obtained of SS Cyg and of U Gem (during the second observing interval) is shown in figure 4. Like U Gem, the spectrum of SS Cygni is complex and it is difficult to separate the continuum and the lines. There are however a number of "emission features" which are seen in both spectra. The "color" temperature of SS Cyg appears to be considerably higher than in U Gem., There is essentially no radiation detected from SS Cyg at wavelengths longer than 130 Å. It is not entirely clear whether this difference is due to absorption in the ISM or the vicinity of the source or due to the underlying source spectrum. Many of the features in the spectrum of SS Cyg are also seen in U Gem.

Mauche et al. (1995) have also parameterized the spectrum in terms of a blackbody with cold intervening absorption. Estimates of N_H from measurements of interstellar absorption lines 3.5×10^{19} cm^{-2} are not not greatly different from those obtained for U Gem (Mauche, Raymond & Córdova 1988), and yield a best 'fit' temperature of 37 ± 2 eV ($\sim$430,000 K), 3–4 times that of U Gem in outburst. If one uses a blackbody parameterization of the system, and assumes a temperature of 20 eV, then $N_H \sim 7.0 \times 10^{19}$ cm^{-2}, $L \sim 2 \times 10^{33}$ ergs s^{-1} and the size of the emitting region is 3×10^{-3} that of the WD surface. If T = 30 eV, then $N_H \sim 4.4 \times 10^{19}$ cm^{-2}, $L \sim 5 \times 10^{32}$ ergs s^{-1} and the size of the emitting region is 1×10^{-4} that of the WD surface. The accretion disk luminosity in SS Cyg is 3×10^{34} ergs s^{-1}. If *EUVE* is actually measuring the bulk of the boundary layer luminosity, which is not clear, then the ratio L_{bl}/L_{disk} is ≤ 0.07 in SS Cyg, while in U Gem it appears to be closer to 1.

4. Summary

EUVE has been used obtain high S/N spectra of U Gem and SS Cyg (and VW Hyi). There are very significant differences in the spectra. It seems quite possible the boundary layer has been unmasked in both systems. Of the two dwarf novae which have been studied in detail, U Gem seems very consistent with the standard theory, while SS Cyg appears to provide a challenge.

None of the *EUVE* observations would have been possible were it not for the dedication of the observers of the AAVSO and of the mission support teams at Berkeley and at GSFC. I have benefitted greatly from discussions with Chris Mauche and John Raymond. My work on the EUV properties of dwarf novae has been supported generously by NASA's *EUVE* guest observer program (NAG5-2572).

REFERENCES

CANNIZZO, J. K., & MATTEI, J. A. 1992, On the Long-Term Behavior of SS Cygni, ApJ, 401, 642

CANNIZZO, J. K., & WHEELER, J. C. 1984, The Vertical Structure and Stability of Alpha-Model Accretion Disks, ApJS, 55, 367

CÓRDOVA, F. A., CHESTER, T. J., MASON, K. O., KAHN, S. M., & GARMIRE, G. P. 1984, Observations of Quasi-Coherent Soft X-ray Oscillations in U Geminorum and SS Cygni, ApJ, 278, 739

CÓRDOVA, F. A., & MASON, K. O. 1984, X-ray Observations of a Large Sample of Cataclysmic Variable Stars Using the Einstein Observatory, MNRAS, 206, 879

HIROSE, M., OSAKI, Y., & MINESHIGE, S. 1991, Three Dimensional Structure of Accretion Disks in Close Binary Systems, Pub. Astr. Soc. of Japan, 43, 809

HOLM, A. V., & POLIDAN, R. S. 1988, SS Cygni in Quiescence, in A Decade of UV Astronomy with the IUE Satellite, ESA SP-281, 179

JONES, M. H. & WATSON, M. G. 1992, The EXOSAT, Observations of SS Cygni, MNRAS, 257, 633

LONG, K. S., BLAIR, W. P., BOWERS, C. W., DAVIDSEN, A. F., KRISS, G. A., SION, E. M., & HUBENY, I. 1993, Observations of the White Dwarf in the U Geminorum System with the Hopkins Ultraviolet Telescope, ApJ, 405, 327

LONG, K. S., MAUCHE, C. W., SZKODY, P., & MATTEI, J. A. 1995, EUVE, Observations of U Gem, in Proceedings of the Padova-Abano Conference on Cataclysmic Variables, in press

LONG, K. S., SION, E. M., HUANG, M., & SZKODY, P. 1994, Cooling of the White Dwarf in U Geminorum Between Outbursts, ApJL, 424, L49

MARSH, T. R., HORNE, K., SCHLEGEL, E. M., HONEYCUTT, R. K., & KAITCHUCK, R. H. 1990, ApJ, 364, 637

MASON, K. O., CÓRDOVA, F. A., WATSON, M. G., & KING, A. R. 1988, The Discovery of Orbital Dips in the Soft X-ray Emission of U Gem During an Outburst, MNRAS, 232, 779

MAUCHE, C. W. 1991, private communication

MAUCHE, C. W. 1995, EUVE, Photometry of SS Cygni: Dwarf Nova Outbursts and Oscillations, these proceedings

MAUCHE, C. W., RAYMOND, J. C., & CÓRDOVA, F. A. 1988, Interstellar Absorption Lines in High Resolution IUE, Spectra of Cataclysmic Variables, ApJ, 335, 829

MAUCHE, C. W., RAYMOND, J. C., & MATTEI, J. A. 1995, EUVE, Observations of the Anomalous 1993, Outburst of SS Cygni, ApJ, 446, in press

MEYER, F., & MEYER-HOFMEISTER, E. 1992, Vertical Structure of Accretion Disks, A&A, 106, 34

NOUSEK, J. A., BALUTA, C. J., CORBET, R. H. D., MUKAI, K., OSBOURNE, J. P. 1994, ASCA, Observations of SS Cygni During an Anomalous Outburst, ApJL, 436, L19

OSAKI, Y. 1974, An Accretion Model for the Outbursts of U Geminorum Stars, Pub. Astr. Soc. of Japan, 26, 429

PANEK, R. J., & HOLM, A. V. 1984, Ultraviolet Spectroscopy of the Dwarf Nova U Geminorum, ApJ, 277, 700

PRINGLE, J. E. 1977, Soft X-ray Emission from Dwarf Novae, MNRAS, 178, 195

RAPPAPORT, S., CASH, W., DOXSEY, R., McCLINTOCK, J., & MOORE, G. 1974, Possible Detection of Very Soft X-rays from SS Cygni, ApJL, 187, L5

RITTER, H. 1990, Catalogue of Cataclysmic Binaries, Low-Mass X-ray Binaries & Related Objects, A&AS, 85, 1179

SION, E. M., LONG, K. S., SZKODY, P., & HUANG, M. 1994, Hubble Space Telescope, Goddard High Resolution Spectrometer Observations of U Geminorum in Quiescence: Evidence for a Slowly Rotating White Dwarf, ApJL, 430, L53

Extreme Ultraviolet Spectroscopy of Magnetic Cataclysmic Variables

FRITS PAERELS,[1] MIN YOUNG HUR,[1]
AND CHRISTOPHER W. MAUCHE[2]

[1]Space Sciences Laboratory and Department of Physics,
University of California, Berkeley, CA 94720-7300

[2]Lawrence Livermore National Laboratory,
L-41, P.O. Box 808, Livermore, CA 94550.

A longstanding problem in the interpretation of the X-ray and extreme ultraviolet emission from strongly magnetic cataclysmic variables can be addressed definitively with high resolution EUV spectroscopy. A detailed photospheric spectrum of the accretion-heated polar cap of the white dwarf is sensitive in principle to the temperature structure of the atmosphere. This may allow us to determine where and how the bulk of the accretion energy is thermalized. The *EUVE* data on AM Herculis and EF Eridani are presented and discussed in this context.

1. Introduction

When matter accretes onto a white dwarf star with a strong ($B > 10^7$ G) magnetic field, the accretion flow near the white dwarf will be dominated by the field geometry, and quasi-1D "column" accretion onto a small polar cap ensues. Such is the case in the "polar" (after the strong polarization of their optical emission), or AM Herculis subclass of cataclysmic variables (CVs). In these objects, the field is known to be strong enough to actually dominate the accretion flow all across the binary, and an accretion disk does not form (see Cropper 1990 for a general review). Spin-orbit synchronization seems to have taken place in all but a few of the members of the class.

Since particles falling freely onto a white dwarf do not acquire enough kinetic energy to penetrate the stellar atmosphere, the bulk of the available kinetic energy must be thermalized and radiated above the atmosphere under optically thin conditions (see Frank, King, & Raine 1990 for a general treatment of accretion physics). In a simple model, the supersonically falling accretion flow therefore encounters a strong shock just above the stellar surface, at which the gas is heated to roughly the virial temperature $kT \sim GMm_p/R \sim 100$ keV, with M and R the stellar mass and radius, and m_p the proton mass. The shock-heated gas cools by hard X-ray emission, roughly half of which strikes the stellar surface. The hard X-ray irradiation heats the atmosphere to soft X-ray/EUV emitting temperatures ($kT \sim$ tens of eV), as is easily shown by a blackbody argument applied to an estimate of the size of the accretion-heated polar cap. The polars should be naturally strong EUV sources.

The same simple argument also predicts that the observed hard and soft X-ray luminosities should be roughly comparable. But while the X-ray spectral characteristics of AM Her stars generally conform to the predictions of the simple model outlined above, the luminosity balance does not. In most objects, there appears to be a large relative overluminosity in soft X-rays/EUV. In AM Her, the first discovered and brightest polar, this problem was already found with *HEAO-1* (Rothschild et al. . 1981). The most recent and most convincing demonstration of the problem is provided by a large homogeneous dataset on numerous polars obtained with the *ROSAT* PSPC (Beuermann & Schwope 1993).

S. Bowyer and R. F. Malina (eds.), Astrophysics in the Extreme Ultraviolet, 309–316.
© *1996 Kluwer Academic Publishers. Printed in the Netherlands.*

2. The Radiative Energy Balance and EUV Spectroscopy

Numerous suggestions have been made as to the cause of the overluminosity in the photospheric radiation. The interesting point here is that each of the various mechanisms implies a detectable effect on the temperature structure of the atmosphere, depending on where and how the bulk of the accretion energy is thermalized. This suggests that high-resolution spectroscopy of the white dwarf photosphere in the EUV band, where the energy distribution peaks, should hold the key to a solution of the "soft X-ray (or EUV) problem." A brief overview of the various suggested mechanisms follows.

The simplest idea is that steady nuclear burning supplies a large fraction of the soft X-ray luminosity—for white dwarfs, the efficiency of nuclear burning for converting mass to energy is much larger than the maximum accretion conversion energy. Existing calculations (Papaloizou, Pringle, & McDonald 1982) indicate, however, that the burning would not be stable for the mass-transfer rates and stellar masses appropriate to AM Her stars.

Kuijpers & Pringle (1982) suggested that dense blobs in the accretion stream would avoid being shocked above the stellar surface if their density contrast is high enough. Instead, they would penetrate to large optical depths into the atmosphere, and deposit their kinetic energy at large optical depths. Since the thermalization occurs at such large depths, the temperature structure of the atmosphere should look like that of an "ordinary" atmosphere with no external heat source. In a further development of this idea however, it has been found that the blobs probably create big "splashes" and may thermalize a significant fraction of their kinetic energy under optically thin conditions (Hameury & King 1988). In that case, the atmosphere, and hence its emission spectrum, should bear the signature of the presence of this external heat source at small optical depths.

The simple calculations for the energy balance of the shocked gas assume instantaneous collisional coupling between ions and electrons. More recently, the structure of the accretion shock has been examined in more detail (Thompson & Cawthorne 1987; Woelk & Beuermann 1992). The length scale associated with energy exchange between ions and electrons can be a significant fraction of the shock height, and a large fraction of the accretion energy can be radiated away across the finite shock thickness if cyclotron cooling is important. This description of the accretion region is valid only if collective effects do not effectively increase the coupling between ions and electrons. Since optically thick cyclotron cooling dominates the radiative energy balance, the electron temperature remains at a fraction of the virial temperature, and the hard bremsstrahlung emission is suppressed.

Woelk & Beuermann (1992) have performed detailed calculations for the accretion region. Their calculation shows that given the right conditions (density in the accretion stream and surface magnetic field strength) the electron temperature in the accretion region does indeed remain relatively low ($\sim$ few keV), and that the hard X-ray emission is indeed suppressed. The idea is consistent with a correlation observed between the soft/hard flux ratio, and the surface magnetic field strength of the white dwarf: objects with stronger magnetic fields have relatively larger soft/hard ratios, implying more efficient cyclotron cooling (Beuermann & Schwope 1993). The one notable exception to this trend is AM Her—with a low surface field strength of only $\sim$ 15 MG (Schmidt, Stockman, & Margon 1981; Latham, Liebert, & Steiner 1981) it nevertheless has one of the largest soft/hard flux ratios.

Here again, there is a heat source at small enough optical depth such that one expects a signature in the photospheric EUV spectrum, which may allow for an independent verification of this explanation for the observed soft/hard flux ratios.

Finally, we point out that an X-ray heated atmosphere, as is present in the "classical" model for the EUV emission from polars, will also bear the mark of heating at small optical depth. In the next section, we will briefly discuss a calculation for the spectrum of an irradiated atmosphere, as an example of the diagnostic power of the photospheric EUV spectrum in polars.

3. EUV Emission from X-ray Irradiated Atmospheres

Any external source of heat in the white dwarf atmosphere, provided it is at sufficiently small depth, will produce a temperature and ionization structure that is very different from the structure of an "ordinary" atmosphere. Specifically, the temperature distribution will be much flatter in the outer parts of the atmosphere, and a temperature inversion will develop in the transition region to the optically thin, hot parts of the accretion region. Assuming LTE for the moment, these properties will qualitatively lead to a larger range of ionization states of a given element being visible in the photospheric spectrum, as compared to the spectrum of an ordinary atmosphere, and a reduced contrast at the absorption edges due to the flat temperature structure. The increased ionization and reduced opacity will also produce a relatively harder spectrum. The temperature inversion may produce a characteristic angular intensity pattern as a function of frequency that would reveal itself in phase-resolved spectroscopy of the accretion spot as limb darkening and brightening. Finally, at the top of the atmosphere in the transition region and above, the conditions are such that, at the right viewing angle, emission features may develop.

All these features are present in calculations for hard X-ray irradiated atmospheres performed by van Teeseling, Heise, & Paerels (1994). These are LTE calculations, for an atmosphere of cosmic abundances (the accreting material maintains a steady supply of metals into the photosphere). The irradiation effect is incorporated by specifying the irradiating radiation field as the upper boundary condition, and solving the transfer equation. In Figure 1, we show an example of the spectrum of an atmosphere at $T_{\rm eff} = 240,000$ K, $\log g = 8$, with and without irradiation by hard thermal bremsstrahlung.

Calculations for the spectrum of the accretion region cooling by optically thick cyclotron emission (Woelk & Beuermann 1992) generally show the same qualitative atmospheric structure. So far, models have been computed for pure hydrogen atmospheres only, which yield roughly the correct optical/UV continuum spectrum and the cyclotron emission spectrum. At higher energies, the opacity is dominated by highly ionized metals, whose presence completely dominates the shape of the EUV/soft X-ray spectrum. Hence, these models cannot yet be directly compared with EUV spectroscopic observational data.

4. EUV/Soft X-ray Emission from AM Her

As mentioned before, the brightest polar, AM Herculis, has a very large soft/hard flux ratio. Its low surface magnetic field strength of ~ 15 MG probably indicates that this large flux ratio cannot be explained by including the effects of cyclotron cooling on the structure of the accretion region. Nevertheless, the emission spectrum shows definite signs of external heating of some sort, as was first found in spectroscopic observations with the diffraction grating spectrometers on *EXOSAT* (Paerels, Heise, & van Teeseling 1994). The white dwarf photospheric emission is definitely much harder than predicted for an ordinary atmosphere. In addition, there was evidence for the presence of absorption

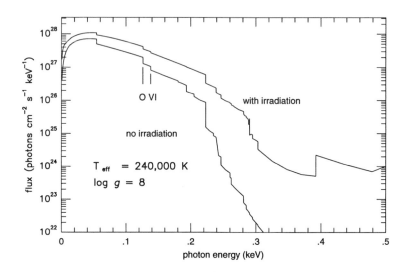

FIGURE 1. Flux spectrum of an atmosphere of cosmic abundances at $T_{\text{eff}} = 240,000$ K, $\log g = 8$. The lower curve is the spectrum for an atmosphere without irradiation, while the upper curve is for the same atmosphere with hard X-ray irradiation. The position of the O VI $2s, 2p$ edges has been indicated. Note the generally reduced contrast at absorption edges and the harder spectral shape for the irradiated atmosphere.

edges due to a wide range of ionization stages of oxygen and neon; this is illustrated in Figure 2, where we show the orbital average spectrum of AM Her as observed with *EXOSAT* in 1983.

The *EXOSAT* data had only moderate sensitivity to absorption features—the spectral resolution was about 3 Å at 100 Å with the high dispersion grating. Instead, these data cover a wide energy band (0.1–2 keV). This source therefore was a natural target for observations with *EUVE*, to obtain high resolution spectroscopy over a narrower band.

We observed AM Her with *EUVE* for a total of 102,000 s effective exposure time over the period 1993 September 23–28, covering a total of 36.9 binary orbits of 3.09 hrs each (Paerels et al. . 1995). We detected the source in the 75–120 Å band. The orbital average spectrum is shown in Figure 3. We positively detect the Ne VI ground state and first excited state absorption edges at 78.5, 85.3 Å. In addition, we detect a number of absorption lines which can be identified with a series of $n = 2$–3,4 transitions in Ne VIII. The simultaneous presence of these two ions with very different ionization potentials ($\chi = 158, 239$ eV, respectively) in the EUV photosphere is qualitatively consistent with external heating of the atmosphere.

Surprisingly, there is no evidence for the O VI $2s, 2p$ edges at 89.8, 98.3 Å. This is again qualitatively consistent with heating at small depths, which tends to reduce the contrast at absorption edges (cf., § 3). However, it remains puzzling why there should be Ne VI absorption, since O VI and Ne VI have similar ionization potentials. An anomalously low oxygen abundance in the accreting gas seems excluded by observations of strong O

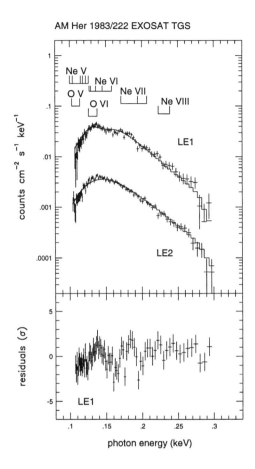

FIGURE 2. Orbital average photospheric spectrum of AM Her as observed with the *EXOSAT* Transmission Grating Spectrometers, with the high dispersion (LE1) and low dispersion (LE2) gratings. The solid line is a simple blackbody fit at $kT_{BB} = 28$ eV, $N_H = 6 \times 10^{19} \text{cm}^{-2}$. Residuals to the LE1 fit are shown in the lower panel. The positions of a number of absorption edges from highly ionized O and Ne has been indicated. Note the presence of discrete structure in the spectra and the residuals, and the fact that the spectrum extends all the way out to the C K edge (280 eV) (from Paerels, Heise, & van Teeseling 1994).

line emission in the far UV (O VI $2s$–$2p$, Raymond et al. 1995) and the optical (O III Bowen fluorescence, Schachter et al. 1991), both arising in the photoionized accretion stream.

There is evidence for a slight hardening of the spectrum at those orbital phases when we view the accretion spot at very shallow angles. This may indicate the presence of a temperature inversion in the photosphere.

All these facts are qualitatively consistent with the spectrum arising in an X-ray ir-

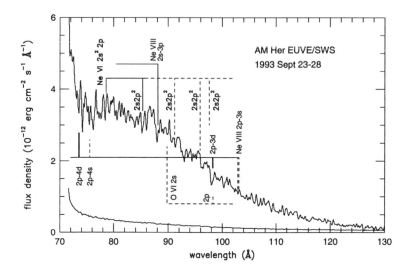

FIGURE 3. Orbital average spectrum of AM Her as observed with *EUVE*. The spectrum has been background subtracted; the lower curve represents the size of the error bars, based on counting statistics. We detect the Ne VI ground state and first excited state edges and Ne VIII $n = 2$–3,4 absorption lines (indicated with solid lines above and below the spectrum).

radiated atmosphere. A detailed spectroscopic comparison with models for irradiated atmospheres to substantiate these suggestions is in preparation. Further details of the observation and the phase dependence of the spectrum may be found in Paerels et al. (1995).

5. EUV Spectroscopy of EF Eri

EF Eridani is another relatively bright polar. Contrary to AM Her, its soft/hard luminosity ratio is closer to that expected for a purely radiatively heated accretion spot (Beuermann, Stella, & Patterson 1987; Beuermann & Schwope 1993). In addition, on at least a number of occasions, the source also exhibited another feature predicted by the classical irradiation model: if the white dwarf atmosphere is heated by hard X-rays from the post-shock gas, any variability in hard X-rays should be tracked by the soft emission. Beuermann, Stella, & Patterson (1987) found such behavior in *Einstein* IPC, data, although the correlation is absent in the recent *ROSAT* data (Beuermann, Thomas, & Pietsch 1991).

We observed EF Eri with *EUVE* for 80,500 s effective exposure time on 1993 September 5–9. The orbital average spectrum is shown in Figure 4. We detect a very strong Ne VI ground state absorption edge at 78 Å. There is evidence for an edge at 98 Å, which could be due to either O VI $2p$ (but no ground state edge seen at 89 Å), or Ne V $2s^22p^2$ (ground state). Another edge near ~ 113 Å may be Mg IV $2s^22p^5$ (ground state). The spectrum is unfortunately too noisy to reveal any absorption lines.

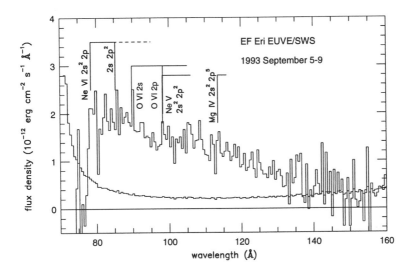

FIGURE 4. Orbital average spectrum of EF Eri as observed with *EUVE*. The spectrum has been background subtracted; the lower curve represents the size of the error bars, based on counting statistics. We detect the Ne VI ground state edge. There is evidence for either O VI $2p$ or Ne V ground state absorption at 98 Å, and possibly Mg IV ground state absorption at $\sim$ 113 Å (indicated with solid lines above the spectrum).

The presence of a strong Ne VI edge indicates lower ionization in EF Eri than in AM Her, consistent with the detection of the lower ionization species Ne V and Mg IV. It is tempting at this stage to interpret the presence of such a strong edge as evidence for the absence of irradiation (cf., § 3), but in view of the limited bandwidth over which we observe the spectrum a detailed comparison with model atmospheres will be necessary to decide whether this suggestion is correct.

6. Conclusions

Detailed spectroscopy of the EUV spectrum of the accretion heated atmosphere of the white dwarf in magnetic CVs can help determine where and how the bulk of the accretion energy is thermalized. Spectroscopy with *EUVE* of AM Her and EF Eri confirms the potential of this diagnostic.

Preliminary investigation of the spectra of AM Her and EF Eri has already produced suggestive spectroscopic evidence for external heating of the white dwarf atmosphere in AM Her, while EF Eri may show the spectrum of an ordinary hot atmosphere, indicating perhaps that at the time of the *EUVE* observation most of the accretion energy was being deposited at large optical depths.

Finally, we mention the fact that *EUVE* observations of two other polars have also provided interesting spectroscopic information, although these objects are weaker, and the statistical quality of the spectra is somewhat limited. Vennes et al. (1995) report the detection of weak O VI absorption edges in the spectrum of VV Pup, and find that they need to assume an anomalously high oxygen abundance in the atmosphere in order to fit the spectrum, while Rosen et al. (*this volume*) detected Ne VI and Ne VIII absorption in RX J1938-461.

F.P. and M.Y.H. were supported by NASA grant NAG-5-2378. C.W.M.'s contribution to this work was performed under the auspices of the US Department of Energy by Lawrence Livermore National Laboratory under contract No. W-7405-Eng-48.

REFERENCES

BEUERMANN, K. STELLA, L., & PATTERSON, J. 1987, Einstein, observations of EF Eridani 2A 0311, the textbook example of AM Herculis type systems, ApJ, 316, 360

BEUERMANN, K., THOMAS, H. -C., & PIETSCH, W. 1991, Short time-scale X-ray variability in the AM Her type binary EF Eridani, A&A, 246, L36

BEUERMANN, K, & SCHWOPE, A. D. 1993, AM Herculis Binaries, in Interacting Binary Stars, ed. A. W. Shafter, ASP Conf. Ser., 56, 119

CROPPER, M. 1990, The polars, Space Sci. Rev, 54, 195

FRANK, J., KING, A., & RAINE, D. 1990, Accretion power in astrophysics 2nd Ed., Cambridge University Press

HAMEURY, J. M., & KING, A. R. 1988, The X-ray light curves of AM Herculis systems, A&A, 235, 433

KUIJPERS, J., & PRINGLE, J. E. 1982, Comments on radial white dwarf accretion, A&A, 114, L4

LATHAM, D. W., LIEBERT, J., & STEINER, J. E. 1981, The 1980, low state of AM Herculis, ApJ, 246, 919

PAERELS, F., HEISE, J., & VAN TEESELING, A. 1994, Simultaneous soft and hard X-ray spectroscopy of AM Herculis with EXOSAT: discovery of photospheric absorption features, ApJ, 426, 313

PAERELS, HUR, M. Y., MAUCHE, C. W., & HEISE, J. Extreme ultraviolet spectroscopy of the white dwarf photosphere in AM Herculis, ApJ, submitted

PAPALOIZOU, J. C. B., PRINGLE, J. E., & McDONALD, J. 1982, Steady nuclear burning on white dwarfs, MNRAS, 198, 215

RAYMOND, J. C., MAUCHE, C. W., BOWYER, S., & HURWITZ, M. 1995, ORFEUS, observations of AM Herculis, ApJ, 440, 331

ROTHSCHILD, R. ET AL. 1981, The X-ray spectrum of AM Herculis from 0.1 to 150 keV, ApJ, 250, 723

SCHACHTER, J., FILIPPENKO, A. V., KAHN, S. M., & PAERELS, F. 1991, Bowen fluorescence in AM Herculis stars, ApJ, 373, 633

SCHMIDT, G. D., STOCKMAN, H. S., & MARGON, B. 1981, A direct measurement of the magnetic field in AM Herculis, ApJ, 243, L157

THOMPSON, A. M., & CAWTHORNE, T. V. 1987, Cyclotron emission from white dwarf accretion columns, MNRAS, 224, 425

VAN TEESELING, A., HEISE, J., & PAERELS, F. 1994, X-ray irradiation of white dwarf atmospheres: the soft X-ray spectrum of AM Herculis, A&A, 281, 119

VENNES, S., SZKODY, P., SION, E. M., & LONG, K. 1995, Extreme ultraviolet spectroscopy and photometry of VV Puppis during a high accretion state, ApJ, 445, in press

WOELK, U., & BEUERMANN, K. 1992, Particle heated atmospheres of magnetic white dwarfs, A&A, 256, 498

EUVE Photometry of SS Cygni: Dwarf Nova Outbursts and Oscillations

CHRISTOPHER W. MAUCHE

Lawrence Livermore National Laboratory, L-41, P.O. Box 808, Livermore, CA 94550, USA

I present *EUVE* Deep Survey photometry and AAVSO optical measurements of the 1993 August and 1994 June/July outbursts of the dwarf nova SS Cygni. The EUV and optical light curves are used to illustrate the different response of the accretion disk to outbursts which begin at the inner edge and propagate outward, and those which begin at the outer edge and propagate inward. Furthermore, we describe the properties of the quasi-coherent 7–9 s sinusoidal oscillations in the EUV flux detected during the rise and plateau stages of these outbursts.

1. Introduction

Simple theory predicts that nonmagnetic cataclysmic variables (CVs; nova-like variables and dwarf novae) should be bright EUV sources. As material sinks through the accretion disk of a CV, half of its specific gravitational potential energy is converted into heat, which is radiated away locally, and half is converted into orbital kinetic energy via the balance of gravitational and centrifugal forces. Consequently, the material in the disk orbiting just above the surface of the white dwarf has a specific kinetic energy equal to $GM_{wd}/2R_{wd}$, where M_{wd} and R_{wd} are the mass and radius of the white dwarf. Unless the white dwarf is rotating near breakup, the material in the disk must dissipate this amount of energy in the boundary layer between the disk and the surface of the white dwarf before accreting onto the star. The luminosity of the boundary layer will be $GM_{wd}\dot{M}/2R_{wd} \approx 20\,L_\odot$ for a $1\,M_\odot$ white dwarf accreting at a rate of $\dot{M} = 10^{-8}\,M_\odot\,\mathrm{yr}^{-1}$. For the expected size of the boundary layer, this luminosity will be radiated at EUV to soft X-ray energies.

Although the luminosities of the boundary layers in VW Hyi (Mauche et al. 1991) and SS Cyg (Mauche, Raymond, & Mattei 1995) appear to be lower than predicted by theory by roughly an order of magnitude, these systems are still sufficiently luminous, sufficiently close ($d \lesssim 100$ pc), and have sufficiently low interstellar column densities ($N_H \lesssim 3 \times 10^{19}$ cm^{-2}) to be observed as bright EUV sources. Indeed, dwarf novae have proven to be particularly rewarding targets for *EUVE*: their EUV flux varies on time scales ranging from seconds to days, they produce high signal-to-noise ratio spectra, and their spectra are bewilderingly complex. On the down side, dwarf novae can only be observed as targets of opportunity, and so require significant and often urgent effort to observe. The success of the existing *EUVE* observations of dwarf novae is due to the efforts of the members, staff, and director of the American Association of Variable Star Observers (AAVSO); *EUVE* Deputy Project Scientist Ron Oliversen; and the staffs of the Center for EUV Astrophysics (CEA), the *EUVE* Science Operations Center at CEA, and the Flight Operations Team at Goddard Space Flight Center.

To date, *EUVE* has observed the nova-like variable IX Vel (van Teeseling et al. 1995) and the dwarf novae SS Cyg (twice), U Gem (Long et al. 1995), and VW Hyi. Having discussed at length the spectroscopic results from the first *EUVE* observation of SS Cyg (Mauche, Raymond, & Mattei 1995), I present here a fuller, though preliminary and largely descriptive, discussion of the photometric results from the Deep Survey (DS) instrument from both *EUVE* observations of SS Cyg.

S. Bowyer and R. F. Malina (eds.), Astrophysics in the Extreme Ultraviolet, 317–324.

2. Observations

EUVE observed SS Cyg in outburst in 1993 August and 1994 June/July. The observations cover the intervals MJD 9216.6 to 9223.1 and MJD 9526.7 to 9536.9 (with a break between MJD 9529.8 and 9532.5). Figure 1 shows the optical light curves of these outbursts. On both occasions, the optical flux was above $V = 10$ for ≈ 16 days, but the 1993 outburst was anomalous in that it took ≈ 5 days for the light curve to reach maximum, whereas typical outbursts (such as the 1994 outburst) reach maximum in 1 to 2 days. Overplotted on Figure 1 is the log of the count rate of the DS instrument for these outbursts. The DS count rate is seen to increase by a factor of $\gtrsim 330$ for the 1993 outburst, and by a factor of ≈ 1000 for the 1994 outburst; at the peak of the outbursts, the count rate in the DS instrument was ≈ 5 counts s^{-1}. Figure 1 demonstrates that the EUV light curve of the 1993 outburst rose more quickly than the optical, and that the rise of the EUV light curve of the 1994 outburst was delayed by ≈ 1 day relative to the optical. It is this delay between the optical and EUV light curves which allows *EUVE* observations of dwarf novae in outburst to be triggered by "amateur" optical observers.

3. Dwarf Nova Outbursts

The first application of the *EUVE* photometry of SS Cyg deals with the relative shapes of the optical and EUV light curves shown in Figure 1 and the implications for the processes responsible for dwarf nova outbursts. For an excellent review of the physics of dwarf nova outbursts, see Cannizzo (1993), from which the following synopsis was prepared.

The vertical structure (the run of pressure, temperature, and flux) of an accretion disk is determined by the equations of hydrostatic equilibrium, energy transport by radiation and convection, and, unlike a stellar atmosphere, energy generation via viscous dissipation of orbital shear. Using standard assumptions, it is found that the effective temperature $T_{\rm eff}$ of the disk as a function of surface density Σ is double-valued: in the Σ-$T_{\rm eff}$ plane, the thermal stability of the disk is described by an "S-shaped" curve. That the form of this relationship leads to an instability in the thermal state of the disk can be understood as follows. Imagine that the surface density of a given annulus increases because mass is added to the annulus from "above" (from higher in the gravitational well) faster than it is removed from "below." Initially, the temperature of the annulus will increase monotonically. However, when the surface density reaches some value $\Sigma_{\rm max}$ when the temperature reaches ~ 6000 K, hydrogen begins to become ionized, causing a sharp increase in both the opacity and the specific heat of the gas. These microscopic changes in the gas drive macroscopic changes in the vertical structure of the disk as radiation replaces convection as the main source of the transport of energy out of the disk. To cope with these changes, the temperature of the annulus must increase discontinuously to remain in thermal equilibrium. On the upper branch of the S-shaped curve, the viscous dissipation is higher because the temperature and hence the pressure is higher. Therefore, it is possible for the surface density of the annulus to decrease as mass drains out of the annulus faster than it is added from "above." Initially, the temperature of the annulus will decrease monotonically. However, when surface density reaches some value $\Sigma_{\rm min}$ when the temperature reaches $\sim 10^4$ K, hydrogen begins to recombine, causing a sharp decrease in both the opacity and the specific heat of the gas. The gas again falls out of thermal equilibrium, this time cooling significantly at fixed surface density.

To describe the response of the full disk to the above instability, it is necessary to specify the dependence of the critical surface densities $\Sigma_{\rm min}$ and $\Sigma_{\rm max}$ on radius, and to

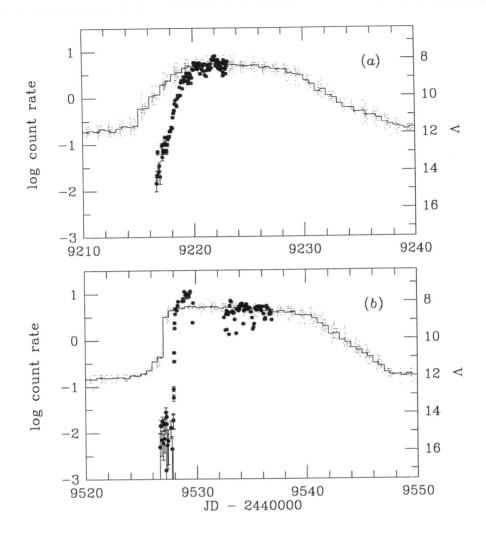

Figure 1. DS count rate (*filled circles with error bars*) and visual magnitude (*small dots and histogram*) as a function of time from (*a*) 1993 August and (*b*) 1994 June/July.

solve the time-dependent hydrodynamic equations which derive from the conservation of mass, energy, and angular momentum. These calculations demonstrate that when a given annulus reaches the value of Σ_{max} appropriate to its radius and makes the transition to the high state, it generates a heating wave which transforms the entire disk to the high state. This disturbance can begin in the inner disk and sweep outward, or begin in the outer disk and sweep inward, generating, respectively, inside-out and outside-in outbursts. Since mass is predominantly transferred inward through the disk by viscous dissipation, the minimum surface density Σ_{min} is reached first in the outer disk. When that annulus makes the transition to the low state, it generates a cooling wave which sweeps inward, transforming the entire disk to the low state.

Due to the radial dependence of the surface density of the disk in quiescence and in outburst, the rise of inside-out and outside-in outbursts is very different. In quiescence, the surface density of the disk is constrained by the critical surface densities Σ_{min} and Σ_{max} to increase with radius; in outburst, material flows toward the white dwarf, causing the surface density to decrease with radius. As the heating wave sweeps inward through the disk in an outside-in outburst, it transforms successively smaller annuli to the high state. The surface density of annuli behind the heating wave quickly increases as material flows in from larger radii. In an inside-out outburst, the heating wave transforms successively larger annuli to the high state, but the surface density of annuli in the inner disk does not increase significantly until the heating wave has swept through the entire disk and material is able to flow into the inner disk from the outer disk.

The differences between the inside-out and outside-in outbursts qualitatively describe the differences in the optical and EUV light curves of the outbursts shown in Figure 1. The 1993 outburst was of the inside-out variety. As the heating wave swept outward though the disk, it transformed successively larger annuli to the high state, causing a constant increase in the rate at which material would reach the inner disk and boundary layer and there radiate at EUV wavelengths. The 1994 outburst, on the other hand, was of the outside-in variety. News of the outburst arrived at the inner disk and boundary layer with the arrival of the heating wave, just ahead of the nearly steady flow of material which would power the boundary layer. The delay of ≈ 1 day between the rise of the optical and EUV light curves of the 1994 outburst is the length of time required for the heating wave to travel the length of the accretion disk.

This delay of ≈ 1 day between the EUV and optical light curves is longer than the delay of ≈ 0.5–0.75 day between the FUV (950–1150 Å) and optical light curves measured by Cannizzo, Wheeler, & Polidan (1986) for the 1980 May outside-in outburst of SS Cyg. The length of the FUV delay is shorter than the EUV delay because the disk itself produces the FUV flux and the heating wave travels a shorter distance before heating an annulus of the disk to the temperature ($\sim 30,000$ K) required for it to radiate in the FUV. This difference between the FUV and EUV light curves of SS Cyg highlights the importance of the above EUV light curves: the optical through FUV light curves measure the response of the disk to the outburst; the EUV light curves measure the rate at which material arrives at the boundary layer. The combination of the EUV and optical, UV, and/or FUV light curves makes for a powerful diagnostic of the mechanisms responsible for dwarf nova outbursts. To follow up these results, we plan to observe VW Hyi this summer simultaneously with *Voyager* and *EUVE*.

4. Dwarf Nova Oscillations

Superposed on the long-term photometric variations associated with its dwarf nova outbursts, SS Cyg exhibits quasi-coherent photometric oscillations ("dwarf nova oscillations," see, e.g., Patterson 1981; Warner 1995) on a time scale measured in seconds. Optical oscillations were detected by Patterson, Robinson, & Kiplinger (1978) with a period of 9.74 s, by Horne & Gomer (1980) with periods of 8.23 and 8.50 s, and by Patterson (1981) with periods of 10.72, 10.90, and 8.9 s. Hildebrand, Spillar, & Stiening (1981) tracked the oscillation over an interval of 6 days and observed its period fall from 7.53 s to 7.29 s and then rise again to 8.54 s. At soft X-ray energies, Córdova et al. (1980) and Córdova et al. (1984) detected oscillations in *HEAO 1* LED 1 data at a period of

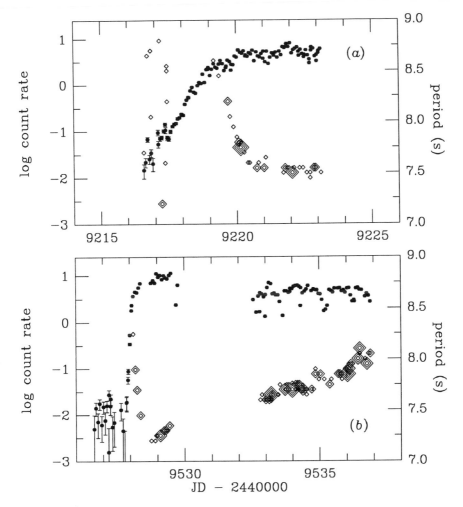

FIGURE 2. DS count rate (*filled circles with error bars*) and pulsation period (*diamonds*) as a function of time from (*a*) 1993 August and (*b*) 1994 June/July. The size of the diamonds corresponds to the relative size of the χ^2 statistic (see text).

9 s and 11 s, respectively. Jones & Watson (1992) detected oscillations in *EXOSAT* LE data at periods between 7.4 and 10.4 s.

To search for oscillations in the EUV flux measured by the DS instrument, we proceed as follows. First, for each of the i satellite orbits, we construct DS count rate light curves with 1 s time resolution. Next, we phase-fold these light curves using a range of trial periods $P_j = P_1 + [j(P_2 - P_1)/j_{max}]$ for $j = 0, 1, \ldots, j_{max}$. Finally, we test for an oscillation with the given period during the given orbit by means of the χ^2 statistic: $\chi^2_{i,j} = \sum_{k=1}^{k_{max}}(r_k - \langle r \rangle)^2/\sigma_k^2$, where r_k is the count rate in phase bin k, $\langle r \rangle = \sum_{k=1}^{k_{max}} r_k/k_{max}$, and $\sigma_k = \sqrt{r_k}$. For an oscillation to be detected by this means, it must have a period within the range of trial periods, it must have a significant amplitude, and it must be coherent in phase and period over the ≲ 2000 s interval during each orbit when the source is visible to the satellite.

To present the results of these calculations, we show in Figure 2 the χ^2 statistic in the form of a pseudo-contour diagram along with the log of the DS count rate. It is apparent from this figure that coherent oscillations exist in the DS count rate for both outbursts. For the 1993 outburst, random peaks in the χ^2 surface are present very early in the outburst, but the oscillation does not stabilize until ≈ MJD 9219 when the count rate is ≈ 2.8 counts s^{-1}. The period of the oscillation is initially ≈ 8.6 s, falls over the next day or so to ≈ 7.8 s, and then asymptotically approaches ≈ 7.5 s over the next few days. For the 1994 outburst, the oscillation turns on during the fast rise to outburst on ≈ MJD 9528 when the count rate is ≈ 2.3 counts s^{-1}. The period of the oscillation is initially ≈ 8.9 s, falls to ≈ 7.2 s (the shortest period ever observed in SS Cyg, or in any other dwarf nova) within a day, and rebounds to ≈ 7.4 s by the end of the observation on MJD 9529.8. When observations resume ≈ 3 days later, the period of the oscillation is ≈ 7.6 s and rises slowly over the next few days to ≈ 8.0 s.

Other quantitative aspects of the oscillations are as follows. First, as is evident from Figure 2, the period of the oscillation correlates with the count rate, being long when the count rate is low and short when the count rate is high. Furthermore, it appears from the 1994 outburst that at a given count rate the period on the rise to outburst is to first order equal to the period on the decline. This is true despite the fact that the period derivative $\dot{P}$ is significantly higher on the rise compared to that on decline. Evidently, the period of the oscillation does not depend to any significant degree on $\dot{P}$, but only on the DS count rate, and, by inference, only on the mass-accretion rate through the boundary layer. Second, the amplitude of the oscillation correlates with the count rate, being high (≈ 100%) when the count rate is low, and moderate (20% ± 10%) when the count rate is high. Furthermore, the amplitude of the oscillation is equal at a given count rate between the 1993 and 1994 outbursts. This fact is particularly striking in the interval log DS count rate = −1.5 to 0, which is traversed slowly on the rise of the 1993 outburst, and very rapidly on the rise of the 1994 outburst. Finally, based on the shape of the phase-folded light curves and on the absence of power at any of the harmonics in power spectra of the light curves (see Fig. 3), the oscillations are purely sinusoidal.

What is the physical mechanism responsible for these oscillations? For a review of this subject, refer to Warner (1995). The low period stability of the oscillations rules out the rotating white dwarf (the DQ Her mechanism) as well as non-radial pulsations of the white dwarf as the cause of the oscillations; pulsations are observed in high inclination systems, ruling out the eclipse by the white dwarf of luminous blobs of material in the inner disk and boundary layer; r-modes and trapped g-modes fail because more than one mode would be excited; oscillations of the accretion disk fail because they are not confined to a particular annulus and hence to a particular period. Viable mechanisms are more difficult to construct. Molteni, Sponholz, & Chakrabarti (1995) argue that shocks are an inevitable consequence of gas flow near the inner edge of an accretion disk, and that shock oscillations lead to a cycling of the accretion luminosity when the cooling time of the shocked gas is comparable to the radial infall time. Warner & Livio (1995) propose that the oscillations are due to the combined action of the differentially rotating surface layers of the white dwarf and magnetically controlled accretion.

The *EUVE* observations of the 1994 outburst of SS Cyg severely constrain the Warner & Livio model. The time derivative of the kinetic energy of the rotating surface layers of the white dwarf is $\dot{E}_{\rm K} = 4\pi^2 I\dot{P}/P^3$ where $I = MR^2$ is the moment of inertia of

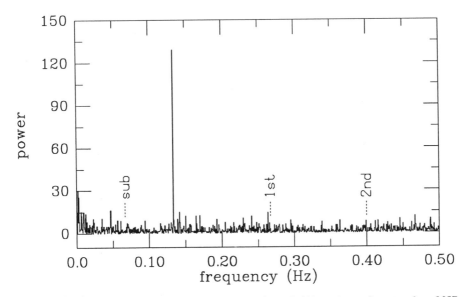

FIGURE 3. Power spectrum of the DS count rate from the half-hour interval centered on MJD 9222.06. The fundamental lies at 7.50 s and the positions of the first and second harmonics and the first subharmonic are indicated.

the surface layers (modeled as a thin-walled hollow cylinder of mass M and radius R). During the rise of the 1994 outburst, $\dot{P} \approx 2\,\text{s}/0.5\,\text{day} \approx 5 \times 10^{-5}\,\text{s s}^{-1}$ and $P \approx 8\,\text{s}$, hence $\dot{E}_K \approx 2 \times 10^{35}\,(M/10^{-10}\,M_\odot)(R/5.5 \times 10^8\,\text{cm})^2\,\text{erg s}^{-1}$. This value is $\gtrsim 100$ times the measured peak boundary layer luminosity (Mauche, Raymond, & Mattei 1995), implying that the mass of the rotating surface layers must be significantly less than $10^{-12}\,M_\odot$.

The author is pleased to acknowledge the invaluable contributions to this research by the members, staff, and director, J. Mattei, of the AAVSO. The author greatly benefited by conversations with J. Cannizzo and B. Warner concerning the dwarf nova outburst mechanism and dwarf nova oscillations. B. Warner, M. Livio, and S. Chakrabarti generously provided copies of material prior to publication. This work was performed under the auspices of the US Department of Energy by Lawrence Livermore National Laboratory under contract No. W-7405-Eng-48.

REFERENCES

CANNIZZO, J. K. 1993, The Limit Cycle Instability in Dwarf Nova Accretion Disks, in Accretion Disks in Stellar Systems, ed. J. C. Wheeler, World Scientific, 6,

CANNIZZO, J. K., WHEELER, J. C., & POLIDAN, R. S. 1986, Dwarf Nova Burst Asymmetry and the Physics of Accretion Disks, ApJ, 301, 634

CÓRDOVA, F. A., CHESTER, T. J., MASON, K. O., KAHN, S. M., & GARMIRE, G. P. 1984, Observations of Quasi-Coherent Soft X-ray Oscillations in U Geminorum and SS Cygni, ApJ, 278, 739

CÓRDOVA, F. A., CHESTER, T. J., TUOHY, I. R., & GARMIRE, G. P. 1980, Soft X-ray Pulsations from SS Cygni, ApJ, 235, 163

HILDEBRAND, R. H., SPILLAR, E. J., & STIENING, R. F. 1981, Observations of Fast Oscillations in SS Cygni, ApJ, 243, 223

HORNE, K., & GOMER, R. 1980, Phase Variability in the Rapid Optical Oscillations of SS Cygni, ApJ, 237, 845

JONES, M. H., & WATSON, M. G. 1992, The EXOSAT Observations of SS Cygni, MNRAS, 257, 633

LONG, K. S., MAUCHE, C. W., SZKODY, P., & MATTEI, J. A. 1995, EUVE Observations of U Gem, in Proc. of the Padova-Abano Conference on Cataclysmic Variables, in press

MAUCHE, C. W., RAYMOND, J. C., & MATTEI, J. A. 1995, EUVE Observations of the Anomalous 1993, August Outburst of SS Cygni, ApJ, 446, in press

MAUCHE, C. W., WADE, R. A., POLIDAN, R. S., VAN, DER, WOERD, H., & PAERELS, F. B. S. 1991, On the X-ray Emitting Boundary Layer of the Dwarf Nova VW Hydri, ApJ, 372, 659

MOLTENI, D., SPONHOLZ, H., & CHAKRABARTI, S. K. 1995, Resonance Oscillations of the Radiative Shocks in Accretion Disks around Compact Objects, ApJ, submitted

PATTERSON, J. 1981, Rapid Oscillations in Cataclysmic Variables. VI. Periodicities in Erupting Dwarf Novae, ApJS, 45, 517

PATTERSON, J., ROBINSON, E. L., & KIPLINGER, A. L. 1986, Detection of a 9.735 Second Periodicity in the Light Curve of SS Cygni, ApJ, 226, L137

VAN, TEESELING, A., DRAKE, J. J., DREW, J. E., HOARE, M. G., & VERBUNT, F. 1995, ROSAT and EUVE Observations of the Nova-like Variable IX Velorum, A&A, submitted

WARNER, B. 1995, Cataclysmic Variable Stars, Cambridge: Cambridge Univ. Press

WARNER, B., & LIVIO, M. 1995, A Model for Dwarf Nova Oscillations, ApJ, submitted

The EUV Excess in Magnetic Cataclysmic Variables

JOHN K. WARREN[1] AND KOJI MUKAI[2]

[1]Space Sciences Laboratory, University of California, Berkeley, CA 94720, USA

[2]Code 668, NASA/Goddard Space Flight Center, Greenbelt, MD 20771, USA

We present preliminary analysis of *EUVE* pointed data of 8 magnetic cataclysmic variables. Blackbody temperatures, luminosities, and interstellar columns have been better constrained. Using these luminosities we look for correlations between the EUV excess (over optical and hard X-rays) and various system parameters. While it appears there is no correlation between the EUV excess and the system inclination and orbital period, correlations are suggested between the EUV excess and the longitude of the accretion spot, the colatitude of the accretion spot, the white dwarf magnetic field, and the magnetic capture radius.

1. Introduction

Magnetic cataclysmic variables, or polars, are cataclysmic variables (CVs) with a white dwarf magnetic field of $\sim$ 10–60 MG. They have no accretion disk and the accretion column lands directly on the white dwarf in a small region(s) near one or both of its magnetic poles. See Cropper (1990) for a review. In the standard (outdated) model the accreting material passes through a standoff shock just above the white dwarf, cooling via optically thin thermal bremsstrahlung hard X-rays (HXR) and optical gyrocyclotron radiation. Half of this radiation is reradiated by the white dwarf in the extreme ultraviolet (EUV) and soft X-rays (SXR). This model of magnetic CV accretion predicts that the ratio R of observed EUV luminosity to the sum of the observed HXR and optical luminosities is obeys the simple relation $R \sim 0.5$, where

$$R = \frac{L_{euv}}{L_{hxr} + L_{opt}} = \frac{L_{hxr,0}\left(1 - a_x\right)}{L_{hxr,0}\left(1 + a_x\right) + L_{opt}} \sim \frac{1 - a_x}{1 + a_x} \qquad (1.1)$$

L_{euv}, L_{hxr}, and L_{opt} are the EUV, HXR and optical luminosities, respectively, $L_{hxr,0}$ is the HXR luminosity directed toward the white dwarf, and $a_x \sim .3$ is the X-ray albedo of the white dwarf. The above approximation, in which the optical luminosity is small enough to be neglected, applies in many systems.

Since the days of *HEAO-1* it has been known that R far exceeds the expected value of 0.5 in many systems. This has been known historically as the SXR excess, but which we call here the EUV excess. Several theories have been proposed to explain the EUV excess, the most promising of which is direct mechanical heating of the white dwarf atmosphere by dense filaments in the accretion stream (Kuijpers & Pringle (1982), Frank, King, & Lasota (1988)). Some as yet unknown mechanism is necessary to separate the flow (at the threading region or perhaps the L1 point) into segments which will then be turned into dense filaments by being pinched by the converging magnetic field lines and stretched by white dwarf tidal forces. No extra pressure or heat is needed to create these segments.

The EUV characteristics of CVs have been difficult to determine, because of their distances ($\geq$ 65 pc) and the high attenuation (a factor of 0.16 at 100 pc and a column of $5 \times 10^{19} \mathrm{cm}^{-2}$) suffered by the EUV photons traversing these distances. In polars, these characteristics include the size, shape, and temperature of the accretion region(s), the number of such regions, and the intervening column. New data from *EUVE* allow tighter

S. Bowyer and R. F. Malina (eds.), Astrophysics in the Extreme Ultraviolet, 325–329.

TABLE 1. Log of Observations

Star Name	Start Date	End Date	Eff. Exp. (ks)
V834 CEN	1993 May 28.1	May 29.6	42.7
EF ERI	1993 Sep. 5.6	Sep. 9.7	120.7
UZ For	1995 Jan. 15.9	Jan. 19.7	120.8
AM HER	1993 Sep. 23.7	Sep. 28.5	143.8
VV PUP	1993 Feb. 7.9	Feb. 9.3	46.0
RE J1149+284	1993 Feb. 22.8	Feb. 25.3	81.0
RE J1844-741	1994 Aug. 17.6	Aug. 24.0	150.0
RX J1938-461	1992 July 8.2	July 9.7	37.3
RX J1938-461	1993 Aug. 16.1	Aug. 17.1	29.1
RX J1938-461	1993 Oct. 6.3	Oct. 10.7	143.2
MR SER	1993 June 1.2	June 2.8	48.7
AN UMA	1993 Feb. 27.9	Feb. 1.2	40.1
EK UMA	1994 Dec. 14.3	Dec. 16.0	50.0

constraints on these characteristics and enable tighter constraints on the EUV excess. In this work we search for correlations between the EUV excess as observed by *EUVE* (Bowyer & Malina (1991)) and other missions, and system parameters. *EUVE* derived parameters are in boldface in the figures.

2. Observations

To date, 11 polars have been observed by *EUVE* in pointed observations: V834 Cen, EF Eri, UZ For, AM Her, VV Pup, RE J1149+284, RE J1844-741, RX J1938-461, MR Ser, AN UMa, and EK UMa. A summary of these observations is provided in Table 1. The data of RE J1844-741 are not publicly available, and two of the other 10—MR Ser and EK UMa—were undetected. The 8 remaining spectra were detected, as expected, only in the short wavelength channel (70–190 Å); no flux was detected in the medium wavelength (140–380 Å) or long wavelength (280–760 Å) channels. Blackbody fits were performed on these spectra, yielding T_{euv}, N_H, & L_{euv}, i.e., the blackbody temperature, the interstellar column, and the bolometrically corrected blackbody luminosity, respectively. These, along with values from other missions (see Ramsay et al. (1994)), are plotted in Figure 1.

3. Results and Discussion

In addition to the *EUVE* observations, we made use of the values compiled by Ramsay et al. (1994) of the HXR and optical luminosities of these 8 systems, as well as the blackbody parameters and HXR and optical luminosities of several other magnetic CVs. Ratios R of observed EUV luminosity to the sum of the (non-simultaneously) observed HXR and optical luminosities were computed, according to Equation 1.1. We then plotted these values of R against various system parameters: orbital inclination i, period P, longitude ψ of the accretion region, colatitude β of the accretion region, white dwarf

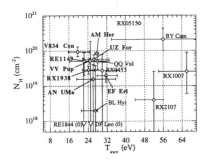
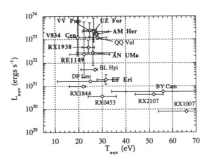

FIGURE 1. Plots of EUV blackbody parameters T_{euv}, N_H, and L_{euv}. EUV Results are in boldface.

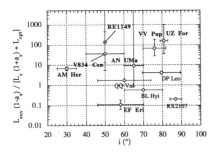
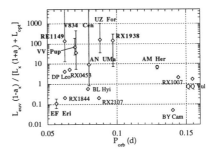

FIGURE 2. EUV excess vs. inclination i and orbital period P.

magnetic field B, and magnetic capture radius R_c. R_c is given by the equation of the ram pressure of the stream with the magnetic pressure (Mukai (1988)):

$$R_c = 1.48 \times 10^{10} B_7^{\frac{4}{11}} R_9^{\frac{12}{11}} D_9^{\frac{4}{11}} \dot{M}_{16}^{\frac{-2}{11}} M_1^{\frac{-1}{11}} \text{ cm},\qquad (3.2)$$

where B_7 is the magnetic field in units of 10^7 gauss, R_9 is the white dwarf radius in units of 10^9 cm, D_9 is the stream diameter in units of 10^9 cm, $\dot{M}_{16}$ is the mass transfer rate in units of 10^{16} gm cm^{-1}, and M_1 is the mass of the white dwarf in units of $M_\odot$.

We did not see a correlation between the EUV excess R and the inclination i, indicating the absence of beaming effects. No correlation is observed between R and the orbital period P, which is related to the secondary mass (Figure 2). However, there seems to exist a correlation between R and the longitude ψ of the accretion region, and between R and the colatitude β of the accretion region (Figure 3).

There may be a marginal correlation between R and the magnetic field B of the white dwarf. Associated with this there may also be a correlation between R and the magnetic capture radius R_c (Figure 4), indicating that in stars where the material is captured farther out the EUV excess is higher. The asynchronous system BY Cam is an exception in both plots. In any case, an *anti*-correlation between R and R_c is expected from the

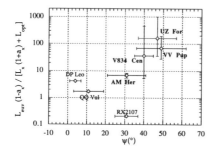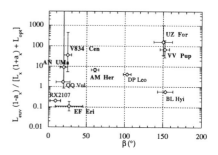

Figure 3. EUV excess vs. longitude ψ and colatitude β of accretion spot.

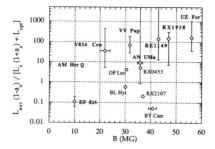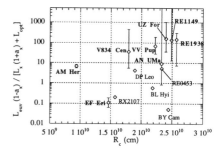

Figure 4. EUV Excess vs. white dwarf magnetic field B and magnetic capture radius R_c.

theory that the Kelvin-Helmolz instability has more time to shred the filaments into fine droplets in streams that are captured farther out (Cropper (1990)). However, this anti-correlation is not observed (see also Ramsay et al. (1994)).

Further analysis is in progress, including the exploration of deviations from blackbody spectral shape in the brighter systems, the results of which will be the subject of a forthcoming paper.

This work was supported by NASA grant NAG 5-2641, and by NASA contract NAS-5-32490.

REFERENCES

Bowyer, S. & Malina, R. F. 1991, The *EUVE* Mission, in Extreme Ultraviolet Astronomy, ed. R. F. Malina & S. Bowyer, New York: Pergamon Press, 397

Cropper, M. S. 1990, The Polars, Space Sci. Rev., 54, 195

Frank, J., King, A. R. & Lasota, J. -P. 1988, The soft X-ray excess in accreting magnetic white dwarfs, A&A, 193, 113

KUIJPERS, J. & PRINGLE, J. E. 1982, Comments on radial white dwarf accretion, A&AL, 114, L4–L6

MUKAI, K. 1988, Accretion streams in AM Her type systems., MNRAS, 232, 175

RAMSAY, G., MASON, K. O., CROPPER, M., WATSON, M. G. & CLAYTON, K. L. 1994, *ROSAT* observations of AN UMa and MR Ser: the status of the soft X-ray excess in AM Her stars., MNRAS, 270, 692

EUVE Spectrophotometry of QS Tel: The Second Pole Becomes Active

S. R. ROSEN,[1] J. P. D. MITTAZ,[2] D. A. H. BUCKLEY,[3] A. LAYDEN,[4] C. McCAIN,[5] J. P. OSBORNE,[1] AND M. G. WATSON[1]

[1] Department of Physics, University of Leicester, University Rd., Leicester, LE1 7RH, UK

[2] Mullard Space Science Laboratory, Holmbury St. Mary, Surrey, RH5 6NT, UK

[3] South African Astronomical Observatory, PO Box 9, Observatory 7935, South Africa

[4] Cerro Tololo Inter-American Observatory, Casilla 603, La Serena, Chile

[5] Mt. Stromlo and Siding Spring Observatories, Weston Creek, PO, ACT 2611, Australia

We present results of *EUVE* spectrophotometry of the EUV luminous polar, QS Tel (RE1938-461), together with contemporaneous optical photometry and spectroscopy. In marked contrast to the *ROSAT* survey observations, the *EUVE* light curve shows two flux maxima per orbital cycle, implying that both magnetic poles were active. A deep, narrow dip is observed during one of the two flux maxima, exhibiting a complex morphology which includes pronounced flickering behaviour. Although this feature is probably caused by stream occultation of the emission region, the apparent lack of spectral hardening at this time disfavours photoelectric absorption by cold gas as the dominant source of opacity. Whilst the overall *EUVE* spectrum can be characterized by a low temperature ($\sim$15eV) blackbody, implying a large soft/hard component flux ratio ($\sim$50), tentative evidence of an absorption edge from NeVI at 85A and lines due to NeVIII and NeVII at 98Å and 116Å respectively indicate that more sophisticated models must be employed. Quasi-simultaneous optical photometry shows a substantial change in the light curve over an interval of just 3 days and little evidence of correlated behaviour with the EUV flux. We consider the implications of these results on the accretion geometry and the structure of the accretion flow.

1. Introduction

QS Tel (RE1938-461) was the brightest of seven new polars (AM Her stars) discovered via the *ROSAT* WFC survey (Pounds et al. 1992). Polars are magnetic cataclysmic variables (CVs) comprising a low mass star donating material to a strongly magnetic ($B \sim 10^7$G), synchronously rotating white dwarf (see Cropper 1990 for an overview). On average, these seven new systems showed a larger EUV/optical flux ratio than previously known polars detected in the WFC survey (e.g., Watson et al. 1993) and it was suspected that they might also be characterized by large soft/hard X-ray flux ratios. Such EUV bright polars may improve our understanding of the homogeneity of the accretion flow and its interaction with the magnetosphere (e.g., mechanical heating of the white dwarf's atmosphere by dense filaments of material in the inflow has been advanced as a mechanism to explain systems with large EUV excesses—Frank, King & Lasota (1988) and references therein). Apart from its brightness, initial interest in QS Tel (Buckley et al. 1993) arose from both its orbital period of 2.33 hr, which places it inside the well known CV period gap, and its simple EUV bright-faint light curve which is symptomatic of a single, small, active pole which passes behind the limb of the white dwarf for a fraction of the rotation cycle. QS Tel has since been found to possess the largest magnetic field yet measured in a polar (Schwope et al. 1995).

S. Bowyer and R. F. Malina (eds.), Astrophysics in the Extreme Ultraviolet, 331–335.
© *1996 Kluwer Academic Publishers. Printed in the Netherlands.*

2. Observations

Two *EUVE* observations of QS Tel were made starting on 1993 Aug 16 (∼30 ks exposure) and on Oct 6 (∼70 ks). The source was only significantly detected in the (70–200Å) short wavelength spectrometer (SWS) and Deep Survey (DS) instruments. Contemporaneous fast (10s) white light optical photometry was obtained from the SAAO on Aug 16. Further fast B band photometry from CTIO was secured on Aug 19/20, whilst time-resolved Hα and Hβ spectroscopy was obtained on the nights of Aug 17 and 18 with the 2.3m ANU telescope at the Siding Springs Observatory.

3. The *EUVE* Light Curves

The *EUVE* SWS and DS light curves of QS Tel were folded on the linear orbital ephemeris of Schwope et al. (1995) whose epoch (phase 0.0) is believed to correspond to inferior conjunction of the companion star. The *EUVE* data reveal two key properties (see Figure 1). Firstly, unlike the bright-faint morphology of the *ROSAT* WFC survey data, the *EUVE* light curve contains two prominent flux maxima per cycle, demonstrating that both magnetic poles were active at that time. This underlying modulation is accompanied by a hardening of the spectrum during the maxima. Secondly, a deep, narrow dip is evident (phase 0.97–1.10) during the maximum that corresponds most closely with that observed during the WFC survey.

The SWS and DS data show that the mean dip profile comprises a slow (∼300s) ingress (phase 0.97–1.01), a broad minimum (phase 1.01–1.09) which appears to contain an interval of enhanced emission between phase 0.04 and 0.08, and a rapid egress (lasting <40s) at phase 1.10 (see Figure 1). Residual flux (≥2%) is detected at all phases in the dip. However, exploiting the sensitivity of the DS instrument allowed us to examine the individual dips in greater detail. This revealed that the enhancement during dip minimum is in fact resolved into pronounced flaring activity. Flaring behaviour is also evident during dip ingress, partially explaining the wide range of ingress profiles observed. In contrast, the epochs of the six observed dip egresses differ by no more than 25s. By dividing the SWS data into two energy bands, we tested the data during the dip for evidence of spectral changes. Although we are hampered by the weak signal at this time, we find no convincing evidence of spectral hardening during dip ingress or dip minimum. Since we believe that the dip is caused by stream occultation of the white dwarf's emission site (see § 4), if we assume that the mid-dip flares are due to intrinsic fluctuations at the emission source, as viewed through the stream, the enhanced S/N available during such flares can be exploited to more sensitively test for hardness ratio changes. None are found, suggesting that photoelectric absorption in a cold, homogeneous medium is not the dominant absorbing mechanism.

The mean *EUVE* spectrum of QS Tel has been fit with simple blackbody models. The best fit yields a temperature of 15 eV and a column of 4.4×10^{19} cm^{-2}. Taken at face value, the data imply a bolometric flux for the EUV/soft X-ray spectral component of about 10^{-9} ergs s^{-1} cm^{-2}. A simple scaling of the estimated hard component flux deduced from an earlier *ROSAT* pointed observation then suggests that the soft/hard component flux ratio may be as high as ∼50 although the lack of *ROSAT/EUVE* simultaneity and the weak constraints on the hard component spectrum mean that this ratio is somewhat uncertain. However, the *EUVE* spectrum (see Figure 2) shows tentative evidence of an absorption edge at 85Å, probably due to NeVI and possible absorption lines at 98Å and 116Å, perhaps due to NeVIII and NeVII respectively. Confirmation of the presence of these and perhaps other absorption features, probably achievable via

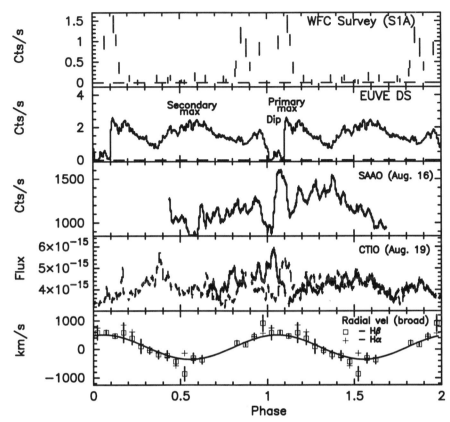

FIGURE 1. From top to bottom: The simple, single pole orbital light curve of QS Tel observed during the *ROSAT* WFC survey: the *EUVE* DS light curve showing the double-peaked (two pole mode) morphology: the SAAO optical light curve: the CTIO photometry, split into two sections and overlayed (solid and dashed lines): the radial velocity motion of the broad component of the optical emission lines (squares = Hβ, crosses = Hα) which reaches maximum redshift near the epoch of the EUV dip. Two cycles are shown for clarity except for the SAAO data where the full dataset is displayed unfolded and the CTIO data where two sections of a continuous timeseries are plotted, overlayed

further, higher quality, dithered mode *EUVE* observations, will confirm the need for more sophisticated modelling of the spectrum, an aspect currently being pursued.

4. Optical Results

The SAAO photometry of QS Tel taken quasi-simultaneously with the first *EUVE* run shows a complex orbital light curve (figure 1). Given that the EUV and optical fluxes likely arise in distinct emission regions, subject to physically and geometrically different orbital effects, it is perhaps not unsurprising that the EUV and optical light curves are essentially uncorrelated. A possible, shallow optical counterpart to the EUV dip might be present although assessment of its intrinsic depth may be affected by a

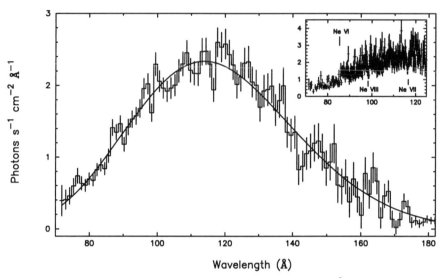

FIGURE 2. The average *EUVE* spectrum of QS Tel accumulated into 1Å bins. The solid line represents the best fitting blackbody curve (kT=15 eV, $N_H = 4.4 \times 10^{19}$ cm^{-2}). The inset shows the more finely resolved data, highlighting a possible absorption edge at 85Å (NeVI) and lines at 98Å (NeVIII) and 116Å (perhaps NeVII)

flare-like event during its latter stages (the EUV and optical dips were not observed simultaneously). The CTIO observation, obtained just 3 days later, shows a somewhat different morphology to the SAAO data and again shows no obvious connection with the *EUVE* light curve—there is also no conspicuous dip that aligns with the EUV feature. The lack of a deep (>25%) optical dip corresponding to the EUV event, together with the slow EUV ingress and EUV flaring behaviour at mid-dip, suggests that the dip is not due to an eclipse by the companion star. The ANU spectra provided important radial velocity information. The optical emission lines contain three components, as previously noted by Schwope et al. (1995). We can only usefully determine the motion of the broad component in our data, finding that it reaches maximum redshift at about the time of the EUV dip. This indicates that the accreting material is flowing most directly away from us (i.e., that the flow is passing through inferior conjunction) at about this time, which is at least qualitatively consistent with the notion that it is the accretion flow that shadows the accretion site. We note that although phase zero supposedly represents inferior conjunction of the secondary star (Schwope et al., 1995), neither the association of the narrow emission line component with the secondary star, or its phasing are yet sufficiently secure to contradict the stream occultation scenario as the cause of the EUV dip.

5. Discussion

The pronounced changes in the EUV light curve of QS Tel presumably reflect alterations in its accretion geometry. In conjunction with a pointed *ROSAT* observation (Clayton et al. 1995; Rosen et al., 1995) which found it in a single pole mode and the optical data of Schwope et al. (1995) which contained evidence of cyclotron emission from two poles, it appears that such changes occur frequently in QS Tel. The most likely causes are 1) increases in the accretion rate which allow material to penetrate more deeply into the magnetosphere and hence reach the second pole, 2) alteration of the balance between the instabilities at the magnetosphere which dictate whether the material is predominantly clumpy or homogeneous when it enters the magnetosphere or 3) asynchronous rotation of the white dwarf relative to the secondary star. The first two possibilities are related to the accretion rate. In either of these cases, with further observations we might expect to recognize a correlation between accretion mode (one or two pole) and the luminosity. The *EUVE* observations presented above apparently observed the system in a bright state (in contrast, the PV phase *EUVE* observation (Warren et al. 1994) found the star in a deep low state, its light curve being dominated by two flare events). If, on the other hand, asynchronism is the cause of the transformation in the accretion geometry, we anticipate that further observations will eventually unveil an evolution of the light curve which is recurrent on the beat period between the binary and white dwarf rotation periods. With its apparently frequent alterations of accretion geometry, QS Tel offers an excellent opportunity to probe the origin of these changes.

The dip observed in the *EUVE* data is structurally complex. Current evidence suggests that this feature arises from occultation of the white dwarf's emission region by the accretion flow rather than via an eclipse by the companion star. The lack of an accompanying hardness ratio change during the dip probably disfavours photoelectric absorption as the dominant absorbing mechanism. Alternative possibilities include electron scattering or partial covering, perhaps by a clumpy medium. Electron scattering would require a column density $\sim 5 \times 10^{24}$ cm^{-2} and a corresponding accretion rate probably in the range form 4×10^{16} gs^{-1} to 4×10^{17} gs^{-1}, to explain the depth of the dip. The inhomogeneous picture is at least qualitatively in agreement with blob bombardment scenario proposed to explain AM Her systems with large EUV excesses. Finally, the observed dip profile might arise from obscuration by two parts of the accretion flow—we doubt that even the latter part of the dip ($0.08 \leq \phi \leq 0.10$) could be explained by an eclipse by the companion star since even at 30–40 s, the duration of dip egress is probably too slow to represent the uncovering of a small ($r \ll R_{WD}$) emission region. If an eclipse is ruled out, the flaring during ingress and around mid-dip might respectively represent the covering and uncovering of the source by that part of the stream that first shadows the source (probably close to the white dwarf), the latter part of the profile being caused as the trailing part of the stream (further out) shadows the source. This would require some curvature or splitting of the stream but this is not perhaps unreasonable. For example, such a trajectory might ensue if the dipole were not centred and/or if there is continued azimuthal drift of the material relative to the field lines once it has begun to thread.

REFERENCES

BUCKLEY, D. A. H., ET AL. 1993, MNRAS, 262, 93

CLAYTON, K. L., ET AL.   1995, Cape Workshop on Magnetic Cataclysmic Variables, ed. D. Buckley, & B. Warner, in press

CROPPER, M. S. 1990, Space Sci. Rev., 54, 195

FRANK, J., KING, A. R., & LASOTA, J. -P. 1988, A&A, 193, 113

POUNDS, K. A., ET AL.   1992, MNRAS, 260, 77

ROSEN, S. R., ET AL.   1995, in prereparation

SCHWOPE, A., ET AL. 1995, A&A, 293, 764

WATSON, M. G. 1993, Adv. Space Res., 13(12), 125

Non-Magnetic Cataclysmic Variables in the *ROSAT* WFC Survey

PETER J. WHEATLEY

Astronomical Institute, Utrecht University, Postbus 80 000, 3508 TA Utrecht, NL.

Six non-magnetic cataclysmic variables were detected during the *ROSAT* WFC survey; four dwarf novae and two nova-like variables. In two dwarf novae (VW Hyi & SS Cyg) the flux evolution through outburst was followed across a broad wavelength range. The two other detections (Z Cam & RX J0640–24) also suggest the presence of distinct luminous EUV emission components; supporting the view that such components are a ubiquitous feature of dwarf nova outbursts. Two bright nova-like variables were detected, but these detections are found to be consistent with the soft tail of the X-ray emission.

1. Introduction

1.1. *Cataclysmic Variables (CVs)*

X-rays in non-magnetic cataclysmic variables are thought arise in the boundary layer between the disk and white dwarf; where material settles onto the surface from its Keplarian velocity. Assuming the white-dwarf is not rotating close to breakup, one half of the total accretion luminosity must be released in this transition. If the boundary layer is thin, i.e. the transition from Keplarian to white-dwarf velocity is sudden, one would expect the entire boundary-layer luminosity to be emitted as X-rays.

In most non-magnetic CVs the boundary layer emission is optically thin and hard ($kT \sim 1$–$10\,\mathrm{keV}$); as such they are not promising EUV sources. The exceptions are systems with a high accretion rate: dwarf novae in outburst and nova-like variables. In these the boundary layer density becomes high enough to make it at least partly optically thick to its own radiation (Pringle & Savonije 1979); the hard X-rays are thermalised and emitted in the EUV ($kT_{eff} \sim 10$'s eV). Pre-ROSAT these soft emission components had been observed only in three systems, all bright dwarf novae in outburst: SS Cyg, VW Hyi and U Gem (Jones & Watson 1992; Pringle et al. 1987; Mason et al. 1988).

EUV emission should be a ubiquitous feature of high $\dot{M}$ CVs if this simple picture is correct. The luminosity should also be comparable with that of the disk (optical/UV). Van Teeseling & Verbunt (1994) show that there is a deficiency of boundary-layer luminosity (hard X-rays compared with optical/UV) which is an increasing function of $\dot{M}$; perhaps indicating the increasing importance of an unobserved EUV component. However, several observations now suggest that the luminosity of the EUV component, where detected, is also lower than that of the disk.

1.2. *ROSAT WFC Survey*

The *ROSAT* Wide Field Camera (WFC) surveyed entire sky in two EUV bandpasses† during the second half of 1990 (Sims et al. 1990). Of the 382 sources detected in the initial survey analysis (Pounds et al. 1993), 17 were identified as CVs. A systematic search for all magnetic CVs, dwarf novae in outburst and the brightest nova-like variables revealed a further four high significance detections (Wheatley 1995a,b).

Six non-magnetic systems were detected, of which four are dwarf novae: VW Hyi, SS Cyg, Z Cam and RX J0640–24; and two are nova-like variables: IX Vel and V3885 Sgr.

† S1a, 90–206 eV; S2a 62–110 eV

S. Bowyer and R. F. Malina (eds.), Astrophysics in the Extreme Ultraviolet, 337–341.

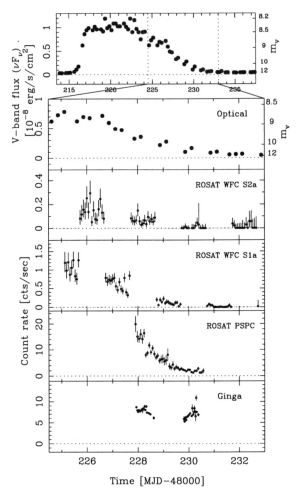

FIGURE 1. Multi-wavelength lightcurves of an outburst of the dwarf nova SS Cygni.

2. Dwarf Novae

Of the four WFC detections of dwarf novae, three were made at the time of outburst (SS Cyg, VW Hyi & Z Cam, Figs. 1–3). There is no simultaneous optical light curve of RX J0640-24, but EUV/optical ratios strongly suggest that it too was detected in outburst.

SS Cyg and VW Hyi are the brightest dwarf novae, in optical and X-rays, and have both been seen as strong EUV sources during outburst. As such they were chosen as targets for a detailed multi-wavelength study, made at the time of the *ROSAT* surveys.

2.1. SS Cyg

SS Cyg was observed during decline from ordinary outburst (Fig. 1). The *ROSAT* WFC and PSPC light curves clearly show a bright EUV component ($kT \sim 20\,$eV), which declines more quickly than the optical (e-folding timescales of one and three days respectively). The luminosity of this component is $\sim 10^{33}\,\mathrm{d}^2_{100\,\mathrm{pc}}\,\mathrm{erg\,s}^{-1}$, which is at

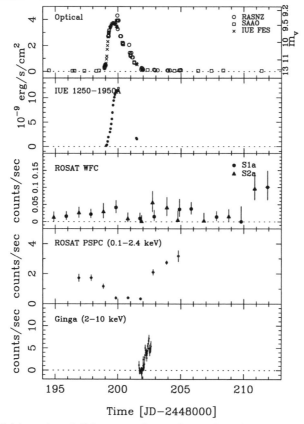

Time [JD−2448000]

FIGURE 2. Multi-wavelength lightcurves of an outburst of the dwarf nova VW Hydri.

least an order of magnitude less than that of the disk. The hard X-ray light curve, measured with *Ginga* (2–20 keV), remains constant throughout this decline, and at a level below that of an observation of quiescence (40 cts/s). Ponman et al. (1995) discuss these observations in detail.

2.2. *VW Hyi*

The multi-wavelength observations of VW Hydri (Fig. 2) are discussed by Wheatley et al. (1995). Here the suppression of the hard X-rays during outburst is well defined by the *ROSAT* PSPC and *Ginga* light curves. Crucially, the *Ginga* observation catches the recovery to the quiescent level; placing it firmly at the end of the optical outburst. This unambiguously associates the X-ray emission with the boundary layer, since models agree that the accretion disk returns to quiescence from the outside in. This probably holds for SS Cyg, since the hard X-rays remain at their apparently suppressed level until very late in the outburst (Fig. 1).

The WFC light curves show no evidence for an EUV component replacing the hard X-rays, but at outburst maximum the survey scans were made with the hard filter (S1a). *EXOSAT* LE observations, in which an EUV component was detected, suggest a best blackbody temperature around 90 000 K (Van Teeseling et al. 1993), which would have been detected with the soft filter (S2a). The non-detection one day later with S2a

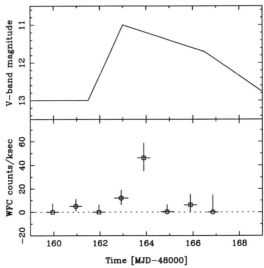

FIGURE 3. Optical and WFC light curves of Z Cam through an outburst (S1a: o, S2a: □).

suggests that the EUV component declines more quickly than the optical in VW Hyi, as in SS Cyg. Upper limits to WFC countrates during outburst limit the blackbody luminosity to $10^{33} \, d^2_{65 \, pc} \, erg \, s^{-1}$ for any temperature; which is substantially below that of the disk ($> 6 \times 10^{33} \, d^2_{65 \, pc} \, erg \, s^{-1}$).

2.3. Z Cam

Z Cam is the third brightest of the eleven other dwarf novae caught in outburst during their survey observations, and is the only one detected. Figure 3 shows a clear EUV enhancement at outburst maximum. The outburst evolves more rapidly than the WFC filters were changed, so little spectral information is available, but for the S2a count rate at MJD 48164 to be so much higher than S1a at MJD 48163 requires either a very soft spectrum or that S1a caught the EUV rise. The low EUV flux at the beginning of the optical decline (S1a at MJD 48165 and S2a at MJD 48166) clearly indicates that the EUV component declines more quickly than the optical, as is seen in SS Cyg and VW Hyi.

2.4. RX J0640–24

RX J0640–24 was discovered through the optical follow-up to the PSPC galactic-plane survey (Beuermann & Thomas 1993). It was detected only in S1a and with a low count rate (10 ± 3 cts/ks) and the light curve is consistent with that of a constant source. It was discovered as a 15th magnitude object, making it the faintest dwarf nova yet detected in the EUV. The discovery magnitude implies an EUV/optical ratio two orders of magnitude greater than that of SS Cyg, so it is likely that RX J0640–24 was in outburst during the WFC survey observation and in quiescence when identified optically.

3. Nova-like Variables

Pounds et al. (1993) identified IX Vel, the brightest UX UMa-type nova-like, in their initial WFC survey analysis. A search in the survey database for the next two brightest UX UMa-types revealed a detection of V8335 Sgr and a non-detection for RW Sex. No systematic search has been made for non-magnetic nova-like variables, which may be expected to have the same EUV properties as dwarf novae in outburst; Van Teeseling showed that they too are underluminous hard X-ray sources.

3.1. *Spectra*

IX Vel is an extremely soft WFC source. Its mean S1a and S2a count rates were 9 ± 3 and 29 ± 5 cts/ks respectively. This suggests a strong EUV emission component, capable of carrying the missing boundary layer luminosity. However, Beuermann & Thomas (1993) report a temperature of $kT \sim 1$ keV with the *ROSAT* PSPC during the survey. At these temperatures optically-thin emission in the WFC bandpass is dominated by line emission, and strong lines in the S2a band give the impression a soft spectrum. Folding a 1 keV "Mewe" spectrum through the responses of the WFC and PSPC demonstrates that this single component can account for the fluxes and colours in both instruments. It was also found that the detection of V3885 Sgr is consistent with an extrapolation of the PSPC spectrum presented by Van Teeseling & Verbunt (1994). The WFC-survey presents no evidence of distinct EUV emission in nova-like variables.

P.J.W. is supported by the NWO under grant PGS 78-277. Much of this work was carried out during a SERC/PPARC studentship with M.G.Watson at the University of Leicester, UK.

REFERENCES

Beuermann, K. & Thomas, H. -C. 1993, The ROSAT view of the cataclysmic variable sky, Adv. Sp. Sci., 13(12), 115

Mason, K. O., Cordova, F. A., Watson, M. G., & King. A. R. 1992, The discovery of orbital dips in the soft X-ray emission of U Gem during an outburst., MNRAS, 232, 779

Jones, M. H. & Watson, M. G. 1992, The EXOSAT observations of SS Cygni, MNRAS, 257, 633

Ponman, T. J., Belloni, T., Duck, S. R., Verbunt, F., Watson, M. G. & Wheatley, P. J. 1995, The EUV/X-ray spectrum of SS Cygni in outburst, MNRAS, in press

Pounds, K. A. et al. 1993, The ROSAT Wide Field Camera all-sky survey of extreme-ultraviolet sources. I - The Bright Source Catalogue, MNRAS, 260, 77

Pringle, J. E., Bateson, F. M., Hassall, B. J. M., Heise, J. & Van, der, Woerd, H. 1987, Multiwavelength monitoring of the dwarf nova VW Hydri. I - Overview., MNRAS, 225, 73

Pringle, J. E. & Savonije, G. J. 1979, X-ray emission from dwarf novae, MNRAS, 212, 519

Sims, M. R. et al. 1990, XUV Wide Field Camera for ROSAT, Optical Engineering, 26(9), 649

Van, Teeseling A. & Verbunt, F. 1994, ROSAT observations of ten cataclysmic variables, A&A, 292, 519

Van, Teeseling, A., Verbunt, F. & Heise, J. 1993, The nature of the X-ray spectrum of VW Hydri, A&A, 270, 159

Wheatley, P. J. 1995a, Cataclysmic variables in the extreme ultraviolet, Ph.D. Thesis, Leicester Univ.

Wheatley, P. J. 1995b, Magnetic cataclysmic variables in the ROSAT WFC survey, Proc. of the Cape workshop on Magnetic Cataclysmic Variables, ASP Conf. Series, ed. D.A.H. Buckley & B. Warner

Wheatley, P. J., Verbunt, F., Belloni, T., Watson, M. G., Naylor, T., Ishida, M., Duck, S. R. & Pfeffermann, E. 1995, The X-ray and EUV spectrum of the dwarf nova VW Hyi in outburst and quiescence, A&A, in press, http://stkwww.fys.ruu.nl:8000/cgi-bin/preprints/preprints

The Three-Dimensional Structure of EUV Accretion Regions of AM Her Stars: Analysis of *EUVE* Light Curves

MARTIN M. SIRK[1] AND STEVE B. HOWELL[2]

[1]Center for EUV Astrophysics, 2150 Kittredge St.,
University of California, Berkeley, CA 94720–5030, USA

[2]Planetary Sciences Institute, Astrophysics Group,
620 North Sixth Ave., Tucson, AZ 85705, USA

We present EUV light curves for a number of AM Her systems observed either as guest observer targets or with the *EUVE* Right Angle Program. We have formed light curves for eight AM Her stars and show in our presentation the similarities and differences present. We draw some conclusions by grouping the systems by inclination, magnetic field strength, and accretion region geometry. In order to understand the physical structures responsible for the EUV emissions, we have developed a software model to generate synthetic light curves. We find that the EUV accretion regions in the systems UZ For, VV Pup, and AM Her cannot be fit with a flat spot confined to the white dwarf surface, regardless of its shape or brightness profile. Rather, a small, symmetric, raised spot is the only shape consistent with the data. The light curves for systems EF Eri, RE1149+28, AN UMa, and V834 Cen show evidence for additional structure that precedes the primary accretion region in phase. Our model indicates that a large portion of the light curve in each system is seen in absorption. Finally, in three systems, we detect a very gradual rise and fall in the EUV flux (<5% of the peak flux) at phases when the spot is completely obscured behind the white dwarf (WD) limb. We attribute this detection to emission from the accretion column that decreases exponentially with distance from the WD and detect emission at heights up to 15% of the WD radius.

1. Introduction and Data

AM Her stars are a class of interacting binaries consisting of a highly magnetic (order $10 - 60$ MG) white dwarf (WD) primary and a red dwarf secondary that fills its Roche Lobe. Stellar material flows through the inner Lagrangian point and falls toward the primary forming an accretion stream. In the absence of a magnetic field, an accretion disk normally forms. However, in AM Her systems, the strong magnetic field captures the ionized material and channels it directly toward one or both of the magnetic poles of the WD primary, forming a hot accretion spot. The interaction of the accretion stream with the magnetic field circularizes the orbits and synchronizes the WD rotation period with the binary orbital period. When viewed from the rotating frame, the primary, secondary, and magnetic field all appear static. The only motion is that of the in-falling material, its free-fall time being about one fourth the orbital period.

Conversion of the kinetic energy of the stream manifests itself in many forms. Electrons in the stream near the spot spiral around the field lines and emit highly polarized cyclotron radiation at optical and near infrared wavelengths. X-rays arise from the post shock region where the supersonic stream meets the WD photosphere. Extreme ultraviolet (EUV) radiation is detected on, or very close to, the heated WD surface; the consequence of reprocessed X-rays and the direct mechanical heating of the WD photosphere by the impacting material. See Liebert & Stockman (1985) and Cropper (1990) for reviews on AM Her systems.

Observations in the EUV are particularly well suited to probing the accretion region

S. Bowyer and R. F. Malina (eds.), Astrophysics in the Extreme Ultraviolet, 343–348.

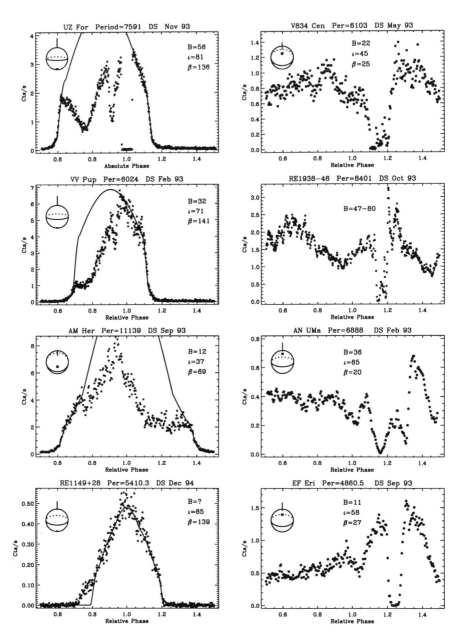

FIGURE 1. Phase folded light curves of eight AM Her systems. Geometric models are fit to the stars in the left column with the derived inclinations ι and spot colatitudes β indicated. Models have not yet been fit to the stars in the right column since their light curves show additional structure. Here, ι and β are from Cropper (1990), as are all the periods (seconds) and magnetic fields B (MG).

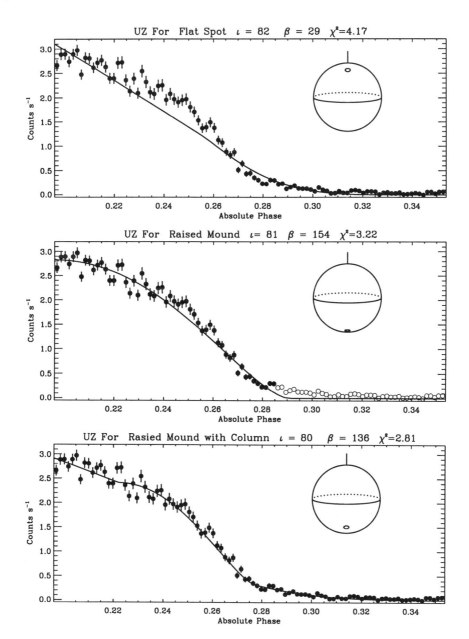

FIGURE 2. Enlargement of the EUV fall phase of UZ For comparing three different models. Only the raised-mound model with luminous accretion column shows no systematic residuals (bottom panel).

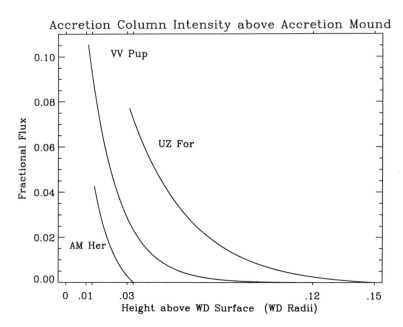

FIGURE 3. Accretion column intensity as a function of distance from the WD surface for UZ For, VV Pup, and AM Her. The column intensities are normalized to the peak intensity for each star.

since the source of radiation is often confined to a very small region on, or near, the WD surface. Furthermore, the systems are intrinsically bright around 100 Å.

Analysis of *Extreme Ultraviolet Explorer* (*EUVE*; Bowyer & Malina 1991) photometry of UZ For (Warren, Sirk, & Vallerga 1995) showed a very small, symmetric EUV accretion spot that is raised a few percent above the WD surface. Vennes et al. (1995) determined a similar height for the accretion spot of VV Pup based on model atmosphere calculations. Vertical extension of the accretion region profoundly affects the EUV light curves, both by lengthening the duration of the bright phase (when the spot is not behind the WD limb), and steepening the rise and fall phases (when the spot rotates into and out of view from behind the WD limb). Understanding the physical processes of AM Her systems requires accurate knowledge of the system geometry, namely the orbital inclination (ι) and the colatitude (β) of the accretion spot. These parameters are often determined from optical light curves. Assuming a flat accretion spot may result in incorrect values of ι and β. Thus, we decided to model the EUV accretion regions, allowing the inclination, spot colatitude, spot size, and spot height to all be free parameters.

Utilizing our own *EUVE* guest observer data (Howell et al. 1995), data from the *EUVE* public archive, and data from the *EUVE* Right Angle Program (McDonald et al. 1994) we construct high time-resolution, phased-folded light curves for eight AM Her systems obtained by the Deep Survey and scanner telescopes. All data are in the Lexan/boron passband ($\geq$ 10% transmission from 67 to 178 Å, peak at 91 Å; Malina et al. 1994). We present the light curves as Figure 1.

Table 1. System Parameters

System	Period (min)	ι (°)	β (°)	r (WD)	h (WD)	h_{col} (WD)	ψ (°)	B (MG)
UZ For$_D$	126.5	*80.5*	*136.1*	*.059*	*.030*	*.15*	*49*	56
UZ For$_{Scan}$	126.5	80.5	*116.2*	*.078*	*.019*	*.14*	*55*	56
VV Pup	100.4	*70.6*	*141.2*	*.022*	*.010*	*.12*	49	32
AM Her	185.6	*36.9*	*68.6*	*.045*	*.013*	*.033*	31	12
RE1149+28	*90.17*	65	139	.063	.016	—	—	—
EF Eri	81.01	74	74	.061	.015	—	—	11

NOTES: The system inclination is ι. The angle between the rotational pole and the EUV accretion spot is β. The radius and the height of the accretion spot are r and h, respectively, in units of WD radii. The maximum height of the accretion column above the WD surface that shows significant flux is h_{col} in units of WD radii. The longitude of the accretion spot is ψ, and the field strength of the primary magnetic pole is B. Italic entries for ι, β, r, h, and h_{col} are the parameter solutions derived from the model fits (light face entries indicate that that particular parameter was fixed, not free). Other italic entries are values determined from *EUVE* data. The remaining light face entries are from Cropper's (1990) review article, *The Polars*.

2. EUV Accretion Spot Models

Three symmetric spot models were attempted: a flat circular spot, a cone with circular base, and a raised mound (sector of rotation). The synthetic light curves from a flat spot closely resemble a cosine curve, and thus, completely fail to account for the steep rise and fall phases of the EUV light curves. The conical spot proved only slightly better. The raised mound, however, fit the data very nicely. We started with UZ For since its inclination and spot colatitude were accurately determined from optical eclipse photometry by Bailey & Cropper (1991), and the EUV spot size from eclipse timings (Warren et al. 1995). To test the applicability of the model, the inclination and the spot colatitude were allowed to be free parameters, even though they were already known. Fitting UZ For, we find a spot radius and spot height of .06 and .03 WD radii, respectively, and an inclination and colatitude of 80 and 136 degrees, in excellent agreement with Bailey & Cropper (1991). The same model is applied to VV Pup, AM Her, and RE1149+28. The derived parameters are listed in Table 1 in italic bold face.

Our raised-mound model fits the steep rise and fall regions of the light curves quite well. The central parts of each light curve are subject to eclipses by the secondary star, eclipses by the accretion stream, and absorption by material very close to the accretion spot, and thus, are excluded from the model fits. A careful study of the residuals of the four fit stars showed a systematic and symmetric underestimate of the flux at the earliest and latest segments of the rise and fall phases, respectively. Rather than invoking a complicated horizontal brightness profile of the accretion spot that symmetrically leads and follows the spot center, we attribute the residual flux to EUV radiation emanating from the accretion column immediately above the accretion spot proper (the residual flux is 1%–2% of the peak flux). Only two additional parameters (a brightness scale factor, and an exponential decay constant) are required to adequately model the light curves. The maximum height of the EUV luminous accretion column may be measured directly from the light curves once ι and β are known. Figure 2 compares the fits of a flat spot, a raised mound, and a raised mound with luminous accretion column to the fall phase of UZ For (top, middle, and bottom panels, respectively). The flat spot fit shows large residuals and places the accretion spot 130° away (in the opposite hemisphere) from its known location. The raised-mound fit does well in the steep fall phase but systematically misses the last portion of the fall phase (plotted as open circles in the middle plot). The final panel includes the accretion column as a source of EUV flux that decays exponentially with height above the WD surface. All points in the rise and fall phase are fit by this model.

In Figure 3 we plot the derived accretion column intensity as a function of distance above the WD for three systems, normalized to the peak intensity measured for each system. UZ For shows column flux at a maximum height of .15 the WD radius.

3. Conclusions

AM Her stars that show symmetric EUV light curves in their rise and fall phases are consistent with small, symmetric accretion regions that are typically 2%–6% of the WD radius in width and 1%–3% of the WD radius in height. These systems do not require additional structures, such as second emitting poles, auroral arcs, or complicated horizontal brightness profiles. Flux detected at the very early rise and very late fall phases may be explained by a luminous accretion column that decreases exponentially in intensity with distance from the WD. The system inclination and accretion spot colatitude may be directly determined from model fits of the light curves. Conversely, if these last two parameters are already known from other sources, the models are constrained, improving the estimates of the three dimensional structure of the EUV accretion region.

The sections of the light curves where the geometric models fail (in phases during which the spot is seen mainly face on) *are* indicative of additional structure. Analysis of phased resolved spectra will address the question of whether the "missing" flux is the result of absorption, or genuine structure in the accretion region when viewed face on. We are refining our techniques in effort to model the systems of the right side of Figure 1, and are investigating the time variability of the EUV light curves.

We thank the CEA Guest Observer Center support personnel. This work is funded in part by NASA contract NAS5-30180. SBH acknowledges support from NASA grant NAG5-2902.

REFERENCES

BAILEY, J. & CROPPER, M. S. 1991, MNRAS, 253, 27

BOWYER, S. & MALINA, R. F. 1991, in Extreme Ultraviolet Astronomy, ed. R. F. Malina & S. Bowyer, New York: Pergamon Press, 397

CROPPER, M. S. 1990, Space Sci. Rev., 54, 195

HOWELL, S. B., SIRK, M. M., MALINA, R. F., MITTAZ, J. P. D., & MASON, K. O. 1995, ApJ, 439, 991

LIEBERT, J. & STOCKMAN, H. S. 1985, in Cataclysmic Variables and Low-Mass X-ray Binaries, ed. D. Q. Lamb & J. Patterson, Dordrecht: Reidel, 151

MALINA, R. F., ET AL. 1994, AJ, 107, 751

MCDONALD, K., ET AL. 1995, AJ, 108, 5

VENNES, S., SZKODY, P., SION, E. M., & LONG, K. S. 1995, ApJ, 445, 921

WARREN, J. K., SIRK, M. M., & VALLERGA, J. V. 1995, ApJ, 445, 909

EUVE and *VLA* Observations of the Eclipsing Pre-Cataclysmic Variable V471 Tauri

S. L. CULLY,[1,2] J. DUPUIS,[2] T. RODRIGUEZ-BELL,[1] G. BASRI,[3]
O. H. W. SIEGMUND,[1,2] J. LIM,[4] AND S. M. WHITE[5]

[1] Experimental Astrophysics Group, Space Sciences Laboratory,
University of California, Berkeley, CA 94720–7450 USA

[2] Center for Extreme Ultraviolet Astrophysics, 2150 Kittredge Street,
University of California, Berkeley, CA 94720–5030 USA

[3] Department of Astronomy, University of California Berkeley, CA 94720–3411 USA

[4] Institute of Astronomy and Astrophysics, Academia Sinica PO Box 1-87,
Nankang, Taipei, Taiwan 115, ROC

[5] Department of Astronomy, University of Maryland College Park, MD 20742 USA

We present observations of the eclipsing binary V471 Tauri by the *Extreme Ultraviolet Explorer* (*EUVE*) and the Very Large Array (*VLA*). The EUV spectrum is dominated by the continuum of the hot white dwarf and the time-averaged spectrum is fitted by a $33.1 \pm 0.5 \times 10^3$ K pure hydrogen white dwarf atmosphere assuming $\log g = 8.5$. An ISM hydrogen column density of $1.5 \pm 0.4 \times 10^{18}$ cm^{-2} is required to explain the attenuation of the white dwarf spectrum thus setting the H I column in the line of sight of the Hyades cluster. The He II λ304 Å line is in emission and varies over the orbital period of V471 Tauri following a sinusoidal modulation with the maximum reached when the K star is at inferior conjunction. Transient dips are detected at orbital phase -0.12 in the SW and MW spectrometers integrated lightcurves but are notably absent in the LW lightcurve indicating the occulting material is ionized. The *VLA* observation suggest the presence of a K star coronal magnetic loop between the two stars reconnecting with the white dwarf magnetic field. Such a structure could be the occulting source needed to explain the dips seen in the lightcurves of V471 Tauri in the EUV.

V471 Tauri is an astrophysically important eclipsing binary (K2V+DA2, $P = 12.5$ hr, $d = 50$ pc, 80° inclination) whose understanding is crucial for theories of binary stars evolution. We have observed V471 Tauri with the *EUVE* observatory with the goal of using the hot white dwarf component as a probe of the intra-system material to gain more insight on the nature of the interaction between the two components of the system. V471 Tauri was observed by *EUVE*, the *VLA* and 5 optical telescopes as part of a coordinated campaign from 1994 November 28–December 3 covering approximately 6 orbits of the system. We present light curves and spectra from the *VLA* and *EUVE* portions of the observation using the *EUVE* DS/S Lex/B (40 – 190 Å) and the *EUVE* SW (70 – 190 Å), MW (140 – 380 Å) and LW (280 – 760 Å) spectrometers.

A spectrum of V471 Tauri was acquired with the *EUVE* spectrometers from 1994 November 28 to December 2, with a total exposure time of about 100,000 s. We show in Figure 1 the time-averaged EUV photon flux spectrum of V471 Tauri made by combining the SW, MW, and LW spectra. In the top part of Figure 1, it is clear the spectral distribution (histogram) is characteristic of that of hot DA star. As expected, some emission lines are detected with He II λ304 Å being the most prominent emission line. Superposed to the spectrum is a fit of pure hydrogen WD model spectra from Vennes (1992) accompanied with the 1, 2, and 3 σ contours in the bottom half of Figure 1. For this fit, $\log g$ is fixed to 8.5 in agreement with the determination of Vennes (1992; based

349

S. Bowyer and R. F. Malina (eds.), Astrophysics in the Extreme Ultraviolet, 349–354.
© 1996 *Kluwer Academic Publishers. Printed in the Netherlands.*

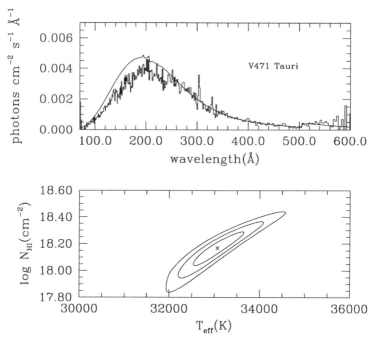

FIGURE 1. (a) Fluxed *EUVE* spectrum of V471 Tauri (histogram) and best fit pure hydrogen model white dwarf model spectrum with gravity fixed to $\log g = 8.5$. (b) Chi-square contours (1, 2, and 3 σ) for the 2-parameters fit (T_{eff}-$\log g$) shown in Figure 1a.

on *IUE* and *EXOSAT* data) and of Kidder (1991). We obtain an effective temperature of $33.1 \pm 0.5 \times 10^3 K$ and an H I interstellar column density of $1.5 \pm 0.4 \times 10^{18}$ cm^{-2} by fitting the spectrum in the $340 - 600$ Å range. Note that the 504 Å photoionization edge of He I is detected from which we measure an ISM column density of $\sim 1.3 \times 10^{17}$ cm^{-2}. Some of the discrepancies at shorter wavelengths are probably due to unaccounted opacities in the atmosphere of the white dwarf that could originate from accretion of material ejected from the K star during flares. Some of the stronger emission lines may be caused by other structures within the system.

In order to understand the origin of the He II λ304 Å emission, we have computed a lightcurve of this line taking special care to subtract the WD continuum. As shown in Figure 2, the lightcurve is then folded to the orbital period of the system with the hope of finding a correlation between the strength of the line and the orbital phase. The data shows a simple sinusoidal modulation consistent with the maximum emission occurring when the K star is at superior conjunction. An interpretation of this phenomenon is that the emission consists of reprocessed EUV radiation in the cool star hemisphere illuminated by the hot white dwarf (Thorstensen et al. 1978). This model also explains the variability in V471 Tauri's Hα emission (Young, Skumanich, & Paylor 1988).

V471 Tauri was monitored simultaneously by the *EUVE* Deep Survey (DS) broadband photometer in the Lexan/Boron band ($40 - 190$ Å). We show in Figure 3 the DS light curve of the entire observation phased on the system orbital period and binned in intervals of 0.01ϕ. The data is dominated by the signal from the white dwarf which can be seen

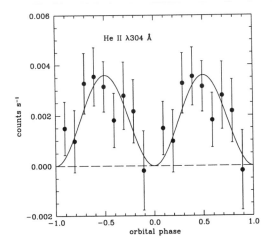

FIGURE 2. He II 304 Å phase binned light curve with fitted sinusoidal modulation

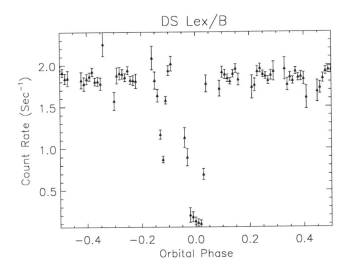

FIGURE 3. *EUVE* DS Lex/B (40 − 190 Å) phase binned light curve. Note dip at phase −0.12.

in eclipse at phase 0.0. The width of the eclipse is approximately 0.066ϕ consistent with the optical eclipse and the K2V star radius of 0.85 $R_\odot$ (Young & Nelson 1972). There is a small residual signal within the eclipse which decreases as the eclipse progresses indicating that approximately 10% of the emission from the system is due to the K star.

A strong dip in the light curve is also seen at phase −0.12 lasting for about 0.05ϕ. The dip appeared in the 3rd orbit of the system and lasted until the 6th orbit. At its deepest, in the third and fourth orbits, approximately 90% of the light from the white dwarf was lost. Figure 4 shows the three *EUVE* integrated spectrometer light curves over the entire

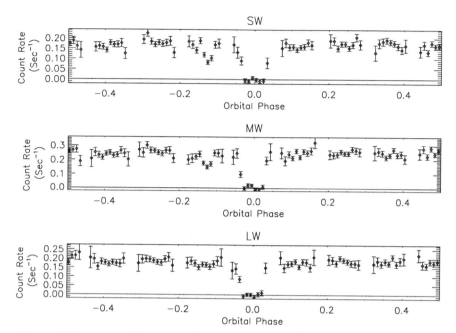

FIGURE 4. Phase binned light curves for the SW $(70 - 190\ \text{Å})$, MW $(140 - 380\ \text{Å})$ and LW $(280 - 760\ \text{Å})$ spectrometers. Note the absence of a dip at phase -0.12 in the LW light curve.

observation. The lack of any discernible dip at -0.12ϕ in the LW light curve indicates that the occulting material is heavily ionized since the opacity of neutral hydrogen increases with increasing wavelength in the EUV (Rumph, Bowyer & Vennes 1994). The occurrence of strong dips in both the SW and MW light curves suggests that occulting gas with a temperature of approximately 10^6 K scatters light from the white dwarf out of the line of sight.

Similar transient dips were reported by Jensen et al. (1986) at phases $0.15, 0.18$, and 0.85 in an *EXOSAT* observation of this system. The white dwarf line of sight passes by the stable Lagrange points of the system at phases 0.17 and 0.83. We believe the occulter is a large coronal loop from the K star with length comparable to the separation of the 2 stars $(3.1\ R_\odot$ Young & Nelson (1972)) located between the stars.

This interpretation is also consistent with the 2.12 hr *VLA* observations (Fig. 5) taken immediately after the *EUVE* observation which show relatively sharp increase/decrease in the 3.6 cm radio emission between $0.15\phi - 0.75\phi$. Lim, White & Cully 1995 have interpreted this increase as being due to an optically thick extended radio source between the stars with a size comparable to the stellar separation. They have also speculated that the source of the radio emission may be due to energetic electrons in the K star coronal loop(s) accelerated by magnetic reconnection between the white dwarf and the K star coronal magnetic fields at the top(s) of the loop(s). The same source could also provide substantial heating of the gas within the loop(s).

It is believed that material is somehow accreted onto the magnetic poles of the white dwarf as evidenced by the 9.25 minute period modulation of the light curve seen by Jensen et al. (1986) and in our observation. This modulation is thought to be due to

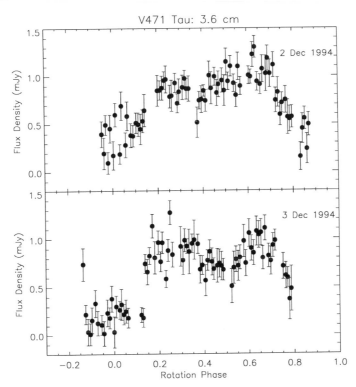

FIGURE 5. Phased binned 3.2 cm light curve taken by the VLA on the two nights following the *EUVE* observation (1994 Dec 2 and Dec 3)

dark spots formed by the accretion of metals on to the magnetic poles of the white dwarf which rotate in and out of view (Clemens et al. (1992)). Since the K star in V471 Tauri does not fill its Roche lobe (Young & Nelson (1972)), no direct mass transfer is possible. However, if large coronal loops from the K star can reconnect with the magnetic field of the white dwarf inside the white dwarf's Roche lobe, accretion of gas from the loop to the white dwarf will result (Lim, White & Cully 1995).

We thank Stéphane Vennes for providing us with white dwarf model spectra. We also thank Martin Sirk for computing deadspot corrections for the DS lightcurve. This research was funded by NASA grant NAG5-2641. S.L.C. acknowledges support from the NASA GSRP. J.D. is also funded by the NASA *EUVE* contract NAS5-30180.

REFERENCES

CLEMENS, J. ET AL. 1992, Whole earth telescope observations of V471 Tauri—The nature of the white dwarf variations, ApJ, 391, 773

JENSEN, K. A, SWANK, J. H., PETRE, R., GUINAN, E. F., SION, E. M., & SHIPMAN, H. L. 1986, EXOSAT observations of V471 Tauri: A 9.25 minute white dwarf pulsation and orbital phase dependent X-ray dips, ApJ, 309, L27

KIDDER, K. M. 1991, Ph.D. Thesis, University of Arizona

LIM, J., WHITE, S. & CULLY, S. 1995, The eclipsing radio emission of the pre-cataclysmic binary V471 Tau, ApJ, submitted

RUMPH, T., BOWYER, S. & VENNES, S. 1994, Interstellar medium continuum, autoionization and line absorption in the extreme ultraviolet, AJ, 107, 2108

THORSTENSEN, J. R., CHARLES, P. A., MARGON, B., & BOWYER, S. 1978, ApJ, 223, 260

VENNES, S. 1992, The constitution of the atmospheric layers and the extreme ultraviolet spectrum of hot hydrogen-rich white dwarfs, ApJ, 390, 590

YOUNG, A. & NELSON, B. 1972, Analysis of the white dwarf eclipsing binary BD+16°516, ApJ, 173, 653

YOUNG, A., SKUMANICH, A., & PAYLOR, V. 1988, Fluorescence-induced chromospheric $H\alpha$ emission from the K dwarf component of V471 Tauri. I. The 1983, epoch, ApJ, 334, 397

Model Spectra for Accretion Disks Truncated at the Inner Edge

RICHARD A. WADE,[1] MARCOS DIAZ,[1] AND IVAN HUBENY[2]

[1] Department of Astronomy & Astrophysics, 525 Davey Lab,
Pennsylvania State University, University Park, PA 16802

[2] Goddard Space Flight Center, Code 681, Greenbelt, MD 20771

Attempts to model the FUV spectra of accretion disks in cataclysmic variables have suggested that in some cases the innermost part of the disk is "missing". We present computed spectra of accretion disks with various amounts of the inner disk removed arbitrarily, to illustrate whether the absence of the inner disk can be detected *via* the profiles of strong lines. Results are presented for lines in the UV and FUV for selected disk parameters and inclinations. UV/FUV line diagnostics can in principle establish whether the hot inner disk is present as expected from standard steady-state disk theory. The presence or absence of the inner disk has direct impact on the interpretation of the EUV spectra of cataclysmic variables, in particular the boundary layer emission.

1. Introduction

The visible, UV/FUV, and EUV light from a cataclysmic variable binary (CV) arises from an accretion structure which is inherently multi-temperature. The EUV spectrum of an accretion disk around a white dwarf is expected to be dominated by emission from the boundary layer (BL), rather than from the disk proper, but analysis of several systems in recent years has suggested that the BL luminosity is less than predicted. To make progress on this question, an accurate model of the *disk* emission in the EUV, based on UV/FUV and visible light data (where the BL contribution should be much less), is needed. But how well can we model the UV/FUV emission from CVs?

2. IX Velorum

Long, et al. (1993) discussed the FUV spectrum of the novalike variable IX Velorum, observed by the *Hopkins Ultraviolet Telescope* (*HUT*) during the ASTRO-1 mission. IX Vel contains a cool secondary star, which feeds an accretion disk around an 0.8 $M_\odot$ white dwarf. The disk is viewed at an angle 60 degrees from face-on. IX Vel is about 95 pc distant. (Data from Beuermann & Thomas 1990.)

The IX Vel spectrum showed a strong, blue continuum upon which was superposed a pattern of resonance lines (P Cygni profiles or blue shifted absorption profiles) from the wind. The Lyman series of hydrogen and some other features were present in absorption, kinematically blended due to the Keplerian orbital motions of gas in the disk.

Attempts to model the FUV spectrum of IX Vel suggested that the innermost part of the disk is "missing". Steady-state models of the disk, assumed to extend inward to the white dwarf surface, and with the mass transfer rate $\dot{M}$ chosen to match the overall observed UV flux, did not match the shape of the observed spectrum particularly well. An improvement was possible, however, if the inner portion of the disk was "removed" from the model flux calculation, with a compensating adjustment of $\dot{M}$ to maintain the observed overall flux level. The hottest regions of the disk were removed by this treatment, resulting in a change in the shape of the disk spectrum, especially near the Lyman limit.

355

S. Bowyer and R. F. Malina (eds.), Astrophysics in the Extreme Ultraviolet, 355–359.

3. Line Diagnostics of a Missing Inner Disk

Based only on the continuum shape, a conclusion that the inner disk is missing cannot be considered to be firmly established. Does interpretation of line profiles support this conclusion? The presence of a strong wind in IX Vel makes use of lines difficult for that system, but other CVs, observed at high S/N and adequate spectral dispersion, might prove amenable to analysis. We present some illustrative line profile calculations, to explore whether non-standard disk structures can be detected *via* line profiles.

In a standard steady-state disk, the effective temperature and orbital velocity of the disk vary with radial distance from the central white dwarf, with the inner disk being the hottest region:

$$T_{eff}(r) \propto r^{-3/4} \times \left[1 - (r_{wd}/r)^{1/2}\right]^{1/4}$$

The Doppler shifts will be $v_{proj}(r, \theta) \propto r^{-1/2} \sin i \cos \theta$, where θ is an azimuthal angle in the disk. The use of kinematic broadening of lines that form at different characteristic temperatures allows one in principle to detect whether the inner disk is present. This hope might be frustrated by too severe blending of lines, or by the formation of lines over sufficiently large ranges of disk radius that the line profile "tool" is effectively blunted.

4. Choice of Models

We adopt a white dwarf mass of $0.9 \, M_\odot$ and radius of $r_{wd} = 6.2 \times 10^8$ cm. We consider two accretion rates, $\dot{M} = 2 \times 10^{-9}$ and $1 \times 10^{-8} \, M_\odot \, \mathrm{yr}^{-1}$. We construct models of the whole disk, or of "evacuated" disks, where $r_{in} > r_{wd}$. The disk is placed at 100 pc, and viewed from several inclination angles, i.

Note that we make an *ad hoc* truncation of the inner disk, without considering changes in the remaining disk structure $T(r)$. This is not entirely self-consistent, but is justified in a preliminary exploration.

The local vertical structure of the disk at each radius is computed using the code TLUSDISK (Hubeny 1988, 1991), and the emerging spectrum (flux and specific intensity) is calculated using the codes SYNSPEC (Hubeny, Lanz & Jeffery 1994: rest frame) and DISKSYN (Keplerian motion and integration over the disk surface).

5. Results

The figures show sample calculated disk spectra in the regions near C IV 1548,1551 Å and the H Lyman lines. Each region contains one or more features that are formed chiefly in the outer disk, which are little altered by the removal of the inner disk and can be used as a reference for comparison.

Each sample is shown at two inclinations, to illustrate the strong effect of viewing angle. Each panel shows a single $\dot{M}$ with various amounts of inner disk removed (r_{in} parameter).

The C IV 1550 Å doublet in Figure 1 is formed mainly at $r < 3r_{wd}$, and the equivalent width of the line is reduced nearly to zero as r_{in} increases. At the higher $\dot{M}$ of the models in Table 2, the same is true for the He II line at 1640 Å (not shown), for which useful "outer disk" lines are Fe II 1611, 1612 Å, C I 1657 Å, and Al II 1671 Å. Both the C IV and He II lines may be strongly affected by a wind from the disk, however.

The Lyman line region (Figure 2) contains features near 1085 Å (He II + N II) and 1066 Å (C II) whose ratio varies by a factor of 2–3 as r_{in} changes in the higher $\dot{M}$ model. This region may be free of excessive contamination by wind lines. The profiles of the Lyman lines themselves (blended with He II) also vary significantly. However, Long et al.

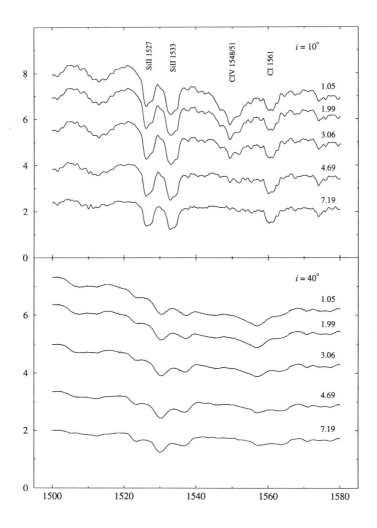

FIGURE 1. The region near C IV for the models specified in Table 1. The upper panel shows spectra for a disk inclination of $i = 10°$, and the lower panel shows $i = 40°$. Numbers at the right end of each spectrum specify the ratio r_{in}/r_{wd}. Wavelengths are in Å units. Flux is given in units of 10^{-12} erg cm^{-2} s^{-1} Å^{-1} for a distance of 100 pc, with no geometric foreshortening taken into account.

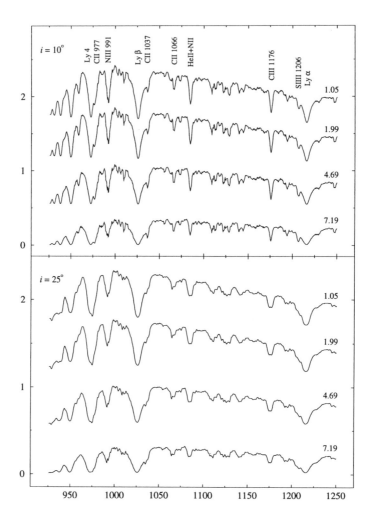

FIGURE 2. The H Lyman line region for the models specified in Table 2. The upper panel shows spectra for a disk inclination of $i = 10°$, and the lower panel shows $i = 25°$. Wavelengths are in Å units. Flux is given in units of 10^{-10} erg cm^{-2} s^{-1} Å^{-1} for a distance of 100 pc, with no geometric foreshortening taken into account. Successively higher spectra in each panel are displaced upwards by 0.5 units to avoid overlap.

TABLE 1. Model details for $M = 0.9$ $M_\odot$, $\dot{M} = 2 \times 10^{-9}$ $M_\odot$ yr^{-1}

r_{in}/r_{wd}	$T(r_{in})$	$v(r_{in})$ $i = 10°$	$v(r_{in})$ $i = 40°$
1.05	40,800 K	740 km s^{-1}	2750 km s^{-1}
1.99	47,000 K	540 km s^{-1}	2000 km s^{-1}
3.06	37,500 K	440 km s^{-1}	1610 km s^{-1}
4.69	28,800 K	350 km s^{-1}	1300 km s^{-1}
7.19	21,700 K	280 km s^{-1}	1050 km s^{-1}

TABLE 2. Model details for $M = 0.9$ $M_\odot$, $\dot{M} = 10^{-8}$ $M_\odot$ yr^{-1}

r_{in}/r_{wd}	$T(r_{in})$	$v(r_{in})$ $i = 10°$	$v(r_{in})$ $i = 25°$
1.05	61,000 K	740 km s^{-1}	1800 km s^{-1}
1.99	70,300 K	540 km s^{-1}	1320 km s^{-1}
4.69	43,200 K	350 km s^{-1}	860 km s^{-1}
7.19	32,500 K	280 km s^{-1}	690 km s^{-1}

suggested that blends with wind lines (e.g., O VI, N IV, S VI) or geocoronal lines (Ly α) limit their individual usefulness.

We conclude that for some disk inclinations and mass transfer rates, a combination of line shape and equivalent width for lines formed at different characteristic temperatures may be useful in assessing whether the disk is truncated from the inside. A complete modeling of the UV/FUV spectrum must of course take account of the shape and level of the continuum as well. If the disk is truncated, the predicted EUV spectrum will be severely altered from the standard (intact disk) model.

The equivalent width diagnostic becomes more difficult but not impossible to use at higher inclination, due to severe blending and spreading of the line profiles. Accurate photometric calibration and adequate signal to noise are essential.

Supported by NASA grant NAGW-3171.

REFERENCES

BEUERMANN, K. & THOMAS, H. -C. 1990, Detection of emission lines from the secondary star in IX Velorum = CPD -48°1577, A&A, 230, 326

HUBENY, I. 1988, A computer program for calculating non-LTE model stellar atmospheres, Comp. Phys. Comm., 52, 103

HUBENY, I. 1991, Model atmospheres for accretion disks, In Structure and Emission Properties of Accretion Disks, ed. C. Bertout et al., IAU Colloquium, Editions Frontières, 129, 227,

HUBENY, I., LANZ, T., & JEFFERY, C. S. 1994, TLUSTY & SYNSPEC—a user's guide, Collaborative Computational Project 7, documentation file

LONG, K. S., WADE, R. A., BLAIR, W. P., DAVIDSEN, A. F., & HUBENY, I. 1993, Observations of the bright nova-like variable IX Velorum with the Hopkins Ultraviolet Telescope., ApJ, 426, 704

The Unusual UV Spectra of
EUV-Discovered AM Herculis Stars

ADRIENNE E. HERZOG,[1] STEVE B. HOWELL,[1]
AND KEITH O. MASON[2]

[1] Planetary Science Institute, Astrophysics Group, 620 North Sixth Ave.,
Tucson AZ, 85705, USA

[2] Mullard Space Science Laboratory, University College London, Holmbury St. Mary,
Dorking, Surrey RH5 6NT, UK

International Ultraviolet Explorer observations of the *ROSAT* WFC-discovered AM Hers RE2107-05, RE1938-46, RE1833-74, RE1149+28, and RE0751+14 reveal UV spectra (1200 − 2000 Å) which are unique when compared with previously studied systems. Specifically, the intensities of individual emission lines due to C IV, C II, N V, He II, and Si IV, are 2 − 4 times stronger then typically seen. However, the equivalent width ratios of these same emission lines, (C II/C IV, N V/C IV, He II/C IV, and Si IV/C IV), are similar to those in typical AM Hers. These results indicate that the EUV-discovered AM Hers must have larger UV emitting volumes. Using additional observations of these stars made by EUVE, we speculate on the cause of their larger UV line fluxes.

1. Introduction

During its EUV all-sky survey, the wide field camera (WFC) on the *ROSAT* satellite discovered several new AM Herculis type stars. In this paper, we present observations of RE0751+14, RE1149+28, RE1844-74, RE1938-46, and RE2107-05. Using the *International Ultraviolet Explorer* (*IUE*), spectra were obtained of these systems over a band pass of 1200–3400 Å. These AM Her stars, discovered in the EUV, and faint at optical wavelengths, proved be bright UV emission line sources. The intensities of the individual emission lines due to C IV, C II, N V, He II, and Si IV are 1.5–3 times stronger then usually seen. However, the equivalent width ratios of these same emission lines are similar to those seen in typical AM Hers. Using additional observations of these stars in the EUV, we compare the continuum flux of the new AM Hers at optical, UV, and EUV wavelengths to previously studied AM Hers, and discuss some of the implications of these data.

2. Ultraviolet Spectroscopy

The *International Ultraviolet Explorer* (*IUE*) satellite was used to obtain ultraviolet spectra of five new AM Hers. The short wavelength prime (SWP) camera was used to make observations during November of 1993, and March and June of 1994. The long wavelength prime (LWP) camera was used to make observations of two of these stars, RE2107 and RE0751, during November of 1993, and June of 1994. Table 1 correlates the individual stars with their relevant observational data. Due to the length of the exposures, which were greater than one orbital period in all cases, phase dependent effects were averaged over.

Figure 1 shows the SWP UV spectra of the 5 new AM Hers. From this figure, it is clear that the UV spectra of these AM Hers are dominated by the extremely strong emission lines. The emission lines are identified as N V (1240), Si III (1297), C II (1333),

S. Bowyer and R. F. Malina (eds.), Astrophysics in the Extreme Ultraviolet, 361–365.
© *1996 Kluwer Academic Publishers. Printed in the Netherlands.*

TABLE 1.

Star	Camera	Exp (min)	Date	$P_{orb}{}^a$ (min)	B^b (MG)	$Dist^a$ (pc)
RE1149	SWP 49336	435	Nov 93	90	—	—
RE2107	SWP 49337	180	Nov 93	125.02	36	190
	LWP 28365	120	Jun 94			
RE1844	SWP 47791	410	Mar 94	89.91	—	—
RE1938	SWP 51048	189	Jun 94	140	56	300
RE0751	SWP 49329	180	Nov 93	340	8–18	—
	SWP 50159	155	Mar 94			
	LWP 26811	115	Nov 93			

Notes: ^{a} Ritter & Kolb 1993; ^{b} Warner 1995.

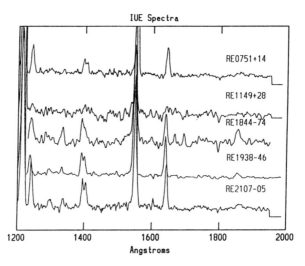

FIGURE 1. *IUE* spectra of RE0751+14, RE1149+28, RE1844-74, RE1938-46, and RE2107-05.

Si IV (1400), C IV (1550), He II (1640), and Fe I (1858). Most likely, the Lyman Alpha line present is contaminated by geocoronal emissions. The LWP data obtained showed no obvious features due to their poor quality.

3. Data Analysis

The spectra of the five new AM Hers were extracted and reduced using *IUE* standard processing. With the wavelength scale and fluxes calibrated, we then used IRAF/splot routines to analyze the spectra. Compared to previously known AM Hers, the emission line equivalent widths are 1.5–3 times stronger, however, the line ratios are similar to those seen in other AM Hers. (See Tables 2 and 3). Figure 2 shows RE2107 and MR Ser plotted together to show the greater dominance of the emission lines in the newly discovered AM Hers. We then compared the fluxes from these sources in the optical,

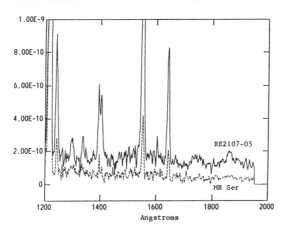

FIGURE 2. Compared to a typical AM Her, the emission line equivalent widths are 1.5–3 times stronger in the EUV discovered stars.

UV, and EUV. Since the EUV data was only obtained during high states, and both our *IUE* observations and those from the *IUE* archives we believe were also obtained at high states, archival V magnitudes at maximum were used to obtain optical flux values. Our UV fluxes came directly from measurements of an average value across the line-free region 1450–1525 Å. Using catalogued EUV fluxes from the EUV all-sky survey (Malina et al. 1994), and the assumptions that the objects have low column densities, blackbody-like spectra near 89 Å, and a constant effective area of the detector, we could calculate EUV flux estimates.

In Figures 3 and 4, we show the spectral energy distributions for the five new AM Hers plus those of two previously studied AM Hers for comparison. The plots were separated onto two graphs for clarity. RE1149 has had five EUV measurements made (Oct 1990, Dec 1992, Feb 1993, Mar 1993, Dec 1994) which are plotted to show its variability. RE0751, a suspected DQ Herculis star, shows a comparatively flat spectral energy distribution, while RE1844 shows a very steep slope between the UV and optical data points. The EUV/UV/Optical flux distributions of the other stars are similar to and generally equivalent with previously known AM Hers.

4. Conclusions

The major differences in the ROSAT/WFC discovered sources appear to be in their larger UV emission line strengths. Although the variability of these sources is quite apparent even in the "high" states, the overall spectral energy distributions, which are similar between the new AM Hers and typical AM Hers, do not change. These results indicate that the EUV discovered AM Hers are likely to have a larger UV emitting volume. The emission line strengths do appear to show a correlation with the magnetic field strength, becoming stronger up to approximately 35 MG and weaker thereafter. This may be related to the weakening of bremsstrahlung emission for fields above 35 MG as noted by Beuermann & Schwope (1994). Full details of this ongoing project will be published elsewhere.

AEH acknowledges support from NSF/REU grant #AST9217971. SBH acknowledges

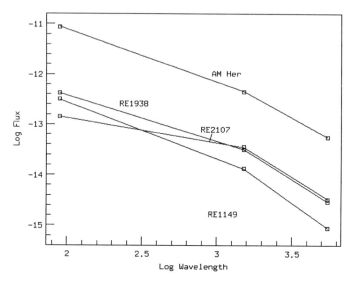

FIGURE 3. The spectral energy distributions for RE1938, RE2107, RE1149, and AM Her. RE1149 has had 5 EUV measurements made which are plotted to show its variability.

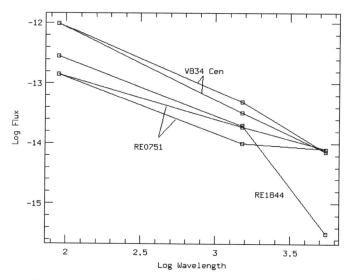

FIGURE 4. The spectral energy distributions for RE0751, RE1844, and V834 Cen.

support from NASA/EUVE grant NAG5-2902. We would like to thank Liz Puchnarewicz for her help with some of the *IUE* data.

TABLE 2. Emission line equivalent widths (Å)

Star	N V	Si III	C II	Si IV	C IV	He II	Fe I
RE2107	41	8	7	44	122	46	17
RE1938	20	6	4	36	154	45	13
RE0751	24	6	3	19	112	39	11
	19	4	4	22	102	38	8
RE1844	64	8	39	63	175	69	37
RE1149	27	—	—	—	227	94	—
V834 Cen[a]	4	—	4	14	53	11	—
V834 Cen[c]	7	—	5	51	67	18	—
AM Her[c]	17	3	5	23	96	25	—
MR Ser[b]	28	13	5	19	48	19	—

Notes: [a]Maraschi & Treves 1984; [b]Szkody et al. 1985; [c]Nousek & Pravdo 1983.

TABLE 3. Ratios of emission line equivalent widths

Star	C II/C IV	N V/C IV	C II/He II	He II/C IV	Si IV/N V	Si IV/C IV	Si IV/He II
RE2107	0.05	0.34	0.14	0.38	1.07	0.36	0.96
RE1938	0.03	0.13	0.10	0.29	1.81	0.23	0.80
RE0751	0.03	0.22	0.08	0.35	0.78	0.17	0.49
	0.03	0.19	0.10	0.36	1.15	0.22	0.62
RE1844	0.22	0.36	0.56	0.40	0.99	0.36	0.90
RE1149	—	0.12	—	0.41	—	—	—
V834 Cen	0.08	0.08	0.36	0.21	3.50	0.26	1.27
V834 Cen	0.07	0.10	0.28	0.27	7.29	0.76	2.83
AM Her	0.05	0.18	0.20	0.26	1.35	0.24	0.92
MR Ser	0.10	0.58	0.26	0.40	0.36	0.21	0.53

REFERENCES

BEUERMANN, K. & SCHWOPE, A. 1994, AM Herculis Binaries, ASP Conference Series, 56, 119

MALINA, ET AL. 1994, EUVE Bright Source List, AJ, 107, 731

MARASCHI, L. & TREVES, A. 1984, Coordinated UV and Optical Observations of the AM Herculis Object E1405 in the High and Low States, ApJ, 285, 214

NOUSEK, J. A., & PRAVDO, S. H. 1983, IUE Observations of E1405: A New AM Herculis type Cataclysmic Variable, ApJ, 266, L39

RITTER, H. & KOLB, U. 1993, A Compilation of Cataclysmic Variables with Known or Suspected Orbital Periods, Cambridge: Cambridge University Press

SZKODY, P., LIEBERT, J., & PANEK, R. 1985, IUE Results on the AM Herculis Stars CW 1103, E1114, and PG 1550, ApJ, 293, 321

WARNER, B. 1995, Cataclysmic Variable Stars: A Review of Observation Properties and Physical Structures, Cambridge: Cambridge University Press

EUV Radiation from B Stars: The Broad Implications for Stellar and Interstellar Astronomy

JOSEPH P. CASSINELLI

Astronomy Department, University of Wisconsin, 475 N. Charter St., Madison, WI 53706, USA

Observations made with the *Extreme Ultraviolet Explorer* (*EUVE*) of the two bright stars ϵ CMa (B2 II) and β CMa (B1 II-III) are discussed. The photospheres show excess EUV radiation. The wind of ϵ CMa exhibits the Bowen Fluorescence mechanism, along with high ionization stages that help explain the nature of the wind shocks. The pulsation and beat phenomena exhibited by the variable star β CMa suggest that deposition of residual pulsation energy might heat and modify the structure of the atmospheres of early-type stars near the β Cephei strip. The possibility that many other B stars show a large excess Lyman continuum radiation is considered as a possible source of the ionization of the warm ionized medium (WIM) in the galactic ISM.

1. Introduction

A major surprise from *EUVE* is that the brightest stellar sources in the 500 to 700 Å range are two B giant stars, ϵ CMa (B2 II) and β CMa (B1 II-III). The stars are bright in the EUV for two reasons. a) The stars lie in a tunnel in the ISM that has a very low HI column density, ($N_H \approx 1 - 2 \times 10^{18}$ cm^{-2}) even though they are at distances of $\approx$ 200 pc. b) The photospheric flux from ϵ CMa and (to a lesser extent) β CMa exceed the predictions of the model atmospheres that adequately fit visual and UV energy distributions.

The first year of *EUVE* observations (AO1) of ϵ CMa are described by Cassinelli et al. (1995, Paper I). Here we present AO2 observations of ϵ CMa and results from both AO1 and AO2 observations of β CMa. The β CMa results will be discussed in an upcoming paper by Cassinelli et al. (1995, Paper II).

Several papers at this meeting are also concerned with the interpretation of the *EUVE* data from ϵ CMa and β CMa. The goal is of this review is to provide a broad overview of the significance of the findings to the the fields of stellar atmospheres, pulsation theory, stellar winds and the galactic interstellar medium.

2. The *EUVE* Spectra of B Stars from 300 to 730 Å

Figures 1A and 1B show a comparison of the *EUVE* LW spectra of the two stars. The results for ϵ CMa in Figure 1A were surprising for several reasons. The photospheric Lyman continuum radiation is a factor of about 30 higher than predicted from model atmospheres. The photospheric radiation is seen on both sides of the He I ionization edge at 504 Å. The star shows strong emission at both the He II Lyα at 304 Å and the resonance line of O III at 374 Å. Also, as is discussed in paper I, emission lines from high ion stages from Fe IX to Fe XVI are present in the SW and MW spectra. The presence of these high ionization species provides new information regarding the wind shocks and X-rays from this star. Figure 1B shows the LW spectrum of β CMa. The spectrum longer than λ 504 is similar to that of ϵ CMa, but the count rate is somewhat smaller because the star has a larger column of interstellar attenuation, twice that of ϵ CMa

367

S. Bowyer and R. F. Malina (eds.), Astrophysics in the Extreme Ultraviolet, 367–374.
© *1996 Kluwer Academic Publishers. Printed in the Netherlands.*

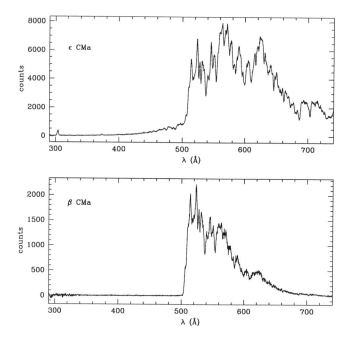

FIGURE 1. Shows *EUVE* spectra in counts (per 0.27 Å bin) versus wavelength as obtained with the long wavelength (LW) spectrometer on *EUVE*. (A) The photospheric spectrum of ε CMa obtained during AO2 during an observation with an exposure time of 122,244 seconds is shown in the top panel. Stellar wind emission lines of He II at 304 Å and of O III at 374 Å can also be seen. (B) The photosphere of β CMa as observed with an exposure time of 104,575 seconds is shown in the bottom panel.

and the star has a smaller photospheric EUV excess. The star β CMa does not show any photospheric radiation shortward of the He I edge, mostly because of the additional Helium ISM attenuation in its direction. The spectra of this star are especially interesting because of their variability.

Figure 2A and 2B show portions of the LW spectra for the 2 stars. There are important contribution to the absorption line spectra from both the photospheres and from the stellar winds. The broad absorption troughs are from photospheric line blends. The lines formed in the winds of hot stars tend to be of higher ionization stages because of the X-rays formed in shocks embedded in the winds, and in both spectra, we see absorption by the resonance line of OV at 630 Å. Since the resolution of the LW spectrometer of about 1200 km/s , is larger than the observed terminal speed of the wind, we should not expect to see a wind P-Cygni profile. The spectrum of β CMa is very similar to that from ε CMa but somewhat noisier because of its lower count rate. Both of the spectra shown in Figure 2 were obtained using the dithering procedure to reduce fixed pattern noise.

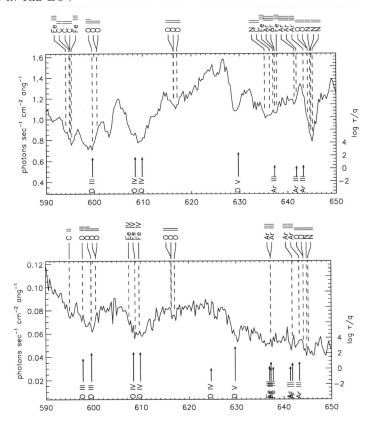

FIGURE 2. The absorption line spectra of (A) ϵ CMa and (B) β CMa in the wavelength band 590 to 650 Å. On each figure are two sets of line identifications derived from models. On the top of each figure are shown model photosphere line identifications. On the bottom the arrows show lines that may be prominent in the winds of the stars.

3. The Photospheric EUV Flux

Hubeny is presenting a review comparing the observed EUV continua with B star models, so here I will just make comments needed for our ISM and pulsation effects discussions. In the case of ϵ CMa, the EUV photospheric radiation exceeds the predictions of model atmospheres by more than an order of magnitude. In the case of β CMa, the result is less straight forward, but if we accept the the effective temperature derived from a model that fits the UV and visual flux distribution, then this star also has an EUV excess (by a factor of about 5). Since the vast majority of B stars have their effective temperatures determined from fits to observable flux distributions, I am choosing to consider in this review that both ϵ CMa and β CMa show an excess. The derived effective temperatures of the two stars are 21000 K for ϵ CMa, and 23250 K for β CMa.

In paper I, we considered two possible causes of the EUV excess in ϵ CMa: it is a non-LTE population effect, or it is caused by an enhancement in the temperature in the outermost regions of the photosphere where the EUV radiation forms. We concluded in favor of the higher temperature explanation because the star also shows a infrared continuum excess in the near 10-25 μm spectral region. Both infrared radiation in that band and the Lyman continuum radiation form at about the same heights in the atmosphere. However, in the IR the opacity is free-free absorption and the source function is the Planck function, and hence the IR flux is sensitive to the temperature, but not to NLTE effects. The temperature must be about 1500 K higher in the continuum formation region of ϵ CMa to explain both the factor of 30 excess in the EUV flux and the roughly 13% excess at 12 μm . The reason the temperature has a largely different effect on the emergent fluxes in the two spectral regions is that the EUV is on the Wien side of the spectrum, while the IR is on the Rayleigh Jeans side. So, for ϵ CMa, the EUV flux varies roughly as T^{13}, while the IR flux is approximately proportional to T. Because both stars have an IR excess (to a lesser extent in β CMa) we are currently investigating the possible causes of heating in the outer photospheres of B stars.

4. Line Emission and Fluorescence in B Star Winds

Over the past several years a new picture has been developed regarding the structure of the winds of hot stars. Line driven winds are unstable and a sinusoidal variation imposed on the velocity distribution will grow to form shock structures (Owocki et al. 1988). As has recently been shown by Cooper (1994), the shocks generated from the instabilities produce a range of temperatures in the wind and the hottest gas gas gives rise to observable X-ray emission. The heated regions emit most of their energy in the form of spectral lines, many of which occur in the EUV. X-rays from ϵ CMa are detected by ROSAT, (Drew et al. 1994) and we see in our spectra of the star, lines from Fe IX (171 Å), Fe XI (181 Å), Fe XV (284 Å), Fe XVI (335 Å). From an analysis of both the X-ray counts and the EUV line radiation, Cohen et al. (1995) conclude the following: the EUV lines confirm that the X-rays are produced thermally in the stellar winds as opposed to non-thermally; The X-ray and especially the EUV photons are subject to wind attenuation; the X-ray energy distribution and the EUV lines from high ions, are consistent with a power law distribution of the emission measure versus temperature; and the power law is of the form expected from the shocked wind theory of Cooper (1994). The simultaneous analysis of X-ray and EUV data provides much more detailed information about shocks and their hard radiation than would be obtainable with either spectral range alone.

MacFarlane et al. (1994) have shown that the X-rays and EUV high ion line emission from shocks have a major effect on the ionization structure in B star winds. In particular they find from a model of ϵ CMa that the ionization fraction of He^{++} changes from being a trace ion with abundance of 10^{-4} of all Helium, to being a major ion species with a $\approx$ 0.3 fractional abundance. With He^{++} being abundant, the origin of the λ 304 Å Lyα line can be explained as from the recombination of He^{++} in the wind. The line is very thick in the wind, and in the multiple scattering transfer of the line radiation it resonantly excites O III. The cascade from the excited level produces O III lines at 3444 Å, 3760 Å, and the final step leads to the resonance line of O III at 374 Å line, which we see in our spectra of ϵ CMa. This is of special interest because we are seeing for the first time in a stellar spectrum the dominant transitions of the Bowen Fluorescence mechanism. Bowen (1935) showed that the fluorescence with He II Lyα could explain the anomalously large strengths of O III lines at 3444 Å and 3760 Å seen in planetary nebulae spectra. In MacFarlane's paper at this meeting, he shows that the λ 304 and λ 374 lines, are formed by a degradation of the X-ray and EUV high ion radiation in much the same way that Hα is produced by the degradation of stellar Lyman radiation in a gaseous nebula. The *EUVE* observations are thus providing fascinating new insight regarding the nebular aspects of winds.

5. Ionization of the Galactic ISM with B Star Emission

Using the estimates of effective temperatures derived from UV and visual continuum studies, we find that both of the B stars bright enough to be detected by the *EUVE* LW spectrometer show an excess. Although the sample is small, we should ask what would be the effect if all B stars had an average of a factor of 10 excess in their EUV continuum?

A possible application concerns the ionization of the warm ionized medium (or WIM). Reynolds (1991a,b) observes diffuse radiation from the galaxy at both Hα and at He I, 5876 Å. Although O stars in the galaxy emit sufficient radiation to produce the observed WIM ionization, they suffer from two problems. The WIM radiation comes from gas with a mean height above the galaxy of 1 kpc. There is some doubt that O stars could be the source of the ionization of the gas at that height above the plane, basically because the O stars remain close to the material from which they formed so that too little of their radiation would penetrate to the WIM region. Secondly, even if the radiation could do so, it may be too hard to explain the WIM ionization. Tufte & Reynolds (1995) find that the He II/ H II recombination radiation of the WIM derived from (λ 5876 / Hα), is only 1/4th the ratio observed from HII regions around O stars. According to Reynolds (private communication) if B stars emitted a factor of 10 more EUV radiation than model atmospheres predict it would go a long way toward solving the WIM problem. The B stars are not so tightly confined to their natal clouds and the ratio of He II / H II produced in B star circumstellar nebulae is more like that seen in the WIM. Having an excess EUV luminosity by a factor of 10 would not affect the effective temperatures of B stars significantly, because the radiation emitted in the EUV region is a negligible fraction of the total luminosity of a B star.

Since we are not likely to find more B stars with the *EUVE* that show the strong EUV continua of ϵ CMa and β CMa, how can we determine if other B stars emit an excess of EUV radiation? Traditionally the EUV radiation from hot stars is studied by observing the ionization of circumstellar HII regions. Kutyrev & Reynolds (1995) have compared the radiation from HII regions around B stars with model predictions and have found that two stars (Spica and 139 Tau) out of eight show too large an ionized emission measure. The other six stars could also have an excess if they are in regions that are density bounded, i.e., with too little mass to absorb all the light from the central star. As a alternative approach, we planning to use the spectrometers on the ISO satellite to search for IR excesses in B stars. Recall that the IR continuum and EUV continuum form at about the same heights in the atmosphere, so if there is an IR excess, a larger EUV excess accompanies it.

6. The β Cephei Phenomenon As Seen in the EUV Radiation of β CMa

A long standing puzzle in interiors theory has been the cause of the pulsation seen in stars within the narrow region parallel to the main sequence called the β Cephei strip. The pulsation has recently been explained. New opacities calculated by Iglesias et al. (1992), showed a peak at a temperature of about 200000 K. Moskalik et al. (1992) and Kiriakides et al. (1992) applied these opacities to β Cephei models and found that the pulsational variability of this class of star could be driven by the classical Kappa

Mechanism. Interior theorists work mostly with mode analyses that predict the periods and modes of oscillation. In the subject of stellar atmospheres there is also interest in the effects of the propagation of the waves through the atmosphere. For example, the propagation of the waves in the cool Mira variables is responsible for the levitation of the atmospheres that is crucial for the formation of grains, and radiative acceleration of the grains then produces the massive dust driven winds from the stars (Bowen & Willson 1991).

Our pair of B stars is ideal for studying the atmospheres and envelopes of the β Cephei class of variables. The star β CMa is the brightest of the β Cephei stars and at a temperature of 23250 K, it lies near the red side of the narrow β Cephei strip in the HR diagram. The star ϵ CMa lies just beyond the redward edge of the strip. The two stars have nearly identical properties in regards to their luminosities, X-ray emission, and wind properties. Yet one pulsates and the other is not reported to be a pulsating variable, (although we find ϵ CMa to show non-periodic variability in our EUV observations). As β CMa is the brightest of the β Cephei stars, it was the first of the class to be well studied, and was the first to show beat phenomena (Meyer 1934). It has 3 periods, at P = 6.00, 6.03, and 5.74 hours, and as seen in the amplitude of the variations the pulsations can show constructive and destructive interference, with a period of 49 days in the case of the 6.0 and 6.03 hour periods. The observations of the pulsations are shown in paper II and in contribution to this meeting by Cohen et al. A portion of the β CMa observations is shown in Figure 3. The bottom section of the figure illustrates the beat phenomenon as it can be seen in the EUV. The periods of pulsation that were derived from observations at visual and UV wavelengths have been detected in the EUV. Future *EUVE* observations are scheduled to better determine the beat phenomenon seen in Figure 3.

The EUV is an ideal spectral range for studying the β Cephei phenomenon. This is because the variation that is seen in the light curve is mostly caused by a variation in the effective temperature of the star, and as we noted earlier, the EUV spectral range is especially sensitive to the radiation temperature since it lies on the Wien side of the spectrum for B stars. Therefore the flux amplitude should be greatly enhanced in the EUV. For the UV studies of β CMa by Beeckmans & Burger (1977), the temperature amplitude of the variation was deduced to be $\Delta T = 180\,^\circ \pm 130\,^\circ$. From our EUV studies, Cohen et al. (these proceedings) have derived $\Delta T = 108\,^\circ \pm 31\,^\circ$. The pulsation is thus much better defined. The EUV results are *consistent* with the UV results, but mostly because of the large error bars in the UV determination. If higher quality UV or visible data were available, a disagreement between the temperature amplitudes in the UV and the EUV might in fact be expected! This is because the EUV continuum is formed at several density scale heights higher in the atmosphere than is the UV continuum. If the temperature amplitude in the EUV would exactly match that in the UV it would mean that there is no damping of the pulsation between the heights of formation in the atmosphere of the two spectral bands. Using the range of the EUV band alone, Cohen et al. find that there may be a somewhat larger amplitude variation at $\lambda < 550$ Å than there is at $\lambda > 550$. The shorter wavelength region forms deeper in the atmosphere so the difference is in the sense that the pulsational amplitude is smaller at larger heights. On the basis of this, we speculate that there may be some damping of the pulsational wave energy in the outer atmosphere. This could presumably lead to an extra source of heating in the case of β CMa. What about for ϵ CMa where there is an even clearer need for additional heating? In the standard picture for explaining pulsations, the pulsation is driven in the region of the envelope with enhanced opacity, where positive work is added to the star during one pulsation cycle. Whether the pulsation is observable at the

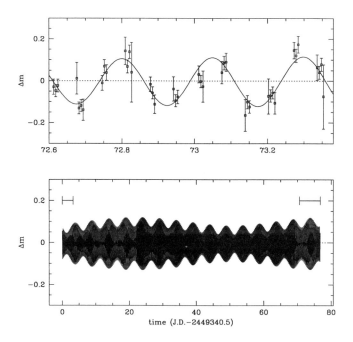

FIGURE 3. The top panel shows the EUV magnitude variation of β CMa versus time during a interval near day 73 in our two part set of observations. In the bottom is shown the beat phenomenon that is derived from the *EUVE* observations, produced by the three periods of oscillation of this star. The time from one minimum to the next in the interference is about 49 days. The two intervals of our observations are indicated by the two bars at the top of the figure.

top depends to a certain extent on the damping of the pulsation in the $\gamma = 5/3$ zone above the driving zone. (where γ is the ratio of specific heats c_P/c_V). Perhaps there is a residual flux of pulsational energy propagating upward also in the case of ϵ CMa, and it becomes preferentially deposited in the region of formation of the EUV continuum. The possibility that pulsational energy plays a role in the heating of the outer photospheres of these hot stars will be a topic of much further study.

7. Conclusions

The goal of this paper has been to highlight the many ways that stellar and interstellar astronomy are affected by the *EUVE* observations of the anomalously bright B giant stars.

Starting from the interior and progressing outward to the ISM we see the following consequences: In the pulsations of β CMa, we now have data for analyzing the possibility that pulsations, driven in a zone in the envelope where T= 200,000 K, can lead to a change in the structure of the outer atmospheres of stars that lie in or near the β Cephei strip. The atmospheres of both ϵ CMa and β CMa produce significantly more EUV radiation than would be predicted from models that are calibrated by visual and UV energy distributions. From the presence of excess of free-free radiation in the infrared, we have found that the enhanced EUV radiation appears to be due to heating of the outer atmospheres. This diagnostic also suggests that IR observations will be useful to survey other B stars for the presence EUV excesses. The wind of ϵ CMa shows a broad range of phenomena not expected before the *EUVE* launch. There is now a clear link between the production of shocks in the wind with the formation of the EUV lines of high ionization stages. The radiation from these lines and the X-ray emission greatly enhance the abundance of doubly ionized helium, which on recombining leads to the Bowen fluorescence mechanism. This fluorescence had not been seen in a stellar spectrum before. The interstellar medium is also affected by the B star EUV radiation. In the case of ϵ CMa (or *Adhara*, as it is referred to in the review by Vallerga and Welsh), the star is responsible for maintaining the ionization of the ISM near the Sun. From a broader perspective the B star EUV radiation may play a role in the ionization of the WIM. The observed HeI(λ 5876) / Hα line ratio can be produced by B star radiation, but probably not by that from O stars. Although the sample of B stars that can be observed with *EUVE* is small, their spectra are leading to an exciting range of investigations.

I thank David Cohen, Joe MacFarlane, Ivan Hubeny and Ron Reynolds for helpful discussions regarding various sections of this review. The research is being supported by NASA grant NAG5-2282.

REFERENCES

Beeckmans, F., & Burger, M. 1977, A&A, 61, 815

Bowen, G. H. & Willson, L. A. 1991, ApJ, 375, L53

Bowen, I. S. 1935, ApJ, 81, 1

Cassinelli, J. P., Cohen, D. H., MacFarlane, J. J., Drew, J. E., Lynas-Gray, A. E., Hoare, M. G., Vallerga, J. V., Welsh, B. Y., Vedder, P. W., Hubeny, I., & Lanz, T. 1995, ApJ, 438, 932. (paper I)

Cassinelli, J. P., Cohen, D. H., Drew, J. E., Lynas-Gray, A. E., Hoare, M. G., Hubeny, I., MacFarlane, J. J., Vallerga, J. V., Welsh, B. Y. 1995, ApJ, in preparation (paper II)

Cohen, D. H., Cooper, R. G., Wang, P., MacFarlane, J. J., Owocki, S. P., & Cassinelli, J. P. 1995, ApJ, in preparation

Cooper, R. G. 1994, Ph.D. Thesis, University of Delaware

MacFarlane, J. J., Cohen, D. H., & Wang, P. 1994, ApJ, 437, 351

Drew, J. E., Denby, M., & Hoare, M. G. 1994, MNRAS, 266, 917

Iglesias, C. A., Rogers, F. J., & Wilson, B. C. 1992, ApJ, 397, 717

Kiriakidis, M., El, Eid, & M. F., Glatzel, W. 1992, MNRAS, 255, 1P

Kutyrev, & Reynolds, 1995, preprint

Meyer, W. F. 1934, Publ. A.S.P, 46, 202

Moskalik, P., & Dziembowski, W. A. 1992, A&A, 256, L5

Owocki, S. P., Castor, J. I., & Rybicki, G. B. 1988, ApJ, 235, 914

Reynolds, R. J. 1991a, in The Interstellar Disk- Halo Connection in Galaxies, Dordrecht: Kluwer, 67

Reynolds, R. J. 1991b, ApJ, 372, L17

Tufte, S. & Reynolds, R. J. 1995, in preparation

Ionization in the Winds of Early-B Stars: Constraints Imposed by *EUVE*

J. J. MacFARLANE,[1,2] D. H. COHEN,[1]
J. P. CASSINELLI,[1] AND P. WANG[2]

[1] Department of Astronomy, University of Wisconsin,
475 N. Charter St., Madison, WI 53706, USA

[2] Fusion Technology Institute, University of Wisconsin,
1500 Johnson Drive, Madison, WI 53706, USA

Extreme Ultraviolet Explorer (*EUVE*) spectral observations of ε CMa (B2 II) provide significant new information about the ionization dynamics of its wind, its mass loss rate, and the interaction of its wind with the soft X-ray/EUV radiation field. We present results from wind ionization calculations which show how EUV radiation, emitted predominantly in the form of Fe IX-XVI lines from plasma with $T \sim 1 - 3 \times 10^6$ K, significantly alters the ionization state of its wind. EUV photons from the hot plasma photoionize He II in the cool portion of ε CMa's wind, producing anomalously high abundances of He III. The subsequent recombination from He III results in He II Lα (304 Å) and Lβ (256 Å) lines which are observed by EUVE. Also observed is O III $\lambda\lambda$ 374, which results from the O III $\lambda\lambda$ 304 multiplet being pumped by He II Lα — the Bowen mechanism (Cassinelli et al. 1995). We report on initial results from numerical simulations which show the effect of the Bowen mechanism on the O III 374 Å line emission.

1. Introduction

Ultraviolet observations of hot stars (Snow & Morton 1976; Lamers & Morton 1976) have shown that they suffer significant mass loss through high velocity winds. For early-O stars, radio (Bieging, Abbott, & Churchwell 1989) and Hα (Lamers & Leitherer 1993) observations provide evidence for mass loss rates $\sim 10^{-6} - 10^{-5}$ M$_\odot$/yr, while wind velocities can be 3,000 km/s or more. Observational determination of mass loss rates for late-O and early-B stars becomes more difficult, however, because their winds are much weaker; i.e., they tend to exhibit lower densities and lower velocities. Fortunately, *EUVE* observations of the B2 II star ε CMa (Cassinelli et al. 1995) have provided valuable new data for studying the winds of early-B stars. Emission lines between 150 Å and 400 Å have been observed from both the hot X-ray emitting region (Fe IX–XVI) and the cool wind (He II and O III). Below, we present results from wind ionization calculations which show the influence of the soft X-ray/EUV radiation on the ionization state of the wind, and the effect of the Bowen mechanism on the O III 374 Å line intensity.

ROSAT observations of ε CMa (Drew, Denby, & Hoare 1993) showed that ε CMa, like most other hot stars, has a significant X-ray luminosity (L$_x$ $\sim 1 \times 10^{-7}$ L$_{bol}$). Two-temperatures fits to the *ROSAT* data suggest a relatively strong low-temperature component ($T \sim 1 - 2 \times 10^6$ K), and a weaker high-temperature component ($T \sim 8 \times 10^6$ K). Plasmas at these temperatures also exhibit strong Fe line emission between 150 Å and 400 Å, which was observed by *EUVE* for ε CMa. Recent studies (MacFarlane, Cohen, & Wang 1994) have predicted that the soft X-ray/EUV radiation of early-B stars can significantly alter the bulk ionization state of their winds. This is unlike the case of early-O star winds, where X-rays tend to produce only a small (but detectable) perturbation on their ionization distribution (Cassinelli & Olson 1979).

Because *EUVE* has detected both the hot plasma Fe line emission and cool wind He II

S. Bowyer and R. F. Malina (eds.), *Astrophysics in the Extreme Ultraviolet*, 375–379.

and O III line emission, it presents a rare opportunity to study the interplay of the hot plasma radiation field and the cool wind.

In this paper, we present several results from numerical simulations describing these effects. More detailed descriptions of this work will be presented elsewhere (Cohen et al. 1995, MacFarlane et al. 1995).

2. Models

Detailed descriptions of our wind ionization code have been presented elsewhere (MacFarlane et al. 1993, 1994); only a brief summary is presented here. The wind is assumed to be spherically symmetric with a velocity which increases monotonically with radius. The density is specified by its mass loss rate and assuming, the mass flux is constant with radius. The radiation field includes contributions from: (1) the photosphere, based on Mihalas' (1972) non-LTE models longward of 912 Å and a fit to the $EUVE$ data for 350 Å $< \lambda < 730$ Å; (2) diffuse radiation in the wind; and (3) EUV/X-ray radiation from a high-temperature plasma, which had a frequency-dependence given by XSPEC models (Raymond & Smith 1977) and a power-law temperature distribution determined from the combined analysis of $ROSAT$ and $EUVE$ data (Cohen et al. 1995). Emission from the hot plasma was assumed to be distributed throughout the wind (shock model), with a peak in the differential emission measure at $r = 1.5 R_*$ ($\alpha = -3$ in Eqn. (10) of MacFarlane et al. 1993). Radiative transfer effects for all radiation sources were computed using a multi-ray impact parameter model.

Non-LTE atomic level populations were computed by solving multilevel statistical equilibrium equations self-consistently with the radiation field. A total of 81 levels for H, He, and O were considered in our atomic model. It was necessary to consider fine-structure splitting of the O III levels so that the Bowen photoexcitation mechanism (Bowen 1935) could be accounted for. This mechanism arises due to overlap of the He II Lα line at 303.78 Å and the O III $1s^2 2s^2 2p^2$ ^{3}P - $1s^2 2s^2 2p3d$ ^{3}P^0 multiplet (see Figure 1). When He III recombines to the $n = 2$ state of He II, the Lα photons emitted can be absorbed, producing an anomalously high population of the O III 3d ^{3}P^0 level. This level subsequently decays to the 3s ^{3}P^0 level, resulting in the spontaneous emission of the $\lambda\lambda$ 374 multiplet, which is observed as a single emission feature by $EUVE$ (Cassinelli et al. 1995). The interaction of overlapping lines is modeled using a generalized Sobolev method with multiple velocity surfaces (Rybicki & Hummer 1978; Puls, Owocki, & Fullerton 1993; see also MacFarlane et al. 1995).

3. Results

We first examine the effect of soft X-ray/EUV radiation on the He ionization balance in the wind of ε CMa. Figure 2 shows computed ionization distributions for He I–III from two calculations: one using a distributed X-ray emission source (solid curves) and the other with no X-ray source (dotted curves). In each case the mass loss rate was 0.8×10^{-8} M$_\odot$/yr. Clearly, the population of doubly ionized He (circles) is dominated by photoionization due to the hot plasma radiation, which dominates the radiation field at $h\nu > 54$ eV. At relatively low velocities ($v \lesssim 0.4 v_\infty$ and $r < 1.5 R_*$) the He III population is lower because of the higher wind densities and the fact that the X-ray differential emission measure peaks at $r = 1.5 R_*$ in this model. Note that in the absence of X-ray-induced photoionization, the He III fraction is $\lesssim 10^{-4}$ throughout the wind. The distribution of He III in the wind is important because its recombination is the source of photons which are ultimately observed by $EUVE$ at 304 Å and 374 Å.

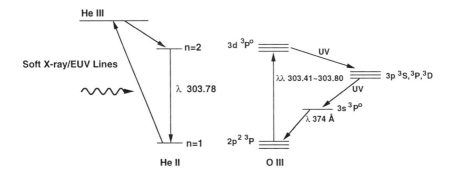

FIGURE 1. Schematic energy level diagram illustrating radiative processes in the wind of
ε CMa.

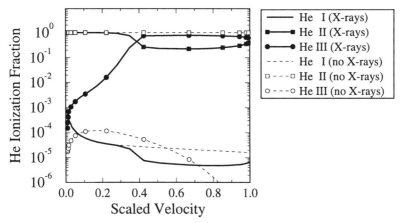

FIGURE 2. Helium ionization distributions calculated with a distributed hot plasma radiation
source (solid curves) and with no X-ray radiation source (dashed curves).

The effect of the Bowen mechanism (photopumping of the O III λ 304 multiplet by
He II Lα) on the O III $\lambda\lambda$ 374 flux can be seen in Figure 3, where the 374 Å profile
is shown from two calculations. The solid curve shows the profile from a calculation
in which line overlap effects on the atomic populations were included, while the dashed
curve is from a calculation in which photoexcitation due to overlapping lines was neglected. (Note, however, that the emergent spectra, which are computed after the level

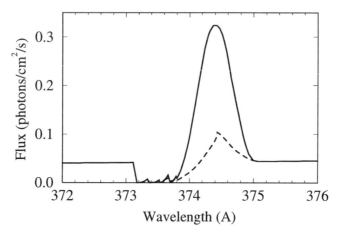

FIGURE 3.   Calculated O III $\lambda\lambda$ 374 line profiles for ε CMa wind ionization models with $\dot{M} = 0.8 \times 10^{-8}$ M$_\odot$/yr. Dashed curve: atomic populations computed without line overlap effects; solid curve: with line overlap effects.

populations are determined, *do* include overlap effects. This gives rise to the structure seen in the absorption component of the profile.) By including the Bowen effect in the calculation, the emission from the 374 Å multiplet is seen to increase substantially. Without the Bowen effect, the $\lambda\lambda$ 374 profile has the characteristic P-Cygni shape, with blue-shifted absorption and red-shifted emission; that is, scattered photospheric radiation with no additional source of photons. Preliminary results suggest that the best agreement with the *EUVE* data for the 304 Å and 374 Å lines is obtained for mass loss rates of $\sim 1 - 2 \times 10^{-8}$ M$_\odot$/yr. Final results and a complete description of this analysis will be presented in a forthcoming paper.

This work was supported in part by NASA grants NAGW-2210 and NAGW 5-2282.

REFERENCES

Bieging, J. H., Abbott, D. C. & Churchwell, E. 1989, ApJ, 340, 518

Bowen, I. S. 1935, ApJ, 81, 1

Cassinelli, J. P. & Olson, R. E. 1979, ApJ, 229, 304

Cassinelli, J. P., Cohen, D. H., MacFarlane, J. J., Drew, J. E., Lynas-Gray, A. E., Hoare, M. G., Vallerga, J. V., Welsh, B. Y., Vedder, P. W., Hubeny, I., & Lanz, T. 1995, ApJ, 438,932

Cohen, D. H. et al. 1995, ApJ, to be submitted

Drew, J. E., Denby, M. & Hoare, M. G. 1994, MNRAS, 266, 917

Lamers, H. J. G. L. M. & Morton, D. C. 1976, ApJS, 32, 715

Lamers, H. J. G. L. M. & Leitherer, C. 1993, ApJ, 412, 771

MacFarlane, J. J., Cohen, D. H., & Wang, P. 1994, ApJ, 437, 351

MacFarlane, J. J., Waldron, W. L., Corcoran, M. F., Wolff, M. J., Wang, P., & Cassinelli, J. P. 1993, ApJ, 419, 813

MacFarlane, J. J., Cohen, D. H., Cassinelli, J. P. & Wang, P. 1995, ApJ, to be submitted

Mihalas, D. 1972, NCAR Tech. Note, STR-76

PULS, J., OWOCKI, S. P., & FULLERTON, A. W. 1993, A&A, 279, 457

RAYMOND, J. C. & SMITH, B. W. 1977, ApJS, 35, 419

RYBICKI, G. B. & HUMMER, D. G. 1978, ApJ, 219, 654

SNOW, T. P. & MORTON, D. C. 1976, ApJS, 32, 429

EUV Radiation from Hot Star Photospheres: Theory Versus Observations

IVAN HUBENY AND THIERRY LANZ

Universities Space Research Association, NASA Goddard Space Flight Center,
Code 681, Greenbelt, MD 20771, USA

The only stars other than white dwarfs whose photospheric extreme ultraviolet radiation has been detected are ϵ and β CMa. It is therefore of considerable theoretical interest to compare the *EUVE* observations of these two giant B stars to predicted spectra. However, both LTE and non-LTE very sophisticated line blanketed model atmospheres fail to match the observed flux. This failure leaves the stellar photosphere theory, even for seemingly "simple" objects as normal B giants were believed to be, in a rather dubious position. This paper briefly summarizes possible reasons for the failure of existing models to describe the *EUVE* observations of hot stars. In particular, we discuss the effects of uncertainties in the line blanketing, and the effects of the photosphere–wind interaction.

1. Introduction

The *EUVE* observations of hot stars present new challenges for the stellar atmospheres theory. It is well known from the theory that the EUV spectra of hot stars may be significantly influenced by departures from the local thermodynamic equilibrium (the so-called non-LTE effects). However, the model implications could not be directly tested by fitting real stellar spectra before we were actually able to observe this spectral region.

It was therefore of considerable theoretical interest to compare the predicted spectra to observations of two giant B stars, ϵ and β CMa, which are the only non-white dwarfs stars for which the Lyman continuum flux can be observed. The fact that LTE models failed to reproduce the observed spectra was not so surprising. However, the real surprise was that even very sophisticated non-LTE line blanketed model atmospheres were similarly unsuccessful. This fact is very disturbing, because the B stars were believed to be relatively simple objects to model. Indeed, they do not possess strong winds; there is no convection in the photosphere; and the continuum opacity is largely dominated by hydrogen and helium. It is fair to say that if we cannot trust computed B star model atmospheres, the models for other stellar types should be trusted even less.

Apart from the alarming consequences for the stellar atmosphere theory *per se*, this failure has profound implications for many branches of astrophysics. For instance, without reliable model atmosphere predictions, we are not able to determine the number of ionizing photons, which consequently leads to significant uncertainties in the interstellar matter modeling.

In this paper we will briefly summarize possible reasons for the failure of existing models to reproduce the *EUVE* observations of hot stars, and outline the way how this situation may be improved.

2. Observations Versus LTE Models for ϵ and β CMa

The detection of ϵ CMa as the brightest EUV source was first presented by Vallerga et al. (1993), and EUV spectrum subsequently analyzed in detail by Cassinelli et al. (1995a; see also Cassinelli, this volume). An analysis of β CMa is in progress (Cassinelli et al. 1995b).

S. Bowyer and R. F. Malina (eds.), Astrophysics in the Extreme Ultraviolet, 381–388.

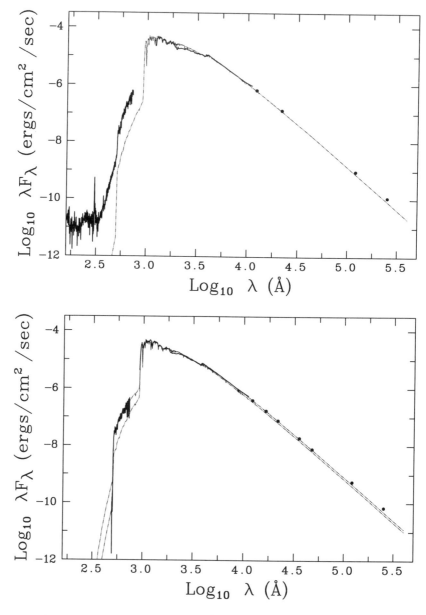

FIGURE 1. Continuous energy distribution from EUV to mid-IR wavelengths, compared with LTE model atmospheres: (a) ϵ CMa and model with $T_{\text{eff}} = 21,000$ K; $\log g = 3.2$; (b) β CMa and models with $T_{\text{eff}} = 23,250$ K; $\log g = 3.5$, and $T_{\text{eff}} = 24,800$ K; $\log g = 3.7$.

In Fig. 1, we present the spectra of both stars together with Kurucz LTE model atmospheres, using the effective temperature and surface gravity determined from the total flux and the UV and optical continuum (for details, see Cassinelli et al. 1995a, b). The effective temperature for ϵ CMa, $T_{\text{eff}} = 21,000$ K, was determined using the measured total flux and angular diameter (Hanbury Brown et al. 1974).

It is clearly seen that the LTE model atmosphere flux falls significantly short of the observed Lyman continuum flux for ϵ CMa. The observed flux is about 30 times higher than predicted! For the other star, β CMa, the situation is not so clear. When we use the measured angular diameter and the total flux, we obtain $T_{\text{eff}} = 24,800$ K. The model flux is then consistent with the UV and even EUV flux, but predicts the V-band flux about 10% higher that observed; such a discrepancy is unacceptably large. If one instead uses the V-band and the total flux but disregards the measured angular diameter, one obtains a lower effective temperature, $T_{\text{eff}} = 23,250$ K — for details refer to Cassinelli et al. (1995b). The corresponding model atmosphere yields a good agreement in the optical and the near UV region, but the flux in the Lyman continuum is by factor of 4 lower than the observed one.

Regardless of the magnitude of discrepancy between observations and models for β CMa, the failure of LTE models to describe the observed Lyman continuum for ϵ CMa is undisputable. Therefore, we have to ask: Is LTE a satisfactory approximation? And, if not, are we able to compute sophisticated enough non-LTE models? And, still if so, do non-LTE provide the remedy of the situation? And, if not, what does? The rest of the paper is devoted to discussing these questions.

3. May NLTE Help?

From the general point of view, LTE is an approximation which may, but does not necessarily need to, be applicable. The so-called non-LTE (or NLTE) approach, which allows for departures from LTE, is a fundamentally better approximation to reality and therefore should be preferable. However, the computational effort needed to calculate such models is by orders of magnitude higher than for LTE models. Moreover, since the problem is enormously complicated, one has to sacrifice many features that can be included in LTE but cannot be handled within the framework of NLTE models. A typical example was, till recently, a treatment of the metal line blanketing–see 3.2.

Therefore, from the practical point of view, one has to ask: do we specifically need a NLTE description for predicting an EUV spectrum? And, do we compute NLTE models accurately enough?

3.1. Why NLTE in EUV?

There are two basic reasons why one expects that departures from LTE play important rôle in the EUV spectrum region.

First, the EUV spectrum region is the region of very high opacity. Indeed, the opacity in the Lyman continuum, He I resonance continuum, and He II Lyman continuum is the strongest opacity source for all hot stars. Since the opacity is large, the observed spectrum is formed high in the atmosphere where the material density is low. A low density is one of the most important reasons for the breakdown of the LTE approximation, because the collision rates (which tend to maintain the Boltzman-Saha LTE statistical equilibrium) become much smaller than the radiative rates. Since the radiation intensity may strongly deviate from its equilibrium, Planckian distribution, the low density implies that a non-equilibrium radiation will cause a non-equilibrium atomic level population distribution.

The second reason is the extreme sensitivity of the Planck function to small changes

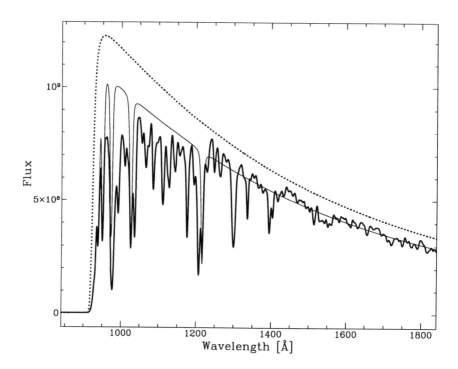

FIGURE 2. Theoretical flux from the unblanketed (thin line) and blanketed (thick lines) NLTE model atmosphere with $T_{\text{eff}} = 21,000$ K; $\log g = 3.2$. To demonstrate the importance of line blanketing, the theoretical continuum for the blanketed model is also displayed (dotted line).

in temperature on the short-wavelength side of its maximum. The source function, and therefore the radiation field, are proportional to the Planck function. Consequently, small differences in temperature in the atmosphere, together with departures from LTE, which may produce relatively modest effects in the optical or near-UV spectrum, translate to large differences in the predicted EUV flux.

3.2. Why Line Blanketing?

By the term "line blanketing" we understand the effect of thousands to millions of metal lines on the atmospheric structure and predicted emergent spectrum. In the case of early B stars, most lines are located in the traditional UV region ($\lambda 900 - 2500$ Å). The fact that the line blocking is largest in the UV does not mean that line blanketing is unimportant in the EUV region. As it is well known, line blanketing influences not only the emergent spectrum (the so-called line blocking), but also the atmospheric structure (the back-warming and the surface cooling). The effect of metal lines is illustrated on Fig. 2, where we plot the predicted flux for an atmosphere with $T_{\text{eff}} = 21,000$ K; $\log g = 3.2$; both for a simple NLTE hydrogen-helium model, as well as for a H-He-Fe line blanketed model. Therefore, a proper treatment of the UV line blanketing is crucial for understanding the EUV spectrum.

Since departures from LTE and line blanketing are both important, we have to compute fully blanketed NLTE models to resolve the dilemma. Until about a decade ago,

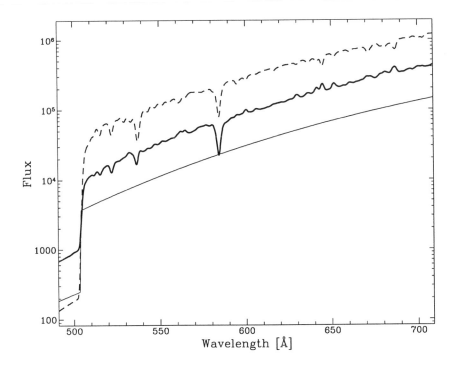

FIGURE 3. Theoretical flux in the Lyman continuum: NLTE unblanketed (thin line) and blanketed (thick line) models, and LTE blanketed model (dashed line), for $T_{\rm eff} = 21,000$ K; $\log g = 3.2$.

such an endeavor was viewed as hopeless. However, with the advent of fast and ingenious numerical methods for solving the radiative transfer equation, even this barrier is being tackled. Anderson (1985, 1989) was the first who calculated NLTE line-blanketed model atmospheres. Werner and colleagues (Werner 1986; Dreizler & Werner 1993) have developed a very sophisticated code based on the accelerated lambda iteration (ALI) method and computed a number of models for very hot stars. Recently, we have developed an hybrid Complete Linearization/ALI method (Hubeny & Lanz 1995), and incorporated it to our computer code TLUSTY (Hubeny 1988; Hubeny & Lanz 1992).

   The first *EUVE* observations of ϵ CMa serendipitously came just when we finished debugging the line-blanketed version of TLUSTY, and applied it by computing several NLTE line blanketed models for hot, metal-rich white dwarfs (Lanz & Hubeny 1995). We have therefore immediately calculated a sophisticated model for ϵ CMA, which took into account all lines of hydrogen and He I, and about 23,000 lines of Fe III and 8000 lines of Fe IV, all in NLTE (some results are presented in Hubeny & Lanz 1995 and Cassinelli et al. 1995). To our surprise and dismay, the NLTE models produced an even lower flux in the Lyman continuum than the LTE models! In Fig. 3, we plot the Lyman continuum flux for the Kurucz LTE model, along with two NLTE models, a simple H-He model, and a line-blanketed H-He-Fe model, for $T_{\rm eff} = 21,000$ K; $\log g = 3.2$ (similarly as in Fig. 2).

   In the next Section, we explain the behavior of the emergent flux in the Lyman continuum in more detail. This analysis also allows us to draw conclusions about what has to be done to improve the model atmospheres.

4. Lyman Continuum Flux

According to the Eddington-Barbier relation (see, e.g., Mihalas 1978), the emergent flux in the Lyman continuum is roughly given by

$$F_\nu \approx S_\nu(\tau_\nu = \frac{2}{3}),$$

where S_ν is the source function, and τ_ν the monochromatic optical depth at frequency ν.

Provided that the photoionization from the ground state of hydrogen is the dominant opacity source in the Lyman continuum ($504 < \lambda < 911$ Å), the source function is given by

$$S_\nu \approx B_\nu(T)/b_1,$$

where $B_\nu(T)$ is the Planck function at temperature T, and b_1 is the NLTE departure coefficient for the hydrogen ground state, $b_1 = n_1/n_1^*$, n_1^* being the LTE population.

The predicted flux in the Lyman continuum may be increased either (i) by increasing $B_\nu(T(\tau_\nu = 2/3))$, or, (ii) by decreasing b_1. The former can be achieved either by keeping the temperature structure unchanged, but moving the point where $\tau_\nu = 2/3$ to the physical layers in the atmosphere where the temperature is higher; or by a genuine increase of T in the continuum-forming layers.

Cassinelli et al. (1995a) suggested that the first possibility, a genuine temperature increase in the Lyman continuum forming region, is more likely. This suggestion was based on the observational fact that the far-IR flux at 12μm is of about 17% higher than predicted. This spectrum region is formed at about the same depth in the atmosphere as the Lyman continuum. However, the IR continuum, which is dominated by H I free-free process, cannot deviate very much from LTE. Rising artificially the temperature in the appropriate layers in the atmosphere by about 16% (i.e. by about 2000 K) would fit not only the IR continuum, but also the Lyman continuum without invoking any NLTE effects. But the question of course remains to know what causes such a temperature increase.

The second possibility cannot be realized in the static atmosphere, because the Lyman continuum behaves similarly as a resonance line in a two-level atom, for which the lower level always ends overpopulated with respect to LTE ($b_1 > 1$–see Mihalas 1978). On the other hand, a stellar wind can in principle decrease b_1 below unity by desaturating the Lyman-α line. Such a mechanism was recently suggested by Najarro et al. (1995).

5. Reasons for the Failure of Current Models

5.1. *Uncertainties in the Photospheric Models*

First, the photospheric models may still be not accurate enough. The most significant uncertainty in this respect is the treatment of line blanketing, in the sense that the total blanketing effect was underestimated due to the following reasons:

• We have taken into account only Fe III and Fe IV. Although these ions likely provide most of the line opacity, we have to explore the effects of considering lines of other species, at least in LTE.

• We have considered only about 30,000 iron lines originating between the *observed* energy levels. Although these lines are presumably the strongest ones, they form only a small fraction of the total number of Fe III and Fe IV lines predicted by Kurucz (1991).

This is potentially the most important source of missing opacity in our models. Our experience from white dwarf models (Barstow et al., this volume) shows that the agreement between observations and theory improves considerably when including *all* iron lines.

• We used hydrogenic photoionization cross sections for Fe III. Again, our experience from white dwarf models (Lanz & Hubeny 1995) shows that using the Opacity Project cross-sections for Fe IV, Fe V, and Fe VI (kindly provided to us by Anil Pradhan and Sultana Nahar) has a significant effect on the predicted spectrum.

• We use very rough estimates of collisional rates for excitation and ionization of iron.

• When constructing the Opacity Distribution Functions (ODF) for individual iron superlevels (see Hubeny & Lanz 1995), we have assumed a microturbulent velocity equal to 0 or 2 km s^{-1}. However, the microturbulent velocity may be larger, which would increase the line opacity. This point needs also to be explored in future models.

5.2. *Photosphere-Wind Interaction*

Alternatively, the theory of photospheres itself is more or less correct (in other words, the above listed possibilities will turn out to yield no significant effects), and the solution of the Lyman continuum problem lies in neglecting a photosphere–wind interaction. Indeed, we know that both ϵ and β CMa possess a weak wind ($\dot{M} \approx 1 \times 10^{-8} M_\odot \text{yr}^{-1}$– Cassinelli et al. 1995a, b). Thus, the wind may in principle influence the photospheric radiation, via the two following mechanisms:

• Depopulation of the ground state of hydrogen via desaturation of the Lyman-α line due to the velocity field at the base of the wind. This point was discussed in detail by Najarro et al. (1995), who conclude that this effect may partially explain the observed Lyman continuum discrepancy. However, their predictions show that the flux enhancement in the Lyman continuum for the observed value of the mass loss rate is about a factor of 0.5 dex, while the total discrepancy is roughly 1.5 dex.

• Irradiation of the photosphere by X-rays which originate in the wind. X-rays from ϵ CMa were detected by *Rosat* (Drew et al. 1994); Cassinelli et al. (1995a) showed that the *EUVE* observations of the He II Lyman-α line for ϵ CMa may be accounted for by recombination of He^{++} which is produced by these X-rays. However, it is questionable whether the X-rays will also increase the Lyman-continuum radiation. Work on this problem is in progress, but the preliminary results indicate that this is not very likely because the X-rays do not seem to provide enough energy.

5.3. *UV Flux Discrepancy*

It may well happen that the solution of the ϵ CMa puzzle will be provided by a combined effect of all the above mechanisms. However, an important piece of evidence that something on the purely photospheric level is not working well is provided by the flux around $\lambda 2200$ Å. There is a clear discrepancy here (see Fig. 1), the observed flux being significantly lower than predicted. An analogous flux discrepancy for other stars is routinely attributed to the well-known $\lambda 2200$ Å interstellar feature. But the *EUVE* observations clearly and undisputably demonstrate that the interstellar column density towards ϵ CMa is very low, $N_H \approx 1 \times 10^{18}$ (Cassinelli et al. 1995a), and therefore the $\lambda 2200$ feature cannot be of interstellar origin! Moreover, this part of the spectrum is formed very deep in the photosphere (even deeper than the optical flux), so one cannot invoke a photosphere–wind interaction. The most likely reason is, again, a missing photospheric opacity.

6. Conclusions

We have shown that neither LTE, nor the current most sophisticated non-LTE line blanketed model atmospheres are able to predict the observed flux in the Lyman continuum for ϵ CMa and likely also for β CMa. This finding serves as a sobering reminder that the stellar atmosphere theory has still a long way to go. Importantly, this failure tempers all far-reaching conclusions about the structure of interstellar matter which are based on *theoretical* estimates of the number of ionizing photons (i.e. the Lyman continuum flux) provided by stars.

We have discussed various possible reasons for the discrepancy, and identified several mechanisms which are likely responsible for the model deficiency. We claim that the failure of the present models is not a fundamental failure; it is just an indication that something in the models is still not being done properly.

The story of ϵ CMa nicely illustrates that history repeats itself. Again and again in modern astrophysics, once a new spectral window opens, new observations disagree with predictions based on old models, and consequently an improved theory has to be developed. We were prepared to accept that in the case of hot white dwarfs and coronae of cool stars, but we were caught in surprise in the case of relatively normal B stars. The case of ϵ and β CMa is even more interesting and important, because these are the only two mildly hot and presumably well understood stars which will ever be observed in the Lyman continuum. Their *EUVE* observations are therefore an invaluable piece of information for the development of the stellar atmosphere theory as a whole.

We would like to acknowledge a fruitful collaboration and numerous conversations with Joe Cassinelli, David Cohen, Janet Drew and Tony Lynas-Gray. The work on this project was supported in part by NASA grant NAGW-3834.

REFERENCES

ANDERSON, L. S. 1985, ApJ, 298, 848

ANDERSON, L. S. 1989, ApJ, 339, 588

CASSINELLI, J. P. ET AL. 1995a ApJ, 438, 932

CASSINELLI, J. P. ET AL. 1995b ApJ, to be submitted

DREIZLER, S., & WERNER, K. 1993, A&A, 278, 199

DREW, J. E., DENBY, M., & HOARE, M. G. 1994, MNRAS, 266, 917

HANBURY, BROWN, R., DAVIS, J., & ALLER, L. R. 1974, MNRAS, 167, 121

HUBENY, I. 1988, Comput. Phys. Commun., 52, 103

HUBENY, I., & LANZ, T. 1992, A&A, 262, 501

HUBENY, I., & LANZ, T. 1995, ApJ, 439, 875

KURUCZ, R. L. 1991, in Stellar Atmospheres: Beyond Classical Models, ed. L. Crivellari, I. Hubeny, & D. G. Hummer, NATO ASI Series C 152, Dordrecht: Kluwer, 441

LANZ, T., & HUBENY, I. 1995, ApJ, 439, 905

MIHALAS, D. 1978, Stellar Atmospheres, San Francisco: Freeman

NAJARRO, F. ET AL. 1995, A&A, in press

VALLERGA, J. V., VEDDER, P. W., & WELSH, B. Y. 1993, ApJ, 414, L65

WERNER, K. 1986, A&A, 161, 177

Photospheric Variability in *EUVE* Observations of β Canis Majoris (B1 II–III)

DAVID H. COHEN, JOSEPH J. MacFARLANE,
AND JOSEPH P. CASSINELLI

Department of Astronomy, University of Wisconsin at Madison,
475 No. Charter Street, Madison, Wisconsin, 53706, USA

The only pulsating early-type star observed by *EUVE*, the β Cephei variable β CMa, displays periodic variability in its Lyman continuum which is basically consistent with the long-known optical and UV variability. The amplitude of the primary pulsation component is significantly larger in the EUV than in the optical or UV. This is consistent with a temperature change being the explanation for the variability. It is notable that the pulsations have been detected in the Lyman continuum because this part of the spectrum is formed in a much higher layer of the photosphere than either the UV or optical continua.

1. β Canis Majoris

The brightest β Cephei variable, β CMa, lies in the region of unusually low column density known as the Canis Major Tunnel. Due partly to the lack of serious interstellar extinction and partly to its intrinsic brightness β CMa is the second brightest star in the *EUVE* long wavelength (LW) spectrometer. Variability has been observed in the optical since 1908 and in the UV over the past two decades. Three pulsation periods have consistently been identified in the optical data. The amplitudes (in the optical) and the periods are so similar that there is a strong beat phenomenon, with beat periods $B_{12} = 49.2$, $B_{13} = 5.4$, and $B_{23} = 4.9$ days.

The dominant period has an inverse amplitude-wavelength relationship that can be explained as a change in effective temperature. Beeckmans & Burger (1977) use the variability at 1810 Å to deduce the change in effective temperature of 180 ± 130 K for the first period and 180 ± 140 K for the second. It should be kept in mind that this temperature change occurs at a Rosseland optical depth of two-thirds. In contrast, the Lyman continuum is formed at a Rosseland optical depth of less than 0.01.

2. β Cephei Variables

The β Cephei variables lie in a small region of the HR diagram above the main sequence. All of the known β Cephei variables in our galaxy have spectral types in the range B0 to B3 and luminosity classes in the range IV to II. These stars show brightness, radial velocity, and line profile variability. The pulsations have generally small amplitudes ($\Delta m_v < 0.2$) and the pulsation periods are typically 4 to 8 hours. Many are "beat" Cepheis with two or more slightly different periods. Increasing variability toward shorter wavelengths indicates that pulsations cause primarily a temperature change (Sterken & Jerzykiewicz 1993).

The β Cephei instability strip includes the end of the core H-burning phase and may extend through shell H-burning. No period-luminosity relationship has been proven. The exact pulsation mechanism has only recently been identified. It is a pure kappa-mechanism due to a bump in the iron opacity near 200,000 K that is now included in the OPAL opacity calculations (Kiriakidis, El Eid, & Glatzel 1992; Moskalik & Dziembowski 1992; Dziembowski & Pamyatnykh 1993).

S. Bowyer and R. F. Malina (eds.), Astrophysics in the Extreme Ultraviolet, 389–394.
© *1996 Kluwer Academic Publishers. Printed in the Netherlands.*

Table 1. Stellar Data

D (pc)	206
m_v	1.97
Sp. Type	B1 II–III
T_{eff} (K)	23250
$\log g$	3.5
N_H (cm^{-2})	2×10^{18}

Table 2. Optical and UV Pulsation Periods

	Optical (Shobbrook 1973)	UV (Beeckmans & Burger 1977)
P_1 (days)	$0.2512985 \pm .0000003$	0.2512985
P_2 (days)	$0.25003 \pm .00004$	0.2500225
P_3 (days)	$0.23904 \pm .00004$	

Table 3. Best-Fit Light Curve Model Parameters

	Best Model			Range (low:high)		
	Period (d)	Δmag	ΔT_b (K)	Period (d)	Δmag	ΔT_b (K)
P_1	0.251065	0.084	108	0.2509955 : 0.2511318	0.059 : 0.108	76 : 139
P_2	0.249952	0.020	26	0.249609 : 0.250082	0.011 : 0.046	14 : 59
P_3	0.239035	0.022	28	0.23904	0.022	28

3. Observations

The primary goal of this study is to determine if the pulsations propagate to the very high atmospheric layers where the Lyman continuum is formed and to accurately measure the temperature change in those layers. To do this we use two, long-duration ($\sim$ 50 ks and $\sim$ 100 ks) $EUVE$ LW spectrometer observations made in the first two cycles of $EUVE$ general observing. The elapsed observing times for the two observations are 3 and 6 days, and they are separated by 70 days.

We first analyzed the light curve composed of the source counts on the interval 504 Å to 700 Å (see Figure 1). The hypothesis of a constant source was ruled out at a very high significance level. The power spectrum revealed a highly significant signal near 6 hours. But the known periods could not be resolved; the data set is not long enough. We then fit models to the light curve. We used the same multiple sinusoidal model that Beeckmans & Burger (1977) used to fit the UV data. Two periods gave a better fit to the data than a single period, and three improved the fit still more. We used the $\Delta\chi^2$ criterion of Lampton, Bowyer, & Margon (1976) to define the 68% confidence limits of parameter space. The best fit and limits are shown in Table 3. The parameters describing the third period were held constant for the error analysis because of the insensitivity of the parameter values describing the first and second periods to the values of those describing the third.

Comparing our results to those from the UV data we see that the trend of increasing variability with decreasing wavelength for the first period continues into the EUV, but that the second period has a relatively constant amplitude (as it does within the UV). We find a temperature change consistent with that found by Beeckmans & Burger (1977) but better constrained. The folded light curve (on P_1) is shown in Figure 2. The contributions from P_2 and P_3 have been subtracted from the data. The fact that the temperature change in the EUV is consistent with that in the UV indicates that the

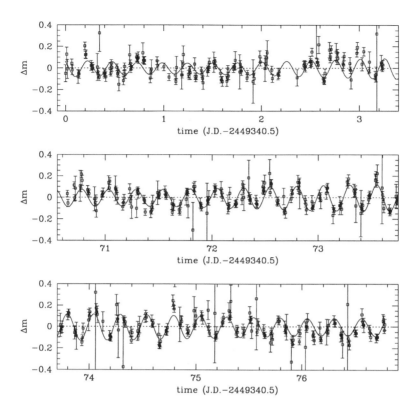

FIGURE 1. The observed light curve with the best-fit three-component sinusoidal model. Note in that the amplitude of the model changes throughout the observation. This is evidence of the beat period(s).

pulsations are not significantly damped as they traverse more than six density scale heights in the photosphere.

Even the best three-component fit is less than ideal. To investigate the cause, we subjected the residuals of the best fit to the same analysis procedure we applied to the data. We found no significant residual power in any one or several frequencies in the power spectrum of the residuals. It seems that either there is some stochastic variability superimposed on the periodic variability or there are significant systematic errors which we have not been able to account for.

In order to investigate the wavelength dependence of the variability within the *EUVE* bandpass we first created two separate light curves—one restricted to counts below 550 Å and the other to counts above 550 Å. We fit these with the same model that was fit to the original light curve. Not only is the light variation larger in the short wavelength band, but the temperature variation is marginally greater in the shorter wavelength band. Because of the $\sim \lambda^3$ dependence of the continuum opacity, the shorter wavelengths in the Lyman continuum are formed at lower layers. Since there is marginally less variability in

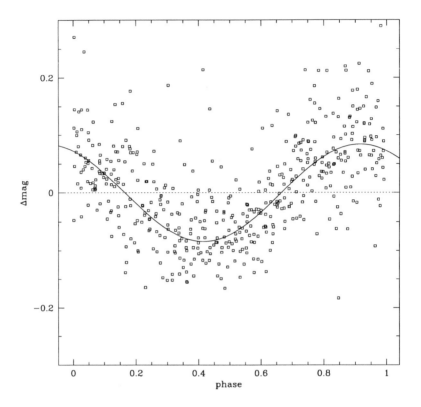

FIGURE 2. The observed light curve, with the best-fit model for periods 2 and 3 subtracted, folded on the best-fit primary period.

the radiation coming from the higher layers of the Lyman continuum formation zone, it is possible that we are seeing pulsations begin to dissipate within the upper photosphere.

Another method to examine the wavelength dependence of the variability is to weight the counts in the data by the sine of the phase of the principle period. The resulting spectrum represents the difference between the spectrum in the high and low states (see Figure 3). There is no way to account for the second and third periods in this analysis however, so the temperature derived from fitting the phase-weighted spectrum will not be as accurate as that derived from the light curve. However the high resolution may provide some information about the lines. Although the data are noisy, it appears that some of the high ion lines such as O V 630 and O IV 554 do not vary as much as the surrounding continuum. This could also be understood in terms of the dissipation of pulsations because the line centers are formed in significantly higher layers than is the surrounding continuum.

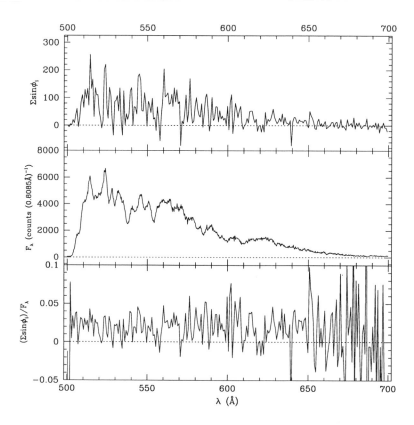

FIGURE 3. The phase-weighted spectrum is shown in the top panel along with the full spectrum in the middle panel and, in the third panel, the ratio of the two is shown.

4. Conclusions

Periodic variability, consistent with β Cephei pulsations are observed in the bright B giant β CMa. The three periods seen in the optical are detected in the *EUVE* data. The change in temperature associated with the primary period is consistent with that seen in the UV, but better constrained (108 ± 31 K). This indicates that the pulsations propagate relatively intact from $\tau_R \approx 2/3$ to $\tau_R \approx 0.01$.

We wish to thank Chris Johns for the use of his period analysis software, and Keivan Stassun and Ken Nordsieck for fruitful discussions.

REFERENCES

BEECKMANS, F., & BURGER, M. 1977, A&A, 61, 815

DZIEMBOWSKI, W., & PAMYATNYKH, A. A. 1993, MNRAS, 264, 204

KIRIAKIDIS, M., EL, EID, M. F., & GLATZEL, W. 1992, MNRAS, 225, 1p

LAMPTON, M., MARGON, B., BOWYER, S. 1976, ApJ, 208, 177

MOSKALIK, P., & DZIEMBOWSKI, W. 1992, A&A, 256, L5

SHOBBROOK, R. R. 1973, MNRAS, 101, 257

STERKEN, C., & JERZYKIEWICZ, M. 1993, β Cephei stars from a photometric point of view, Space Science Reviews, 62, 95

EUV/Soft X-ray Emission from Classical Novae

JAMES MacDONALD

Department of Physics and Astronomy, University of Delaware, Newark, DE 19716, USA

Multiwavelength observations have shown that, after optical decline, the stellar remnants of classical nova outbursts evolve at constant, near-Eddington, bolometric luminosity to high effective temperature ($> 2 \; 10^5$ K), before turning off. Here we briefly review the observations of classical novae in this phase of evolution and show that, in principle, EUV and soft X-ray observations can be used to determine the mass of the underlying white dwarf and place limits on the rate of mass loss by stellar winds and the rate of mass gain due to accretion from the stellar companion. We also describe our model atmosphere calculations of EUV/Soft X-ray emission from hot, high gravity stars and their application to the *EUVE* all-sky survey detection of Nova Cygni 1992 (V1974 Cyg).

1. Introduction

A classical nova outburst occurs when a white dwarf in a close binary system accretes sufficient hydrogen-rich material from a non-degenerate companion to trigger a thermonuclear runaway (Starrfield et al. 1974). The ensuing rapid release of nuclear energy causes the photospheric luminosity to increase to greater than the Eddington limit, L_{Edd}, and the white dwarf envelope, of mass M_{env}, to expand to red giant dimensions on a dynamical timescale, ejecting a mass, M_{ej}, of between 10^{-6} and 10^{-4} $M_{\odot}$. Early hydrodynamic simulations found that only a fraction of the envelope is ejected in the initial explosion. The remaining envelope quickly returns to hydrostatic equilibrium and enters a steady nuclear burning phase with a constant bolometric luminosity close to L_{Edd}.

Observationally the recognition that a constant bolometric luminosity phase is a common characteristic of classical novae followed upon a combination of ultraviolet and infrared observations of Nova FH Serpentis 1970 (Hyland & Neugebauer 1970; Geisel, Kleinmann & Low 1970; Gallagher & Code 1974; Gallagher & Starrfield 1976). For this moderately fast nova, the visual luminosity fell quite rapidly, with the dominance of the UV luminosity after the first week reflecting the rapidly increasing photospheric temperature. The fall in the UV after approximately 60–80 days and the simultaneous emergence of the infrared is due to absorption and re-emission by dust grains in the ejecta. During the first 100 days, the sum of UV, visual and IR luminosities are roughly constant (Gallagher & Starrfield 1978). The subsequent non-detection of FH Ser in the UV by Gallagher & Holm (1974) can be interpreted as a hardening of the spectra to wavelengths shorter than 912 Å.

Although estimates of the time to turn-off of the nuclear burning were uncertain, it was immediately clear that evolution of the stellar remnant on a purely nuclear time scale was too slow to be consistent with the observations. Turn-off requires some mechanism of mass loss (Starrfield 1979; MacDonald, Fujimoto & Truran 1985, hereafter MFT). In the next section, we briefly review the theory of the turn-off of classical novae. In particular, the stellar remnants are expected to evolve to high effective temperatures ($T_{\mathrm{eff}} > 2 \; 10^5$ K) and be copious emitters of EUV/soft X-ray photons. We argue that measurement of the maximum T_{eff}, attained at turn-off, can be used to determine the mass of the white dwarf, and that the time to reach maximum T_{eff} constrains the rate of mass loss from

S. Bowyer and R. F. Malina (eds.), Astrophysics in the Extreme Ultraviolet, 395–400.

the stellar remnant and also the rate of any accretion from the companion star. This is followed by discussion and interpretation of the EUV/soft X-ray observations of classical novae, including analysis of the *EUVE* all-sky survey detection of Nova Cyg 1992 (V1974 Cyg).

2. The Evolution to Turn-Off

In discussing the evolution of the stellar remnant, it is useful to divide the nova outburst into three phases: A) Envelope expansion in response to the thermonuclear runaway, B) Common envelope evolution and C) Steady nuclear burning at constant bolometric luminosity, with possible wind mass loss or accretion from an irradiated companion.

Of crucial importance to the life-time of phase B is the amount of envelope material remaining on the stellar remnant after phase A. Recent hydrodynamic calculations of phase A differ significantly in the fraction of envelope material that is ejected. For example, Starrfield et al. (1992) in simulations of Nova Her 1991 find M_{ej}/M_{env} in the range 0.09–0.67 whereas Prialnik & Shara (1995) find M_{ej}/M_{env} between 0.92 and 1.00. We suspect that this significant disagreement is due in part to Prialnik & Shara (1995) completely suppressing convective energy transport in hydrodynamic phases of evolution and to their method of removal of mass zones that are escaping from the star (Prialnik & Kovetz 1995). Both these effects work in the direction of increasing the amount of ejected material.

If, as the calculations of Starrfield and collaborators suggest, a significant fraction of the envelope remains after phase A, then the stellar remnant is larger than the orbital separation and phase B is entered. As shown by MacDonald (1980) and MFT, dynamical friction quickly ejects almost all of the material lying outside the orbit at about the escape speed in the vicinity of the orbit. For this phase to occur, M_A, the envelope mass at the end of phase A, must satisfy (MFT)

$$\frac{M_A}{M_\odot} \gtrsim 2 \ 10^{-5} \left(\frac{M_{WD}}{M_\odot}\right)^{-4.5} \left(\frac{S}{10^{11}cm}\right)^{0.3}$$

where M_{WD} is the white dwarf mass and S is the orbital separation. For a typical S of 10^{11} cm, this is satisfied for all the simulations of Starrfield et al. (1992). Indirect observational support for the existence of phase B in classical novae comes from comparison with very slow novae such as RR Tel and RT Ser in which the companion is a giant star (Feast & Glass 1974) and the orbital separation too large for a significant common envelope phase. The very slow novae have long turn-off times and more importantly very low expansion velocities of order < 100 km s^{-1} whereas classical novae have expansion velocities of order the orbital velocity of the companion star or greater (MacDonald 1980).

As shown by MFT, the life times of phases A and B are much shorter than that of phase C and hence it is the lifetime of phase C that determines t_{rem}, the time for turnoff of classical nova remnant. In the absence of wind mass loss, the only process to consume envelope material is nuclear burning. On the basis of polytropic models for the envelope, MFT then estimate, for a composition representative of novae ejecta, that

$$t_{rem} \simeq 20 \left(\frac{M_{WD}}{M_\odot}\right)^{-6.3} yr$$

which has a very strong dependence on white dwarf mass. However it is unrealistic to assume that the hot remnant with a luminosity close to the Eddington limit will not have a wind, whether optically thick (Bath 1978; Kato 1983; Kato & Hachisu 1994) or line-driven (MacDonald 1984; MFT). Furthermore for systems with small orbital periods, irradiation of the companion by the hot remnant may induce accretion onto the white dwarf prolonging phase C (Ögelman et al. 1993). If an independent estimate of M_{WD} can be obtained then comparison of t_{rem} with the nuclear burning time scale in principle gives the rate of mass loss or mass gain, which can be compared to the theoretical prediction for e.g., optically winds. Determinations of M_{WD} by standard spectroscopic techniques have large errors, primarily because the radial velocity of the white dwarf cannot be directly measured and must instead be inferred from accretion disk emission lines (see Warner 1990 for a review) Attempts to obtain M_{WD} from a mass-luminosity relation similar to that for AGB stars are hampered by uncertainties in the distance. However, MFT argue that T_{max}, the maximum effective temperature reached by the remnant before turning-off mainly depends on M_{WD} and their polytropic model gives a dependence

$$t_{max} \approx 6.6 \ 10^5 \left(\frac{M_{WD}}{M_\odot} \right)^{1.6} K$$

Hence determination of T_{max} from fitting stellar atmosphere models to the results of a campaign of monitoring classical novae at soft X-ray/EUV energies has the potential of determining M_{WD}. To achieve accurate estimates, it is expected that the simple polytropic model of MFT will need to be replaced by detailed stellar evolution calculations of phase C and beyond.

3. The EUV/Soft X-ray Observations

The detection of soft X-rays from three classical novae, Nova GQ Muscae 1983, Nova PW Vulpeculae 1984 and Nova QU Vulpeculae 1984, at times ranging from 100 to 1000 days after optical maximum with the *EXOSAT* satellite (Ögelman, Beuermann, & Krautter 1984; Ögelman, Krautter & Beuermann 1987, hereafter OKB) has provided observational verification that the constant bolometric luminosity phase does extend to high effective temperatures ($> 2 \ 10^5$ K), as originally suggested by MFT.

Since the majority of the *EXOSAT* observations were made with only the thin Lexan filter on the low energy telescope, there is, in general, insufficient data in the X-ray measurements alone to discriminate between emission from a hot photosphere and emission from shocked circum-stellar gas. OKB also find that the single observation of Nova Vulpeculae 1984 # 2 with the Boron filter, 307 days after optical maximum, is not significant enough to exclude or support a thermal bremsstrahlung spectrum. However, they favor emission from a hot photosphere on the basis of consideration of plasma cooling times and the absence of coronal [Fe XIV] λ 5303 lines that should be emitted from shocked gas (Kurtz et al. 1972). Krautter & Williams (1989) rule out shocked circumstellar gas by considering the degree and temporal evolution of the ionization of the ejecta of GQ Muscae and argue that it is consistent with photo-ionization from a hot radiation source.

EXOSAT observations made two months after the January 1985 outburst of the recurrent nova RS Ophiuchi (Mason et al. 1987) have been interpreted as due to X-rays from a shock front moving into a circumstellar envelope formed by the pre-outburst wind of the red giant companion (O'Brien, Kahn & Bode 1987). Even so, the X-ray emission 200 days after outburst is consistent with an origin on the surface of a white dwarf (Mason et al. 1987).

The greater energy resolution of the *ROSAT* PSPC permits a clearer distinction between a hot photosphere and shocked gas. The X-ray emission from GQ Mus in February 1992, nine years after optical maximum, has a very soft spectrum and is consistent with a photospheric origin, at near L_{Edd} (Ögelman et al. 1993). *ROSAT* spectra of Nova Hercules 1991 (Lloyd et al. 1992), and Nova Cygni 1992 (Krautter, Ögelman & Starrfield 1992), 5 and 60 days after optical maximum respectively are hard and the implied luminosities are below 10^{-2} L$_{Edd}$, perhaps indicating an origin in shocked gas.

GQ Mus was observed twice again with the *ROSAT* PSPC in 1993 (Shanley et al. 1995). The count rates fell sharply from February 1992 to January 1993 and GQ Mus was not detected in August/September 1993, indicating that GQ Mus had turned-off 10 years after optical maximum. Nova Cygni 1992 (V1974 Cyg) was monitored with *ROSAT* throughout its active phase in X-rays and was found to turn off 18 months after optical maximum (Starrfield et al. 1995).

In anticipation of these *ROSAT* detections of classical novae, MacDonald & Vennes (1991, hereafter MV) performed model atmosphere calculations of the X-ray emission from the photospheres of hot, high gravity stars, using representative values for the white dwarf mass and luminosity of $M_{WD} = 1.2$ $M_\odot$ and $L_{WD} = 4.2$ 10^4 L$_\odot$. Stellar atmospheres in local thermodynamic equilibrium were calculated for a range of effective temperatures from 10^5 to 10^6 K (the corresponding range in log g is from 4.85 to 8.85) and two compositions representative of the ejecta of CNO novae and ONe novae. The code used was developed for LTE modeling of the atmospheres of hot white dwarfs, and is based on that described by Mihalas, Auer, & Heasley (1975). It uses a complete linearization method coupled to a Feautrier elimination scheme. This powerful method allows successful convergence of model atmospheres very close to the Eddington limit, as are needed for classical nova stellar remnants. In addition to free-free and electron scattering opacity, all the important photoionization edges of H, He and the ions of C, N, O, and Ne that are present for T_e between 10^5 and 10^6 K (i.e., C IV–C VI, N IV–N VII, O IV–O VIII, and Ne IV–Ne X) were included.

An important result of this study is that black body fits to the X-ray spectra can grossly underestimate, by as much as a factor of 2, the true effective temperature. Indeed for GQ Mus, Ögelman et al. (1993) found the best black body fit temperature of 2.5 10^5 K gave a bolometric flux that was about a factor 100 too high to be consistent with the Eddington limit. However, within the 2 and 3 σ error limits, they could find a black body fit with higher temperature ($\simeq 3.5$ 10^5 K) and lower bolometric flux. We estimate that use of MV model atmospheres (see Fig. 3 of MV) would give $T_{\rm eff} = 4$ 10^5 K. Furthermore, Starrfield et al. (1995) find that blackbody fits to the V1974 Cyg data give super-Eddington fluxes, and that this problem is cured by use of the MV fluxes.

V1974 Cyg was also detected 280 days after visual maximum by *EUVE* in the all-sky survey. The count rates were 0.108 ± 0.010 s^{-1} and 0.091 ± 0.010 s^{-1} in the Lexan/Boron and Al/C/Ti filters respectively (Bowyer et al. 1994). A preliminary analysis of this data was given by MacDonald & Vennes (1994). Shown in Figure 1 are contours of $T_{\rm eff}$ (labelled in units of 10^3 K) and log column density (in cm^{-2}) plotted as functions of the predicted MV count rates for an assumed distance of 2 kpc. Also shown are the data for V1974 Cyg taken at face value (diamonds, full line) and corrected for a possible Al/C/Ti X-ray leak (diamonds, dashed line). $T_{\rm eff}$ was found to be between 3 10^5 and 3.5 10^5 K and the column density about 1.8 10^{21} cm^{-2}. Stringfellow & Bowyer (1994) have performed a similar analysis using better calibrated effective area curves for the two filters. They find $T_{\rm eff} = 4.0 \pm 0.5$ 10^5 K and $N_H \sim 3 10^{21}$ cm^{-2}. We note that the temperatures found for these two objects and also for QU Vul from analysis of the *EXOSAT* data by MV are remarkably similar.

Using the above values for T_{max}, our estimate for M_{WD} is 0.75 $M_\odot$ for both stars. Furthermore, to get the short turnoff times requires mass loss at rates of 8 10^{-6} $M_\odot$ yr^{-1} for GQ Mus and 5 10^{-5} $M_\odot$ yr^{-1} for V1974 Cyg. The estimated mass is much lower than thermonuclear runaway theory requires for classical nova outbursts (MacDonald 1983). A possible resolution of this discrepancy is that there is an optically thick outflow

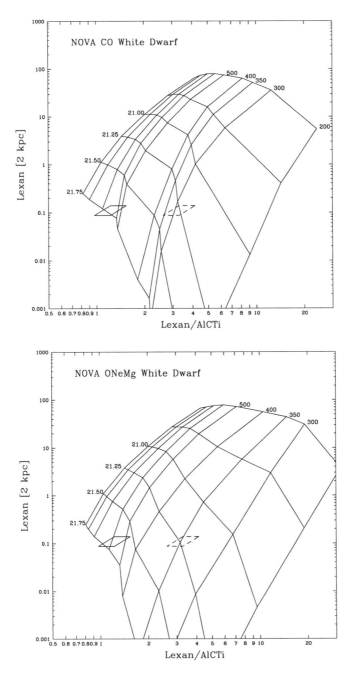

FIGURE 1. Analysis of the V1974 Cyg *EUVE* date.

from the surface which results in a more extended envelope and lower T_{eff} than for a hydrostatic envelope. Alternatively, the simple polytropic model of MFT may be too crude to accurately predict the effective temperature.

REFERENCES

Bath, G. T. 1978, MNRAS, 182, 35

Feast, M. W. & Glass, I. S. 1974, MNRAS, 167, 81

Gallagher, J. S. & Code, A. D. 1974, ApJ, 189, 303

Gallagher, J. S. & Holm, A. V. 1974, ApJ, 189, L123

Gallagher, J. S. & Starrfield, S. 1976, MNRAS, 176, 3

Gallagher, J. S. & Starrfiled, S. 1978, ARAA, 16, 171

Geisel, S. L., Kleinmann, D. E. & Low, F. J. 1970, ApJ, 161, L101

Hyland, A. R. & Neugebauer, G. 1970, ApJ, 160, L177

Kato, M. 1983, PASJ, 35, 507

Kato, M. & Hachisu, I. 1994, ApJ, 437, 802

Krautter, J., Gelman, H. & Starrfield, S. 1992, IAUC, 5550

Krautter, J. & Williams, R. E. 1989, ApJ, 341, 968

Kurtz, D. W., Vanden, Bout, P. A., & Angel, J. R. P. 1972, ApJ, 178, 701

Lloyd, H. M., et al. 1992, Nature, 356, 222

MacDonald, J. 1980, MNRAS, 191, 933

MacDonald, J. 1983, ApJ, 267, 732

MacDonald, J. 1984, ApJ, 283, 241

MacDonald, J., Fujimoto, M. Y., & Truran, J. W. 1985, ApJ, 294, 263

MacDonald, J. & Vennes, S. 1991, ApJ, 373, L51

MacDonald, J. & Vennes, S. 1994, AAS meeting Minneapolis, MN, 1994 May 29–June 2

Mason, K. O., Córdova, F. A., Bode, M. F. & Barr, P. 1987, in RS Ophiuchi, 1985, and the Recurrent Nova Phenomenon, ed. M.F. Bode, Utrecht: VNU Science Press, 167

Mihalas, D., Auer, L. H., & Heasley, J. N. 1975, NCAR Tech. Note, TN/STR-104

Mukai, K., et al. 1985, Sp Sci Rev, 40, 151

O'Brien, T., Kahn, F. D. & Bode, M. F. 1987, in RS Ophiuchi, 1985, and the Recurrent Nova Phenomenon, ed. M.F. Bode, Utrecht:VNU Science Press, 177

Ögelman, H., Beuermann, K. & Krautter, J. 1984, ApJ, 287, L31

Ögelman, H., Krautter, J. & Beuermann, K. 1987, A&A, 177, 110

Ögelman, H., Orio, M., Krautter, J. & Starrfield, S. 1993, Nature, 361, 331

Prialnik, D. & Shara, M. 1995, preprint

Prialnik, D. & Kovetz, A. 1995, ApJ, 445, 789

Shanley, L., Ögelman, H., Gallagher, J. S., Orio, M., & Krautter, J. 1995, ApJ, 438, L95

Starrfield, S. 1979, in White Dwarfs & Variable Degenerate Stars, ed. H.M. Van Horn & V. Weidemann, 274

Starrfield, S., Truran, J. W., Sparks, W. M. & Kutter, G. S. 1974, ApJ, 176, 169

Starrfield, S., et al. 1992, ApJ, L71

Starrfield, S., et al. 1995, these proceedings

Stringfellow, S. G. & Bowyer, S. 1994, to appear in Interacting Binary Stars, ed. A.W. Shafter

Warner, B. 1990, in IAU Colloq. 122, Physics of Classical Novae, ed. A. Cassatella, Berlin: Springer-Verlag, 24

Modelling the Soft X-Ray and EUV Emission in Classical Novae: *EUVE* and *ROSAT* Observations of V1974 Cygni

GUY S. STRINGFELLOW[1,2] AND STUART BOWYER[1]

[1] Center for EUV Astrophysics, University of California, 2150 Kittredge Street, Berkeley, CA 94720–5030, USA

[2] Present Address: Department of Astronomy and Astrophysics, The Pennsylvania State University, 525 Davey Lab, University Park, PA 16802–6305, USA

We have conducted an extensive analysis of the observability of Classical Novae with the *EUVE* Lex/B and Al/Ti/C detectors. Predicted count rates have been computed using optically thin, isothermal plasma models for solar and metal-rich compositions, and hot ONeMg white dwarf model atmospheres. We find *EUVE* to be quite sensitive to both the EUV and soft X-ray emission emitted by the underlying hot white dwarf during novae outbursts, except for the coolest temperatures with very high intervening hydrogen column density. These results are used to interpret the emission detected during the *EUVE* all-sky survey of Nova Cygni 1992 ($\equiv$ V1974 Cyg), 279–290 days after visual maximum. We find the best fit to the observed emission from V1974 Cyg arises from a hot ONeMg white dwarf with surface temperature $\sim 4 \times 10^5$ K and a mass of $\sim 1.2~M_\odot$, and derive an interstellar hydrogen column density of $\sim 3 \times 10^{21}$ cm^{-2}. Virtually all this emission arises from supersoft X-rays rather than the EUV. We also report the detection of V1974 Cyg with the *EUVE* Deep Survey detector at 549 days after visual maximum. This observation is compatible with the above properties, indicating that the mechanism responsible for the soft X-ray emission, connected with the underlying white dwarf, had not yet entirely turned off. We also present analysis of a *ROSAT* PSPC observation which is contemporaneous with the *EUVE* survey observations; this independently confirms the high column density we derived from the *EUVE* survey observation. Light curves for the *EUVE* and *ROSAT* observations are presented. Statistical tests for variability show that *all* of these observations are indeed highly variable over various time scales. The *EUVE* survey data shows one day variations, the *EUVE* DS data show $\sim$30 minute fluctuations, while the *ROSAT* data vary rapidly on time scales of seconds. The *EUVE* data shows no periodic variability on any time scale. The implications of the rapid variability are briefly discussed.

1. Introduction

Nova Cygni 1992 (V1974 Cyg) was the optically brightest nova in nearly two decades, having been discovered on 19.1 Feb 1992 (Collins 1992), three days before reaching maximum visual brightness at $m_V \approx 4.4$ on 22.1 Feb. It was subsequently classified as an ONeMg nova (e.g., Hayward et al. 1992). We previously reported the detection of V1974 Cyg with the *Extreme Ultraviolet Explorer* (*EUVE*) during the all-sky survey (Stringfellow & Bowyer 1993a,b), after the nova had entered its nebular stage at about 285 days after maximum visual brightness ($V_{\max}$). A wide variety of distance estimates have been employed, yielding a range of 1.4–3.2 kpc (Chocol et al. 1993; Stringfellow & Bowyer 1994; Paresce et al. 1995; and references therein), which suggests the interstellar hydrogen column density, $N_{\rm H}$, should be $\geq 10^{21}$ cm^{-2}. In this paper we show that the *EUVE* detection arises predominantly from supersoft X-rays ($\lambda < 44$ Å), in a regime where the *EUVE* effective area curves have less sensitivity, but where they are nonetheless still capable of detecting very bright sources. The emission detected during the all-sky sur-

S. Bowyer and R. F. Malina (eds.), Astrophysics in the Extreme Ultraviolet, 401–405.
© *1996 Kluwer Academic Publishers. Printed in the Netherlands.*

vey is best described by radiation emitted by a white dwarf with surface temperature ~ 400000 K, and $N_{\mathrm{H}} \sim 3 \times 10^{21}$. This conclusion is in general agreement with the analysis of *ROSAT* observations during the *EUVE* survey observation, an example of which is discussed below. *EUVE* and *ROSAT* observations during Nov–Dec 1992 are analyzed for variability, along with a subsequent *EUVE* pointed observation made with the Deep Survey detector 549 days after V_{max}. We find all these lightcurves to be highly variable.

2. Modelling the *EUVE* Survey Observations

Stringfellow & Bowyer (1993ab, 1994) reported detection of V1974 Cyg by the *EUVE* during the course of the all-sky survey, scanned between 27 Nov and 8 Dec 1992. The mean count rates in the Lex/B and Al/Ti/C filters are 0.112 ± 0.010 counts s^{-1} and 0.102 ± 0.014 counts s^{-1}, corresponding to 18 σ and 9 σ detections, respectively. The *EUVE* scanner effective areas are sensitive down into the supersoft X-ray domain and clearly the flux detectable in both supersoft X-rays and EUV depends critically on the precise value of N_{H}.

Several potential emission mechanisms could be operating in V1974 Cyg simultaneously. The clumpy nova ejecta model discussed by Williams (1992) and Saizar & Ferland (1994) is illustrative of the physically distinct components and their interactions, although the properties of each component could vary widely in each nova, and indeed during the evolution of any given nova. A general composite model will contain the hot ($> 10^5$ K) underlying white dwarf, while the ejecta contains warm ($10^4 - 10^5$ K) clumps and a hot (10^6 K), interclump plasma filling a large volume. Shocks arising through interactions with various components may also exist, although perhaps only during specific evolutionary stages during the outburst. Although there have been no reported observations indicating the presence of dust in the ejecta of V1974 Cyg thus far, another fast ONeMg nova, Nova Her 1991 ($\equiv$ V838 Her), has been shown to form dust early, within a few days after outburst (Harrison & Stringfellow 1994). Obviously, each nova is unique, even within a specific class, and their ejecta environment are complex and variable.

We have computed an extensive grid of isothermal, optically thin plasma models for both solar and metal-rich compositions, the latter representative of the ejecta of ONeMg novae. Generally speaking, these models alone are unable to account for the observed *EUVE* emission as the dominant component. Hence, we forgo discussing details of these models here; early results were presented by Stringfellow & Bowyer (1994), and full details of these and the white dwarf modelling will appear elsewhere.

Figure 1 shows our derived *EUVE* count rates for a 1.2 $M_\odot$ ONeMg white dwarf with surface temperatures ranging from $10^5 - 10^6$ K (MacDonald & Vennes 1991). The models span the wavelength range from $10 - 300$ Å. The curves correspond to various values of N_{H} at a distance of 1 kpc, but these can be scaled simply to any distance. Note the *EUVE* count rates are generally quite high, implying that the hottest ONeMg novae with massive white dwarfs are easily detectable by *EUVE* out to large distances and high N_{H}. Observability of the coolest ONeMg novae ($\sim$ few $\times 10^5$ K) are most strongly dependent upon the intervening N_{H}. The two horizontal lines in Figure 1 are the fitted count rates of the models at 1.4 kpc (lower line) and 3.0 kpc (upper line), which bound the range of permitted models for the range in derived distances to V1974 Cyg.

The observed *EUVE* count rates imply a white dwarf temperature between $3-5 \times 10^5$ K, whereas the modelled ratio of the count rates in the two filters demands that $N_{\mathrm{H}} > 10^{21}$ cm^{-2}; at $N_{\mathrm{H}} = 3 \times 10^{21}$ cm^{-2}, the ratio of the Al/Ti/C to Lex/B count rates become insensive (i.e., flat) to the WD effective temperature, and lower than the observed ratio by about a factor of 4. This simply reflects the ratio of the effective area curves of the

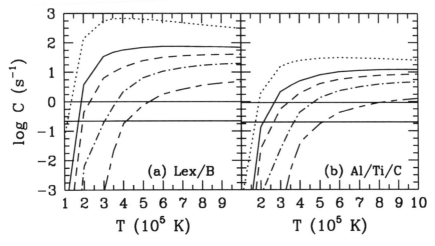

FIGURE 1. Computed *EUVE* count rates for a 1.2 $M_\odot$ ONeMg white dwarf at 1 kpc. The various curves are the computed count rates for model white dwarfs with corresponding surface temperatures attenuated by the ISM for values of log N_H of 21.5, 21.0, 20.7, 20.5, and 20.0 cm^{-2} from the bottom up, respectively. The *EUVE* filters employed in the computations of the model count rates are noted in each panel. The two horizontal lines correspond to the *EUVE* observed count rates for V1974 Cyg in each filter if it lies at a distance of 1.4 kpc (*lower*) or 3 kpc (*upper*).

two filters below 44 Å (the last wavelength at which ground based calibrations exist for both filters). The Lex/B filter effective area curve has been adjusted in this regime to match the *EUVE* survey count rate observed for Sco X-1, using its well known X-ray modelled parameters, but no such secondary calibration has yet been performed for the Al/Ti/C filter. Therefore, we attribute the factor of 4 discrepancy between the modelled and observed count rate in the Al/Ti/C filter to this 'secondary calibration' problem. Inspection of the effective area curves for the two filters indicates that the Al/Ti/C count rate should always be at least a factor of ~10 lower than that in the Lex/B filter between 44 − 160 Å. Al/Ti/C-to-Lex/B count rate ratios near unity are permitted only for sources with EUV-dominated flux in the narrow range 160 − 180 Å, or according to our analysis presented herein, when the radiation arises below 44 Å. Thus, we conclude from the analysis of the *EUVE* survey data that essentially all of the emission observed from V1974 Cyg arises from supersoft X-rays. Note that the observed count rate ratio for Sco X-1 is in accord with the above arguments. *Analysis of EUVE sources with observed Al/Ti/C-to-Lex/B count rate ratios greater than 0.1 could yield a population of (previously unidentified?) supersoft X-ray candidates.*

3. Rapid Soft X-ray Variability in V1974 Cyg

The *EUVE* survey light curve for the Lex/B filter is shown in Figure 2a. The count rates correspond to actual scanned times over 12^h time intervals. We also report for the first time the *EUVE* detection of V1974 Cyg with the Deep Survey detector (DS) during a ~38 ks pointed observation in August 1993, 549 days after V_{max}. This latter light curve is displayed in Figure 2b. The DS data have been corrected for the loss of sensitivity resulting from lying on the detector "dead spot." We have performed a chi-

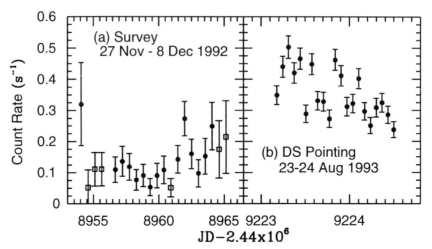

FIGURE 2. Lightcurves of V1974 Cyg observed by *EUVE* during (a) the Lex/B filter during the all-sky survey, where the open squares denote $< 3\ \sigma$ significance, and (b) a pointed observation with the Deep Survey detector. Each survey bin represents the actual scanned time during a 12 hour interval (typically $\sim 100 \pm 50$ s), whereas the DS data are the continuous pointed time per *EUVE* orbit (roughly 30.8 min).

square analysis on both data sets and find each to be highly variable, with confidence levels of 99% for the survey data and $\gg$99.9% for the DS data.

ROSAT has followed the temporal evolution of V1974 Cyg at irregular intervals with the PSPC, starting at 60 days after V_{max}. One such observation occurred on 6 Dec 1992, 288 days after V_{max}, which overlaps with the observational period during the *EUVE* survey data; it corresponds to a low-state in the survey light curve (fifth point from the left in Fig. 2a). This lightcurve is shown in Figure 3, where the data have been barycenter corrected, and binned into 10 s intervals. While the mean count rate over this continuous observation is $31.4\ \mathrm{s}^{-1}$, visual inspection clearly shows rapid fluctuations by as much as a factor of two in less than 20 s. The data span a range of $18 - 55$ counts s^{-1}. We have performed a Kolmogorov-Smirnov (KS) and Cramer-von Mises statistical tests, and find the data to be *inconsistent* with a constant source intensity at an extremely high confidence level, $\gg$99.9%. Other *ROSAT* PSPC data sets we have analyzed also exhibit such behavior as that in Figure 3. Discovery of such rapid variability, on time scales of seconds, implies (a) a compact region for the source of the supersoft X-rays, and (b) undoubtedly accretion instabilities are responsible. Whether this corresponds to an accretion stream/hotspot or related instead to reformation of an accretion disk, is presently unclear and requires further analysis which is presently underway. It may even be possible to account for the rapid X-ray variability by invoking clumpy accretion of the remaining ejecta which did not reach, or maintain, escape velocity; this material could be "raining" back onto the white dwarf surface. The concept of stable (i.e., constant) thermonuclear burning on the white dwarf surface requires review in light of these results.

We wish to thank the CEA science and DASS teams for useful discussions and support, Martin Sirk for assistance with the *EUVE* analysis software, and Stephane Vennes for providing the white dwarf atmosphere models. This research has been supported in part by NASA contract NAS5-30180.

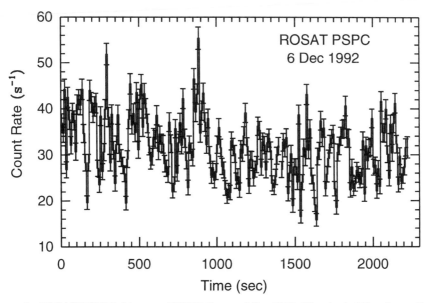

FIGURE 3. *ROSAT* PSPC lightcurve of V1974 Cyg on 6 Dec 1992. The start of the observation corresponds to JD = 2448962.9360987. The data have been binned into 10 s intervals, and are clearly highly variable.

REFERENCES

COLLINS, P. 1992, IAU Circular No. 5454

CHOCOL, D., HRIC, L., URBAN, Z., KOMZIK, R., GRYGAR, J., & PAPOUSEK, J. 1993, A&A, 277, 103

HARRISON, T. E. & STRINGFELLOW, G. S. 1994, ApJ, 437, 827

HAYWARD, T. L., GEHRTZ, R. D., MILES, J. W., & HOUCK, J. R. 1992, ApJ, 401, L101

MACDONALD, J. & VENNES, S. 1991, ApJ, 373, L51

PARESCE, F., LIVIO, M., HACK, W., & KORISTA, K. 1995, A&A, in press

SAIZAR, P. & FERLAND, G. J. 1994, ApJ, 425, 755

STRINGFELLOW, G. S. & BOWYER, S. 1993a, IAU Circular No. 5803

STRINGFELLOW, G. S. & BOWYER, S. 1993b, BAAS, 25, 1250

STRINGFELLOW, G. S. & BOWYER, S. 1994, In Interacting Binary Stars, ed. A. W. Shafter, ASP Conference Series, 37, 315

WILLIAMS, R. E. 1991, ApJ, 392, 99

Novae and Helium Novae As Bright EUV Sources

MARIKO KATO

Department of Astronomy, Keio University, Hiyoshi, Kouhoku-ku,
Yokohama, 223, Japan

I present theoretical light curves of novae and helium novae in EUV and visible bands derived from optically thick wind theory. The EUV light curves are very useful in determining the white dwarf mass and the distance to the star. Helium novae are bright EUV sources but much less luminous in the optical. Identification of helium novae will be observational evidence for the growth of the white dwarf, which will eventually become a Type Ia supernova. A semi-detached system may be a new progenitor model of Type Ia supernovae, which is a very bright EUV source with a faint optical counterpart.

1. Classical Nova

Novae are a thermonuclear runaway event on a white dwarf of a close binary system. A hydrogen shell flash on the white dwarf causes an outburst in which the star quickly brightens up and the hydrogen-rich envelope greatly extends. Most of the envelope is eventually blown off. In the H-R diagram, the star moves from the lower left part (accreting white dwarf) to the upper right (red giant region), undergoing strong mass loss. After the optical luminosity peak, the star moves blueward on the horizontal track with constant bolometric luminosity.

The evolutional track of the decay phase of classical novae is shown in Figure 1 (for more details see Kato & Hachisu 1994). The timescale of such a decay phase of classical novae is summarized in Table 1. The optically thick wind mass loss (the dashed part) lasts until the effective temperature exceeds $2-3 \times 10^5$ K. After that the star continuously moves leftward until it reaches the turning point where the hydrogen burning extinguishes around the point of maximum effective temperature in Figure 1. As the EUV and soft X-ray luminosity quickly drop around this point, the total duration in Table 1 gives the turn off time of EUV and soft X-ray after the optical maximum. Then the star returns to the pre-outburst magnitude.

Soft X-ray emission from GQ Mus has been observed to drop in 1993. Its turn off time of 9 yrs indicates the white dwarf mass of 0.68 $M_\odot$ from this table. This value is consistent with the value 0.5–0.6 $M_\odot$ obtained by the light curve analysis (Kato 1995b).

Theoretical light curves are calculated using the optically thick wind theory (Kato 1983; Kato & Hachisu 1994) which is so far the only method in reproducing nova light curves. Figure 2 shows the light curve fitting for classical nova V1668 Cyg (Nova Cyg 1978). Theoretical light curves of massive white dwarfs show a rapid evolution because of the small envelope mass. Both the optical and UV data show a good agreement with the model of a 1.0 $M_\odot$ white dwarf. The distance to the star can also be estimated from the comparison between observed apparent magnitude and the theoretical absolute magnitude and turns out to be 2.9 kpc (UV) and 3 kpc (optical). The theoretical expansion velocity at the photosphere is also consistent with *IUE* data. Details are published by Kato (1994) and Kato & Hachisu (1994).

The thick, solid curve starts at $t = 780$ days and is the EUV light curve (100–912 Å)

407

S. Bowyer and R. F. Malina (eds.), Astrophysics in the Extreme Ultraviolet, 407–411.

TABLE 1. X-ray turn-off time of classical novae

M_{WD} $(M_\odot)$	t (wind) (yr)	t (static) (yr)	t (total) (yr)
1.33	0.11	0.038	0.148
1.2	0.26	0.15	0.41
1.1	0.39	0.34	0.73
1.0	0.61	0.74	1.35
0.9	0.87	1.4	2.27
0.8	1.4	2.7	4.1
0.7	2.4	5.3	7.7
0.6	4.0	8.9	12.9
0.5	7.4	16.8	24.2

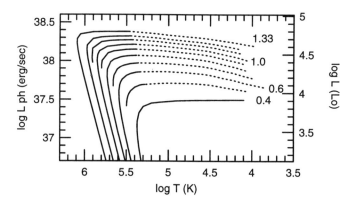

FIGURE 1. The evolutionary tracks of the decay phase of classical novae are plotted in the theoretical H-R diagram. Each curve corresponds to different white dwarf mass, i.e., 1.33 $M_\odot$, 1.2 $M_\odot$, 1.1$M_\odot$, 1.0 $M_\odot$, 0.9 $M_\odot$, 0.8 $M_\odot$, 0.7 $M_\odot$, 0.6 $M_\odot$, 0.5 $M_\odot$, and 0.4 $M_\odot$. Dashed part of the curve denotes the wind phase.

expected for 1.0 $M_\odot$ white dwarf model. This curve shows that the classical nova is a bright EUV source and becomes bright after the optical magnitude begins to decrease. This is because the effective temperature increases with time, and the wavelength at the spectral maximum shifts from the optical to the UV and EUV region. As the EUV curve has a different shape from the others, observational data will be useful in determining the white dwarf mass and the distance to the star with high accuracy.

2. Recurrent Novae

A recurrent nova is also a thermonuclear runaway event on a massive white dwarf. It brightens within a few days and fades in the optical with a short decline time of 10 days to several months. The outburst repeats every few decades. No heavy element enhancement has been reported.

The example of light curve fitting of recurrent novae, U Sco, is shown in Figure 3. The best fit curves for U Sco are the models of mass $M_{WD} \sim 1.38$ $M_\odot$ with heavy element abundance $0.01 < Z < 0.03$ as shown in this figure. Very similar results are also obtained

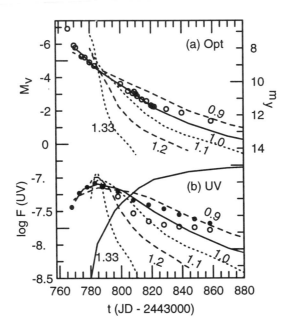

FIGURE 2. Theoretical and observational light curves for the decay phase of classical novae. Theoretical ones are denoted by lines. White dwarf mass is attached to each curve. (a) Optical light curves. The left ordinate shows the absolute visual magnitude for the theoretical curves. Observational data of Nova Cygni 1978 (Gallagher et al. 1980) are shown by open circles. Its apparent y-magnitude is written on the right ordinate. (b) UV light curves. Open circles denote the UV flux (1140–3290 Å) and filled circles do the summation of the UV and IR fluxes (> 12000 Å) in the unit of erg cm^{-2} s^{-1} (Stickland et al. 1981). Theoretical UV flux, $F = L_{\mathrm{UV}}/4\pi D^2$, is also shown by lines, where D is the distance to the star and assumed to be 2.88 kpc. The solid curve rising at $t = 780$ days is the EUV light curve (100–912 Å) for 1.0 $M_\odot$ model.

in V394 CrA and T CrB, and $1.35 \leq M_{\mathrm{WD}} \leq 1.38$ $M_\odot$ with $0.004 < Z < 0.02$ for V745 Sco (Kato 1995b). These white dwarf masses are near the Chandrasekhar limit and very close to the critical mass, i.e., 1.38 $M_\odot$, for accreting hot white dwarfs to become a Type Ia supernova or a neutron star triggered by accretion-induced collapse.

As recurrent novae decline almost linearly, fitting of the theoretical light curves have an ambiguity. Multiwavelength observations, such as EUV or FUV, are very helpful for accurate fitting of light curves to estimate the parameters such as the distance to the star.

3. He Nova—As an Evidence of Mass-Growing WD

When a part of the accreted matter remains after the hydrogen shell flash, the white dwarf develops a helium layer under the hydrogen burning zone. This helium layer will grow in each hydrogen shell flash. When the mass of the helium layer reaches a critical value, an unstable helium burning triggers a nova-like phenomenon.

Figure 4 shows light curves of helium flashes. This light curve is similar to that of a typical nova, but the development is very slow. The optical magnitude drops in the

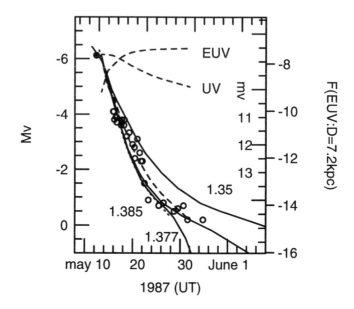

FIGURE 3. Light curve fittings of recurrent nova U Sco in 1987 outburst. Observational optical data are taken from Sekiguchi et al. (1988). These data are well fitted by theoretical models denoted by thick curves, i.e., 1.377 $M_\odot$ white dwarf with chemical comosition (X,Z) = (0.1, 0.01), (0.7, 0.02), (0.1, 0.02) and 1.385 $M_\odot$ with (0.1, 0.01). Dashed curves in the upper part of this figure denote theoretical light curves in EUV ($100 - 912$ Å) and UV ($1200 - 2000$ Å) bands of the best fitted model 1.377 $M_\odot$ with (0.1, 0.01).

later phase, but magnitudes of short wavelength are still large. In this phase, the helium nova will be observed as a very bright EUV source with a faint optical counterpart. The chemical composition of the star is highly hydrogen-deficient. Identification of such an object is very important, because it would connect recurrent novae to Type Ia supernovae (Kato 1995a), i.e., it will be the first observational evidence of mass-increasing white dwarfs which will eventually become Type Ia supernovae.

4. A New Model of Type Ia Supernovae

Hachisu & Kato (1995) proposed a new scenario of Type Ia supernovae. They considered a binary initially consisting of a white dwarf and a main-sequence star. When the secondary (main-sequence star) evolves toward the red-giant and fills its Roche lobe, mass transfer from the secondary begins. The mass transfer rate is high enough to cause a stable H burning at the WD surface. The optically-thick wind occurs at the surface of the WD envelope which takes the angular momentum, as well as mass, away from the binary. They followed the binary evolution for various sets of the binary parameters and found that binaries of initial mass ratio $q < 1.15$ experience stable mass transfer from the companion to the white dwarf, which can become a Type Ia supernova. This scenario predicts a progenitor of binaries consisting of an accreting hot white dwarf with an optically thick wind and a dwarf/semi-dwarf companion. Therefore, this progenitor

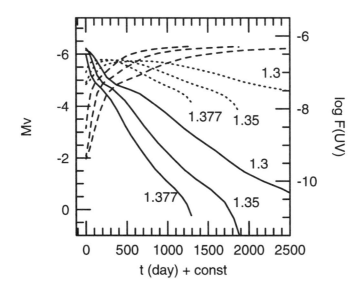

FIGURE 4. Optical, UV, and EUV light curves for He nova: dashed line is EUV (100 − 912 Å); solid line is optical; dotted line is FUV (912 − 2000 Å).

can be observed as a very bright EUV source whereas it is very faint in the optical band because its effective temperature is as high as $\log T \sim 5.2$. Observational identification is needed to examine this scenario.

REFERENCES

GALLAGHER, J. S., KALER, J. B., OLSON, E. C., HARTKOPF, W. I., & HUNTER, D. A. 1980, PASP, 92, 46

HACHISU, I. & KATO, M. 1995, ApJ, in preparation

KATO, M. 1983, PASJ, 35, 507

KATO, M. 1994, A&A, 281, L49

KATO, M. 1995a, in Lecture Notes in Physics, Proc. of IAU Colloq. 151 on Flares and Flashes, Berlin: Springer, in press

KATO, M. 1995b, in proceedings of Padova conference on Cataclysmic variables, Dordrecht: Kluwer, in press

KATO, M. & HACHISU, I. 1994, ApJ, 437, 802

SEKIGUCHI, K., FEAST, M. W., WHITELOCK, P. A., OVERBEEK, M. D., WARGAU, W., & SPENCER, JONES, J. 1988, MNRAS, 234, 281

STICKLAND, D. J., PENN, C. J., SEATON, M. J., SNIJDERS, M. A. J., & STOREY, P. J. 1981, MNRAS, 197, 107

The Hot Winds of Novae

PETER H. HAUSCHILDT,[1]
S. STARRFIELD,[1] E. BARON,[2] AND F. ALLARD[3]

[1] Department of Physics & Astronomy, Arizona State University, Tempe, AZ 85287

[2] Department of Physics & Astronomy, University of Oklahoma, Norman, OK 73019-0225

[3] Department of Physics, Wichita State University, Wichita, KS 67260-0032

We discuss the physical effects that are important for the formation of the late wind spectra of novae. Nova atmospheres are optically thick, rapidly expanding shells with almost flat density profiles, leading to geometrically very extended atmospheres. We show how the properties of nova spectra can be interpreted in terms of this basic model and discuss some important effects that influence the structure and the emitted spectrum of nova atmospheres, e.g., line blanketing, NLTE effects, and the velocity field. Most of the radiation from hot nova winds is emitted in the spectral range of the *EUVE* satellite. Therefore, we present *predicted EUVE* spectra for the later stages of nova outbursts. Observations of novae with *EUVE* could be used to test our models for the nova outburst.

1. Introduction

The modeling and analysis of *early* nova spectra has progressed significantly during the last 2–3 years by the construction of detailed model atmospheres and synthetic spectra for novae by Hauschildt et al. (1992, 1994ab, 1995) . In the early stages of the nova outburst, the spectrum is formed in an optically *very* thick shell (in both lines and continua) with a flat density profile, leading to very extended continuum and line forming regions (hereafter, CFR and LFR, respectively). Because of the large variation of the physical conditions inside the spectrum forming region, the classical term "photosphere" is not appropriate for novae. The large geometrical extension leads to a very large electron temperature gradient within the CFR and LFR, allowing for the observed simultaneous presence of several ionization stages of many elements. Typically, the relative geometrical extension R_{out}/R_{in} of a nova atmosphere is $\approx 100\dots1000$, which is much larger than the geometrical extensions of hydrostatic stellar atmospheres (even in giants R_{out}/R_{in} is typically less than 2) or supernovae (SNe). Nova atmospheres are also very different from SN atmospheres with respect to their energy balance. Whereas SNe spectra show constantly decreasing color temperatures and decreasing bolometric luminosities, the color temperatures of nova atmospheres generally increase with time and their bolometric luminosity is constant. This is caused by the presence of a central energy source in novae (the hot white dwarf) which is missing in SNe (where the only sources of energy are the radioactive decays of the Ni and Co nuclei).

The electron temperatures and gas pressures typically found in nova photospheres lead to the presence of a large numbers of spectral lines, predominantly Fe-group elements, in the LFR and a corresponding influence of line blanketing on the emergent spectrum. The situation is complicated significantly by the velocity field of the expanding shell which leads to an enhancement of the overlapping of the individual lines. This in turn makes simplified approximate treatments of the radiative transfer by, e.g., the Sobolev-approximation, very inaccurate and more sophisticated radiative transfer methods, which treat the overlapping lines and continua simultaneously, must be used in order to obtain reliable models. The line blanketing also leads to strong wavelength redistribution of

S. Bowyer and R. F. Malina (eds.), Astrophysics in the Extreme Ultraviolet, 413–417.

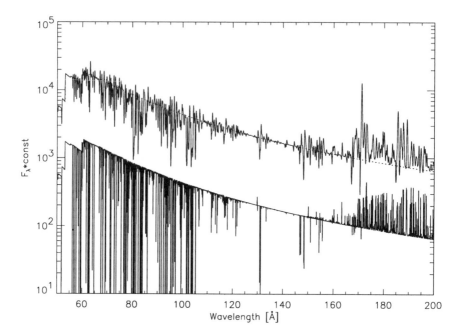

FIGURE 1. Experimental nova spectra for $T_{\text{eff}} = 250000$ K and a density law of the form $\rho \propto r^{-3}$ in the EUV spectra range. The upper spectrum is a synthetic spectrum computed with an expansion velocity of $v_{\max} = 1000 \, \text{km s}^{-1}$. The dotted line is the corresponding continuum spectrum (all spectral lines neglected). For comparison we show in the lower spectrum (shifted by -1 dex with respect to the upper spectrum for clarity) the results for a *static* atmosphere with otherwise the same structure.

the radiative energy, therefore, the temperature structure of the shell must be calculated including the effects of the line blanketing.

The situation is further complicated by the fact that the (electron) densities of the CFR and LFR in a nova atmosphere are very low compared to classical stellar atmospheres. This leads to an overwhelming dominance of the radiative rates over the collisional rates. In addition, the radiation field is very non-grey. These two effects lead to large departures from local thermodynamic equilibrium (LTE) in the CFR and LFR. Therefore, NLTE effects must be included self-consistently in the model construction, in particular in the calculation of the temperature structure and the synthetic spectra. As discussed in the previous paragraph, the effects of the line blanketing on the radiative rates requires a careful treatment of the radiative transfer in the NLTE calculations and simple approximations can lead to wrong results.

In this paper, we discuss the formation of the later wind spectrum of a nova. In the later phases of the nova outburst, the spectrum of the extremely hot white dwarfs can be observed directly only in the EUV spectral range. In both the optical and UV spectral ranges the radiation from the hot wind is re-processed by the nebular around the nova. In addition, the effective temperatures of the wind are so high, in excess of $250,000$ K, that most of the energy is emitted in the EUV range.

2. Results

In order to extract detailed quantitative information on velocities, densities, temperatures, and compositions, and constrain theoretical explosion models, it is necessary to model the nova spectrum in detail. We use our computer code **PHOENIX** (Version 5.5) to compute model atmospheres and synthetic spectra for novae. This is an updated version of the code used for the analyses of the early spectra of Nova Cygni 1992 (Hauschildt et al. , 1994a) and SN 1993J (Baron et al. , 1994), a more detailed description can be found in Hauschildt et al. (1995) and Allard & Hauschildt (1995).

The low densities and complicated radiation fields in the atmosphere require that the most important species be treated using multi-level NLTE. At present, we can treat H I (15 levels), He I (11 levels), He II (15 levels), Na I (3 levels), Mg II (18 levels), Ca II (5 levels), Ne I (26 levels), and Fe II (617) in NLTE. The Fe II 617 level model atom includes 13675 primary transitions that are self-consistently included in the radiative transfer and statistical equilibrium calculations and an additional (up to) 1.2 million secondary weak transitions that are included as background line opacity (see Hauschildt & Baron 1995 for details).

Our NLTE model atmospheres show that the early evolution of a nova shell, the "optically thick" phase, can be divided into at least 3 very different epochs. The first and very short-lived stage is the "fireball" stage, first detected in the infrared by Gehrz (see Gehrz, 1988) and analyzed in the UV by Hauschildt et al. (1994a). In this stage, the density gradient in the nova atmosphere is high, $N \approx 15$ and the effective temperatures are dropping from ≈ 15000 to < 10000 K (lower T_{eff} are probable but have not yet been observed). In this stage, the nova spectrum resembles that of a SN with low velocities ($v_{\mathrm{max}} \approx 4000 \, \mathrm{km \, s^{-1}}$ for Nova V1974 Cygni 1992).

As the density and temperatures of the expanding fireball drop, the material becomes optically thin and deeper layers become visible. In this stage, the "optically thick wind phase," the atmosphere evolves to a very flat density profile, $N \approx 3$, and the LFR, as well as the CFR, has a very large geometrical extension, values of $\Delta R/R \approx 100$ in the LFR are common. The large geometrical extension causes a very large temperature gradient in the LFR, typically the electron temperatures range between $1,100$ K and $50,000$ K for a model with $T_{\mathrm{eff}} = 6,500$ K and from 40000 K to $> 1,000,000$ K for a model with $T_{\mathrm{eff}} = 200,000$ K (we emphasize that all of the regions can be visible simultaneously in the emitted spectrum). This explains the observed fact that nova spectra can show, simultaneously, lines from different ions of the same element. The structure of the atmospheres and the calculated spectra are very sensitive to changes of the abundances of iron, carbon, nitrogen, and oxygen, which makes abundance determinations of these elements possible. However, for reliable abundance determinations, NLTE effects need to be included self-consistently for the elements under consideration. Consistent with the results of Pistinner et al. (1995) we also find that the synthetic spectra are insensitive to large changes in the luminosity. However, the spectra are *very* sensitive to changes in the *form* of the velocity profile inside the atmosphere.

In terms of the classical scheme of nova spectrum classification (McLaughlin, 1960), the increase in the effective temperature during the wind stage corresponds to the transition from the "premaximum spectrum" to the "diffuse enhanced spectrum" and later into the "nebular spectrum."

The EUV spectra of novae are dominated by the line blanketing of many thousands of overlapping spectral lines. The line blanketing changes the structure of the atmosphere, and the emitted spectrum, so radically that it must be included *self-consistently* in the models in order to be able to reliably compare the results to observed spectra. In Fig. 1

we display experimental synthetic spectra for a model with $T_{\text{eff}} = 250,000\,\text{K}$, $\rho \propto r^{-3}$, and a linear velocity law with $v_{\text{max}} = 1,000\,\text{km}\,\text{s}^{-1}$. The plot also shows the continuum spectrum obtained by neglecting all lines (dotted curve). Line blanketing has a significant impact on the spectrum emitted by the model, in particular in the EUV. Many lines in the EUV are *absorption* lines, only a relatively small fraction of the lines show P Cygni profiles. This is typical for nova spectra and is more pronounced in models with lower effective temperatures. The impact of the velocity field on the spectrum is demonstrated by comparing it to a spectrum emitted by a *static* atmosphere with otherwise identical parameters (lower spectrum in Fig. 1).

The strong coupling between continua and lines, as well as the strong overlap of lines, requires a unified treatment of the radiative transfer to (at least) full first order in v/c (i.e., including advection and aberration terms, our calculations are done using a fully relativistic radiative transfer). Methods that separate continua and lines (e.g., the Sobolev approximation) give unreliable results. The "density" of lines in the EUV is so high that practically the complete wavelength range is affected by lines, even for very high effective temperatures. For slightly cooler models, most of the observed "emission lines" are just gaps between "clusters" of lines, thus they are regions of transparency which have *less* opacity than the surrounding wavelength regions. In addition, the radiation fields inside the nova continuum and line forming regions nowhere resemble blackbody or grey distributions. This condition, combined with the low densities, causes very large departures from LTE. The NLTE radiative transfer and rate equations *must* be solved self-consistently including the effects of line blanketing in the UV and optical spectral regions. The effects of neglecting the line blanketing on the departure coefficients are enormous, as shown by Hauschildt & Ensman (1994) for the case of SN model atmospheres.

The large deviation from a blackbody energy distribution and the extreme non-gray spectrum cause a large effect of the lines on the temperature structure of nova atmospheres. The effects of line cooling (in the optically thin regions) and back-warming (in the inner regions) must be included in a proper analysis of nova spectra. The temperature changes introduced by line blanketing can amount to more than $10000\,\text{K}$ in the CFR and LFR, thus changing the synthetic spectra significantly. This demonstrates that a nova model atmosphere must include line blanketing self consistently in order to derive parameters, in particular elemental abundances, otherwise the results are unreliable.

3. Summary

The agreement between our synthetic spectra and observed *early* nova spectra is very good from the UV to the near IR. We are now able to reproduce the UV and optical spectra of very different types of classical novae. The basic physics and the modeling of early nova atmospheres is now well understood but the later nova wind models are still in an experimental stage. We are currently working on improvements of the later stage models (more NLTE species). Another important step will be to investigate the effects of density inversions (clumps) on the emitted spectrum and a more systematic investigation of the effects of different velocity fields on nova spectra is also required. Furthermore, we are currently working on a detailed treatment of the "pre-nebula" phase, i.e., the phase where allowed, semi-forbidden and sometimes forbidden lines are simultaneously present in the observed nova spectra.

It is a pleasure to thank H. Störzer, J. Krautter, G. Shaviv and S. Pistinner for stimulating discussions. This work was supported in part by a NASA LTSA grant to Arizona

State University, by NASA grant NAGW-2999; as well as grants to G. F. Fahlman and H. B. Richer from NSERC (Canada). Some of the calculations presented in this paper were performed at the San Diego Supercomputer Center (SDSC), supported by the NSF, and at the NERSC, supported by the US DoE, we thank them for a generous allocation of computer time.

REFERENCES

ALLARD, F. & HAUSCHILDT, P. H. 1995, ApJ, in press

BARON, E., HAUSCHILDT, P. H., & BRANCH, D. 1994, ApJ, 426, 334

GEHRZ, R. D. 1988, AR&AA, 26, 377

HAUSCHILDT, P. H. & BARON, E. 1995, JQSRT, submitted

HAUSCHILDT, P. H. & ENSMAN, L. M. 1994, ApJ, 424, 905

HAUSCHILDT, P. H., STARRFIELD, S., AUSTIN, S. J., WAGNER, R. M., SHORE, S. N., & SONNEBORN, G. 1994a, ApJ, 422, 831

HAUSCHILDT, P. H., STARRFIELD, S., SHORE, S. N., ALLARD, F., & BARON, E. 1995, ApJ, in press

HAUSCHILDT, P. H., STARRFIELD, S., SHORE, S. N., GONZALES-RIESTRA, R., SONNEBORN, G., & ALLARD, F. 1994b, AJ, 108, 1008

HAUSCHILDT, P. H., WEHRSE, R., STARRFIELD, S., & SHAVIV, G. 1992, ApJ, 393, 307

McLAUGHLIN, D. B. 1960, in Stellar Atmospheres, edited by Greenstein, J. L., number VI in Stars and Stellar Systems, page 585, University of Chicago Press

PISTINNER, S., SHAVIV, G., HAUSCHILDT, P. H., & STARRFIELD, S. 1995, ApJ, in press

The X-ray and EUV Turnoff of GQ Mus and V1974 Cyg

S. STARRFIELD,[1] J. KRAUTTER,[2] S. N. SHORE,[3] I. IDAN,[4]
G. SHAVIV,[4] AND G. SONNEBORN[5]

[1] Dept. of Physics & Astronomy, Arizona State University, Tempe, AZ 85287, USA

[2] Landessternwarte, Königstuhl, D69117, Heidelberg, Germany

[3] Dept. of Physics and Astronomy, Indiana University at South Bend,
1700 Mishawaka Ave, South Bend, IN 46634, USA

[4] Dept. of Physics, The Technion, 32000 Haifa, Israel

[5] Lab. for Astr. and Solar Phys., Code 681, NASA Goddard Space Flight Center,
Greenbelt, MD 20771, USA

Both GQ Mus and V1974 Cyg were observed to turnoff in X-rays by ROSAT. The turnoff of
V1974 Cyg was also observed with EUVE. GQ Mus was observed near the beginning of its
outburst with *EXOSAT* and then 7 years later by *ROSAT* in the all-sky survey. Later *ROSAT*
PSPC observations showed that its X-ray intensity was slowly declining with time and it was
not detected in the last pointing that occurred in August 1993. We observed GQ Mus with *IUE*
over the entire active phase of its outburst and found a change in the slope of the UV continuum
around the time that the X-rays turned off. V1974 Cyg was observed by *ROSAT* throughout
its entire active phase in X-rays which lasted about 18 months. V1974 Cyg was detected in the
EUVE all-sky survey, but not in pointed observations that occurred in August 1993 (and June
and November 1994). We use the measured times of the active phases to determine important
properties of these two novae. For example, for V1974 Cyg we predict that more than 10^{-5} $M_\odot$
of helium rich material was left on the white dwarf when it returned to quiescence. For GQ Mus,
$\sim 10^{-4}$ $M_\odot$ was left on the white dwarf. These results imply that a significant amount of the
helium seen in nova ejecta was produced in outbursts prior to the one that was just observed.
They also imply that the mechanism which mixes core material into the ejecta must be efficient.

1. Introduction

It is now commonly accepted that the cause of the classical nova outburst is a thermonuclear runaway (hereafter, TNR) in the accreted hydrogen-rich envelope on the white
dwarf (hereafter, WD) component of a close binary system. The secondary star in the
system fills its Roche lobe and loses mass through the inner Lagrangian point into the
lobe surrounding the WD. The material enters the Roche lobe of the WD with the angular momentum of the secondary and, therefore, creates an accretion disk before falling
onto the WD. Hydrodynamic studies have shown that the consequence of this accretion
is a growing layer of hydrogen-rich gas on the WD. When both the WD luminosity and
the rate of mass accretion onto the WD are sufficiently low, so that the deepest layers of
the accreted material become electron degenerate, a TNR occurs in the accreted layers.
The TNR heats the accreted material ($\sim 10^{-4}$ $M_\odot$ to $\sim 10^{-6}$ $M_\odot$ depending on WD
mass) until an explosion occurs. This leads to the ejection of a fraction of the envelope
(Starrfield 1989, and references therein).

S. Bowyer and R. F. Malina (eds.), Astrophysics in the Extreme Ultraviolet, 419–424.
© 1996 *Kluwer Academic Publishers. Printed in the Netherlands.*

2. The Constant Bolometric Luminosity Phase

Hydrodynamic simulations have predicted, and X-ray, EUV, and IR observations have confirmed, that there is a phase of constant bolometric luminosity and high T_{eff} following the initial outburst (Starrfield et al. 1972; Shore et al. 1993, 1994). The cause is that only a fraction of the accreted envelope is ejected during the initial phase of the explosion. The remaining mass (anywhere from 10% to 90%) ultimately relaxes to quasistatic equilibrium, with a radius in the range from $\sim 10^{10}$ to $\sim 10^{11}$ cm (Truran 1982; Starrfield 1989). Because the shell source is still burning in the envelope, the luminosity of the rekindled WD stabilizes at the core-mass luminosity (Paczynski 1971; Iben 1982), which is close to L_{Edd} for the most massive WDs. Such a high luminosity can drive mass loss in a wind and cause the radius of the pseudo-photosphere, which occurs at some point in the expanding material, to decrease (MacDonald, Fujimoto, & Truran 1986). Since the pseudo-photosphere shrinks in radius at constant luminosity, the T_{eff} of the WD increases to values exceeding 10^5 K and produces a phase of EUV or soft X-ray emission as was observed for both GQ Mus and V1974 Cyg (Ögelman et al. 1993; Krautter et al. 1995; Shanley et al. 1995; Stringfellow and Bowyer 1995). Once nuclear burning ends, this material collapses back onto the WD and cools (Starrfield et al. 1991).

One important question about the nova explosion is the length of the active phase - how long does it take for nuclear burning to end and the nova to return to quiescence? Over the past three years, *ROSAT* has observed both the turn-on and turn-off of V1974 Cyg (Krautter et al. 1995) and the turn-off of GQ Mus (Shanley et al. 1995). We now, therefore, have measured the length of the active phase for two novae (V1974 Cyg: 18 months; GQ Mus: 9 years) plus, a measure of both the cooling time and the rate of cooling after maximum (Idan et al. 1995; Shanley et al. 1995).

3. The X-ray and UV Evolution of GQ Mus

GQ Mus 1983 was observed in the optical and UV near its maximum optical brightness (Krautter et al. 1984). It was also detected by *EXOSAT* (Ögelman, Krautter, and Beuermann 1987). During the 1980's it's optical spectrum indicated that there was a hot, photoionizing source in the system (Krautter and Williams 1989; Péquignot et al. 1993). While it was detected in 1990 during the *ROSAT* All Sky Survey(Ögelman et al. 1993), *ROSAT* PSPC observations showed that it was turning off in X-rays (Shanley et al. 1995). This was corroborated by optical observations which indicated that the nebular ionization was dropping (Hamuy et al. 1994).

GQ Mus was observed by *IUE* throughout its outburst. Although it was initially bright enough to obtain both SWP(1200 Å to 2000 Å) and LWP (2000 Å to 3200 Å) spectra, in the past few years we have only obtained long exposure SWP observations (Idan et al. 1995). Because the nova was faint, it was necessary to correct for the residual background and camera artifacts that dominate the weakest spectra and we have developed techniques to treat these problems (Idan et al. 1995; see also Crenshaw et al. 1990). Each spectrum was then dereddened to $E(B - V) = 0.45$ (Krautter et al. 1984) using the standard Galactic extinction law (Savage & Mathis 1979). We performed pseudo-filter photometry on our spectra using a uniform weighting with 50 Å bandpasses.

We chose three wavelength intervals which were found to be relatively free of emission lines. The results are shown in Figure 1, which displays the binned flux at 1375 Å divided by the binned flux at 1575 Å versus the binned flux at 1575 Å divided by the binned flux at 1825 Å. Hereafter, we refer to the plotted points as the UV colors. The data for GQ Mus are marked by asterisks. We have also plotted the UV colors for hot WDs such as

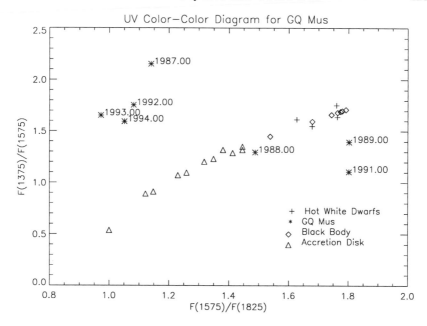

FIGURE 1. This figure displays the UV colors for GQ Mus from 1987 to 1994. The data for GQ Mus are marked by asterisks. We have also plotted the UV colors for hot white dwarfs such as PG1159 and H1504 (+), black-bodies ranging in temperature from 5×10^4 to 10^6 K (◊), and theoretical predictions for the colors of accretion disks (△). The errors are about 10%.

PG1159 and H1504 (+), black-bodies ranging in temperature from 5×10^4 to 10^6 K (◊), and theoretical predictions for the colors of accretion disks (△) (Idan et al. 1995). The errors on each point for GQ Mus are probably ~10%. We do not plot data from before 1987 because of the presence of strong emission lines which affected the determination of the continuum.

It seems clear that the colors of GQ Mus are well separated from those of either hot WDs, black-bodies, or accretion disks. Note, however, that while the point for 1988 does fall close to those for accretion disks (for the highest mass accretion rates), that may only be a coincidence since it is evolving to the right in the diagram. We also mention that the UV colors of other novae fall in the lower-left part of this diagram and nowhere near the position of GQ Mus at any time (Idan et al. 1995). It is also interesting that the colors of this nova do not fall close to those of single WD's with $T_{\rm eff} \sim 150,000$ K. We would have expected the continuum for the radiating WD in the GQ Mus system to have approximately the same colors as these stars if the accretion disk has not reformed.

The change in the colors from 1987 to the years 1988 through 1991 is real and indicates that the WD was increasing in temperature. This is confirmed by comparison with contemporaneous optical observations of GQ Mus which showed that the strength of [Fe X] 6374 Å was increasing over this time period and in 1988 its strength exceeded Hα (Krautter & Williams 1989; Péquignot et al. 1993). During this interval its UV colors fell neither in the region where we predict that novae with accretion disks should lie nor did optical spectra show the emission lines and continuum characteristic of an accretion disk. We suggest that none was present in the system.

The UV color-color diagram also shows that between 1991 and 1992 the nova rapidly evolved to a region of lower temperature. It was over this same interval that pointed *ROSAT* observations (PSPC) showed that the X-rays were turning off and it was not detected by the PSPC in late 1993 (Shanley et al. 1995). We have, therefore, determined from both UV and X-ray observations that the end of the active period of GQ Mus was in 1992 which is about 9 years after the explosion was first seen in the optical.

4. The Importance of Measuring the Turnoff Time

In this section we use the fact that we have obtained, for the first time, the length of the active phase for two novae. We assume that the soft X-ray emission is the result of ongoing nuclear burning in the remnant envelope on the WD which is responsible for the constant bolometric luminosity phase of the outburst (Starrfield et al. 1972, 1974).

We also need the mass of the white dwarf but this has been done neither for GQ Mus nor for V1974 Cyg from radial velocity studies. We can, however, use the observed length of the active phase to estimate the mass of the white dwarf. We do this following the discussion in Starrfield et al. (1991). We can estimate the envelope mass necessary to initiate a TNR and both calculations and analytic estimates show that that mass is a function of white dwarf mass (Starrfield 1989; Starrfield et al. 1991). We can then use the luminosity determined from the core-mass luminosity relationship (Paczynski 1971; Iben 1982) in combination with the theory of radiation pressure driven mass loss to calculate the mass loss rate as a function of luminosity and, hence, as a function of WD mass. Combining the two relations, allows us to obtain the length of the active phase as a function of white dwarf mass.

Using the measured intervals of the active phases for these two novae implies that the mass of the WD in GQ Mus is $\sim 1.2\ M_\odot$ and the mass of the WD in V1974 Cyg is $\sim 1.3\ M_\odot$. The concomitant luminosities are $\sim 5 \times 10^4\ L_\odot$ for V1974 Cyg and $\sim 4 \times 10^4\ L_\odot$ for GQ Mus. The value for V1974 Cyg is in reasonable agreement with the value determined by Shore et al. (1993, 1994). The value determined by Hassell et al. (1990) for GQ Mus was $\sim 2.5 \times 10^4\ L_\odot$ which is close enough to our estimate for the purposes of this paper. For the rest of this section, however, we shall assume a 1.25 $M_\odot$ WD and a luminosity of $4.5 \times 10^4\ L_\odot$ (Iben 1982).

The analyses of the X-ray emission from these novae imply effective temperatures for the WD $\sim 3 \times 10^5$ K. Because these values depend on black-body fits to the observed X-ray PSPC energy distributions, they are not well determined. Nevertheless, a hot, luminous 1.25 $M_\odot$ WD radiating at the core-mass luminosity with an effective temperature of 3×10^5 K has a radius of 5.6×10^9 cm. This value is 15 times larger than the equilibrium radius of a 1.25 $M_\odot$ WD (3.8×10^8 cm). Therefore, at the beginning of the X-ray decline phase, the configuration of the remnant was that of a WD with a hot, extended envelope.

We now assume that the final decline is caused by the WD consuming all available hydrogen and the energy radiated during the decline comes only from gravitational contraction of the hot, extended, *helium-rich layers*. We can then use the Kelvin-Helmholtz time scale (Cox & Giuli 1967) to obtain the envelope mass required to produce the observed energy (Krautter et al. 1995). We use a time for the collapse of the envelope of 6 months for V1974 Cyg and other parameters for a 1.25 $M_\odot$ WD to arrive at a mass for the helium layer of $\sim 10^{-5}\ M_\odot$. This is an upper limit to the remnant envelope mass. Assuming a lower mass for the WD in GQ Mus and a cooling time scale of ~ 3 years (Shanley et al. 1995), implies that it ended its outburst with more than $10^{-4}\ M_\odot$ of helium on the surface. These values are significantly larger than the amount of mass thought to be left on the WD after the outburst (Starrfield 1989). In addition, the value for V1974 Cyg is a large fraction of the ejected mass ($> 10^{-4}$ $M_\odot$: Shore et al. 1993).

The fact that we observed the entire nuclear burning time for their outbursts allows us to determine additional parameters. We again assume constant bolometric luminosity phases of 18 months and 9 years. Therefore, their total radiated luminosities were $\sim$ 9×10^{45} ergs (V1974 Cyg) and $\sim 5 \times 10^{46}$ ergs (GQ Mus). V1974 Cyg ejected $\sim 10^{-4}$ $M_\odot$ at a speed of $\sim 2 \times 10^8$ cm s^{-1} (Shore et al. 1993) for a total kinetic energy of $\sim 4 \times 10^{45}$ ergs. GQ Mus ejected $\sim 10^{-4}$ $M_\odot$ at a speed of $\sim 10^8$ cm s^{-1} (Krautter et al. 1984; Hassell et al. 1990) for a total kinetic energy of $\sim 10^{45}$ ergs. These results imply that $\sim 10^{-6}$ $M_\odot$ of hydrogen was burnt to helium during their outbursts. This is less than 1% of the mass of the ejecta and suggests that the *observed* enrichment of helium in nova ejecta (Hassell et al. 1990; Péquignot et al. 1993; Andreä et al. 1994; Austin et al. 1995) does *not* come from nuclear burning but from mixing with material that was already present on the surface of the WD.

5. Discussion

We have assumed that the X-ray decline marked the end of the nuclear burning phase when all the hydrogen had been fused to helium. One of the consequences of the outburst on both these novae, therefore, must be a massive layer of helium on the surface of the WD. Once accretion onto the WD is again initiated, the infalling material must penetrate the helium-rich layer in order to reach to the CO or ONeMg core material which is seen in nova ejecta. This helium, however, must also be ejected because the observations also show large enrichments of helium in nova ejecta. This claim is supported by the helium enrichment found by Austin et al. (1995) for V1974 Cyg (He/H $\sim$5 times solar). This result also implies that efficient mixing must occur between core and accreted material prior to the TNR. Therefore, the large amounts of helium observed in nova outbursts is not produced by the current outburst but in previous outbursts. In addition, if the WD is as massive as we have found in our calculations, then the accretion time scale is relatively short and the mixing mechanism must be significantly more efficient than diffusion (Prialnik and Kovetz 1984).

In summary, we have used the observed turn-off times of V1974 Cyg and GQ Mus to determine important properties of the nova outburst.

REFERENCES

ANDREÄ, J. , DRECHSEL, H. , & STARRFIELD, S. 1994, A&A, 291, 869

AUSTIN, S. J., WAGNER, R. M., STARRFIELD, S., SHORE, S. N., SONNEBORN, G., & BERTRAM, R. 1995, ApJ, in preparation

CRENSHAW, D. M., NORMAN, D. J., & BRUEGMAN, O. W. 1990, PASP, 102, 463

HAMUY, M., PHILLIPS, M., & WILLIAMS, R. E. 1994, IAUC #5953

HASSELL, B. J. M. et al., 1990, in Physics of Classical Nova, ed. A. Cassatella & R. Viotti, Berlin: Springer-Verlag, NY, 202

IBEN, I. JR. 1982, ApJ, 259, 244

IDAN, I., STARRFIELD, S., SHORE, S., KRAUTTER, J., SONNEBORN, G., & SHAVIV, G. 1995, ApJ, in preparation

KRAUTTER, J., et al., 1984, A&A, 137, 307

KRAUTTER, J., ÖGELMAN, H., STARRFIELD, S., WICHMANN, R. & PFEFFERMAN, R., 1995, ApJ, in press

KRAUTTER, J. & WILLIAMS, R. E. 1989, ApJ, 341, 968

MacDONALD, J., FUJIMOTO, M. Y., & TRURAN, J. W. 1985, ApJ, 294, 263

ÖGELMAN, H., KRAUTTER, J., & BEUERMANN, K. 1987, A&A, 177, 110

ÖGELMAN, H., ORIO, M., KRAUTTER, J., & STARRFIELD, S. 1993, Nature, 361, 331

PACZYNSKI, B. 1971, Acta Astron., 21, 271

PÉQUIGNOT, D., PETITJEAN, P., BOISSON, C., & KRAUTTER, J. 1993, A&A, 271, 219

PRIALNIK, D. & KOVETZ, A. 1984, ApJ, 281, 367

SAVAGE, B., & MATHIS, J. 1979, ARAA, 17, 73

SHANLEY, L., ÖGELMAN, H., GALLAGHER, J. S., ORIO, M., & KRAUTTER, J. 1994, ApJL, 438, L95

SHORE, S. N., SONNEBORN, G., STARRFIELD, S., GONZELEZ-RIESTRA, R., & AKE, T. B. 1993, AJ, 106, 2408

SHORE, S. N., SONNEBORN, G., STARRFIELD, S., GONZELEZ-RIESTRA, R., & POLIDAN, R. S. 1994, ApJ, 421, 344

STARRFIELD, S. 1989, in Classical Novae, ed. M. Bode and A. N. Evans, Wiley Press, New York, 39

STARRFIELD, S. 1992, in Variability in Stars and Galaxies: Reviews in Modern Astronomy, 5, ed. G. Klare, Heidelberg: Springer-Verlag, 73

STARRFIELD, S., SPARKS, W. M., & TRURAN, J. W. 1974, ApJS, 28, 247

STARRFIELD, S., TRURAN, J. W., SPARKS, W. M., & KUTTER, G. S. 1972, ApJ, 176, 169

STARRFIELD, S., TRURAN, J., SPARKS, W. M., & KRAUTTER, J. 1991, in Extreme Ultraviolet Astronomy, ed. R. Malina, & S. Bowyer, 168

STRINGFELLOW, G., & BOWYER, C. S. 1995, in Proceedings EUVE IAU Colloquium, ed. S. Bowyer, Berkeley, CA, these proceedings

TRURAN, J. W. 1982, in Essays in Nuclear Physics, ed. C. A. Barnes, D. D. Clayton, and D. N. Schramm, Cambridge, University Press, 467

EUV Constraints on Models of Low Mass X-Ray Binaries

D. J. CHRISTIAN,[1] J. E. EDELSTEIN,[2] M. MATHIOUDAKIS,[1]
K. McDONALD,[1] AND M. M. SIRK[1]

[1] Center for EUV Astrophysics, 2150 Kittredge St. Berkeley, CA, 94720-5030, USA

[2] Department of Astronomy, Space Sciences Lab., University of California Berkeley, 94720, USA

We present *EUVE* survey results for moderate column directions containing known low-mass X-ray binaries (LMXB). We derive Lexan band (100 Å) count rates and upper limits for nearly 40 LMXB chosen generally with $E_{B-V} \leq 0.3$. Detections include Sco X-1, Her X-1, and the GRO transient CJ0422+32. Super soft sources in the LMC yield 3 σ upper limits of ≤ 10 counts ks-1. The extrapolation of two component spectral models (such as blackbody plus thermal bremsstrahlung), are in agreement with the survey upper limits. Contemporary LMXB spectral models, which involve Comptonization in an inner disk corona, predict a large flux of EUV photons. If the above model is correct in the EUV, such a component could be detected in source with low column densities. We argue that additional intrasystem column hampers its detection.

1. Introduction

Low Mass X-ray Binaries (LMXB) with their near Eddington luminosities, X-ray bursts, and dipping behavior provide the opportunity to study accretion onto neutron stars. They are best classified on the basis of their color-color diagrams, which show a Z pattern for the persistently bright sources and less defined patterns (Atoll) for the burst sources (Hasinger & van der Klis 1989; van der Klis 1989). This spectral behavior is correlated with the short time-scale variability and often show broad features or quasiperiodic oscillations (QPO) in their power spectra. Constructing a physical model of LMXB from the spectral data has been made difficult, due to the limited sensitivity and band-passes of X-ray instrumentation. Models involving a boundary layer plus an accretion disk (Mitsuda 1984) are indistinguishable from Comptonization models (White, Stella, & Parmar 1989) for many X-ray instruments. Contemporary models have sought to build a self-consistent model to explain both the spectral and QPO behavior. Such models often require a large input of soft (≤ 1 keV) photons (Lamb 1989; Schulz & Wijers 1993).

2. Observations and Analysis

EUVE conducted an all-sky survey between July 1992 and January 1993 with three scanning telescopes. Scanners A and B each had two Lexan/boron ("Lexan") quadrants with a bandpass covering 58–174 Å and two Al/Ti/C filters with a bandpass covering 156–234 Å. The bandpasses of scanner C covered 345–605 Å for the Ti/Sb/Al filter, and 500–740 Å for the Sn/SiO filter (Bowyer et al. 1996).

2.1. *Skymaps and Pigeonholes*

We calculated count rates from the *EUVE* all-sky survey for J2000 positions taken from the White, Giommi, & Angelini catalogue for known LMXB generally with E_{B-V} less

S. Bowyer and R. F. Malina (eds.), Astrophysics in the Extreme Ultraviolet, 425–429.

than 0.3 (van Paradijs 1993). However, a few globular cluster sources and blackhole candidates with higher E_{B-V} were included. This software produces skymaps, which are a binned distribution of EUV photons for a chosen circular region of the sky. A maximum likelihood technique is used to test for variations above the expected background (Lewis 1993; Bowyer et al. 1996). To verify count rates greater than 3 σ in the Lexan band skymaps (since we do not expect any real flux above 100 Å for most LMXB) we constructed "pigeonholes" (lists of photon events), with 24' radius centered on the position of the source. Results from the all-sky survey are shown in Table 1. The likelihood significance quoted in the table is related to the square of the Gaussian significance, σ. Sources with a downward arrow in Table 1 denote 3 σ upper limits.

3. Results

Sco X-1 and Her X-1, which did appear in the first *EUVE* All Sky survey (Bowyer et al. 1994) are also reported here showing large Lexan and Al/C count count rates. The high column density of Sco X-1 cuts off photons with wavelengths longer than ~64 Å, therefore the Al/Ti/C result is an X-ray leak. The Lexan count rate for Sco X-1 is some fraction of X-ray leak, since the bandpass does extend down to 30 Å (discussed below). The Lexan count rate for Her X-1 is not suspect, because of its low interstellar column. Her X-1 (although an accreting X-ray pulsar) has been the only X-ray binary *EUVE* has been able to study in any detail (Vrtilek et al. 1994). The hard transient, GRO CJ0422+32 (Nova Persei 1992) is detected in Lexan at a 4.5 σ level. Since the first all-sky source catalog use a 6 σ detection threshold, CJ0422+32 was not included in that catalog. Super-soft source in the LMC (Greiner, Hasinger, & Kahabka 1991), such as Cal 83, Cal 87, RX J0439.8-6809, and RX J0527.8-695 have upper limits of $\leq$15 counts per ks.

We have predicted the Scanner Lexan count rates based on X-ray model fits from simultaneous *Einstein* solid state spectrometer (SSS; 0.5-4.5 keV) and monitor proportional counter (MPC; 1.2-20.0 keV) data (Christian 1993). Best fitting spectral models from the *Einstein* LMXB survey included a form of unsaturated Comptonization (USC; A $E^{-\Gamma}$exp(-E/kT), where Γ is the spectral index), and a blackbody plus thermal bremsstrahlung (BB+TB). We folded these models through the Lexan effective area. The majority of predictions are an order of magnitude lower than the survey upper limits. Less than 10% of the observed Lexan flux is expected for the X-ray model of Sco X-1. A large fraction of the counts are from wavelengths shorter than 30 Å. The unfolded SSS+MPC model extrapolated to the Lexan bandpass is shown in Figure 1.

4. Discussion

Many of the present models of LMXB are based on a possibly non-unique model fits of the X-ray spectra (e.g., Vacca et al. 1987) We have learned models of LMXB must be able to account for QPO, spectral-temporal correlations, and the observed spectra. Contemporary Comptonization models of LMXB have attempted this (Lamb 1989; Ponman, Foster, & Ross 1990; Schulz & Wijers 1993). Such models consider a cocoon of Comptonizing material surrounding the neutron star and assume an input spectrum, such as a cutoff power-law or blackbody. The emerging spectrum is a function of the scattering optical depth, which is very sensitive to the mass accretion rate. Such models can produce the observed spectra with a similar number of free parameters as two component models (Ponman, Foster, & Ross 1990). Phenomenological models of the LMXB spectra

TABLE 1. EUVE Survey Observations of Low Mass X-ray Binaries.

Source	Name	E_{B-V}	S	Lexan Time	counts ks^{-1}	S	Al/Ti/C Time	counts ks^{-1}
(1)	(2)	(3)	(4)	(5)	(6)	(7)	(8)	(9)
0042+327	4U0042+327	0.2	0.9	1225	22 ↓	0.4	1217	28 ↓
0422+32	CJ0422+32	0.4	22.3	702	51±15	7.3	694	87 ↓
0439.8-65	RXJ0439.8-6	0.1	6.9	8376	14 ↓	0.4	8041	17 ↓
0512-401	NGC1851	0.1	0.0	1349	15 ↓	6.5	1343	53 ↓
0521-720	LMC X-2	0.1	1.30	9218	10 ↓	1.2	8714	16 ↓
0527.8-69	RXJ0527.8-61	0.1	0.0	14201	3 ↓	2.5	13633	16 ↓
0543-682	CAL 83	0.1	0.9	31555	7 ↓	0.0	30498	21 ↓
0547-711	CAL 87	0.1	1.8	13366	8 ↓	3.5	13381	16 ↓
0614+091	4U0614+091	0.3	1.0	988	31 ↓	3.4	942	83 ↓
0620-003	V616 Mon	0.4	5.6	1188	42 ↓	0.5	1144	54 ↓
0748-767	EXO0748-676	0.42	6.7	6495	15 ↓	6.0	6284	26 ↓
0918-549	4U0918-549	0.3	2.0	2386	19 ↓	4.7	2308	47 ↓
0921-63	A0921-630	0.2	1.1	3128	15 ↓	0.4	2951	25 ↓
1124-684	Nova Mus 91	0.25	0.3	1782	22 ↓	0.9	1738	54 ↓
1254-69	4U1254-690	0.35	6.7	508	72 ↓	0.3	491	67 ↓
1617-155	Sco X-1	0.3	792	2024	284±16^a	273	1010	248±23^a
1627-673	4U1627-673	0.1	1.3	889	30 ↓	5.1	603	74 ↓
1656+354	Her X-1	≤ 0.05	611	1912	180 ± 13	7.0	1907	42 ↓
1704+24	A1704+240	0.3	3.9	1144	40 ↓	8.2	676	91 ↓
1728-169	GX 9+9	0.3	6.2	574	75 ↓	5.1	558	104 ↓
1735-44	V926 Sco	0.15	0.2	649	39 ↓	3.8	638	80 ↓
1820-303	NGC6624	0.3	1.9	1209	28 ↓	3.4	1154	75 ↓
1822-371	V691 CrA	0.15	0.0	1232	15 ↓	0.6	1160	55 ↓
1832-33	NGC6652	0.1	0.3	1032	28 ↓	0.8	970	74 ↓
1908+005	Aql X-1	0.4	7.9	1083	51 ↓	2.1	1049	71 ↓
1916-053	V1405 Aql	0.2	4.5	1045	45 ↓	2.0	1018	66 ↓
1957+115	4U1957+115	0.4	6.4	773	59 ↓	0.0	712	64 ↓
2127+119	M15	≤ 0.06	0.9	1088	30 ↓	3.0	1023	82 ↓
2129+47	V1727 Cyg	0.5	1.7	1578	24 ↓	0.2	1498	31 ↓
2142+38	Cyg X-2	0.45	6.9	1430	38 ↓	0.3	1352	44 ↓

Col 1–Source
Col 2–Alternate name
Col 3–E_{B-V} adapted from van Paradjis 1993.
Col 4&7–S is the likelihood significance equal to the square of the Gaussian significance σ.
Col 5&8–Exposure Time in seconds.
Col 6–The Lexan/boron count rate per ks. The down arrow (↓) indentifies a 3 σ upper limit.
Col 9–Same as Column 6, but for the aluminum/titanium/carbon filter.
^{a} X-ray leak.

generally predict 1 to 2 orders of magnitude less EUV luminosity as compared to the X-ray luminosity. This comparison is shown in Table 2.

The Lamb model predicts a large flux of soft photons, possibly from high-harmonic cyclotron emission. If we assume soft photons are produced near the Eddington luminosity for a 1.4 $M_{\odot}$ neutron star and a typical LMXB power-law (with photon index of 2), we would expect only ∼2 counts ks^{-1} for a source with a column of 3×10^{21} cm^{-2} in the Lexan bandpass. For a column which allows photons in the *EUVE* Lexan bandpass (e.g., ∼ 5×10^{20} cm^{-2}) we would expect ∼100 counts ks^{-1}, which would have been

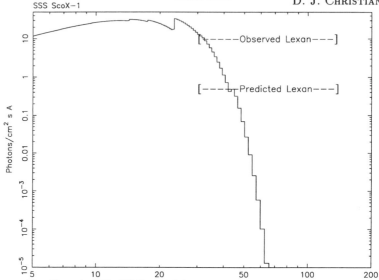

FIGURE 1. The *Einstein* SSS+MPC model spectrum of Sco X-1 extraplolated into the Lexan bandpass for comparison. Predicted and Observed Lexan fluxes calculated using the Lexan effective area are indicated.

TABLE 2. Comparison of Lexan and X-ray Luminosities

Name	Model	Luminosity[a] Lexan 10^{38} ergs s^{-1}	Luminosity 0.5–20.0 keV 10^{38} ergs s^{-1}
(1)	(2)	(3)	(4)
CJ0422+32	CompST[b]	2e-4	1.4
NGC1851	USC	0.003	0.02
LMC X-2	USC	0.1	2.0
CAL 83	BB[c]	0.3	0.002
4U0614+091	BB+USC	7e-4	0.02
A0921-630	USC	1e-4	0.01
Sco X-1	BB+TB	0.03	0.7
Her X-1	$BB + PL^d$	0.001	0.01
NGC6624	USC	0.03	0.6
X1822-37	BB+TB	9e-6	0.008
M15	USC	0.005	0.02
4U 2129+47	BB+TB	2e-5	0.001
Cyg X-2	BB+TB	0.07	1.4

Col 1–Source
Col 2–The best fitting model from SSS+MPC spectral fits 0.5–20.0 keV (Christian 1993) unless otherwise noted.
Col 3–The predicted Lexan Luminosity in units of 10^{38} ergs s^{-1}.
Col 4–The 0.5–20.0 keV Luminosity in units of 10^{38} ergs s^{-1}.
[a] Luminosites based on distances as compiled in Christian & Swank 1995.
[b] Derived from Griener, Hasinger, & Thomas 1994
[c] Sunyaev and Titarchuk (1980) form of Comptonization from Pietsch et al. 1993.
[d] From Vrtilek et al. 1994

detectable in the all-sky survey. Sources like X2127+119 in M15 and X1822-37 fulfill the column density requirement, but are known to have large intrasystem absorption that would absorb most EUV photons. However, the column for X2127+119 does vary with orbital phase (Hertz & Grindlay 1983), and it could be detectable in the EUV when at the low column phase (0.0). Callanan (this proceedings) report a possible detection of M15 from a pointed *EUVE* deep survey observation.

This research was supported under NASA contract NAS5-30180.

REFERENCES

BOWYER, S. ET AL. 1994, ApJS, 93, 569

BOWYER, S. ET AL. 1996, ApJS, in press

CHRISTIAN, D. J. & SWANK, J. H. 1995, ApJ, submitted

CHRISTIAN, D. J. 1993, Ph.D. Thesis, U. of Maryland

GREINER, J., HASINGER, G., & THOMAS, H. C. 1994, A&A, 281, L61

HERTZ, P. & GRINDLAY, J. E. 1983, ApJ, 275, 105

HASINGER, G., & VAN DER KLIS, M. 1989, A&A, 255, 79

LAMB, F. K. 1989, Proc. 23rd ESLAB Symp. 1, 215

LEWIS, J. W. 1993, JBIS, 46, 346

MITSUDA, K. 1984, Ph.D. Thesis, U. of Tokyo

MORRISON, R., & McCAMMON, D. 1983, ApJ, 270, 119

PONMAN, T. J., FOSTER, A. J., & ROSS, R. R. 1990, MNRAS, 246, 287

SCHULZ, N. S. & WIJERS, R. A. M. J. 1993, A&A, 273, 123

SUNYAEV, R. A., & TITARCHUK, L. G. 1980, Sov. Astr. L., 12, 117

VACCA, W. D., SZTANJO, M, LEWIN, W. H. G., TRUMPER, J., VAN PARADIJS, J., & SMITH, A. 1987, A&A, 172, 143

VAN PARADIJS, J. ET AL. 1990, PASJ, 42, 633

VAN PARADIJS, J. 1993, X-ray Binaries, ed. W. H. G. Lewin & E. P. J. van den Heuvel, Cambridge Univ. Press

VAN DER KLIS, M. 1989, ARA&A, 27, 517

VRTILEK, S. D. ET AL. 1991, ApJS, 76, 1127

VRTILEK, S. D., ET AL. 1994, ApJL, 436, L9

WHITE, N. E., PEACOCK, A., & TAYLOR, B. G. 1985, ApJ, 296, 475

WHITE, N. E., STELLA, L., & PARMAR, A. 1988, ApJ, 324, 363

An *EUVE* Detection of a Low-Mass X-ray Binary? AC211 in M15

PAUL J. CALLANAN,[1] JAY BOOKBINDER,[1]
JEREMY J. DRAKE,[2] AND ANTONELLA FRUSCIONE[2]

[1] Harvard-Smithsonian Center for Astrophysics, Cambridge 60 Garden Street, Cambridge, MA 02138, USA; and Eureka Scientific Inc., Oakland, CA, USA

[2] Center for EUV Astrophysics, 2150 Kittredge Street, University of California, Berkeley, CA 94720–5030, USA

We have observed M15 with EUVE, in an attempt to detect AC211, the well know globular cluster Low Mass X-ray Binary (LMXB). Our observations are part of an attempt to characterize the *EUVE* properties of LMXBs by looking at those towards which the extinction is extremely low (as it is for M15): these observations are impossible for the vast majority of Galactic LMXBs. *EUVE* successfully detected this cluster, at a countrate of 0.002 counts s^{-1}. In this paper we discuss the association of this source with AC211, and alternative scenarios for *EUVE* emission from M15.

1. Introduction

Low Mass X-ray Binaries (LMXBs) are systems where a neutron star or black hole primary accretes material (via an *accretion disk*) from a low mass ($M \leq 1 M_\odot$) secondary. As such, they provide a unique opportunity for the study of compact objects and their interaction with their companion stars (e.g., see Bhattacharaya & van den Heuvel 1991 for a review). Although they have been studied extensively from radio to $\sim$MeV energies, little is known of their EUV (10–100 eV) spectral characteristics. This is primarily because most LMXBs are relatively distant (>1 kpc), and lie behind >0.5 magnitudes of optical extinction, dramatically reducing any EUV flux. It is only with the launch of *EUVE* that observations can be performed which even approach the required sensitivity, and even then only for the relatively unobscured systems.

2. The LMXB in M15

There is a class of LMXBs toward which the reddening is accurately known—the globular cluster LMXBs. Of these the clusters with the lowest reddening are M15 ($E_{B-V} \sim 0.05$) and NGC 1851 ($E_{B-V} \sim 0.02$; Djorgovski 1993). These are the lowest measured towards any LMXB and hence these systems provide a unique opportunity to study the EUV characteristics of LMXBs in general. Furthermore, the low metallicity of M15 (only 0.01 Solar) implies that the effects of photoelectric absorption within this system are likely to be substantially reduced, further increasing the chances of detecting any EUV component.

The optical counterpart of the M15 LMXB, AC211, is the best studied cluster LMXB at optical wavelengths (e.g., Ilovaisky 1989). The optical light curve of AC211 shows an eclipsing modulation strongly indicative of a high inclination system: however, the X-ray light curve is only weakly modulated at the orbital period of 17.1 hrs (Auriere et al. 1993). The large degree of X-ray heating inferred from the optical colors and luminosity, coupled with the relatively modest X-ray luminosity of $\sim 5 \times 10^{36}$ ergs s^{-1} (e.g., Hertz & Grindlay 1983), implies that most of the emitted radiation from this system is unobserved. This is either because it is obscured, or because the bulk of the emission lies in a hitherto

S. Bowyer and R. F. Malina (eds.), Astrophysics in the Extreme Ultraviolet, 431–435.
© 1996 *Kluwer Academic Publishers. Printed in the Netherlands.*

unobserved spectral region. Although the former hypothesis was favored initially (e.g., Fabian, Guilbert, & Callanan 1987), it has since been shown to be untenable (see Dotani et al. 1990 for more details).

Hence we conclude that circumstantial evidence exists for an unobserved spectral component: could this be in the form of EUV emission? We obtained an *EUVE* observation of M15 in an attempt to constrain the contribution of such a component to the overall energetics of the system.

3. Observations and Data Reduction

The *EUVE* instrumentation has been described in detail by Bowyer & Malina (1991) and by Welsh et al. (1990). In brief, the *EUVE* Deep Survey and Spectrometer telescope (DSS) feeds an EUV spectrometer, which covers the wavelength range from 70–760 Å in three bands, and the Deep Survey photometer (DS). About half of the on-boresight photons are not intercepted by the dispersive elements of the spectrometer and pass through a boron coated Lexan filter (Vedder et al. 1989) to a focus on the DS detector. This telescope, filter and detector combination has significant transmission between about 65 and 170 Å, peaking at approximately 83 Å with an effective area of approximately 12 cm^2. The effective area curve is illustrated in Bowyer et al. (1994).

M15 was acquired by the *EUVE* Deep Survey and Spectrometer telescope (DSS) on UT 1994 June 6 13:55, and was observed until UT 1994 June 8 06:08.

We reduced the data corresponding to satellite night-time, which comprises about 25 minutes or so of each *EUVE* 90 minute orbit. Although the DS and SW spectrometer detectors remained switched on throughout the observation (except for episodes of very high geocoronal and particle background when count rates exceeded detector safety thresholds), the higher daytime geocoronal background can dominate the relatively faint flux from M15, and the daytime data are not useful. M15 was detected in the DS instrument only after a significant fraction of the photons from the entire observation were co-added. The average count rate in the DS was approximately 0.0017 count s^{-1}. As expected, this count rate is much too low to yield any appreciable signal in the spectrometer.

In order to produce light curves from the DS data, we first constructed photon event lists from the satellite telemetry, mapped in celestial coordinates, in the QPOE format. The QPOE event lists included sufficient sky area so as to facilitate accurate background subtraction. Episodes of high geocoronal and particle background were first filtered out, and light curves were then calculated from the photon event lists using an "aperture photometry" method for time bin sizes of 10,000 s. Shorter time bins resulted in rather noisier data. The image is illustrated in Figure 1, and the light curve is plotted in Figure 2. The light curve appears to be variable, albeit only at the 96% confidence level. There is no obvious modulation on the orbital ephemeris of AC211, but this does not rule against an origin in the LMXB because of (a) the shallow nature of the 1–10 keV modulation and (b) the low signal-to-noise of our data.

4. Discussion

4.1. *AC211 As the EUV Source*

Assuming an effective temperature of 50–100 eV (the range for which the *EUVE* DS photometer is most sensitive for the reddening discussed above), the observed *EUVE* countrate corresponds to a luminosity of $\geq 10^{35}$ ergs s^{-1} (see Callanan et al. 1995 for further details).

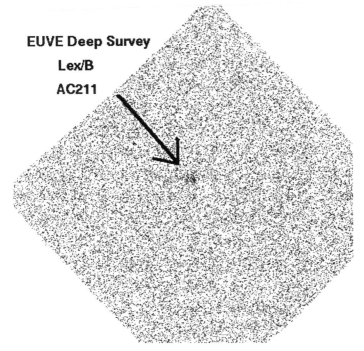

EUVE Deep Survey

Lex/B

AC211

FIGURE 1.

If this flux originates from AC211, then the true EUV luminosity will be even higher because of the column intrinsic to the binary. For example, a column of $\sim 9 \times 10^{20}$ cm^{-2} (well within the range reported by Callanan et al. 1987) yields an intrinsic luminosity of $\sim 10^{38}$ ergs $^{-1}$. Hence, without a simultaneous measurement of EUV flux and X-ray column, we can only constrain the EUV flux of AC211 to be 10^{35}–10^{38} ergs s^{-1}.

EUV emission from accreting neutron stars has previously been observed from X-ray pulsars (e.g., Her X-1 and RX J0059.2-7138) and the so called "Extreme Ultra Soft" sources (Shulman et al. 1985: Hughes 1994; Hasinger 1994). The former may be explained by emission from the neutron star magnetosphere (McCray & Lamb 1976). The most plausible explanation for the latter may be steady thermonuclear burning on the surface of a white dwarf (van den Heuvel et al. 1992). However, as AC211 is a bursting neutron star (e.g., Dotani et al. 1990), neither of these explanations can be valid in this case. However, a more precise estimate of the AC211 EUV luminosity is probably required before more detailed modelling is warranted.

5. Alternative Scenarios

The central star (L$\sim$3000L$_\odot$) of the well known planetary nebula K 648 in M15 might be expected to be a significant EUV source. However, the effective temperature is thought to be 40,000 K (Adams et al. 1984) and as such the system is beyond the sensitivity limit of *EUVE* (given the 10.5 kpc distance to M15).

However, it is more difficult to exclude a population of optically faint planetary nebulae/post-AGB stars. For example, systems such as K1-16 (Fruscione et al. 1995)

AC211 EUVE DS Light Curve

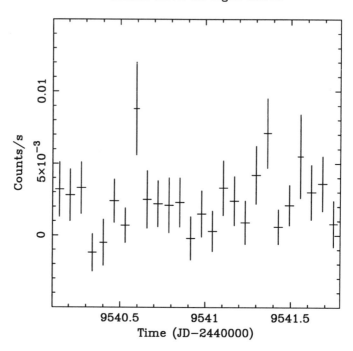

FIGURE 2.

would be detectable at the distance of M15, and could make a significant contribution to the EUV flux. However, preliminary calculations based on the current M15 mass function show that a sufficiently large number of such stars are unlikely exist in the cluster core (Callanan et al. 1995).

Conclusive evidence for the association of our EUV flux with AC211 is only likely to come if significant variability can be found in the EUV lightcurve: a longer observation is required to establish this with certainty.

6. Conclusions

We have detected EUV emission from the globular cluster M15. Although it is likely to be associated with AC211, we cannot as yet exclude the possibility that some of the emission is due to a population of post-AGB/planetary nebula stars. If the emission is indeed dominated by AC211, we can constrain the LMXB EUV luminosity to $\sim 10^{35}$–10^{38} ergs s^{-1}. It is clear that further observations of this and other low column LMXBs (i.e., that in NGC 1851) are required to establish the true origin of the M15 EUV emission, and the ubiquity of LMXB EUV emission in general.

REFERENCES

ADAMS, S., SEATON, M, J., HOWARTH, I. D., AURIERE, M. & WALSH, J. R. 1984, MNRAS, 207, 471

Auriere, M. et al. 1984, A&A, 138, 415

Bhattacharaya, D. & E. P. J. & van den Heuvel 1991, Phys. Reports, 203, 1

Bowyer, S., Lieu, R., Lampton, M., Lewis, J., Wu, X., Drake, J. J., Malina, R. F. 1994, ApJS, in press

Bowyer, S., Malina, R. F. 1991, in Extreme Ultraviolet Astronomy, ed. R.F. Malina & S. Bowyer, New York: Pergamon Press, 397

Callanan, P. J. et al. 1987, MNRAS, 224, 781

Callanan, P. J., Bookbinder, J., Drake, J. J. & Fruscione, A. 1995, ApJ, in preparation

Djorgovski, S. 1993, in Structure and Dynamics of Globular Clusters, ed. S. Djorgovski & G. Meylan, ASPCS 50, 1993

Dotani, T., et al. 1990, Nature, 347, 534

Fabian, A. C., Guilbert, Callanan, P. J. 1987, MNRAS, 225, 29

Fruscione, A., et al. 1995, ApJ, 441, 726

Hasinger, G. 1994, in The Evolution of X-ray Binaries, AIP Conf. Proc. 308, ed. S.S. Holt & C.S. Day

Hertz, P. & Grindlay, J. E. 1983, ApJ, 275, 105

Hughes, J. P. 1994, ApJL, in press

Ilovaisky, S. 1989, in Proc. 23rd ESLAB Symposium, ESA SP-296, 151

McCray, R. A. & Lamb, F. K. 1976, ApJ, 204, L115

Schulman, S. et al. 1975, ApJ, 199, L101

van Paradijs, J. 1993, in X-ray Binaries, ed. W.H.G. Lewin, J. van Paradijs, & E.P.J. van den Heuvel, Cambridge: Cambridge Univ. Press

Vedder, P. W., Vallerga, J. V., Siegmund, O. H. W., Gibson, J., & Hull, J. 1989, in Proc. SPIE, Space Sensing, Communications, and Networking, ed. M. Ross & R.J. Temkin, 1159, 392

Welsh, B. Y., Vallerga, J. V., Jelinsky, P., Vedder, P. W., & Bowyer, S. 1990, Opt. Eng., 29, 752

Extreme Ultraviolet Emission from Neutron Stars

R. S. FOSTER,[1] J. EDELSTEIN,[2] AND S. BOWYER[2]

[1]Remote Sensing Division, Code 7210, Naval Research Laboratory,
Washington, DC 20375, USA

[2]Center for EUV Astrophysics, 2150 Kittredge Street,
University of California, Berkeley, CA 94720–5030, USA

We summarize the detections of extreme ultraviolet (EUV) emission from neutron stars. Three firm detections have been made of spin-powered pulsars: the aged millisecond pulsar PSR J0437−4715, the middle-aged X-ray pulsar Geminga, and the radio pulsar PSR B0656+14. These observations allow us to evaluate both power-law and thermal-law emission models as the source of the EUV flux. For the case of PSR B0656+14 the lack of flux modulation with pulse period argues that the EUV radiation originates from the cooling neutron star surface rather than from a hot polar cap. If the emission is from a thermalized neutron star surface, then limits can be placed on the surface temperature. For the case of Geminga we can explain the observed EUV flux using thermal models that are consistent with standard neutron cooling scenarios. We also have a weak indication that the EUV emission from Geminga is pulsed in a manner consistent with the lowest energy channel observed with *Rosat*. For the case of the millisecond pulsar PSR J0437−4715 standard neutron star cooling models require surface re-heating. We compare different heating models to the data on this object. We rule out re-heating by crust–core friction, and find that models for the accretion from the interstellar medium, accretion from the white dwarf companion and a particle-wind nebula do not account for the EUV luminosity. Models of pulsar re-heating by magnetic monopole catalysis of nucleon decay are used to establish new limits to the flux of monopoles in the Galaxy. A single power-law source with properties derived from X-ray data cannot explain the EUV flux from PSR J0437−4715. The strongest model for explaining the EUV emission consists of a large ∼ 3 km^2 polar cap heated from particle production in the pulsar magnetic field. We consider the prospects for detecting other neutron stars in the extreme ultraviolet.

1. Introduction

The detection of extreme ultraviolet (EUV) emission from neutron stars using the *Extreme Ultraviolet Explorer* (*EUVE*) represents a challenging observation that requires both long integration times on the sources and a population of nearby neutron stars. Most neutron stars that we know about in the Galaxy are at distances between 0.5 and 10 kpc away. The closest known radio pulsar is ∼ 100 pc distant. High column densities of neutral hydrogen, particularly in the plane of the Galaxy, introduce a large photoelectric absorption in the *EUVE* instrumental bandpass that strongly attenuates any EUV emission. Theoretical models for neutron star cooling predict that neutron stars with ages exceeding ∼ 10$^{6.5}$ years will have very low or non-existent fluxes in the EUV or soft X-ray from thermal surface emission as they will have surface temperatures well below 10^5 K. The detection of any neutron stars as EUV sources is certainly a surprise.

As of March 1995 there are at least 706 known radio pulsars (Taylor et al. 1995) representing a 27% increase in the number of known spin-powered neutron stars since the previous published catalogue (Taylor, Manchester, & Lyne 1993). A number of these new pulsars were found as a result of several sensitive searches for millisecond period

S. Bowyer and R. F. Malina (eds.), Astrophysics in the Extreme Ultraviolet, 437–442.
© *1996 Kluwer Academic Publishers. Printed in the Netherlands.*

TABLE 1. *EUVE* neutron star detections

Source	Count (ct/s)	Age (yrs)	DM (pc cm^{-3})	Distance (pc)
PSR J0437−4715	0.0143	5×10^9	2.65	140
PSR B0656+14	0.024	1×10^5	14	260
Geminga	0.008	5×10^5	—	< 500
Her X-1	0.18*	$< 1 \times 10^7$	—	< 5000

* Lexan sky survey detection of an accretion powered source (Vrtilek et al. 1994).

pulsars at high Galactic latitudes. One of these recently discovered millisecond pulsars is PSR J0437−4715 (Johnston et al. 1993) and is located at a distance of only about 140 pc away and is the closest known millisecond pulsar. Edelstein, Foster, & Bowyer (1995) and Shemi (1995) argue that a number of the closest neutron stars might be EUV sources. Examination of the *EUVE* and *Rosat* databases for unidentified sources may lead to the discovery of additional pulsars from the very local population of isolated neutron stars.

2. Observations

Three spin-down powered pulsars have been detected with *EUVE*. They are the pulsars PSR B0656+14 and Geminga, plus the millisecond pulsar PSR J0437−4715. Table 1 gives the count rate, spin-down age, dispersion measure, and distance. We also list the accretion powered neutron star Hercules X-1, however this EUV source is probably not the neutron star but the system's accretion disk. The *EUVE* sources are all coincident with the radio or X-ray source positions within the 1 arcminute uncertainty of the *EUVE* pointing.

All three detected spin-powered pulsars were observed with the 100 Å (Lexan) filter on the Deep Survey Instrument. The Geminga source has the longest integration of 250,458 seconds between 1994 January 14 to 24. A total of 2306 counts were received in the detector which resulted in a net count of 2032 after subtracting background. The pulsar PSR B0656+14 had a shorter exposure of 91,021 seconds with 3358 photons counted during the observation. After removing the background count rate a net count of 2109 photons were detected during the observation.

The millisecond pulsar PSR J0437−4715 was observed between 1994 January 31 and February 4 for a total of 71,886 s. The photon count rate after correcting for instrumental vignetting and telemetry losses was 0.0143 ± 0.0008 counts s^{-1}. Using an assumed ionization fraction $X = N_e/N_H \sim 0.1$ we can estimate the neutral hydrogen column density of $\sim 7 \times 10^{19}$ cm^{-2} from the pulsar dispersion measure of 2.65 pc cm^{-3}. This source was observed by Becker & Trumper (1993) with *Rosat* and shown to have pulsed X-ray emission commensurate with the pulsars 5.75 ms rotational period.

The source flux for each pulsar is derived from the count rates using the Deep Survey effective area of an assumed hydrogen column N_H and a spectral model of the instrumental response. The Deep Survey response function is highly sensitive to the absorbing column densities between 10^{19} cm^{-2} and 10^{21} cm^{-2}. Detection of pulse modulation was

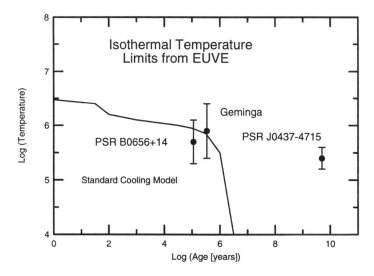

FIGURE 1. Isothermal temperate versus pulsar age for the three spin-powered pulsars detected with *EUVE*. Also plotted is a standard neutron star cooling model (e.g., Page & Applegate 1992).

not possible for PSR J0437−4715 due to limited time resolution in the mode used for these observations.

3. Discussion

The *EUVE* data allow us to discriminate between different power-law models and blackbody source models for emission from nearby neutron stars. In the case of PSR J0437−4715, we found that the power-law model based upon the *Rosat* observations by Becker & Trumper (1993) cannot explain the observed EUV emission. We found that the EUV data can be explained by extrapolations of the *Rosat* power-law model only for the case where the total absorbing column of neutral hydrogen exceeds the measured hydrogen column out of the Galaxy in this direction.

The companion star to the millisecond pulsar has been detected optically (Bell, Bailes, & Besell 1993) and identified as a cool white dwarf with a color temperature of 4000 ± 350 K. The star is too cool and too far (65 to 160 pc distant) to be the source of the EUV emission.

It is possible that the flux we are observing from PSR J0437−4715 is thermal in origin. Edelstein, Foster, & Bowyer (1995) argued the the EUV flux and X-ray emission are both consistent with a single blackbody with a temperature of $\sim 5.7 \times 10^5$ K from an emitting region of ~ 3 km^2 with a Galactic absorbing column of $N_{\mathrm{H}} \sim 5 \times 10^{19}$ cm^{-2}. If the emission is from the entire surface of the neutron star, then we can place a lower limit of the surface temperature of $1.6 \times 10^5 d_{140}^{-0.5}$ K for a distance of 140 pc and a radius of 10 km. An upper limit constrains the temperature to be less than $4.0 \times 10^5 d_{140}^{-0.5}$ K.

Applying similar isothermal models to the observed emission from the two younger neutron stars we can derive temperature limits as a function of spin-down age. We

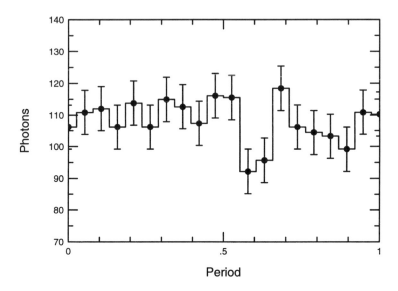

FIGURE 2. Extreme ultraviolet photon counts versus pulse phase for the spin-power pulsar Geminga. The pulse profile is computed using the spin-down model from the X-ray timing observations of Halpern & Holt (1992) and Halpern & Ruderman (1993).

plot these limits in Figure 1. Also in Figure 1 is a standard neutron star cooling curve following Page & Applegate (1992). The two 10^5 year old neutron stars fit the cooling curve while the 5×10^9 year old millisecond pulsar shows excess emission compared to standard cooling models. This leads us to speculate on the source of the neutron star surface re-heating that must be going on, if the EUV emission is thermal in origin.

Edelstein, Foster, & Bowyer (1995) considered five possible mechanism for thermal re-heating in old neutron stars including crust-core friction from pulsar spin-down, Bondi-Hoyle type accretion from the ISM, accretion from the companion star, a particle-wind nebula generated from the neutron star, and magnetic monopoles. All of these suggestions are difficult to justify observationally, although the particle-wind nebula mechanism following Arons & Tavani (1993) might explain the emission within an order of magnitude.

It has been suggested that neutron stars could be heated by magnetic "monopole-catalyzed" nucleon decay (Freese, Turner, & Schramm 1983 and Kolb & Turner 1984). Superheavy magnetic monopoles ($\leq 10^{21}$ GeV) that hit the surface of a neutron star will lose sufficient energy through electronic interactions to be captured by the star. These monopoles accumulate inside the neutron star at a rate proportional to the monopole flux. The estimated upper limit to the Galactic magnetic monopole flux is

$$F_G \leq 8.3 \times 10^{-25} \left(\frac{\tau_9}{5}\right)^{-1} (\sigma_{-28}\, v)^{-1}\, R_{10}^4\, \beta_{-3}^2\, M_{1.4}^{-2}\, \mathrm{cm}^{-2}\, \mathrm{sr}^{-1}\, \mathrm{s}^{-1}\,,$$

where τ_9 is the neutron star age in units of 10^9 yr, σ_{-28} is the interaction cross section in units of 10^{-28} cm^2, v is the relativistic velocity difference between the monopole and interacting nucleon assumed in units of the speed of light $c = 3 \times 10^{10}$ cm s^{-1}, R_{10} is the neutron star radius in units of 10 km, β_{-3} is the monopole's velocity far from the star in units where $\beta_{-3} = 10^3 \times vc^{-1}$ (the viral velocity of a monopole in the Galaxy is $\sim 10^{-3}c$), and $M_{1.4}$ is the neutron star mass in solar units. The upper limit to the surface temperature of PSR 0437$-$4715 and the object's extreme age results in a value for F_G that is 3 orders of magnitude lower than the upper limit derived from PSR 1929+10 (Freese, Turner, & Schramm 1983).

The photon arrival times for the data from the millisecond pulsar PSR J0437$-$4715 have insufficient time resolution to identify pulsed emission. Future observations using a high time resolution mode on the *EUVE* spacecraft will allow for the identification of any possible pulsed emission in the photon count statistics. The photons collected from both PSR B0655+14 and Geminga were folded using the spin-down models with differing results. No pulsed emission was identified from PSR B0655+14 while a weak indication of pulsed emission exists for Geminga (see Figure 2). An inverted pulse (or dip) near the phase 0.5 is similar to the profile seen in the lowest energy 0.07$-$0.28 keV band from the *Rosat* data (Halpern & Holt 1992 and Halpern & Ruderman 1993).

4. Summary

Three spin-powered pulsars have been detected with *EUVE*. All three sources have previously been detected in soft X-rays with *Rosat*. The two pulsars with spin-down ages less than 10^6 years have EUV emission that fit standard thermal models for neutron star cooling. The apparent excess emission from the millisecond pulsar PSR 0437$-$4715 seen in both the extreme ultraviolet and soft X-rays are not consistent with standard models describing thermal flux from this object. However, a large polar cap heated from the back flow of relativistic particles produced in the magnetic beam might be producing a ~ 3 km^2 region that would explain both the EUV and X-ray observations below 0.4 keV. Alternatively, if the EUV emission is due to a thermalized surface on the neutron star then we place a limit on the surface temperature of the neutron star of $1.6 - 4.0 \times 10^5$ K. Some surface re-heating mechanism would be required to explain this temperature according to the standard models for neutron star cooling.

The discovery of a number of new nearby radio pulsars and the large list of still unidentified EUV sources provides an opportunity for the detection of additional neutron stars in the EUV. Further study of this population may provide clues to the nature of the EUV emission from neutron stars.

We acknowledge useful conversations in the preparation of this work with C. Dermer, J. Arons, and M. Tavani. We also thank P. Ray and D. Chakrabarty for their assistance in providing software for the ephemeris conversions. Basic research in precision pulsar astrophysics at the Naval Research Laboratory is supported by the Office of Naval Research. The *Extreme Ultraviolet Explorer* is supported by NASA contract NAS5-30180.

REFERENCES

ARONS, J. & TAVANI, M. 1993, High-energy emission from the eclipsing millisecond pulsar PSR 1957+20, ApJ, 403, 249

BECKER, W. & TRÜMPER, J. 1993, Detection of pulsed X-rays from the binary millisecond pulsar J0437$-$4715, Nature, 356, 528

EDELSTEIN, J., FOSTER, R. S., & BOWYER, S. 1995, Extreme Ultraviolet Emission from the Millisecond Pulsar J0437$-$4715, ApJ, in press

FREESE, K., TURNER, M., & SCHRAMM, D. N. 1983, Monopole catalysis of nucleon decay in old pulsars, Phys. Rev. L., 51, 1625

HALPERN, J. P. & HOLT, S. S. 1992, Discovery of soft X-ray pulsations from the gamma-ray source Geminga, Nature, 357, 222

HALPERN, J. P. & RUDERMAN, M. 1993, Soft X-ray properties of the Geminga pulsar, ApJ, 415, 286

JOHNSTON, S., ET AL. 1993, Discovery of a very bright, nearby binary millisecond pulsar, Nature, 361, 613

KOLB, E. & TURNER, M. 1984, Limits from the soft X-ray background on the temperature of old neutron stars and on the flux of superheavy magnetic monopoles, ApJ, 286, 702

PAGE, D. & APPLEGATE, J. H. 1992, The cooling of neutron stars by the direct Urca process, ApJL, 394, L17

SHEMI, A. 1995, Are Unidentified EUV Sources the Closest Neutron Stars?, MNRAS, submitted

TAYLOR, J. H., MANCHESTER, R. N., & LYNE, A. G. 1993, Catalog of 558 pulsars, ApJS, 88, 529

TAYLOR, J. H., MANCHESTER, D. N., LYNE, A. G., & CAMILO, F. 1995, in preparation

VRTILEK, S. D. ET AL. 1994, Multiwavelength observations of Hercules X-1/HZ Hercules, ApJL, 436, L9

EUV/Soft X-ray Spectra for Low B Neutron Stars

ROGER W. ROMANI,[1] MOHAN RAJAGOPAL,[1]
FORREST J. ROGERS,[2] AND CARLOS A. IGLESIAS[2]

[1]Department of Physics, Stanford University, Stanford, CA 94305-4060, USA

[2]Lawrence Livermore National Laboratory, Livermore, CA 94550, USA

Recent *ROSAT* and *EUVE* detections of spin-powered neutron stars suggest that many emit 'thermal' radiation, peaking in the EUV/soft X-ray band. These data constrain the neutron stars' thermal history, but interpretation requires comparison with model atmosphere computations, since emergent spectra depend strongly on the surface composition and magnetic field. As recent opacity computations show substantial change to absorption cross sections at neutron star photospheric conditions, we report here on new model atmosphere computations employing such data. The results are compared with magnetic atmosphere models and applied to PSR J0437-4715, a low field neutron star.

1. Introduction

Thermal radiation from neutron star surfaces is of continuing interest to the X-ray community because it provides important clues to the compact objects' evolution and to the physics of the high density interior. Model atmosphere calculations for low field neutron stars (Romani 1987) with effective temperatures $5.0 < \log(T_{\text{eff}}) < 6.5$ and a variety of surface compositions showed that the emergent spectra could differ strongly from a Planck function. Extension of these results to atmospheres with strong ($> 10^{12}$G) fields has proved possible for pure H models (Pavlov et al. 1994). Satisfactory opacities in strong fields (Miller 1992) are not yet available for heavy elements, but the effects should be significantly weaker for a given field than in the H atmosphere case.

In the last few years there has been a dramatic improvement in our ability to probe neutron star thermal fluxes with a number of strong *ROSAT* detections (Ögelman 1994) and, more recently, some detections with *EUVE* (Edelstein, Foster & Bowyer 1995). These results include measurements of millisecond pulsars and other low field neutron stars. With the availability of flux ratios and some low resolution spectra it is important to compare the observations with models of varying surface composition; iron is of particular interest since it is the surface element for a BPS composition run. Recently it has also been realized that previous opacity computations, e.g., the Los Alamos Opacity Library results, based on DCA (detailed configuration accounting) were not sufficiently accurate. In particular, new opacities by the OPAL (Rogers & Iglesias 1994) and OP (Seaton et al. 1994) groups, including configuration term structure, have significantly modified the contribution of Fe and other heavy elements. The most important changes are due to M shell transitions near 100eV, which provide large increases (up to four-fold) in solar abundance opacities. These modifications have already solved a number of outstanding stellar physics puzzles. For the neutron star thermal spectrum problem the new opacities are even more significant, since these objects can have pure Fe atmospheres, with strong surface gravities and temperatures $\log(T_{\text{eff}}) \sim 5 - 6$ ensuring that the emergent spectrum forms precisely in the region where the opacity changes have a maximal effect.

These two developments have prompted us to recompute atmosphere structures and

S. Bowyer and R. F. Malina (eds.), Astrophysics in the Extreme Ultraviolet, 443–447.
© *1996 Kluwer Academic Publishers. Printed in the Netherlands.*

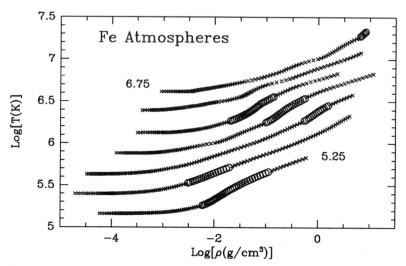

FIGURE 1. Fe model atmosphere structures. Convectively unstable regions are shown with open circles.

emergent spectra for weakly magnetized neutron stars. Here we report on preliminary results using OPAL H and Fe opacities. Improved equation of state data also allow us to investigate the atmospheric stability. We compare the new Fe spectra with emergent spectra for light elements and show how the sensitivity of recent observations can constrain the atmospheric properties of some pulsars.

2. Model Atmosphere Computations

For the iron atmospheres a pure Fe equation of state table was computed in the range $4.5 < \log(T) < 8$, $-14 < \log(\rho) < 5$, covering $\log(R = \rho/T^3) < -18$. The upper cut-off in density was imposed to keep relativistic corrections small. For this pure Fe gas 'monochromatic' opacities were computed as 10^4 Rosseland group means spaced linearly over $0 < u = \frac{E}{k_B T} < 20$. To extend to high energies in the final emergent spectrum, this was supplemented by 10^4 Planck mean opacities spaced linearly in energy to 10 keV. As a representative light element, pure H equations of state and Planck mean 'monochromatic' opacities extending to 10keV were also computed. We have grids for solar abundance compositions as well; computations of these atmospheres are in progress.

The atmosphere calculations were started from models obtained by imposing hydrostatic equilibrium on the exact gray atmosphere solution. Typically 100 grid zones were utilized, extending to a Rosseland mean optical depth $\tau_R \sim 300$. Mean opacities at $\sim 10^3$ frequencies were computed and the transfer equations were solved at each energy group to compute the atmospheric flux. At large monochromatic τ_ν, the Milne integrals were replaced by diffusion to preserve convergence to the exact grey solution. The atmospheric temperature runs were corrected using an adaptive Lucy-Unsöld procedure; convergence to the radiative zero solution within 1% was achieved throughout. Final emergent spectra were then computed from $> 10^4$ Planck mean subgroups and binned down to the desired resolution.

Atmospheric stability was checked by computing the energy gain ratio for a convection

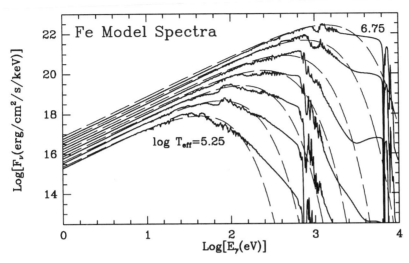

FIGURE 2. Emergent spectra for 10km, $1.4M_\odot$, Fe atmosphere neutron stars computed from
OPAL opacities. Blackbody spectra are shown with dashed lines.

cell along the radiative solution, $dT/dz|_{rad}$:

$$\Gamma \approx \frac{\kappa_{cm}k_B^2}{24\,\sigma\mu^2 m_H^2}\rho(dT/dz|_{rad} - dT/dz|_{ad})^{1/2}C_P/(T\,g)^{3/2} \qquad (2.1)$$

where the local adiabatic gradient was determined from the EOS computations. Although
the atmospheres become formally convective at large depth the very large surface gravity
ensures that convection is weak in most cases, since $\Gamma \propto 1/g$. For the coldest iron
atmospheres $\log(T_{\rm eff})\lesssim5.25$ the convective flux transport contributes only a few percent
change to $T(\rho)$; for the other models corrections are smaller (Fig. 1). For H, convection
can be significant at large depth for $\log(T_{\rm eff})\lesssim5.5$, although atmospheric B fields will
provide some further suppression. Spectral modifications due to photospheric convection
are not included here; these decrease the flux above the thermal peak slightly.

Our products are grids of Fe and H atmospheres computed with the surface gravity
of a $1.4M_\odot$, 10 km neutron star (e.g., Fig. 2). Emergent spectra (with zero redshift)
are computed for $5 < \log(T_{\rm eff}) < 6.75$. To compare with observations these must be
redshifted to infinity; spectra for various neutron star radii and masses are adequately
represented by varying this redshift. These spectra are available for comparison with
EUV/soft X-ray observations of neutron stars. (Contact: rwr@astro.stanford.edu)

3. Application to PSR J0437-4715

Using the *ROSAT* PSPC Becker and Trümper (1993) have studied the nearby (140pc),
low field (B=2.5×10^8G) millisecond pulsar PSR J0437-4715, detecting pulsed X-rays in
the 0.1-2.4 keV band. They find that the time averaged spectrum is fit by a composite
model of a powerlaw plus blackbody. The pulse fraction peaks at ~ 1 keV, indicating
that pulsation is due to variation in the 'blackbody' component.

We have re-analyzed the PSPC data, phasing to the radio pulse ephemeris (Manchester,
private comm.). We find that the X-ray pulse arrives in phase with the radio emission

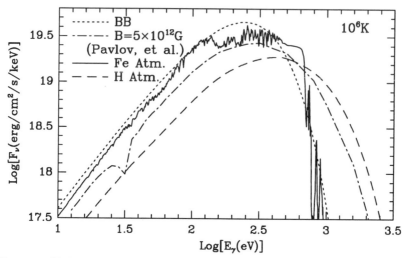

FIGURE 3. Variation with surface composition and magnetic field at T=10^6 K. H and Fe spectra are from the present work; the H atmosphere with B=5×10^{12}G is from Pavlov et al. 1994. Note that all spectra depart significantly from the equivalent blackbody.

TABLE 1. Fits to PSR J0437-4715 PSPC Data

Model	$T_{\rm eff}(10^6$ K)	$A_{\rm eff}({\rm cm}^2)$	$\chi^2/{\rm DOF}$
BB	1.4	9.8×10^9	1.33
H	0.56	4.0×10^{11}	0.85
Fe	1.0	3.8×10^{10}	2.0

(residual $ROSAT$ clock errors, however, allow a phase error $\Delta\phi \sim 0.1$), supporting the idea that the thermal pulsed component is associated with a heated polar cap. We therefore extract the off-pulse spectrum and find the best fit powerlaw to represent the unpulsed, presumably plerionic, emission. This fit is acceptable, but shows significant residuals. To test the effect of surface composition on the polar cap flux, we then fit the on-pulse spectrum by adding a thermal component. Results are summarized in Table 1.

While the blackbody is statistically acceptable, the implied polar cap area is very small. The Fe atmospheres, having a sharp L edge at $\sim$0.9keV, cannot provide a good fit to the data. Light element, e.g., H, atmospheres produce a significantly better fit than the best blackbody due to the excess above the thermal peak (Figure 3). The large emission region implied by the lower temperature is in better agreement with theoretical expectations as well; the cap radius for an aligned vacuum dipole is $r_{cap} = (2\pi r_{NS}^3/Pc)^{1/2}$ giving an area 1×10^{11}cm^2 while the cap size inferred for surface core emission ($r_{cap} \sim 0.043 r_{NS}P^{-1/2}$, Rankin 1990) gives an area of 9×10^{11}cm^2.

Similarly, the presence of a light element atmosphere simplifies interpretation of the recent detection of PSR J0437-4715 in the 100Å filter of the $EUVE$ deep survey instrument (Edelstein, Foster & Bowyer 1995). The $EUVE$ flux for an absorbing column of $N_H \sim 10^{20}$cm^{-3}pc is in agreement that expected from an H atmosphere with an effective temperature of $T_{\rm eff} \sim 6 \times 10^5$K and an emitting area of $2 - 3 \times 10^{11}$cm^2.

4. Conclusions

Computations of Fe and H atmospheres with revised opacity and equation of state data provide a baseline set of atmospheres to compare with observed spectra of neutron stars. While for most young pulsars strong $\sim 10^{12}$G magnetic fields ensure that magnetic atmospheres are needed for detailed comparison, the gross differences between the H and Fe spectra in the present spectra still provide a useful probe of the surface composition. Moreover, these spectra are directly applicable to millisecond pulsars and other low field neutron stars.

For the Fe atmospheres, improvements to the atmospheric computations and the OPAL opacity data have resulted in significant changes to the absorption edge and line strengths in the emergent spectra, especially at energies in the EUV/Soft X-ray range, $\sim 50 - 300$ eV. These new spectra should be used to illustrate the effects of heavy element atmospheres. For the H atmospheres computation of stability criteria along the radiative zero solution indicates that there can be significant convective energy transport for low T_{eff}, although the high surface gravity and the presence of magnetic fields suppresses the convection. In the light element atmospheres the trend towards 'harder' spectra for a given T_{eff} with excess emission above the Wien peak is clear. For high magnetic fields this excess is somewhat suppressed (Pavlov et al. 1994).

Because of the strong broad-band features in these models, phase resolved X-ray spectral data, even of low resolution, can constrain the composition of neutron star surfaces. Comparison of our models with PSR J0437-4715 PSPC observations suggests that, as for other soft X-ray pulsars, 'thermal' spectra harder than the equivalent blackbody provide the best description of the data; a light element surface can provide such a spectrum. The change in the inferred T_{eff} has significant effect on the thermal history and on the inferred EOS at supernuclear densities. By inference, application of magnetic H atmospheres to high-field young neutron stars may be appropriate, as well. This is fortunate, since magnetic opacities for heavy elements will be difficult to compute.

Thus phase resolved soft X-ray spectra can isolate thermal surface emission from pulsar magnetospheric or plerionic flux. Phase resolved fluxes from $EUVE$ images can also provide important constraints. We note that spectral results from $EUVE$ would be even more revealing if moderate age ($\log(T_{\text{eff}}) \gtrsim 5.5$) pulsars can be found with weak interstellar absorption $N_H \lesssim 10^{19}$cm^{-3}pc. In this case measurement of spectral lines and breaks can strongly probe the surface composition and redshift. Future AXAF and XMM observations should be able to effect such tests for more distant objects.

REFERENCES

BECKER, W. & TRÜMPLER, J. 1993, Nature, 365, 528

EDELSTEIN, J., FOSTER, R., & BOWYER, S. 1995, ApJ, submitted; see also Foster, these proceedings

MILLER, M. C. 1992, MNRAS, 255, 129

ÖGELMAN, H. 1994, In Lives of Neutron Stars, ed. J. van Paradijs & A. Alpar,. Dordrecht: Kluwer

PAVLOV, G. G., ET AL. 1994, In Lives of Neutron Stars, ed. J. van Paradijs & A. Alpar,. Dordrecht: Kluwer

RANKIN, J. M. 1990, ApJ, 352, 247

ROGERS, F. J. & IGLESIAS, C. A. 1994, Science, 263, 50

ROMANI, R. W. 1987, ApJ, 313, 718

SEATON, M. J. ET AL. 1994, MNRAS, 266, 805

EUV Studies of Solar System Objects: A Status Report

SUPRIYA CHAKRABARTI[1] AND G. RANDALL GLADSTONE[2]

[1]Center for Space Physics, Boston University,
725 Commonwealth Avenue, Boston, MA 02215, USA

[2]Southwest Research Institute, PO Drawer 28510, San Antonio, TX 78228, USA

EUV studies have contributed substantially to our understanding of the physical and chemical properties the Sun, planets, and their satellites. Although the spectroscopic data set is limited to Venera 11/12, Voyager 1/2, Astro 1/2, EUVE, Galileo, and a handful of sounding rocket experiments, these data have provided important insights regarding the atmospheres and surfaces of several planets and satellites to the point where rudimentary comparative planetology can be conducted. In this paper we highlight some of these results.

1. Introduction

The EUV spectral region is rich in emission features from the most common planetary atmospheric species. Recently, it has also been proposed to be useful for studies of the surfaces of solar system bodies without atmospheres. Spectrographs aboard planetary spacecraft, most notably the Voyagers, have established the general EUV/FUV spectral characteristics of the upper atmospheres and plasma environments of most of the planets. The *EUVE* mission continues to provide new insights into solar system objects. The space-shuttle-based Astro missions and sounding rocket experiments have also added to our understanding of the solar system through their EUV observations.

The richness of the EUV spectral region for planetary studies has been discussed by Feldman & Bagenal (1991) (see their Table 1). Extreme ultraviolet emission features include transitions of atoms and their ions (e.g., H, He, He$^+$, Ne, Ar, O, O$^+$, O^{++}, N, N$^+$, N^{++}, S$^+$, S^{++}, S^{+++}) as well as molecules (e.g., H$_2$ and N$_2$). Through occultation studies, one can also infer the altitude distribution of abundant EUV-absorbing molecules (e.g., O$_2$ and N$_2$ on Earth, or hydrocarbons on the giant planets).

In this paper we provide a summary of EUV observations of solar system objects. The discussions will be broken into observations of the study of the Moon, the terrestrial planets, the jovian planets, the interplanetary medium, and comets.

2. The Moon

There have been only a small number of studies on EUV emissions from the Moon. An EUV spectrophotometer on the Mariner 10 spacecraft observed the Moon in the 550–1250 Å range and obtained a geometric albedo of 2–10% at a solar phase angle of 74°(Wu & Broadfoot 1977).

Prompted by the *ROSAT* soft X-Ray observation of the Moon (Schmitt et al. 1991), Edwards et al. (1991) simulated lunar EUV emissions from L- and M-shell X-Ray fluorescence. The authors predicted much higher emission peaks in the 90–500 Å range from this process than from the reflected solar spectrum.

During the all-sky survey phase of the *EUVE* operations, *EUVE* scanned the Moon on several occasions. Gladstone et al. (1994) used this data to estimate the solar EUV

S. Bowyer and R. F. Malina (eds.), Astrophysics in the Extreme Ultraviolet, 449–456.
© *1996 Kluwer Academic Publishers. Printed in the Netherlands.*

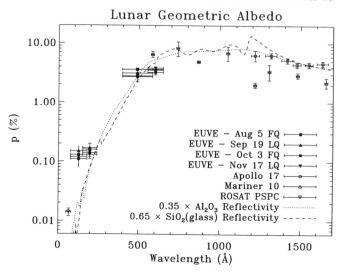

FIGURE 1. EUV albedo of lunar surface obtained by *EUVE* and other experiments.

irradiance and its variation. The authors used the bulk reflectivities of SiO_2 and Al_2O_3 (Phillip 1985 and Gervais 1991) to explain the photometric observations (Figure 1). The average geometric albedos obtained in the 150–650 Å range varied from 0.15% to 3.5% with an upper limit of 0.13% in the 75–180 Å range. The authors found that the primary features are consistent with reflected sunlight, rather than the X-ray fluorescence emissions suggested by Edwards et al. (1991). Subsequently, Gladstone (1994) used the *EUVE* spectrometer to obtain the EUV spectrum of the Moon, which indicated that the observed signal is primarily due to reflected sunlight.

3. Terrestrial Planets

Although there have been numerous EUV observations of Earth's upper atmosphere, only a single EUV observation of Mars and a handful for Venus have been reported (e.g., Krasnopolsky et al. 1994; Bertaux et al. 1981; Hord et al. 1991). As a result, EUV observations of terrestrial planets remain an almost unexplored domain. Even the first round of discovery missions will not fill this gap.

Terrestrial EUV observations started with broadband photometric measurements from sounding rockets and satellites which obtained the intensity distributions of the bright features from the geocorona (H I Lyman α and He I 584 Å), nighttime ionosphere (O I 911 Å, 1304 Å, and 1356 Å features), plasmasphere (He II 304 Å), and the interplanetary medium (He I 584 Å and He II 304 Å) [for an excellent review of Earth's EUV airglow see Meier (1991)]. Detailed experimental and modeling studies have shown evidence for the upflow of ionospheric ions in the magnetosphere (Chiu et al. 1986), detected the presence of a non-thermal population of atomic oxygen (Cotton et al. 1993), inferred solar line center flux and line width from He I 584 Å airglow (Bush & Chakrabarti 1995), and obtained the ionospheric temperature from the shape of O I recombination continuum (Feldman et al. 1992).

The atmospheres of Mars and Venus have been observed in the FUV from a number of U.S. and Soviet spacecraft. He I 584 Å emission from Venus was observed by the Ven-

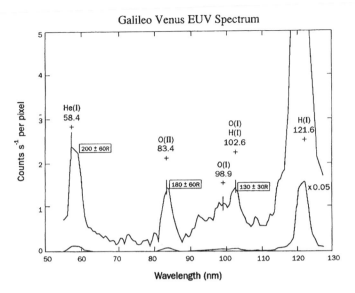

FIGURE 2. Venus airglow spectrum obtained by the Galileo UV spectrograph.

era 11/12 spacecrafts (Chassefiere et al. 1986). O II 834 Å emissions and, surprisingly, argon emissions at 869 Å and 1048 Å were also recorded by Venera missions. These observations were conducted by broadband instruments, however, which makes unambiguous determination of atmospheric composition a difficult task (Bertaux et al. 1981). In particular, the argon detections are especially suspect. On its trip to Jupiter, the Galileo spacecraft obtained an EUV spectrum of Venus (Hord et al. 1991), which includes several features due to O, O$^+$, He and H (Figure 2).

EUV spectroscopy of Mars and Venus can provide important clues for understanding not only the present structure of their upper atmospheres, but also their outgassing history. The first step towards that goal was achieved by the *EUVE* mission in January 1993, when it observed the He I 584 Å feature on Mars (Krasnopolsky et al. 1994). The observed spectrum shows the presence of martian helium emissions above the geocoronal background. The measured brightness of the He I 584 Å emission was 43 Rayleighs, which results in an inferred lifetime of 5×10^4 years for helium in the martian atmosphere.

A recent sounding rocket experiment by S. A. Stern and colleagues obtained an EUV spectrum of Venus, and the *HUT* experiment aboard the Astro-2 mission obtained EUV spectra of Venus and Mars. These results, although they have not yet appeared in the literature, will undoubtedly provide new insights and improve our understanding of the upper atmospheres of the terrestrial planets.

4. Jovian Planets

Of the Jovian planets, Jupiter remains the most extensively studied at EUV wavelengths. The UVS experiments on the Voyager 1 and 2 spacecraft have provided the primary EUV spectroscopic data set for the giant planets and their satellites. The EUV spectra of the Jupiter, Saturn, and Uranus, when properly scaled by their heliocentric distances, appear very similar (Yelle et al. 1987). They do differ, however, above 1100 Å, and the UVS data at these longer EUV wavelengths have been used to infer upper atmospheric composition.

The highest spectral resolution observations of Jupiter in the EUV were obtained by the *HUT* spectrograph aboard the Astro-1 (Feldman et al. 1993) and Astro-2 missions. These measurements clearly indicate the hydrogen Lyman series of lines as well as fluorescence emissions excited by solar Lyman β.

The *HUT* observations at superior spectral resolution have provided important new insight into the controversy over H_2 band airglow emissions seen on Jovian planets. Feldman et al. (1993) noted that an electron impact excitation source due to atmospheric dynamo, such as that proposed by Clarke et al. (1987), is required to explain the data. This source is needed in addition to solar fluorescence (Yelle et al. 1987), which accounts for up to 22%, and photoelectrons (Waite et al. 1983).

Recently, the *EUVE* mission observed the Jupiter and Io system in conjunction with the comet Shoemaker-Levy 9 impact. The data show an increase of Jupiter's He I 584 Å brightness by about a factor of 1.5 during the comet impact over its pre-impact value. Gladstone et al. (1995) surmise that the most likely explanation for the increase is the rising of helium in the impact-generated plumes from the well-mixed lower atmosphere into the depleted upper atmosphere. The extra helium in the upper atmosphere is then able to scatter more sunlight at 584 Å. An investigation of the EUV light curves obtained by *EUVE* shows the He I 584 Å enhancement during individual comet fragment impacts for the larger fragments (G, H, and K). The 1–2 hour delay of the brightening of the signal indicates the time required for substantial lateral expansion of the plumes as the impact sites rotate toward the direction of Earth.

Strong EUV emissions from the Io Plasma Torus (IPT) were discovered by the UVS instrument during the flyby of Voyager 1 (Broadfoot et al. 1979). The IPT is the only substantial emitter of EUV radiation (rather than simply being a reflector of sunlight) in the solar system. Sulfur and oxygen ions in a relatively dense plasma torus ($\sim 10^3$ cm^{-3}) surrounding the orbit of Io are excited by electron impact to radiate at several characteristic EUV wavelengths. Most of the energy for the torus comes from Jupiter's rotation. As neutral sulfur and oxygen atoms escape from Io through sputtering or other mechanisms, they are likely to become ionized and join the torus population. As they do so, they are swept up by Jupiter's corotating magnetic field which, since it moves much faster than Io's orbital velocity around Jupiter, gives the newly-formed torus sulfur and oxygen ions an energy boost of about 525 eV and 262 eV, respectively.

High-resolution EUV spectra of the IPT were obtained by the Astro-1 and Astro-2 missions. These spectra are totally dominated by emissions from the oxygen and sulfur ions (Moos et al. 1991). The *EUVE* spectrographs have also been used to study the IPT, and they have the advantage of being able to map out the spatial distribution of several of the EUV emissions over the entire torus (Hall et al. 1994). In addition to a strong dusk/dawn brightness asymmetry, the *EUVE* IPT data of the O II 539 Å feature also seem to indicate the presence of other torus-like features inside the orbit of Io (Figure 3). *EUVE* spectrometer data of the Io Plasma Torus before, during, and after the SL-9 impacts show no significant change in their relative brightnesses (Hall et al. 1995).

For the other outer planets, the Voyager UVS (Broadfoot et al. 1977) made extensive use of solar and stellar occultations to obtain atmospheric composition and temperature structure. This is possible due to the fact that the opacity of the atmosphere in different EUV wavelength regions can differ significantly (Broadfoot et al. 1989).

EUVE Jupiter System O⁺ Images

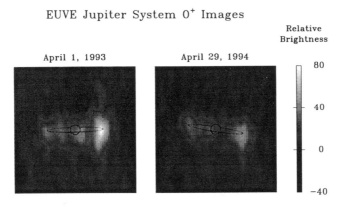

FIGURE 3. Images of the Jupiter (circle at the center) and Io Torus (dashed line) obtained at 539 Å by EUVE. Note the presence of torus-like features inside Io's orbit.

5. Satellites

While passing Neptune, the Voyager 2 UVS experiment obtained an EUV spectrum of Triton (Broadfoot et al. 1989), which showed that the primary features are due to various band systems of N_2. The other primary emission feature include the N II 1085 Å and H I 1216 Å lines. The authors noted that the data do not include positive detections of CO, Ar and Ne. From the absolute brightness and day/night asymmetry of the observed intensity of H I Lyman α, the authors concluded that photoelectrons and magnetospheric electrons might be responsible for some of the features.

The extended atmosphere of Triton, which resembles that of Pluto (and also somewhat that of Titan), also contributes to the formation of a proposed "plasma arc" (partial torus) around Neptune. Broadfoot et al. 1989 used this model to explain the presence of H I Lyman β emissions in the aurora and the relative absence of H I Lyman α.

6. Comets

A sounding rocket observation (Green et al. 1991) of Comet Austin in the 910–1180 Å band at approximately 3 Å resolution revealed an emission feature at 1128 Å which the authors attributed to a forbidden line of atomic oxygen. Such a feature has not been observed in the laboratory or in the terrestrial atmosphere, nor in any other comet spectrum (Feldman et al. 1991), and has raised an interesting controversy regarding atomic spectroscopy [see, for example, Slanger (1991) and its response]. Green et al. 1991 argue that the optically thick conditions that might explain the presence of 1128 Å emission might be present due to the large extent of the coma in a dynamically new comet.

The EUV data from this same sounding rocket flight were used to obtain limits on the relative abundances of cometary helium and argon (Stern et al. 1992). Although only upper limits to the abundances were possible, this was the first constraint of He and Ar abundances in comets. The authors found that the relative abundance of He/O in comet Austin is 1.5×10^4 depleted and no more than 30 times enriched in Ar/O compared to solar abundances. The He/O depletion can be used to infer that the cometary nuclear ices are not always maintained within a few degrees of absolute zero.

7. The Interplanetary Medium

There have been only limited observations of the interplanetary medium in the EUV. The Voyager UV spectrometer, with a 1.5×10^6 seconds exposure, has collected a spectrum of the sky in the direction of the North Galactic Pole (Holberg 1986). This observation, clearly contains the He I 584 Å feature produced by resonance scattering of solar emissions by neutral helium atoms in the heliosphere. The interpretation of this emission has subsequently raised the possibility that the solar 584 Å line is redshifted (Chassefiere et al. 1988).

The presence of 0.02 Rayleigh (1.6×10^3 photons cm^{-2} sec^{-1} ster^{-1}) of He II 304 Å emissions was observed in the photometric data obtained during the Apollo-Soyuz mission in the Earth's shadow cone (Paresce et al. 1981). These data were interpreted as due to multiple scattering of solar emissions, either by plasmaspheric He$^+$ ions or by He$^+$ in the interstellar medium produced by photoionization of He. A more recent analysis of *EUVE* survey observations obtained brightnesses of 1.30 Rayleigh, 0.040 Rayleigh, and 0.029 Rayleigh for the He I 584 Å, 537 Å, and the He II 304 Å emissions, during an exposure of 575,232 seconds looking down Earth's shadow (Jelinsky et al. 1995). As with the Apollo-Soyuz data, it is unclear what portion of these signals is due to multiple scattering in Earth's geocorona and plasmasphere.

On its way to Jupiter the Galileo spacecraft has made routine spectroscopic observations of the interplanetary medium (Hord et al. 1991). Although only the hydrogen Lyman α observations have been analyzed to date, He I 584 Å data also exist and are undergoing detailed analysis.

8. Summary

As EUV instruments have become more sensitive, with higher spectral and spatial resolution than the early exploratory ones, newer insights are being obtained regarding planetary atmospheres and their surfaces. Already several questions raised by the Voyager mission have been addressed by the Astro-1 and *EUVE* missions. The Astro-2 mission and several sounding rocket experiments have obtained new data on several solar system objects. Results from these data have not been reported yet. We also note that neither *EUVE* nor *HUT* is ideal for solar system studies, since *EUVE* was designed to either avoid detection or otherwise suppress many spectral features found in planetary atmospheres, and *HUT* lacked imaging capability. Progress in EUV instruments and availability of space flight opportunity will undoubtedly continue to increase our knowledge and understanding of the solar system.

We acknowledge the support of NASA grant NAG5-2260.

REFERENCES

Bertaux, J. -L., Blamont, J. E. Lepine, V. M, Kurt, V. G., Romanova, N. N. & Smirnov, A. S. 1981, Venera 11 and Venera 12 Observations of EUV Emissions from the Upper Atmosphere of Venus, Planet. Space Sci., 29, 149

Broadfoot, A. L. 1977, Ultraviolet Spectrometer Experiment for the Voyager Mission, Space Sci. Rev., 21, 183

Broadfoot, A. L., et al. 1979, Extreme Ultraviolet Observations from Voyager 1 Encounter with Jupiter, Science, 204, 979

Broadfoot, A. L., et al. 1989, Ultraviolet Spectrometer Observations of Neptune and Triton, Science, 246, 1459

BUSH, B. C. & CHAKRABARTI, S. 1995, Analysis of Lyman α and He I 584 Å Airglow Measurements using a Spherical Radiative Transfer Model, J. Geophys. Res., in press

CHASSEFIERE, E., BERTAUX, J. -L., KURT, V. G. & SMIRNOV, A. S. 1986, Venus EUV Measurements of Helium at 58.4 nm from Venera 11 and Venera 12 and Implication of Outgassing History, Planet. Space Sci., 34, 585

CHASSEFIERE, E., DALAUDIER, F., & BERTAUX, J. L. 1988, Estimate of Interstellar Helium Parameters from Prognoz 6 and Voyager 1/2 EUV Resonance Glow Measurements Taking into Account a Possible Redshift in the Solar Line Profile, A&A, 201, 113

CHIU, Y. T., ROBINSON, R. M., SWENSON, G. R., CHAKRABARTI, S., & EVANS, D. S. 1986, Imaging the Outflow of Ionospheric Ions in the Magnetosphere, Nature, 322, 441

CLARKE, J. T., HUDSON, M. K., & YUNG, Y. L. 1987, The Excitation of Far Ultraviolet Electroglow Emissions on Uranus, Saturn and Jupiter, J. Geophys. Res., 92, 15,139

COTTON, D. M., GLADSTONE, G. R., CHAKRABARTI, S., & LINK, R. 1993, Sounding Rocket Observations of a Hot Atomic Oxygen Geocorona, J. Geophys. Res., 98, 21,651

EDWARDS, B. C., PRIEDHORSKY, W. C. & SMITH, B. H. 1991, Expected Extreme Ultraviolet Spectrum of the Lunar Surface, Geophys. Res. L, 18, 2161

FELDMAN, P. D., & BAGENAL, F. 1991, EUV Planetary Astronomy, Extreme Ultraviolet Astronomy, R. F. Malina and S. Bowyer Eds, Pergamon, 252

FELDMAN, P. D., & MORRISON, D. 1991, The Apollo 17 Ultraviolet Spectrometer: Lunar Atmosphere Measurements Revisited, Geophys. Res. L., 18, 2105

FELDMAN, P. D. ET AL. 1991, Observations of Comet Levy 1990XX, with the Hopkins Ultraviolet Telescope, The ApJ, 379, L37

FELDMAN, P. D. ET AL. 1992, The Spectrum of Tropical Oxygen Nightglow Observed at 3 Å Resolution with the Hopkins Ultraviolet Telescope, Geophys. Res. L., 19, 453

FELDMAN, P. D., McGRATH, M. A., MOOS, H. W., DURRANCE, S. T., STROBEL, D. F., & DAVIDSEN, A. F. 1993, The Spectrum of Jovian Dayglow Observed at 3Å with the Hopkins Ultraviolet Telescope, ApJ, March 93, 279

GERVAIS, F. 1991, Aluminum Oxide Al_2O_3, in Handbook of Optical Constants of Solids II, ed. E. D. Palik, New York: Academic Press, 761

GLADSTONE, G. R. 1994, Extreme Ultraviolet Observations of the Moon, Proceedings of the Tenth Thematic Conference on geologic Remote Sensing, I, 275, Environmental Research Institute of Michigan, Ann Arbor

GLADSTONE, G. R., HALL, D. T., & WAITE, J. H., JR. 1995, EUVE Observations of Jupiter During the Impact of Comet Shoemaker-Levy 9, Science, 268, 1595

GLADSTONE, G. R., McDONALD, J. S., BOYD, W. T., & BOWYER, S. 1994, EUVE Photometric Observations of the Moon, Geophys. Res. L., 21, 461

GREEN, J. C., CASH, W., COOK, T. A., & STERN, S. A. 1991, The Spectrum of Comet Austin from 910 to 1180 Å, Science, 251, 408

HALL, D. T., ET AL. 1994, Extreme Ultraviolet Explorer Satellite Observation of Jupiter's Io Plasm Torus, ApJ, 426, L51

HALL, D. T., ET AL. 1995, Io Torus EUV Emissions During the Comet Shoemaker-Levy/9 Impacts, Geophys. Res. L., submitted

HOLBERG, J. B. 1986, Far-ultraviolet Background Observations at High Galactic Latitude, II, Diffuse Emission, ApJ, 311, 969

HORD, C. W., ET AL. 1991, Galileo Ultraviolet Spectrometer Experiment: Initial Venus and Interplanetary medium Cruise Results, Science, 253, 1548

JELINSKY, P., VALLERGA, J. V., & EDELSTEIN, J. 1995, First Spectral Observations of the Diffuse Background with the Extreme Ultraviolet Explorer, ApJ, 442, 653

KRASNOPOLSKY, V. A., BOWYER, C. S., CHAKRABARTI, S., GLADSTONE, G. R., & McDONALD, J. S. 1994, First measurement of Helium on Mars: Implication for the Problem of Radiogenic Gases on the Terrestrial Planets, Icarus, 109, 337

MEIER, R. R. 1991, Ultraviolet Spectroscopy and remote Sensing of the Upper Atmosphere,

Space Sci. Rev., 58, 1

MOOS, H. W., ET AL. 1991, Determination of Ionic Abundances in the Io Torus Using the Hopkins Ultraviolet Telescope, ApJ, 382, L105

OGAWA, H., PHILLIPS, E., & JUDGLE, D. L. 1984, Line Width of the Solar EUV He I Resonance Emissions at 584 and 537 Å, J. Geophys. Res., 89, 7537

PHILIPP, H. R. 1985, Silicon Dioxide SiO_2, Glass, in Handbook of Optical Constants of Solids, ed. E. D. Palik, New York: Academic Press, 749

PARESCE, F., FAHR, H, & LAY, G. 1981, A Search for Interplanetary He II 304 Å emission, J. Geophys. Res., 86, 10038

SCHMITT, J. H. M. M., SNOWDEN, S. L., ASCHENBACH, B., HASINGER, G., PFEFFERMANN, E., PREDEHL, P., & TRÜMPER, J. 1991, A Soft X-Ray Image of the Moon, Nature, 349, 583

SLANGER, T. G. 1991, The Spectrum of Comet Austin, Science, 253, 452

STERN, S. A., GREEN, J. C., CASH, W., & COOK, T. A. 1992, Helium and Argon Abundance Constraints and the Thermal Evolution of Comet Austin 1989c1, Icarus, 95, 157

WAITE, J. H., ET AL. 1983, Superthermal Electron Processes in the Upper Atmosphere of Uranus: Aurora and Electroglow, J. Geophys. Res., 88, 6143

WU, H. H., & BROADFOOT, A. L. 1977, The Extreme Ultraviolet Albedos of the Planet Mercury and of the Moon, J. Geophys. Res., 82, 759

YELLE, R. V., MCCONNELL, J. C., SANDEL, B. L., & BROADFOOT, A. L. 1987, The Dependence of Electroglow on the Solar Flux, J. Geophys. Res., 92, 15110

Three-Dimensional Modelling of *EUVE* Observations of the Io Plasma Torus

N. THOMAS,[1] D. E. INNES,[1] AND R. LIEU[2]

[1]Max-Planck-Institut für Aeronomie, D-37191 Katlenburg-Lindau, Germany

[2]Center for EUV Astrophysics, 2150 Kittredge St.,
University of California, Berkeley, CA 94720-5030, USA

First results from a 3-D model of *EUVE* observations of the Io Plasma Torus are reported. The semi-empirical model calculations follow a method previously used to describe visible and near-UV emissions. The extension to EUV wavelengths is described. Several EUV emissions have been successfully modelled although some discrepancies remain at this stage. Most EUV emissions peak at a jovicentric distance of $\approx$ 5.8 R_J. The observed dawn-dusk asymmetry of the torus was well fitted with a shift parameter (ϵ) of 0.03. The modelling also indicates that optical depth effects need to be considered for several EUV emission lines.

1. Introduction

The first observations of the Io plasma torus (IPT) at extreme ultraviolet wavelengths (EUV) were made by the ultraviolet spectrometer (UVS) instrument on Voyager 1 (Broadfoot et al., 1979). The data showed strong emission from a ring of ionized sulphur and oxygen species which completely surrounded Jupiter at a distance of $\approx$ 5.9 R_J (1 Jupiter radius (R_J) = 7.14 10^7 m). The spectra were obtained at rather poor resolution ($\lambda/\delta\lambda \approx 25$) and were severely blended. However, emissions at 685 Å (S^{2+}), 833 Å (O^+ and O^{2+}), 910 Å (S^+), 1020 Å (S^{2+}), and 1070 Å (S^{3+}) could be identified.

Recent developments in EUV instrumentation have resulted in observations of the IPT at spectral resolutions roughly 10 times superior to that of the Voyager UVS instrument. The *Extreme Ultraviolet Explorer* (*EUVE*) spacecraft has achieved a resolution of $\lambda/\delta\lambda \approx$ 200 in the range 400–750 Å in addition to providing 2-D images of the IPT in each line (Hall et al. 1994). While the fine structure of the EUV emissions remains blended, the higher resolution allows a comparison of the strengths of groups of lines within the multiplet. Table 1 shows the total emitted power from the IPT within several wavelength bands from a 6 10^4 s *EUVE* exposure in March 1993 (Hall et al. 1994).

Hall et al. (1994) have generated a fit to the spectrum (integrated over the entire IPT) using a spectral fitting algorithm. By assuming constant electron temperature, they derived a composition. This paper reports the results of our initial attempts to model these data using a 3-D semi-empirical description of the IPT previously used to fit emissions at optical wavelengths (Thomas 1992; 1993; 1995). We concentrate, in particular, on four aspects of the observations; the possible influence of self-absorption, the observed east-west asymmetry, the spatial distribution of the emission, and the total emitted power. The results must be considered preliminary at this stage because further checking of the model (in particular the EUV rate coefficients) is required. We begin by describing briefly the model.

2. The Model

The semi-empirical model of the IPT, which has previously been described by Thomas (1992), solves the diffusive equilibrium equation for a multispecies plasma following the method of Bagenal (1985), giving the ion distribution along each field line based on the

457

S. Bowyer and R. F. Malina (eds.), Astrophysics in the Extreme Ultraviolet, 457–464.

TABLE 1. Observed and modelled photon fluxes for several torus species

Wavelength (Å)	EUV Emitted Power Observed (W)	EUV Emitted Power Modelled (W)	Difference (%) (W)	Maximum Modelled Optical Depth	East-West Asymmetry
OII 426–434	$2.1\ 10^9$	$3.1\ 10^9$	48	0.43	2.7
OIII 505–512	$2.2\ 10^9$	$5.1\ 10^9$	132	0.02	2.0
OII 536–543	$0.98\ 10^{10}$	$1.1\ 10^{10}$	12	0.26	2.5
SII 639–643	$4.0\ 10^9$	$2.4\ 10^{10}$	500	0.95	2.4
SIV 654–665	$2.7\ 10^{10}$	$4.3\ 10^{10}$	59	0.36	2.3
SIII 671–685	$1.0\ 10^{11}$	$2.3\ 10^{11}$	130	1.32	2.3
SIII 697–705	$7.6\ 10^{10}$	$6.9\ 10^{10}$	−9	0.30	2.3
SIII 724–731	$8.7\ 10^{10}$	$8.6\ 10^{10}$	−1	0.27	2.3

plasma sciences (PLS) in situ measurements from Voyager I (Bagenal 1994). An offset tilted dipole (OTD) approximation of Jupiter's magnetic field was assumed. The ion temperatures were also obtained from PLS data ($T_\| = T_\perp$ was assumed) while the electron temperature (T_e) was taken from Sittler & Strobel (1987). Suprathermal electrons have not been included. Taylor et al. (1995) have shown they contribute only a minor fraction to EUV emissions. For forbidden optical transitions, the emission from each volume element along the line of sight was computed by using a classical five-level equilibrium calculation following Osterbrock (1989). The integrated brightness was then computed by numerical integration assuming azimuthal symmetry about the OTD.

The model has been extended to include EUV emissions by incorporating higher levels in the model ions. Each model ion contains data for transitions between 15 to 35 levels. Rather than listing all levels and rate data, Table 2 simply gives the number of levels included for each ion and references to the transition probability and collision strength data for the observed EUV lines. In cases where data are only available for rates between LS levels, the fine structure rates for electric-dipole allowed transitions have been obtained by scaling the total rate by the relative line strength according to the rules of Russell-Saunders coupling (Condon & Shortly 1970) and the fine-structure rates for forbidden transitions have been obtained by scaling by the state degeneracies.

3. Optical Depth Effects in the EUV

Strobel & Davis (1980) noted that several of the transitions making up the EUV spectrum of the IPT are short-lived and that self-absorption needs to be considered. They suggested that optical depths as high as 3 could occur in the IPT. Shemansky & Smith (1981) pointed out that whilst the O^+/O^{2+} transitions at 833 Å should not be influenced by self-absorption, it was possible that the S^{2+} transitions near 680 Å might be affected. However, they suggested that the absence of measurable changes in the 680/833 Å line ratio scanning across the IPT (Shemansky & Sandel 1981) provided an upper limit for both self-absorption ($< 10\%$) and the column density of S^{2+} ($5.5\ 10^{12}$ cm^{-2}). The limit on the column density was made assuming an ion temperature of 26 eV. The question of optical depth effects has received almost no attention since these early papers.

The model can be used to determine the column density of S^{2+} in the IPT. The value of this quantity at the western ansa when the torus is seen edge-on is $1.97\ 10^{13}$ cm^{-2} exceeding the limit set by Shemansky & Smith (1981) by 4.

TABLE 2. Atomic data

Ion	No. of levels	λ	A–value	Ω
OII	19	426–434	WSG	LM
		536–543	HH1	I
OIII	25	505–512	NS	I
SII	31	639–643	BD	M
SIII	27	671–685	S	HH2
		697–705	HH2	HH2
		724–731	HH2	HH2

REFERENCES— WSG = Wiese et al. (1966); LM = Landini & Monsignori Fossi (1990); HH1 = Ho & Henry (1984a); I = Itikawa et al. (1983); NS = Nussbaumer & Storey (1981); BD = Bates & Damgaard (1949); M = Mewe (1972); RRC = Ryan et al. (1989); HH2 = Ho & Henry (1984b); S = Sultana (1993).

In order to assess the influence of self-absorption, we have included this effect in the model. Even if transitions are optically thick in the plane of the torus, the optical depth out of the plane is always small, so that resonance line trapping will not affect level populations. The derived populations are then used to obtain the emissivity, η, and the opacity, χ. The final intensities are obtained by integrating the radiation transfer equation along the line of sight. Between any two depths, τ_1 and τ_2, the intensity, I, at frequency, ν, is related by

$$I(\tau_1, \nu) = I(\tau_2, \nu)e^{\tau_2(\nu)-\tau_1(\nu)} + \int_{\tau_1(\nu)}^{\tau_2(\nu)} S(\nu, t)e^{t-\tau_1(\nu)}dt \qquad (3.1)$$

where $S(\nu)$ is the source function, $\eta(\nu)/\chi(\nu)$.

The S^{2+} 678 Å transition between the 3p3d $^3D_1^o$ and the ground state, 3p^{2}2 $^3P_0^e$ has the largest collision strength and transition probability ($8.9\ 10^9$ s^{-1}) in the EUV wavelength range. The optical depth at line centre is

$$\tau_0 = 2.6(\frac{N_i}{10^{13}[cm]})(\frac{10^5[K]}{T_i})^{1/2} \qquad (3.2)$$

where N_i and T_i are the S^{2+} column density and ion temperature, respectively.

Figure 1 shows the optical depth of all S^{2+} lines between 670 Å and 710 Å at the point of maximum brightness on the western ansa of the torus. The observer sees the torus edge-on. An optical depth of 1.3 in the S^{2+} 678 Å transition is evident. Several other transitions are also affected. The total flux from the 680 Å multiplet is reduced by 15% at this point in the torus compared to the optically thin case. From Figure 1 that the 680/700 Å line ratio must also be reduced because of the lower optical depth of the 700 Å multiplet.

To illustrate this further, simulated images of the torus have been produced at several wavelengths. Five examples are shown in Figure 2. The two lower panels (d and e) show the appearance of the torus at 680 Å including and ignoring the effect of optical depth (all lines in this multiplet have been summed). Visually, the general appearance of the torus remains unaltered but quantitatively there are some changes. Over a projected area of $3.25\ 10^{11}$ km^2 centred on Jupiter, the observed intensity in the 675–686 Å band

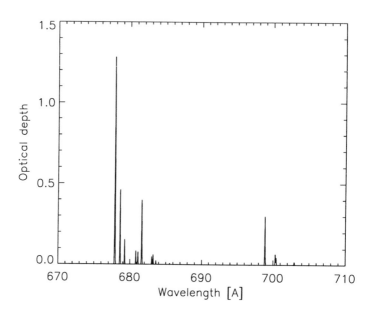

FIGURE 1. Optical depth of SIII lines near 700 Å

is reduced by 7% because of optical depth effects. The effects are most noticeable at the ansae. The 697–705 Å band is reduced by 0.5% for the same observing geometry. It may be interesting to look for spatial and temporal variations in the 680/700 Å ratio and whether they correlate with total SIII intensity.

Although these calculations indicate that self-absorption will be most significant for the S^{2+} 680 Å line, Table 1 shows that lines within other multiplets can also reach high values. Caution should be exercised when treating the SII 641 Å and the OII 430 Å multiplets.

4. The East-West Asymmetry

Sandel & Broadfoot (1982) noted that UVS data showed IPT EUV emissions to be asymmetric about Jupiter, being brighter at the western ansa than at the eastern. Barbosa & Kivelson (1983) and Ip & Goertz (1983) suggested that this was the result of a dawn-dusk electric field. The field shifts the centre of the IPT towards the east as seen from the Earth leading to compression of the plasma at the dusk ansa and expansion at the dawn ansa. The expansion cools the plasma at the dawn ansa (relative to Voyager 1 in situ measurements at the dusk ansa) giving lower volume emission rates.

The model includes the effect of the dawn-dusk electric field following the method described in Smyth & Combi (1988). Schneider & Trauger (1995) have presented evidence that this description is not adequate to explain the short timescale oscillation of the western ansa. However, the temporal resolution of *EUVE* is insufficient to resolve this motion and therefore the formulation of Smyth & Combi is adopted as a first approach.

The method allows the determination of first order corrections to the ion and electron densities and temperatures using a single independent variable ϵ, the so-called "shift parameter," which describes the ratio of the dawn-dusk electric field strength to that of the corotation field. Smyth & Combi (1988) relate the densities and temperatures to the western (dusk) ansa which was investigated by the Voyager 1 in situ measurements.

The shift of the IPT with respect to Jupiter and the difference in brightness between the two ansae as seen from the Earth, have been measured on several occasions at optical wavelengths (e.g. Morgan, 1985). Values in the range $\epsilon = 0.03\pm0.01$ have been found.

The influence of the dawn-dusk electric field is clearly evident in $EUVE$ images. Hall et al. (1994) found observed brightness ratios of 2.3±0.3 and 2.7±0.5 for SIII 680 Å and OII 539 Å, respectively, for the regions between 5 and 6 R_J on either side of the planet. Using $\epsilon = 0.03$, the model predicts brightness ratios of 2.4 (SIII 680 Å) and 2.5 (OII 539 Å), in excellent agreement with observation.

The slight difference in the magnitude of the asymmetry for the two lines is the result of the slightly higher excitation energy of OII. Optical depth effects are not major contributors to the observed difference in the ratio between the two species. This is, in part, because the reduction in the ion density in the eastern ansa is partially compensated for by the decrease in ion temperature (see Eq. (2)).

Figure 2 shows the appearance of the IPT in both OII 539 Å and SIII 680 Å. In all cases, the IPT is seen edge-on. Table 1 gives the predicted east-west asymmetry for 8 EUV lines based on $\epsilon = 0.03$. It should be noted that the modelled brightness ratio in the EUV is highly sensitive to ϵ and that values outside the range $0.025< \epsilon < 0.035$ are clearly incompatible with the east-west ratios reported by Hall et al. (1994).

5. The Spatial Distribution of Emission

The steep rise in T_e with radial distance in the IPT (Sittler & Strobel 1987) has led to a misconception that the maximum emission at EUV wavelengths must come from regions well outside Io's orbit and up to 7 R_J from Jupiter (e.g. McGrath et al., 1995). Figure 2 shows that the maximum emission from the IPT actually occurs close to 6 R_J for the major species. This is simply because the effect of the rise in T_e with radial distance is more than compensated for by the sharp decrease in density. Thus, despite the fact that only the tail of the electron distribution excites EUV emissions, density effects dominate. The rise in T_e does push the maximum emission out slightly beyond the position of the maximum at visible wavelengths (the "ribbon"). However, $EUVE$, which has a spatial resolution of about 1 R_J, should not be able to resolve the difference.

An exception to this behaviour is the OIII 509 Å emission which peaks at 7.2 R_J in the current model. The extended width of the OIII emission should, in principle, be detectable by $EUVE$ and will be the subject of further investigations. The appearance is caused by the increasing mixing ratio of OIII with jovicentric distance (see Bagenal et al., 1992). This remarkable extension of the emission also leads to an unusual modelled east-west asymmetry ratio at this wavelength because of the restriction of the calculation to the region between 5 and 6 R_J (see Table 1).

6. The Total Emitted Power

The model has been used to compute the total emitted power from the IPT within 8 wavelength bands which include major emission features. The results are compared to the observed values in Table 1 and the difference between the predicted and observed intensities is given as a percentage of the observed. The results show significant similarities and discrepancies. It should be noted at this point that while errors in the observed fluxes are, in most cases, less than 10%, the errors in the emission rate coefficients for EUV

Models of IPT EUV emissions

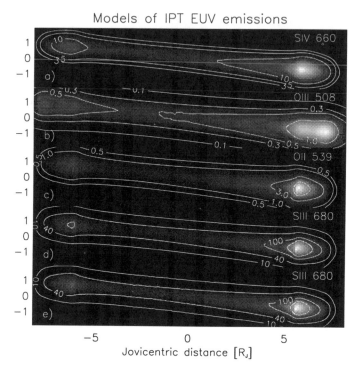

FIGURE 2. Models of the spatial distribution of IPT EUV emissions: (a) SIV 660 Å, (b) OIII 508 Å, (c) OII 539 Å, (d) SIII 680 Å excl. optical depth effects, (e) SIII 680 Å incl. optical depth effects.

transitions often far exceed this. Hence, it is not straightforward to determine whether the IPT was either (a) strictly "Voyager-like" at the time of the *EUVE* observations, (b) significantly different from Voyager times or (c) diagnosed incorrectly from Voyager measurements.

Bearing this point in mind, the most satisfactory result is for OII 539 Å for which observed matches predicted at the 10% level. OII is now thought to be the most abundant species in the IPT. The Voyager 1 PLS experiment was unable to differentiate between OII and SIII because of their identical mass per charge ratios while the UVS experiment could not separate OII from OIII because of the almost identical wavelengths of their emissions at 833 Å. The excellent agreement between the current Voyager model and the *EUVE* data suggests that this difficulty has been correctly resolved. There remains, however, the OII 430 Å multiplet which is significantly weaker than predicted although the collision strengths for this line are poorly known.

Two of the three SIII multiplets are well modelled. However, the model of the 680 Å emission is more than a factor of two too high. The similarity in the excitation energies of these three multiplets indicates that these lines should be in a simple ratio determined by their thermally averaged collision strengths. This, in turn, implies that if the observed ratios for the three SIII multiplets are correct, the atomic data source for the model must be wrong. It should be noted that we have used theoretical calculations for these lines while the COREQ program used to analyse Voyager UVS data (e.g. Shemansky, 1988) scales theoretical values to match experimental data. The resolution of this problem remains the subject of further work.

The 60% difference between model and observation for the SIV emission at 660 Å is considered to be a good result given the uncertainty in the mixing ratio of this minor species. This is also true to a lesser extent of the OIII emission at 509 Å. The model suggests that emission at high jovicentric distance dominates at this wavelength. This clearly needs to be investigated but it appears that a reduced mixing ratio near 7 R_J would bring the model into closer agreement with observation.

The SII 641 Å emission shows by far the largest disagreement probably because of the assumed emission rate coefficients which have only been crudely estimated so far.

7. Conclusion

The good agreement in several areas between observations and the model shows this to be a good starting point for the detailed modelling of *EUVE* observations of the IPT. There remain points where further work needs to be done. In particular, verification and checking of the emission rate coefficients is urgently required, especially for SII. However, the successful description of the east-west asymmetry shows that this approach will be very useful in deriving the properties of the Io plasma torus.

N. Thomas would like acknowledge useful discussions with D.T. Hall, M.A. McGrath, and M. Taylor during the International Jupiter Watch Torus Discipline workshop in Las Cruces, New Mexico, April 26–27, 1995. R. Lieu acknowledges the support of NASA contract NAS5-30180.

REFERENCES

BAGENAL, F. 1985, Plasma conditions inside Io's orbit: Voyager measurements, J. Geophys. Res., 90, 311

BAGENAL, F. 1994, Empirical model of the Io plasma torus: Voyager measurements, J. Geophys. Res., 99, 11,043

BAGENAL, F., ET AL. 1992, The Abundance of O^{++} in the Jovian Magnetosphere, Geophys. Res. L., 19, 79

BARBOSA, D. D. & M. G. KIVELSON 1983, Dawn-dusk electric field asymmetry of the Io plasma torus, Geophy. Res. L., 10, 210

BATES, D. R. & A. DAMGAARD 1949, The calculation of the absolute strengths of spectral lines, Phil. Trans. Roy. Soc. London, 242, 101

BROADFOOT, A. L., ET AL. 1979, Extreme ultraviolet observations from Voyager I encounter with Jupiter, Science, 204, 979

CONDON, E. U. & G. H. SHORTLY 1970, The theory of atomic spectra, Cambridge: Cambridge University Press

HALL, D. T., ET AL. 1994, Extreme Ultraviolet Explorer satellite observation of Jupiter's Io plasma torus, ApJL, 426, L51

HO, Y. K. & R. J. W. HENRY 1984a, Oscillator strengths for OII ions, J. Quart. Spectrosc. Radiat. Transfer, 31, 57

HO, Y. K. & R. J. W. HENRY 1984b, Oscillator strengths and collision strengths for S III, ApJ, 282, 816

IP, W. -H. & C. K. GOERTZ 1983, An interpretation of the dawn-dusk asymmetry of UV emission from the Io plasma torus, Nature, 302, 232

Itikawa, Y., S. Hara, T. Kato, S. Nakazaki, M. S. Pindzola, & D. H. Crandall 1983, Recommended data on excitation of carbon and oxygen ions by electron collisions, Institute of Plasma Physics, Nagoya University

Landini, M. & B. C. Monsignori, Fossi 1990, The X–UV spectrum of thin plasmas, A&AS, Ser., 82, 229

McGrath, M. et al. 1995, Response of the Io Plasma Torus to Comet Shoemaker–Levy 9, Science, 267, 1313

Mewe, R. 1972, Interpolation Formulae for the Electron Impact Excitation of Ions in the H-, He-, Li-, and Ne-Sequences, A&A, 20, 215

Morgan, J. S. 1985, Temporal and spatial variations in the Io torus, Icarus, 62, 389

Nussbauer, H. & P. J. Storey 1981, O III: Intercombination and Forbidden Lines., A&A, 99, 177

Osterbrock, D. E. 1989, Astrophysics of Gaseous Nebulae and Active Galactic Nuclei, Mill Valley, Calif.: Science Books

Ryan, L. J., L. A. Rayburn, & A. J. Cunningham 1989, Measurements of oscillator strengths for EUV emissions of ionized oxygen, nitrogen and sulfur, J. Quant. Spectros. Radiat. Transfer, 42, 295

Sandel, B. R. & A. L. Broadfoot 1982, Io's hot plasma torus—A synoptic view from Voyager, J. Geophys. Res., 87, 212

Schneider, N. M. & J. T. Trauger 1995, The Structure of the Io Torus, ApJ, in press

Shemansky, D. E. 1988, Energy branching in the Io plasma torus: The failure of neutral cloud theory, J. Geophys. Res., 93, 1773

Shemansky, D. E. & B. R. Sandel 1981, The injection of energy into the Io plasma torus, J. Geophys. Res., 87, 219

Shemansky, D. E. & G. R. Smith 1981, The Voyager 1 EUV spectrum of the Io plasma torus, J. Geophys. Res., 86, 9179

Sittler, E. C. & D. F. Strobel 1987, Io plasma torus electrons: Voyager I., J. Geophys. Res., 92, 5741

Smyth, W. H. & M. R. Combi 1988, A general model for Io's neutral gas clouds II: Application to the sodium cloud, ApJ, 328, 888

Strobel, D. F. & J. Davis 1980, Properties of the Io plasma torus inferred from Voyager EUV data., ApJL, 238, L49

Sultana, N. Nahar 1993, Transition Proabilities for Dipole Allowed Fine Structure Transitions in Si–like Ions: Si I, S III, Ar V and Ca VII, Physica Scripta, 48, 297

Taylor, M. H. et al. 1995, A comparison of the Voyager 1 UVS and PLS measurements of the Io Plasma Torus, J. Geophys. Res., submitted

Thomas, N. 1992, Optical observations of Io's neutral clouds and plasma torus, Surv. Geophys., 13, 91

Thomas, N. 1993, Detection of [OIII]λ5007 emission from the Io plasma torus, ApJL, 414, L41

Thomas, N. 1995, Ion temperatures in the Io plasma torus, J. Geophys. Res., 100, 7925

Wiese, W. L., M. W. Smith, & B. M. Glennon 1966, Atomic Transition Probabilities, 1, NSRDS–NRS 22

ALEXIS Lunar Observations

B. C. EDWARDS, J. J. BLOCH, D. ROUSSEL-DUPRÉ,
T. E. PFAFMAN, AND SEAN RYAN

Mail Stop D436, Los Alamos National Laboratory, Los Alamos, NM, USA

The *ALEXIS* small satellite was designed as a large area monitor operating at extreme ultraviolet wavelengths (130 − 190 Å). At these energies, the moon is the brightest object in the night sky and was the first source identified in the *ALEXIS* data. Due to the design of *ALEXIS* and the lunar orbit, the moon is observed for two weeks of every month. Since lunar emissions in the extreme ultraviolet are primarily reflected solar radiation these observations may be useful as a solar monitor in the extreme ultraviolet. The data show distinct temporal and spectral variations indicating similar changes in the solar spectrum. We will present a preliminary dataset of lunar observations and discussions covering the variations observed and how they relate to the solar spectrum.

1. Introduction

The *Alexis* experiment consists of three pairs of co-aligned extreme ultraviolet telescopes. Each telescope has a 33 degree field of view, narrow energy band response centered on 133, 174, or 188 Å and one half degree spatial resolution. The *ALEXIS* satellite is spin-stabilized with the telescopes pairs pointed 107 degrees (pair 3), 137 degrees (pair 2), and 167 degrees (pair 1) from the sun direction. In this orientation the telescopes observe half the sky during each spin period of roughly 50 s. (For details on the *Alexis* experiment and satellite see Bloch et al. (1990) and Bloch this conference) This mission design provides a good platform for semi-continuous monitoring of extreme ultraviolet phenomenon.

In the case of the moon, *Alexis* semi-continually monitors its extreme ultraviolet emissions for two weeks around the full moon every lunar cycle. Since the lunar emission is primarily reflected solar radiation (Edwards et al. 1991; Gladstone et al. 1993) it provides a direct monitor of solar activity in the extreme ultraviolet. The primary solar lines that contribute to this part of the spectrum are from Fe X (primarily between 174 and 177 Å) and Fe XI (primarily between 180 and 190 Å) (Heroux 1974) which originate in the solar corona and photosphere. In addition to the reflected solar spectrum some of the lunar emission may be due to L- and M-shell fluorescence from the lunar surface (Edwards 1991).

2. Discussion

In determining the count rates associated with the moon we first converted the observed events into lunicentric coordinates and flagged events within 1.5 degree of the moon's position. We calculated the count rates by integrating the counts inside a 1 degree radius circle around the moon's position and using the remainder of the flagged events to determine the background rate (Figure 1). These rates were then corrected for the *Alexis* vignetting function. Complications in this calculation arose due to errors in the photon flagging routine and the aspect determination. Because these errors defocused the lunar image we chose an integration box much larger than the half degree resolution of *Alexis* would warrant. This insured all lunar photons were collected but it also reduced the signal to noise of our results. Future processing will correct these problems. For this processing we have also ignored variations in the intensity due to different phases of the

S. Bowyer and R. F. Malina (eds.), Astrophysics in the Extreme Ultraviolet, 465–470.

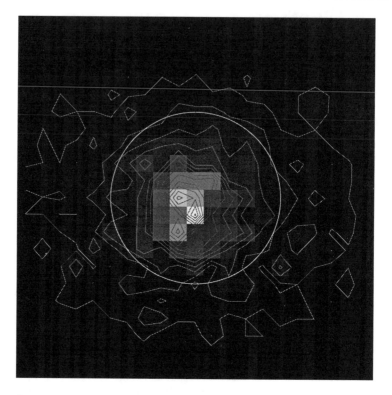

FIGURE 1. Image of the moon in lunicentric coordinates. The grayscale and contours show the relative counts observed. The circle shown is the 1° radius area used to calculate the lunar count rate. Counts outside this circle were used to calculate the background rate.

moon. This will affect each pair of telescopes differently. In telescope pair three this will cause errors of less than 14 percent in the absolute rates but will not affect our rate ratio discussions.

The observed count rates varied from 0 to several counts per second depending on the telescope and time. Figures 2a, 2b, and 2c show averaged rates for over one year of *Alexis* observations. Each of the plotted data points represents the roughly 20 minutes of data obtained for each orbit of *Alexis* (see Figure 3a and 3b). The data have high statistics in some cases and show count rate variations of greater than two on the scale of hrs and greater than five on the scale of days. These rate variations can be compared to solar 10.7 cm indices (Figure 4) but no strong correlation is seen.

The ratio of the 188 and 174 Å bands on telescope pair 3 can also be examined (Figure 5). The ratio is seen to vary primarily between 0.5 and 2 but statistically significant values are seen at all values. The ratio is not well correlated with the solar 10.7 cm indices (Figure 5). Since the 188 Å band is primarily Fe XI and the 174 Å band is primarily Fe X, the ratio of these bands could be a very sensitive monitor of the relative Fe ion states that exist in the solar corona. The ion state and thus our rate ratio is a very sensitive solar corona temperature monitor in the range between 1 million and 1.5 million degrees.

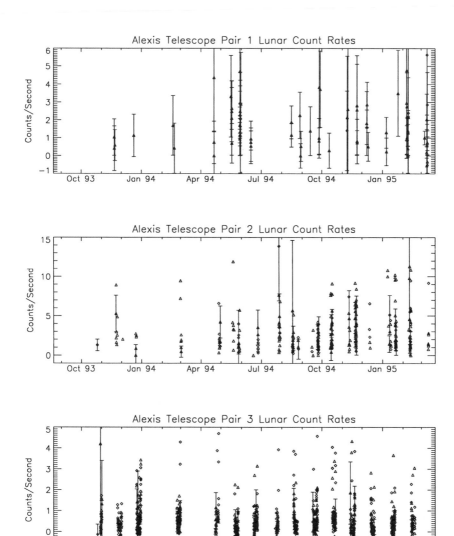

FIGURE 2. Lunar count rates observed by Alexis. a) Telescope pair 1, b) Telescope pair 2, c) Telescope pair 3. In all plots the diamonds are for the A telescopes (1A:133 Å, 2A:133 Å, 3A:174 Å) and the triangles are for the B telescopes (1B: 174 Å, 2B: 188 Å, 3B:188 Å). Several representative error bars are shown.

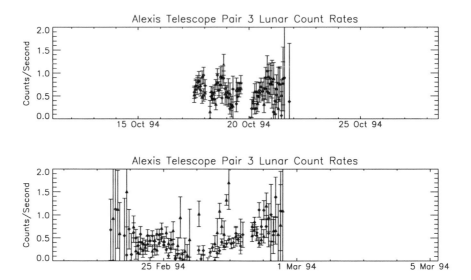

FIGURE 3. Lunar count rates observed by *Alexis* for several representative periods. The diamonds represent the 3A telescope and the triangles represent the 3B telescope. The count rates while in the center of the telescopes are observed to have good statistics. The rates determined for observations on the edge of the field of view rise due to error in the background determination.

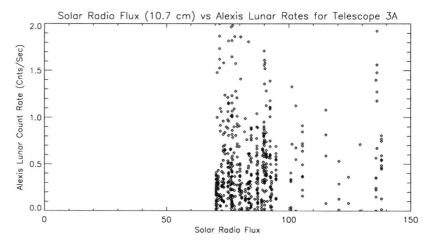

FIGURE 4. Solar 10.7 cm emission vs. *Alexis* lunar count rates (3A). Error bars are the same as plotted in Figures 1 and 2.

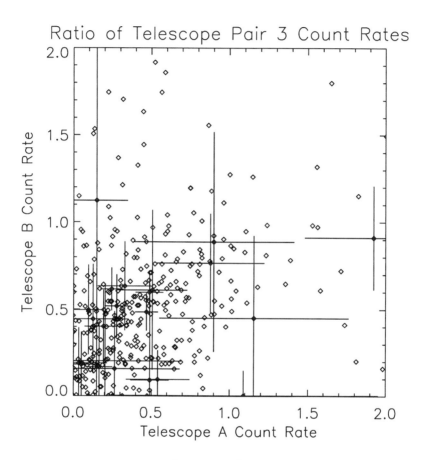

FIGURE 5. Rate ratio displayed as 3A:174 Å vs. 3B:188 Å. Only points where both 3A and 3B had overlapping observations are plotted. Several representative error bars are shown.

3. Conclusions

Alexis observations of the moon have high statistics and show interesting temporal and spectral variations. This data directly relates to processes occurring in the outer layers of the sun and may be a unique monitor of their temperature variations. Future work is required to improve the data quality and determine its value as a solar monitor.

The authors would like to thank the entire *Alexis* team for their hard work and dedication. This work is supported by the Department of Energy.

REFERENCES

BEARMAN, P. W. & GRAHAM, J. M. R. 1980, Vortex shedding from bluff bodies in oscillating flow: A report on Euromech 119, J. Fluid Mech., 99, 225

BLOCH, J. J., ET AL. 1990, Design, Performance, and Calibration of the ALEXIS Ultrasoft X-ray Telescopes, SPIE 1344, EUV, X-Ray, and Gamma-Ray Instrumentation for Astronomy

EDWARDS, B., PRIEDHORSKY, W. C., & SMITH, B. W. 1991, Expected Extreme Ultraviolet Spectrum of the Lunar Surface, Geophys. Res. Lett., 18, 2161

GLADSTONE, G. R., McDONALD, J. S., BOYD, W. T., & BOWYER, S. 1993, EUVE Photometric Observations of the Moon, Geoph. Res. Lett., 21(6), 461

HEROUX L., COHEN, M., & HIGGINS, J. E. 1974, Electron Densities Between 110 and 300 km Derived From Solar EUV Fluxes of August 23 1972, J. Geophys. Res., 29(34)

RE J1255+266—Detection of an Extremely Bright EUV Transient

MICHAEL DAHLEM

Space Telescope Science Institute†, 3700 San Martin Drive, Baltimore, MD 21218, USA

During a pointed *ROSAT* observation in the direction of the Coma cluster of galaxies an exceptionally bright EUV source, RE J1255+266, was detected serendipitously. The source is located close to the Galactic North pole, at $b_{II} \simeq 89°$. Its observed EUV flux (62–110 eV) at the time of the detection was of order 7×10^{-9} ergs s^{-1} cm^{-2}, making RE J1255+266 temporarily one of the brightest EUV sources in the sky.

The EUV flare of RE J1255+266 has a light curve with a decay time of about 0.86 days. With respect to earlier non-detections, the source brightened by a factor of > 7000. Such a behavior has not been observed before. Thus, it is unclear what type of source RE J1255+266 might be. The most likely optical counterpart is a faint ($V \sim 18.5$ mag) object with a blue spectrum (taken from an objective-prism Schmidt plate). For more details on the optical identification see the paper by J. Pye (this conference).

Simultaneous observations with CGRO/BATSE resulted in non-detections of the source in the 8–50 keV energy range.

1. Introduction

I report here on the serendipitous detection of a transient EUV source with the *ROSAT* Wide-Field Camera (WFC). The WFC is an EUV camera with its own optics onboard *ROSAT*. It is an independent instrument, aligned to the optical axis of the X-ray telescope (XRT). The WFC field of view (FOV) is 5° across, compared to 2° (PSPC) and 40′ (HRI). It works in the energy range of 17–210 eV, depending on the selection of one out of four filters. For more details see Barstow & Willingale (1988) and Briel et al. (1994).

2. Detection with the *ROSAT* WFC

RE J1255+266 was detected during a *ROSAT* pointed observation near the eastern boundary of the Coma cluster of galaxies on June 25–July 7, 1994. The source was seen during several satellite orbits. The total integration time of the pointing is 16.74 ks. At an off-axis angle of 2°.2 the WFC registered a very bright source at $\alpha, \delta(2000) = 12^h 55^m 07^s.6, +26° 41′ 21″ \pm 1′$ which would have been missed with the primary detector used for the pointing, the *ROSAT* High-Resolution Imager. Since the source could not be identified with any known object we named it—according to the IAU convention—RE J1255+266 (Dahlem & Kreysing 1994).

3. Other Observations

Simultaneous observations with the Compton Gamma-Ray Observatory BATSE instruments (Fishman et al. 1989) resulted in non-detections by both the Large Area Detectors and the spectrometers, which sets upper limits to the 8–50 keV flux of RE J1255+266 during the EUV outburst (Dahlem et al. 1995; D95).

Follow-up radio observations with the VLA at 1.4 and 4.9 GHz on September 29, 1994, also resulted in non-detections. The same holds for earlier observations toward RE

† Affiliated with the Astrophysics Division in the Space Science Department of ESA

S. Bowyer and R. F. Malina (eds.), Astrophysics in the Extreme Ultraviolet, 471–474.
© 1996 *Kluwer Academic Publishers. Printed in the Netherlands.*

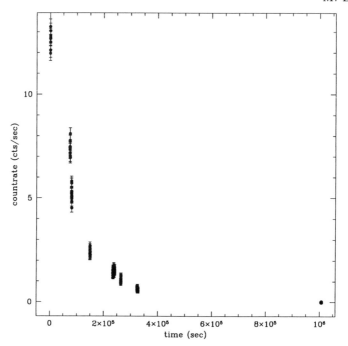

FIGURE 1. EUV light curve of RE J1255+266, binned into 100 second intervals. The observations started on June 25, 1994, 12:17 UT, and ended on July 7, 1994, 04:31 UT.

J1255+266 (D95), including the all-sky surveys of both *ROSAT* (WFC and PSPC) and *EUVE* (Pounds et al. 1993; Voges 1992; Bowyer et al. 1994).

A deep objective-prism Schmidt plate from the Hamburg Quasar Survey (HQS; Engels et al. 1988) shows a source with a blue spectrum about $1'.3$ NE of the WFC position, spatially coincident with a very faint pointlike object on the POSS plate at $\alpha, \delta(2000) = 12^h\ 55^m\ 10^s.7, +26° \ 42'\ 28'' \pm 1''$. D95 found this object to be the most likely optical counterpart of RE J1255+266. Follow-up observations by Watson (1995) corroborate this initial assumption. His results are summarized by Pye (this conference).

4. The Unusual Properties of RE J1255+266

Besides the fact that it is the first EUV transient—and the only one observed so far with good signal-to-noise—RE J1255+266 exhibits several characteristics which have not been observed before in transients that were detected in other wavebands.

RE J1255+266 was detected when its count rate in the 62–110 eV band (WFC filter S2; Pounds et al. 1993) was about 14 counts s^{-1}. At this time it was one of the brightest EUV sources on the sky. The sensitivity of the WFC at that time with respect to the all-sky survey in 1990 was 0.187. Thus, at the time of the sky survey, the maximum count rate would have been 76.5 counts s^{-1}, i.e., twice as bright as HZ 43, the brightest source at this energy during the WFC survey (Durisen et al. 1976; Pounds et al. 1993). Lacking spectral information we must adopt a spectral slope, assuming a thermal plasma of, e.g., $T = 2 \times 10^5$ K and an absorbing column density of $N(H) = 10^{19}$ cm^{-2} for the conversion of count rates to flux units. At the time of the all-sky survey a count rate of $1\ s^{-1}$ corresponded to $5 - 7 \times 10^{-11}$ ergs s^{-1} cm^{-2}. Scaling this by the sensitivity loss since then $(1/0.187)$, the flux at the time of the detection was roughly

$$f_{EUV} = 0.3 - 1.1 \times 10^{-8}\ [\text{ergs s}^{-1}\text{cm}^{-2}]. \qquad (4.1)$$

Taking the mean value, this very high flux leads to a luminosity of

$$L_{EUV} \sim 8.6 \times 10^{29}\ \frac{D^2}{[\text{pc}^2]}\ [\text{ergs s}^{-1}]. \qquad (4.2)$$

For $N(H) = 10^{18}$ cm^{-2} the flux would be a factor of 2 to 3 lower. The $N(H)$ values adopted above are only fractions of the total column density measured in HI line emission, $N(H)_{Gal} = 8.676 \times 10^{19}$ cm^{-2} (Hartmann & Burton 1995).

Figure 1 shows the unusual light curve of RE J1255+266. The half-light time of the (exponential) decay is $t_{1/2} = 0.86 \pm 0.04$ days (D95). This is much shorter than the decay times of novae or dwarf novae (e.g., Kwok & Leahy 1984, Richter 1992). In comparison with long-duration stellar flares of up to a few hrs (cf. Schmitt 1994) the decay time of the outburst observed in RE J1255+266 is much longer. Decay times of order one day are only observed in RS CVn systems.

Also the brightening factor of > 7000 (D95) is unprecedented. This value we obtained by comparing our WFC pointing with non-detections during an earlier deep *ROSAT* pointing and the WFC and *EUVE* all-sky surveys.

These characteristics of RE J1255+266 do not resemble the properties of any known class of objects. Thus, further observations or theoretical calculations are needed in order to clarify the nature of this bizarre source.

It is a pleasure to thank H.-C. Kreysing, S. M. White, C. Kouveliotou, and D. Engels for their contributions to this paper.

REFERENCES

BARSTOW, M. A. & WILLINGALE, R. 1988, Journal of the Brit. Interplan. Soc., 41, 345

BOWYER, S., LIEU, R., LAMPTON, M., LEWIS, J., WU, X., DRAKE, J. J., & MALINA, R. F. 1994, ApJS, 93, 569

BRIEL, U. G. ET AL. 1994, The ROSAT Users' Handbook, MPE Garching

DAHLEM, M. & KREYSING, H. -C. 1995, IAU Circ., 6085

DAHLEM, M. ET AL. 1995, A&A, 295, L13; D95

DURISEN, H. R., SAVEDOFF, M. P., & VAN, HORN, H. M. 1976, ApJ, 206, L149

ENGELS, D. GROOTE, D., HAGEN, H. -J., & REIMERS, D. 1988, Optical Surveys for Quasars, ed. P. S. Osmer et al., Astr. Soc. of the Pacific Conf. Ser. 2, 143

FISHMAN, G. J. ET AL. 1989, Proc. Gamma Ray Observatory Science Workshop, NASA, Washington, 2

HARTMANN, D. & BURTON, W. B. 1995, The Leiden-Dwingeloo Atlas of Galactic Neutral Hydrogen, Cambridge Univ. Press in preparation

KWOK, S. & LEAHY, D. A. 1984, ApJ, 283, 675

POUNDS, K. A. ET AL. 1993, MNRAS, 260, 77

PYE, J. 1995, This proceedings.

RICHTER, G. A. 1992, Reviews in Modern Astronomy, 5, 26

SCHMITT, J. H. M. M. 1994, ApJS, 90, 735

VOGES, W. 1992, Space Science with Particular Emphasis on High Energy Astrophysics, Proc. of the European International Space Year Meeting, Munich, 9

WATSON, M. G. 1995, Proc. of the Cape Workshop on Magnetic Cataclysmic Variables, Cape Town, January 23 submitted

The Secrets of EUVE J2056-17.1

M. MATHIOUDAKIS,[1] J. J. DRAKE,[1] N. CRAIG,[1]
D. KILKENNY,[2] J. G. DOYLE,[3] M. SIRK,[1] J. DUPUIS,[1]
A. FRUSCIONE,[1] C. A. CHRISTIAN,[1] AND M. J. ABBOTT[1]

[1]Center for EUV Astrophysics, 2150 Kittredge Street, University of California,
Berkeley, CA 94720-5030, USA

[2]South African Astronomical Observatory, PO Box 9, Observatory 7935, South Africa

[3]Armagh Observatory, Armagh BT61 9DG, Northern Ireland

EUVE J2056-17.1 is one of the brightest sources in the First *EUVE* Source Catalog with 0.24 counts s^{-1} in the Deep Survey Lexan/B band. We present optical and EUV results that show this source is one of the most active late-type dwarfs. *EUVE* observed a large flare with energy in excess of 10^{35} ergs in its Lexan/B band. The quiescent optical spectrum of the source reveals strong hydrogen Balmer and Ca II H and K emission. A strong Li I 6707 Å line is also present in the spectrum. We have estimated a Li abundance of log N(Li) = 2.5±0.4. The high Li abundance and the high flare activity favors an interpretation where the enhanced Li is sustained by spallation reactions.

1. Introduction

The *Extreme Ultraviolet Explorer* (*EUVE*) is the first mission fully dedicated to extreme ultraviolet (EUV, 60–740 Å) astronomy. The First *EUVE* Source Catalog (FEC) contains 410 EUV sources (Bowyer et al. 1994). The sources included in the FEC include late-type stars, white dwarfs, early-type stars, cataclysmic variables, X-ray binaries, novae, extragalactic sources and several unidentified objects. Almost 50% of the objects detected during the all-sky survey are late-type stars, whereas in the deep survey 70% of the objects are late-type stars. This comparison shows that late-type stars are exposure time limited as compared to the remaining objects that are further away and therefore limited by absorption from the interstellar medium. These facts would suggest that a large fraction of the unidentified objects are late-type stars. The unidentified objects may constitute a potentially interesting class of objects since they are bright in the EUV and faint in optical wavelengths.

We present EUV and optical results on the source EUVE J2056-17.1, which is one of the brightest unidentified objects in the FEC. We identify the object as a new, very active flare star. An enormous flare was detected during the *EUVE* observations. Follow up optical spectroscopy revealed strong Hα and Ca II H and K emission. The Li I 6707.8 Å line is also present in the spectrum.

2. Observations and Data Reduction

2.1. *EUV Observations*

The *EUVE* all-sky survey was conducted using three coaligned scanning telescopes, which point at right angles to the satellite spin axis. A fourth telescope, the Deep Survey (DS) telescope, was aligned along the spin axis and, while pointing to the anti-sun direction, advanced at ∼ 1° per day along the ecliptic equator. The filter on the DS telescope comprises three sections. The center section is the Lex/B (60–200 Å) with two panels of Al/C (170–360 Å) on either side. In survey mode, as the spacecraft rotated around its spin axis, a source drifted into the field of view of the DS telescope and was observed in

475

S. Bowyer and R. F. Malina (eds.), Astrophysics in the Extreme Ultraviolet, 475–479.

the Lex/B and Al/C filters. The source makes a full revolution through the two filters during an orbital night and traces a circle passing from the Lex/B into the Al/C filter. Useful data are collected only during orbital night, which lasts for about 1900 s. The combination of higher effective areas and longer exposure times gave the DS Lex/B a higher sensitivity by approximately a factor of 10 compared to the all-sky survey.

EUVE J2056-17.1 was observed with the *EUVE* DS instrument from approximately 1992 August 2 11:00 UT to August 4 22:00 UT. The orientation of the *EUVE* telescopes and the geometry of the survey was such (see Haisch et al. 1993) that the three scanning telescopes observed the same position in the sky approximately three months later. Initially one of the most puzzling results was that this very bright DS source was not detected during the all-sky survey. Analysis of the DS Lex/B count rate as a function of time showed that the source gave a very large outburst during the observations.

2.2. *Optical Observations*

A search of various astronomical catalogs and databases revealed no known source at the position of EUVE J2056-17.1. An examination of the HST Guide Star Catalogue finding chart shows an object with $V \approx 10.5$ associated with the *EUVE* position. Ground-based spectroscopic observations of this source were obtained with the 2.1 m telescope at the Kitt Peak National Observatory and the 1.9 m telescope at the South African Astronomical Observatory. The optical spectrum of the source is characterized by strong Hα and Ca II H and K emission. The deep molecular bands of TiO (4760 Å), MgH (4780 Å), and CaOH (4780) allow us to determine a spectral type in the range of dK7e–dM0e, which would suggest that the source is at a distance of 50 ± 12 pc.

3. Results and Discussion

3.1. *The EUV Flare*

The lightcurve of the flare is shown in Figure 1. The flare is characterized by a sharp rise lasting less than 4,000 s reaching a peak count rate of 0.42 ± 0.02 counts s^{-1}. This is about 1 order of magnitude higher than the quiescent state. The main characteristics of the decay phase are an initial fast decay followed by a much longer tail. The total duration of the flare was ~ 1.1 days. The flare was not detected in the DS Al/C. The non-detection could be partly attributed to the broad wavelength coverage of the Al/C filter, which includes a high background contribution from the He II 304 Å geocoronal emission, and partly to considerable absorption by the interstellar medium. We have determined the flare energy using the Monsignori-Fossi & Landini (1994) line emissivities for an average coronal temperature of 10^7 K. *EUVE* spectroscopic observations of flares from active stars such as HR 1099 and AU Mic have shown that the wavelength range covered by the Lex/B band is indeed dominated by lines of Fe XVIII–Fe XXIII formed in the temperature range $10^{6.7}$–$10^{7.2}$ K (Brown 1994; Cully et al. 1994). The interstellar medium attenuation was taken into account using the Rumph et al. (1994) photoionization cross sections for a hydrogen column of $N_{H\,I} = 3 \times 10^{19}$ cm^{-2}. We used a He I / H I ratio of 0.1. Using these parameters, we estimate the total energy of the flare in the Lex/B band to be in excess of 10^{35} ergs.

The flare on EUVE J2056-17.1 is one of the most energetic flares reported on a late-type dwarf. Its lightcurve is similar to the flare observed on the dMe star AU Mic (Cully et al. 1994). The AU Mic flare lasted for about 1.5 days and was characterized by an impulsive rise to the peak, followed first by a fast decay, and then by a much longer tail. The observed time scales were much longer than the radiative and cooling time scales; this led Cully et al. (1994) to conclude that the long decay phase could be explained by a

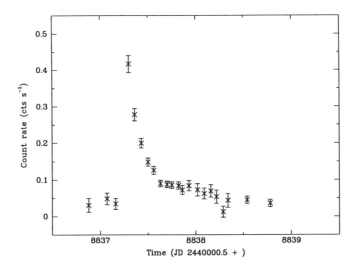

FIGURE 1. The lightcurve of the EUV flare in the DS Lex/B band (60–200 Å). Time is in Julian Dates since JD 2440000.5. We have used one time bin per *EUVE* orbit for the flare. The "quiescent" data have been averaged over two orbits.

model of rapid expansion causing the plasma to become tenuous sufficiently quickly that it avoids catastrophic radiative cooling. However, in this model no additional heating is assumed during the decay phase of the event. It remains to be investigated whether the long lightcurve could be fit by a model in which the plasma cools by radiation and/or conduction with continual heating taking place throughout the duration of the event.

The EUV flares observed to date indicate that the EUV carries a large fraction of the flare's radiative losses. The high energy and long duration of this particular event being similar to the largest flares seen on RS CVn binaries (Haisch et al. 1991). The "quiescent" EUV luminosity of EUVE J2056-17.1 is $L_{\text{Lex/B}} = 10^{30}$ ergs s^{-1}.

3.2. Detection of Lithium

One of the most conspicuous lines in the optical spectrum of this source is the Li I 6707.8 Å resonance line. The Li abundance of EUVE J2056-17.1 was estimated using the spectrum synthesis program MOOG, together with a model atmosphere generated using the MARCS program for solar metallicity (Gustafsson et al. 1975). An effective temperature of 4000 K and surface gravity of $\log g = 4.6$ was used in the computation. These parameters were determined from the optical spectrum. The Li abundance derived this way is $\log N(\text{Li}) = 2.5 \pm 0.4$. Lithium is rarely seen on late-type dwarfs with spectral types later than G5, and this is attributed to the fact that deep convection transports Li to high temperatures where it is destroyed. *Could EUVE J2056-17.1 be a weak T-Tauri star?* The chromospheric activity levels of EUVE J2056-17.1 are considerably lower than those of pre–main-sequence stars, they are, in fact, consistent with those of very active late-type dwarfs. The source is not associated with any clouds or star forming regions and has no entry in the *IRAS* Point Source Catalog. It would have been extremely interesting if EUVE J2056-17.1 turned out to be a weak T-Tauri star. However, we have no additional information at this point to support such a suggestion.

A possible interpretation that could explain the excess Li may be related to the high

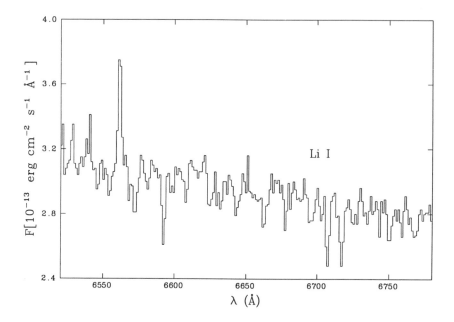

FIGURE 2. An optical spectrum of EUVE J2056-17.1 centered in the Hα region with a spectral resolution of ~4 Å; note the presence of Li I 6707.8 Å.

flare activity of the object. Li can be produced during energetic flares by spallation reactions (Canal et al. 1980). For a Li abundance similar to that of EUVE J2056-17.1, Schramm et al. (1990) predict that the energy required for producing Li by spallation is ≈ 10^{46}–10^{47} ergs. The "quiescent" EUV luminosity of the source implies a time-averaged flare energy of $L_{\text{tot}}^* \approx 10^{30}$ ergs s^{-1}. Assuming that the age of the star is similar to that of nearby flare stars, ~ 10^9 years, the energy produced is 3×10^{46} ergs. If the above assumption is correct, the energy released by flares is sufficient to sustain a high Li abundance on this source. Both the high Li abundance and strong coronal emission would suggest that EUVE J2056-17.1 is a young object that recently arrived on the main sequence.

4. Conclusions

We have presented optical and extreme ultraviolet results showing that the previously unidentified source EUVE J2056-17.1 is a new late-type star with a spectral type in the range dK7e–dM0e. The "quiescent" chromospheric and coronal emission of the object show that this is one of the most active late-type dwarfs. A flare with energy in excess of 10^{35} ergs and duration of ~1.1 days was observed in the *EUVE* DS Lex/B band. The optical spectrum shows a strong Li I 6707.8 Å line. The estimated Li abundance is log N(Li) = 2.5±0.4. We have discussed the possibility of Li production in the atmosphere of EUVE J2056-17.1 by spallation reactions during flares. Production of Li by such a process could delay the fast destruction occurring in the convection zone and therefore sustain an unusually high Li abundance.

This work has been supported by NASA Contract NAS5-30180. Research at Armagh Observatory is grant-aided by the Dept. of Education for Northern Ireland.

REFERENCES

BOWYER, S., LIEU, R., LAMPTON, M., LEWIS, J., WU, X., DRAKE, J. J., & MALINA, R. F. 1994, ApJS, 93, 569

BROWN, A. 1994, in Proc. Eighth Cambridge Workshop on Cool Stars, Stellar Systems and the Sun, ed. J.P. Caillault, Astron. Soc. Pac. Conf. Ser., 64, 23

CANAL, R., ISERN, J., & SANAHUJA, B. 1980, ApJ, 235, 504

CULLY, S. L., FISHER, G. H., ABBOTT, M. J., & SIEGMUND, O. H. W. 1994, ApJ, 435, 449

GUSTAFSSON, B., BELL, R. A., ERIKSSON, K., & NORDLUND, A. 1975, A&A, 42, 407

HAISCH, B. M., STRONG, K., & RODONO, M. 1991, ARAA, 29, 275

HAISCH, B. M., BOWYER, S., & MALINA, R. F. 1993, JBIS, 46, 331

MONSIGNORI-FOSSI, B., & LANDINI, M. 1994, Solar Physics, 152, 81

RUMPH, T., BOWYER, S., & VENNES, S. 1994, AJ, 107, 2108

SCHRAMM, D. N., STEIGMAN, G., & DEARBORN, D. S. P. 1990, ApJ, 359, L55

Searching *EUVE* Data for Transient/Flaring Extreme Ultraviolet Sources

J. LEWIS, S. BOWYER, M. LAMPTON, X. WU, AND M. MATHIOUDAKIS

Center for EUV Astrophysics, 2150 Kittredge Street,
University of California, Berkeley, CA 94720-5030, USA

The *Extreme Ultraviolet Explorer (EUVE)*, because of its sky survey strategy, performed two observations of each point along a 180° by 1° strip of the ecliptic during the initial survey phase of the mission. One observation used the deep survey telescope, and another, 90 days earlier or later, used the all sky scanner telescopes, two of which have a nearly identical passband to that of the ecliptic deep survey. Since the completion of the initial sky survey, *EUVE* has been used to carry out deep, pointed observations of selected targets. Many areas of the sky have therefore been observed two or more times, allowing us to compare count rates for some objects over a long temporal baseline. Objects with significantly varying count rates for widely separated times are of particular astrophysical interest.

With this technique, we have discovered one such object, which appears in the First *EUVE* Source Catalog as EUVE J2056-171. We present upper and lower limits on how frequently other highly-variable objects will be detected by *EUVE* in future observations.

1. *EUVE* Survey Strategy

The *EUVE* mission began with a six-month survey period, consisting of an all-sky survey in four EUV bandpasses (100 Å, 200 Å, 400 Å, and 600 Å) and a simultaneous deep survey in the 100 Å and 200 Å bandpasses, covering a 1° by 180° strip along the ecliptic. The deep survey telescope was pointed in the anti-sun direction for the duration of the survey phase, while the all-sky scanning telescopes, pointing 90° away from the deep survey axis, swept out great circles along lines of ecliptic longitude at a rate of about one revolution per orbital night. After six months, a narrow 180° strip along the ecliptic had been observed by the deep survey telescope, with average integration times on the order of 10000 s, and the all-sky survey telescopes had observed about 90% of the sky, with total integration times ranging from about 500 s along the ecliptic plane to 20000 s near the ecliptic poles (Bowyer et al. 1994; Bowyer et al. 1996).

Since the initial survey, *EUVE* has been used to conduct many deep, pointed observations of selected targets using the deep survey/spectrometer telescope. The all-sky survey telescopes are also used to perform concurrent observations at 90° angles from the spectrometer pointing via the *EUVE* Right Angle Program (RAP; McDonald et al. 1994).

Since the entire sky was covered in the initial *EUVE* survey, any detection in deep survey, guest observer (GO), or RAP data can be compared to the all-sky survey maps. Our approach is to compare each pointed observation with the all-sky survey data for the same field, looking for discrepancies between count rates of any sources detected in that region during any phase of the mission. Since the effective exposure times of pointed observations are typically several orders of magnitude longer than the all-sky survey exposure for most parts of the sky, the pointed data will generally give a tight upper limit on any quiescent emission from the possible source.

S. Bowyer and R. F. Malina (eds.), Astrophysics in the Extreme Ultraviolet, 481–484.

2. Detection Criteria for EUV Transients

Our goal is to determine whether or not there is a population of objects which are normally faint in the 100 Å EUV band, but are subject to occasional significant outbursts which could be detected by *EUVE*, and whether any of our catalogued detections which lack known optical counterparts could be due to these EUV transients.

We invoke four criteria for candidate EUV transient detections. First, the object must be detected at the 5 σ confidence level in one or more observation. Second, the object must brighten or dim by a factor of 5 or more between observations. Third, the 3 σ confidence limits on the count rate for each observation must not overlap, after correcting for the different effective areas of the all-sky and deep survey telescopes. Fourth, the object has no well-known optical or X-ray counterpart.

3. Data Selection

We chose to study a group of detections that met the detection criteria for the Second *EUVE* Source Catalog, but were not yet optically identified. Some of these objects were observed only during the all-sky survey; since no RAP, GO, or deep survey data were available for comparison purposes, they were excluded from this study. The remaining 22 objects with multiple observations were reprocessed to recover the unbinned photon data, which includes enough timing information to produce light curves. The reprocessed data was analyzed to obtain detection significance, maximum likelihood count rate, and 3 σ confidence intervals for the count rates.

One of these objects, EUVE J2056-171, underwent a large flare during a deep survey observation. This source was not detected in the all-sky survey, and was the only object from our sample to meet all the criteria for a significant transient.

4. Flaring and Quiescent Observations of EUVE J2056-171

A light curve for the deep survey observation of EUVE J2056-171 was generated by splitting the approximately 2.5 days worth of view time into individual orbits, each consisting of approximately 2000 s of uninterrupted view time. These data are shown in Figure 1. The light curve shows a significant flare with a peak count rate of about 0.55 counts s^{-1}. It is not clear from this data whether the object was ever in a truly quiescent state during the deep survey observation; most of the data preceding and following the main flare are consistent with a count rate of approximately 0.1 counts s^{-1}.

Three months later, the neighborhood of EUVE J2056-171 was observed by the scanning telescopes during the all-sky survey. The total exposure time was about 500 s, spread out over about five days. The object was not detected during this set of observations ($\chi^2 = 2.1$, compared to our nominal detection threshold of 25); the 3 σ confidence interval for the count rate includes (after correcting for the different effective areas of the two telescopes) the 0.1 counts s^{-1} "quiescent" rate derived from the deep survey data, and is also consistent with zero counts s^{-1}.

EUVE J2056-171 has been identified with a *HST* Guide Star Catalog object GSC_06349_00200 by Mathioudakis et al. (1995). Follow up optical and ultraviolet spectroscopy of this object suggests a dwarf star with spectral type between dK7e and dM0e, at a distance of 50 $\pm$ 12 pc.

The flare on EUVE J2056-17.1 has several similarities to the AU Mic flare reported by Cully et al. (1994), where it was suggested that the AU Mic flare was associated with a large coronal mass ejection. Very energetic flare events from low mass stars

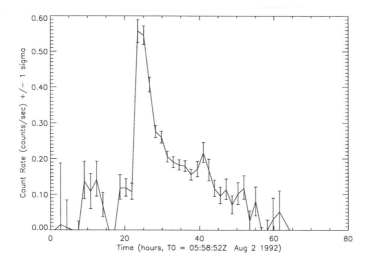

FIGURE 1. Light curve of EUVE J2056-171

may constitute the most efficient way by which angular momentum is lost by the star's atmosphere, leading to rotational spin down and consequently affecting the dynamo generated atmospheric heating process.

5. Statistical Limits on Frequency of EUV Transients

From the single confirmed transient observed during the deep survey phase of the mission, we can calculate upper and lower limits on the frequency of detection of similar events. It is assumed that the events are uniformly distributed over the sky, and that the number of events in a given time interval can be modeled as a Poisson random variable.

We treat the entire deep survey data set as a single "bin" in which one event was observed. The maximum likelihood estimate of the Poisson parameter λ is exactly one event per bin, as observed. To obtain the upper and lower limits of a 90% confidence interval around this value, we use a procedure similar to that used in for $EUVE$ source detection and count rate estimation:

Let N be the number of observed events in our single bin, and λ_{exp} be the number of expected events. We vary λ_{exp} until the statistic

$$C = 2(\lambda_{exp} - N \log \lambda_{exp}) \qquad (5.1)$$

is minimized. In our case, $\lambda_{min} = N = 1$, and $C_{min} = 2.0$. The quantity $(C - C_{min})$ for other values of λ follows a χ^2 distribution about λ_{min} with 1 degree of freedom. A 90% confidence interval for λ corresponds to a ΔC of 2.71, giving a range from approximately 0.11 to 3.6 events per "bin."

It is of interest to determine the size of the search space identified here as a "bin." The field of view of the deep survey 100 Å filter is about 2 square degrees and the observation period spanned 6 months, therefore one "bin" roughly corresponds to 1 year-degree^2 of sky coverage. In principle, the expected number of flare detections would be proportional to the observation time and field of view (all else remaining equal); in practice, however, the actual number of detections is also extremely sensitive to the survey strategy, making extrapolation from the deep survey result difficult.

For example, the all-sky survey instrument has a field of view about 12 times larger than that of the deep survey 100 Å filter. However, if EUVE J2056-171 had undergone an identical flare during the all-sky survey observation, only about 15 photons would have been received during the 24 hrs of maximum brightness. The flare would not have been detected, even if we somehow knew when and where to look for it. The detection efficiency for equivalent flares is therefore almost zero when the all-sky survey data is used in isolation, negating any benefit from the larger field of view.

If the survey telescopes had been pointed at EUVE J2056-171 for the same 24 hours, the flare very likely would have been detected. This suggests that the RAP and GO pointed data are likely to contain detectable transient events, although background, effective area, and point spread function differences may degrade the detection efficiency from that expected by extrapolating from the deep survey flare detection rate.

6. Conclusions

The design of the *EUVE* survey mission allows us to compare views of the same object at different times to detect variability. The most significant transient event detected so far appears to be a stellar flare. We estimate (based on limited statistics) that the frequency of events of this magnitude is approximately one per year per square degree. In the existing database of EUV objects, there are a number of flare stars, so there is a significant population of candidate objects that are observable at EUV wavelengths. We expect that continuing guest observer pointings and Right Angle Program pointings will reveal further objects undergoing outburst.

This work has been supported by NASA contract NAS5-30180.

REFERENCES

BOWYER, S., LIEU, R., LAMPTON, M., LEWIS, J., WU, X, DRAKE, J. J., & MALINA, R. F. 1994, ApJS, 93, 569

BOWYER, S., LAMPTON, M., LEWIS, J., WU. X, JELINSKY, P., & MALINA, R. F. 1996, ApJS, in press

CULLY, S. L., FISHER, G. H., ABBOTT, M. J., & SIEGMUND, O. H. W. 1994, ApJ, 439, 449

MATHIOUDAKIS, M., ET AL. 1995, The Secrets of EUVE J2056.1, these proceedings

MCDONALD, K., CRAIG, N., SIRK, M. M., DRAKE, J. J., FRUSCIONE, A., VALLERGA, J. V., & MALINA, R. F. 1994, Astron. J., 108, 1843

The *ALEXIS* Point Source Detection Effort

DIANE ROUSSEL-DUPRÉ,[1] JEFF BLOCH,[1] SEAN RYAN,[1]
BRADLEY EDWARDS,[1] TIMOTHY PFAFMAN,[1]
KERI RAMSEY,[2] AND STEVE STEM[1]

[1]Astrophysics and Radiation Measurements Group, Los Alamos National Laboratory,
P.O. Box 1663, MSD436, Los Alamos, New Mexico 87545 USA

[2]Space Data Systems Group, Los Alamos National Laboratory, P.O. Box 1663, MSD440,
Los Alamos, New Mexico 87545 USA

Los Alamos National Laboratory's *ALEXIS* satellite (a wide area EUV monitoring instrument) was launched April 25, 1993. Due to the damage sustained at launch by the satellite, the *ALEXIS* project team has had to spend over a year devising new methods to determine spacecraft attitude knowledge, essential for putting photons back on the sky correctly. These efforts have been successful and currently the *ALEXIS* attitude solutions are precise to better than 0.5 degree close to the original 0.25 degree pre-flight specification. This paper will discuss the number and types of point sources that have been revealed in the *ALEXIS* data to date. We will also discuss *ALEXIS* observations of the June, 1994 super outburst of the Cataclysmic Variable VW Hyi, a program to look for simultaneous EUV emission from Gamma Ray Bursts, as well as an effort to detect EUV transients with a 12 − 24 hour response time.

1. Introduction

Los Alamos National Laboratory's *ALEXIS* satellite, containing six EUV telescopes was launched April 25, 1993. *ALEXIS* is a sky monitor/survey experiment, and with each 50 second rotation, the satellite's six wide field of view telescopes scan nearly half of the sky. The telescopes have narrow energy responses centered at 66,73 and 93 eV. They are arranged in binocular pairs which are arranged to look (1) along, (2) 45° to and (3) 90° to what was to suppose to be the spin axis. The six telescopes are tuned to 3 different wavelengths.

2. Attitude Solutions and Point Source Calibration

Due to the damage sustained to the magnetometer at launch which invalidated most of the preflight analysis software and on-board attitude control algorithms (Bloch 1995), the *ALEXIS* project team has had to (1) learn to steer the satellite from the ground and (2) spend over a year devising new methods to determine spacecraft attitude knowledge, essential for putting photons back on the sky correctly. Part of this effort was to use the data from bright sources (HZ43 primarily) to determine the individual bore sight corrections required for each new version of the attitude solutions. These efforts have been successful and currently the *ALEXIS* attitude solutions are precise to better than 0.5 degrees, which is the minimum requirement to be able to detect weak sources in the sky maps. The first *ALEXIS* sky maps that contained substantial amounts of co-added data were produced on November 4, 1994, which revealed several point sources in addition to the bright sources HZ43 and the Moon which had been detected in earlier, less sophisticated processing.

The first effort in identifying point sources entailed using our first generation processed sky maps to validate the attitude solutions. To do this, we made 10 day composite sky maps, looked for point sources within these maps and then plotted as a function of time

S. Bowyer and R. F. Malina (eds.), Astrophysics in the Extreme Ultraviolet, 485–489.
© 1996 *Kluwer Academic Publishers. Printed in the Netherlands.*

TABLE 1. Previously identified sources found in the *ALEXIS* data

Name	RA (°)	Dec (°)	Spec. Type
WD0050-332	12.74	−33.29	DA1
RE2214-491	33.64	−49.19	DA
Feige 24	38.44	3.58	DA0
V471 Tau	57.54	17.18	DA2
VW Hyi	61.60	−71.05	CV
MCT0455-2812	73.97	−28.05	DA
G191-B2B	76.41	52.94	DA0
WD0549+158	88.21	15.67	DA1
WD0642-166	101.12	−16.86	DA2
U Gem	118.05	22.00	CV
RE1032+532	157.80	53.41	DA
WD1254+223	194.27	22.01	DA1
HZ43	199.23	29.19	DA1
WD1501+55	222.54	66.39	DZ0
WD2111+498	318.04	49.94	DA2
RE2156-543	329.62	−54.35	DA
WD2309+10	348.04	10.98	DA0
WD2309+105	348.07	11.0	DA0

the positional offsets looking for systematics. We have done this for HZ 43, G191-B2B, RE2214-49, and MCT0455-2812. Only HZ43 and G191-B2B were visible in all three telescope pairs. For the initial data processing, the histogram distribution of offsets was double peaked with a peak at 0.25 and 0.45 degrees. The peak at 0.45 was due to under determined bore sight correction terms for telescope pair 2 (TP2) and TP3 which have recently been corrected with the addition of several months more data. Currently, point sources observed in TP2 and TP3 have offsets of order 0.25 or less.

To determine point source locations, we first create standard deviation maps and Lampton fluctuation maps (Lampton 1994), which improves detection at low count rate levels, and search for point sources using the IDL "FIND" routine, and/or more recently use a parameter estimation procedure developed by Cash (1978). The on-line data bases for *EUVE*, *ROSAT*, Yale Bright Star, Gliese, and White Dwarf catalogs are then searched for likely candidates. Two of the detector filters have small pinhole leaks that allows O and B star UV radiation through, therefore, the Yale Bright Star catalog must be searched to assure that the source is a real EUV source. Recent changes to the data processing pipeline eliminate events associated with known pinholes. If these on line catalogs do not yield a likely candidate, we then use the WEB SKYVIEW page to search other catalogs.

We compiled a list of all sources found in the 10 day maps. We then added all files for 1 year into composite maps for each telescope and then all maps for each energy band and searched these for point sources. Lists of the point sources found are located in the following tables. The quick first pass through the data has yielded eighteen bright previously identified sources (Table 1) and five transients, three of which are unique to the *ALEXIS* data set (Table 2). It is also possible that the detection of V471 Tau should be included in the transient list, but more work will be required to determine when it was observed. Several unknown sources were also identified as having a lampton fluctuation significance of greater than 10^{-5} probability of random fluctuation which are potentially other transients. They will need further study before they are formally included in the final *ALEXIS* source catalog.

TABLE 2. Observed *ALEXIS* transients

Name	RA (°)	Dec (°)	Spec. Type	Telescope
VW Hyi	61.60	−71.05	CV	(1B−73 eV)
U Gem	118.05	22.00	CV	(2A−93 eV)
ALEXIS J1114+430	168.66	42.73	CV	(1B−73 eV)
				(1A/2A−93 eV)
				(2B−66 eV)
ALEXIS J1139-685	174.51	−69.9	unknown	(1B−73 eV)
ALEXIS J1644-302	251.23	−3.21	unknown	(2B−66 eV)

TABLE 3. *ALEXIS* pre-flight and observed count rate

Telescope	Pre-flight Estimates (c/s)	Observed (c/s, Jan 94)	Observed (c/s, Mar 95)
HZ43			
1A	0.12	0.02 ±0.015[†]	0.025±0.015
1B	0.49	0.46 ±0.05	0.55±0.05
3A	0.18		0.26±0.03
3B	0.13		0.18±0.02
G191-B2B			
1B	0.057		0.05±0.015

[†] due to mirror response much narrower than initially predicted

Since HZ43 has been observed for more than a year, it can be use to determine the long term stability of the *ALEXIS* telescope response. Table 3 contains the HZ43 predicted and observed count rates. As can be seen, the recent TP1 count rates are consistent to within the 1 sigma error bars of the observations made more than a year ago. In addition, the observed count rates for HZ43 and G191-B2B are consistent with preflight estimates. Thus, we conclude that the *ALEXIS* telescopes have retained full preflight sensitivity two years after launch.

3. Transient Search

Three methods are currently being used to detect EUV transients in the *ALEXIS* data: (1) archival search, (2) daily automated sky map searches, and (3) manual inspection of daily sky map. The archival searches to date have primarily used the 10 day sky maps. This study has produced the list of sources found in Table 3 as well as the super outburst of Cataclysmic Variable (CV) VW Hyi observed May/June 1994. Searches for transients on shorter time scales will be done as soon as the detector backgrounds are better understood. Real time transient searches were initially done daily by hand. ALEXIS J1114-430 was discovered in this manner in November, 1994. Starting 1 January, 1995, these searches were automated, so now the 12, 24, and 48 hour sky maps are searched for point sources. Point source locations found in the data by search algorithms are sent via e-mail to the science team for evaluation. U Gem and *ALEXIS* 1139-685 (Roussel-Dupré et al. 1995a) were discovered in this mode. In addition, manual searches of the data are done to spot check the e-mail point source lists which recently produced ALEXIS J1644-032 (Roussel-Dupré et al. 1995b) a system that was just below threshold on 24 and 48 hour maps, but exceeded threshold in the 36 hour map.

Of the transients observed by *ALEXIS*, three have been CV's and two are of unknown origin. Two of the three CV's, VW Hyi and U Gem, have been previously studied, but *ALEXIS* affords a detailed complete light curve unlike some previous observations.

The complete super outburst light curve of VW Hyi was observed by *ALEXIS* May 30–June 6, 1994. The optical light curve can be classified as a Bateson type S6 due to the decrease of 0.5 mag a 1.5 days after peak brightness was observed. The *ALEXIS* observations show the EUV radiation lagging the optical enhancement by 0.75 days similar to previous super outbursts (Pringle et al. 1987). C. Mauche of LLNL obtained *EUVE* target of opportunity observations during the outburst, however, 24 hrs of *EUVE* observations were not started until 2.5 days after initial optical enhancement. The EUV enhancement observed by *ALEXIS* continued exhibiting large fluctuations until 6 June at which time it turned off rapidly. The *ALEXIS* light curve will be combined with the *EUVE* results to try to piece together as much information as possible about the super outburst.

ALEXIS J1114-430 was first observed in outburst on 28 November, 1994 as the satellite field of view (FOV) drifted over the location of the source. The outburst lasted until 13 December, 1994 at which time the source dropped below the detector threshold. The light curve was quite flat topped throughout most of the outburst until the final decay very reminiscent of outbursts from SS Cyg (Jones & Watson 1992) and U Gem (Long et al. 1995). A literature search revealed a recently identified CV, AR UMa, within the error box. AR UMa, previously identified as an oxygen-rich, semi-irregular system with period of 69 days (Jura & Kleinmann 1992), has recently been reclassified by (Remillard et al. 1994) as an AM Her type system. Fortuitously, Remillard (private communication) obtained optical spectra of this source on 13 December, 1994 which showed enhanced emission thus confirming our identification of this source in outburst. The optical light curve from the roboscope at Indiana University (Honeycutt, Turner & Robertson, 1995) for this time showed AR UMA to be in a 130+ day enhancement; the EUV outburst occurred in the middle of this enhancement. The first EUV observations occur eight days after a short duration 0.5 magnitude optical enhancement. Because the telescope drifted onto the AR UMa location after the optical enhancement there is no way of knowing what the maximum EUV brightness was at the time of optical peak. The optical emission from AR UMa was essentially constant during the *ALEXIS* observed outburst.

ALEXIS J1139-685 and ALEXIS J1644-032 both have similar time signatures and look to be potentially a new type of transient. Both systems were observed to be in outburst for of order 24–36 hours. Target of opportunity observations were requested of *EUVE* for these systems and in both cases, *EUVE* was able to slew to target and observe these systems starting less than 24–36 hrs after maximum light. But in both cases, 24 hrs of observations failed to detect the transients. Other potential transients have also been observed in the data with time scales of 12–36 hours. These systems are different than the *ROSAT* Wide Field Camera transient, 2RE J1255+266 Dahlem et al. (1995); if ALEXIS J1139-685 and ALEXIS J1644-032 had similar decay rates as 2RE J1255+266, *EUVE* should have been able to easily detect the transients in 24 hrs of observation. Searches of the error boxes for counterparts have failed to produce good candidates for these systems. *ARIEL V* reported the identification of fast-transient X-ray sources in their archive data (Pye & McHardy 1983). They observed the transients to be essentially isotropic in distribution and predict that the rate should be about one per 3 days. The average duration of the X-ray fast transients was typically a few hours. Although preliminary estimates for the number of fast transients observed by *ALEXIS* appear to be similar to those observed by *ARIEL V*, the *ALEXIS* transients are of longer duration than those observed by *ARIEL V* and thus might constitute a different class of objects.

An effort is also underway to search the *ALEXIS* archive for EUV detections of gamma ray bursts (GRB). An automated program takes the *ALEXIS* photon list and checks to see if the detectors were on, at the correct high voltage setting, not looking at the earth, etc. during times of the BACODINE GRB triggers. The majority of the time the detectors have been off, but for a hand full of events for which the telescopes were in proper configuration, no clear signature of GRB events were present in the data. This effort continues with searches pre and post burst for weak emission within the error boxes.

4. Conclusion

As *ALEXIS* begins its third year on orbit with all of the problems of the launch failure now resolved, the full potential of the monitoring capability of *ALEXIS* is starting to be realized. Especially exciting is the complete coverage of CV outburst light curves and the detection of fast transients.

This effort is supported by the US Department of Energy, but would not be possible without the dedicated efforts of the entire *ALEXIS* team.

REFERENCES

BLOCH, J. J. 1995, EUV Astrophysics with ALEXIS: The Wide View, these proceedings

CASH, W. 1978, Parameter Estimation in Astronomy through Application of the Likelihood Ratio, ApJ, 228, 939

DAHLEM, M., KRYSING, H. -C., WHITE, S., ENGELS, D., CONDON, J. J., & VOGES, W. 1995, RE J1255+266: Detection of an extremely bright EUV transient, A&A, 295L, 13

HONEYCUTT, K., TURNER, G. W., & ROBERTSON, J. 1995, RoboScope's Data Archive, World Wide Web http://www.astro.indiana.edu/apt/

HWANG, L. -S. & TUCK, E. O. 1992, The EXOSAT Observations of SS Cygni, MNRAS, 257, 633

JURA, M. & KLEINMANN, S. G. 1992 1992, Oxygen-Rich Semiregular and Irregular Variables, ApJS, 83, 329

LAMPTON, M. 1994, Two-sample Discrimination of Poisson Means, ApJ, 436, 784

LONG, K. S., MAUCHE, C. W. SZKODY, P., & MATTEI, J. A. 1995, EUVE Observations of U Gem, in Proc. of the Padova-Abano Conference on Cataclysmic Variables, in press

PRINGLE, J. E., BATESON, F. M., HASSALL, B. J. M., HEISE, J., VAN DER WOERD, H., HOLBERG, J. B., POLIDAN, R. S., VAN AMERONGERN, S., VAN PARADIJS, J., & VERBUNT, F. 1987, The Ariel V sky survey of fast-transient X-ray sources, MNRAS, 225, 73

PYE. J. P., & McHARDY, I. M. 1983, The Ariel V sky survey of fast-transient X-ray sources, MNRAS, 205, 875

REMILLARD, R. A., SCHACHTER, J. F., SILBER, A. D., & SLANE, P. 1994, 1ES 1113+432: Luminous, Soft X-ray Outburst from a Nearby Cataclysmic Variable AR Ursae Majoris, ApJ, 426, 288

ROUSSEL-DUPRÉ, D., BLOCH, J. J., EDWARDS, B. C., PFAFMAN, T. E., PRIEDHORSKY, W. C., RYAN, S., SMITH, B. W., SIEGMUND, O. H. W., CULLY, S., RODRIGUEZ-BELL, T., VALLERGA, J., & WARREN, J. 1995a, ALEXIS J1139, IAU Circ. No. 6152

ROUSSEL-DUPRÉ, D. & THE ALEXIS TEAM 1995b, ALEXIS J1644, IAU Circ. #6170

The *EUVE* Optical Identification Campaign II: Late-Type and White Dwarf Stars

N. CRAIG,[1] A. FRUSCIONE,[1] J. DUPUIS,[1] M. MATHIOUDAKIS,[1]
J. J. DRAKE,[1] M. ABBOTT,[1] C. CHRISTIAN,[1] R. GREEN,[2]
T. BOROSON,[2] AND S. B. HOWELL[3]

[1] Center for EUV Astrophysics, 2150 Kittredge Street, University of California,
Berkeley, CA 94720–5030, USA

[2] NOAO, P.O. Box 26732, 950 N. Cherry, Tucson AZ 85726–6732, USA

[3] Planetary Science Institute, 620 N. Sixth Ave., Tucson, AZ 85705–8331, USA

We present optical identifications of nine previously unidentified extreme ultraviolet (EUV) sources discovered during the *Extreme Ultraviolet Explorer* (*EUVE*) satellite surveys. The all-sky survey detected four of the sources and the more sensitive deep survey detected the other five sources. Three of the four all-sky survey sources, EUVE_J1918+59.9, EUVE_J2249+58.5, and EUVE_J2329+41.4, are listed in present catalogs as having possible associations with optical counterparts but without spectral class. The first two of these sources are hot DA white dwarfs showing an optical spectrum with broad Balmer lines. The source EUVE_J2329+41.4 is listed as having a possible association with an unclassified M star. We show that a pair of dMe stars are actually optical counterparts located within the error circle of the *EUVE* source position. The EUVE_J2114+503 remains unidentified even though all the possible candidates have been studied. Based on the count rates we predict a fainter white dwarf or a cataclysmic variable counterpart for this candidate. All five sources discovered with the *EUVE* deep survey, EUVE_J0318+184, EUVE_J0419+217, EUVE_J2053-175, EUVE_J2056-171 and EUVE_J2233-096, have been identified as late-type stars. The spectral classes, distances, visual magnitudes, and estimated hydrogen column densities for these *EUVE* sources are presented.

1. Introduction

Opening new windows to the electromagnetic spectrum for astronomical study has traditionally resulted in exciting discoveries of new classes of astrophysical objects. The new types of objects are often relatively inconspicuous at optical and other wavelengths. Examples include the X-ray binaries and the radio galaxies. Consequently, the promise of such rewarding discoveries has formed a large part of the scientific motivation for building instruments to observe in hitherto unobserved bandpasses. The last major remaining window to the electromagnetic spectrum opened for astronomers was the extreme ultraviolet (EUV), a region roughly identified with the wavelength range 50–900 Å.

The *Extreme Ultraviolet Explorer* satellite (*EUVE*) has recently completed an all-sky survey between 60 and 740 Å and a deep survey (DS) between 67 and 364 Å in a 2° × 180° swath along the ecliptic (Bowyer & Malina 1991). The spectral region longward of 200 Å was surveyed for the first time by *EUVE*. To date, the surveys and subsequent observations have resulted in three published catalogs of approximately 500 EUV sources: the Bright Source List (BSL) of Malina et al. (1994); the First *EUVE* Source Catalog (FESC) of Bowyer et al. (1994); and the McDonald et al. (1994) catalog of serendipitous EUV sources detected in the first year of the Right Angle Program (RAP). It was found that the counterparts to about 80% of the sources in these catalogs could be readily identified from existing astronomical catalogs by searching for objects

491

S. Bowyer and R. F. Malina (eds.), Astrophysics in the Extreme Ultraviolet, 491–496.

thought to be EUV-bright and that coincided with the observed source positions (e.g., see Bowyer et al. 1994).

The EUV range provides coverage of a critically important and very wide temperature range, and contains plasmas with temperatures ranging from several 10^4 K to $\sim 20 \times 10^6$ K that produce photons in the EUV from many different elements in a wide variety of ionization stages. Consequently, about 50% of the EUV sources detected by *EUVE* were found to be active late-type stars (G, K, and M dwarfs and tidally interacting binaries) whose coronae have typical temperatures of $\sim 1 - 10 \times 10^6$ K (e.g., Pallavicini 1989). The other $\sim$50% comprise primarily hot white dwarfs, with a smaller number of early-type stars, active galactic nuclei (AGN), planetary nebulae, cataclysmic variables (CVs), X-ray binaries, and even one or two novae and thermally emitting neutron stars.

In this paper, second in the *EUVE* Optical Identification Campaign series (c.f., Craig et al. 1995 and Mathioudakis et al. 1995b), we present optical identifications for nine of the *EUVE* sources that currently do not have identified optical counterparts.

1.1. *EUVE_J0318+184*

Six possible optical counterparts were observed in the field of this NOID type EUVE_J0318+184. The source which is 35" from the *EUVE* position displays the spectrum the chromospheric activity indicators of Hα emission in (6563 Å) as well as Ca II H & K emission (3963 and 3934 Å, respectively). The spectral type of the counterpart is estimated as late K (dK5e–dK7e).

1.2. *EUVE_J0419+217*

EUVE_J0419+217 is detected in the DS Lexan filter (60−200 Å) and is listed as a possible counterpart in the BSL with spectral type A0sp and visual magnitude $m_v = 5.38$ with identification 56 Tau. The identification quality is Q $= 2$ indicating that this A0 type star is within the error circle. As the general criterion, if a detected object has a spectral class earlier than A0 and is brighter than $m_v = 5$, the emission could be a UV leak (Bowyer et al. 1994). Further investigation of this DS Lexan/B detection (Fig. 2b of McDonald et al. 1994) with 13 counts ks^{-1} is consistent with what one expect s from a UV leak for the given visual magnitude, $m_v = 5.38$.

Another possibility is that the A0 star 56 Tau has a visually unresolved hot white dwarf companion as the EUV source. This star is similar in brightness to several A stars found by *ROSAT* (Fleming et al. 1991), e.g., A3V star β Crt, and *EUVE*, and which were later identified as hot whit e dwarfs through *IUE* followup observations.

We have observed all the sources within the error circle as well as the brighter candidates outside the 1′ error circle in this field. Two candidates display signatures of high chromospheric activity. From the measurements of TiO (4670 Å) and MgH (4780 Å) bands we derived dM2e, dM3e, type spectra, respectively.

In order to examine whether the DS Lexan/B count rate can be attributed to the dMe stars we used the Monsignori Fossi & Landini (1994) line emissivities for a coronal temperature of log $T = 6.8$ and determined the EUV flux. This flux is consistent with the EUV fluxes of the most active dMe stars presented by Mathioudakis et al. (1995a).

Because the active late-type stars lie within the error circle, it appears that the two counterparts, are the likely contributors to the EUV emission and that "Possible ID" 56 Tau (Malina et al. 1994) is not the *EUVE* source but a UV leak also contributing to the total counts.

1.3. EUVE_J1918+599 and EUVE_J2249+585

EUVE_J1918+599 and EUVE_J2249+585 are detected by both the *EUVE* survey (Bowyer et al. 1994) and by the *ROSAT* WFC (Pounds et al. 1993). The optical counterparts are published as a "blue star," LB342, (Dixon & Sonnenborg 1980) and "UV-emission source," LAN 23, respectively (Lanning 1973). No spectral type is given in the literature for these sources. Within the EUV error circle, the candidates were observed and the candidates for EUVE_J1918+599 and EUVE_J2249+585 clearly reveal white dwarf spectra that exhibit broad H lines as illustrated in Figure 1.

The quasi-equivalent widths were measured in Å, based upon truncating the line profiles at 50 Å relative to the line center. The residual intensity at these points defines the baseline of the integration. This definition of width has the advantage of avoiding uncertainties caused by the continuum placement, which can be quite large because of the broad wings. The equivalent widths of Hβ and Hγ of both of the white dwarfs were fitted against the line blanketed model atmospheres of Wesemael (1980) for log $g = 8$. The accuracy of the equivalent widths are within few percent in each case and will contribute about 10% error to the temperature determination. The errors caused by the uncertainty of the choice of log g is also in about the 10% range for EUVE_J1918−599 but closer to 20% for EUVE_J2249+585. We obtain a range of $T_{\rm eff}$ = 30000–35000 for EUVE_J1918+599 and of 55000–60000 for EUVE_J2249+585. From the $T_{\rm eff}$ values we obtained the M_v values given by Liebert (1988). We estimate M_v in the range of $9.7 - 10.1$ for EUVE_J1918+599 while for EUVE_J2249+585 a range of $8.9 - 9.2$. These values reflect the uncertainties in the equivalent widths and the model log g value. From the absolute visual magnitude, M_v, and estimated visual magnitude, m_v, from our KPNO spectra (which are also in good agreement with the SIMBAD m_v values), we derive distances to these white dwarfs. Distances of 95 and 110 pc are derived for EUVE_J1918−599 and EUVE_J2249+585. The errors in distances are on the order of 10% to 20%. Based on preliminary follow-up results conducted by Vennes, Thejll, & Dupuis (1995), EUVE_J2249+585 and EUVE_J1918+599 are DA1.0 and DA1.5 temperature class white dwarfs, which agree with the results of this work.

1.4. EUVE_J2053−175

The spectrum of the brightest object within the error circle, has been obtained with the Perkins Telescope of the Lowell Observatory. Two other brighter candidates, 7 and 13, outside the error circle, were obtained at KPNO. Bright candidate reveals a spectral type late K. However, it does not show the clear evidence for chromospheric activity such as Balmer emission. It is premature to comment on the Ca emission since the spectral region does not extend to the 3900 Å range. The EUV light curve was constructed from the DS data to examine the source variability in the EUV, and no variability was detected. No conclusive spectral type nor definitive optical identification can be attributed to this field. However the source fits the description of "late-type star with no documented activity." Thus, this candidate is a possible counterpart or, in Bowyer et al.'s scheme, has identification quality Q = 2.

1.5. EUVE_J2056−171

EUVE_J2056−171 is one of the brightest DS *EUVE* sources. This object has recently been identified as a very active late-type dwarf. The spectrum exhibits the Hα and Ca II H and K emission lines and is given securely as type dK7e–dM0e. A large flare with an energy in excess of 10^{35} ergs was observed during the *EUVE* observations in the Lexan/B band. The optical spectrum of the source shows strong Hα and Ca II H and K emission. The Li I 6707.8 Å absorption line is al so present in the spectrum. The Li I

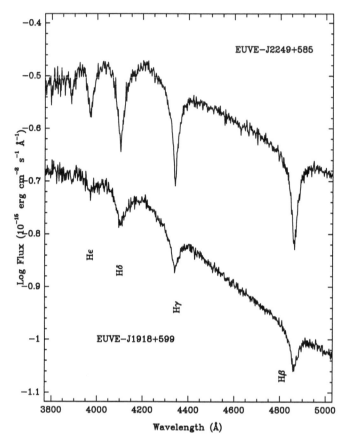

FIGURE 1. Spectra of the optical counterparts of EUVE_J1918+59 and EUVE_J2249+585

equivalent width implies a Li abundance of log N(Li) $= 2.5 \pm 0.4$. A detailed analysis of EUVE_J2056−171 has been presented by Mathioudakis et al. (1995b)

1.6. *EUVE_J2114+503*

The field of EUVE_J2114+503 is a very reddened field in the Galactic plane. Six candidates were observed at Kitt Peak in 1993, but none of the spectra indicated obvious EUV emission signatures. In 1994 additional bright candidates outside the 1′ radius were observed at Lowell Observatory. Again we do not detect any likely EUV candidates. All the remaining possible candidates in this field have magnitudes less than $m_v = 14$. This fact suggests that it is a very bright object in the EUV and relatively faint in optical wavelengths. Moreover, because this source lies in the Galactic plane, it must be relatively nearby—within 100 pc or so. If we plot this source on the count rate vs. visual magnitude diagram in Figure 5 of the BSL (Malina et al. 1994), with Lexan/B counts 60 counts ks^{-1} and $m_v = 14$ and fainter, it falls into either the class of white dwarf stars or cataclysmic variables. Both possibilities fit the observational evidence.

1.7. *EUVE_J2233−096*

EUVE_J2233−096 is detected in the DS Lexan/B filter and cataloged in the FESC with count rate 11 counts ks^{-1}. Previous work by Stephenson (1986) and Gliese & Jahreiss (1991), reported an M3 star within the error circle at this position. However, no references to any activity was given. Spectra of the two brightest stars within the error circle were obtained. A candidate which is 18″ from the *EUVE* source position, shows a spectrum characteristic of type dMe with Hβ and Hγ prominent in emission. From the TiO band measurements, we estimate a spectral type of dM3.5e.

1.8. *EUVE_J2329+414*

Currently EUVE_J2329+414 is identified as a "possible" ID, as an M star (G 19 0−28) in the FESC and as G 190−27 in the BSL. They are a pair of dMe stars, both located within the 1′ error circle of the *EUVE* position. Spectra show Balmer emission lines in both counterparts 6A and 6B. Both stars are likely to contribute to the observed EUV emission. The Hα, Hβ, Hγ, and Hδ flux measurements and comparison with the libraries of stellar spectra of Jacoby et al. (1984), Pettersen & Hawley (1989), and Mathioudakis & Doyle (1991) suggest the spectral types are dM4.5e and dM3.5e, respectively. Visual magnitudes of 11.87 and 12.44 are derived for these candidates. From the TiO band we derive the absolute magnitudes $M_v = 12$ and 11.97, respectively, and a distance of 14.2 pc for the pair. Note that the Gliese Catalog lists these objects as M2 and M3 stars and gives a distance of 14.8 pc, which is in a good agreement with our findings.

This work has been supported by NASA contract NAS5−29298.

REFERENCES

Bowyer, S., & Malina, R. F. 1991, in Extreme Ultraviolet Astronomy, ed. R. F. Malina & S. Bowyer, New York: Pergamon, 397

Bowyer, S., Lieu, R., Lampton, M., Lewis, J., & Wu, X. 1994, ApJ, 93, 569

Craig, N., et al. 1995, AJ, in press

Dixon, R. S., & Sonneborn, G. 1980, A Master List of Nonstellar Optical Astronomical Objects, Columbus: Ohio State U. Press

Fleming, T. A., Schmitt, J. H. M. M., Barstow, M. A., Mittaz, J., P., D. 1991, A&A, 246L, 47

Gliese, W. 1991, Catalog of Nearby Stars IDL Database Library

Jacoby, G. H., Hunter, D., & Christian, C. A. 1984, ApJS, 56, 257

Lanning, H. H. 1973, PASP, 85, 70

Liebert, J., Dahn, C. C., & Monet, D. G. 1988, ApJ, 332, 891

Malina, R. F., et al. 1994, AJ, 107, 751

Mathioudakis, M., Fruscione, A., Drake, J. J., McDonald, K., Bowyer, S., & Malina, R. F. 1995a, A&A, in press

Mathioudakis, M., et al. 1995b, A&A, in press

Mathioudakis, M. & Doyle, J. G. 1991, A&A, 244, 409

McDonald, K., Craig, N., Sirk, M. M., Drake, J. J., Fruscione, A., Vallerga, J. V., & Malina, R. F. 1994, AJ, 108, 1843

Monsignori-Fossi, B. & Landini, M. 1994, Solar Physics, 152, 81

Pallavicini, R. 1989, A&AR, 1, 177

Pettersen, B. R., & Hawley, S. L. 1989, A&A, 217, 187

Stephenson, C. B. 1986, AJ, 91, 144

Vennes, S., Thejll, X., & Dupuis, J. 1995, in prepreparation

Wesemael, F., Auer, L. H., Van, Horn, H. M., & Savedoff, M. P. 1980, ApJS, 43, 59

An Optical Study of the Field of EUVE J1027+323: Discovery of a QSO and a Hidden Hot White Dwarf

RICARDO GÉNOVA,[1,2] STUART BOWYER,[1]
STÉPHANE VENNES,[1] RICHARD LIEU,[1] J. PATRICK HENRY,[3]
JOHN E. BECKMAN,[2] AND ISABELLA GIOIA[3,4]

[1] Center for EUV Astrophysics, 2150 Kittredge Street,
University of California, Berkeley, CA 94720-5030, USA

[2] Instituto de Astrofísica de Canarias, E–38200 La Laguna, Tenerife, Spain

[3] Institute for Astronomy, University of Hawaii,
2680 Woodland Drive, Honolulu, HI 96822, USA

[4] Istituto di Radioastronomia del CNR, Via Gobetti 101, I–40129 Bologna, Italy

We have carried out optical and Far UV studies of the field around the EUV source EUVE J1027+323. We find two sources which contribute to this flux which are spatially unresolvable with *EUVE*. One is a non-cataloged QSO and one is a "hidden" hot white dwarf.[†] Reasonable scenarios ascribe the majority of the flux to the white dwarf.

1. Introduction

The EUV source EUVE J1027+323 was detected in the course of the all-sky survey conducted from the *EUVE* satellite and has been listed as "not identified" in the First *EUVE* Source Catalog (Bowyer et al. 1994). A detailed optical study of the field reveals the presence of two objects which contribute to the EUV flux in the region: a high redshift QSO and a "hidden" hot white dwarf companion to a main-sequence star.

2. The Region Around EUVE J1027+323

The *EUVE* source J1027+323 was detected by *EUVE* within a 60″ radius circle centered at $(\alpha, \delta)=(10^{\mathrm{h}}27^{\mathrm{m}}08^{\mathrm{s}}, +32°23'18'')$, a line of sight that was scanned between 15 and 21 of November, 1992.[‡] The Bell Laboratories H I survey (Stark et al. 1992) show that the total neutral hydrogen column density out of the Galaxy in this direction is $\leq 2 \times 10^{20}$ cm^{-2}.

The Simbad database shows two sources within 3′ of the EUV source: RE 1027+322, located within a 39″ radius circle centered at $(\alpha, \delta)=(10^{\mathrm{h}}27^{\mathrm{m}}11^{\mathrm{s}}.7, +32°23'24''.0)$, detected by the *ROSAT* satellite Wide Field Camera (WFC) and listed as "not identified" in the First *ROSAT* Catalog (Pounds et al. 1993) and in the Optical Atlas of *ROSAT* Sources (Shara et al. 1993); the second source is *IRAS* 10245+3238, detected by the *IRAS* satellite toward $(\alpha, \delta)=(10^{\mathrm{h}}27^{\mathrm{m}}23^{\mathrm{s}}.73, +32°22'52''.09)$. Given the small angular separation between the *ROSAT*, *IRAS*, and *EUVE* detections we believe that these detections correspond to the same object.

CCD images of the field obtained with the University of Hawaii 2.2 m telescope on Mauna Kea and the IAC–80 telescope on Observatorio del Teide show within the error

---

† Based in part on observations obtained at the Michigan–Dartmouth–MIT Observatory.

‡ All coordinates in this paper are J2000.0

S. Bowyer and R. F. Malina (eds.), Astrophysics in the Extreme Ultraviolet, 497–502.

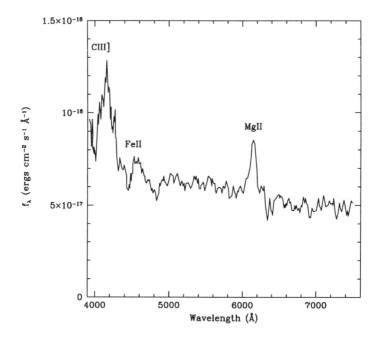

FIGURE 1. Optical spectrum of the faint object detected in the field. C III 1909 Å and Mg II 2798 Å lines seen in smission indicate a redshift of this object of 1.18. The spectral features are typical of QSOs.

circle of the *EUVE* detection a relatively bright object that we identify with the star GSC 02511 00033, included in the *Hubble Space Telescope* Guide Star Catalog, with magnitude 13.1.

Another fainter object, which we identify below as extragalactic, lies close to this star with an angular separation of 13″ that is unresolvable with the *EUVE* or WFC instrumentation.

3. Observations

3.1. *Visual and Far UV Observations*

An optical spectrum of the faint object that lies close to the main-sequence star in the field is shown in Figure 1. The spectrum shows characteristics typical of QSOs. Prominent in the spectrum are strong emission lines which we identify as redshifted C III 1909 Å, Fe II 2080 Å, and Mg II 2798 Å. From the observed wavelengths of C III and Mg II lines at 4135 Å and 6131 Å we deduce a redshift of 1.18. From the flux observed at 5500 Å we deduce an apparent visual magnitude of the QSO of 19.5 ± 0.2.

We obtained an optical spectrum of the relatively bright main-sequence star. The spectrum shows strong Ca II H and K lines in absorption and other spectral features typical of main-sequence G stars, comparison with the spectra published by Jacoby et al. (1984) leads us to classify the star as G2($\pm$2)V.

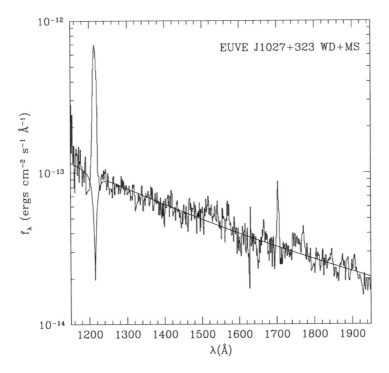

FIGURE 2. FUV spectrum of the bright star present in the field. A white dwarf synthetic spectrum (T_{eff} = 40,000 K, log g = 8.0) is superposed on the spectrum and testifies to the presence of a white dwarf star "hidden" by the bright object.

An *IUE* short wavelength spectrum (SWP49778) taken in the direction of the main-sequence star is shown in Figure 2.† The spectrum is characterised by rapidly diminishing flux with increasing wavelength. Geocoronal Lyman α line emission is superposed on the corresponding stellar line whose blue wing is seen blueward of that emission. A relatively strong emission feature that does not match any known line is seen at ~1700 Å and could be due to stray light from the QSO: at a redshift of 1.18, the rest wavelength of this emission would be 783 Å, a broad Ne VIII λ774 Å emission has been observed in the spectra of QSOs (Hamann et al. 1995; Cohen et al. 1995). The overall shape of the spectrum is typical of a white dwarf star. Superposed on the spectrum shown in Figure 2 is a synthetic spectrum of a DA white dwarf computed using the code developed by Vennes (1992) assuming a canonical value of the white dwarf gravity, log g = 8.0, and an effective temperature of 40,000 K. Hotter models which include heavy metal opacities would also be consistent with the *IUE* data.

3.2. *X-ray Observations*

The source EUVE J1027+323 was observed during the *ROSAT* survey using the Position Sensitive Proportional Counter (PSPC; Pfefferman et al. 1986). Shortward of 44.3 Å,

---

† Obtained from the Regional Data Analysis Facility at Goddard Space Flight Center

the flux of this source is too low to be detected with an upper limit of 0.006 counts s^{-1} in the band from 5 to 44 Å. The count rate in the 44 to 100 Å band is 0.073 ± 0.016 counts s^{-1} (Schmitt 1995).

3.3. EUV Observations

The source EUVE J1027+323 was found in the Lexan/B band of *EUVE*. Bowyer et al. (1994) estimate the count rate from the source in this band to be 0.040 counts s^{-1}. At longer wavelengths the source was not detected.

The source was also detected by the Wide Field Camera (WFC) of *ROSAT* (Wells et al. 1990). Pounds et al. (1993) estimated the photon count rate toward the source to be 0.017 ± 0.005 counts s^{-1} in the C/Lexan/B S1 filter and ≤ 0.022 counts s^{-1} in the Be/Lexan S2 filter.

4. Discussion

Both QSOs and hot white dwarfs are known EUV and X-rays emitters. We consider two hypotheses for the objects in the field of EUVE J1027+323. For the first hypothesis we derive an *EUVE* effective area from the instrumental area given in Bowyer et al. (1995) combined with the intervening interstellar medium absorption. For the interstellar medium absorption we combine the interstellar cross sections of Rumph et al. (1994), with an estimate of $N(\text{H\,I})$ to the sources. Later in this paper we obtain a distance to the white dwarf of ~400 pc. From Fruscione et al. (1994), we find $N(\text{H\,I})$ to stars in this direction are 3×10^{19} (300 pc) and 6×10^{20} (350 pc). From the Bell Lab Survey the total Galactic $N(\text{H\,I})$ in this direction is $\sim 2 \times 10^{20}$. We adopt $N(\text{H\,I}) = 2 \times 10^{20}$ as a reasonable estimate for both the Galactic and extragalactic sources in this field. Following Vennes et al. (1993), we assume He I/H I=0.07 and He II/H I=0.03. From the count rates in § 3.2 we find an EUV flux, λf_λ, equal to 1.1×10^{-11} erg cm^{-2} s^{-1}.

Using the same value for the ISM absorption in combination with the *ROSAT* PSPC effective area (Pfefferman et al. 1986) and the photon fluxes measured we obtain fluxes $\leq 1.2 \times 10^{-13}$ erg cm^{-2} s^{-1} in the 5–44 Å band, and 4.6×10^{-13} erg cm^{-2} s^{-1} in the 44–100 Å band.

4.1. Scenario A

We assume that most of the flux detected in the X-rays emanates from the QSO and that both the white dwarf and the QSO contribute to the flux observed in the EUV. At high energies QSO spectra follow a power law of the form $f \propto \nu^{-\alpha}$ with ν ranging between ~1.3 (Mushotzky et al. 1993) and ~1.03 (Maccacaro et al. 1988). We adopt a value of 1.2 for the spectral index and use the flux observed in the 44 to 100 Å band to predict a flux from the QSO in the 5 to 44 Å wavelength range; this is 1.0×10^{-13} erg cm^{-2} s^{-1}. This value is less than the upper limit measured with the *ROSAT* PSPC. In the EUV we predict a flux of 7.1×10^{-13} erg cm^{-2} s^{-1}; this is 6% of the observed EUV flux. In this case the white dwarf will have a nearly pure hydrogen atmosphere and will produce the remainder of the EUV flux.

4.2. Scenario B

We now consider the case in which the EUV and X-ray flux is essentially due to the white dwarf. We use the photon fluxes measured by *EUVE* and *ROSAT* and hot white dwarf model atmospheres to set limits to the effective temperature of the star and the column density of the intervening interstellar medium. In Figure 3 we present the *EUVE* and *ROSAT* count rates in the $T_{\text{eff}} - \log(N_\text{H})$ plan. Assuming a canonical value of the stellar

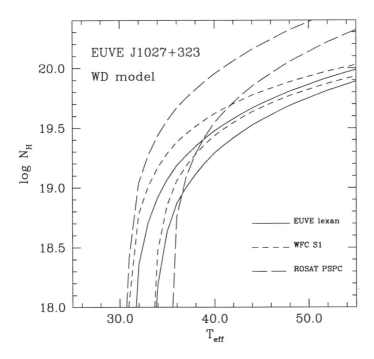

FIGURE 3. Range of T_{eff} vs. log N_{H} values derived from EUV and X–ray measurements. The count rates measured in these bands are analyzed assuming that log $g = 8.0$.

gravity, log $g = 8.0$, the count rates measured restrict the values of T_{eff} and $N(\text{H\,I})$ to 37,000±5,000 K and $\leq 5 \times 10^{19}$ cm^{-2} respectively. In this case there would be virtually no metal opacity in the atmosphere of the white dwarf.

5. Conclusions

Our analysis of the field of the source EUVE J1027+323 has shown the presence of two previously uncataloged objects: a white dwarf hidden by a main-sequence star and a QSO. Detection of these objects had been hampered by their small angular separation from the visually brighter main-sequence star.

From the synthetic spectrum fit to its *IUE* spectrum we deduce an apparent visual magnitude of the white dwarf star of 17.2±0.2. The effective temperature that we have used, 40,000 K, is the lower limit set by the absence of red wing in the stellar Lyman α line; this result is consistent with the EUV result. With this value and the absolute magnitude of the white dwarf star deduced from its FUV flux, we find the distance modulus is within the range from 7.3 to 8.3. For the main-sequence star we find the distance modulus to be within the range from 7.9 to 8.7. These values give heliocentric distances from 380 to 550 pc to the main-sequence star, and from 290 to 460 pc to the white dwarf star. Given the overlap in distance in combination with the angular separation, lower than 10″, we believe it is highly likely that the objects are binary companions. In this case the distance to the system falls in the range from 380 to 460 pc.

If the white dwarf atmosphere is essentially pure hydrogen and is at a temperature of 40,000 K, the object would join a subclass of DA white dwarfs similar to GD 153 or HZ 43 which display pure hydrogen EUV continua in *EUVE* spectroscopy (Dupuis & Vennes 1995). A review by Chayer et al. (1995) present evidence for the presence of a weak mass loss which would deplete the atmospheres of these objects from heavier elements in a few Myr after the onset of the cooling sequence. If the white dwarf is found to be in a close binary rather than in a wide binary, the question of its surface chemical composition would place constraints on the efficiency of accretion mechanisms and mass loss rate from the main-sequence companion.

John Thorstensen shared with us part of the time he had allocated at MDM Observatory. This research has been supported by a grant to RG from Gobierno Autónomo de Canarias, and by NASA contract NAS5-30180. JPH and IG wish to thank the NSF for support under grant AST91-19216.

REFERENCES

BOWYER, S., LAMPTON, M., LEWIS, J., WU, X., JELINSKY, P., LIEU, R. & MALINA, R. F. 1995, ApJS, submitted

BOWYER, S., LIEU, R., LAMPTON, M., LEWIS, J., WU, X., DRAKE, J. J., & MALINA, R. F. 1994, ApJS, 93, 569

CHAYER, P., FONTAINE, G., & WESEMAEL, F. 1995, ApJ, in press

COHEN, R. D., ET AL. 1995, in prepreparation

DUPUIS, J., & VENNES, S. 1995, in White Dwarfs, ed. K. Werner & D. Koester, Springer, Lecture Notes in Physics, 443, 323

FRUSCIONE, A., HAWKINS, I., JELINSKY, P., & WIERCIGROCH, A. 1994, ApJS, 94, 127

JACOBY, G. H., HUNTER, D. A., & CHRISTIAN, C. A. 1984, ApJS, 56, 257

HAMANN, F., ZUO, L., & TYTLER, D. 1995, ApJ, in press

MACCACARO, T., GIOIA, I. M., VOLTER, A., ZAMORANI, G., & STOCKE, J. T. 1988, ApJ, 26, 680

MUSHOTZKY, R. F., DONE, C., & POUNDS, K. A. 1993, ARA&A, 31, 717

PFEFFERMAN, E., ET AL. 1986, Proc. SPIE, 733, 519

POUNDS, K. A., ET AL. 1993, MNRAS, 260, 77

RUMPH, T., BOWYER, S., & VENNES, S. 1994, AJ, 107, 2108

SCHMITT, J. H. M. M. 1994, private communication

SHARA, M. M., SHARA, D. J., & McLEAN, B. 1993, PASP, 105, 387

STARK, A. A., GAMMIE, C. F., WILSON, R. W., BALLY, J., LINKE, R. A., HEILES, C., & HURWITZ, M. 1992, ApJS, 79, 77

VENNES, S. 1992, ApJ, 390, 590

VENNES, S., DUPUIS, J., RUMPH, T., DRAKE, J., BOWYER, S., CHAYER, P., & FONTAINE, G. 1993, ApJL, 410, L119

WELLS, A., ET AL. 1990, Proc. SPIE, 1344, 230

Recent Advances in EUV Solar Astronomy

G. A. DOSCHEK

Code 7670, E.O. Hulburt Center for Space Research,
Naval Research Laboratory, Washington, DC 20375-5320, USA

I discuss recent advances in EUV solar astronomy. The new work is primarily a result of current solar space missions such as the *Yohkoh* high energy solar physics mission, as well as upcoming space missions such as the ESA *Solar and Heliospheric Observatory (SOHO)*. I discuss spectroscopic and atomic physics work, and new results concerning solar flares that are directly relevant to stellar research.

1. Introduction

The Sun is the only astrophysical object for which the electrodynamics and plasma physics of high temperature plasmas can be studied in detail. Amazingly, plasmas that reach equivalent kinetic temperatures of more than 100 million degrees can occur by natural processes in the solar atmosphere. Accelerated particles in these plasmas are sufficiently energetic to produce nuclear reactions and consequent gamma ray continuum and line emission in a relatively low density medium. These energetic phenomena are believed to be the result of the conversion of magnetic energy into kinetic energy, but the precise physical processes are not known. The physics of solar activity is intrinsically fascinating, even if the rest of the Universe did not exist.

There are many examples in non-solar astronomy that involve complex electrodynamical processes, e.g., astrophysical jets, magnetic fields in the interstellar medium, stellar atmospheres and violent processes in these atmospheres, compact objects with accretion disks, etc. In most of these examples the sources cannot be spatially resolved, and therefore electrodynamical theories involving these source must by necessity rely on considerable conjecture. Understanding electrodynamical processes in the solar atmosphere can replace at least some of the conjectures with documented physical insight based on physical understanding achieved through solar physics.

In a less sophisticated sense, the Sun serves as a source for high resolution emission line spectroscopy of cosmic plasmas. Because solar activity produces plasmas that span a very wide temperature range, from a few thousand to millions of degrees, spectral lines characteristic of all these temperature regions can be recorded by high resolution solar spectroscopic experiments, and in particular, very high resolution can be achieved because the line intensities are obviously much higher than in other cosmic plasmas. High resolution solar spectroscopy experiments have been carried out for about 30 years, and have been a great stimulus for the study of the spectroscopy and atomic physics of highly ionized ions. Because of solar spectroscopy, UV, EUV, and X-ray emission line spectra of cosmically abundant elements are now well-understood in terms of spectral classification and plasma diagnostic potential for deriving important physical parameters such as electron temperature and density. For example, most of the atomic physics and plasma diagnostics necessary for the analysis of emission line spectra from the current *EUVE* and *ASCA* missions already exists, and is continually being improved upon because of new solar space experiments, either existing or planned. Laboratory research involving tokamak plasmas has also helped with EUV and X-ray atomic physics and spectroscopy, but the main driver for the cosmically abundant elements has been, and will continue to be, the Sun.

S. Bowyer and R. F. Malina (eds.), Astrophysics in the Extreme Ultraviolet, 503–510.
© *1996 Kluwer Academic Publishers. Printed in the Netherlands.*

In this paper I highlight some recent results in EUV and X-ray spectroscopy, and discuss some surprising new flare observations that indicate the occurrence of some counterintuitive processes in flaring magnetic flux tubes.

2. Current and Future Solar Spectroscopic Space Missions

There is no current solar EUV spectroscopic orbiting spacecraft experiment. However, there have been several recent flights of a high spectral and spatial resolution rocket experiment, designed and built by the solar group at Goddard Space Flight Center. This instrument, the Solar EUV Rocket Telescope and Spectrograph (SERTS), has obtained calibrated high resolution spectra of solar quiet and active regions, and an extensive active region line list with intensities has been published by Thomas & Neupert (1994). The last previous well-calibrated line list for EUV wavelengths greater than about 100 Å was published by Malinovsky & Heroux (1973), based on a rocket experiment. For EUV wavelengths less than 100 Å, the most recent line list with calibrated intensities is the Acton et al. (1985) list, again based on a rocket spectrograph experiment.

Major new high resolution EUV and UV results can be expected from the ESA cornerstone mission, the *Solar and Heliospheric Observatory (SOHO)* spacecraft, to be launched around the end of 1995. The *SOHO* spacecraft will be in orbit around the L1 Lagrangian point and will have an unobstructed view of the Sun. The payload consists of 12 instruments furnished by different consortia. Although *SOHO* is an ESA mission, NASA is supplying three solar experiments and the spacecraft operations and science analysis centers will be at GSFC.

SOHO contains two major experiments devoted to EUV and UV spectroscopy. The lead institution for the EUV instrument, the Coronal Diagnostics Spectrometer (CDS), is the Rutherford Appleton Laboratory. The spectrometer contains both normal and grazing incidence optics, and covers selected wavelength regions between 155 and 787 Å. The spectral resolving power is wavelength dependent but is very high, ranging between 3,500 and 12,000. The spatial resolution along the slit is also quite high, about 2 arcseconds over a 4 arcminute field of view. The lead institution for the UV spectrometer, called Solar Ultraviolet Measurement of Emitted Radiation (SUMER), is the Max Planck Institute for Aeronomy. This instrument will cover the wavelength region from 500 to 1600 Å, with a spectral resolving power > 18,000 and a spatial resolution along the slit of about 1 arcsecond.

Primarily because of the reflectance properties of mirrors, the simultaneous measurement of both UV and EUV lines with high resolution has not been done before in solar astronomy, and therefore many promising new plasma diagnostics can be used for the first time in the analysis of solar spectra obtained by CDS and SUMER. For example, intersystem lines of many transition region ions such as O IV and O V fall in the UV, but their resonance lines fall in the EUV. Combining the UV and EUV lines of these ions provides both temperature and density sensitive line ratios.

The *SOHO* spectroscopy experiments have formed the basis for a concentrated effort by solar atomic physicists to make a quantum leap improvement in EUV and UV atomic data, and the application of these data to plasma diagnostics. The goal is to make analysis of CDS and SUMER data straightforward for solar researchers who are not trained in spectroscopy. This requires a systematic approach to atomic physics calculations and their adaptation to plasma diagnostics. These atomic physics efforts are discussed in detail elsewhere in this Proceedings.

Although there are no current EUV orbiting spacecraft solar missions, the Japanese *Yohkoh* high energy solar physics mission contains four X-ray Bragg crystal spectrome-

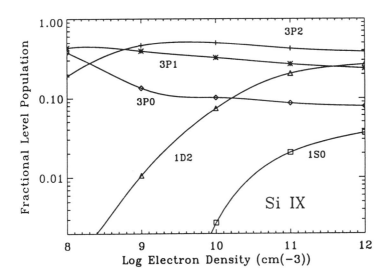

FIGURE 1. Ground configuration fractional level populations

ters. These spectrometers, in combination with the soft and hard X-ray telescopes on *Yohkoh*, have revealed some surprising properties of solar flares. Although X-ray observations, the results concern the physics of flares which transcends wavelength ranges. Some of the results are discussed in Section 4.

3. Atomic Physics and Plasma Diagnostics

The new atomic physics calculations carried out in support of future space missions in astronomy can be compared with the recent SERTS results (Thomas & Neupert 1994) and the earlier results from Malinovsky & Heroux (1973). As an illustration, consider Si IX of the C I isoelectronic sequence. The ground configuration, $2s^22p^2$, consists of five levels, $^3P_{0,1,2}$, $^1D_2$, and $^1S_0$. The populations of these levels vary with electron density, which leads to density sensitivity of some of the lines at typical stellar atmosphere densities.

The theoretical level populations of the ground configuration are shown in Figure 1, and theoretical relative line intensities are given in Table 1, compared with the SERTS and Malinovsky & Heroux (1973) data. The theoretical intensities are from Bhatia & Doschek (1993), assuming a temperature of about 1.3×10^6 K for the formation of Si IX. The best density sensitive line is the line at 258.095 Å, which terminates on the $^1D_2$ level. In general, the theoretical relative line intensities (arbitrarily with respect to the line at 341.974 Å) agree reasonably well with the data. The density sensitive ratio indicates a density of about 10^{10} cm^{-3} for the SERTS active region, which seems reasonable. Although there is general agreement between theory and experiment, there are some problems, such as the 296.137 Å line ratio, for which SERTS data and theory do not agree very well.

In most calculations, the relative line intensities generally agree, but sometimes there are disturbing exceptions. These exceptions may result from unresolved line blends,

TABLE 1. Si IX Line Intensity Ratios

Wave			Intensity Ratio (ergs)				
	SERTS	MH[†]	10^8	10^9	10^{10}	10^{11}	10^{12}
55.3[‡]	...	3.24	1.60	3.41	4.03	4.06	4.08
223.75	...	1.34	0.294	0.633	0.743	0.735	0.727
225.033[‡]	...	2.51	0.874	1.89	2.22	2.19	2.16
227.01	...	4.41	1.50	3.30	4.40	5.86	6.90
258.095	1.69	...	5.99(-2)	0.252	1.30	3.88	5.49
290.693	1.13	1.17	0.537	0.776	0.859	0.857	0.851
292.801	2.40	(2.40)	1.18	2.37	2.68	2.65	2.63
296.137	7.08	3.52	1.40	3.67	4.51	4.50	4.50
341.974	1.0	...	1.0	1.0	1.0	1.0	1.0
344.958	0.588	...	0.570	0.571	0.571	0.570	0.570
345.130	2.41	...	0.993	2.08	2.26	2.23	2.20
349.872	4.76	...	0.697	2.98	3.93	3.96	3.98

[†] MH solar data from Malinovsky & Heroux. The MH data are forced to agree with SERTS for the 292.801 Å line.

[‡] blend

instrumental calibration problems, or inaccurate atomic physics. They may also result from nonequilibrium conditions in the plasma; equilibrium is usually assumed, but not demonstrated. One of the perplexing problems concerns lines of Fe IX. The Fe IX resonance line at 171.075 Å is one of the strongest lines in the coronal solar spectrum, and it is a strong line in stellar spectra of stars such as Procyon. At longer wavelengths there are two other weaker Fe IX lines at 241.739 Å and 244.912 Å due to intersystem transitions to the ground state. The ratio of these two intersystem lines is an excellent electron density diagnostic for solar and stellar flare densities (e.g., Feldman, Doschek, & Widing 1978). However, the calculated ratio of the intersystem line intensities to the resonance line intensity is too large by about a factor of 5 to 10 in solar spectra. The reason for this discrepancy is unclear.

C I isoelectronic sequence ions are relatively simple from the atomic physics standpoint compared to the ions, Fe IX - Fe XVI, which produce strong line emission in the EUV, and which are strong in *EUVE* spectra of stars such as Procyon. For these ions, better atomic data are needed; however, some data exist and have been used in the analysis of *EUVE* and solar spectra. There are some excellent density diagnostics for stellar atmospheres. For example, Bhatia & Doschek (1995) have recently completed calculations for Fe X, adopting a 54 level ion. Fe X provides the excellent density diagnostic ratio, 175.266 Å/174.534 Å. This ratio has been used by Young et al. (1995) to derive solar densities from SERTS spectra, and by Drake, Laming, & Widing (1995) to derive a density for the atmosphere of Procyon using *EUVE* spectra.

It is important for the user of atomic data to be aware of the approximations that are adopted in different atomic physics calculations. In the Fe X example discussed above, it is important to include resonance contributions in order to obtain an accurate density diagnostic. Resonance excitation of a particular excited level is produced by dielectronic capture onto an ion, followed by autoionization leaving the ion in the particular excited state, or in an excited state that can decay into the level in question. This process can be a very important excitation mechanism in some cases. For Fe X, this is discussed and illustrated by Young et al. (1995). Neglecting resonances in the Fe X line ratio mentioned above would lead to an overestimate of density by about a factor of 2.5 over the most sensitive range of the line ratio.

4. New Solar Flare Results from *Yohkoh*

The *Yohkoh* spacecraft contains high spatial and time resolution soft and hard X-ray telescopes, along with a set of wide band X-ray spectrometers and four highly sensitive Bragg crystal spectrometers. Details of the instrumentation can be found in *Solar Physics* (vol 136, no. 1, 1991). The combination of high spatial and spectral resolution X-ray instrumentation, coupled with high time resolution and sensitivity, has not occurred previously in solar flare space missions. I will briefly discuss three areas of solar flare research: (1) the dynamics of multimillion degree plasmas, (2) the association of high speed soft X-ray upflows with hard X-ray emission, and, (3) the morphology of soft X-ray flare loops.

In flare numerical simulation models, the soft X-ray flare is formed by the ablation of hot plasma from the chromosphere (chromospheric evaporation). This evaporation might be caused by beams of high energy nonthermal electrons or protons that are accelerated in flaring magnetic flux loops, or by a conduction front due to plasma heating in the loops. In either case, the models predict that X-ray spectral line profiles should be primarily Doppler blueshifted at flare onset, if the flare occurs well within the solar disk (in order to observe the Doppler shift). The Bragg crystal spectrometer (BCS) package on *Yohkoh* was specifically designed for high sensitivity, in order to obtain statistically meaningful line profiles at flare onset and thereby provide an observational test of the models. (The chromospheric evaporation model is not universally accepted, e.g., see Feldman 1991).

A study by Mariska, Doschek, & Bentley (1993) of 219 flares observed by BCS has shown that intense blueshifted line components (relative to a stationary spectral component) occur in only about 10 percent of all flares. In the majority of flares, a relatively intense stationary spectral component exists even at flare onset. The presence of this component is difficult to understand in terms of the numerical simulation models. The result does not imply that chromospheric evaporation is not a source of any coronal mass supply, but it does imply that some other physical processes are also at work and that these have not been properly accounted for in the detailed modelling.

In the subset of flares that do show intense blueshifted components, these components first appear at flare onset and disappear or become quite weak at flare maximum and in the decay phase. Therefore, to observe solar-type dynamical activity in stellar spectra, high time resolution over the rise phase of a stellar flare is needed. Ca XIX profiles, and the integrated Ca XIX spectral intensity, are shown in Figure 2 for a solar-type flare exhibiting intense blueshifts at flare onset. The profiles are normalized to the same peak intensity in order to facilitate comparison of line shapes. The histogram spectrum is a typical decay phase spectrum without a blueshifted or significantly Doppler broadened component. These profiles can be easily scaled to EUV wavelengths in order to infer the spectral and time resolution needed to observe upflowing plasma with future EUV experiments. In the beam model of solar flares, the chromospheric heating resulting in evaporated plasma is produced by high energy accelerated particles that heat the chromosphere by Coulomb collisions. Electrons in these beams also produce hard X-ray bursts by thick target Bremsstrahlung. This implies a close correlation between the onset times of hard X-ray bursts and upflowing plasma that emits soft X-rays. These times have been known to be correlated from previous *Solar Maximum Mission* observations, but with *Yohkoh* it has been possible to investigate the correlation with much better time resolution. The onset times of the hard X-ray bursts and high speed upflows, i.e.,

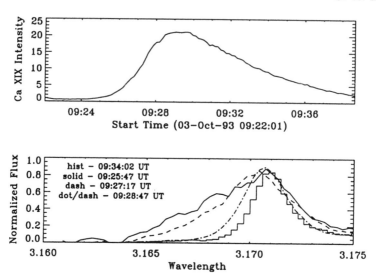

FIGURE 2. Ca XIX resonance line profiles and integrated light curve

plasma with speeds greater than about 400 km s^{-1}, are very well-correlated, to within a few seconds, as illustrated in Figure 3 (Bentley et al. 1994). Investigation of *Yohkoh* soft X-ray images reveals areas with light curves similar to the hard X-ray bursts (Hudson et al. 1994), and these occur at the footpoints of loops. Although some aspects of beam model physics appear verified by these observations, there are some significant problems. In some cases, the footpoint locations do not seem well-connected to the brightest soft X-ray features, making it difficult to understand how evaporation from these footpoints could have filled the soft X-ray loop. An examination of the entire soft X-ray line profile at flare onset (including speeds less than 400 km s^{-1} reveals in some cases blueshifts that can occur up to 100 s before the soft X-ray flare (Plunkett & Simnett 1994). This may imply that protons, in addition to electrons, are an important source of energy during the impulsive phase of flares (Plunkett & Simnett 1994). Perhaps most importantly, a region of apparent *confined* high energy particles, close to, but outside the soft X-ray emitting loop or loops, has been observed for some flares (Masuda et al. 1994). It is difficult to explain confined high energy particles in the corona, since it always seems possible for the particles to expand in some direction or directions along magnetic field lines.

There is a similar confinement problem with the soft X-ray flare. One of the most perplexing results from *Yohkoh* concerns the morphology of soft X-ray flare loops. Because a flare loop is essentially a magnetic tube with the field lines oriented in the direction of the major loop axis, models of gas behavior in loops predict a smooth distribution of plasma, i.e., constant pressure along the loop axis except for short time periods in which nonequilibrium dynamical effects may exist. However, observations by the *Yohkoh* soft X-ray telescope (SXT) of many flare loops (see Figure 4) show that a confined bright region is often found at the tops of flare loops (Acton et al. 1992; Feldman et al. 1994; Doschek, Strong, & Tsuneta 1995), and these regions exist from near flare onset to late in the flare decay phase. Measurements of temperature and emission measure along the flare loop axis indicate that these bright regions represent a region of enhanced pressure

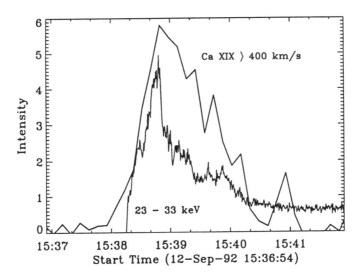

FIGURE 3. Ca XIX and hard X-ray light curves. The Ca XIX curve represents plasma moving at Doppler speeds greater than 400 km s^{-1}

FIGURE 4. SXT LDE image (2 Nov. 1992, 10:13:36 UT)

in the loop. It is difficult to understand how such a region can exist for long time periods, and explaining these structures is now a current and important problem in flare modelling.

Some indication of these bright regions was noticed in previous *Skylab* observations of long duration events (LDEs) (Kahler 1977; Vorpahl, Tandberg-Hanssen, & Smith 1977), which are very large flares, spatially, and last for many hours. They were detected because of the size and longevity of LDEs, and because the bright regions are particularly noticeable in these events. The dramatic illustration of a bright knot shown in Figure 4 is an SXT image of an LDE flare. However, the *Yohkoh* observations show that these bright regions are common properties of compact flares as well.

In summary, the observations from *Yohkoh* support some features of solar models, in the sense that evidence for high energy particle beams and evaporating plasma is seen in flares. But the details of the images and spectra conflict strongly in many cases with the details of the models, and the apparent longevity of confined bright regions, observed in both soft and hard X-ray images, appears to require significant revision or additions in the physics of flare models.

The author was supported by a NASA Grant from the Astrophysics Data Program. The author thanks J. Mariska for assistance with manuscript preparation.

REFERENCES

ACTON, L. W., BRUNER, M. E., BROWN, W. A., FAWCETT, B. C., SCHWEIZER, W., & SPEER, R. J. 1985, ApJ, 291, 865

ACTON, L. W. ET AL. 1992, Publ. Astron. Soc. Japan, 44, L71

BENTLEY, R. D., DOSCHEK, G. A., SIMNETT, G. M., RILEE, M. L., MARISKA, J. T., CULHANE, J. L., KOSUGI, T., & WATANABE, T. 1994, ApJL, 421, L55

BHATIA, A. K. & DOSCHEK, G. A. 1993, Atomic Data and Nuclear Data Tables, 55, 281

BHATIA, A. K. & DOSCHEK, G. A. 1995, Atomic Data and Nuclear Data Tables, in press

DOSCHEK, G. A., STRONG, K. T., & TSUNETA, S. 1995, ApJ, 440, 370

DRAKE, J. J., LAMING, J. M., & WIDING 1995, ApJ, 443, 393

FELDMAN, U. 1991, Flare Physics in Solar Activity Maximum 22, ed. Y. Uchida, R.C. Canfield, T. Watanabe, & E. Hiei, Springer, Lecture Notes in Physics, 387, 146

FELDMAN, U., DOSCHEK, G. A., & WIDING, K. G. 1978, ApJ, 219, 304

FELDMAN, U., SEELY, J. F., DOSCHEK, G. A., STRONG, K. T., ACTON, L. W., UCHIDA, Y., & TSUNETA, S. 1994, ApJ, 424, 444

HUDSON, H. S., STRONG, K. T., DENNIS, B. R., ZARRO, D., INDA, M., KOSUGI, T., & SAKAO, T. 1994, ApJL, 422, L25

KAHLER, S. 1977, ApJ, 214, 891

MALINOVSKY, M. & HEROUX, L. 1973, ApJ, 181, 1009

MARISKA, J. T., DOSCHEK, G. A., & BENTLEY, R. D. 1993, ApJ, 419, 418

MASUDA, S., KOSUGI, T., HARA, H., TSUNETA, S., & OGAWARA, Y. 1994, Nature, 371, 495

PLUNKETT, S. P. & SIMNETT, G. M. 1994, Solar Phys., 155, 351

THOMAS, R. J. & NEUPERT, W. M. 1994, ApJS, 91, 461

VORPAHL, J. A., TANDBERG-HANSSEN, E., & SMITH, J. B., JR. 1977, ApJ, 212, 550

YOUNG, P. R., MASON, H. E., BHATIA, A. K., DOSCHEK, G. A., & THOMAS, R. J. 1995, these proceedings

Looking for the FIP Effect in EUV Spectra: Examining the Solar Case

BERNHARD HAISCH,[1] JULIA L. R. SABA,[1,2]
AND JEAN-PAUL MEYER[3]

[1] Lockheed Solar and Astrophysics Laboratory, Dept. 91-30, Bldg. 252, 3251 Hanover St., Palo Alto, CA 94304, USA

[2] stationed at Solar Data Analysis Center, Code 682.2, NASA Goddard Space Flight Center, Greenbelt, MD 20771, USA

[3] Service d'Astrophysique, CEA/DSM/DAPNIA, Centre d'Etudes de Saclay, 91191 Gif-sur-Yvette, France

Systematic differences between elemental abundances in the corona and in the photosphere have been found in the Sun. The abundance anomalies are correlated with the first ionization potentials (FIP) of the elements. The overall pattern is that low-FIP elements are preferentially enhanced relative to high-FIP elements by about a factor of four; the transition occurs at about 10 eV. This phenomenon has been measured in the solar wind and solar energetic particle composition, and in EUV and X-ray spectra of the corona and flares. The FIP effect should eventually offer valuable clues into the process of heating, ionization and injection of material into coronal and flaring loops for the Sun and other stars. The situation for the Sun is remarkably complex: substantial abundance differences occur between different types of coronal structures, and variations occur over time in the same region and from flare to flare. Anomalies such as enhanced Ne/O ratios, distinctly at odds with the basic FIP pattern, have been reported for some flares. Are the high-FIP elements underabundant or the low-FIP elements overabundant with respect to hydrogen? This issue, which has a significant impact in physical interpretation of coronal spectra, is still a subject of controversy and an area of vigorous research.

1. Introduction

The ability of the *Extreme Ultraviolet Explorer* to carry out coronal spectroscopy has opened a number of opportunities to make progress in solar-stellar astrophysics, e.g., the determination of differential emission measures (Mewe et al. 1996), the measurement of coronal densities (Brickhouse et al. 1996; Schmitt, Haisch, & Drake 1994). These capabilities are beginning to constrain how solar outer atmospheric 0 may be scaled to various stellar conditions. However there is also an interesting new opportunity for EUV spectroscopy to contribute in the other direction by shedding light on a solar phenomenon which is not yet well understood: the First Ionization Potential (FIP) effect. Although originally discovered in cosmic rays, the FIP effect of interest here is an empirical relationship between abundance anomalies of heavy elements in highly ionized states in the solar corona and the ionization potentials of those elements in their neutral states (anomalous in comparison to the photospheric composition). This could yield valuable clues into the process of heating, ionization and injection of material into coronal and flaring loops, as well as into more open coronal field structures. Work by Drake, Laming, & Widing (1995a, 1996) and Drake et al. (1995) has begun to explore the evidence of a FIP effect in stellar coronae. Other similar investigations can be anticipated. A concise overview of the solar situation should be useful, given that the solar FIP literature could be characterized as not an easy read for the non-specialist.

511

S. Bowyer and R. F. Malina (eds.), Astrophysics in the Extreme Ultraviolet, 511–518.

TABLE 1. Photospheric Abundance of Major Elements and Their First Ionization Potentials

Atomic Number	Element	log Phot. Abund.	FIP (eV)
1	H	12	13.6
2	He	10.99 ± 0.04	24.6
6	C	8.55 ± 0.05	11.3
7	N	7.97 ± 0.07	14.5
8	O	8.87 ± 0.07	13.6
10	Ne	8.09 ± 0.10	21.6
11	Na	6.32 ± 0.03	5.1
12	Mg	7.58 ± 0.02	7.6
13	Al	6.48 ± 0.02	6.0
14	Si	7.55 ± 0.02	8.1
16	S	7.24 ± 0.06	10.4
18	Ar	6.56 ± 0.10	15.8
20	Ca	6.35 ± 0.02	6.1
26	Fe	7.51 ± 0.01	7.9
28	Ni	6.25 ± 0.02	7.6

2. Brief History

One of the assumptions of classical stellar atmospheres theory has been that the composition does not spatially vary in the upper layers of a star. (Thermonuclear-process gradients of course exist in the interior.) The first evidence for solar abundance anomalies were acquired in UV and EUV spectrograms taken during a series of sounding rocket flights between March 1959 and May 1963. These were analyzed by Pottasch in several papers, but in particular we point to Pottasch (1964) which contains a listing (in his Table IV) of the 14 most abundant elements comparing their chromospheric and transition region abundances to the then-standard photospheric ones; this paper is also well-known as the origin of the differential emission measure technique. The elements Mg, Al and Si were found to be three times more abundant in the upper atmosphere than in the photosphere; Fe was about ten times more abundant. Further references to other papers from that era and into the 1970s may be found in the introductory discussion of the recent Drake et al. (1995) paper.

The key discovery was the recognition of a pattern not in solar atmospheric abundances, but in the galactic cosmic ray (GCR) composition. Cassé & Goret (1978) noted the correlation between GCR abundances and the FIP of heavy elements. Table 1 lists the abundances of the major elements along with the potentials, in eV, required to the neutral species (Anders & Grevesse 1989; Grevesse & Noels 1993). Elements with FIP's < 10 eV were found to be enhanced relative to those with higher FIP.

Cassé and Goret specifically considered the possibility that flare stars might be significant contributors to the supply of GCR's. The prevailing view had been that supernovae were the primary, if not exclusive, source of cosmic rays. In the case of the Sun, solar energetic particle (SEP) events certainly feed material into interplanetary space, most of which will ultimately flow out into the interstellar medium. Analysis of the solar data was a major undertaking in which Meyer (1985a) analyzed all existing spacecraft observations of SEP events and came to the conclusion that: "All data show the imprint of an ever-present basic composition pattern... that differs from the photospheric composition by a simple bias related to first ionization potential." The SEP study, accumulated solar wind data, and evidence for coronal gas abundance anomalies all of a similar sort led to the proposal by Meyer (1985b) that GCR's originate not in supernova ejecta, but in late-type stellar coronal material. The complexity of the data and the interpretation in this fashion is well illustrated in the Meyer (1985b) article: It is a rare example of an ApJ Supplement paper with its own table of contents preceding the abstract.

3. FIP Observations of the Solar Corona

In its simplest form, the observation is that in the solar corona, heavy elements with low-FIP (< 10 eV) appear preferentially enhanced by about a factor of four relative to high-FIP (> 10 eV) ones. Figure 1 illustrates this. It shows the abundances and their uncertainties for the photosphere (light box symbols) vs. the corona (heavy-lined dotted symbols). There is a clear segregation above 10 eV. The data have been normalized for Si. This highlights one of the major problems: It is not yet settled whether the high-FIP elements are underabundant in the corona with respect to hydrogen, as the figure makes things look, or whether the low-FIP elements are overabundant, which the Si-normalization would thus misrepresent.

Actually, this composition bias is not found uniformly over the surface of the Sun and the range can be much broader than a factor of four (e.g., Feldman 1992; Meyer 1993a,b). According to Athay (1994): "In the transition region and corona the abundance of low-FIP elements relative to their photospheric values may vary by a factor of 1 to 15 depending on the region's characteristics." EUV observations of many localized features on the surface of the Sun (mainly more or less compact, active region or flaring loops), first gave the impression that, the more open the magnetic field structure, the larger the FIP effect (Widing & Feldman 1989; Sheeley 1995). This apparent positive correlation between FIP effect and opening of the field lines seemed dramatically confirmed by the very large low-FIP element enhancements found in specific diffuse, open-field structures within coronal holes, the polar plumes, as well as in other diverging field structures: Mg/Ne ratios 10 to 20 times as large as in the photosphere were observed (whereas the relative abundances of Ca, Na and Mg in the plume are photospheric within a factor of 2) (Widing & Feldman 1989, 1992). However it now appears that a specific (diffusion) mechanism (by Marsch and collaborators; Laming, private communication) may account for the plume abundances; the large open coronal holes have a more photospheric composition.

Of special interest in the context of EUV astronomy are the *Skylab* observations analyzed by Widing & Feldman (1989) since they involve spectroheliograms in the 315–625 Å regime. In particular they examined the ratios of of Ne VI to Mg VI line emission around 400 Å (see also Sheeley 1995). The range of observed Mg/Ne ratios extended over a factor of 20.

On a larger scale, by contrast, the FIP effect is most prominent over predominantly closed-field regions, which can include both quiet Sun and active regions. In the large open-field regions, primarily coronal holes with their associated cooler, high-speed streams, the composition is closer to photospheric (Meyer 1993a,b; von Steiger et al. 1995). Such a trend is drawn from a variety of data ranging from EUV and X-ray spectroscopy to in-situ measurements of the solar wind and interplanetary energetic particles, from observation of stable coronal structures to flares and flare-associated events. In the stellar case, the only available measurement is spectroscopy, but this does not necessarily mean that the non-spectroscopic solar data are irrelevant, since the relative contributions to a single disk-integrated stellar coronal spectrum may involve a different mix of structures than for the Sun.

Regarding active region and flare observations, a closer investigation suggests that they might be understood in terms of emerging, new, photospheric, non-FIP-biased material,

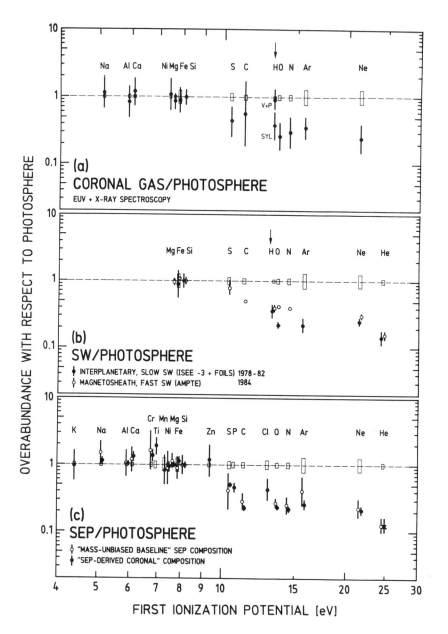

FIGURE 1. Change of elemental abundance as a function of first ionization potential between the photosphere (light box symbols) vs. the corona (heavy-lined dotted symbols) for coronal structures, the solar wind (SW) and solar energetic particles (SEP). Normalization has been forced for Si. The issue remains whether the high-FIP elements are underabundant or the low-FIP elements overabundant with respect to hydrogen. Evidence favors the latter; see Figure 2 (from Meyer 1993a).

which progressively changes its composition as it rises into the corona, over time scales of a day. As the material expands, the scale size of the magnetic field grows accordingly. It now seems most likely that the magnetic field opening just accompanies the rise of the gas, but is not the cause for its change in composition. So, the correlation between composition and opening of the field within active regions and flares, while real, is probably irrelevant (Meyer 1993a,b; Sheeley 1995). As for the extreme contrast between the polar plume and the wider coronal hole FIP effects, it is not understood. Maybe, all this just tells us that the magnetic field geometry is not the most crucial parameter controlling the intensity of the FIP effect.

One way around such apparent contradictions is to bypass the poorly understood details and let the Sun provide its own average FIP effect by examining a full-disk spectrum. The photoelectric recording of the 50–300 Å solar spectrum obtained with a rocket on 1969 April 4 by Malinovsky & Heroux (1973) provides such data. Although no flares took place during the few minute exposure, solar activity was near a maximum: the sunspot number had peaked in November 1968, but the flare maximum was still a year away (see Table 1 of Haisch, Antunes, & Schmitt 1995). With ~ 0.25 Å resolution and this wavelength coverage, it is quite similar to *EUVE* stellar spectra. This stellar-like solar spectrum has been thoroughly reanalyzed using modern atomic data by Laming et al. (1995). Their conclusion is that the canonical factor of $\sim 3 - 4$ relative enhancement for low-FIP elements appears provided one uses lines formed at $T \geq 10^6$ K; whether or not the FIP effect disappears below that temperature is not well determined by this spectrum since the major transition region lines are at longer wavelengths. Nevertheless, such a break in abundance pattern would be consistent with the fact that the chromospheric network disappears in spectroheliograms originating at $\sim 10^6$ K. This leads to the suggestion that the discrepancies between FIP effect in various discrete solar features may be resolved by considering the height of formation of the observed feature since different types of structures dominate above vs. below 10^6 K. One must also keep in mind that this spectrum is after all a single snapshot of the Sun, which may, or may not, be representative of the Sun at different times or in different activity states.

Two further complexities regarding the solar FIP observations are that variations occur from flare to flare and even over the course of time for a given active region; and that there are non-negligible changes in the ratios of elements *within* either of the groups, e.g., an enhancement of the Ne/O ratio—both being high FIP elements—in some flares (Schmelz 1993) and active regions (Strong, Lemen, & Linford 1991). Sylwester, Lemen, & Mewe (1984) were the first to find spectroscopic evidence for the variation of the coronal Ca abundance in high-temperature solar flare plasmas. On the basis of over 200 spectra taken by the Solar Maximum Mission Flat Crystal Spectrometer, Strong, Lemen, & Linford (1991) found that the relative abundance of Fe/Ne can vary by as much as a factor of about 7 and could change on timescales of less than 1 h. Good reviews of the solar EUV and X-ray spectroscopic results can be found in Feldman (1992) and Saba (1995).

4. Absolute Calibration of the FIP Effect

The overall FIP pattern of relative enhancements is now well established but the key question is not resolved: Are the high-FIP elements underabundant or the low-FIP elements overabundant with respect to hydrogen? Absolute coronal abundances were first derived from flare observations involving both X-ray lines and the continuum by Veck & Parkinson (1981). In the absence of hydrogen lines, the continuum provides the necessary reference point. The following absolute abundances were reported: Si

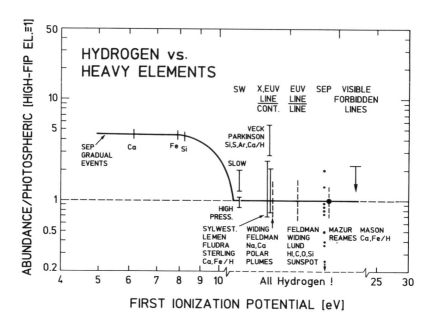

FIGURE 2. Compilation of data attempting to anchor the FIP pattern to hydrogen. The evidence favors the interpretation that low-FIP elements are genuinely overabundant (from Meyer 1993a).

$(7.7^{+0.2}_{-0.3})$, Ca $(6.5^{+0.1}_{-0.2})$, S $(6.9^{+0.1}_{-0.3})$, and Ar $(6.4^{+0.2}_{-0.3})$. Comparison to Table 1 shows somewhat better agreement with the interpretation that the high-FIP elements S and Ar are underabundant than that the low-FIP elements Si and Ca are enhanced.

Since then, a wide variety of observations with entirely different techniques, ranging from EUV and X-ray spectroscopy to in-situ measurements of solar wind and energetic particles, have pointed to the opposite conclusion: they seem to converge on an absolute enhancement of the low-FIP elements; high-FIP elements appear to have roughly photospheric abundances relative to H, or to be only slightly depleted (Meyer 1993a,b; Mazur et al. 1993; Reames 1995; von Steiger et al. 1995). Note that such a behaviour, if confirmed, is not surprising, since H itself is a high-FIP element, which is neutral or ionized at about the same temperatures as other high-FIP elements.

Very recent X-ray studies, a line-to-continuum study by Fludra & Schmelz (1995), and a line-to-line study by Phillips et al. (1994, 1995) seem, however, to support the earlier analysis of Veck & Parkinson (1981) indicating a depletion of high-FIP elements. A discussion of some of the problems involved in all EUV and X-ray determinations of absolute abundances can be found in Saba (1995). As an illustration of how difficult this type of determination is, we examine here the procedure followed by Phillips et al. (1994) in analyzing X-ray flare data provided by the Bragg Crystal Spectrometer onboard the *Yohkoh* satellite. Two very different Fe lines lie close together in the X-ray spectrum and can thus be observed by the same instrument with good relative calibration: the Fe XXV resonance line at 1.850 Å, and the Fe Kα and Kβ fluorescence line at 1.936/1.940 and 1.757 Å respectively. The former is the usual type of collisionally excited line, in this case that of He-like Fe. The K lines involve inner shell transitions. If one of the innermost electrons (in the K-shell) is removed by photoionization, this vacancy will be quickly filled in by an electron from one of the next higher levels. The most common involves a transition from the L-shell, which would yield the Fe $K\alpha$ lines, but transitions from the M-shell are also possible, and these give rise to the $K\beta$ line. The K lines can be formed in more than one ionization stage of Fe, since the energy level structure for these innermost shells changes little as a function of the number of outermost filled shells. The strength of both the Fe XXV and the Fe K lines depend on nearly the same emission measure of hot ($\sim$ 20 MK) material during a flare: the former because that material is the source of the collisional excitation, the latter because irradiation from the material gives rise to the photoionization leading to fluorescence. The line ratio during a flare is thus a function of known atomic physics parameters, a common hot coronal emission measure, and the ratio of coronal [Fe/H] (see eqn. 7 of Phillips et al. 1994). Their conclusion was that the abundance of Fe, a low-FIP element, was not more than a factor of 2 larger in the corona than in the photosphere, again suggesting a high-FIP element underabundance (see Phillips et al. 1995 for the $K\alpha$ line discussion).

So, while most data still converge towards an overabundance of low-FIP elements relative to hydrogen, the high-FIP element abundances being photospheric or only slightly depressed, the question of this absolute calibration cannot be considered entirely settled.

At this time it remains difficult to synthesize the diverse and contradictory solar observations into a single coherent picture. As variable as it appears to be, there is a FIP effect, and it is encouraging that the one attempt to analyze sun-as-a-star data (Laming, Drake, & Widing 1995) did succeed in observing this.

This work was supported in part by the Lockheed Solar and Astrophysics Laboratory Independent Research Program.

REFERENCES

ANDERS, E. & GREVESSE, N. 1989, Abundances of the Elements: Meteoritic and Solar, Geochim. Cosmochi. Acta, 53, 197

ATHAY, R. G. 1994, Separation of Low First Ionization Potential Ions From High First Ionization Potential Neutrals in the Low Chromosphere, ApJ, 423, 516

BRICKHOUSE, N. S. 1995, Dissecting the EUV Spectrum of Capella, these proceedings

CASSÉ, M. & GORET, P. 1978, Ionization models of cosmic ray sources, ApJ, 221, 703

DRAKE, J. J., LAMING, J. M., & WIDING, K. G. 1995a, Stellar Coronal Abundances. II. The First Ionization Potential Effect and Its Absence in the Corona of Procyon, ApJ, 443, 393

DRAKE, J. J., LAMING, J. M., & WIDING, K. G. 1996, The FIP Effect and Element Abundance Anomalies in Late-Type Stellar Coronae, these proceedings

DRAKE, J. J., LAMING, J. M., WIDING, K. G., SCHMITT, J. H. M. M., HAISCH, B. M., & BOWYER, S. 1995, The Elemental Composition of the Corona of Procyon: Evidence for the Absence of the FIP Effect, Science, 2677, 1470

FELDMAN, U. 1992, Highly Ionized Atoms in Space; or Highly Ionized Atoms, What Are They Teaching Us About the Solar Coronal Heating Problem? Physica Scripta, 46, 202

FLUDRA, A. & SCHMELZ, J. T. 1995, Absolute Abundances of Flaring Coronal Plasma Derived from SMM Spectral Observations, ApJ, 447, 936

GREVESSE, N. & NOELS, A. 1993, Cosmic Abundances of the Elements, in Origin and Evolution of the Elements, ed. N. Prantzos, E. Vangioni-Flam, & M. Cassé, Cambridge Univ. Press, 15

HAISCH, B., ANTUNES, A., & SCHMITT, J. H. M. M. 1995, Solar-Like M-Class X-ray Flares on Proxima Centauri Observed by the ASCA Satellite, Science, 268, 1327

LAMING, J. M., DRAKE, J. J., & WIDING, K. G. 1995, Stellar Coronal Abundances. III. The Solar First Ionization Potential Effect Determined from Full-Disk Observations, ApJ, 443,

416

MALINOVSKY, M. & HEROUX, L. 1973, An Analysis of The Solar Extreme-Ultraviolet Spectrum between 50 and 300 Å, ApJ, 181, 1009

MAZUR, J. E., MASON, G. M., KLECKER, B., & McGUIRE, R. E. 1993, The Abundances of Hydrogen, Helium, Oxygen, and Iron Accelerated in Large Solar Events, ApJ, 404, 810

MEYER, J. -P. 1985a, The Baseline Composition of Solar Energetic Particles, ApJS, 151

MEYER, J. -P. 1985b, Solar-Stellar Outer Atmospheres and Energetic Particles, and Galactic Cosmic Rays, ApJS, 173

MEYER, J. -P. 1993a, Element Fractionation at Work in the Solar Atmosphere, in : Origin and Evolution of the Elements, ed. N. Prantzos, E. Vangioni-Flam, & M. Cassé, Cambridge Univ. Press, 26

MEYER, J. -P. 1993b, Elemental Abundances in Active Regions, Flares and the Interplanetary Medium, Adv. Space Res., 13(9), 377

MEWE, R., VAN, DEN, OORD, G. H. J., SCHRIJVER, C. J., & KAASTRA, J. S. 1996, DEM Analysis with the Utrecht Plasma Code, these proceedings

PHILLIPS, K. J. H., PIKE, C. D., LANG, J., WATANABE, T., & TAKAHASHI, M. 1994, Iron Kβ Line Emission in Solar Flares Observed by Yohkoh and the Solar Abundance of Iron, ApJ, 435, 888

PHILLIPS, K. J. H. ET AL. 1995, Evidence for the Equality of the Solar Photospheric and Coronal Abundance of Iron, Adv. Space Res., 15(7), 33

POTTASCH, S. 1964, On the Interpretation of the Solar Ultraviolet Emission Line Spectrum, Space Science Revs., 3, 816

REAMES, D. V. 1995, Coronal Abundances of O, Ne, Mg, and Fe in Solar Active Regions, Adv. Space Res. 15(7), 41

SABA, J. L. R. 1995, Spectroscopic Measurements of Element Abundances in the Solar Corona: Variations on the FIP Theme, Adv. Space Res., 15(7), 13

SABA, J. L. R. & STRONG, K. T. 1993, Coronal Abundances of O, Ne, Mg, and Fe in Solar Active Regions, Adv. Space Res. 13(9), 391

SCHMELZ, J. T. 1993, Elemental Abundances of Flaring Solar Plasma: Enhances Neon and Sulfur, ApJ, 408, 373

SCHMITT, J. H. M. M., HAISCH, B. M., & DRAKE, J. J. 1994, A Spectroscopic Measurement of the Coronal Density of Procyon, Science, 265, 1420

SHEELEY, N. R. 1995, A Volcanic Origin for the Material in the Solar Atmosphere, ApJ, 440, 884

STRONG, K. T., LEMEN, J. R., & LINFORD, G. A. 1991, Abundance variations in solar active regions Adv. Space Res., 11, 151

SYLWESTER, J., LEMEN, J. R., & MEWE, R. 1984, Variation in observed coronal calcium abundance of X-ray flare plasmas Nature, 310, 665

VECK, N. J., & PARKINSON, J. H. 1981, Solar Abundances from X-ray Flare Observations, MNRAS, 197, 41

VON, STEIGER, R., WIMMER, SCHWEINGRUBER, R. F., GEISS, J., & GLOECKLER, G. 1995, Abundances Variations in the Solar Wind, Adv. Space Res. 15(7), 3

WIDING, K. G. & FELDMAN, U. 1989, Abundance Variations in the Outer Solar Atmosphere Observed with Skylab Spectroheliograms, ApJ, 344, 1046

WIDING, K. G. & FELDMAN, U. 1992, Element Abundances and Plasma Properties in a Coronal Polar Plume, ApJ, 392, 715

The Sun in Time: Evolution of Coronae of Solar-Type Stars

MANUEL GÜDEL[1] AND EDWARD F. GUINAN[2]

[1]Paul Scherrer Institut, CH-5232 Villigen PSI, Switzerland, and
JILA, University of Colorado, Boulder, CO 80309-0440

[2]Department of Astronomy and Astrophysics, Villanova University, Villanova, PA 19085

We report on the results of a multi-frequency program to study coronal X-ray, EUV, and microwave activity of solar-type (G0-5 V) stars of greatly different ages. These stars are of interest as proxies for the Sun for ages from 50 Myr to 10 Gyr. Coronal temperatures decrease with X-ray/EUV luminosity and with increasing age and rotation period according to power-law relations. The young Sun had an extremely luminous corona with $L_{\mathrm{EUV,X}} = 100 - 600 L_{\mathrm{EUV,X}}$ (present Sun) and temperatures up to 10 MK; this information is pivotal for the study of the young planetary atmospheres. The findings further suggest an intimate connection between the presence of very hot coronal plasma and non-thermal radio-emitting particles.

1. A Diary of Solar Life: An Age Sequence of Solar Proxies

The Sun's magnetic activity and consequently its chromospheric and coronal emissions are expected to have declined steadily to present levels as the solar rotation slowed due to magnetic braking (Skumanich 1972). Despite the fact that evolutionary models indicate that the Sun was less luminous in its early ZAMS phase ($L_{\mathrm{ZAMS}} \approx 0.76 L_{\odot}$), Canuto et al. (1982) have estimated that in the Sun's pre-main-sequence T Tauri phase, solar X-ray, EUV, and UV emissions were enhanced by up to *hundreds of times present levels*. Direct observation of high-energy emissions from younger solar-type stars is the only means of establishing the level of the solar short wavelength flux at earlier epochs.

Table 1 presents our sample of solar proxies (spectral types G0-G5 V), selected mainly for their well established rotation periods (as chiefly measured by optical photometry) and reliably determined ages (partly supported by cluster or moving group memberships, age-rotation relations, and isochrone ages). The ages of the targets range from approximately 70 Myr (near-zero-age main-sequence = ZAMS; EK Dra) to 8.8 Gyr (terminal-age main-sequence = TAMS; β Hydri). For most of these stars, we obtained *ROSAT* PSPC observations in the Guest Observer Program; further, *ASCA* and *EUVE* data are being collected, and a vast amount of VLA, *IUE* and optical data obtained over the last decade clearly makes this an exciting moment to assemble bits and pieces of information into a more coherent picture on the long-term evolution of solar magnetic activity.

2. Dissecting the Sun's Activity in Time: Previous Findings

Coronal parameters are expected to vary with the efficiency of the magnetic dynamo. For example, Güdel et al. (1995a) found rotational modulation in the cooler X-ray component of the young EK Dra; through geometric modeling, they determined that

• the cool plasma of ~ 2 MK is confined to heights (above the photosphere) similar to active regions in the quiet solar corona, i.e., $\sim 0.1 R_{\odot}$. Its coronal filling factor is moderate, showing that young solar proxies possess well localized, magnetically confined structures that we identify with "nonflaring stellar active regions".

• The electron density of the cool component exceeds $4 \cdot 10^{10}$ cm^{-3}.

519

S. Bowyer and R. F. Malina (eds.), Astrophysics in the Extreme Ultraviolet, 519–524.

TABLE 1. Coronal properties of the Sun in time (X-ray data from present observations, or Güdel et al. 1995b or Dorren & Guinan 1994; stellar 8.5 GHz radio data from Güdel et al. 1994, 1995ab; L_X for β Hyi scaled to one solar radius)

Star	Spectr. Type	P_{rot} (d)	Age (Gyr)	$\mathrm{Log}L_X$ (erg s^{-1})	T_{hot}/T_{cool} (MK)	$\mathrm{EM}_{hot}/\mathrm{EM}_{cool}$	$\mathrm{Log}L_R$ (erg s^{-1}Hz^{-1})
EK Dra	G0 V	2.75	0.07	29.92	$9.33^{+.67}_{-.78}/2.29^{+.34}_{-.29}$	≥ 1.16	14.1 ± 0.5
π^1 UMa	G1.5 V	4.68	0.3	29.09	$6.35^{+.71}_{-.71}/1.41^{+.14}_{-.12}$	1.15	≤ 13.37
HN Peg	G0 V	4.86	0.3	28.95	...	...	≤ 13.38
χ^1 Ori	G1 V	5.08	0.3	29.13	$7.65^{+.86}_{-.55}/1.56^{+.27}_{-.18}$	0.65	...
HD 1835	G2 V	7.65	0.6	29.13	$6.52^{+.98}_{-1.2}/1.62^{+.26}_{-.37}$	0.79	...
VB 64	G2 V	8.7	0.6	29.41	...	...	...
κ^1 Cet	G5 V	9.2	0.7	28.82	$7.21^{+1.6}_{-1.9}/1.74^{+.20}_{-.25}$	0.45	...
β Com	G0 V	12.4	1.6	28.11	$3.66^{+1.6}_{-.57}/1.24^{+.11}_{-.11}$	0.45	...
15 Sge	G5 V	13.5	1.9	28.05	$3.68^{+1.9}_{-2.0}/1.06^{+.38}_{-.36}$	0.60	...
Sun	G2 V	25.4	4.6	27.3	$3.03^{+.56}_{-.32}/1.22^{+.15}_{-.11}$	0.74	10.7
α Cen A	G2 V	~ 30	5-6	27.11	...	...	...
β Hyi	G2 IV	~ 45	9	27.08	$2.09^{+.31}_{-.18}$	...	...

• Thus, a significant portion of the high (non-flaring) X-ray luminosity L_X in EK Dra is *not* due to large volumes but rather due to high plasma density. This in turn requires, for plasma confinement, stronger average coronal magnetic fields than on the Sun.

EK Dra is the first solar proxy discovered as a non-thermal radio source (Güdel, Schmitt, & Benz 1994), evident both in quiescent and flaring emission (Güdel et al. 1995a). This shows that the young Sun was a source of copious acceleration of non-thermal particles, producing a radio luminosity L_R more than 3 orders of magnitude higher than the quiet Sun's (which is thermal emission from lower atmospheric levels).

Slightly older solar-type stars continue to keep their strong non-thermal radio emission for some time. The rapidly rotating G5V star Gliese 755, estimated to be approximately 200 Myr old, was detected at $\mathrm{log}L_R = 14.03$ ($\mathrm{log}L_X = 29.52$), while, on the other hand, HN Peg and π^1 UMa remained undetected ($\mathrm{log}L_R \leq 13.37$), indicating a time of the order of 300 Myr for the decay of steady radio emission to below $\mathrm{log}L_R = 13.5$.

3. Coronal Cooling with Age

The coronal L_X of a GV star is well correlated with its rotation period P_{rot} and consequently with its age, with $L_X \propto P_{rot}^{-2}$ (Dorren, Guinan, & DeWarf 1994).

In Fig. 1a, an age sequence of *ROSAT* spectra of G stars is shown, with each *ROSAT* spectrum renormalized to a constant amplitude of the lower-energy peak. The quiet solar corona was modeled using the continuous emission measure (EM) distribution given in Raymond & Doyle (1981) between 1 and 5 MK (supplemented with realistic noise). All spectra were fitted with two Raymond-Smith type thermal plasma components. Notice that the Sun's L_X is close to β Hyi's, despite the latter's significantly higher age (8.8 Gyr; see Dorren, Güdel, & Guinan 1995). This is because β Hydri has, as a consequence of its leaving the main sequence, expanded to a radius of $1.6R_\odot$. Its surface X-ray flux is, however, $\sim 2\frac{1}{2}$ times smaller than the Sun's.

In our stars, the low-energy peaks contain photons from both plasma components (typically with comparable contributions), while the bump around 0.8 keV is largely

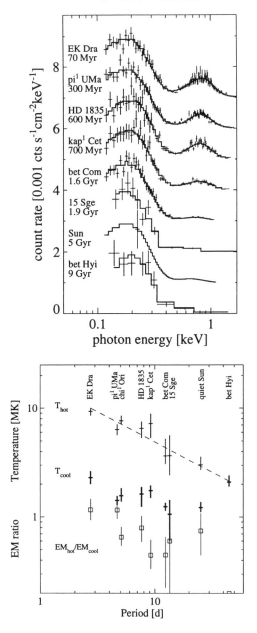

FIGURE 1. a (top): *ROSAT* PSPC pulse-height spectra of the solar proxy age series, normalized such that the left (softer) peak is about equal for all spectra. Individual spectra have been shifted by multiples of 0.001 cts s^{-1}cm^{-2}keV^{-1} for illustration. Age monotonically increases from top (EK Dra: 70 Myr) to bottom (β Hydri: 8.8 Gyr). The consistent decrease of hard emission in the peak around 0.8 keV indicates a decrease in EM at high T. Crosses are observations, while histograms represent fit (2-T Raymond-Smith). b (bottom): Coronal T_X and EM ratios for the same stars, plotted versus P_{rot}. *Thin*: Hotter PSPC temperature; dashed line represents a power-law. *Bold*: Cooler PSPC temperature. *Quadrangles*: EM ratio of hot and cool temperature.

determined by hot (5–15 MK) plasma. Quite evidently, the relative amount of hot plasma and therefore hard radiation diminishes with age.

Fig. 1b shows the numerical fit results. The higher temperature decreases from around 10 MK near the ZAMS age ($P_{rot} \approx 2$ d) to $\sim 2 - 3$ MK beyond the age of the Sun. For T_{hot}, we thus find a decay law

$$T_{hot} = 1.74 \cdot 10^7 P_{rot}^{-0.55 \pm 0.03} \text{ [K]}, \tag{3.1}$$

$$T_{hot} = 4.7 \cdot 10^6 t^{-0.31} \text{ [K]}, \tag{3.2}$$

where P_{rot} is in days and t, the stellar age, is in Gyr. The second equation has been derived using a rotation-age relation for G stars, $\log t = 1.75 \log P_{rot} - 1.81$ (t in Gyr, P_{rot} in d; Dorren et al. 1994).

The nearly consistent decrease of the EM ratios for the hotter stars (i.e., EM_{hot}/EM_{cool}; quadrangles in Fig. 1b) further suggests that the hotter plasma not only becomes cooler with age, but also that *it becomes less abundant*, and thus that the *distribution of plasma falls off increasingly steeper with age* (starting from nearly flat in young stars). Further, the variation of T_{cool} along with T_{hot} is indicative of a *continuous EM distribution*: apparently, the whole distribution shifts toward cooler temperatures with increasing age, although the *ROSAT* PSPC spectral fits are not sensitive enough to reveal more than two plasma components. From Fig. 2a, T_{hot} is correlated with the *total* L_X:

$$L_X \approx 84 T_{hot}^{3.97 \pm 0.23} \approx 50 T_{hot}^4 \text{ [erg s}^{-1}] \tag{3.3}$$

(T in K). Remarkably, this remains true if we add close binary systems consisting of two almost identical solar-type G stars each: ER Vul (G1V+G1V), HR 8358 (G5V+G5V), and TZ CrB (G0V+G0V). The data for these binaries are from Dempsey et al. (1993). To account for binarity, their L_X were divided by two. Similar relations have been reported for less restricted star samples (e.g., Schrijver, Mewe, & Walter 1984). For a theoretical discussion of such relations, see Jordan & Montesinos (1991).

Since $L_R \propto L_X$ in active G stars (Güdel et al. 1995b), we find $L_R \approx 1.6 \cdot 10^{-14} T_{hot}^4$. It implies that microwave luminous stars are those with hot coronae, and particle acceleration is intimately connected to the production of very hot (thermal) plasma.

4. EUV Properties: Their Relevance to Planetary Atmospheres

We summarize EUV properties of solar analogs as obtained from ROSAT/Wide Field Camera Survey observations (Fig. 2b). The EUV luminosity roughly fulfills a relation similar to the X-rays (we give both relations here),

$$L_{EUV} \approx 5.3 \cdot 10^{29} P_{rot}^{-2.06 \pm 0.15} \approx 3.8 \cdot 10^{27} t^{-1.17} \text{ [erg s}^{-1}] \tag{4.4}$$

$$L_X \approx 8.4 \cdot 10^{30} P_{rot}^{-2.43 \pm 0.15} \approx 2.6 \cdot 10^{28} t^{-1.39} \text{ [erg s}^{-1}] \tag{4.5}$$

(P_{rot} in days, t in Gyr). The much increased EUV and X-ray emissions of the "younger Suns" are of primary relevance for the formation of planetary ionospheres. In particular, they yield information on the young Sun's influence on the upper terrestrial atmosphere. Similarly, the enhanced UV emission is crucial for our understanding of the photochemistry in the young Earth's atmosphere.

5. Summary and Conclusions

The Sun's evolution from ZAMS to TAMS involves spin-down due to angular momentum loss in a magnetized wind. The feedback to the magnetic activity itself via an

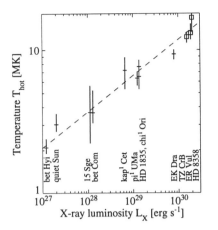 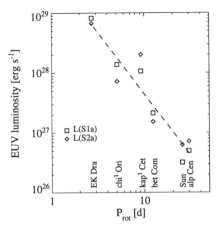

FIGURE 2. **a (left):** The coronal T_{hot} versus L_X in the [0.1,2.4] keV energy range, clearly showing a strong correlation. The three stars indicated with quadrangles are three close binary systems consisting of two identical G dwarfs each (see text). **b (right):** ROSAT/WFC (Survey) luminosities for a sample of solar-like G stars. Quadrangles and diamonds indicate luminosities in the S1 and in the S2 filters, respectively. The diagonal is a power-law (Eq. 4.4).

internal dynamo is theoretically poorly understood, but can be readily studied with the currently growing data pool of X-ray, EUV, UV, optical, and radio data of solar proxies at different ages. We find that the X-ray/EUV luminosity of solar-like stellar coronae decreases with the rotation period as $\sim P^{-2\ldots-2.4}$, an that they become cooler with increasing P_{rot} and thus age, with $T_{hot} = 1.74 \cdot 10^7 P^{-0.55}$ [K] and $T_{hot} = 4.7 \cdot 10^6 t^{-0.31}$ [K]. By implication, $L_X \approx 50 T_{hot}^4$ [erg s^{-1}].

Very hot quiescent plasma ($\gtrsim 10^7$ K) is restricted to *young* solar proxies. The same appears to be true for the persistent presence of a large amount of non-thermal radio-emitting electrons detectable in microwaves. We thus find a three-fold tight correlation between L_R, L_X, and T_{hot}, indicating that particle acceleration is intimately related with the production of *very hot* coronae.

The *ROSAT* project has been supported by the Bundesministerium für Forschung und Technologie (BMFT) and the Max-Planck-Gesellschaft (MPG). This research has been supported by NASA *ROSAT* grants NAG5-1662 and NAG5-1703, and by NASA *IUE* grants NAG5-382 and NAG5-1703. We also acknowledge the support of NSF grant AST 86-16362. MG has been supported by the Swiss National Science Foundation (grant 8220-033360) and by NASA grant NAG5-1887 to the University of Colorado.

REFERENCES

CANUTO, V. M., LEVINE, J. S., AUGUSTSSON, T. R. & IMHOFF, C. L. 1982, UV radiation from the young Sun and oxygen and ozone levels in the prebiological paleoatmosphere, Nature, 296, 816

DEMPSEY, R. C., LINSKY, J. L., SCHMITT, J. H. M. M. & FLEMING, T. A. 1993, The ROSAT All-Sky Survey of active binary coronae II: Coronal temperatures of the RS Canum Venaticorum Systems, ApJ, 413, 333

Dorren, J. D. & Guinan, E. F. 1994, The Sun in Time: Detecting and Modelling Magnetic Inhomogeneities on Solar-type Stars, in The Sun as a Variable Star, IAU Colloq. 143, ed. J. M. Pap et al., Cambridge, 206

Dorren, J. D., Guinan, E. F. & DeWarf, L. E. 1994, The Decline of Solar Magnetic Activity with Age, in The 8th Cambridge Workshop on Cool Stars, Stellar Systems, and the Sun, ed. J.-P. Caillault, ASP Conf. Ser., 64, 399

Dorren, J. D., Güdel, M., & Guinan, E. F. 1995, X-Ray Emission from the Sun in its Youth and Old Age, ApJ, in press

Güdel, M., Schmitt, J. H. M. M. & Benz, A. O. 1994, Discovery of microwave emission from four nearby solar-type G stars, Science, 265, 933

Güdel, M., Schmitt, J. H. M. M., Benz, A. O. & Elias, N. M. II 1995a, The corona of the young solar analog EK Draconis, A&A, in press

Güdel, M., Schmitt, J. H. M. M. & Benz, A. O. 1995b, Microwave emission from X-ray bright solar-like stars: The F-G main-sequence and beyond, A&A, in press

Jordan, C. & Montesinos, B. 1991, The dependence of coronal temperature son Rossby numbers, MNRAS, 252, 21P

Raymond, J. C. & Doyle, J. G. 1981, Emissivities of strong ultraviolet lines, ApJ, 245, 1141

Schrijver, C. J., Mewe, R. & Walter, F. M. 1984, Coronal activity in F-, G-, and K-type stars II: Coronal structure and rotation, A&A, 138, 258

Skumanich, A. 1972, Time-scales for Ca II emission decay, rotational braking, and lithium depletion, ApJ, 171, 565

Fe XIII Emission Lines Observed by *EUVE* and the S082A Instrument On-Board *Skylab*

F. P. KEENAN,[1] J. J. DRAKE,[2] V. J. FOSTER,[1] C. J. GREER,[1] S. S. TAYAL,[3] AND K. G. WIDING[4]

[1] Department of Pure and Applied Physics,
The Queen's University of Belfast, Belfast BT7 1NN, N. Ireland

[2] Center for EUV Astrophysics, 2150 Kittredge Street,
University of California, Berkeley, CA 94720-5030, USA

[3] Department of Physics and Center for Theoretical Studies of Physical Systems,
Clark Atlanta University, Atlanta, GA 30314, USA

[4] Code 4174W, E. O. Hulburt Center for Space Research,
Naval Research Laboratory, Washington DC 20375, USA

Recent R-matrix calculations of electron impact excitation rates for Fe XIII are used to derive the theoretical electron density sensitive emission line ratios R_1 = I(318.12 Å)/I(320.80 Å) and R_2 = I(256.42 Å)/I(251.95 Å), which are found to be up to 50% different from earlier diagnostics. A comparison of the current line ratios with both solar flare and active region observations, obtained by the Naval Research Laboratory's S082A spectrograph on board *Skylab*, reveals generally good agreement between densities deduced from Fe XIII and those estimated from diagnostic line ratios in species formed at similar temperatures. This provides experimental support for the accuracy of the line ratio calculations, and hence the atomic data adopted in their derivation. In *Extreme Ultraviolet Explorer* satellite (*EUVE*) spectra the Fe XIII emission lines are found to be severely blended. However, an analysis of these lines measured in the spectra of Procyon and α Cen demonstrates that they still allow very approximate values of the electron density to be inferred. Moreover, it should be possible to increase the accuracy of the measured line fluxes, and hence of the inferred densities, if longer exposures of the stars concerned can be obtained.

1. Introduction

Emission lines arising from $3s^2 3p^2 - 3s 3p^3$ transitions in Fe XIII have been frequently detected in solar EUV spectra (Dere 1982), while more recently they have been measured in *Extreme Ultraviolet Explorer* (*EUVE*) satellite observations of late-type stars, such as Procyon (Drake, Laming & Widing 1995). The diagnostic potential of these lines was first noted by Flower & Nussbaumer (1974), and since then several authors have produced theoretical Fe XIII line ratios, the most recent being those of Brickhouse, Raymond & Smith (1995), which employ the electron excitation rates of Fawcett & Mason (1989). However very recently, Tayal (1995) has used the R-matrix code to calculate electron rates for transitions in Fe XIII, which include the effects of resonances converging to the $3s^2 3p^2$, $3s 3p^3$ and $3s^2 3p 3d$ states.

In this paper we use the Tayal (1995) atomic data to derive diagnostic line ratios for Fe XIII, and compare them both with previous calculations, and with solar and stellar observations from *Skylab* and *EUVE*.

2. Theoretical Ratios

The model ion for Fe XIII has been discussed by Keenan et al. (1995), where details of the line ratio calculations may be found. In Figure 1 we plot the ratio R_1 = I($3s^2 3p^2$

S. Bowyer and R. F. Malina (eds.), Astrophysics in the Extreme Ultraviolet, 525–529.
© 1996 *Kluwer Academic Publishers. Printed in the Netherlands.*

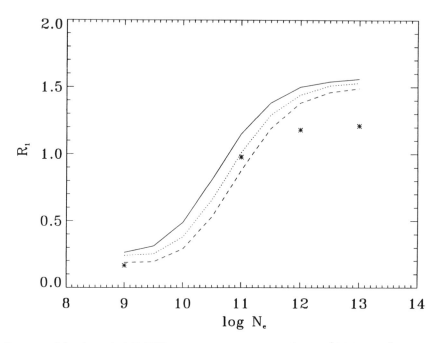

FIGURE 1. The theoretical Fe XIII emission line ratio $R_1 = I(318.12\ \text{Å})/I(320.80\ \text{Å})$, plotted as a function of electron density at electron temperatures of log T_e = 5.9 (*solid line*); 6.2 (*dotted line*); 6.5 (*dashed line*). The calculations of Brickhouse et al. (1995) at log T_e = 6.2 are shown as stars.

^{1}D– $3s3p^3\ ^1$D)/I($3s^23p^2\ ^3$P$_2$ – $3s3p^3\ ^3$P$_2$) = I(318.12 Å)/I(320.80 Å) as function of electron density at the temperature of maximum Fe XIII fractional abundance in ionisation equilibrium, log T_{max} = 6.2 (Arnaud & Raymond 1992), plus ±0.3 dex about this value. Also shown in the figure are the calculations of Brickhouse et al. (1995) at log T_{max}. Our line ratios are up to 50% different from those of Brickhouse et al. , which is principally due to the adoption of improved electron excitation rate data in the present paper. We note that a similar figure for the ratio $R_2 = I(3s^23p^2\ ^1$D– $3s3p^3\ ^1$P)/I($3s^22p^2\ ^3$P$_2$ – $3s3p^3\ ^3$S) = I(256.42 Å)/I(251.95 Å) may be found in Keenan et al. (1995).

3. Observational Data

The Fe XIII 251.95, 256.42, 318.12 and 320.80 Å emission lines discussed in § 2 have been extensively observed in the solar spectrum by the Naval Research Laboratory's S082A spectrograph on board *Skylab*. This instrument operated in the 171–630 Å wavelength range in two sections (171–350 Å and 300–630 Å), and produced dispersed images of the Sun on photographic film with a spatial resolution of 2″ and a maximum spectral resolution of ∼0.1 Å. It is discussed in detail by Dere (1978). In Table 1 we summarize measurements of R_1 and R_2 for two active regions and two flares. Unfortunately, R_2 could not be determined in most of the features due to blending of the 256.42 Å line with the very strong He II transition at 256.32 Å. As the R_1 and R_2 ratios involve lines which

TABLE 1. Fe XIII emission line ratios and derived logarithmic electron densities.

Feature	R_1	R_2	log N_e (R_1)	log N_e (R_2)	log N_e (other)
(i) Skylab S082A observations[a]					
Active region McMath 12390	0.31	...	9.7	...	9.5[b]
Active region McMath 12375	0.25	...	9.5	...	9.4[b]
1973 Dec 17 flare, 0044 UT	0.50	0.62	10.3	10.5	10.4[b]
1973 Dec 17 flare, 0045 UT	0.82	...	10.8	...	10.6[b]
1974 Jan 21 flare, 2346 UT	1.03	...	11.0	...	10.8[c]
(ii) EUVE observations					
Procyon[d]	0.70±0.50	...	≤11.3	...	9.5[b]
α Cen[e]	0.78±0.41	...	10.6±0.8	...	9.2[b]

NOTES:
[a]These line ratio data should be accurate to ±20%, and hence the derived values of log N_e to ±0.2 dex.
[b]Determined from the I(219.12 Å)/I(211.32 Å) line ratio in Fe XIV.
[c]Determined from the I(215.97 Å)/I(191.29 Å) line ratio in S XI.
[d]Observed between 1993 Jan 11, 2219 UT and 1993 Jan 15, 0028 UT (100,000 sec total exposure time).
[e]Observed between 1993 May 29, 1355 UT and 1993 May 31, 0836 UT (45,000 sec total exposure time).

are close together in wavelength and are approximately equal in intensity, we estimate that the data listed in Table 1 should be accurate to approximately ±20%.

The *EUVE* instrumentation has been described in detail elsewhere. In summary, the spectrograph covers the wavelength range 70–760 Å in three bandpasses, commonly referred to as "Short", "Medium" and "Long", or SW, MW and LW, where the bandpasses are as follows: SW 70–190 Å; MW 140–380 Å; LW 280–760 Å, and the resolution in each bandpass varies from $\lambda/\Delta\lambda = 190$ (blue end) to 290 (red end). The best stellar candidates for studying EUV emission lines due to Fe XIII are Procyon (F5 IV) and α Cen (G2 V + K0 V). However as the resolution of the MW spectrometer near 256 Å is about 0.8 Å; we did not therefore expect to be able to extract fluxes for Fe XIII 256.42 Å, since it is blended with the much stronger He II line at 256.32 Å. Hence we have concentrated on the Fe XIII lines around ~318–320 Å. The intensities of the Fe XIII lines were estimated for each star by fitting Gaussian profiles using a least squares technique (see Keenan et al. 1995), the profile fit for α Cen being shown in Figure 2. The resultant R_1 ratios are summarised in Table 1, where uncertainties have been calculated from the error matrix in the least squares fitting procedure.

4. Results and Discussion

The logarithmic electron densities deduced from the observed values of R_1 and R_2 in the solar features are summarised in Table 1. In view of the observational uncertainties in the line ratios (see § 3), the derived values of log N_e should be accurate to approx-

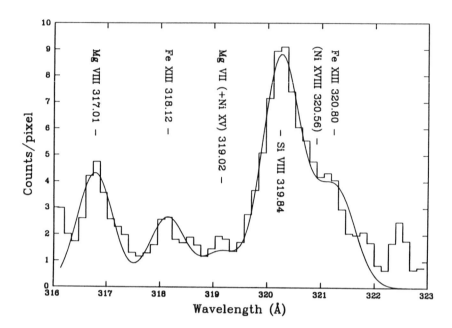

FIGURE 2. *EUVE* spectrum of α Cen in the 316–323 Å wavelength range, showing the results of a profile fitting procedure for the emission lines indicated.

imately ±0.2 dex. Also given in the table are the densities estimated from either the I(219.12 Å)/I(211.32 Å) emission line ratio in Fe XIV or the I(215.97 Å)/I(191.29 Å) ratio in S XI; as these ions are formed at electron temperatures close to that of Fe XIII, densities deduced from their line ratios should be similar to those estimated for the Fe XIII emitting region of the plasma.

An inspection of Table 1 reveals that, for the 1973 Dec 17 flare at 0044 UT, the densities derived from R_1 and R_2 are compatible, with a difference of only 0.2 dex. In addition, these values of N_e, and those estimated from R_1 in the other events, are in good agreement with densities found from line ratios in species formed at similar temperatures to Fe XIII, with discrepancies that average only ~0.1 dex. These results provide experimental support for the accuracy of the current line ratio calculations, and hence the atomic data adopted in their derivation.

The observational uncertainty of the R_1 ratio for Procyon in Table 1 implies that no firm estimate of the electron density is possible from Figure 1, with only an upper limit of $\log N_e \leq 11.3$ being inferred. However in the case of α Cen, the error in the ratio is somewhat smaller, and hence the density may be constrained to $\log N_e \simeq 10.6\pm0.8$. As for the solar observations, these values may be compared with the densities in Table 1 determined from line ratios in Fe XIV, namely $\log N_e = 9.5$ and 9.2 for Procyon and α Cen, respectively (see Keenan et al. 1995). Although at face value these densities are not strictly compatible with the Fe XIII results, in reality the true error in disentangling the Fe XIII line fluxes from those of the blending species is larger than the simple Poisson error. It is clear therefore that these lines do not currently provide any reliable diagnostic information on the Fe XIII emitting regions of stellar atmospheres, due to the large observational uncertainties in the line ratios. This is caused in part by blending, but primarily by the lack of adequate signal-to-noise in the spectral data to allow detailed profile fitting to be performed, and hence the contribution of blends to be properly assessed and removed. Clearly, higher quality observations of the ~316–323 Å wavelength region in Procyon and α Cen are required in order for reliable profile fitting to be attempted, and we hope to obtain these during future allocations of observing time with *EUVE*.

We would like to thank the Principal Investigators S. Bowyer and R.F. Malina and the *EUVE* Science Team for advice and support, and N. Brickhouse for a copy of her work in advance of publication. JJD was supported by NASA contract NAS5–30180 during the course of this work, while VJF acknowledges financial support from the Particle Physics and Astronomy Research Council of the United Kingdom. This work was also supported by the Royal Society and NATO travel grant CRG.930722.

REFERENCES

ARNAUD, M. & RAYMOND, J. C. 1992, ApJ, 398, 394

BRICKHOUSE, N. S., RAYMOND, J. C. & SMITH, B. W. 1995, ApJS, in press

DERE, K. P. 1978, ApJ, 221, 1062

DERE, K. P. 1982, Solar Phys., 77, 77

DRAKE, J. J., LAMING, J. M. & WIDING, K. G. 1995, ApJ, in press

FAWCETT, B. C. & MASON, H. E. 1989, Atom. Data Nucl. Data Tables, 43, 245

FLOWER, D. R. & NUSSBAUMER, H. 1974, Astr. Astrophys., 31, 353

KEENAN, F. P., FOSTER, V. J., DRAKE, J. J., TAYAL, S. S. & WIDING, K. G. 1995, ApJ, submitted

TAYAL, S. S. 1995, ApJ, in press

Solar EUV Rocket Telescope and Spectrograph (SERTS) Observations of Fe XII Emission Lines

F. P. KEENAN,[1] R. J. THOMAS,[2] W. M. NEUPERT,[2]
V. J. FOSTER,[1] C. J. GREER,[1] AND S. S. TAYAL[3]

[1]Department of Pure and Applied Physics,
The Queen's University of Belfast, Belfast BT7 1NN, N. Ireland

[2]Laboratory for Astronomy and Solar Physics, Code 680,
NASA/Goddard Space Flight Center, Greenbelt, MD 20771, USA

[3]Department of Physics and Center for Theoretical Studies of Physical Systems,
Clark Atlanta University, Atlanta, GA 30314, USA

Theoretical electron density sensitive emission line ratios involving transitions in the 186–383 Å wavelength range are compared with observational data for a solar active region and a subflare, obtained by the *Solar EUV Rocket Telescope and Spectrograph (SERTS)*. Electron densities derived from the majority of the ratios are consistent with one another, and are also in good agreement with the values of density estimated from diagnostic lines in other species formed at similar temperatures to Fe XII. These results provide observational support for the general accuracy of the diagnostic calculations. In addition, our analysis indicates that a line at 283.70 Å in the active region spectrum is the $3s^2 3p^3 \, ^2D_{3/2} - 3s3p^4 \, ^2P_{1/2}$ transition in Fe XII, the first time (to the best of our knowledge) that this line has been identified in the solar spectrum. Several of the line ratios considered are predicted to be relatively insensitive to the adopted electron temperature and density, and the generally good agreement found between theory and observation for these provides evidence for the reliability of the *SERTS* instrument calibration. The application of the Fe XII diagnostics to *EUVE* observations of the F5 subgiant Procyon is briefly discussed.

1. Introduction

Emission lines due to $3s^2 3p^3 - 3s3p^4$ and $3s^2 3p^3 - 3s^2 3p^2 3d$ transitions in Fe XII have been frequently observed in solar spectra (see, for example, Kastner & Mason 1978). Dere et al. (1979) noted that the ratios of these lines are very sensitive to electron density, and hence are potentially very useful N_e–diagnostics for the emitting plasma. These authors plotted several theoretical Fe XII line ratios as a function of density, but restricted their results to diagnostics involving either $3s^2 3p^3 - 3s3p^4$ or $3s^2 3p^3 - 3s^2 3p^2 3d$ transitions only. This was primarily because the S082A spectrograph on board *Skylab*, from which Dere et al. obtained their observational data, detected solar spectra in two wavelength settings, making the measurement of ratios involving lines in both wavelength regions difficult to determine. More recently however, the *Solar EUV Rocket Telescope and Spectrograph (SERTS)* has obtained solar spectra over the full wavelength range of 170–450 Å in a single exposure (Thomas & Neupert 1994), thereby allowing the development and testing of Fe XII diagnostics involving both $3s^2 3p^3 - 3s3p^4$ and $3s^2 3p^3 - 3s^2 3p^2 3d$ transitions. This is potentially very important, as both the *Extreme Ultraviolet Explorer* satellite (*EUVE*) and the *Coronal Diagnostic Spectrometer* on board the *Solar and Heliospheric Observatory (SOHO)* mission, should detect many of the Fe XII EUV lines, which fall within the 186–383 Å wavelength range. In this paper we therefore compare theoretical Fe XII line ratios with the *SERTS* observations, and hence investigate their usefulness as electron density diagnostics.

S. Bowyer and R. F. Malina (eds.), Astrophysics in the Extreme Ultraviolet, 531–536.
© 1996 Kluwer Academic Publishers. Printed in the Netherlands.

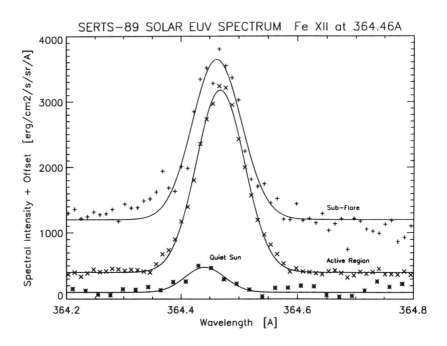

FIGURE 1. Plot of the active region and subflare spectra obtained with *SERTS* in the wavelength interval 364.2 − 364.8 Å, containing the Fe XII 364.46 Å line. Also shown in the figure are the *SERTS* Quiet Sun observations in this wavelength range.

2. Observational Data

The solar spectra analysed in the present paper are those of an active region and a small subflare, recorded on Eastman Kodak 101–07 emulsion by *SERTS* during a rocket flight on 1989 May 5 at 17:50 UT (Neupert et al. 1992). The observations cover the wavelength region 235.46–448.76 Å in first order and 170–224.38 Å in second, with a spatial resolution of about 7 arc sec and a spectral resolution of better than 80 mÅ (FWHM). The active region measurements used here were spatially averaged over the central 4.6 arc min of the spectrograph slit, whereas the subflare results come from a 22 arc sec portion of the same dataset.

The Fe XII transitions identified in the *SERTS* spectra are listed in Table 1, where we note that we have detected the 196.62 and 283.70 Å lines, which were not listed by Thomas & Neupert (1994) in their summary of the 1989 May 5 observations. Our tenative measurement of the $3s^2 3p^3$ $^2D_{3/2}$ $-3s3p^4$ $^2P_{1/2}$ line at 283.70 Å is of particular importance, as to the best of our knowledge this transition has not previously been detected in solar spectra. Intensities of the Fe XII lines were determined by fitting gaussian profiles to microdensitometer scans of the recorded spectra. The intensities of the 364.46 Å line in the two solar features are listed in Table 1; the observed intensities of the other Fe XII transitions may be inferred from these using the line ratios given in the table.

The quality of the observational data are illustrated in Figure 1, where we plot the active region and subflare spectra between 364.2–364.8 Å. Also shown for comparison in

TABLE 1. Fe XII transitions in the 1989 May 5 *SERTS* observations.

| Transition | $\lambda(\text{Å})$ | $R = I(\lambda)/I(364.46 \text{ Å})$ | | Ratio |
		Active region[a]	Subflare[b]	designation
$3s^2 3p^3\ ^2D_{5/2} - 3s^2 3p^2(^3P)3d\ ^2F_{7/2}$	186.88	5.73+0[c]	1.04+1	R_1
$3s^2 3p^3\ ^4S - 3s^2 3p^2(^3P)3d\ ^4P_{1/2}$	192.37	1.02+1	1.18+1	R_{11}
$3s^2 3p^3\ ^4S - 3s^2 3p^2(^3P)3d\ ^4P_{3/2}$	193.51	5.49+0	5.35+0	R_{12}
$3s^2 3p^3\ ^4S - 3s^2 3p^2(^3P)3d\ ^4P_{5/2}$	195.12	5.24+0	6.62+0	R_{13}
$3s^2 3p^3\ ^2D_{5/2} - 3s^2 3p^2(^1D)3d\ ^2D_{5/2}$	196.62[d]	7.95−1	−	R_2
$3s^2 3p^3\ ^2P_{3/2} - 3s^2 3p^2(^1D)3d\ ^2S$	200.41	1.57+0	2.93+0	R_9
$3s^2 3p^3\ ^2P_{3/2} - 3s^2 3p^2(^1D)3d\ ^2P_{3/2}$	201.13	1.69+0	1.48+0	R_3
$3s^2 3p^3\ ^2D_{5/2} - 3s^2 3p^2(^3P)3d\ ^2P_{3/2}$	219.43	5.83−1	7.77−1	R_4
$3s^2 3p^3\ ^2D_{3/2} - 3s3p^4\ ^2P_{1/2}$	283.70[d]	7.85−2	−	R_8
$3s^2 3p^3\ ^2D_{5/2} - 3s3p^4\ ^2P_{3/2}$	291.01	4.69−1	4.42−1	R_5
$3s^2 3p^3\ ^2D_{3/2} - 3s3p^4\ ^2D_{3/2}$	335.04	5.59−2	7.08−2	R_6
$3s^2 3p^3\ ^2D_{5/2} - 3s3p^4\ ^2D_{5/2}$	338.27	3.30−1	5.27−1	R_7
$3s^2 3p^3\ ^4S - 3s3p^4\ ^4P_{1/2}$	346.86	2.88−1	2.72−1	R_{14}
$3s^2 3p^3\ ^4S - 3s3p^4\ ^4P_{3/2}$	352.11	6.22−1	7.54−1	R_{15}
$3s^2 3p^3\ ^4S - 3s3p^4\ ^4P_{5/2}$	364.46	1.00+0	1.00+0	−
$3s^2 3p^3\ ^2P_{3/2} - 3s3p^4\ ^2D_{5/2}$	382.86	3.06−2	3.38−2	R_{10}

[a] $I(364.46 \text{ Å}) = 288.0 \text{ erg cm}^{-2}\,\text{s}^{-1}\,\text{sr}^{-1}$.
[b] $I(364.46 \text{ Å}) = 260.0 \text{ erg cm}^{-2}\,\text{s}^{-1}\,\text{sr}^{-1}$.
[c] $A \pm B$ implies $A \times 10^{\pm B}$.
[d] Line not listed by Thomas & Neupert (1994).

the figure are the *SERTS* Quiet Sun data in this wavelength range, which we note are of too low a quality for reliable Fe XII line ratios to be derived.

3. Theoretical Ratios

The model ion adopted for Fe XII has been discussed by Keenan et al. (1995), where details of the line ratio calculations may be found. In Table 1 we list the electron density sensitive emission line ratios R_1 through R_{10}; plots of these ratios as a function of electron density may be found in Keenan et al. (1995).

4. Results and Discussion

In Table 2 the electron densities derived from the observed values of R_1 through R_{10} in Table 1 are summarised. An inspection of the table reveals that the logarithmic electron densities deduced from R_1, R_3 and R_9 in the active region are all ≥ 11.0, which are much higher than those previously estimated for active regions at electron temperatures where the Fe XII lines are formed in ionisation equilibrium, $T_{max} = 1.4 \times 10^6$ K (Arnaud & Raymond 1992). For example, Brickhouse, Raymond & Smith (1995) derived log $N_e \simeq$ 9.5 for the *SERTS* active region from line ratios in Fe XIII, which is formed at a similar electron temperature to Fe XII (T_{max}(Fe XIII) $= 1.6 \times 10^6$ K; Arnaud & Raymond 1992). The very large Fe XII densities deduced here probably arise from overestimates

TABLE 2. Fe XII logarithmic electron densities.

Ratio	Log $N_e(R)$ Active region	Subflare
R_1	11.0	H[a]
R_2	9.6	–
R_3	11.2	11.0
R_4	9.6	9.8
R_5	10.1	10.1
R_6	9.2	9.3
R_7	10.1	10.3
R_8	9.7	–
R_9	11.5	H
R_{10}	9.5	9.5
Mean density[b]	9.7±0.3	9.8±0.4

[a]Indicates that the observed ratio is larger than the theoretical high density limit.
[b]Average density excluding results from the R_1, R_3 and R_9 ratios.

TABLE 3. Fe XII density insensitive ratios.

Ratio	Observed Active region	Subflare	Theoretical[a]
R_{11}	1.02+1[b]	1.18+1	2.55+0
R_{12}	5.49+0	5.35+0	6.78+0
R_{13}	5.24+0	6.62+0	1.12+1
R_{14}	2.88−1	2.72−1	3.67−1
R_{15}	6.22−1	7.54−1	6.99−1

[a]Ratios calculated for $N_e = 10^{10}$ cm^{-3}; $T_e = T_{max} = 1.4 \times 10^6$ K (Arnaud & Raymond 1992).
[b]A±B implies A $\times 10^{\pm B}$.

of the relevant line ratios due to blending, a conclusion which is also supported by the fact that R_1 and R_9 in the subflare are larger than the theoretical high density limits.

Electron densities deduced from the remaining Fe XII line ratios in Table 2 are consistent, with discrepancies of typically ≤0.3 dex with the mean values, $\overline{log\ N_e}$ = 9.7±0.3 and 9.8±0.4 for the active region and subflare, respectively. These mean densities are in excellent agreement with the values determined from diagnostic ratios in species formed at similar temperatures to Fe XII. For example, as noted above, Brickhouse et al. (1995) found log $N_e \simeq 9.5$ for the active region from Fe XIII, while for the subflare we derive log $N_e \simeq 9.5$ from the I(219.12 Å)/I(211.32 Å) ratio in Fe XIV (Keenan et al. 1991), which has $T_{max} = 1.9 \times 10^6$ K (Arnaud & Raymond 1992). In addition, the fact that the R_8 ratio leads to an electron density in good agreement with those inferred from the other Fe XII diagnostics confirms our tentative identification of the 283.70 Å feature as the $3s^2 3p^3\ ^2D_{3/2} - 3s3p^4\ ^2P_{1/2}$ line. These results provide experimental support for the theoretical Fe XII line ratios (and hence the atomic data used in their derivation), and implies that they may be applied in the future to high resolution observations from the *Coronal Diagnostic Spectrometer* on the *SOHO* mission (Harrison 1993).

We note that the line ratios R_{11} through R_{15} in Table 1 are predicted to be relatively insensitive to the adopted plasma parameters (temperature and density), and hence may be useful in investigating either blending or the reliability of the instrument calibration (Neupert & Kastner 1983). The observed values of R_{11} through R_{15} are therefore summarised in Table 3, along with the calculated ratios. An inspection of the table reveals excellent agreement between theory and observation for R_{12}, R_{14} and R_{15}, with discrepancies of $\leq 25\%$ in all cases, once again providing support for the theoretical ratios. Additionally, as R_{12} contains the 364.46 and 193.51 Å transitions, measured in first and second order, respectively, the good agreement also provides verification of the *SERTS* instrument calibration between the two orders. However agreement between theory and observation for R_{11} and R_{13} is very poor, which is probably due to a combination of blending in the observational data and errors in the theoretical line ratios (see Keenan et al. 1995 for more details). More theoretical work on the Fe XII spectrum is clearly needed.

Several of the Fe XII lines discussed in this paper have been observed in *EUVE* spectra of the F5 subgiant Procyon (Drake, Laming & Widing 1995). Unfortunately, these authors do not list an intensity for the 364.46 Å line (probably due to blending with nearby strong features, such as Mg IX 367.34 Å), and only give data for the 186.88, 192.37, 193.51 and 195.12 Å transitions. However the ratio I(186.88 Å)/I(192.37 Å) is predicted to be N_e–sensitive, and the Drake et al. value of ~ 0.6 implies $\log N_e \simeq 9.5$. This is in excellent agreement with the densities estimated for Procyon from species formed at similar temperatures to Fe XII, including Fe XIII and Fe XIV, which give $\log N_e \simeq 9.5$–9.7 (Drake et al.). We note that although the 192.37 Å line, observed in second order by *SERTS*, is blended with the first order Mn XV 384.75 Å transition in this dataset (Thomas & Neupert 1994), this problem does not of course apply to the *EUVE* observations. In addition, although 186.88 Å is blended with S XI in *SERTS*, it appears to be resolved by *EUVE* (Drake *et el.*).

Both the I(193.51 Å)/I(192.37 Å) and I(195.12 Å)/I(192.37 Å) ratios are predicted to be insensitive to N_e and T_e, with theoretical values of 2.6 and 4.3, respectively. The *EUVE* measurement of the former, I(193.51 Å)/I(192.37 Å) $\simeq 1.7$, is in reasonable agreement with theory, but the observed value of I(195.12 Å)/I(192.37 Å) $\simeq 2.1$ is smaller than theory predicts. This problem is also found in the *SERTS* observations (see Table 3), and is probably due to errors in the adopted atomic data (see Keenan et al. 1995).

Clearly, more work on the Fe XII emission lines in *EUVE* observations is required, especially regarding their usefulness as N_e–diagnostics. We plan to undertake this research in the near future (Drake & Keenan 1995).

VJF is grateful to PPARC for financial support, while CJG acknowledges the award of a research studentship from the Department of Education for N. Ireland. We are also grateful to Jeremy Drake for a copy of his paper on *EUVE* observations of Procyon in advance of publication. This work was supported by NATO travel grant CRG.930722 and the Royal Society. The *SERTS* rocket program was funded under NASA RTOP 879-11-38.

REFERENCES

ARNAUD, M. & RAYMOND, J. C. 1992, ApJ, 398, 394

BRICKHOUSE, N. C., RAYMOND, J. C. & SMITH, B. W. 1995, ApJS, in press

DERE, K. P., MASON, H. E., WIDING, K. G. & BHATIA, A. K. 1979, ApJS, 40, 341

DRAKE, J. J. & KEENAN, F. P. 1995, ApJ, in preparation

DRAKE, J. J., LAMING, J. M. & WIDING, K. G. 1995, ApJS, in press

HARRISON, R. A. 1993, The Coronal Diagnostic Spectrometer for SOHO, RAL Report, SN–93–0007

KASTNER, S. O. & MASON, H. E. 1978, A&A, 67, 119

KEENAN, F. P., DUFTON, P. L., BOYLAN, M. B., KINGSTON, A. E. & WIDING, K. G. 1991, ApJ, 373, 695

KEENAN, F. P., FOSTER, V. J., BROWN, P. J. F., THOMAS, R. J., NEUPERT, W. M. & TAYAL, S. S. 1995, MNRAS, submitted

NEUPERT, W. M., EPSTEIN, G. L., THOMAS, R. J. & THOMPSON, W. T. 1992, Solar Phys., 137, 87

NEUPERT, W. M. & KASTNER, S. O. 1983, A&A, 128, 181

THOMAS, R. J. & NEUPERT, W. M. 1994, ApJS, 91, 461

Skylab Observations of Temperature and Density Sensitive Emission Line Ratios in Ne VI

C. J. GREER,[1] V. J. FOSTER,[1] F. P. KEENAN,[1] R. H. G. REID,[2]
J. G. DOYLE,[3] H. L. ZHANG,[4] AND A. K. PRADHAN[4]

[1] Department of Pure and Applied Physics, The Queen's University of Belfast,
Belfast BT7 1NN, Northern Ireland

[2] Department of Applied Mathematics and Theoretical Physics,
The Queen's University of Belfast, Belfast BT7 1NN, Northern Ireland

[3] Armagh Observatory, Armagh BT61 9DG, Northern Ireland

[4] Department of Astronomy, Ohio State University, Columbus, Ohio 43210, USA

Recent calculations of electron and proton impact excitation rates in Ne VI are used to derive the intensity ratios of lines in the $\sim$402–1006 Å wavelength range as a function of electron temperature (T_e) and density (N_e). These results are presented in the form of ratio–ratio diagrams, which should in principle allow both N_e and T_e to be deduced for the Ne VI line emitting region of a plasma. Electron temperatures and densities derived from ratio–ratio diagrams involving the 562.7, 997.4, 999.6 and 1006.1 Å lines, in conjunction with observational data for a sunspot obtained with the Harvard S-0555 spectrometer on board Skylab, are found to be compatible, and in good agreement with plasma parameters determined using other methods. This provides some support for the diagnostic calculations presented in this paper, and hence the atomic data used in their derivation. However agreement between theory and observation is very poor for other Ne VI lines in the sunspot spectrum, and for most transitions observed in S-0555 active region and flare data, which is probably due to blending with lines from N III, Mg VI and Mg VII. The application of the calculations to non-solar EUV sources is discussed.

1. Introduction

Emission lines arising from $2s^2 2p$–$2s 2p^2$ transitions in boron-like ions have often been identified in solar UV and EUV spectra. Flower & Nussbaumer (1975), first noted the diagnostic potential of these lines presenting electron density and temperature sensitive emission line ratios for O IV calculated using electron impact excitation rates derived in the Distorted-Wave approximation (Eissner & Seaton 1972). Zhang, Graziani & Pradhan (1994) have calculated electron excitation rates for Ne VI using the R-matrix method of Burke & Robb (1975) and these results are used here to derive solar plasma diagnostics for this ion. An assessment of the validity of these diagnostics is made by way of a comparison with observations from the S-0555 instrument on board Skylab.

2. Atomic Data and Theoretical Line Ratios

The Ne VI model ion is discussed in detail by Keenan et al. (1995) where details of the ratio calculations may be found. The line ratios considered by us in the present paper include the following:

S. Bowyer and R. F. Malina (eds.), Astrophysics in the Extreme Ultraviolet, 537–541.

$$R_1 = I(2s^2 2p\ ^2P_{3/2}-2s2p^2\ ^4P_{3/2})/I(2s^2 2p\ ^2P_{1/2}-2s2p^2\ ^4P_{1/2})$$
$$= I(1006.1\ \text{Å})/I(997.4\ \text{Å}),$$

$$R_2 = I(2s^2 2p\ ^2P_{3/2}-2s2p^2\ ^4P_{5/2})/I(2s^2 2p\ ^2P_{1/2}-2s2p^2\ ^4P_{1/2})$$
$$= I(999.6\ \text{Å})/I(997.4\ \text{Å}),$$

$$R_3 = I(2s^2 2p\ ^2P_{3/2}-2s2p^2\ ^4P_{3/2})/I(2s^2 2p\ ^2P_{3/2}-2s2p^2\ ^2D_{3/2,5/2})$$
$$= I(1006.1\ \text{Å})/I(562.7\ \text{Å}),$$

$$R_4 = I(2s^2 2p\ ^2P_{3/2}-2s2p^2\ ^2S)/I(2s^2 2p\ ^2P_{3/2}-2s2p^2\ ^2D_{3/2,5/2})$$
$$= I(435.7\ \text{Å})/I(562.7\ \text{Å}),$$

$$R_5 = I(2s^2 2p\ ^2P_{1/2}-2s2p^2\ ^2S)/I(2s^2 2p\ ^2P_{3/2}-2s2p^2\ ^2D_{3/2,5/2})$$
$$= I(433.2\ \text{Å})/I(562.7\ \text{Å}),$$

$$R_6 = I(2s^2 2p\ ^2P_{3/2}-2s2p^2\ ^2P_{1/2})/I(2s^2 2p\ ^2P_{3/2}-2s2p^2\ ^2D_{3/2,5/2})$$
$$= I(403.3\ \text{Å})/I(562.7\ \text{Å}),$$

and
$$R_7 = I[(2s^2 2p\ ^2P_{1/2}\ -2s2p^2\ ^2P_{1/2}) + (2s^2 2p\ ^2P_{3/2}-2s2p^2\ ^2P_{3/2})]/I(2s^2 2p\ ^2P_{3/2}-\ 2s2p^2$$
$$^2D_{3/2,5/2})$$
$$= I(401.9\ \text{Å})/I(562.7\ \text{Å}).$$

Under solar conditions, the above ratios are usually sensitive to variations in both the electron temperature and density. Hence in principle they should only be used to determine N_e or T_e when the other plasma parameter has been independently estimated. For example, Figure 1 is the ratio–ratio diagram of R_1 vs R_3 for a grid of (log N_e, log T_e) values. Using such figures it is possible to simultaneously determine both the electron temperature and density from the measured values of the ratios.

3. Observational Data

The Ne VI lines in the wavelength interval 401.9–1006.1 Å discussed in § 2 have been identified in solar spectra obtained with the Harvard S-0555 EUV spectrometer on board *Skylab*. This instrument, which covered the wavelength region 280–1350 Å, observed a spatial area of 5 × 5 arcsec with a spectral resolution of ~1.5 Å (FWHM) using an integration time of 0.04 s and a step length of 0.2112 Å. It is discussed in detail by Reeves, Huber & Timothy (1977) and Reeves et al. (1977). We have determined Ne VI line strengths, and hence ratios, by using the STARLINK reduction package DIPSO (Howarth & Murray 1991) to fit Gaussian profiles to the S-0555 spectra; such profiles were found to give acceptable fits to the observational data. Line intensities derived from the profile fitting should be accurate to typically ±20%, implying that the resultant line ratios have an uncertainty of ±30%. The derived values of R_1 to R_7 are summarised in Table 1 for a sunspot located close to disk centre recorded on August 29, 1973 (discussed by Doyle et al. 1985), an active region observed at the limb on December 16, 1973 (Doyle, Mason & Vernazza 1985), and a large two-ribbon flare observed on September 7, 1973 at 12:55, 14:03 and 15:52 UT (Doyle 1983).

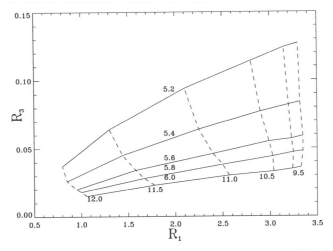

FIGURE 1. Plot of the theoretical Ne VI emission line ratio $R_1 = I(2s^2 2p\ ^2P_{3/2}-2s2p^2\ ^4P_{3/2})/I(2s^2 2p\ ^2P_{1/2}-2s2p^2\ ^4P_{1/2}) = I(1006.1\ \text{Å})/I(997.4\ \text{Å})$ against $R_3 = I(2s^2 2p\ ^2P_{3/2}-2s2p^2\ ^4P_{3/2})/I(2s^2 2p\ ^2P_{3/2}-2s2p^2\ ^2D_{3/2,5/2}) = I(1006.1\ \text{Å})/I(562.7\ \text{Å})$, where I is in energy units, for a range of logarithmic electron temperatures ($\log T_e = 5.2$–6.0; T_e in K) and logarithmic electron densities ($\log N_e = 9.5$–12.0; N_e in cm^{-3}). Points of constant T_e are connected by solid lines, while those of constant N_e are joined by dashed lines. Note that the curves for $\log N_e < 9.5$ are coincident with that for $\log N_e = 9.5$.

4. Results and Discussion

In Table 2 we summarise the logarithmic electron densities and temperatures derived from the observed values of R_1 to R_7 in conjunction with the diagnostic calculations. For the sunspot, the plasma parameters estimated from (R_1, R_3) and (R_2, R_3) are consistent, with discrepancies of ~0.1 dex in both $\log N_e$ and $\log T_e$. These differences correspond to only a $\sim10\%$ change in the line ratios, which is well within both the theoretical ($\pm20\%$) and observational ($\pm30\%$) uncertainties. However the ratios R_4 to R_7 in the sunspot, R_3 to R_6 in the active region and R_3 to R_7 in the flare, are all much larger than the theoretical low temperature limits. Such discrepancies are too great to be ascribed to theoretical or observational errors. In some of these instances, we have at least been able to estimate a $\log N_e$ range from the R_1 ratio portions of the diagnostic curves, by assuming that the temperatures of the Ne VI emitting regions lie within the range plotted ($\log T_e = 5.2$–6.0), where the fractional abundance of Ne VI in ionisation equilibrium is $\geq 10^{-3}$ (Arnaud & Rothenflug 1985). The general disagreement between theory and observation for the R_3 to R_7 ratios is most likely due to blending. However for the flare data there may be additional problems, as typically ~100 seconds were needed by the S-0555 instrument to scan through the spectrum from the long wavelength lines at ~1000 Å to the transitions between ~400–563 Å, during which time there may have been significant variations in the intensities of the flare lines, which would hence affect any determinations of R_3. We note that in the few instances where we have been able to measure reliable Ne VI line ratios, and hence determine plasma parameters, these are in good agreement with other independent estimates. For example, Keenan et al. (1994) derived $\log N_e = 11.1$ and 10.4 for the flare at 14:03 and 15:52 UT, respectively, using line ratios in O V, which is formed at a similar electron temperature to Ne VI ($\log T_{\max}(\text{O V})$

Table 1. Observed Ne VI emission line ratios

Solar feature	R_1	R_2	R_3	R_4	R_5	R_6	R_7
(a) Sunspot	2.85	5.31	5.64×10^{-2}	1.06	1.17	1.86	4.21
(b) Active Region	...	...	2.01×10^{-1}	4.60	2.82	1.98	1.71
(c) Flare, 12:55 UT	2.43	...	5.73×10^{-1}	...	1.01	...	4.37
(d) Flare, 14:03 UT	1.83	...	1.07	...	1.89	...	7.34
(e) Flare, 15:52 UT	3.25	...	4.69×10^{-1}	...	1.92	...	5.53

Table 2. Ne VI logarithmic electron densities and temperatures (log N_e, log T_e) derived from ratio–ratio diagrams (L = observed line ratio is larger than the theoretical low temperature limit)

Feature	(R_1,R_3)	(R_2,R_3)	(R_1,R_4)	(R_1,R_5)	(R_1,R_6)	(R_1,R_7)
(a) SS	10.6,5.60	10.7,5.52	...,L	...,L	...,L	...,L
(b) AR	...,L	...,L	...,L	...,L	...,L	...,5.51
(c) Flare	10.8–11.1,L	...,L	......	10.8–11.1,L	......	10.8–11.1,L
(d) Flare	11.2–11.4,L	...,L	......	11.2,11.4,L	......	11.2–11.4,L
(e) Flare	9.6–10.0,L	...,L	......	9.6–10.0,L	......	9.6-10.0,L

= 5.40; Arnaud & Rothenflug 1985). These densities are compatible with those listed in Table 2. Hence there is limited observational support for the accuracy of the theoretical diagnostics presented in this paper, and hence the atomic data used in their derivation.

Clearly, higher spectral resolution observations of the Ne VI lines in the 400–1000 Å wavelength range are required, in order to remove the problems of blending, and hence reliably determine emission line ratios which may be used to determine both the electron temperature and density of the Ne VI emitting region of the solar atmosphere. Such observations should be possible in the future using the *Coronal Diagnostic Spectrometer* (*CDS*) and *Solar Ultraviolet Measurements of Emitted Radiation* (*SUMER*) instruments on the upcoming *Solar and Heliospheric Observatory* (*SOHO*) mission, due for launch in 1995. CDS covers the 150–800 Å wavelength range which will allow measurements of the R_4 to R_7 ratios in Ne VI, while SUMER will obtain spectra between 500–1600 Å permitting determinations of R_1 to R_3. However for other astronomical sources the situation is more problematical. Although the *Extreme Ultraviolet Explorer* (*EUVE*) satellite observes the 280–760 Å wavelength region in the LW passband (Bowyer & Malina 1991), which contains the Ne VI 401.9–562.7 Å transitions, these data are only obtained at a resolution of ~ 2 Å. As a result, the only Ne VI line that has been reliably observed to the best of our knowledge, is 401.9 Å (Drake, Laming & Widing 1995). However, the *Lyman/FUSE* mission planned for the end of the decade will obtain spectra in the 912–1500 Å region at a resolution of ~ 0.04 Å (Linsky 1993). This should be sufficient to accurately determine values of the R_1 and R_2 ratios in Ne VI, and hence investigate their usefulness as plasma diagnostics for astronomical sources.

VJF is grateful to the Particle Physics and Astronomy Research Council of the United Kingdom for financial support, while HLZ and AKP acknowledge partial support from the US National Science Foundation (PHY 91–15057) for the Iron Project. Research at Armagh Observatory is grant-aided by the Department of Education for Northern Ireland. This work was also supported by the Royal Society and NATO grant CRG.930722.

REFERENCES

Arnaud, M. & Rothenflug, R. 1985, A&A, 60, 425

Bowyer, S. & Malina, R. F. 1991, in Extreme Ultraviolet Astronomy, ed. R.F. Malina & S. Bowyer, New York: Pergamon Press, 397

Burke, P. G. & Robb, W. D. 1975, Adv. Atom. Molec. Phys., 11, 143

Doyle, J. G. 1983, Solar Phys., 89, 115

Doyle, J. G., Mason, H. E. & Vernazza, J. E. 1985, A&A, 150, 69

Doyle, J. G., Raymond, J. C., Noyes, R. W. & Kingston, A. E. 1985, ApJ, 297, 816

Drake, J. J., Laming, J. M. & Widing, K. G. 1995, ApJS, in press

Eissner, W. & Seaton, M. J. 1972, J. Phys. B, 5, 2187

Flower, D. R. & Nussbaumer, H. 1975, A&A, 45, 145

Howarth, I. D. & Murray, M. J. 1991, Starlink User Note, 50.13

Keenan, F. P., Warren, G. A., Doyle, J. G., Berrington, K. A. & Kingston, A. E. 1994, Solar Phys., 150, 61

Keenan, F. P., Foster, V. J., Reid, R. H. G., Doyle, J. G., Zhang, H. L. & Pradhan, A. K. 1995, A&A, in press

Linsky, J. L. 1993, Mem. S. A. It., 64, 323

Reeves, E. M., Huber, M. C. E. & Timothy, J. G. 1977, Appl. Opt., 16, 837

Reeves, E. M., Timothy, J. G., Huber, M. C. E. & Withbroe, G. L. 1977, Appl. Opt., 16, 849

Zhang, H. L., Graziani, M. & Pradhan, A. K. 1994, A&A, 283, 319

The Arcetri Spectral Code for Optically-Thin Plasmas

BRUNELLA C. MONSIGNORI FOSSI[1] AND MASSIMO LANDINI[2]

[1] Arcetri Astrophysical Observatory, Florence, Italy

[2] Department of Astronomy and Space Science, University of Florence, Italy

The Arcetri-95 spectral code for optically thin plasmas computes the continuum and line emission of the ions of the most abundant elements. It includes the most updated atomic models and the main atomic processes for ions from Fe IX to Fe XXIII and for ions of the Be-like, C-like, and N-like isoelectronic sequences. The power emitted per unit emission measure is produced as a function of temperature and density.

Comparison with observations requires the knowledge of the differential emission measure (DEM) as a function of temperature. A numerical code evaluates the best DEM distribution that satisfies observations and theoretical predictions. The spectral code together with the DEM code allows to compute synthetic spectra for any specified temperature distribution model of the plasma.

1. Introduction

The X-ray–UV emission from optically thin astrophysical plasmas has been measured by a large number of space programs, with the aim to study both solar and celestial sources. Future space missions are achieving high spectral resolution and sensitivity. A number of theoretical evaluations of emission spectra have been made over the temperature range 10^4 to 10^8 K (Landini & Monsignori Fossi 1970; Mewe 1972; Raymond & Smith 1977; Kato 1976; Stern et al. 1978; Gaetz & Salpeter 1983; Mewe et al. 1985; Landini & Monsignori Fossi 1990). In the large majority of cases the line emissivity has been computed using the semi-empirical formula of the effective Gaunt factor which relates the collision strength to the oscillator strength, but now a large number of accurate theoretical computations for electron excitation rate of ions exists and the updated theoretical spectral codes are produced using the most accurate atomic data and complete evaluation of the level population for each ion (Doyle & Keenan 1992; Monsignori Fossi & Landini 1994a; Brickhouse et al. 1995b).

An upgrading of the Arcetri X-ray–EUV spectral code (Landini & Monsignori Fossi 1990) is now available and allows to evaluate continuum and line emission. Because a large number of lines pertaining to iron falls in the X-ray–EUV spectral range we have started to review the atomic data for ions of iron and, at this time, we have added the Be-like, C-like, and N-like isoelectronic sequences. A critical assessment of the evaluation of atomic models and main atomic processes (collisional excitation rates, radiative decays, ...) has been performed. Stationary balance has been assumed to compute the number density population for each ion and the intensity of the emission lines has been evaluated as a function of temperature and density. The detections of Be-like, C-like, N-like and iron lines in X-ray–EUV spectral range provide inputs for spectral diagnostics in very broad temperature ($5.0 \leq \log T \leq 8.0$) and density ($10^6 \leq \log N_e \leq 10^{15}$ cm^{-3}) regimes and allow investigation of astrophysical plasmas in very different physical conditions.

S. Bowyer and R. F. Malina (eds.), Astrophysics in the Extreme Ultraviolet, 543–552.

2. The Atomic Processes

2.1. *Ionization Balance*

In stationary conditions the degree of ionization of an element is obtained by equating the ionization and recombination rates that relate successive stage of ionization. The evaluation of the rates is performed following approaches similar to those used by Arnaud & Rothenflug (1985) and is described in Landini & Monsignori Fossi (1990, 1991). For the iron ions the computation of Arnaud & Raymond (1992) is used.

2.2. *The Radiative Emission*

The computer code evaluates continuum and line emission. The continuum emission includes free-free, free-bound and two-photons decay from H-like ions (Landini & Monsignori Fossi 1990). The line emission includes a large number of radiative transitions of all the ions of the most important elements.

The bound-bound emissivity (power per unit volume) is given by:

$$P_{i,j} = N_j(X^{+m})A_{j,i}\frac{hc}{\lambda_{i,j}} \qquad \text{erg cm}^{-3}\,\text{s}^{-1}$$

where $A_{j,i}$ (s^{-1}) is the *Einstein* spontaneous emission coefficient; $N_j(X^{+m})$ is the number density of the level j of the ion $(+m)$ of the element X.

In low density plasmas the collisional excitation processes are generally faster than ionization and recombination timescales, therefore the collisional excitation is dominant over ionization and recombination in producing excited states. The number density population of level j must be calculated by solving the statistical equilibrium equations and including all the important collisional and radiative excitation and de-excitation mechanisms:

$$N_j(N_e\Sigma_i C^e_{j,i} + N_p\Sigma_i C^p_{j,i} + \Sigma_{i<j} A_{j,i}) = \Sigma_i N_i(N_e C^e_{i,j} + N_p C^p_{i,j}) + \Sigma_{i>j} N_i A_{i,j}$$

with $C^e_{j,i}$ and $C^p_{j,i}$ the electron and proton collisional excitation rates $(\text{cm}^{-3}\,\text{s}^{-1})$, $C^e_{i,j}$ and $C^p_{i,j}$ the electron and proton collisional de-excitation rates $(\text{cm}^{-3}\,\text{s}^{-1})$. The collisional excitation rate for a Maxwellian electron velocity distribution is given by:

$$C^e_{i,j} = \frac{8.63 \times 10^{-6}}{T_e^{1/2}}\frac{\Upsilon_{i,j}(T_e)}{\omega_i}\,exp\left(\frac{-\Delta E_{i,j}}{kT_e}\right)$$

where ω_i is the statistical weight of level i, k is the Boltzmann constant and $\Upsilon_{i,j}$ (*the effective collision strength*) is the thermally-averaged collision strength:

$$\Upsilon_{i,j}(T_e) = \int_0^\infty \Omega_{i,j}\,exp\left(-\frac{E}{kT_e}\right)d\left(\frac{E}{kT_e}\right)$$

where $\Omega_{i,j}$ is the collision strength, which is related to the electron excitation cross section, and E is the energy of the scattered electron.

In many cases the proton collisional excitation and de-excitation processes must be included. They become comparable with electron collisional processes only for transitions where $\Delta E_{i,j} \ll kT_e$. This happens, for instance, for transitions between fine structure levels at high temperatures. In our code, till now, the proton collisional excitation and de-excitation rates are neglected.

The solution of the electron-ion scattering problem is complex and takes a great deal of computing resources (for an overview see Mason & Monsignori Fossi 1994). The main approximations used for electron-ion scattering are *Distorted Wave* (DW), *Coulomb Bethe* (CBe) and the more elaborate *Close-Coupling* (CC) approximation. In general DW is thought to be accurate to about 25% and CC to better than 10%. Resonance structures can contribute significantly to the excitation rates, particularly for forbidden and intersystem lines. A lot of effort has recently been put into the assessment of published atomic data, in particular electron excitation rates (Lang 1994). Here the evaluation of the effective collision strengths over a Maxwellian distribution has been performed following a slight modification of the method suggested by Burgess & Tully (1992) for critically examining the electron scattering data calculations (for details see Monsignori Fossi & Landini 1994b).

2.3. The Data Base

An extensive review of the available atomic data has been performed following the suggestions of the Abingdon meeting (Lang 1994); the most updated atomic models have been selected for iron ions, Be-like, C-like, and N-like ions. When available, the observed values for the energy of the levels have been used in order to make an easier comparison with the observations; the radiative decay probabilities and electronic collision strengths or effective collision strengths have been collected for any pair of levels. In the next sections we give a brief description of the Arcetri data base.

2.3.1. Iron Ions

Fe IX: The atomic model consists of 17 levels ($3p^6\ ^1S$, $3p^53d\ ^3P$, ^3F, ^1D, ^3D, ^1F, ^1P, $3p^54s\ ^3P$, ^1P). The collision strengths are computed by Fawcett & Mason (1989), the observed energies and the radiative probability are from Flower (1977b) and Fawcett & Mason (1991).

Fe X: The atomic model consists of 31 levels ($3s^23p^5\ ^2P, 3s3p^6\ ^2S, 3s^23p^43d\ ^4D$, ^4F, ^2P, ^4P, ^2D, ^2F, ^2G, ^2D, ^2S, ^2P, ^2D). The collision strengths are computed by Mason (1975) and the observed energies are from Corliss & Sugar (1985).

Fe XI: The atomic model consists of 47 levels ($3s^23p^4\ ^3P$, ^1D, $^1S, 3s3p^5\ ^3P$, ^1P, $3s^23p^33d\ ^5D$, ^3D, ^3F, ^1S, ^3G, ^1G, ^1D, ^3P, ^1F, ^3S, $^1P, 3p^6\ ^1S$). The collision strengths and the radiative probabilities are computed by Mason (1975), the observed energies are from Behring et al. (1972), Bromage et al. (1977), and Doschek et al. (1976).

Fe XII: The atomic model consists of 29 levels ($3s^23p^3\ ^4S$, $^2D,^2\,P, 3s3p^4\ ^4P,^2\,S$, $^2D,^2\,P$, $3s^23p^23d\ ^4P$, $^2F,^2\,D,^2\,P,^2\,S$). The effective collision strengths are computed by Tayal & Henry (1986), Tayal, Henry & Pradhan (1987), the observed energies and the radiative probabilities are from Flower (1977a). Other radiative probabilities are from Bromage et al. (1978) and Tayal & Henry (1988).

Fe XIII: The atomic model consists of 24 levels ($3s^23p^2\ ^3P$, ^1D, ^1S, $3s3p^3\ ^5S$, ^3D, ^3P, ^1D, ^3S, ^1P, $3s^23p3d\ ^3P$, ^1D, ^3D, ^1P, ^1F). The collision strengths and the observed energies are from Fawcett & Mason (1989). The radiative probabilities are from Mc Kim Melville & Berger (1965), Bromage et al. (1978) and Corliss & Sugar (1985).

Fe XIV: The atomic model consists of 12 levels ($3s^23p\ ^2P$, $3s3p^2$, ^4P, ^2D, ^2S, ^2P, $3s^23d\ ^2D$). The effective collision strengths are from Dufton & Kingston (1991), the observed energies are from Mason (1975). The radiative probabilities are from Froese, Fisher, & Liu (1986).

Fe XV: The atomic model consists of 16 levels ($2s^2\ ^1S, 3s3p\ ^3P, P, 3p^2\ ^1D$, ^3P, ^1S, $3s3d$ $3s4s\ ^3S$, ^1S). The effective collision strengths, radiative probabilities and the observed energies are from Christensen et al. (1985) and Pradhan (1988).

Fe XVI: The atomic model consists of 19 levels ($3s\ ^2S, 3p\ ^2P, 3d$ $^2D, 4s\ ^2S, 4p\ ^2P$, $4d\ ^2D, 4f\ ^2F$, $5s\ ^2S, 5p\ ^2P$, $5d\ ^2D, 5f\ ^2F$). The observed energies are from Corliss & Sugar (1985); the oscillator strengths and collision strengths are from Sampson, Zhang, & Fontes (1990).

Fe XVII: The atomic model consists of 37 levels ($2p^6$ ^1S, $2p^53s$ ^3P, ^1P, $2p^53p$ ^3S, ^3D, ^3P, ^1P, ^1D, ^1S, $2p^53d$ ^3P, ^3F, ^3D, ^1D, ^1F, ^1P, $2s2p^63s$ ^3S, ^1S, $2s2p^63p$ ^3P, ^1P, $2s2p^63d$ ^3D, ^1D). The collision strengths, the radiative probabilities and the observed energies are from Bhatia & Doschek (1992).

Fe XVIII: The atomic model consists of 108 levels ($2p^5$ ^2P, $2s2p^6$ ^2S, $2p^43s$ ^4P, ^2P, ^2D, ^2S, $2p^43p$ ^4P, ^4D, ^2D, ^2P, ^4S, ^2S, ^2F, $2p^43d$ ^4D, ^4F, ^2F, ^2D, ^2G, ^2S, ^2P, ^4P, $2s2p^53s$ ^4P, ^2P, $2s2p^53p$ ^4S, ^4D, ^4P, ^2P, $2s2p^53d$ ^4P, ^4D, ^2P, ^2D, ^2F, ^2S). The collision strengths, the radiative probabilities and the observed energies are from Cornille et al. (1992).

Fe XIX: The atomic model consists of 10 levels ($2s^22p^4$ ^3P, ^1D, ^1S, $2s2p^5$ ^3P, ^1P, $2p^6$ ^1S). The observed energies, the collision strengths and the radiative probabilities are from Loulergue et al. (1985).

Fe XX: The atomic model consists of 13 levels ($2s^22p^3$ ^4S, ^2D, ^2P, $2s2p^4$ ^4P, ^2D, ^2S, ^2P). The observed energies, the collision strengths and the radiative probabilities are from Bhatia & Mason (1980a).

Fe XXI: The atomic model consists of 36 levels ($2s^22p^2$ ^3P, ^1D, ^1S, $2s2p^3$ ^5S, ^3P, ^3D, ^3S, ^1P, ^1D, $2s^22p3s$ ^3P, ^1P, $2s^22p3d$ ^3F, ^1D, ^3D, ^3P, ^1F, ^1P, $1s^22p^4$ ^3P, ^1D, ^1S). The observed energies, the collision strengths and radiative probabilities are from Mason et al. (1979) for the first 31 levels. The effective collision strengths and the observed energies for the configuration $1s^22p^4$ are from Aggarwal (1991).

Fe XXII: The atomic model consists of 20 levels ($2s^22p$ ^2P, ^4P, ^2D, ^2S, $2s2p^2$ ^4P, ^2D, ^2S, ^2D, $2p^3$ ^4S, ^2D, ^2P, $2p^3$ ^4S, ^2D, ^2P, $2s^23s$ ^2S, ^2P, $2s^23p$ ^2P, $2s^23d$ ^2D). The observed energies are from Edlen (1983) and Bhatia et al. (1986), collision strengths and radiative probabilities are from Bhatia et al. (1986).

Fe XXIII: The atomic model consists of 30 levels ($2s^2$ ^1S, $2s2p$ ^3P, ^1P, $2p^2$ ^3P, ^1D, ^1S, $2s3s$ ^3S, ^1S, $2s3p$ ^3P, ^1P, $2s3d$ ^3D, ^1D, $2s4s$ $^3S_1$, ^1S, $2s4p$ ^3P, ^1P, $2s4d$ ^3D, ^1D). The observed energies are from Bhatia & Mason (1981), the collision strengths and the radiative probabilities are from Bhatia & Mason (1986) for the first 20 levels. For the configurations 2s4s, 2s4p 2s4d the atomic data are supplied by Bhatia (private communication).

2.3.2. Be-like Ions

C III: The atomic model consists of 20 levels ($2s^2$ ^1S, $2s2p$ ^3P, ^1P, $2p^2$ ^3P, ^1D, ^1S, $2s3p$ ^3P, ^1P, $2s3d$ ^3D, ^1D). The collisional strengths and the radiative probabilities are from Bhatia & Kastner (1993a); the observed energies are from NIST energy levels database.

O V, Ne VII, Mg IX, Ca VII, Si XI, Ni XXV: The atomic model consists of 10 levels ($2s^2$ ^1S, $2s2p$ ^3P, ^1P, $2p^2$ ^3P, ^1D, ^1S). The collisional strengths and the radiative probabilities for allowed transitions are from Zhang & Sampson (1992). For forbidden and intercombination transitions the radiative probabilities are from Muhlethaler & Nussbaumer (1976) and Nussbaumer & Storey (1979). The observed energies are from NIST energy levels database.

2.3.3. C-like Ions

The atomic model consists of 46 levels ($2s^22p^2$ ^3P, ^1D, ^1S, $2s2p^3$ ^5S, ^3D, ^3P, $2s^22p3d$ ^3F,

O III: The collisional strengths and the radiative probabilities are from Bhatia & Kastner (1993b). The observed energies are from Edlen (1985) and Wiese et al. (1966).

Ne V: The collisional strengths and the radiative probabilities are from Bhatia & Doschek (1993a). The observed energies are from Edlen (1985) and Wiese et al. (1966).

Mg VII: The collisional strengths and the radiative probabilities are from Bhatia & Doschek (1995). The observed energies are from Edlen (1985) and Wiese et al. (1966).
Si IX: The collisional strengths and the radiative probabilities are from Bhatia & Doschek (1993b). The observed energies for the configurations $2s^2 2p^2, 2s2p^3, 2p^4$ are from Edlen (1985).
S XI: The collisional strengths and the radiative probabilities are from Bhatia et al. (1987). The observed energies are from Edlen (1985) and Wiese et al. (1966).
Ca XV: The collisional strengths and the radiative probabilities are from Bhatia & Doschek (1993c) The observed energies for the configurations: $2s^2 2p^2, 2s2p^3, 2p^4$ are from Edlen (1985).
Zn XXV: The collisional strengths and the radiative probabilities are from Bhatia et al. (1987) The observed energies for the configurations: $2s^2 2p^2, 2s2p^3, 2p^4$ are from Edlen (1985).

2.3.4. N-like Ions

The atomic model consists of 13 levels ($2s^2 2p^3\ ^4S,\ ^2D,\ ^2P, 2s2p^4\ ^4P,\ ^2D,\ ^2S,\ ^2P$). The collisional strengths and the radiative probabilities are from Bhatia & Mason (1980b).
Mg VI, Si VII, S X: The observed energies are from Wiese et al. (1966).
Ar XII, Ca XIV: The observed energies are from Edlen (1984).

3. Plasma Diagnostics

The reason for studying astrophysical spectra is to determine the physical parameters of the plasma, such electron density, temperature distribution, element abundances and velocity fields. The first item to be addressed is to identify correctly the lines in the always higher resolution spectra. For this aim it is useful to produce the synthetic spectrum of the source using any temperature distribution model of the plasma.

3.1. Differential Emission Measure Analysis

Physical information on the source model are hidden in the emission measure, which is a quantity that any theoretical model should predict. The aim of temperature diagnostic is to extract such empirical information using a set of measured line intensities.
The following procedure has been developed (Monsignori Fossi & Landini 1991):
- A set of density insensitive lines must be selected.
- The differential emission measure:

$$\mathrm{DEM} = N_e^2 \frac{dV}{dT} = f(T)$$

is defined and assumption is made that it is a function of temperature only.
- The flux measured at Earth for each line j is given by:

$$I_j = \frac{1}{4\pi D^2} \int_V G_j(T) N_e^2\, dV = \int_T G(T) f(T)\, dT$$

where D is the distance from the source to the Earth and G(T) is the so-called *contribution function*, that takes into account the atomic data of the transition. A set of these equations are a system of integral equations of Frehdolm of first order.
- An inversion technique must be applied to evaluate the best $f(T)$ function which satisfies the observed lines.
It is particularly useful to compute the solution, following an iterative procedure in which the $f(T)$ is described by a cubic spline function through a small number (n) of mesh

SERTS−89 Active region

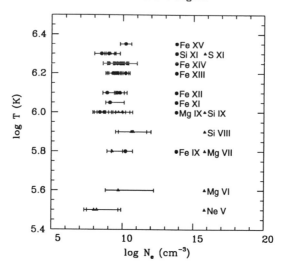

FIGURE 1. The *EUVE* spectrum of AT Microscopii. *Top:* the observed SW spectrum. *Bottom:* the synthetic spectrum (Monsignori Fossi et al. 1995b)

points $(f_i = f(T_i), i = 1, ..., n)$ and f_i are changed until the best fit is obtained. The $(r+1)$ approximation is achieved by the previous one (r) using the recurrent expression:

$$\ln f_i^{r+1} = \ln f_i^r - 2\lambda\Sigma_j \frac{(I_{ex,j} - I_{ob,j})^2}{\sigma_s^2} \frac{\partial I_{ex,s}}{\partial f_i}$$

where $I_{ob,j}$ is the observed signal, $I_{ex,j}$ is the expected signal, computed using the current $f^r(T)$ approximation, and λ is a Lagrange multiplier.

This procedure gives only positive solutions, is usually quickly converging and allows to control oscillation or flattening through the number of mesh points. It can give large indetermination in the temperature intervals where poor constraints are put by the available observations. For this reason it is important to have an appropriate selection of the lines, specially with regard to their sensitivity to different temperatures. Using the measured DEM, the synthetic spectrum can been evaluated and used for a detailed comparison with the observation to get new identifications and a general check on the theory. This technique was applied to study the spectra recorded by the *Extreme Ultraviolet Explorer* (*EUVE*) spectrometers for the active stars AU Mic (Landini & Monsignori Fossi 1993; Monsignori Fossi & Landini 1994c), AT Mic (Monsignori Fossi et al. 1995b), EQ Peg (Monsignori Fossi et al. 1995a). In Figure 1 the result for the SW section spectrum of AT Mic is shown.

3.2. Density Sensitive Lines: Comparison with Experimental Data

The code can evaluate the line emissivities for temperatures between 10^4 and 10^8 K and densities between 10^5 and 10^{16} cm^{-3}. The X-EUV spectral region is crowded by bright iron lines and Be-, C-, and N-like ions lines. Several lines of the most abundant element are useful to perform density diagnostic in astrophysical plasmas.

The electron density may be evaluated from the ratios of proper spectral line pairs pertaining to the same ion. If the source is homogeneous, this procedure does not make

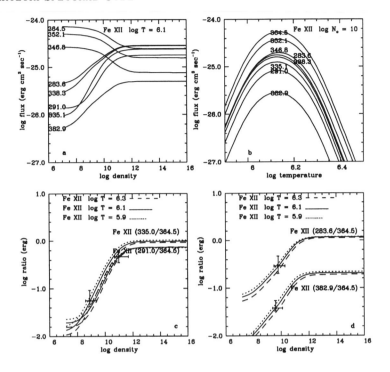

FIGURE 2. Fe XXI ratio useful to determine density in flares.

any assumption about the size of the emitting volume, the relative ion abundance and the elemental abundance.

A number of density sensitive lines were detected by the *EUVE* spectrometers both in the spectra of active stars as Capella, AU Mic, EQ Peg, Algol (Dupree et al. 1993; Monsignori Fossi & Landini 1994c; Monsignori Fossi et al. 1995a; Stern et al. 1995) and in the solar-like stars as αCen, Procion (Mewe et al. 1995; Young et al. 1995). The densities evaluated during the large flare on the star AU Mic, observed by *EUVE* spectrometers, on July 1992 are shown in Figure 2 together with the density deduced in the OSO-5 solar flare. The stellar flare requires at any time density larger than observed in the solar flare.

4. Comparison with Other Spectral Codes

Comparison between different spectral codes for optically thin plasma has been recently performed (Brickhouse et al. 1995a). In the EUV spectral region, the agreement, between the codes, is rather good specially for high temperature regimes, where the main contribution to the emission comes from the high ionization stages of iron ions. The main differences of the examined codes concern the evaluation of the line emissivity and the spectral range. The Brickhouse et al. (1995a) code solves the statistical equations only for the iron ions; the Mewe et al. (1995) code computes the line emissivity with the Gaunt factor approximation and covers the spectral range below 1000 Å; the Masai code (Masai 1994) evaluates the emission only in the X-ray spectral region. The Arcetri-95 code covers the spectral region 1 − 3000 Å. For several ions the line's emissivity in the IR is also evaluated. The Arcetri code, at this time, seems the most complete, because it is updated with the most reliable atomic data for iron ions and Be, C and N isoelectronic sequences ions and produces the power emitted per unit emission measure as function of temperature and density.

5. Conclusions

The best available atomic data for Be, C, and N-like isoelectronic ions and for Fe IX to Fe XXIII have been used to evaluate level population under the balance among electron collisional excitations and de-excitation and radiative decay and inserted in the updated version of the Arcetri spectral code. It produces synthetic spectra of plasma emission for temperatures $(4.0 \leq \log T \leq 8.0)$ and densities $(10^5 \leq \log Ne \leq 10^{15})$. A numerical procedure to evaluate empirical DEM distribution together with the theoretical spectrum allows one to compute the synthetic spectra which recover the observations.

An updated version of Arcetri spectral code (January 1995) in the wavelength range $60 - 760$ Å is available for free distribution via WWW (http://www.arcetri.astro.it). In the future we will provide on WWW the updated version for the soft X-ray spectral region including the B-like ions of the most abundant elements.

We would like to acknowledge Drs. Bhatia and Zhang for providing their computations. The research was supported by the Italian Space Agency (ASI).

REFERENCES

Aggarwal, K. M. 1991, ApJS, 77, 677

Arnaud, M. & Rothenflug, R. 1985, A&AS, 60, 425

Arnaud, M. & Raymond, J. C. 1992, A&AS, 398, 394.

Behring, W. E., Cohen, L. & Feldman, U. 1972, ApJ, 175, 493

Behring, W. E., Cohen, L. Feldman, U., & Doschek, G. A. 1976, ApJ, 203, 521

Bhatia, A. K. & Doschek, G. A. 1992, Atomic Data Nuclear Data Tables, 52, 1

Bhatia, A. K. & Doschek, G. A. 1993a, Atomic Data Nuclear Data Tables, 55, 315

Bhatia, A. K. & Doschek, G. A. 1993b, Atomic Data Nuclear Data Tables, 55, 281

Bhatia, A. K. & Doschek, G. A. 1993c, Atomic Data Nuclear Data Tables, 53, 195

Bhatia, A. K. & Doschek, G. A. 1995, Atomic Data Nuclear Data Tables, in press

Bhatia, A. K., Feldman, U., & Seely, J. F. 1986, Atomic Data Nuclear Data Tables, 35, 319

Bhatia, A. K. & Kastner, S. O. 1993a, ApJ, 408, 744

Bhatia, A. K. & Kastner, S. O. 1993b, Atomic Data Nuclear Data Tables, 54, 133

Bhatia, A. K. & Mason, H. E. 1980a, A&A, 83, 380

Bhatia, A. K. & Mason, H. E. 1980b, MNRAS, 190, 925

Bhatia, A. K. & Mason, H. E. 1981, A&A, 103, 324

Bhatia, A. K. & Mason, H. E. 1986, A&A, 155, 413

Bhatia, A. K., Seely, J. F., & Feldman, U. 1987, Atomic Data Nuclear Data Tables, 36, 453

Brickhouse, N. S., Edgar, R., Kaastra, J., Kallman, T., Liedahl, D., Masai, K., Monsignori Fossi, B. C., Petre, R., Sanders, W., Savin, D. W., & Stern, R. 1995a, Legacy, 6

Brickhouse, N. S., Raymond, J. C., & Smith, B. W. 1995b, ApJ, in press

Bromage, G. E., Cowan, R. D., & Fawcett, B. C. 1977, Phys. Scripta, 15, 177

BROMAGE, G. E., COWAN, R. D., & FAWCETT, B. C. 1978, Monthly Notices Roy. Astron. Soc., 183, 19

BURGESS, A. & TULLY, J. A. 1992, A&A, 254, 436.

CHRISTENSEN, R. B., NORCROSS, D. W., & PRADHAN, A. K. 1985, Physical Review A, 32, 93

CORLISS, C. & SUGAR, J. 1985, in Spectroscopic Data for Iron, ed. W. L. Wiese

CORNILLE, M., DEBAU, J., LOULERGUE, M., BELY-DEBAU, F., & FAUCHER, P. 1992, A&A, 259, 669

DOSCHEK, G. A., FELDMAN, U., VAN HOOSIER, M. E., & BARTOE, J. D. F. 1976, ApJS, 31, 417

DOYLE, J. G. & KEENAN, T. P. 1992, A&A, 264, 173

DUFTON, P. L. & KINGSTON, A. E. 1991, Physica Scripta, 43, 386

DUPREE, A. K., BRICKHOUSE, N. S., DOSCHEK, G. A., GREEN, J. C., & RAYMOND, J. C. 1993, ApJ, 418, L41

EDLEN, B. 1983, Physica Scripta, 28, 483

EDLEN, B. 1984, Physica Scripta, 30, 135

EDLEN, B. 1985, Physica Scripta, 31, 345

FAWCETT, D. R. & MASON, H. E. 1989, Atomic Data Nuclear Data Tables, 43, 245

FAWCETT, B. C. & MASON, H. E. 1991, Atomic Data Nuclear Data Tables, 47, 17

FLOWER, D. R. 1977a, A&A, 54, 163

FLOWER, D. R. 1977b, A&A, 56, 451

FROESE, FISHER, C. & LIU, B. 1986, Atomic Data Nuclear Data Tables, 34, 261

GAETZ, T. J. & SALPETER, E. E. 1983, ApJS, 52, 155

KASTNER, S. O., NEUPERT, W. M., & SWARTZ, M. 1974, ApJ, 191, 261

KATO, T. 1976, ApJS, 30, 394

LANG, J. ED. 1994, Atomic Data Nuclear Data Tables, 57, no 1

LANDINI, M. & MONSIGNORI FOSSI,B. C. 1970, A&A, 6,468

LANDINI, M. & MONSIGNORI FOSSI, B. C. 1990, A&AS, 82, 229

LANDINI, M. & MONSIGNORI FOSSI, B. C. 1991, A&AS, 91, 183

LANDINI, M & MONSIGNORI FOSSI, B. C. 1993, A&A, 275, L17

LOULERGUE, M., MASON, H. E., NUSSBAUMER, H., & STOREY, P. J. 1985, A&A, 150, 246

MASAI, K. 1994, ApJ, 437, 770

MASON, H. E. 1975, MNRAS, 170, 651

MASON, H. E., DOSCHEK, G. A., FELDMAN, U., & BHATIA, A. K. 1979, A&A, 73, 74

MASON, H. E. & MONSIGNORI FOSSI, B. C. 1994, Astron. Astroph. Rev., 6, 123

MC, KIM, MELVILLE, J. & BERGER, R. 1965, Planet. Space Sc., 13, 1131

MEWE, R. 1972, Solar Phys., 22, 459

MEWE, R., GRONESCHILD, E. H. B. M., & VAN DEN OORD, G. H. J. 1985, A&AS, Ser., 62, 197

MEWE, R., KAASTRA, J. S., SCHRIJVER, C. J., VAN DEN OORD, & ALKEMADE, F. J. M. 1995, A&A, in press

MONSIGNORI FOSSI, B. C. & LANDINI, M. 1991, Adv. Space Res, 11(1), 281

MONSIGNORI FOSSI, B. C. & LANDINI, M. 1994a, Solar Phys, 152, 81

MONSIGNORI FOSSI, B. C. & LANDINI, M. 1994b, Atomic Data Nuclear Data Tables, 57, 125

MONSIGNORI FOSSI, B. C. & LANDINI, M. 1994c, A&A, 284, 900

MONSIGNORI FOSSI, B. C., LANDINI, M., FRUSCIONE, A., & DUPUIS, J. 1995a, ApJ, in press

MONSIGNORI FOSSI, B. C., LANDINI, M., DRAKE, J. J., & CULLY, S. L. 1995b, A&A, in press

MUHLETHALER, H. P. & NUSSBAUMER, H. 1976, A&A, 48, 109

Nussbaumer, H., & Storey, P. J. 1979, A&A, 74, 244

Pradhan. A. K. 1988, Atomic Data Nuclear Data Tables, 40, 335

Raymond, J. C. & Smith, B. W. 1977, ApJS, 35, 419

Sampson, D. H., Zhang, H. L., & Fontes, C. J. 1990, Atomic Data Nuclear Data Tables, 44, 210

Stern, R., Wang, R., Bowyer, S. 1978, ApJS, 37, 195

Stern, R. A., Lemen, J. R., Schmitt, J. H. M. M., & Pye, J. P. 1995, ApJ, in press

Tayal, S. S., & Henry, R. J. W. 1986, ApJ, 302, 200

Tayal, S. S., Henry, R. J. W., & Pradhan, A. 1987, ApJ, 319, 951

Tayal, S. S., & Henry, R. J. W. 1988, ApJ, 329, 1023

Thomas, R. J. & Neupert, W. M. 1994, ApJ, 91, 461

Wiese, W. L., Smith, M. W., & Glennon, B. M. 1966, Atomic Trans. Prob., 1 NBS

Young, P. R., Mason, H. E., & Thomas, R. J. 1995, in Proc. 3rd SOHO Workshop, in press

Zhang, H. & Sampson, D. H. 1992, Atomic Data Nuclear Data Tables, 52, 143

DEM Analyses with the Utrecht Codes

R. MEWE,[1] G. H. J. VAN DEN OORD,[2]
C. J. SCHRIJVER,[3] AND J. S. KAASTRA[1]

[1]SRON Laboratory for Space Research, Sorbonnelaan 2, 3584 CA Utrecht, The Netherlands

[2]Sterrekundig Instituut Utrecht, P.O. Box 80 000, 3508 TA Utrecht, The Netherlands

[3]Stanford/Lockheed Inst. for Astrophys. and Space research, LPARL,
Dept. 91-30, Bldg. 252, 3251 Hanover Street, Palo Alto, CA 94304-1191, USA

We address the inversion problem of deriving the differential emission measure (DEM) distribution $\mathbf{D}(T) = n_e n_H dV/d \log T$ from the spectrum of an optically thin plasma. In the past we have applied the iterative Withbroe-Sylwester technique and the Polynomial technique to the analysis of *EXOSAT* spectra of cool stars, but recently we have applied the inversion technique discussed by Craig & Brown (1986) and Press et al. (1992) in the analysis of *EUVE* spectra of cool stars. The inversion problem–a Fredholm equation of the first kind–is ill–posed and solutions tend to show large, unphysical oscillations. We therefore apply a second–order regularization, i.e., we select the specific DEM for which the second derivative is as smooth as is statistically allowed by the data. We demonstrate the importance of fitting lines and continuum *simultaneously*, discuss the effect on the DEM of continuum emission at temperatures where no line diagnostics are available, and address possible ways to check various model assumptions such as abundances and photon destruction induced by resonant scattering.

1. Introduction

The X-ray spectrum of optically thin sources contains contributions from plasmas at different temperatures as is indicated by the presence of lines which have different formation temperatures. The underlying continuum emissivity is a slowly varying function of the plasma temperature and varies as $\sim \exp(-hc/\lambda kT)/\sqrt{T}$. Hence, at temperatures $kT \gtrsim 2hc/\lambda$ the shape of the continuum does not vary significantly making it a poor temperature probe. For the analysis of *EUVE* spectra this causes problems for T $\gtrsim$ 20 MK because at these temperatures only few lines are present, while the continuum cannot be used to constrain T. Though the continuum is of limited use as a diagnostic at high temperatures, it is of paramount importance for the spectral analysis as a whole. Spectral fitting based on only line emission can give erroneous results. We show this with a straightforward example: the effect of non-standard abundances in a source. Suppose an observed spectrum contains a number of dominant Fe lines. When the emissivity of each line is known one can assign an emission measure to each line corresponding to the formation temperature of the line. The result is, however, based on the *assumed* Fe abundance: if this is too high the derived emission measures will be too low. The check which can be made is to see whether the emission measures derived for the lines result in an acceptable fit to the continuum. If the assumed abundances are too high, and the derived emission measure as a consequence too low, the theoretically calculated continuum will be below the observed continuum signalling that there is something wrong with the assumed model. This simple example demonstrates the importance of fitting both lines and continuum simultaneously.

How accurately can we recover the temperature distribution of the source plasma using the spectral lines as temperature probes? The formation process of a spectral line depends on the ionization balance and on the excitation process, which both depend on temperature. As a result, the temperature information is spread out over a typical

S. Bowyer and R. F. Malina (eds.), Astrophysics in the Extreme Ultraviolet, 553–560.

width σ_T over which the line is formed and which is generally roughly a factor two in temperature, i.e., $\log \sigma_T \approx 0.3$. This corresponds to an intrinsic limit for the temperature resolution which can be achieved, regardless of the resolution of the instrument.

In analyzing the temperature structure of a source from the observed spectrum, one usually introduces the concept of the *differential emission measure* (DEM) which is a weighting function measuring the contribution of the plasma at each temperature to the observed spectrum. We point out a fundamental problem concerning the retrieval of the DEM from an observed spectrum. Suppose the spectrum is observed in n wavelength bins. The *observed* counts in the bins form the elements of a vector $\mathbf{g}$. In a similar way the *expected* counts from a spectral model, at a temperature T_j, can be represented by a vector $\mathbf{f}_j$. Determining the DEM comes down to solving over m temperature bins:

$$\mathbf{g} = \sum_{j=0}^{m} D_j \mathbf{f}_j = D_1 \mathbf{f}_1 + D_2 \mathbf{f}_2 + \ldots D_m \mathbf{f}_m,$$

with D_j the differential emission measure at temperature T_j. If the vectors $\mathbf{f}_j$ would constitute an orthogonal set then it would be straightforward to determine the coefficients D_j. Simply taking the dot product of $\mathbf{f}_j$ on both sides of the expression would give $D_j = \mathbf{f}_j \cdot \mathbf{g} / \mathbf{f}_j \cdot \mathbf{f}_j$ since $\mathbf{f}_j \cdot \mathbf{f}_i = 0$ for $i \neq j$. However, in reality the vectors $\mathbf{f}_j$ do not constitute an orthogonal set which means that *no unique solution exists* for the coefficients D_j. Each vector $\mathbf{g}$ can be decomposed into the vectors $\mathbf{f}_j$ in an infinite number of ways. Additional information has to be provided to select a D_j from the infinite set of solutions. Exempli gratia, for a one-temperature fit one provides the information that the DEM is a delta-function which restricts the number of solutions to the m values for D_j determined from $D_j = \mathbf{f}_j \cdot \mathbf{g} / \mathbf{f}_j \cdot \mathbf{f}_j$. Amongst these m values the one is selected which best satisfies a quality criterion for the fit, e.g., χ^2. In general, spectral fitting is based on a combination of assumptions about the expected solution and a quality criterion for the fit.

Since detectors cover only a finite wavelength range one can in principle only obtain information about the differential emission measure in the temperature range for which spectral lines can be observed within the wavelength range of the detector. We note that the *absence* of a specific line in an observed spectrum provides information for the DEM that is as relevant as the *presence* of lines. The continuum, however, contains information from a much broader temperature range than the lines. In other words, the emission measure associated with the continuum does not have to correspond necessarily to the emission measures derived from the lines. This limits the use of the continuum as a check for, e.g., abundance variations. There is a way out: use a complementary instrument which can observe at higher or lower energies to see whether other temperature components are present in a source. These components can then be responsible for an 'excess' continuum at the observed wavelengths but for which no line diagnostics are available. If the presence of other components is not confirmed, the continuum can be used for checking the basic model assumptions.

Various inversion techniques exist to recover a DEM distribution over the range of temperatures for which the instrument is sensitive, using optically thin model spectra, convolved with an appropriate instrument response (cf., Thompson 1991).

In the past we have applied an iterative deconvolution technique that was based on a weighting-factor method originally proposed by Withbroe (1975), modified later by Sylwester et al. (1980): the Withbroe-Sylwester (WS) method. The technique is formulated to exclude a priori negative values for the emission measure. It makes no assumptions about the functional form of the DEM, although the final result is subject to an implicit smoothing, because the weighting functions contain integrals over the entire spectral range. It was applied to the analysis of X-ray spectra of solar flares and to spectra of several cool stars measured with the transmission grating spectrometer (TGS) on board *EXOSAT* (Lemen et al. 1989). Lemen et al. also used the Polynomial (P) method, which assumes that the DEM can be approximated by $D(T) = \alpha \exp[\omega(T)]$, where $\omega(T)$ is a polynomial function of temperature for which a sum of Chebyshev polynomials was chosen allowing any arbitrary functional form for the DEM to be approximated as the number of terms in the sum is increased. In the *EXOSAT* analysis they have used polynomials of order 9. This technique also precludes negative values for the DEM. Recently we have developed the software–built into our new spectral code SPEX–for an inversion technique as discussed by Craig & Brown (1986) and by Press et al. (1992) which uses a matrix inversion by singular-value matrix decomposition combined with second-order regularization and which has been applied to the analysis of *EUVE* spectra of cool stars.

2. Inversion by Statistical Regularization

Spectra emitted by optically thin coronae are linear combinations of isothermal spectra and are interpreted as statistical realizations of models calculated with our code for optically thin plasmas. In our models we assume (thermal) collisional ionization equilibrium, thus ignoring possible transient effects, and we ignore effects that depend on the plasma density, which affect only a few lines. Let an isothermal plasma of temperature T emit a spectrum that, when incorporating interstellar absorption, instrumental efficiencies, and instrumental smoothing, is represented by $\mathbf{f}(\lambda_i, T_j)$ in a linear grid of wavelength bins of width $\Delta\lambda$ (i = 1, ..., n) and a logarithmic grid of temperature intervals of width $\Delta \log(T)$ (j = 1, ..., m), where $\mathbf{f}(\lambda_i, T_j)$ is the plasma emissivity per unit $n_e n_H$. For a composite plasma with temperatures ranging from T_1 up to T_m, the net expected spectrum $\mathbf{g}(\lambda_i)$ is given by:

$$\mathbf{g}(\lambda_i) = \int \mathbf{f}(\lambda_i, T) n_e(T) n_H dV(T) \approx \sum_{j=1}^{m} \mathbf{f}(\lambda_i, T_j) \mathbf{D}(T_j) \Delta \log(T). \qquad (2.1)$$

Eq. (2.1) constitutes a Fredholm equation of the first kind for the differential emission measure $\mathbf{D}(T) = n_e n_H dV/d\log T$ and can be written as a vector equation: $\mathbf{g} = \mathbf{F} \cdot \mathbf{D}$, in which $\mathbf{F}$ is a matrix with m columns and n rows, of which the elements are given by $\mathbf{F}_{ij} = \int_{\lambda_i}^{\lambda_i + \Delta\lambda} \mathbf{f}(\lambda, T_j) d\lambda \Delta \log T \equiv \mathbf{f}_i(T_j) \Delta \log T$. Each column of $\mathbf{F}_{ij}$ consists of a 'spectral' vector containing the discretized spectrum at a certain temperature. The formal least-squares solution of this problem requires an inversion of

$$\mathbf{F}^T \mathbf{F} \cdot \mathbf{D} = \mathbf{F}^T \cdot \mathbf{g}, \qquad (2.2)$$

in which $\mathbf{F}^T$ is the transpose of $\mathbf{F}$. The terms $[\mathbf{F}^T\mathbf{F}]_{ij}$ are proportional to dot products $\mathbf{f}_i \cdot \mathbf{f}_j$ while the terms $[\mathbf{F}^T\mathbf{g}]_j$ are proportional to $\mathbf{f}_j \cdot \mathbf{g}$, in which one recognizes the vector decomposition discussed in the introduction. Let $\tilde{\mathbf{g}}$ represent an observed spectrum which contains noise. The measurement errors are taken into account by introducing a weighting $\tilde{\mathbf{g}}_i \rightarrow s\tilde{\mathbf{g}}_i/s_i$, $\mathbf{F}_{ij} \rightarrow s\mathbf{F}_{ij}/s_i$ with s the geometrical mean of the errors. The common practice is then to solve Eq. (2.2) in the classical generalized least-squares sense, $\|\mathbf{F} \cdot \mathbf{D} - \tilde{\mathbf{g}}\|^2 = \min$, in which one recognizes the χ^2-method. This method does not, however, take into account that information has been lost due to the smoothing action of the kernel in Eq. (2.1). Also, the method forces the solution to fit noisy data points as good as possible, causing high-frequency oscillations in $\mathbf{D}$. The way to remedy these problems is to introduce an additional quadratic constraint $\|\mathbf{R} \cdot \mathbf{D}\|^2$ which reflects some desired property for $\mathbf{D}$. One then minimizes a joint functional comprising the classical χ^2 constraint and the additional constraint: $\|\mathbf{F} \cdot \mathbf{D} - \tilde{\mathbf{g}}\|^2 + \varrho\|\mathbf{R} \cdot \mathbf{D}\|^2 = \min$, where ϱ is a Lagrangian multiplier commonly referred to as the *regularization parameter* $(0 \leq \varrho < \infty)$ which controls the degree of smoothness of the solution. The matrix form of the regularized problem is then given by

$$(\mathbf{F}^T \mathbf{F} + s^2 \,\varrho\mathbf{R}) \cdot \mathbf{D} = \mathbf{F}^T \cdot \tilde{\mathbf{g}}. \tag{2.3}$$

A suitable choice for the quadratic constraint is *second-order regularization*, given by $\mathbf{R} \cdot \mathbf{D} = \mathbf{D}''$, which makes the second derivative as smooth as statistically allowed by the data. The matrix $\mathbf{R}$ has a band diagonal shape over five columns (see Mewe et al. 1995). Because we use $\Delta \log T \approx (\log \sigma_T)/3 \approx 0.1$, the matrix $\mathbf{R}$ smooths the information over a typical line-formation interval. We note that constraints which require $\mathbf{D}$ to be positive are too restrictive when the background-subtracted spectra contain bins with only a few, and possibly, negative counts (see Thompson 1991).

When blends occur in an observed spectrum of lines with different formation temperatures, and when a continuum is present in the spectrum, the photon noise together with the imposed smoothing will cause some degree of cross-talk over a temperature range exceeding σ_T resulting in a coupling between different temperature components of $\mathbf{D}$. However, in the analysis of the *EUVE* spectra such 'contamination' is limited because the continuum bins have relatively low count rates, corresponding to a low statistical weight, while lines with different formation temperatures are generally well separated.

We stress that the above inversion method is *not* an iterative procedure. An iterative method requires an initial *DEM* distribution, and, depending on the details of the distribution, the iteration may not converge to the true best–fit solution, but instead may yield a solution corresponding to a local minimum in the χ^2-space. This problem is avoided by the method discussed in this paper.

3. Applications and Examples

A detailed DEM analysis using the Mewe et al. (1985, 1986) plasma emission code and the Withbroe-Sylwester (WS) and Polynomial (P) methods was performed on the *EXOSAT–TGS* spectra (5–200 Å) of Capella, σ^2 CrB , and Procyon by Lemen et al. (1989). An example of the DEM results for Capella is shown in Fig. 2.

For the analysis of the *EUVE* spectra isothermal equilibrium spectra were calculated using our newly developed SPEX code, which contains the MEKA spectral code, available as a subroutine to the XSPEC package, but extended with spectral lines between 300–2000 Å (cf. Kaastra & Mewe 1993, Mewe & Kaastra 1994). The calculated spectra are modified by interstellar absorption using absorption cross sections of Rumph et al. (1994), adopting abundance ratios He I/H I = 0.1, He II/H I = 0.01, and convolved with the instrument response. Solar photospheric abundances from Anders & Grevesse (1989) are used. We have used a range of temperatures between 3×10^4 K–10^8 K divided in 36 logarithmically spaced temperature values (i.e. $\Delta \log(T) \simeq 0.1$). The lower temperature boundary is determined by the presence of strong He II lines which form $\lesssim 10^5$ K, the upper boundary is chosen well above the highest formation temperature of the lines within the *EUVE* wavelength range. We have chosen a wavelength binning equal to about half the spectral resolution (i.e., 0.25, 0.5, and 1 Å for the SW, MW, and LW, respectively). The emission measure is plotted as $\mathbf{D}(T)\Delta \log T$ vs. $\log T$, with $\Delta \log T = 0.1$. The total emission measure in any temperature range can be obtained by summing the values of $\mathbf{D}(T)\Delta \log T$ over this range.

We have analyzed the spectra of eight late-type stars (cf., Schrijver et al. 1995 and these proceedings), and the pre-main-sequence star AB Dor (cf., Rucinski et al. 1995),

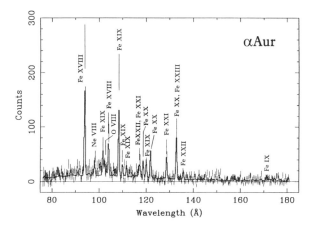

FIGURE 1. The *EUVE* background-corrected spectrum for Capella for the SW passband, binned into intervals of 0.25 Å (from Mewe et al. 1995). The observed spectrum is shown with the uncertainties on the observed number of counts. The thick solid line shows the best–fit spectrum for solar abundances (determined from all three bands), for which the DEM distribution $\mathbf{D}(T)$ is plotted as a thick solid line in Fig. 2.

as observed with the spectrometers on board the *Extreme Ultraviolet Explorer* (*EUVE*) (e.g., Bowyer & Malina 1991) in the short-, medium-, and long-wavelength bands SW (64–202 Å), MW (160–320 Å), and LW (256–808 Å), respectively.

In Fig. 1 we show the SW spectrum of Capella together with the best fit (determined from all three wavelength bands). The corresponding DEM is shown in Fig. 2. The differential emission measure shows a contribution from plasma below 10^5 K which is mainly caused by the He II line at 304 Å. In the formation of this line radiative transfer effects play a rôle so that the optically thin approximation does not apply. The DEM at these temperatures must therefore be considered merely as a *formal* solution. There is little emission from plasma at temperatures in the range $10^5 - 10^6$ K while a significant part of the observed spectrum originates from plasma at temperatures in the range $2 \ 10^6 - 10^7$ K. This plasma component is also found in the analysis of *EXOSAT* data. What is surprising is the presence of a hot 'tail' in the DEM-distribution above 20 MK. A similar tail or hot component was found for all objects except for χ^1 Ori which is a solar-like star. We stress that the presence of this 'superhot' component is not the result of our specific inversion algorithm but of the fact that we fitted the line and continuum simultaneously. Other inversion methods such as the Polynomial method also show the presence of such a hot component in the DEM analysis of *EUVE* data (cf., Stern et al. 1995).

In the following we discuss the origin of the *high-temperature component* for Capella in particular, and other sources in general. In the case of Capella, the hot component is most likely due to a real hot plasma component. The analysis of the *EUVE* spectrum for solar abundances shows that this hot component produces about half of the SW continuum (cf., Fig. 1) and the presence of spectral lines up to Fe XXIV in the Capella (LW) spectrum suggests that at least part of this hot component is real. Observations with instruments looking at harder photons than the *EUVE* should then confirm the presence of a hot plasma component. This is indeed substantiated by the *EXOSAT–TGS*

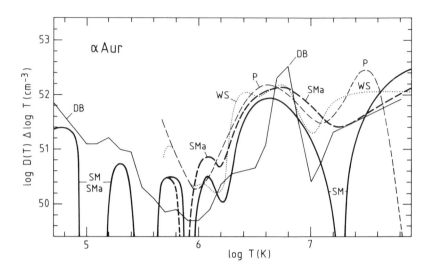

FIGURE 2. DEM curve $\mathbf{D}$(T)$\Delta \log$ T on a log scale derived by Schrijver, Mewe et al. 1995 for the model with solar abundances from the $EUVE$ spectra of Capella (**thick solid** line, labeled **SM**) and for the case with Fe abundance 0.3 × solar (**thick dashed** line, **SMa**). For comparison are also drawn the DEM derived by Dupree, Brickhouse et al. (1993) by a DEM analysis from the Capella $EUVE$ spectrum (**thin solid** line, **DB**) and the $EXOSAT$ results derived by Lemen et al. (1989) (**thin dotted** (**WS**) and **thin dashed** (**P**) lines for the WS and P methods, respectively; original results were multiplied by a factor 2.42 so as to correct for n_H/n_e and different temperature interval). All curves are for a temperature interval $\Delta \log$ T = 0.1.

observations which indicate a hot (20–40 MK) component for Capella (cf., Fig. 2). As we discussed in the introduction, the need to add continuum photons can indicate that a plasma component is present in the source at temperatures where no lines form in the $EUVE$ wavelength range or it signals that something is wrong with the assumed plasma model. In both cases the high-temperature component in the DEM arises because the values of the DEM required to fit the spectral lines do not generate enough associated continuum photons. To make the fit acceptable any inversion algorithm will add plasma at temperatures where the lines are not influenced. For the $EUVE$ this corresponds to temperatures above 20 MK.

More generally, incorrect values for the assumed abundances can be a second cause of a high-temperature component. Line excitation $\propto n_i n_e$ with n_i the density of the emitting ions and n_e the electron density. Continuum emission (predominantly free-free and free-bound emission from H- and He ions) varies proportionally to $n_e n_H$. The line/continuum ratio is therefore proportional to n_i/n_H, the ion abundance, which in turn is proportional to the element abundance. Many of the dominant lines in the $EUVE$ spectra, especially in the SW band are from iron. Reducing the Fe abundance will therefore make the continuum relatively stronger with respect to the lines and making it less necessary for the algorithm to invoke a high-temperature component. To illustrate this effect, we also show in Fig. 2 the DEM for a reduced Fe (0.3 × solar) abundance (thick dashed line). The high-temperature component is, as expected, reduced in strength while the DEM

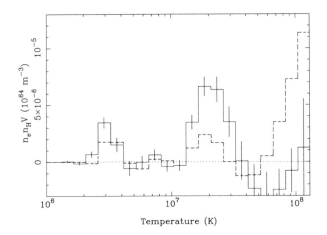

Temperature (K)

Figure 3. Results of a DEM analysis of a simulated *EUVE* spectrum including noise and background subtraction. In the input two-temperature model we take temperatures of 2.9 and 18.4 MK, emission measures of 5 and 15 (10^{58} m^{-3}), d = 12 pc, exposure time 10^5 s and solar abundances except for Fe (0.3 × solar). Fitting with 0.3 × solar Fe abundances recovers the input model (solid line) but when the Fe abundance is incorrectly assumed to be solar (dashed line) a hot component arises above ∼ 20 MK to correct the line/continuum ratio.

becomes larger in the 2 $10^6 - 10^7$ K range, required to fit the Fe lines when the abundance is lower.

Reduced Fe abundances (in the range ∼ 0.2–0.4) are probably needed to reconcile *EUVE* observations on Algol (Stern et al. 1995) and AB Dor (Rucinski et al. 1995 and these proceedings) with *ASCA* data, or to explain the re–analyzed *EINSTEIN* SSS spectrum of Capella and the *ASCA* spectrum of AR Lac (e.g., Drake et al. 1994).

To demonstrate the effect of abundances we have simulated an *EUVE* spectrum with noise and typical background subtraction for a model with two temperature delta-functions and solar abundances, except for Fe where we take 0.3 × solar. If we make a DEM fit with the same model we restore the two components, but when we wrongly assume a solar Fe abundance, indeed a hot component arises above about 20 MK in order to fit the line/continuum ratio with the too large Fe abundance (cf. Fig. 3).

A third explanation for the hot DEM component may be provided by the assumption that we are missing in the spectral code many weaker, closely spaced lines that contribute to and mimic the observed SW continuum such as L-shell lines from the lighter elements (Ne, Mg, Si, S) and Fe M-shell emission (Liedahl, private communication).

A fourth possibility is a non-thermal continuum of comparable strength as the thermal continuum (e.g., Stern et al. 1995) and with a power-law slope similar to a Crab-like spectrum so as to give a flat spectrum on a wavelength scale.

Finally, there is the possibility that the stronger lines are weakened –compared to the optically thin approximation–by the effect of resonant scattering. This effect which is discussed in detail by Schrijver et al. in these proceedings may play a rôle in the coronae of e.g., α Cen and Procyon. When a plasma becomes optically thick (optical depth $\tau \gtrsim 1$) the effect on the intensity of a resonance line is determined by the processes competing with the spontaneous radiative decay to the ground level: (1) radiative branching to other levels, (2) collisional de-excitation, or (3) true absorption in a nearby, dense medium. Suppose that all these processes destroy a fraction f of the photons per absorption. As a photon is absorbed and re-emitted ('scattered') $\sim \tau$ times before it eventually escapes, the plasma can still be considered effectively optically thin as long as $f\tau \ll 1$. The net effect on the line intensity compared to the optically thin case therefore depends on the competing processes and on the geometry. Scattering will only play a rôle in attenuating the photon line flux when there is an asymmetry in the locations of emitting and scattering region. In a stellar corona collisional de-excitation can be neglected, but scattering can re-direct photons towards the underlying, dense chromosphere where they are lost due to absorption, e.g., when the line emission occurs at the base of a scattering corona (or overlying, hot wind). For $\tau > 1$ scattering thus results in a weakening of the line flux and the optically thin model overestimates the line flux, hence underestimates the DEM in the line-forming temperature region, leading to the generation of a hot component in the DEM so as to compensate for the underestimated continuum strength. A complication is that abundance variations and scattering have opposite effects on the line/continuum ratio: an increase in abundance leads to an increase in the line/continuum ratio but also to an increase in optical depth which leads to a reduction of this ratio.

In conclusion, with the advent of high-resolution spectrometers as $EUVE$ and in future $SOHO$, $AXAF$, and XMM, care should be taken in the spectral fitting procedures while lines and continuum should both be used. Every DEM determination should be inspected on its individual merits taking into account (1) all the assumptions made in the spectral model and (2) considering the influence of continuum emission from plasma at temperatures where no line diagnostics is available in the relevant wavelength range.

REFERENCES

Anders, E. & Grevesse, N. 1989, Geochimica et Cosmochimica Acta, 53, 197

Bowyer, S. & Malina, R. F. 1991, in Extreme Ultraviolet Astronomy, ed. R.F. Malina & S. Bowyer, New York: Pergamon Press, 397

Craig, I. J. D. & Brown, J. C. 1986, Inverse Problems in Astronomy., Adam Hilger Ltd.

Drake, S. A., Singh, K. P., White, N. E. & Simon, T. 1994, ApJ, 436, L87

Dupree, A. K., Brickhouse, N. S., Doschek, G. A. et al., 1993, ApJ, 418, L41

Kaastra, J. S. & Mewe, R. 1993, Legacy, 3, 16

Lemen, J. R., Mewe, R., Schrijver, C. J. & Fludra, A. 1989, ApJ, 341, 474

Mewe, R. & Kaastra, J. S. 1994, European Astron. Soc. Newsletter, Issue 8, 3

Mewe, R., Gronenschild, E. & van den Oord, G. 1985, A&AS, 62, 197

Mewe, R., Kaastra, J. S., Schrijver, C. J. et al., 1995, A&A, 296, 477

Mewe, R., Lemen, J. R. & van den Oord, G. H. J. 1986, A&AS, 65, 551

Press, W. H., Teukolsky, S. A., Vetterling, W. T. & Flannery, B. P. 1992, Numerical Recipes, Cambridge University Press

Rucinski, S. M., Mewe, R., Kaastra, J. S. et al., 1995, ApJ, in press

Rumph, T., Bowyer, S., & Vennes, S. 1994, AJ, 107, 2108

Schrijver, C. J., Mewe, R., van den Oord, G. H. J. et al., 1995, A&A, in press

Stern, R. A., Lemen, J. R., Schmitt, J. H. M. M. et al., 1995, ApJL, in press

Sylwester, J., Schrijver, J. & Mewe, R. 1980, Solar Phys., 67, 285

Thompson, A. M. 1991, In Intensity Integral Inversion Techniques: a Study in prereparation for the SOHO Mission, ed. R.A. Harrison & A.M. Thompson,. Rutherford Appleton Lab. Rep. RAL-91, 18

Withbroe, G. L. 1975, Solar Phys., 45, 301

Plasma Emission Codes:
Comparisons and Critiques

HELEN E. MASON

Department of Applied Mathematics and Theoretical Physics, Silver Street, Cambridge
CB3 9EW, UK

A great deal of effort in recent years has gone into the development of spectroscopic techniques to probe the physical parameters of solar, stellar and other astrophysical plasmas. One aspect of this work is the calculation of plasma emission codes which are used to study EUV spectral lines. These codes require the input of a large amount of atomic data. In this paper, we present an overview of the atomic processes involved and an assessment of the accuracy of the parameters which are incorporated into different emission codes.

1. Introduction

The task of this review has been made significantly easier because a workshop on *Plasma Emission Codes* was held in Napa Valley on November 2nd–5th, 1994. The proceedings are being published as a report (Brickhouse et al. 1995). A benchmark comparison was made between different plasma emission codes for wavelength ranges relevant to recent astrophysical observations. In this review, we give an overview of the strengths and weaknesses of the various plasma emission codes; we discuss the problem areas identified at the Napa Valley meeting, in particular wavelength ranges where codes produce different synthetic spectra or where there are substantial discrepancies with observations; we give a *user's guide* to the relevant atomic data; we ask what has been learnt from the past twenty years of solar observations and finally we look to future developments in the plasma codes for solar spectral and astrophysical observations. An extensive review of spectroscopic diagnostics in the UV and EUV wavelength ranges for solar and stellar plasmas has recently been published by Mason & Monsignori Fossi (1994).

2. Napa Valley Workshop

The Napa Valley workshop made benchmark comparisons for the wavelength ranges 0.1–10 KeV (*ASCA*); 40–85 Å (*DXS*); 85–125 Å and 125–180 Å (*EUVE*); in parenthesis are the relevant astrophysical observations. The plasma emission codes which were included in the comparative study were MEKA (Mewe, Kaastra et al.); KM (Kato & Masai); BRS (Brickhouse, Raymond & Smith); MFL (Monsignori Fossi & Landini). These codes have similar features—they all deal with the continuum and line emission from hot, optically thin plasmas. The formulations used for continuum emission (free-free, free-bound, two-photon) are in general agreement, so in this review we shall concentrate on the approximations used to compute the line emission spectrum. The main features to be considered are: the treatment of ionisation and recombination processes; the type of model ion used for the level populations (treatment of *metastable levels* and highly excited levels); the representation of electron collisional excitation rates; inclusion of proton rates; the contribution of di-electronic and inner shell satellite lines. The main features of each code are described below.

561

S. Bowyer and R. F. Malina (eds.), Astrophysics in the Extreme Ultraviolet, 561–568.

2.1. *MEKA, XSPEC, SPEX*

The first version of the MEKA code, published by Mewe (1972), covered the wavelength range 1–300 Å. Updated versions, including improved atomic data were published by Mewe & Gronenschild (1981), Mewe et al. (1985). The electron density dependence of the population for the lower levels are allowed for together with a correction for cascades from higher levels.

The electron excitation rates are represented by an approximation based on the $\bar{g}$ formula (Van Regemorter 1962). This is used by Mewe and other authors to estimate the intensity for the solar lines, so we discuss it here in some detail.

The averaged collision strength, Υ, is defined as

$$\Upsilon_{ij} = \int_0^\infty \Omega_{ij}\ exp(\frac{-E_j}{kT})\ d(E_j/kT) \qquad (2.1)$$

where Ω_{ij} is the collision strength for the transition i to j, E_j is the electron energy relative to the final state j, k is the Boltzmann constant and T is the electron temperature of the plasma.

The electron collision rate is then given by the well known formula

$$q_{ij} = 8.63\ 10^6\ T^{-0.5}\ exp(-E_{ij}/kT)\Upsilon_{ij}/\omega_i \qquad (2.2)$$

The averaged collision strength Υ (which Mewe calls $\bar{\Omega}$) is related to $\bar{g}$ by

$$\Upsilon_{ij} = \frac{8\pi}{\sqrt{3}}\frac{\omega_i f_{ij}}{E_{ij}}\bar{g} \qquad (2.3)$$

where $\omega_i f_{ij}$ is the radiative oscillator strength and E_{ij} is the energy difference between levels i and j.

It should be emphasised that the $\bar{g}$ formulation was developed for dipole transitions. Mewe (1972) extended its use to forbidden transitions. He uses an expression of the form

$$\bar{g} = A + (By - Cy^2 + Dy^3 + E)e^y E_1(y) + (C + D)y - Dy^2 \qquad (2.4)$$

for both allowed and forbidden transitions. The parameters A, B, C, D, and E are derived by fitting to available electron excitation rates. The parameter y is E_{ij}/kT, and $E_1(y)$ is a standard exponential integral. A serious problem arises when electron excitation rates are not available and the parameters must be guessed.

Doschek & Cowan (1984) produced a synthetic spectrum from a composite of observed solar spectra. Mewe et al. (1985) re-normalised their semi-empirical parameters to fit this spectrum better.

The synthetic spectra developed by Mewe and coworkers include di-electronic recombination and inner shell satellite lines. The plasma emission code uses the most recent ionisation and recombination rates and is able to deal with time-dependent processes. Kaastra & Mewe (1993) reported on an updated version of these synthetic spectra called XSPEC. In 1996, a revised and extended version of these codes, called SPEX will be available. This will cover the wavelength range 1–2000 Å, with improved atomic data including new n=2 to n=3, n=4 electron excitation data.

2.2. *KM*

Kato(1976) published a synthetic spectrum covering the wavelength range 1–250 Å, this was later updated and extended by Masai (1984a, 1984b). The treatment was similar to that of Mewe and coworkers, using the $\bar{g}$ formulation for electron excitation and some allowance for cascade effects. Again time dependent ionisation, using standard ionisation and recombination rates was incorporated. The most recent update is by Masai (1994), using the Arnaud & Raymond (1992) ionisation and recombination calculations for the iron ions.

2.3. BRS

The Raymond & Smith (1977, 1988) synthetic spectra covered the wavelength range 1–200 Å. The $\bar{g}$ formulation similar to Mewe's was used for the electron collisional excitation and the metastable levels were also treated in a similar way to Mewe. The formulation for ionisation (+excitation/auto-ionisation) was based on the Exchange Classical Impact Parameter method and the re-combination processes were calculated with a refined form of the Burgess General Formula. The codes can deal with time dependent processes. Recently Brickhouse et al. (1995) have published new synthetic spectra for the iron ions in the wavelength range 70–600 Å using the Arnaud & Raymond (1992) ionisation balance and the best available electron excitation rates. The statistical equilibrium equations are solved for each ion. Future improvements to this work include a study of other elements and an extension of the wavelength range to cover the X-ray wavelength region.

2.4. MFL

Landini & Monsignori Fossi (1972, 1990, 1991) published a synthetic spectrum in the wavelength range 1–2000 Å. These authors used a $\bar{g}$ formulation similar to Mewe and coworkers. Their ionistaion and recombination rates were a little different. Recent work by Monsignori Fossi & Landini (1994) uses the ionisation rates of Arnaud & Raymond (1992) for the iron ions. They solve the statistical equilibrium equations for the low lying levels of each ion and represent the electron excitation rates using the Burgess & Tully (1992) method, which is described in a later section.

2.5. *Problems Identified at the Napa Valley Workshop*

In the X-ray wavelength range, large discrepancies between the spectral codes were found to be due primarily to the use of different formulations for the ionisation balance. In addition, a major problem was identified for the intensity ratios of the transitions between n=2 to n=3 levels and n=2 to n=4 levels of highly ionised systems. In the 40–85 Å range, there were discrepancies between the intensities for the Fe XV and Fe XVI spectral lines. In the 85–125 Å region some discrepancies were identified for the iron ions Fe XVIII–XXII and it was noted that the Ni ions need more theoretical work. The wavelength range >200 Å was not investigated.

3. Accuracy of Atomic Calculations

The atomic data required for the analysis of astrophysical spectra includes: energy levels and radiative data; electron and proton excitation rates; ionisation and re-combination rates. Recent progress in these areas includes laboratory measurements of atomic parameters, for which only theoretical calculations previously existed. Although in several cases the independent theoretical calculations had converged to what was assumed to be a high accuracy result, unfortunately the laboratory measurements give a different value. It is beyond the scope of this review to discuss the details, but simply to sound a note of caution.

3.1. *Solar Context*

In the 1970s, many new atomic physics calculations were prompted by the *Skylab* UV observations and the HRTS (High Resolution Telescope Spectrometer) rocket instrument. The emphasis in the 1980s was on solar flare spectra, stimulated by the success of projects such as the Solar Maximum Mission, SOLEX, SOLFLEX and HINOTORI. These instruments obtained high resolution spectra in the X-ray wavelength regions, below 25 Å as well as the UV (1000–2000 Å). The instruments on the YOHKOH satellite have recently provided some fascinating X-ray images and spectra. Interest in the UV wavelength range has also been revived with the successful flights of the SERTS (Solar EUV Rocket Telescope and Spectrometer) (Neupert et al. 1992; Thomas & Neupert 1994). Much effort has been directed towards a major satellite mission, the Solar Heliospheric Observatory (*SOHO*) to be launched in October 1995. This will carry several UV spectroscopic instruments (160–1600 Å), including the Coronal Diagnostic Spectrometer (CDS) (Harrison & Sawyer 1992) and the Solar Ultraviolet Measurement of Emitted Radiation (SUMER) instrument (Lemaire et al. 1992).

3.2. *Ionisation Balance*

The large discrepancies for the iron ions between Arnaud & Raymond (1992) and Arnaud & Rothenflug (1985) require further investigation. Shifts on the order of 0.1–0.2 dex are found in the temperatures corresponding to peak ion abundances. The peak values of the ionisation ratio typically differ by 10–20%. Detailed studies of the formulations used ionisation and recombination rates can be found in Raymond (1988), Mewe (1990), and the recent paper by Jordan (1995). The main questions to be addressed are the more accurate account of indirect processes in the ionisation rates (Moores & Reed 1995) and the treatment of electron density dependence and alternative decay channels in the di-electronic recombination rates.

3.3. *Electron Excitation Rates*

Accurate electron scattering calculations are now available for many of the ions which are abundant in the solar corona and transition region. These were reviewed at an atomic data assessment workshop held in Abingdon, UK, in 1992 (Lang 1994).

3.3.1. *Fe XV and Fe XVI*

Below we consider the accuracy of the representations of these atomic data in the various plasma emission codes, in particular for Fe XV and Fe XVI which were reviewed by Badnell & Moores (1994).

Sampson & Zhang (1992) studied the use of the Van Regemorter formula for collision strengths. Their abstract states, "It is found to be a very poor approximation, especially for $\Delta n \geq 1$ excitation transitions from levels $l < n - 1$, and the recommendation is made that with recent advances in calculation procedures and available accurate atomic data, use of the Van Regemorter formula be discontinued." This approximation is only applicable to optically allowed (electric dipole) transitions. There is no justification for applying this formulation to other types of transitions.

The ion Fe XVI is abundant in solar active regions and flares. Strong spectral lines from this ion have been observed over a wide wavelength range. The transitions 3s–3p and 3p–3d fall between 250 and 365 Å and lines from transitions between n=3 to n=4, n=3 to n=5 fall between 30 and 80 Å. These lines have also been observed in stellar spectra. Very few solar spectra have been recorded in this wavelength region, however, spectra in the 10–100 Å wavelength region have been obtained for a solar flare with the XSST rocket instrument (Acton et al. 1985). Cornille et al. (1993, 1995) have compared the theoretical and observed intensity ratios for the strong Fe XVI lines which were prominent in these spectra. They note that the spectral lines corresponding to transitions between 3p–4s (62.88 Å, 63.71 Å), 3p–4d (54.13 Å, 54.72 Å, 54.77 Å) and 3d–

4f (66.25 Å, 66.36 Å) are as strong as the spectral lines corresponding to 3s–4p (50.35 Å, 50.56 Å) transitions. There is also an important temperature sensitivity for the spectral line intensity ratios. A similar phenomena is seen for the spectral lines for n=3 to n=5 transitions. Cornille et al. (1995) give a comparison of their theoretical intensity ratios and those obtained using the $\bar{g}$ formulation given in the various plasma emission codes. The closest agreement is found for Doschek & Cowan (1985), which is to be expected since their values are obtained from the solar observations! The synthetic spectra of Mewe et al. (1985) and Landini & Monsignori Fossi (1989) give major discrepancies (up to a factor 10) for the intensity ratios of the Fe XVI lines relative to the transition 3s–4p. The $\bar{g}$ formulation used in these plasma emission codes gives a very poor representation of the Fe XVI spectral lines between 50–60 Å. Very recent versions of these codes have better representations for the Fe XVI excitation rates, but what about other ions, such as Fe XV, for which no accurate atomic data has yet been published?

Electron scattering calculations for Fe XV have recently been completed by Bhatia et al. (1995) including the n=3 to n=4 transitions. They find that the electron collision rates for the monopole and quadrupole transitions are large in comparison to the values for the dipole transition $3s^2$ $^1S_0$–$3s4p$ $^1P_1$ (52.91 Å). They calculate intensity ratios for the Fe XV spectral lines and compare these with the values obtained in the solar flare spectrum from XSST. The agreement between the theoretical and observed values is good. In fact, it is possible to identify the very strong line at 69.65 Å as the Fe XV transition $3s4s$ $^1S_0$–$3s3p$ $^1P_1$. This is almost a factor of 10 stronger than the resonance line at 52.91 Å. The theoretical and observed intensity ratios for these lines are completely different from those estimated by the synthetic spectrum codes, even the most recent ones!

Since the Fe XV and Fe XVI X-ray lines have not been extensively studied with solar spectra, it is perhaps not so surprising that such major inconsistencies have gone unnoticed until very recently, however the plasma emission codes have been used to analyse astrophysical data, for example *EXOSAT* observations of stellar coronae (cf., Schrijver et al. 1989; Lemen et al. 1989); *ROSAT* (Schmitt 1992) and DXS Bragg Crystal Spectrometer (44–83 Å; Sanders 1995). For such analyses, large inaccuracies in the simulated spectra for Fe XVI and Fe XV could lead to a misinterpretation of the distribution of material as a function of temperature.

3.3.2. *Fe XVIII–Fe XXIV*

Extensive electron scattering calculations have been carried out for the iron ions Fe XVIII–XXIII, both within the n=2 configurations and between the n=2 to n=3 configurations. Solar observations in the 90–150 Å wavelength range are sparse, but a very good flare spectrum was obtained with the with OSO-5. The diagnostic potential of these iron lines is discussed in Mason et al. (1984), together with references to atomic calculations. Electron excitation rates have been calculated for iron and other elements (e.g., Ne, Mg, Si, S, Ca) of solar interest by Mason, Bhatia, Doschek, Dubau and colleagues using the University College London distorted wave program. Many calculations have also been carried out with the more sophisticated RMATRX, close coupling program, developed at Queens' University Belfast. The atomic data for the $2s^2 2p^n - -2s2p^{n+1}$ transitions is very accurate and references can be found in the reviews in Lang (1994).

The recent versions of the plasma emission codes are all based on these atomic data. There is no obvious reason for discrepancies between the synthetic spectra. One important process which is not accounted for in these codes is that of proton excitation. It is particularly important for these highly ionised systems and should be included in the statistical equilibrium equations.

The n=2 to n=3 transitions in Fe XVIII–Fe XXIV give rise to spectral lines in the X-ray wavelength range (10–20 Å). These were recorded by a different instrument on OSO-5. These electron scattering calculations are more complex than those within the n=2 configurations, but the accuracy of available atomic data should be good. For several ions electron excitation rates have also been calculated for the n=2 to n=4 transitions.

3.3.3. Fe IX–Fe XIV

The ions Fe IX–Fe XIV give rise to a whole host of spectral lines in the UV wavelength range, of particular interest to the SOHO-CDS instrument. The available atomic data for these ions was reviewed by Mason (1994). These recommended data were used by BRS, MFL. It was pointed out by Mason that some of the available calculations had very limited accuracy. Since that review, new atomic data have been published for several of these ions. For example, the new calculations for Fe X are reviewed by Young et al. (1995) and a comparison is made between theoretical intensity ratios and solar (SERTS) and stellar (*EUVE*) spectra.

An international collaboration called the Iron Project (co-ordinated by D. Hummer) aims to calculate accurate electron excitation data for these and many other ions using the most accurate methods available. A series of papers reporting on this work is being published in Astronomy and Astrophysics. As part of this project, Storey et al. (1995) report on new calculations for the famous Fe XIV green coronal line (5303 Å). They use a very accurate target and include all the important resonance contributions to the collision strength.

3.4. Atomic Data Assessment Program

Burgess & Tully (1992) have developed a graphical procedure to fit the collision strengths and to obtain upsilon (Υ) (OMEUPS). The collision strengths (or Upsilon's) are plotted on a reduced energy (or temperature) scale such that 0 corresponds to threshold (or zero temperature)and 1 corresponds to infinite energy (or temperature). The correct functional behaviour of the collision strength (or upsilon) on energy (or temperature) which depends on the type of transition (i.e., dipole, exchange, forbidden) is taken into account. The high energy (or temperature) limit is obtained from the Bethe or Born approximations. A five point spline is fitted to the reduced data by a least squares procedure. The advantage and beauty of this method is its precision and simplicity. Any errors or problems with the atomic data are immediately obvious. Published data from different authors are all stored in the same format, with only five points required to reproduce the whole energy (or temperature) range. This method is being used by several groups to provide an atomic databank which will be made available to the solar and astrophysics community.

4. Future Directions

4.1. Synthetic Spectrum Project

A new synthetic spectral code (45–2000 Å) is being prepared as part of a collaboration between Dere, Monsignori Fossi, Mason and coworkers using the graphical assessment method of Burgess & Tully (1992). Dere has written a version of the Burgess & Tully assessment method which runs in IDL (Interactive Data Language) which is extensively used for solar analyses. The immediate aim is to prepare for the analyses of spectra from SOHO. It is anticipated that the essential data and basic programs should be ready by the end of 1995. Significant progress has been achieved through the willingness of atomic physicists to make their results available in digital form.

4.2. *CDS and SUMER Scientific Software*

Preparations for *SOHO* have been in progress for several years. The aim is to build a package of analysis software for spectroscopic diagnostics. An important component of this project is the Atomic Data and Analysis Structure (ADAS) which was developed by Summers for fusion research and is now being adapted for solar application. ADAS is a very powerful suite of subroutines which solves the collisional radiative model for equilibrium or non-equilibrium conditions. It contains a comprehensive atomic data analysis, storage and processing procedure. All the recommended atomic data from Lang (1994) and more recent calculations (eg Iron Project) will be incorporated together with atomic data assessed using OMEUPS. ADAS is capable of producing ionisation balance calculations together with low lying level populations. These are being incorporated into solar analysis programs for spectroscopic diagnostics (electron temperature and density determination, emission measure analyses, elemental abundance determinations).

5. Conclusion

Maybe at the end of this review one might expect a league table for the different synthetic spectral codes. This would result in the loss of some good relations and collaborators! The task of assembling the synthetic spectral codes is formidable. Each project has its own strengths and weaknesses. The main conclusion of this review is rather obvious, that the BEST plasma emission are the ones which are able to represent the atomic rates the most accurately. The question still arises with regard to missing atomic data. Fortunately, the atomic physics community is always receptive to suggestions for interesting and useful new computations.

The financial support of PPARC is acknowledged.

REFERENCES

ACTON, L. W., BRUNER, M. E., BROWN, W. A., FAWCETT, B. C., SCHWEIZER, W. & SPEER, R. J. 1985, ApJ, 291, 865

ARNAUD, M. & RAYMOND, J. C. 1992, ApJ, 398, 39

ARNAUD, M. & ROTHENFLUG, R. 1985, A&AS, 60, 425

BADNELL, N. R. & MOORES, D. L. 1994, Atomic Data Nucl. Data Tabl., 57, 329

BHATIA, A. K., MASON, H. E. & BLANCARD, C. 1995, pre-print

BRICKHOUSE, N. S., EDGAR, R., KAASTRA, J., KALLMAN, T., LIEDAHL, D., MASAI, K., MONSIGNORI FOSSI, B., PETRE, R., SANDERS, W., SAVIN, D. W. & STERN, R. 1995, preprint

BRICKHOUSE, N. S., RAYMOND, J. C. & SMITH, B. W. 1995, ApJS, in press

BURGESS, A. & TULLY, J. A. 1992, A&A, A&A, 254, 436

CORNILLE, M., DUBAU, J. A., MASON, H. E., BLANCARD, C. & BROWN, W. A. 1993, in UV and X-ray Spectroscopy of Laboratory and Astrophysical Plasmas, ed. E. Silver & S. Kahn, Cambridge: Cambridge Univ., 101

CORNILLE, M., DUBAU, J. A., MASON, H. E. & BLANCARD, C. 1995, pre-print

DOSCHEK, G. A. & COWAN, R. D. 1984, ApJS, Ser., 56, 67

HARRISON, R. A. & SAWYER, E. C. 1992, in Proc. of the First SOHO Workshop, ESA SP-348, 17

JORDAN, C. 1995, Astron. Soc. of the Pacific, in press

KAASTRA, J. J. & MEWE, R. 1993, Legacy, 3, 16

KATO, T. 1976, ApJS, Ser., 30, 397

LANG, J. 1994, Atomic Data Nucl. Data Tabl., 57

LANDINI, M & MONSIGNORI FOSSI, B. C. 1972, A&AS, Ser., 7, 291

LANDINI, M. & MONSIGNORI FOSSI, B. C. 1990, A&AS, Ser., 82, 229

LANDINI, M. & MONSIGNORI FOSSI, B. C. 1991, A&AS, Ser., 91, 183

LEMAIRE, P., WILHELM, K., AXFORD, W. I., CURDT, W., GABRIEL, A. H., GREWING, M., HUBER, M. C. E., JORDAN, S. D., KUEHNE, M., MARSCH, E., POLAND, A. I., THOMAS, R. J., TIMOTHY, G. J., & VIAL, J. -C. 1992, in Proc. of the First SOHO Workshop, ESA SP-348, p13

LEMEN, J. R., MEWE, R., SCHRIJVER, C. J. & FLUDRA, A. 1989, ApJ, 341, 474

MASAI, K. 1984, Astrophys. and Space Sci., 98, 367

MASAI, K. 1994a, JQSRT, 51, 211

MASAI, K. 1994b, ApJ, 437, 770

MASON, H. E., BHATIA, A. K., KASTNER, S. O., NEUPERT, W. M. & SWARTZ, M. Sol. Phys., 92, 199

MASON, H. E. & MONSIGNORI FOSSI, B. C. 1994, The A&A Rev., 6, 123

MASON, H. E. 1994, Atomic Data Nucl. Data Tabl., 57, 305

MEWE, R. 1972, Sol. Phys., 22, 459

MEWE, R. & GRONENSCHILD, E. H. B. M. 1981, A&AS, Ser., 45, 11

MEWE, R., GRONENSCHILD, E. H. B. M. & VAN, DEN, OORD, G. H. J. 1985, A&AS, Ser., 62, 197

MEWE, R. 1990, in Physical Processes in Hot Cosmic Plasmas, 39, ed. W. Brinkmann et al., Dordrecht: Kluwer Acad. Publ.

MONSIGNORI FOSSI, B. C. & LANDINI, M. 1994, Sol. Phys., 152, 81

MOORES, D. L. & REED, K. J. 1995, Advan. Atomic. Mol. Phys., in press

NEUPERT, W. M., EPSTEIN, G. L., THOMAS, R. J. & THOMPSON, W. T. 1992, Sol. Phys., 137, 87

RAYMOND, J. C. 1988, in Hot Thin Plasmas in Astrophysics, ed. R. Pallavicini, Dordrecht: Kluwer Acad. Publ., 3

RAYMOND, J. C. & SMITH, B. W. 1977, ApJS, 35, 419

SAMPSON, D. H. & ZHANG, H. L. 1992, Phys. Rev. A, 45, 1556

SANDERS, W. 1995, these proceedings

SCHMITT, J. H. M. M. 1992, in The Seventh Cambridge Workshop on Cool Stars, Stellar Systems, and the Sun, ASP Conf. Ser., 26, 83

SCHRIJVER, C. J., LEMEN, J. R. & MEWE, R. 1989, ApJ, 341, 484

STOREY, P. J., MASON, H. E. & SARAPH, H. E. 1995, preprint

THOMAS, R. J. & NEUPERT, W. M. 1994, ApJS, 91, 461

VAN, REGEMORTER, H. 1962, ApJ, 136, 906

YOUNG, P. R., MASON, H. E., BHATIA, A. K., DOSCHEK, G. A. & THOMAS, R. J. 1995, these proceedings

The Opacity Project and the Iron Project

ANIL K. PRADHAN

Department of Astronomy, The Ohio State University,
174 West 18th Avenue, Columbus, OH 43210, USA

Systematic and large-scale calculation of accurate plasma opacities and atomic data by the Opacity Project and the Iron Project has applications in many areas of astrophysics. Analysis of *EUVE* observations using monochromatic opacities of elements calculated by the Opacity Project is described. Theoretical methods and atomic calculations are discussed briefly. Recent work related to ionization balance, photoionization and recombination, and the modeling of plasmas in local thermodynamic equilibrium (LTE) and non-LTE are discussed. New calculations for the important iron ions, under the Iron Project, are also described and their relevance to the more extensive non-LTE calculations is pointed out. The comprehensive radiative and collisional data sets from the Opacity and the Iron Projects should be applicable to a large number of sources in the EUV and other wavelength regions.

1. Introduction

The Opacity Project (Seaton et al. 1994) is an international collaboration of atomic physicists and astrophysicists to calculate stellar opacities for a given chemical composition of astrophysically abundant elements. While the primary goal of the Opacity Project (hereafter OP) is to obtain accurate *mean* opacities for models of stellar interiors, it was found that the *monochromatic* opacities could be employed for the analysis and interpretation of emission spectra from stellar sources such as obtained by the *Extreme Ultraviolet Explorer* (*EUVE*) of objects like the hot photospheres of young white dwarfs at temperatures and densities normally prevalent in stellar envelopes under LTE conditions. While departures from LTE are to expected in certain cases, a detailed comparison of these models with observations should provide the basis for a more precise determination of non-LTE effects, which in turn require an even more extensive set of physical processes and parameters.

The bulk of the opacities calculations consists of atomic calculations using ab init. quantum mechanical methods that yield highly accurate radiative parameters, photoionization cross sections and radiative transition probabilities. Complementing the Opacity Project work on radiative processes, a new project called the Iron Project (Hummer et al. 1993; Bautista et al., in this volume) was initiated to address the collisional processes in astrophysical objects, particularly the spectra of Iron and other Iron peak elements in various ionization stages. These data will be applicable to the interpretation of emission line spectra and NLTE modeling of a variety of EUV sources such as the Interstellar medium, stellar coronae, flares, winds, active galactic nuclei, quasars, novae, and supernovae.

Below, we briefly describe the Opacity Project and the Iron Project (hereafter OP and IP respectively), followed by a discussion of some issues related to LTE and NLTE models and available atomic data sets for astrophysical plasma diagnostics.

2. The Opacity Project

It was noted by Simon (1982) that the then available opacities from the Los Alamos Opacities Library produced theoretical period ratios for Cepheid variables that disagreed widely with observations, and that an arbitrary increase in the metal opacities of factors

569

S. Bowyer and R. F. Malina (eds.), Astrophysics in the Extreme Ultraviolet, 569–576.
© *1996 Kluwer Academic Publishers. Printed in the Netherlands.*

of 2 or 3 appeared to ameliorate the problem. Simon issued what he termed "a plea for re-examination of the metal opacities." Two independent groups responded to the "plea": one at the Lawrence Livermore National Laboratory (Rogers & Iglesias 1992) and the other, an international group of about 30 atomic and astrophysicists called the International Opacity Project (OP). Both groups realized that the atomic physics of the Los Alamos opacities was inadequate and incomplete, and far more radiative data was needed to account for the missing opacity. The methods involved, both in terms of the atomic physics and the equations of state, employed by the two groups were quite different. None the less, at the end of the day (actually a decade: 1983–1994) both groups obtained essentially similar opacities over most of the temperature-density regimes of practical interest. The enhancement in metal opacities in both calculations ranged from factors of 2 to 5 over the Los Alamos opacities, particularly dominated by Iron and its many ions that contributed millions of lines and continua to the overall opacity of stellar interiors. A comparison of the two sets, as well as a complete description of the OP work, is given by Seaton et al. (1994).

The quantity of particular interest in stellar models is the Rosseland mean opacity defined by

$$\frac{1}{\kappa_R} = \frac{\int_0^\infty \frac{1}{\kappa_\nu} g(u) du}{\int_0^\infty g(u) du}, \tag{2.1}$$

where u = hν/kT and g(u) is the weighting function corrected for stimulated emission

$$g(u) = u^4 e^{-u} (1 - e^{-u})^{-2}. \tag{2.2}$$

The Rosseland mean is a difficult quantity to calculate as it involves the reciprocal of the monochromatic opacity which in general is a very complicated function of the photon frequency, involving a large number of bound-bound, bound-free, free-free, and scattering contributions from all constituent atomic species in the plasma. For many complex atomic systems the opacity contribution may be a very large over a wide wavelength range. As an example, Fig. 1 shows the monochromatic opacity spectrum due to one such ion, Fe II (Nahar & Pradhan 1994).

The OP opacities may be computed for an arbitrary mixture of elements from the monochromatic opacities of elements calculated at a wide range of temperature and density, and at a fine mesh of up to 10^5 photon frequencies (with the resolution shown in Fig. 1).

The monochromatic opacities for all elements with nuclear charge Z = 1–14, 16, 18, 20, and 26, in all ionization stages, are computed and pre-tabulated to enable the computation of mixture opacities for any chemical composition.

3. *EUVE* Observations and Modeling

The observations of hot, young white dwarfs, and their relation to stellar evolution, is one of the most important findings of the *EUVE* program (Vennes 1995, in this volume). The analysis of spectra of white dwarfs early in their evolution is essential to the determination of the precise evolutionary path, along the H-R diagram, from the planetary nebula phase down to the stable, final state. As the ejecta from the PN phase is driven by radiation pressure on the one hand, and the gravitational stratification of elements in the remnant core takes place on the other hand, some interesting "inversions" in the photospheres are observed. Iron, owing to the rich and complex atomic structure of its low ionisation stages and its relatively high abundance, is radiatively levitated by the opacity inherent in the millions of possible transitions, primarily in the UV and the EUV.

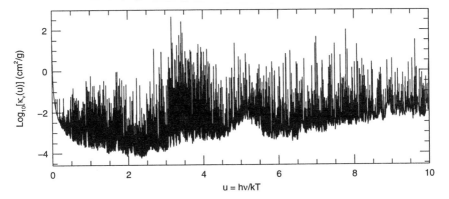

FIGURE 1. Monochromatic opacity of Fe II at Log $T = 4.2$, Log $N_e = 16$.

For example, the *EUVE* observations reveal the EUV spectra to be dominated by Iron in certain wavelength regions. A full-scale theoretical computation therefore depends on an accurate and complete determination of the opacities of constituent elements.

The emissivities of the elements in the emitting photospheres can be derived in LTE. The medium resolution spectra can be modeled using the emissivities of elements in the emitting photospheres calculated in LTE from the monochromatic opacities of elements, with abundances optimised to fit the spectrum—thus determining the composition of the white dwarf photosphere. The observations clearly show the prominent contribution of iron which is radiatively excited in the intense radiation field and emits copiously in the EUV, in particular through the ionization stages Fe IV and Fe V. Several other elements have also been identified. It is necessary to calculate the monochromatic opacities of all of these elements at a T-ρ mesh appropriate to the photospheres of hot white dwarfs: Log $T = 4.5 - 5.5$, Log $N_e = 12 - 20$.

The calculated opacities include all the main atomic processes: bound-bound, bound-free, free-free, electron scattering, and line broadening.

The presence of elements in the white dwarf photospheres may be deduced from spectral observations and certain spectral features. A number of elements have been observed in the *IUE* and the *EUVE* spectra of white dwarfs (Vennes, this volume). Among these are C, N, O, Si, S, and Fe. In Fig. 2 we present the theoretical emissivities, calculated using the monochromatic opacities, as each new element is added to the mixture. The contribution of the metal opacities is clearly seen. While the contribution of silicon and sulfur is small, and manifests only certain detailed spectral features, the dramatic effect of the iron opacity on the entire background in the lower wavelength region is evident. In order to emphasize the contribution of the individual opacities, the computed fluxes in Fig. 2 do not include the effect of the interstellar medium and therefore may not be directly compared to the observations (the spectra have been smoothed to a resolution of 1 Å to match the resolution of the *EUVE*).

A representative model of the theoretically computed flux is compared with the *EUVE* observations of a young white dwarf in Fig. 3. Work is in progress to improve the models by more accurate estimates of abundances, the inclusion of other observed elements such as neon and magnesium, and departures from NLTE (discussed below). In particular the discrepancies between observations and theory may be attributable to the fact that the current OP data for Fe I–V is not very accurate, and is being recalculated to higher ac-

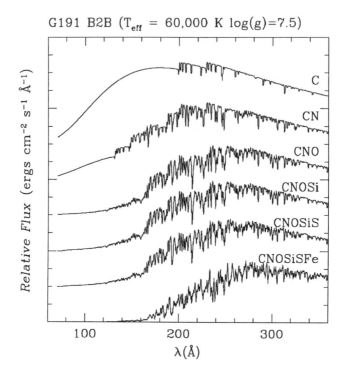

FIGURE 2. Incremental mixture emissivities of C,N,O,Si,S, and Fe. The bottom-most curve demonstrates the dominant effect of iron on the overall, as well as the detailed, opacity of the plasma. All abundance fractions relative to H are 3×10^{-6}.

curacy under the IP (Nahar & Pradhan 1994; Bautista & Pradhan 1995a). For example, work in progress for Fe V shows that the new data will yield significantly higher opacity in the wavelength range < 300 Å.

4. Non-LTE Models: Collisional and Recombination Data

While departures from LTE would not be unexpected, the precise nature of such effects is difficult to determine quantitatively since extensive NLTE calculations are necessary. Such calculations need to include all relevant radiative and collisional processes important in line formation, coupled with an accurate treatment of radiative transfer (e.g., Hubeny & Lanz 1995). A huge amount of accurate atomic data is needed. We describe below two projects that are extensions of the Opacity Project and are devoted to large-scale calculation of collisional and recombination parameters required in NLTE codes.

4.1. The Iron Project

The primary aim of the Iron Project is to compute collisional data for the iron-peak elements in various ionization stages (Hummer et al. 1993). The first two phases of the project deal with fine structure transitions along several isoelectronic sequences and for all iron ions. Some radiative data for low ionization stages of iron ions, Fe I–V, is also being computed to higher accuracy than in the earlier work from the Opacity Project

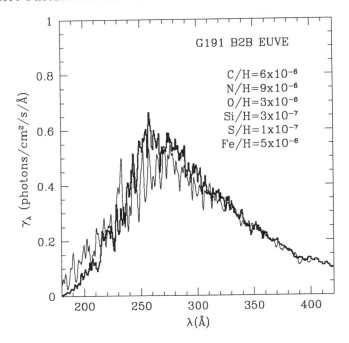

FIGURE 3. Comparison of the theoretical emissivity (thin line) and the observed *EUVE* spectra (thick line). The white dwarf model assumes the given chemical abundance ratios, at $T = 56,000$ K and $\log g = 7.5$.

(e.g., Nahar & Pradhan 1994). Information on the IP and the latest list of publications may be obtained on World Wide Web (URL: http://www.am.qub.ac.uk/projects/iron/).

The IP work has applications in both the LTE and the NLTE modeling of *EUVE* spectra. The new radiative calculations for Fe I and Fe II, and the opacities calculations therefrom, differ greatly from the earlier OP data. For example, Fig. 4 shows the recently calculated photoionization cross section for the ground state of Fe I, which is *up to 3 orders of magnitude* higher than the data currently available (Bautista & Pradhan 1995a).

The NLTE models require collisional cross sections and rates for a large number of transitions in many of the iron ions. Recently Zhang & Pradhan (1995a,b) have calculated electron impact excitation rate coefficients for 10,011 IR, O, and UV transitions in Fe II and 23,871 transitions in Fe III. Similar calculations are in progress for Fe IV, Fe V and Fe VI. New effective collision strengths for Ni II have also been computed recently (Bautista & Pradhan 1995b).

Further details on the IP work are given in the papers in this volume by Bautista et al. and Mason; the latter work deals with the important coronal ions of iron.

4.2. *Unified and Consistent Electron-Ion Recombination Rates*

Electron-ion recombination calculations, consistent with the Opacity Project photoionization data, are being carried out to determine, for the first time, accurate and self-consistent ionization balance in astrophysical sources, as well as for NLTE populations dependent on partial photoionization and recombination cross sections for individual levels.

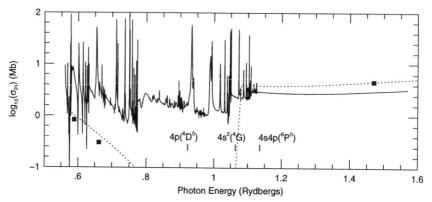

FIGURE 4. Photoionization cross section of the ground state $3d^6 4s^2 (^5D)$ of Fe I; full curve = Bautista & Pradhan (1995a); filled squares = Reilman & Manson (1979); broken curve = Verner et al. (1993). The orders of magnitude enhancement over the previous calculations is primarily due to the coupling among various states dominated by the open $3d$ and the $4s$ subshells, which is also manifest in the autoionization resonances.

Accurate Recombination rates are crucial to the calculation of ionization balance in plasmas; however an equally important point is to obtain both the recombination and the ionization cross sections rates *within the same formulation* in order to be self-consistent. A new method has recently been developed by Nahar & Pradhan (1995, and references therein) to calculate unified electron-ion recombination rates that include both the radiative and the di-electronic recombination (RR and DR) processes in an ab init. manner, giving a single, total recombination rate coefficient $\alpha_R(T)$ for each atom or ion as a function of temperature. These recombination rates are calculated using the same atomic physics (i.e., eigenfunction expansions for the electron+ion system) as in the OP work, and therefore are fully consistent with the new OP photoionization cross sections (which may be obtained from TOPBASE, Cunto et al. 1993). It might be emphasized that the currently available RR and DR rates are calculated in different approximations and are, in general, inconsistent with the photoionization cross sections, possibly leading to considerable loss of accuracy in ionization balance calculations.

Fig. 5 illustrates the unified recombination rate coefficients obtained with the new method for an isoelectronic sequence of ions (carbon-like). Along the sequence the $\alpha_R(T)$ increase with nuclear charge Z in the low temperature region, where the background recombination (i.e., RR type) dominates, but the relative importance of DR diminishes with Z converging to a wide bump in the figure in the high-T region (Log $T > 5.0$). In the low-T region the resonant recombination (DR type) may also contribute for some ions, as indicated by a smaller bump.

4.2.1. *Photoionization and Recombination into Individual States*

In addition to the total photoionization and recombination rates, it is also necessary to compute partial photoionization and recombination rates for the individual levels of an ion populated in NLTE. Whereas some such calculations are in progress, these are extremely difficult for the complex iron ions where the levels of each ion are coupled to the levels of adjacent ions through photoionization and recombination. The calculations under way should yield the partial photoionization and recombination rates for the *metastable* levels of the ions concerned; the partial rates being dependent upon the

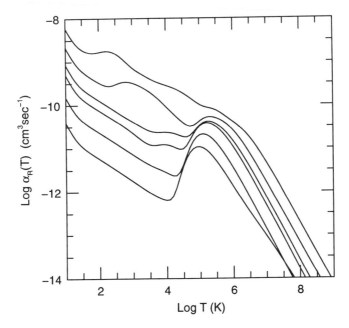

FIGURE 5. Total electron-ion recombination rate coefficients for ions in the carbon sequence; the curves represent recombination rates for C I, N II, O III, F IV, Ne V, and Si IX from bottom to top (Nahar & Pradhan 1995). Applicable to ionization balance calculations, these unified rates include both the radiative and the di-electronic recombination processes and are consistent with the OP photoionization data.

individual level populations which, in turn, may be determined from the solution of a collisional-radiative model. For Fe I–V, for example, a number of such metastable levels need to be considered, dominated by the lowest even parity configurations of the type $3d^n 4s^m$ which dominate the wavefunctions of a relatively large number of terms. Given the large differences in the photoionization cross sections from available data for these ions (e.g. Fig. 4), it is clear that current photoionization models need considerable revision in order to incorporate the new photoionization and recombination data.

5. Conclusion

The Opacity Project and the Iron Project have enabled a far more extensive and detailed treatment of radiative and collisional processes in astrophysical models than heretofore possible. The huge amount of atomic data computed by the Opacity Project and the Iron Project should meet most of the needs of the next generation of LTE and NLTE models. Information on databases and ongoing work on the OP and the IP may be obtained from the author (pradhan@payne.mps.ohio-state.edu).

I would like to thank Stephane Vennes, Sultana Nahar, and Manuel Bautista for some of the figures. This work is partially supported by the US National Science Foundation (PHY-9421898), the NASA LTSA program (NAGW-3315), and the NASA ADP program (NAS 5-32643). The computational work is carried out on the CRAY Y-MP8/64 and the MPP CRAY T3D at the Ohio Supercomputer Center in Columbus, Ohio.

REFERENCES

Bautista, M. A. & Pradhan, A. K. 1995a, Photoionization of neutral Iron, J. Phys. B Lett., in press

Bautista, M. A. & Pradhan, A. K. 1995b, Atomic Data from the Iron Project XIV. Electron excitation rates and emissivity ratios for forbidden transition in Ni II and Fe II, A&A, submitted

Cunto, W., Mendoza, C., Ochsenbein, F., & Zeippen, C. J. 1993, TOPbase at the CDS, A&A, 275, L5

Hummer. D. G., Berrington, K. A., Eissner, W., Pradhan, A. K., Saraph, H. E., & Tully, J. A. 1993, Atomic data from the IRON Project, A&A, 279, 298

Nahar, S. N. & Pradhan, A. K. 1994, Atomic data for opacity calculations: XX. Photoionization cross sections and oscillator strengths for Fe II, J. Phys. B: At. Opt. Phys., 27, 429

Nahar, S. N. & Pradhan, A. K. 1995, Unified electron-ion recombination rate coefficients for Sulfur and Silicon Ions, ApJ, in press

Reilman, R. F. & Manson, S. T. 1979, Photoabsorption cross section for positive atomic ions with $Z \leq 30$, AJSS, 40, 815

Rogers, F. J. & Iglesias, C. A. 1993, Radiative atomic Rosseland mean opacity tables, ApJS, 79, 507

Seaton, M. J., Yu, Y., Mihalas, D., & A. K., Pradhan 1994, Opacities for stellar envelopes, MNRAS, 266, 805

Simon, N. R. 1982, A plea for re-examination of metal opacities, ApJ, 260, L87

Verner, D. A., Yakovlev, D. G., Band, I. M., & Trzhaskovskaya, M. B. 1993, Subshell photoionization cross sections and ionization energies of atoms and ions from He to Zn, At. Data Nucl. Data Tables, 55, 233

Zhang, H. L. & Pradhan, A. K. 1995a, Atomic Data from the Iron Project VI. Collision strengths and rate coefficients for Fe II, A&A, 293, 953

Zhang, H. L. & Pradhan, A. K. 1995b, Relativistic and electron correlation effects in electron impact excitation of Fe^{2+}, J. Phys. B: At. Opt. Phys., submitted

The Iron Project:
Atomic Data for Fe I–Fe VI

MANUEL A. BAUTISTA, SULTANA N. NAHAR,
JIANFANG PENG, ANIL K. PRADHAN, AND HONG LIN ZHANG

Department of Astronomy, Ohio State University,
174 West 18th Avenue, Columbus, OH 43210, USA

Recent progress in large scale computations of photoionization cross sections, radiative transitions probabilities, and collision strengths for the lowest ionization stages of Iron, i.e., Fe I–VI, are reported. These results are part of an international collaboration called the IRON Project. The present results exhibit large differences with respect to data currently in use and is expected to be of importance in the study of astrophysical objects in the UV and the EUV.

1. Introduction

One of the major limitations in the analysis of observations of astronomical objects is the availability of reliable atomic data. Accurate photoionization cross section are necessary in the calculation of the photoionization and emissivity models, the opacities, and recombination rates for the ionization balance of the plasmas considered. Electron excitation cross sections and transition probabilities are required for the interpretation of emission spectra.

The IRON Project (IP) is an international collaboration devoted to large scale computations of electron-impact excitation cross sections, as well as photoionization cross sections and radiative transition probabilities for Iron peak elements and ions (Hummer et al. 1993). These data will complement the extensive radiative computations by the Opacity Project (OP; Seaton et al. 1992) which is currently available with the data server TOPbase (Cunto et al. 1993).

Despite the considerable interest in the low ionization stages of Iron in laboratory and astrophysical plasmas, accurate atomic data for these species was not available. This is because the calculations for these ions are particularly difficult, mainly owing to the complex electron-electron correlation effects involved. Another difficulty is the large number of atomic states, the coupling effects, and the complex resonance structures that need to be considered for an accurate treatment. For example, accurate representation of autoionizing resonances and bound-bound transition in Fe IV and Fe V are necessary in the computation of monochromatic opacities in the EUV spectra of hot, young white dwarfs (Vennes 1995; Pradhan 1995 in this issue).

In the present paper, we report on the latest advances in the IP for the lowest ionization stages of Iron, Fe I–VI, whose lines are very prominent in the UV and EUV spectra of many objects.

2. Radiative Data for Fe I–Fe V

Extensive calculations of photoionization cross sections and transition probabilities for Fe I–VI are being carried out by the Ohio State group. For these calculations, multi-configuration coupled-channel eigenfunction expansions for the electron+ion system are considered using the R-matrix method as developed for the OP and the IP.

Previous calculations for these ions were carried out in the central field type approximation by Reilman & Manson (1979) and Verner et al. (1993); however, these computations

S. Bowyer and R. F. Malina (eds.), Astrophysics in the Extreme Ultraviolet, 577–581.
© *1996 Kluwer Academic Publishers. Printed in the Netherlands.*

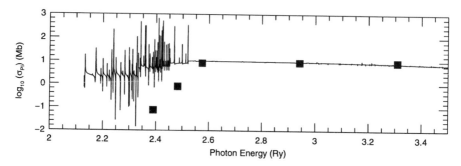

FIGURE 1. Photoionization cross section of the ground state $3d^6$ (^5D) of Fe III: solid curve = present; filled dots = Reilman & Manson (1979).

do not take account for the coupling effects and resonances that are of considerable importance for atomic systems such as the low ionization stages of Iron (e.g. Nahar & Pradhan 1994). Earlier R-matrix calculations for Fe I–VI were also done by Sawey & Berrington (1992) using much smaller close-coupling expansions than what is considered in the present work; this results in important configuration interaction missing from their calculations (for example, their calculations do not obtain the ground state of Fe I, and in general their photoionization cross sections are up to an order of magnitude, or more, lower).

We have recently calculated the photoionization cross sections for the ground state of neutral Iron (Bautista & Pradhan 1995a). A comparison of our results with those reported by Reilman & Manson (1979) and Verner et al. (1993) is shown by Pradhan (1995 this issue). It is found that the previous computations for Fe I do not obtain the dominant contributions to the cross section, owing to the absence of coupling effects, which results in a difference with respect to our results of over *three orders of magnitude* for photon energies of up to about 1 Rydberg. Photoionization cross sections of Fe II have been reported by Nahar & Pradhan (1994), where differences of up to an order of magnitude are found in the ground state cross section with respect to the results by Reilman & Manson (1979) and Sawey & Berrington (1992).

Radiative transition probabilities (f-values) have been calculated for Fe II for 35,941 LS multiplet transitions. For the transitions for which experimental values are available, the calculated f-values agree generally to within 10% (Nahar and Pradhan 1994). Recently, radiative transition probabilities for 21,589 dipole fine structure transitions, and over 19,000 transitions between LS multiplets, in Fe II have been reported by Nahar (1995) using experimentally observed energies and theoretical line strengths. For all transitions compared, the present lifetimes and f-values compare better with the experimentally observed values than the Kurucz (1981, 1991) data for Fe II. Transition probabilities are being calculated for Fe I, Fe III, Fe IV and Fe V.

Photoionization cross sections of Fe III–V are currently in progress. However, ground state cross sections of Fe III and Fe V have been obtained and they exhibit considerable differences with respect to earlier computations. In Fig. 1 we show the ground state photoionization cross section of Fe III, $3d^6$ ^5D, together with the results by Reilman & Manson (1979). It is noticed that their results neglect the important contribution of the resonances near the threshold and underestimate the cross section at low photon energies by several factors; however, the agreement between the background cross sections in the higher energy regions is very good.

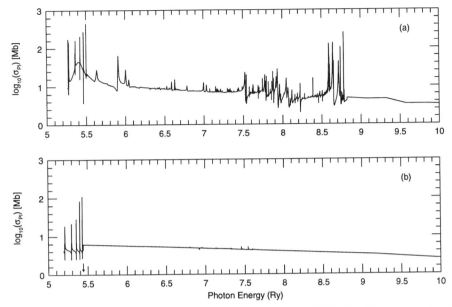

FIGURE 2. Photoionization cross section of the ground state $3d^4$ (^5D) of Fe V: *(a)* present; *(b)* Sawey & Berrington (1992).

The ground state photoionization cross section of Fe V is presented in Fig. 2(a), while the earlier results by Sawey & Berrington (1992) are shown in Fig. 2(b). The large number of additional terms in our close-coupling expansion gives rise to more extensive resonance structures and enhance the background cross section by over a factor of two.

The IP work at Ohio State is now being extended to also calculate not only the total but also the *partial* photoionization cross sections into excited states of the residual ion. These data are required for non-LTE spectral models and the calculation of total and partial electron+ion recombination rate coefficients (e.g. Nahar and Pradhan 1995).

3. Collision Strengths for Fe II–Fe VI

The main emphasis of the IP is on the computation of collision strengths for astronomically important atoms and ions. Among these, the low ionization stages of Iron are of particular importance since their rich atomic structures give rise to a large number of transitions that are seen in many objects, and in almost all the regions of the spectrum. The dipole allowed transitions in these systems in particular correspond to a multitude of observed UV and EUV lines, $\lambda\lambda 3000 - 200$ Å. Accurate collisional data for these ions must include fine structure, and therefore needs to take into account relativistic effects, as well as the extensive electron-electron correlation effects.

Considerable effort has been dedicated to the calculation of collisional data for Fe II, and a data set of Collision strengths and rate coefficients for 142 levels and 10,011 associated fine structure transitions, including many transitions observed in the UV, is currently available (Pradhan & Zhang 1993; Zhang & Pradhan 1995a). These data are expected to be more accurate than earlier results by Nussbaumer & Storey (1980) and

Keenan et al. (1988). Other calculations for some additional transitions (and which provide a check on the accuracy of the present data set) are being carried out (Bautista & Pradhan 1995b).

Collisional data for Fe III have recently been obtained for 219 even parity levels and 23,871 associated fine structure transitions. These data exhibit differences of up to a factor two with respect to earlier calculations by Berrington et al. (1991) for some of the low lying forbidden transitions. A detailed study of relativistic and electron correlation effects has been carried out in these calculations which suggests that the present data should be accurate to about $10 - 30\%$ (Zhang & Pradhan 1995b). A more extensive calculation for this ion which include odd parity terms and most of the transitions observed in the UV and EUV is being carried out.

4. Discussion

The present status of the second phase of the IP regarding collisional and radiative data for the low ionization stages of Iron is reported. Photoionization cross sections and transition probabilities for Fe II are already available and the corresponding data for Fe I and Fe III–V is expected to become available in the near future. Collision strengths for Fe II and Fe III have been calculated and may be used for the analysis of observed spectra. Collisional data for Fe IV–VI is currently being carried out.

The third phase of the IP that deals with Iron peak elements and ions other than Iron is currently in progress and, for instance, the first results for Ni II have recently been obtained (Bautista & Pradhan 1995b). Extensive ab initio calculations for the Iron peak elements and ions is difficult and requires hundreds to thousands of CPU hrs on modern supercomputers. For this reason we are investigating the capabilities of new technologies such as massive parallel processors (MPP). The first of these calculations on the MPP CRAY T3D are reported by Bautista & Pradhan (1995b).

This work is partially supported by the U.S. National Science Foundation (PHY-9115057), NASA LTSA program (NAGW-3315), and NASA ADP program (NAS 5-32643). The computational work is carried out on the CRAY Y-MP8/64 and the MPP CRAY T3D at the Ohio Supercomputer Center in Columbus, Ohio.

REFERENCES

Bautista, M. A. & Pradhan, A. K. 1995a, Photoionization of neutral Iron, J. Phys. B Lett., in press

Bautista, M. A. & Pradhan, A. K. 1995b, Atomic Data from the Iron Project XIV, Electron excitation rates and emissivity ratios for forbidden transition in Ni II and Fe II, A&A, submitted

Berrington, K. A., Burke, P. G., Hibbert, A., Mohan, M., & Baluja, K. L. Electron impact excitation of Fe$^+$ using the R-matrix method incorporating fine-structure effects, J. Phys. B: At. Opt. Phys., 21, 339

Berrington, K. A., Zeippen, C. J., Le Dourneuf, M., Eissner, W., & Burke, P. G. Electron impact excitation of Fe^{2+}: I. A 17-level fine-structure calculation, J. Phys. B: At. Opt. Phys., 24, 3467

Bell, R. A., Paltoglou, G., & Tripicco, M. J. 1994, The calibration of synthetic colours, MNRAS, 268, 771

Cunto, W., Mendoza, C., Ochsenbein, F., & Zeippen, C. J. TOPbase at the CDS, A&A, 275, L5

HUMMER. D. G., BERRINGTON, K. A., EISSNER, W., PRADHAN, A. K., SARAPH, H. E., & TULLY, J. A. 1993, Atomic data from the IRON Project, A&A, 279, 298

KEENAN, F. P., HIBBERT, A., BURKE, P. G., & BERRINGTON, K. A. 1988, Fine-structure populations for the ^6D ground state of Fe II, ApJ, 332, 539

KURUCZ, R. L. 1981, Semiempirical calculation of gf values: Fe II, Smithsonian Astrophysical Observatory Special Report 390, Cambridge

KURUCZ, R. L. 1991, private communication

NAHAR, S. N. 1995, Atomic Data from the Iron Project VII. Radiative dipole transitions probabilities for Fe II, A&A, 293, 967

NAHAR, S. N. & PRADHAN, A. K. 1994, Atomic data for opacity calculations: XX. Photoionization cross sections and oscillator strengths for Fe II, J. Phys. B: At. Opt. Phys., 27, 429

NAHAR, S. N. & PRADHAN, A. K. 1995, Unified electron-ion recombination rate coefficients of Silicon and Sulfur ions, ApJ, in press

PRADHAN, A. K. 1995, these proceedings

REILMAN, R. F. & MANSON, S. T. 1979, Photoabsorption cross section for positive atomic ions with Z≤ 30, AJSS, 40, 815

SAWEY, P. M. J. & BERRINGTON, K. A. 1992, Atomic data for opacity calculations: XV. Fe I-IV, J. Phys. B: At. Opt. Phys., 25, 1451

SEATON, M. J., ZEIPPEN, C. J., TULLY, A. K., PRADHAN, A. K., MENDOZA, C., HIBBERT, A., & BERRINGTON, K. A. 1992, The Opacity project—Computation of atomic data, Rev. Mexicana Astron. Astrof., 25, 19

VENNES, S. 1995, these proceedings

VERNER, D. A., YAKOVLEV, D. G., BAND, I. M., & TRZHASKOVSKAYA, M. B. 1993, At. Data Nucl. Data Tables, 55, 233

ZHANG, H. L. & PRADHAN, A. K. 1995a, Atomic Data from the Iron Project VI. Collision strengths and rate coefficients for Fe II, A&A, 293, 953

ZHANG, H. L. & PRADHAN, A. K. 1995b, Relativistic and electron correlation effects in electron impact excitation of Fe^{2+}, J. Phys. B: At. Opt. Phys., submitted

EUV Line Intensities of Fe X

P. R. YOUNG,[1] H. E. MASON,[1] A. K. BHATIA,[2]
G. A. DOSCHEK,[3] AND R. J. THOMAS[2]

[1] Department of Applied Mathematics and Theoretical Physics,
Silver Street, Cambridge CB3 9EW, UK

[2] Laboratory for Astronomy and Solar Physics, Code 680,
NASA-Goddard Space Flight Center, Greenbelt MD 20771, USA

[3] E.O. Hulburt Center for Space Research, Naval Research Laboratory,
Washington DC 20375-5000, USA

The 4 configuration, distorted wave calculation of Bhatia & Doschek (1995) (hereafter referred to as BD95), together with the ground transition calculation of Pelan & Berrington (1995) are here used to predict the intensities of the Fe X EUV lines, which are then used to derive electron densities from several solar spectra, including the recent SERTS spectra.

1. Introduction

The identification of the red coronal line as being the ground transition of Fe X helped lead to the idea of the corona as a million degree plasma. The importance of Fe X today lies in it being formed over the temperature region 8–12×10^5K, bridging the different morphologies of the transition region and coronal plasmas.

The main EUV lines of Fe X are summarised in table 1: the $3s^2 3p^5\ ^2P$–$3s^2 3p^4 3d\ ^2P$, ^2D, ^2S transitions contribute over 80% of the power output of Fe X, giving rise to a complex of strong lines in the 170–190Å region; the $3s^2 3p^4 3d\ ^4D_{5/2,7/2}$ levels are barely split and decay down to the $3s^2 3p^5\ ^2P_{3/2}$ level via electric dipole and magnetic *quadrupole* transitions, respectively, giving rise to a self-blended line observed at 257.26Å.

2. Atomic Data

2.1. *Radiative Data and Energy Levels*

A 13 configuration model of Fe X was used in the radiative code SUPERSTRUCTURE (Eissner et al. 1974), to produce transition probabilities and energy levels. The transition probabilities were found to be within 5% of those of Fawcett (1991). Observed energy levels from Corliss & Sugar (1982) and Jupén et al. (1994), were used together with scaled SUPERSTRUCTURE energies for the remaining levels.

2.2. *Electron Excitation Data*

Mason (1994) assessed the (then) available electron collisional excitation data for Fe X. Subsequently, a Distorted Wave calculation has been done by BD95 for 5 values of the incoming electron energy for a four configuration model of Fe X. This calculation is compared with two previous calculations in Figure 1 (left frame), where the *reduced*† collision strength is plotted against the reduced energy for the strong transition that gives rise to the 174.53Å line.

The high energy limit point at $E_r = 1$ can be accurately determined for allowed transitions using oscillator strengths from, e.g., SUPERSTRUCTURE via the expression $4\omega f/\Delta E$. This allows a crude comparison of the calculations—the BD95 results are

---

† see Burgess & Tully (1992) for details

S. Bowyer and R. F. Malina (eds.), Astrophysics in the Extreme Ultraviolet, 583–587.
© *1996 Kluwer Academic Publishers. Printed in the Netherlands.*

TABLE 1. Important Fe X lines and their associated transitions. Note that, for Fe X, $^2P_{3/2}$ is the ground level and $^2P_{1/2}$ is the first excited level.

Upper Level		Lower Level		Wavelength
Configuration	Term	Configuration	Term	(Å)
$3s3p^6$	$^2S_{1/2}$	$3s^23p^5$	$^2P_{3/2}$	345.72
$3s^23p^43d$	$^4D_{5/2,7/2}$		$^2P_{3/2}$	257.26
	$^2D_{5/2}$		$^2P_{3/2}$	174.53
	$^2P_{3/2}$		$^2P_{3/2}$	177.24
	$^2S_{1/2}$		$^2P_{3/2}$	184.54
	$^2D_{3/2}$		$^2P_{1/2}$	175.27

$\sim 20\%$ more accurate than the Mann (1983) results due to the inclusion of the $3s3p^53d$ configuration, but a further $\sim 20\%$ improvement is still possible. This can largely be achieved by including the $3s^23p^33d^2$ configuration (Nussbaumer, 1976).

2.3. Resonance Structure in the Ground Transition

BD95 did not include any resonance structure in their model, which is particularly important for the ground $^2P_{3/2}-^2P_{1/2}$ transition. This calculation has been done separately by Pelan & Berrington (1995), and the two are compared in figure 1 (right frame), together with the results of Mohan et al. (1994). Although this latter R-matrix calculation included some resonance structure, the fine-structure cross-section below the first excited state was not calculated (Pelan, private communication).

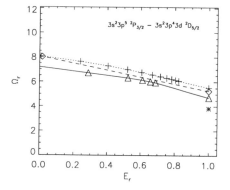 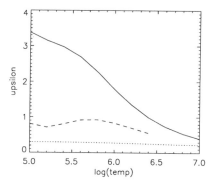

FIGURE 1. The left frame shows the reduced collision strength for the Fe X transition that gives rise to the 174.53 Å line. The three sets of points represent the results of BD95 ($\triangle$), Mason 1975 ($\diamond$) and Mann 1983 ($+$). $*$ represents the accurate high energy limit point, and $E_r = 0$ corresponds to the threshold energy of the transition. The right frame compares the thermally-averaged collision strengths for the ground $^2P_{3/2}-^2P_{1/2}$ transition obtained by: Pelan & Berrington (——); Mohan et al. (- - -); Bhatia & Doschek (.....).

3. Density Diagnostics

In the 170–190Å region, three density diagnostics are formed by taking the ratio of the 174.53, 177.24 and 184.54Å lines relative to the 175.27Å line. The new results for one

TABLE 2. Intensity ratios for the 175.27Å line taken relative to the 174.53, 177.24, 184.54 lines; the figure in brackets is the derived density. The observations used are: [1]Malinovsky & Heroux (1972); [2]Behring et al. (1976); [3]Dere et al. (1979); [4]Drake et al. (1995). (This latter spectrum is of the star Procyon, observed by *EUVE*—the 184.54Å line was observed in a different bandpass and so the derived density is less certain.)

Line	Full Disk Sun[1]	Full Disk Sun[2]	Solar Flare[3]	Procyon[4]
174.53	0.085 (8.7)	0.18 (9.3)	0.5 (10.7)	0.16 (9.2)
177.24	0.16 (8.8)	0.26 (9.2)	0.4 (9.5)	0.30 (9.3)
184.54	0.41 (8.8)	0.6 (9.1)	n/a	0.92 (9.5 ?)

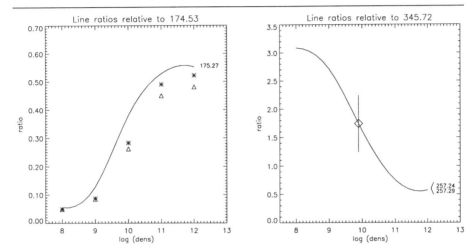

FIGURE 2. The left frame shows the new results presented here for the 175.27/174.53 ratio (solid line). The △ are the same results but with the BD95 collision strength for the ground transition, while the ∗ are the results of Brickhouse et al. (1994), who used the collision strengths of Mann for the transition shown. The right frame shows the 257.26/345.72 ratio with SERTS intensities and error bars marked on (note: theory splits the 257.26 blend into lines at 257.24,.29Å).

of these ratios are presented in figure 2 (left frame), and used to derive the densities in table 2 for various spectra.

The 257.26/345.72 ratio is the only other useful density diagnostic in the EUV for Fe X and was observed by the Solar EUV Rocket Telescope and Spectrograph (SERTS) flight of 1989.

4. Results from SERTS

4.1. *Densities Derived from Fe X*

Figure 2 (right frame) shows the density obtained from the SERTS spectrum of Thomas & Neupert (1994) using the Fe X 257.26 and 345.72 Å lines. However this spectrum has been spatially-averaged over a region containing considerable inhomogeneities.

The variation of the SERTS emission over the slit of the spectrograph for the Fe X 257.26Å, 345.72Å, and Fe XVI 335.40Å lines is shown in figure 3. It can be seen that the subflare (seen clearly near the centre of the slit in the Fe XVI line), gives different

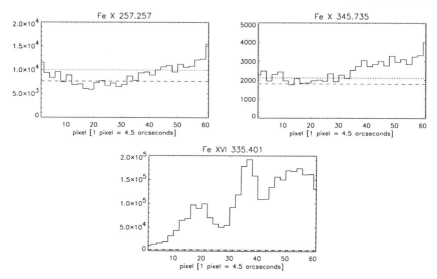

FIGURE 3. Spatial variation across the SERTS slit of Fe X 257.26Å and 345.72Å (seen at 345.735Å by SERTS), and Fe XVI 335.40Å. The dashed line represents the average background, while the dotted line represents the 1 σ variation of the background.

TABLE 3. Densities derived from the SERTS spectrum for various ions formed around 10^6 K. The collisional data for Mg VIII and Si X are from Zhang et al. (1994) and this, together with the radiative data were supplied by K. P. Dere.

Ion	T_{max}	Ratio	Density	Data Source
Mg VIII	5.9	436.73/430.45	5.0×10^8?	present
Si VIII	5.9	276.85/319.84	2.5×10^{10}	Dwivedi (1994)
Fe X	6.0	257.26/345.72	8.9×10^9	present
Si IX	6.0	349.87/345.13	$\gtrsim 10^{10}$	Mason & Bhatia (1978)
Si X	6.1	356.03/347.41	1.3×10^9?	present

contributions to the two Fe X lines. One can estimate the subflare 257.26/345.72 ratio as ≈ 1.2 and the active region loops (to the right of the subflare) ratio as ≈ 2.5, leading to densities $\approx 2.5 \times 10^{10}$ and $\approx 2.0 \times 10^9$, respectively.

4.2. Comparison with Other Ions

An important study for both solar and atomic physics is to check if ions with similar temperatures of peak abundance (T_{max}) give consistent results for the electron density. For Fe X, with T_{max} around 1×10^6 K, we have Si VIII–X and Mg VIII, all with useful density diagnostics lying within the SERTS wavelength range.

The densities predicted by these diagnostics are shown in table 3 where one can see a fair degree of scatter. However the Mg VIII ratio is only sensitive to densities $\lesssim 10^9$, and the error bars are large enough to make this density estimate highly uncertain. The Si X ratio involves the 347.41Å line which was marred by a plate flaw (Thomas & Neupert, 1994), again making the density estimate uncertain.

TABLE 4. Wavelength bands in Å covered by CDS

NIS		308–381		513–633	
GIS	151–221	256–338	393–493	656–785	

5. Future Observations

The SERTS rocket will be flown again in the spring of 1995, with a new multilayer-coated grating tuned to enhance the spectral region between 170–220Å. This should allow the full set of Fe X diagnostics to be observed in that wavelength range for the first time ever from individual solar features.

In addition, the *Solar and Heliospheric Observatory* (*SOHO*), to be launched in October 1995, will contain several instruments including the Coronal Diagnostic Spectrometer (CDS) which will cover the EUV region of the solar spectrum. CDS's two spectrometers (the grazing incidence, GIS, and the normal incidence, NIS) will cover most of the 150–800Å region in the bands shown in table 4.

As can be seen, the 170–190Å lines will all be observed, together with the 257.26 blend and the 345.72 line. However, these latter two lines are covered separately by the GIS and NIS, respectively, and so the ratio will be subject to calibration uncertainties.

The 345.72 and 174.53 lines may be used to help cross-calibrate the GIS and NIS since the 345.72/174.53 ratio is relatively density and temperature insensitive over $logT = 6.0 \pm 0.15$ and $logN_e \sim 8 - 10$, with values 0.02–0.05.

PRY and HEM acknowledge the support of EPSRC and PPARC, respectively.

REFERENCES

BEHRING, W. E., COHEN, L., FELDMAN, U., & DOSCHEK, G. A. 1976, ApJ, 203, 521

BHATIA, A. K. & DOSCHEK, G. A. 1995, ADNDT, in press

BRICKHOUSE, N. S., RAYMOND, J. C., & SMITH, B. W. 1995, ApJS, in press

BURGESS, A. & TULLY, J. A. 1992, A&A, 254, 436

CORLISS, C. & SUGAR, J. 1982, J. Phys. Chem. Ref. Data, 11, 1

DERE, K. P., MASON, H. E., WIDING, K. G., & BHATIA, A. K. 1979, ApJS, 40, 341

DRAKE, J. J., LAMING, J. M., & WIDING, K. G. 1995, ApJ, 443, 393

DWIVEDI, B. N. 1994, Sol. Phys., 153, 199

EISSNER, W., JONES, M., & NUSSBAUMER, H. 1974, Comp. Phys. Comm., 8, 270

FAWCETT, B. C. 1991, ADNDT, 47, 319

JUPÉN, C., ISLER, R. C., & TRABERT, E. 1994, MNRAS, 264, 627

MALINOVSKY, M. & HEROUX, L. 1973, ApJ, 181, 1009

MANN, J. B. 1983, ADNDT, 29, 407

MASON, H. E. 1975, MNRAS, 170, 651

MASON, H. E. 1994, ADNDT, 57, 305

MASON, H. E. & BHATIA, A. K. 1978, MNRAS, 184, 423

MOHAN, M., HIBBERT, A., & KINGSTON, A. E. 1994, ApJ, 434, 389

NUSSBAUMER, H. 1976, A&A, 48, 93

PELAN, J., & BERRINGTON, K. A. 1995, A&AS, in press

THOMAS, R. J. & NEUPERT, W. M. 1994, ApJS, 91, 461

ZHANG, H. L., GRAZIANI, M., & PRADHAN, A. K. 1994, A&A, 283, 319

Diagnostics of EUV Spectral Emission from Boron-like Ions

JIANFANG PENG AND ANIL K. PRADHAN

Department of Astronomy, The Ohio State University, Columbus, OH 43210, USA

Recent developments of EUV diagnostics from transitions in Boron-like ions, C II, N III, O IV, Ne VI, Mg VIII, Al IX, Si X, and S XII, are discussed based on new atomic data (Zhang, Graziani & Pradhan (1994)) and systematic theoretical calculations of line intensity ratios (Peng & Pradhan 1995). The available EUV and UV lines which are good temperature and density diagnostics are summarized. Possible applications of these lines to the investigation of the physical conditions in astrophysical objects are proposed.

1. Introduction

The atomic structure of Boron-like ions affords useful diagnostics of sources in the EUV and UV spectral emissions. Previous studies have focused primarily on low-Z B-like ions C II, N III, O IV, and Ne IV, and no systematic study of the isoelectronic sequence has been done. Previous works include the C II and O IV calculations by Hayes & Nussbaumer (1984, 1983), the C II work by Lennon et al. (1985), the N III calculations by Stafford et al. (1993), and the earlier work by Luo & Pradhan (1990) and Blum & Pradhan (1992) on C II, N III and O IV. By and large, there is good agreement between all of the above calculations that have all been carried out in the close coupling approximation using the R-matrix method. Recently Zhang et al. (1994) have provided extensive results for the collisional rate coefficients for B-like ions, that include the higher members of the sequence, calculated with high precision under the auspices of the Iron Project (Hummer et al. 1993). Some of the new data has been employed in diagnostic studies of the UV lines of O IV observed in gaseous nebulae and solar spectra (O'Shea et al. 1995; Keenan et al. 1993a) and Ne VI line ratios observed from the Sun (Keenan et al. 1993b; Cook et al. 1994; Keenan et al. this meeting).

Recently a comprehensive and systematic study of the line ratios of B-like ions is reported by Peng & Pradhan (1995), using the new Zhang et al. data. Some line ratios for the EUV and UV lines are identified from the point of view of providing clean diagnostics, with the criterion that a temperature sensitive line ratio should be relatively independent of the density effects and vice-versa. While all possible line ratios in the 15-level model were considered, selected results are reported for a few of the ones that showed the best sensitivity in either the temperature or the density. In this report we focus on the EUV wavelength range with lines from B-Like ions while Z > 8. More extensive results of the UV lines from C, N, O ions are given in Peng & Pradhan (1995).

2. The Atomic Data and Collisional-Radiative Model

The atomic data involves the collisional rate coefficients and the radiative decay rates. The collisional rate coefficients are taken from the Zhang et al. (1994) and the radiative decay rates from the Dankworth & Trefftz (1978), Hayes & Nussbaumer (1983, 1984), and Merkelis et al. (1994). In this work B-like ions with Z > 12 are not considered since the collisional rate coefficients by Zhang et al. did not make a complete relativistic treatment and may not be as accurate as for the lower Z ions. The collisional calculations by Zhang et al. (1994) were carried out in a 8-state (15-level) close coupling approximation,

S. Bowyer and R. F. Malina (eds.), Astrophysics in the Extreme Ultraviolet, 589–593.
© 1996 Kluwer Academic Publishers. Printed in the Netherlands.

Table 1. Selected Line Ratios For Density Diagnostics

	$\dfrac{^4P_{5/2} - ^2P^0_{3/2}}{^4P_{3/2} - ^2P^0_{3/2}}$	$\dfrac{^4P_{3/2} - ^2P^0_{3/2}}{^4P_{1/2} - ^2P^0_{3/2}}$	$\dfrac{^4P_{5/2} - ^2P^0_{3/2}}{^4P_{1/2} - ^2P^0_{3/2}}$	$\dfrac{^4P_{3/2} - ^2P^0_{1/2}}{^4P_{1/2} - ^2P^0_{1/2}}$
Ne VI	999.6/1006.1	1006.1/1010.6	999.6/1010.6	993.0/997.4
Mg VIII	773/784	784/791	773/791	764/771
Al IX	692/704	704/713	692/713	681/689
Si X	625/639	639/650	625/650	612/622
S XII	520/539	539/554	520/554	504/516

Notes: Wavelengthes (Å) for Ne VI are in air (Morton 1991). For other ions the wavelength given are approximate.

including the terms: $2s^2 2p(^2P^o_{1/2,3/2})$, $2s2p^2(^4P_{1/2,3/2,5/2}, {}^2D_{3/2,5/2}, {}^2S_{1/2}, {}^2P_{1/2,3/2})$, $2p^3\ (^4S^o_{3/2}, {}^2D^o_{3/2,5/2}, {}^2P^o_{1/2,3/2})$. They also examined the effect of including the n=3 terms, $2s^2 3s(^2S)$, $2s^2 3p(^2P^o)$ and $2s^2 3d(^2D)$, and found these to enhance the excitation rate coefficients for some transitions in Ne VI. The Zhang et al. results are in agreement with those obtained earlier by Hayes (1992) for Ne VI and appear to indicate a general tendency for the resonances converging on to the $n = 3$ thresholds to enhance significantly the excitation of some of the higher $n = 2$ states (i.e. those dominated by $2p^3(^4S^o, {}^2D^o, {}^2P^o)$). Recently, Keenan et al. (1994) used the 11-state data and calculated the Ne VI line ratios for comparison with solar UV observations. In the present work we study similar line ratios for Mg VII and Si X. Our collisional-radiative model includes the 15 levels as mentioned above. The level populations are determined by solving the coupled set of equations for the 15 levels system and intensity ratios are carried out in the optical thin plasma.

3. EUV Line Ratios of Density and Temperature Diagnostics

There are a number of transitions in the EUV that could potentially be employed as temperature or density diagnostics in B-like ions. The present work involved a comprehensive study of all the line ratios along the isoelectronic sequence as functions of temperature and density. Below we present selected results (line ratios for other ions and lines may be obtained on request from the authors).

Fig. 1 shows a typical intensity ratios $I(^4P_{5/2} - ^2P^o_{3/2})/I(^4P_{3/2} - ^2P^o_{3/2})$ for the ions considered in the low density regime. The wavelengths of the UV and the EUV lines in the line ratio $(^4P_{5/2} - ^2P^o_{3/2})/(^4P_{3/2} - ^2P^o_{3/2})$ range from the $\lambda\lambda$ (2327.6, 2329.1) pair in C II, to $\lambda\lambda$ (520.3, 539.3) in S XII. The densities range from 10^2 cm^{-3} for C II up to 10^{12}cm^{-3} for S XII. The criterion used here is that the temperature sensitivity be small and the value of line ratio be not too small or large. Depending on the ion charge, the line intensities involving the ^4P levels are sensitive to wide range of electron densities. Table 1. gives ratios of wavelengths of lines involving the ^4P levels for the various ions that were found to be reasonably temperature-independent so as to serve as good density diagnostics.

Unlike density sensitive lines temperature sensitivity requires that the energy differences between the upper levels should be large enough so as to affect a temperature-dependence through the electron excitation rate coefficients. Further, to minimise density dependence the radiative decay rates should be much larger than the collisional

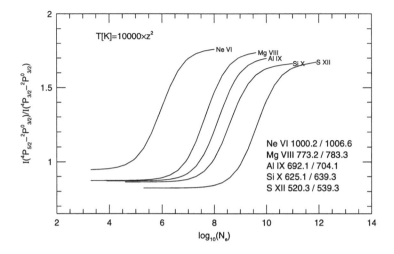

FIGURE 1. Density Diagnostic Line Ratios

TABLE 2. Selected Line Ratios For Temperature Diagnostics

ION	$\dfrac{^2P_J - ^2P^0_{J'}}{^2D_J - ^2P^0_{J'}}$
C II	905/1337
N III	687/992
O IV	556/791
Ne VI	403/563
Mg VIII	317/437
Al IX	287/393
Si X	261/356
S XII	222/300

NOTE: Wavelengthes (Å) are not observed but caculated.

de-excitation or redistribution rates from, or between, the upper levels. The $2s2p^2$ ($^2D_{3/2,5/2}$, $^2S_{1/2}$, $^2P_{1/2,3/2}$) levels, lying above the metastable ^4P term, fulfill these criteria and give rise to a number of lines that could be used as good temperature diagnostics. All possible line ratios were investigated and it was found that the ratios of lines $^2D_J - ^2P^o_{j'}$ and $^2P_J - ^2P^o_{j'}$ are particularly sensitive to electron temperature. Table 2 gives a wavelength ratios for these multiplets in the ions considered. The individual fine structure components are very close in wavelength and most have not been observed.

Fig. 2 shows four ions considered. With increasing ion charge there is a larger spread with density. However, the line ratios given provide an example of the lines that might be used as temperature indicators in the EUV. The temperatures considered are scaled with ion charge as z^2.

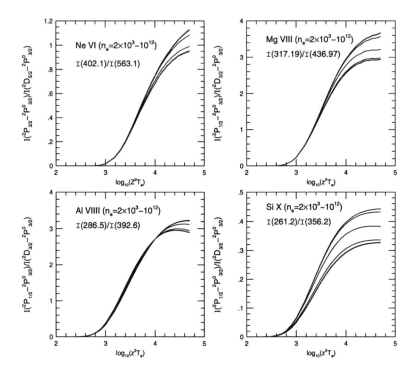

FIGURE 2. Temperature Disagnostic Line Ratios

4. Conclusion

A sample of the extensive study of temperature and density dependent EUV and UV line ratios of B-like ions has been summarized. Line ratios are investigated for pairs of transitions along the isoelectronic sequence, making it possible to employ a given diagnostic ratio for more than one ion in the sequence for the same or similar sources, and to enable a study of the temperature and density dependence as function of ion charge. These line ratios should be applicable to the astronomical objects like QSQs, AGNs, nova and hot stars.

This work was supported by a grant from the U.S. National Science Foundation (PHY-9421898). The computational work was carried out on the Cray Y-MP at the Ohio Supercomputer Center in Columbus, Ohio.

REFERENCES

BLUM, R. D., & PRADHAN, A. K. 1992, ApJS, 80, 425
DANKWORTH, W. & TREFFTZ, E. 1978, A&A, 65,93
HAYES, M. A. & NUSSBAUMER 1984, A&A, 134,193
HAYES, M. A. & NUSSBAUMER 1983, A&A, 124,279
HAYES, M. A. 1992, J. Phys. B, 25,2649

HUMMER, D. G., BERRINGTON, K. A., EISSNER, W., PRADHAN, A. K., SARAPH, H. E. & TULLY, J. A. 1993, A&A, 279, 298

LUO, D. & PRADHAN, A. K. 1990, Phys. Rev. A, 41,165

LENNON, D. J., DUFTON, P. L., HIBBERT, A., KINGSTON, A. E. 1985, ApJ, 294, 200

KEENAN, F. P., CONLON, E. S., BOWDEN, D. A., DWIVEDI, B. N. & WIDING 1994a, Solar Phys., 149, 137

KEENAN, F. P., WARREN, G. A., DOYLE, J. G., BERRINGTON, K. A., KINGSTON, A. E. 1994b, Solar Phys., 150, 61

KEENAN, F. P., FOSTER, V. J., REID, R. H. G., DOYLE, J. G., ZHANG, H. L. & PRADHAN, A. K. 1994, private communication

MERKELIS, G., VILKAS, M. J., GAIGALAS, G. & KISIEULIUS, R. 1994, Physica Scripta, submitted

O'SHEA, E., FOSTER, V. J., KEENAN, F. P., DOYLE, J. G., REID, R. H. G., ZHANG, H. L., & PRADHAN, A. K. 1995, A&AS, submitted

PENG, J. F. & PRADHAN, A. K. 1995, A&AS, submitted

PENG, J. F. & PRADHAN, A. K. 1994, ApJ, 432, L123

ZHANG, H. L., GRAZIANI, M. & PRADHAN, A. K. 1994, A&A, 283, 319

STAFFORD, R. P., HIBBERT, A., & BELL, K. L. 1993, MNRAS, 260, L11

Line Ratio Diagnostics Applicable to Astronomical Spectra in the 50–3000 Å Wavelength Region

F. P. KEENAN

Department of Pure and Applied Physics, The Queen's University of Belfast, Belfast BT7 1NN, N. Ireland

A bibliography has been produced of the most reliable emission and absorption line ratio diagnostic calculations currently available for application to the spectra of astrophysical sources in the UV and EUV wavelength region (50–3000 Å). References are listed containing diagnostics for species in the Li through P isoelectronic sequences, as well as the iron ions Fe II–Fe XXIII and nickel ions Ni XVII–Ni XXV. Also given is the wavelength range for which diagnostic calculations are presented in each reference, along with the type of diagnostic considered. These include, for example, emission line ratios for determining electron temperatures and densities, and absorption line diagnostics for evaluating hydrogen densities.

1. Introduction

Line ratios involving transitions in the ultraviolet (UV) and extreme ultraviolet (EUV) regions of the spectrum frequently provide excellent temperature and density diagnostics for the emitting or absorbing plasma. Over the past 20 years, many such diagnostics have been developed for application to astronomical spectra, such as those of the solar transition region/corona (Dere 1978; Vernazza & Reeves 1978), and stellar and inter-stellar observations from the *International Ultraviolet Explorer* and *Copernicus* satellites (Hyung, Aller & Feibelman 1994; Jenkins & Shaya 1979; Jordan 1988). More recently, new observing opportunities, such as those afforded by the *Hubble Space Telescope* (*HST*), *Extreme Ultraviolet Explorer* (*EUVE*), *Orbiting and Retrievable Far Ultraviolet Telescope* (*ORFEUS*), and *Hopkins Ultraviolet Telescope* (*HUT*), have lead to a large increase in both the quality and quantity of astronomical UV and EUV observations, and there is clearly an urgent requirement for diagnostics which may be applied to the analysis of such data.

There have been several recent reviews on the importance of diagnostics for analysing UV and EUV astronomical spectra (Dwivedi 1994; Feldman & Doschek 1991; Feldman et al. 1992). However of particular importance is that of Mason & Monsignori Fossi (1994), which discusses the spectroscopic techniques used to study astrophysical plasmas, the atomic processes involved, recent observations and plans for future space missions. My aim therefore is to complement the Mason & Monsignori Fossi review by providing a bibliography of the most reliable emission and absorption line ratio diagnostics currently available for transitions observable in the UV and EUV spectra of astronomical sources. Such diagnostics will not only be applicable to observations from, for example, *Skylab, HST, EUVE, ORFEUS*, and *HUT*, but will also be useful for analysing data from upcoming missions such as the *Solar and Heliospheric Observatory* (*SOHO*).

2. Bibliography of Diagnostic Calculations

I have compiled a list of references to the most reliable diagnostic calculations currently available for application to astronomical spectra in the UV and EUV wavelength

S. Bowyer and R. F. Malina (eds.), Astrophysics in the Extreme Ultraviolet, 595–596.

range, 50–3000 Å. Species are listed by isoelectronic sequence, apart from ions of Fe and Ni which are considered separately. After each species, I summarise the wavelength range for which diagnostic calculations are presented in the relevant reference. In some instances, references contain diagnostics involving transitions over a wider wavelength range than that considered here; in these instances I list the full wavelength coverage of the diagnostics. For example, the Fe XIII diagnostic calculations of Young, Mason & Thomas (1995) consider not only the EUV lines at $\sim$202 Å, but also the IR transitions at $\sim$1.08 μm, while Cai & Pradhan (1993) similarly provide data for both UV ($\sim$910 Å) and optical ($\sim$6731 Å) lines in S II. Also given is a list which indicates the type of diagnostic calculated in the relevant reference. The vast majority of the diagnostics are emission line electron temperature or density diagnostics; this is not surprising as most work in this area has been performed in order to analyse solar or nebular emission line spectra.

Omitted from the lists are diagnostics for H-like, He-like and F-like ions, as to the best of my knowledge there are no calculations for these sequences applicable to spectra in the 50–3000 Å wavelength range, apart from electron temperature sensitive emission line ratios involving the He II 256–304 Å lines (Cassinelli et al. 1995) and the Ar X 165–4257 Å transitions (Feldman & Doschek 1977). Nor do I list diagnostics that involve lines from different elements, as these are rather limited in number (see, for example, Keenan, Dufton & Kingston 1990).

I would like to apologize in advance to anyone whose work has been inadvertently omitted from the tables. For the future, I would be grateful if readers would inform me of such omissions, and also of any of their more recent diagnostic work that could be included in future versions of these tables.

Full copies of the bibliography, which includes 15 tables and 156 references, are available from the author either by post or by electronic mail (from F.Keenan@qub.ac.uk on Internet).

I would like to thank all those who sent me reprints/preprints of their work for inclusion in this paper, including L. Aller, A. Bhatia, J. Cassinelli, G. Doschek, B. Dwivedi, U. Feldman, S. Kastner, H. Mason, J. Peng, A. Pradhan and H. Zhang.

REFERENCES

CAI, W. & PRADHAN, A. K. 1993, ApJS, 88, 329

CASSINELLI, J. P. et al., 1995 ApJ, 438, 932

DERE, K. P. 1978, ApJ, 221, 1062

DWIVEDI, B. N. 1994, Space Sci. Rev., 65, 289

FELDMAN, U. & DOSCHEK, G. A. 1977, J. Opt. Soc. Am., 67, 726

FELDMAN, U. & DOSCHEK, G. A. 1991, ApJS, 75, 925

FELDMAN, U., MANDLEBAUM, P., SEELY, J. F., DOSCHEK, G. A. & GURSKY, H. 1992, ApJS, 81, 387

HYUNG, S., ALLER, L. H. & FEIBELMAN, W. A. 1994, PASP, 106, 745

JENKINS, E. B. & SHAYA, E. J. 1979, ApJ, 231, 55

JORDAN, C. 1988, J. Opt. Soc. Am. B, 5, 2252

KEENAN, F. P., DUFTON, P. L. & KINGSTON, A. E. 1990, ApJ, 353, 636

MASON, H. E. & MONSIGNORI, FOSSI, B. C. 1994, A&AR, 6, 123

VERNAZZA, J. E. & REEVES, E. M. 1978, ApJS, 37, 485

YOUNG, P. R., MASON, H. E. & THOMAS, R. J. 1995, Proc. 3rd SOHO Workshop, in press

The *EUVE* Guest Investigator Science Program

KEN ANDERSON AND BRETT STROOZAS

Center for EUV Astrophysics, 2150 Kittredge Street,
University of California, Berkeley, CA 94720-5030, USA

In order to promote research using data from NASA's *Extreme Ultraviolet Explorer* (*EUVE*) satellite, the Center for EUV Astrophysics has implemented a Guest Investigator (GI) Science Program. The purpose of the GI Program is to provide researchers with information, services, and training in the use of public *EUVE* data sets; in effect, it offers to the research community the technical experience and intricate knowledge of the *EUVE* data sets resident at CEA. All interested researchers are encouraged to participate as GIs.

1. Introduction

NASA's *Extreme Ultraviolet Explorer* (*EUVE*) satellite was launched on 7 June 1992. The purpose of the mission is to perform a six-month all-sky survey in the EUV wavelength region (60–740 Å) followed by a multi-year program of guest observer (GO) pointed spectrometer observations (Bowyer & Malina 1991). The survey phase of the mission has been completed and *EUVE* is now in its third year of GO observations.

Proprietary data rights on a large volume of *EUVE* data have already expired and are available to the public. The basic *EUVE* data sets include catalogs, skymaps, and photon event lists from the survey phase of the mission, and multichannel spectra from GO observations. Access to *EUVE* public data (all in the astronomical standard FITS format) as well as to various associated information and services is available via the CEA World Wide Web (WWW) site (http://www.cea.berkeley.edu/).

2. The GI Science Program

To encourage and promote scientific research using these unique data sets, the Center for EUV Astrophysics (CEA) at the University of California, Berkeley, has implemented the Guest Investigator (GI) Science Program. The purpose of the GI Program is to provide researchers with the information and training necessary to use the publicly available *EUVE* data sets. The GI Program is open to all researchers and offers to the research community the technical experience and intricate knowledge of the *EUVE* data that is resident at CEA.

Whether working remotely or visiting at CEA, all GIs receive a variety of benefits and individualized services that include the following:

- easy, quick, and complete access to all public *EUVE* data sets
- dedicated personal support from CEA scientific/technical experts
- information on and training in the use of *EUVE* data and software (i.e., IRAF)
- the use of CEA computing resources

Anyone can become a GI—astronomers doing EUV science, technologists interested in test-bed technology transfer activities, engineers concerned about the affects of long-term exposure of hardware instruments to the Space environment, educators working to establish "hands-on" science lesson plans, and the general public. Some examples of past and current GI support activities include helping researchers to build and analyze EUV

S. Bowyer and R. F. Malina (eds.), Astrophysics in the Extreme Ultraviolet, 597–598.

light curves and spectral data sets; to use public *EUVE* data to prepare, validate, and enhance observing programs on other facilities (e.g., the Hubble Space Telescope); to set up information services on the WWW; and to test tools developed for *EUVE* that may be useful for other missions.

Becoming a GI is extremely simple. Just fill out the *brief* registration form available via the CEA WWW site (http://www.cea.berkeley.edu/) or contact the *EUVE* Science Archive (archive@cea.berkeley.edu; 510-642-3032).

3. Summary

A large volume of *EUVE* science data is now publicly accessible. To promote its use CEA has established the GI Science Program, the purpose of which is to provide to researchers information on and training in the use of publicly available *EUVE* data sets. In effect, the GI Program offers to the research community the technical experience and intricate knowledge of the *EUVE* data that resides at CEA. The Program is open to all interested researchers—astronomers, technologists, engineers, educators, and the general public—and CEA welcomes and encourages your participation.

This program is supported by NASA contract NAS5-29298.

REFERENCES

BOWYER, S. & MALINA, R. F. 1991, The Extreme Ultraviolet Explorer Mission, in Extreme Ultraviolet Astronomy, ed. R. F. Malina & S. Bowyer, New York: Pergamon, 387

The *Extreme Ultraviolet Explorer* Public Right Angle Program

K. McDONALD, N. CRAIG, E. OLSON, AND C. A. CHRISTIAN

Center for EUV Astrophysics, 2150 Kittredge St., University of California,
Berkeley, CA 94720–5030, USA

The new *Extreme Ultraviolet Explorer* (*EUVE*) Public Right Angle Program (RAP) offers the opportunity for researchers to obtain observations with the *EUVE* imaging telescopes, oriented at right angles to the *EUVE* spectrometer used for the NASA Guest Observer Program. Scientists may submit proposals electronically through the World Wide Web or e-mail using a template form to list specific targets and present the scientific motivation for the work. The RAP electronic proposal process is streamlined from proposal submission through data delivery and is a prototype for a system to be used for all guest observers during the *EUVE* extended mission starting in February, 1996.

1. Introduction

The *Extreme Ultraviolet Explorer* (*EUVE*) Public Right Angle Program (RAP) is a new program that makes the observations taken with the *EUVE* scanning telescopes available to scientists around the world. These imaging instruments, primarily used during the *EUVE* all-sky survey (Bowyer et al. 1994; Bowyer et al. 1996), are pointed along a great circle oriented perpendicular to the *EUVE* Deep Survey/Spectrometer used for guest observer observations. The scanners can be used to collect data of high scientific interest simultaneously with the spectroscopy of scheduled GO targets. The RAP was started at CEA as a way to optimize the data collected by specifying particular spacecraft roll angles (subject to spacecraft and instrument pointing constraints) so that targets of astrophysical significance could be observed (McDonald et al. 1994). The goal of the Public RAP is to provide a simple, effective, on-line method to propose targets, schedule observations, and disseminate data products to the scientific community as well as provide a testbed for the *EUVE* extended mission GO program.

2. Proposal Submission Process

RAP proposals are submitted by accessing and completing the proposal form available through the WWW. The ASCII format form is sent via electronic mail and processed through the *EUVE* Project Office at Goddard Space Flight Center. Approved targets are added to the database of targets available for scheduling. If a target satisfies the pointing constraints, it is placed in the upcoming *EUVE* schedule, and the Principal Investigator is notified by electronic means.

3. Data Reduction and Delivery for RAP Targets

When a scheduled RAP target has been observed, the data is reduced using the standard EUV package software in IRAF and FITS formats which preserve the event data. These data are distributed to the Principal Investigator via any of a number of different media (ftp, 8mm tar, CD-ROM, etc.). The Principal Investigator is awarded 6 months proprietary data rights to the data taken during this observation, at which time the data

S. Bowyer and R. F. Malina (eds.), Astrophysics in the Extreme Ultraviolet, 599–600.

will become public domain and accessible to anyone upon requests. Information about observed targets are published in the *EUVE* Electronic Newsletter on a monthly basis. No financial support is provided to investigators for analysis of this data via the Public RAP, but all investigators are encouraged to apply for other means of financial support (such as NASAs Astrophysics Data Program).

4. Application of Public RAP to *EUVE* Extended Mission

In February 1996, the *EUVE* satellite will enter the extended mission phase in which changes to science operations will be implemented. Procedures to streamline all parts of the system, including observing proposals, observation scheduling, data delivery are being tested currently. Since the Public RAP has similar characteristics to those planned for the *EUVE* extended mission, the proposal system of the RAP serves as a working model for the methods to be used throughout the *EUVE* and other NASA satellites.

This work has been supported by NASA contract NAS5-29298.

REFERENCES

Bowyer, S., Lieu, R., Lampton, M., Lewis, J., Wu, X., Drake, J. J., & Malina, R. F. 1994, The First Extreme Ultraviolet Explorer Source Catalog, ApJS, 93, 569

Bowyer, S. & Malina, R. F. 1991, The Extreme Ultraviolet Explorer Mission, in Extreme Ultraviolet Astronomy, ed. R. F. Malina & S. Bowyer, New York: Pergamon Press, 387

Bowyer, S. et al. 1996, ApJS, in press

McDonald, K., Craig, N., Sirk, M. M., Drake, J. J., Fruscione, A., Vallerga, J. V., & Malina, R. F. 1994, Serendipitous EUV Sources Detected During the First Year of the Extreme Ultraviolet Explorer Right Angle Program, AJ, 108, 1843

The Berkeley Spectrometer for *ORFEUS*: Laboratory and In-Flight Performance

MARK HURWITZ AND STUART BOWYER

Center for EUV Astrophysics, 2150 Kittredge Street,
University of California, Berkeley, CA 94720–5030, USA

The Berkeley spectrometer aboard the *ORFEUS* payload achieved a variety of "firsts" during its inaugural mission in September 1993. The instrument utilizes spherical gratings with mechanically ruled varied line-spacing, and curved microchannel plate detectors with delay-line anode readout systems, to cover the $390 - 1200$ Å band at a resolution of $\lambda/5000$. The instrument will be discussed, and its performance illustrated with calibration and in-flight spectra. Science highlights from the ORFEUS-I mission will be presented (oral presentation only). The payload will be available for use by guest investigators during the ORFEUS-II mission currently scheduled for late 1996.

1. Introduction and Background

ORFEUS, the *Orbiting Retrievable Far and Extreme Ultraviolet Spectrometers*, is a joint project of NASA and DARA, the German space agency. It is an outgrowth of a longstanding collaboration between Prof. Bowyer and the Space Astrophysics Group at UC Berkeley, and Prof. Grewing and the Astronomical Institute at the University of Tuebingen (AIT).

The *ORFEUS* telescope is 1 meter in diameter and 4 meters long (Grewing et al. 1991). The Berkeley spectrometer (Hurwitz & Bowyer 1991) sits at the prime focus of the $f/2.4$ normal incidence primary mirror. This instrument is designed to provide high-resolution ($\lambda/5000$) spectroscopy of point sources between 390 and 1200 Å, with an effective area of about $4 - 6$ cm^2. Alternatively, an off-axis paraboloidal mirror can be driven into the light path, collimating the beam and directing it into an Echelle spectrometer provided by AIT and the Landessternwarte Heidelberg (LSW). That instrument is designed to provide $\lambda/10,000$ spectroscopy of point sources between 900 and 1250 Å at somewhat lower sensitivity.

The resolution of the *ORFEUS* spectrometers significantly exceeds that of other instruments with comparable sensitivity at wavelengths of overlap. Below 760 Å, the spectrometers on the *Extreme Ultraviolet Explorer* offer somewhat lower sensitivity (but longer observing times), and a resolution of $\lambda/300$. Between 800 and 1250 Å, the *Hopkins Ultraviolet Telescope* offers higher sensitivity, and comparable observing times, but again the resolution is limited to about $\lambda/300$. The *Hubble Space Telescope* exceeds *ORFEUS* performance above about 1170 Å. A smaller telescope mounted parallel to the *ORFEUS* optical axis is IMAPS, (Jenkins et al. 1988) provided by the Princeton University Observatory. IMAPS offers extremely high resolution ($\lambda/2 \times 10^5$) in the far ultraviolet (FUV), but more limited sensitivity.

ORFEUS flew successfully aboard the German space platform ASTRO-SPAS, which was deployed from the shuttle Discovery in September of 1993 and recaptured some 5 days later. During this inaugural mission (hereafter ORFEUS-SPAS), the Berkeley spectrometer functioned well. More than 100 individual pointings were successfully carried out with the Berkeley spectrometer, including multiple observations of faint sources. This

601

S. Bowyer and R. F. Malina (eds.), Astrophysics in the Extreme Ultraviolet, 601–609.

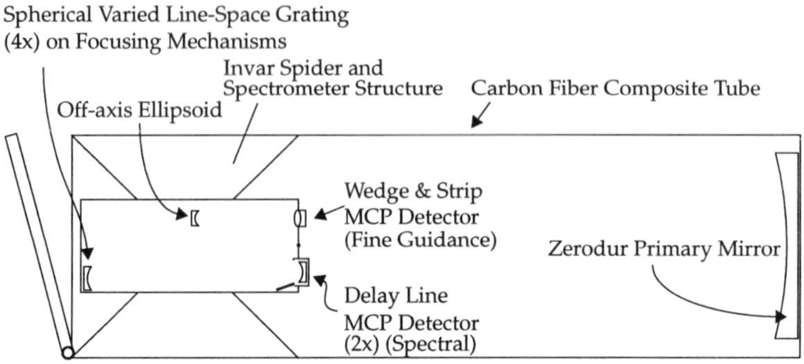

FIGURE 1. Cross-section schematic of *ORFEUS* telescope with Berkeley spectrometer.

total includes targets observed for the Berkeley science program, targets observed for our collaborators at AIT and LSW, and joint target observations.

The spacecraft has flown once since the ORFEUS-SPAS I mission, carrying a cryogenic infrared instrument for study of Earth's atmosphere (CRISTA-SPAS). A reflight of the *ORFEUS* instrument package is scheduled for late 1996. During this mission 50% of the available science time will be devoted to observations led by guest investigators, who may elect to utilize either of the *ORFEUS* spectrometers or IMAPS.

In this paper we discuss the performance of the Berkeley spectrometer in detail. The spacecraft and telescope are discussed insofar as they affect the spectra and the observing program.

2. The Berkeley Spectrometer: General Principles and Design Details

In Figure 1 we show the position of the Berkeley spectrometer within the *ORFEUS* telescope (the Echelle spectrometer is omitted for clarity). The Berkeley spectrometer occupies the inner 50% of the aperture diameter and obscures the inner 25% of the primary mirror area. For reference, the instrument Z axis is parallel to the optical axis; the prime focus defines the plane of $Z = 0$, and the Z coordinate increases toward the top of the telescope.

Shadows from five spider vanes divide the monolithic telescope aperture into five segments of the outer annulus. This segmentation is illustrated in Figure 2, where we show an incoming "photon's view" of the system. Starlight from all five segments comes to a common focus some 240 cm from the primary mirror, then diverges into the Berkeley spectrometer volume.

As the beams from the five segments diverge, each segment of the annulus strikes a distinct optic within the Berkeley spectrometer. The smallest segment, representing a 36° wedge of the annulus, is intercepted by an off-axis ellipsoidal optic near $Z = +60$ cm which images the target and aperture onto a sealed tube microchannel plate detector sensitive to wavelengths near 1500 Å. This fine guidance system is used for initial coalignment of the *ORFEUS* telescope and the external star tracker of the ASTRO-SPAS. It can also be used for postflight spectral reconstruction to correct for target "jitter"

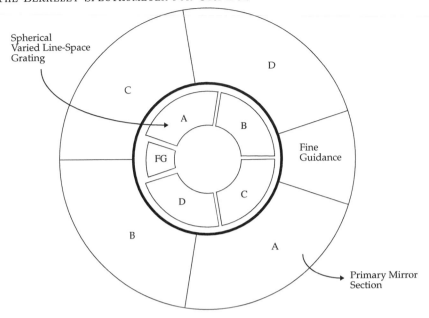

FIGURE 2. Telescope as "seen" by incoming photon. The Z axis is perpendicular to page, pointing at viewer.

during an observation, although this was not necessary for spectra collected during the ORFEUS-SPAS I mission.

Each of the larger segments labeled A through D subtends 81° of the annulus and strikes a distinct diffraction grating. The gratings are located at the extreme end of the spectrometer, near $Z = +100$ cm. The position of the gratings is shown in Figure 2; from this vantage point the viewer is seeing the "back side" of each grating, and of course, in the actual instrument the gratings are obscured by the surrounding spectrometer structure, thermal blankets, etc. Each grating disperses a unique sub-bandpass across one of two spectral detectors near the focal plane of the spectrometer. In Figure 3 we show the layout of detectors, including both the spectral and the fine guidance sealed tube, in the focal plane near $Z = 0$. Note that the scale has been expanded by a factor of 2 in this figure.

The reasons for adopting this overall design are fairly straightforward and are discussed in Hurwitz & Bowyer (1986). The telescope diameter and overall length are the maximum that can be accommodated within the spacecraft and shuttle envelope; a fast primary was a necessity. High-resolution spectroscopy required that either the aperture be subdivided into slower segments, or that a secondary mirror be introduced. The latter would have increased the effective focal length, imposing more severe restrictions on allowable spacecraft jitter. Furthermore, the potential gain in effective area (e.g., utilizing the entire beam vs. only a segment of it) would have been offset by the low reflectivity of normal incidence optics at EUV/FUV wavelengths. And as a practical matter the beam would have required subsequent subdivision in any case, given the very broad spectral bandpass (a factor of 3 in wavelength), the high spectral resolution that was desired, and the limited number of resolution elements provided by even the most advanced detector systems.

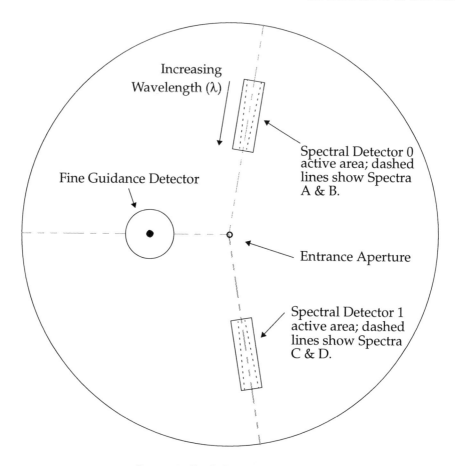

FIGURE 3. Focal plane detector layout.

TABLE 1.

Grating	Central groove dens. (/mm)	Nom. Bandpass (Å)	Coating (First Mission)	(Second Mission)
A	6000.0	389 − 523	Evaporated Ir	Evaporated Ir
B	4550.0	513 − 690	Evaporated Ir	Evaporated Ir
C	3450.4	676 − 910	Evaporated Ir	Sputtered SiC
D	2616.6	892 − 1200	Evaporated Ir	Sputtered SiC

The geometry of each grating/detector system is identical save for rotation about the Z axis and/or reflection through the detector midplane. The incidence angle is 12°. Central groove densities, nominal bandpasses, and coatings are contained in Table 1. The wavelength coverage ($\lambda_{max}/\lambda_{min}$) of each grating is identical, as is the ratio of central wavelength to groove density.

When the design was first proposed, (e.g., Hurwitz & Bowyer 1986) the optics were

TABLE 2.

Dim.	Technique	Length (mm)	Resolution (μm)	Digitization (μm)
X	Delay line	82	30 − 40	3.5
Y	Chg. division	26	200	120

toroidal in surface figure, with uniform line spacing, and followed the classical Rowland circle geometry. It was subsequently realized that higher resolution and reduction of astigmatism was possible using spherical optics with mechanically ruled varied line-spacing (SVLS) in a non-Rowland mounting (Harada et al. 1991). Spectrometers of this type had been proposed earlier (Harada & Kita 1980) but had never before been used in an astronomical application. The geometry enforced by the prime focus spectrograph position imposed unique constraints on the optical design. The adopted SVLS parameters, discussed in Harada et al. (1991), were optimized at Berkeley with an iterative raytracing technique.

The two spectral detectors incorporate curved microchannel plates with delay-line anodes (Siegmund et al. 1991) to encode the X position of the dispersed photons. ORFEUS-SPAS I was the first space flight for detectors of this type, and they performed extraordinarily well. Along the Y axis, imaging is used only to separate the two spectra and to isolate the spectra from detector background; resolution requirements are modest. Y-axis imaging is provided by an interleaved wedge/wedge anode pattern and charge division. Spectral detector parameters are contained in Table 2. Both spectral detectors are housed in a sealed vacuum chamber equipped with doors; both utilize KBr photocathodes deposited on the front microchannel plate surface.

We now present relevant characteristics of other hardware systems before returning to system performance values.

3. Spacecraft, Telescope, and Apertures

The ASTRO-SPAS, fabricated by Daimler-Benz Aerospace (formerly Deutsche Aerospace), offers high scientific performance, but its resources are designed for short duration missions.

Absolute pointing is accurate to within a few arc seconds; compressed gas thrusters provide the necessary torques. An external star tracker provides a reference signal to the attitude/maneuvering control system (AMCS). During ORFEUS-SPAS I, the AMCS successfully placed almost every target within the 20″ diameter aperture; one or two observations failed because of a lack of suitable guide stars.

Some tens of watts of power are provided to the experiments, using energy stored in batteries. Data are primarily archived by on-board tape recorders; about 130 kbits s^{-1} are allocated to the scientific experiments. This high speed (HS) channel records up to about 4000 spectral photon events per second, but cannot be accessed until the spacecraft is recovered and returned to Earth. A much slower (QL or quick-look) telemetry channel allows transmission of some 8 kbits s^{-1} of data to the ground in real time, but only during periods when the ASTRO-SPAS is in contact with the shuttle and the shuttle is in contact with the ground via TDRSS. The ASTRO-SPAS can record the data stream from only a single experiment at any given time. This limitation does not restrict functionality, since only one of the Berkeley or Echelle spectrometers can receive the *ORFEUS* telescope beam, and the targets that can be observed with IMAPS are too bright for either *ORFEUS* spectrometer.

The *ORFEUS* telescope, fabricated by Kayser-Threde, contains three major mechanisms. One operates the large door at the $+Z$ extremum of the telescope. A second drives the collimator mirror used to direct the beam into the Echelle spectrometer. A third actuates an aperture blade at the prime focus. For ORFEUS-SPAS I, the science aperture was a circular hole 20″ in diameter. In the second mission, observations with the Berkeley spectrometer will be carried out with three apertures near the prime focus. A 20″ on-axis hole admits the light from the target and diffuse emission. A second 20″ diameter hole is located off axis and admits diffuse emission only. These holes are displaced perpendicular to the dispersion direction, so their individual spectra will be separated on the spectral detectors, enabling diffuse emission to be subtracted from the target spectrum. A third hole, 60″ in diameter is also located off axis, and is covered by a thin tin filter. In most applications the presence of this aperture can be ignored. The filter will be used only for extreme ultraviolet (EUV) observations of the bright B stars ϵ and β CMa, whose integrated FUV flux could otherwise scatter from the gratings and overwhelm the EUV signal.

For a variety of reasons it was not practical to measure the end-to-end focus of the complete telescope and spectrometer system at EUV/FUV wavelengths prior to launch. The telescope and spectrometer were focused independently, and the overall system focus relied on mechanical mounting tolerances. The telescope tube is fabricated of carbon-fiber composite, and is therefore prone to shrinkage as water vapor is outgassed on orbit. Thermal modeling of the Berkeley spectrometer indicated that our internal Invar structure would offer high dimensional stability, but the uncertainty in the thermal model was difficult to ascertain prior to flight experience. For all these reasons, we equipped the Berkeley spectrometer gratings with independent Z-axis focusing mechanisms, driven by stepper motors which can be actuated on orbit. (These mechanisms also facilitated preflight alignment and internal focusing in our laboratory calibration chamber.) The focusing period was scheduled early in the ORFEUS-SPAS I mission, but this period was occupied by critical but originally unscheduled checkout and alignment activities. Problems then arose in the telescope mechanisms; these too were eventually rectified, but for a time they threatened to end the science portion of the mission. When normal operations were restored, we put a premium on data gathering rather than fine tuning of the spectrometer performance. Evaluation of the real-time spectra verified that the on-orbit focus was at least reasonably close to the laboratory value, so we did not attempt to reinsert a focusing activity in the remaining mission time line. Postflight analysis reveals some defocus; the in-flight resolution was $\lambda/3000$.

4. Performance: Resolution, Effective Area, and Backgrounds

Various terms contribute to the net resolution error budget. Estimates for these terms are found in Table 3, both under ideal and realistic assumptions. Values are FWHM, in micrometers. The "ideal" column corresponds to a resolution of about $\lambda/9500$. The "realistic" column corresponds to a resolution of about $\lambda/5900$.

Preflight laboratory data confirm the realistic resolution to within about 10%. Here jitter should be negligible, but the entrance pinhole is comparatively large (25 μm). In these circumstances we expect a resolution of about $\lambda/5500$. In Figure 4 we show a small section of a laboratory spectrum of argon emission lines near 544 Å (Grating B). The two bright features in this figure are separated by 4.26 Å, and show FWHM values of almost precisely $\lambda/5000$.

In Figure 5 we show a small section of the spectrum of a symbiotic binary star near the far-ultraviolet O VI lines (Grating D) collected during the ORFEUS-SPAS I mis-

TABLE 3.

Term	Ideal	Realistic
Primary mirror spot size	8	12
Grating aberrations (fab. and mounting)	4	15
Detector resolution	30	40
Defocus	0	20
Uncorrected jitter	0	12
TOTAL	31.3	50.1

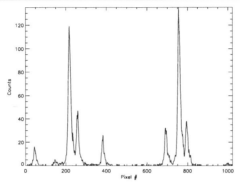

FIGURE 4. Section of a laboratory Grating B spectrum of argon.

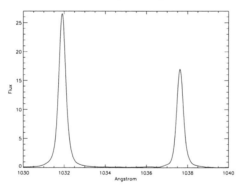

FIGURE 5. Section of a flight Grating D spectrum of a symbiotic binary star.

sion. The FWHM of these features is $\lambda/3000$. Interstellar absorption lines in continuous spectra show this characteristic width, indicating that it is an instrumental limit, not an astrophysical effect. Spacecraft jitter, for which no corrections were applied, would limit the resolution at about $\lambda/5400$. The most natural explanation for the $\lambda/3000$ is a defocus somewhere in the system as discussed above.

The in-flight effective area can be estimated from observations of the well studied hot DA white dwarf G191-B2B (Vennes & Fontaine 1992). In Figure 6 we show the in-flight effective area (solid line); this is somewhat below preflight estimates based on "theoretical" expectations for individual components, but not grossly so. The effective area was about 10% higher upon initial instrument turn-on, but quickly declined to the

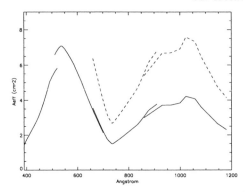

FIGURE 6. Spectrometer effective area during ORFEUS-SPAS I mission (solid), estimated -II mission (dotted).

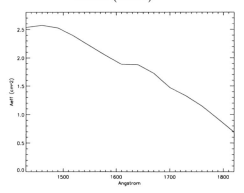

FIGURE 7. Fine Guidance effective area.

value shown here, then stabilized. This decline is presumably associated with an initial outgassing of contaminants, which then condensed on the optics and/or detector.

For the next mission, SiC coatings on Gratings C and D should improve the effective area, as shown by the dashed line in Figure 6. The 10% on-orbit loss may have been recovered by postflight cleaning and recoating of the optics; however, it could easily recur soon after deployment. At the time of writing, the preflight calibration for the next mission has not been carried out.

The effective area of the fine guidance system, again estimated from ORFEUS-SPAS 1 observations of G191-B2B, is shown in Figure 7. The barium fluoride entrance window of the sealed tube reduces the effective area of this system to negligible values below about 1400 Å.

We acknowledge the support of NASA grant NAG5-696. The *ORFEUS* program is supported by DARA grant WE 3 50 OB 8501 3.

REFERENCES

GREWING, M. ET AL. 1991, The ORFEUS Mission, in Extreme Ultraviolet Astronomy, ed. R. F. Malina & S. Bowyer, New York: Pergamon, 437

HARADA, T. & KITA, T. 1980, Appl. Opt., 19, 3987

HARADA, T., KITA, T., HURWITZ, M., & BOWYER, S. 1991, International Conference on the Application and Theory of Periodic Structures, Proc. SPIE, 1545, 2

HURWITZ, M. & BOWYER, S. 1986, A High Resolution Spectrometer for EUV/FUV Wavelengths, Proc. SPIE, 627, 375

HURWITZ, M. & BOWYER, S. 1991, The Berkeley EUV Spectrometer for the ORFEUS Mission, in Extreme Ultraviolet Astronomy, ed. R. F. Malina & S. Bowyer, New York: Pergamon, 442

JENKINS, E. B., JOSEPH, C. L., LONG, D., ZUCCHINO, P. M. CARRUTHERS, G. R. BOTTEMA, M. & DELAMERE, W. A. 1988, Proc. SPIE, 932, 213

SIEGMUND, O. H. W., LAMPTON, M., RAFFANTI, R., & HERRICK, W. 1991, High Resolution Delay Line Readouts for Microchannel Plates, Nuclear Instr. and Methods in Phys. Res., 310, 311

VENNES, S. & FONTAINE, G. 1992, An Interpretation of the Spectral Properties of Hot Hydrogen-Rich White Dwarfs with Stratified H/He Model Atmospheres, ApJ, 401, 288

An Instrument to Study the Diffuse EUV Astronomical Background

S. BOWYER,[1] J. EDELSTEIN,[1] M. LAMPTON,[1] L. MORALES,[2]
J. PEREZ MERCADER,[3] AND A. GIMENEZ[3]

[1] Center for EUV Astrophysics, 2150 Kittredge St.,
University of California, Berkeley, CA 94720–5030, USA
[2] Universidad Autonoma de Madrid, Ciudad Universitaria de Cantoblanco,
28049 Madrid, Spain
[3] LAEFF, Instituto Nacional De Tecnica Aerospacial,
Apartado de Correos 50727, 28080 Madrid, Spain

The extreme ultraviolet (EUV) diffuse background is the most poorly known of any of the diffuse astronomical backgrounds. Only upper limits to this flux exist, obtained with spectrometers with very crude (from $\approx$ 15 to 30 Å) resolution; these limits are generally one to two orders of magnitude larger than the expected sources of cosmic flux. A variety of source mechanisms have been postulated to radiate in this bandpass; the most discussed is the hot phase of the interstellar medium. A speculative possibility is that hot dark matter in the form of massive, radiatively unstable neutrinos in our Galaxy will produce a unique line in this bandpass. We describe an instrument employing a new type of spectrometer which will provide ~5 Å resolution and unprecedented sensitivity for diffuse EUV radiation. The instrument will be carried aboard the newly developed Spanish Minisat satellite.

1. Introduction

Initial investigations of the diffuse astronomical background in the EUV were carried out with broad-band detectors on rockets and short duration orbital flights (Cash, Malina, & Stern 1976; Stern & Bowyer 1979; Bloch et al. 1986). The *Alexis* Satellite developed by the Los Alamos group was designed to provide broad band measurements of the cosmic EUV flux (Bloch et al.); and Lieu et al. (1995) obtained upper limits to this flux with *EUVE*.

A few spectrographic measurements have been made. Holberg (1986) obtained data from 520 to 1100 Å with 30 Å resolution with the Voyager 2 ultraviolet spectrometer. Edelstein & Bowyer (1995) have noted that the Voyager upper limits were overly stringent and have derived more appropriate limits from this data. Labov & Bowyer (1991) flew a grazing incidence spectrometer to measure the cosmic background from 80 to 650 Å with a resolution of 15 Å on a sounding rocket. These authors tentatively identified features which might have been produced by the hot phase of the ISM, but the features noted were close to the limiting sensitivity of the instrument and are now believed to be spurious (Edelstein & Bowyer 1995). Jelinsky et al. (1995) made innovative use of the spectrometers on *EUVE* to obtain astronomically important upper limits to the EUV background. The Wisconsin group has developed an instrument to carry out soft X-ray spectroscopy on the diffuse cosmic background; preliminary results from this experiment are presented elsewhere in this Volume.

2. Diffuse Cosmic Emission Mechanisms

It has been more than thirty years since Spitzer (1956) suggested that hot, million-degree gas pervades our Galaxy and more than twenty years since the soft X-ray background (now generally believed to be produced by a high-temperature component of the

611

S. Bowyer and R. F. Malina (eds.), Astrophysics in the Extreme Ultraviolet, 611–616.
© *1996 Kluwer Academic Publishers. Printed in the Netherlands.*

ISM) was first detected (Bowyer, Field, & Mack 1968). However, models ascribing this emission to a high temperature plasma have been surprisingly unsuccessful (Cox 1995). If the diffuse soft X-ray background is the product of emission from a hot gas, much of the radiated power will be in emission lines from highly ionized atoms radiating at EUV wavelengths.

The actual lines observed from a hot ISM will be strongly dependent upon the temperature and thermal history of this material (Breitschwerdt & Schmutzler 1994). Several temperatures have been suggested for this phase. Soft X-ray data suggest 10^6 K gas (Cox & Reynolds 1987). Absorption line data showing O VI (Jenkins 1978a,b) is often cited in combination with the soft X-ray data as further evidence for a 10^6 K gas, but the peak of the emission curve for O VI is at the substantially lower temperature of $\sim 3 \times 10^5$ K. High ionization absorption lines observed in stellar spectra taken with *IUE* indicate a temperature of 3×10^5 K (Savage 1987) as does the observation of emission lines at far UV wavelengths (Martin & Bowyer 1989). Breitschwerdt & Schmutzler (1995) have suggested that the soft X-ray emission is the product of residual high ionization states and that the actual kinetic temperature of this plasma could be as low as 4×10^4 K.

A speculative possible contributor to the cosmic EUV background is emission from neutrinos in our Galaxy undergoing radiative decay. This scenario has been explored in substantial detail in a series of papers by Sciama and co-workers (Sciama 1994). Evidence for this emission has been searched for but not found in *IUE* data taken on a quasi-stellar object which lies in a cluster of galaxies (Fabian, Naylor, & Sciama 1991), and with the Hopkins Ultraviolet Telescope, which observed a cluster of galaxies on the flight of ASTRO I (Davidsen et al. 1991). In both cases the line would have been sufficiently redshifted to move it into the bandpass observed. Though these results have been generally interpreted as ruling out the Sciama hypothesis, Bowyer et al. (1995) have shown that this is not correct and that Sciama's hypothesis is still viable.

3. The Instrument

We have developed an instrument to measure the diffuse EUV background. We have employed normal incidence rather than grazing incidence optics in this instrument because this allows a folded light path and a much smaller instrument size. Through the use of special coatings, this instrument reaches wavelengths as short as 350 Å.

A spectrometer for diffuse radiation disperses radiation viewed through an aperture to a single location on the detector corresponding to the wavelength of the radiation. The observed intensity depends on the product of the area of the aperture times the solid angle of the sky observed. Increasing the width of the aperture increases this product but will degrade the overall spectral resolution. Increasing the solid angle by increasing the aperture height will also increase this product, but grating aberrations in existing spectrograph designs increase rapidly for off-axis rays, strongly limiting the extent to which this parameter can be increased. For example, a conventional Rowland spectrometer performs well for point sources on axis, but has severe aberrations for radiation as little as 1° off-axis.

To determine potential combinations of grating surface and ruling parameters for an optimum diffuse radiation spectrometer, we developed a general expression describing the optical path for radiation incident upon an arbitrary polynomial surface with variable space diffraction rulings converging to a single point on the detector. We chose plane-cylindrical radiation emanating from a slit aperture as a source. In contrast, the Rowland spectrograph utilizes a spherical source from a point on the slit. Following Fermat's Principle, we minimized variation of the path function over the grating's aperture and found solutions which eliminated aberrations to third order for on-axis illumination and

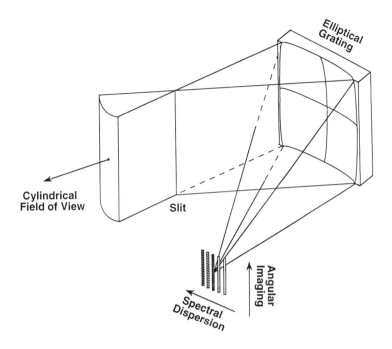

FIGURE 1. Schematic diagram of the optical design of the spectrometer.

retained most of this performance well off-axis. Constant line spacing and rotationally symmetric grating surfaces were then imposed on our solution to simplify the ruling and figuring process, and approximations to the optimum theoretical solution were sought. An elliptical surface was found to provide a solution close to theoretical and this was numerically raytraced over a range of illumination angles and wavelengths to verify the predicted parameters. This design retained at least 80% of its performance to 4° off axis. Moreover, spatial resolution better than 0.1° is achieved along the sky in the direction along the spectrograph slit, which permits radiation from bright stars to be identified and removed. A schematic of the optical design is shown in Figure 1.

A major innovation is the use of a low internal background multichannel plate detector used in connection with wedge and strip encoding (Lampton et al. 1986). At EUV wavelengths a prime contributor to the noise is background in the detector (Lieu et al. 1993). This background consists of two components (a) an internal background due to the radioactive decay of potassium in the microchannel plates, and (b) a charged particle background. We employ special microchannel array plates with no potassium and surround the detector with a charged particle anticoincidence system to significantly reduce both of these backgrounds.

Two of these spectrometers will be flown as a single instrument package (the Espectrografo Ultravioleta extremo para la observacion de la Radiacion Diffusa, or EURD) on the Spanish Minisat satellite in 1996. The gratings on these spectrographs have different

TABLE 1. Key Instrument Parameters

Bandpass:	350–1100 Å
Field of view:	26° × 8°
Grating:	8 cm diameter
	18 cm focal length
	holographically ruled
	2460 lines mm^{-1}
Grating overcoating:	long wavelength: silicon carbide
	short wavelength: boron carbide
Detector:	Multichannel plate with
	wedge and strip encoding
Detector photocathode:	magnesium fluoride
Size (each spectrograph):	40 × 40 × 13 cm
Weight (each spectrograph):	11 kg

coating to optimize the shorter and longer bandpasses of the total bandpass covered. A summary of key instrument parameters is provided in Table 1.

4. Flight Electronics and Software

The first stage detector electronics consists of three ultra low noise charge sensitive amplifiers which receive the microchannel plate signals and provide shaped pulse signals to the downstream analog-to-digital conversion system. These charge amplifiers are especially designed to be free of overload saturation artifacts from cosmic-ray events, and are designed to provide image stability over the three year duration of the mission.

An important part of the detector electronics is an electronic pulse calibration system. Each second, a trio of charge pulses is generated by an onboard quartz-crystal controlled oscillator. The amplitudes of these pulses are controlled by a digitally switched attenuator to produce accurate charge signals for the detector electrodes; these charge amplitude ratios have been chosen to encode positions in the extreme corners of the field of view of each detector. In this way, the stability of the entire detector electronics can be monitored through instrument development, calibration, test, integration, and during the mission.

The digital flight electronics system is centered in a high capability microprocessor with associated ROM, RAM, control logic, and communications chips. The architecture adopted is based on the flight proven ATT DSP32C microprocessor. Although this chip family is radiation tolerant, specific provisions have been taken to assure its survival in case of latchup triggered by a high energy cosmic ray event.

There are two principal functions of the EURD microprocessor. A photon formatting task takes random photoevents in their wedge-strip format and converts them into event (x,y) coordinates using full 32-bit arithmetic in order to avoid introducing computational artifacts into the accumulated images. Each event is flagged with the anticoincidence shield status along with a variety of other information.

The data communications task of the EURD microprocessor executes concurrently with the photon formatting. Data are transferred to and from the spacecraft bus in a high-speed block format. The EURD processor is continuously available to receive a command block, or to dispatch a data block to the onboard data storage system.

5. Overall Performance

In Figure 2 we compare the predicted sensitivity for the EURD spectrometers for 100 hrs and 1000 hrs of observing time with the best upper limits available in the bandpass

TABLE 2. Diffuse Galactic ISM Experiments

	Bandpass	Resolution (E/ΔE)
EUVE (Jelinsky et al. 1995)	190–250 Å	10
	400–460 Å	10
Los Alamos (Bloch et al., 1995)	130–190 Å	10
Wisconsin (Sanders et al., 1995)	40–80 Å	20
Penn State (Burrows et al. 1995)	10–50 Å	40–60
This experiment	350–1000 Å	200

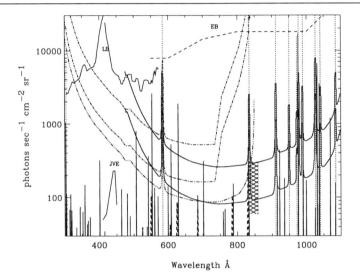

FIGURE 2. Existing upper limits to the diffuse EUV cosmic background. LB are the 15 Å resolution limits of Labov and Bowyer. EB are the 30 Å resolution limits derived by Edelstein and Bowyer from Voyager data. JVE are the limits of Jelinsky et al. The curved, solid, horizontal lines are the flux limits provided by 100 and 1000 hrs of observation with the long wavelength spectrometer. The curved, dot-dashed, hotizontal lines are the same limits for the short wavelength spectrometer with the aluminum filter. The lower dash-triple dot line show similar limits with no filter. The solid vertical lines are the expected ISM emission from a steady-state collisionally ionized plasma. The heavy dashed vertical lines are the intensities from the delayed recombination model of Breitschwerdt and Schmutzler. The dotted vertical lines are expected airglow lines. The cross-hatched region shows the range of the emission predicted by Sciama for a halo of radiatively decaying neutrinos.

covered. Two distinct models of interstellar line emission are also shown. One model assumes collisional ionization equilibrium, using the emissivities of Monsignori-Fossi & Landini (1995) and the emission measure from Bowyer et al. (1995). The other model is the delayed recombination model of Breitschwerdt & Schmutzler (1994 and private communication) whose intensity fits the 0.25 keV soft X-ray background at high galactic latitudes and for which we have assumed attenuation due to the ISM in local cloud of 5×10^{17} (Frisch 1994). The line from decaying neutrinos is from Sciama (1994), and the airglow lines are from Chakrabarti et al. (1984).

In Table 2, we summarize key parameters of instruments which have been, or will soon be used to study the character of the diffuse EUV and soft X-ray background. The EURD instrument has substantial capabilities both in absolute terms and in comparison with these other instruments.

We thank Eric Korpela for help in various aspects of this development and for useful discussions. Gerald Penegor, Josef Dalcolmo, Charles Donnelly, and Ray Chung provided important technical contributions. This work was supported by NASA Grant NRG05-003-450.

REFERENCES

BOWYER, S., FIELD, G., & MACK, J. 1968, Nature, 217, 32

BOWYER, S., LIEU, R., SIDHER, S. D., LAMPTON, M., & KNUDE, J. 1995, Nature, 375, 212

BLOCH, J. J., JAHODA, K., JUDA, M., McCAMMON, D., SANDERS, W. T., & SNOWDEN, S. L. 1986, ApJ, L, 308, L59

BLOCH, J. J. 1995, This volume

BREITSCHWERDT, D. & SCHMUTZLER, T. 1994, Nature, 371, 774

BREITSCHWERDT, D. & SCHMUTZLER, T. 1995, IAU Symposium 171: New Light on Galaxy Evolution, Heidelberg, ed. Ralf Bender & Roger Davies, Dordrecht: Kluwer Academic Publishers

BURROWS, ET AL. 1995, SPIE Proceedings Volume 2518: EUV, X-Ray & Gamma-Ray Instrumentation for Astronomy VI, submitted

CASH, W., MALINA, R., & STERN, R. 1976, ApJL, 204, L7

CHAKRABARTI, S., KIMBLE, R., & BOWYER, S. 1984, J. Geophys. Res., 89, 5660

COX, D. P. & REYNOLDS, R. J., 1987, Ann. Rev. A&A, 25, 303

COX, D. P. 1995, these proceedings

DAVIDSEN, A., ET AL. 1991, Nature, 351, 128

EDELSTEIN, J. & BOWYER, S. 1995, ApJ, submitted

FABIAN, A. C., NAYLOR, T., & SCIAMA, D. W. 1991, MNRAS, 249, 21

FRISCH, P. 1994, Science, 265, 1423

HOLBERG, J. B. 1986, ApJ, 311, 969

JELINSKY, P., VALLERGA, J. V., & EDELSTEIN, J. 1995, ApJ, 442, 653

JENKINS, E. B. 1978a, ApJ, 219, 845

JENKINS, E. B. 1978b, ApJ, 220, 107

LABOV, S. E. & BOWYER, S. 1991, ApJ, 371, 810

LAMPTON, M., SIEGMUND, O. H. W., BIXLER, J., & BOWYER, S. 1986, Proc. SPIE, 627, 383

LIEU, R., BOWYER, S., LAMPTON, M., JELINSKY, P., & EDELSTEIN, J. 1993, ApJ, 417, L41

MARTIN, C. & BOWYER, S. 1989, ApJ, 338, 677

MONSIGNORI-FOSSI, B. & LANDINI, M. 1995, these proceedings

SANDERS, W. T. & EDGAR, R. J. 1995, these proceedings

SAVAGE, B. D. 1987, in Interstellar Processes, ed. Hollenbach & Thronson, Dordrecht: Reidel Publishing

SCIAMA, D. W. 1994, Modern Cosmology and the Dark Matter Problem, New York: Cambridge University Press

SPITZER, L., JR. 1956, ApJ, 124, 20

STERN, R. & BOWYER, S. 1979, ApJ, 230, 755

INDEX

INDEX